Ecological Studies

Analysis and Synthesis

Edited by

W. D. Billings, Durham (USA) F. Golley, Athens (USA)
O. L. Lange, Würzburg (FRG) J. S. Olson, Oak Ridge (USA)

Volume 29

Vegetation and Production Ecology of an Alaskan Arctic Tundra

Edited by
Larry L. Tieszen

With 217 Figures

Springer–Verlag New York Heidelberg Berlin

Library of Congress Cataloging in Publication Data
Main entry under title:

Vegetation and production ecology of the Alaskan
  arctic tundra.

   (Ecological studies; 29)
   Bibliography: p.
   Includes index.
   1. Tundra flora—Alaska—Barrow. 2. Primary
productivity (Biology)—Alaska—Barrow. 3. Tundra
ecology—Alaska—Barrow. 4. Primary productivity
(Biology)—Arctic regions. I. Tieszen, Larry L.
II. Series.
QK146.V43   581.5′264   78-14039

Printed in the United States of America.

9 8 7 6 5 4 3 2 1

ISBN 0–387–90325-9 Springer–Verlag   New York

ISBN 3–540–90325-9 Springer–Verlag   Berlin   Heidelberg

Preface

This volume on botanical research in tundra represents the culmination of four years of intensive and integrated field research centered at Barrow, Alaska. The volume summarizes the most significant results and interpretations of the primary producer projects conducted in the U.S. IBP Tundra Biome Program (1970–1974). Original data reports are available from the authors and can serve as detailed references for interested tundra researchers. Also, the results of most projects have been published in numerous papers in various journals.

The introduction provides a brief overview of other ecosystem components. The main body presents the results in three general sections. The summary chapter is an attempt to integrate ideas and information from the previous papers as well as extant literature. In addition, this chapter focuses attention on processes of primary production which should receive increased emphasis. Although this book will not answer all immediate questions, it hopefully will enhance future understanding of the tundra, particularly as we have studied it in Northern Alaska.

The synthesis of these primary producer papers as well as those of other ecosystem components is being published in "An Arctic Ecosystem: The Coastal Tundra at Barrow, Alaska" (J. Brown et al., Eds.) and can be considered a companion volume. "Truelove Lowland, Devon Island, Canada: A High Arctic Ecosystem" (L. C. Bliss, Ed.) provides an invaluable comparison from the High Arctic, and "Fennoscandian Tundra Ecosystems" (F. E. Wielgolaski, Ed.) summarizes recent studies in northern Europe. Comparative information and the synthesis at the international level can be found in a number of recent books including: "Structure and Function of Tundra Ecosystems" (T. Rosswall and O. W. Heal, Eds.) and "Primary Production and Production Processes, Tundra Biome" (L. C. Bliss and F. E. Wielgolaski, Eds.). "Tundra; Comparative Analysis of Ecosystems" (L. C. Bliss et al., Eds.) is being prepared as an international synthesis.

The rising environmental awareness in the world coincided with the discovery of large oil deposits on the Alaskan Arctic Coastal Plain. The period 1970–1974 was therefore especially interesting as the U.S. Tundra Biome Program served as a scientific focus for the environmental interests of private industry, the U.S. Government and the public. In many ways we have all benefited from this confrontation and perhaps the resolution of future environmental issues will now be easier. The interest, encouragement, and support of these diverse groups is sincerely appreciated. Direct financial support of the program has been derived from three major sources: the National Science Foundation, the State of Alaska through the University of Alaska, and individual companies and members of the petroleum industry. The NSF funding was under the joint sponsorship of the U.S. Arctic Research Program (Office of Polar Programs) and the U.S. International Biological Programme (Ecosystem Analysis).

The Office of Naval Research through its Naval Arctic Research Laboratory (NARL) at Barrow provided the field and laboratory support without which the U.S. Tundra Biome Program would not have been possible. Two former directors of NARL, Dr. Max C. Brewer and Mr. John Schindler, deserve particular credit for facilitating this support. Alpine field support was provided through the University of Alaska's Institute of Arctic Biology for Eagle Summit and through the University of Colorado's Institute of Arctic and Alpine Research at its Mountain Research Station for Niwot Ridge. Administratively, the Tundra Biome Center at the University of Alaska provided vital contractual and support services. Dr. George C. West, Director of the Tundra Biome Center, Dr. Keith Van Cleve, and Mr. David Witt deserve particular credit. The U.S. Army Cold Regions Research and Engineering Laboratory (CRREL) provided overall management, specialized logistic and equipment support and editing services to the program through an interagency agreement with the NSF and in the spirit of Public Law 40–280 which authorized the U.S. IBP. Mr. Stephen L. Bowen of CRREL and his staff, particularly Donna Murphy and Audrey White, were responsible for the preparation of all Biome reports and the final draft of this volume. Mr. Harold Larsen, CRREL illustrator, supervised the preparation of illustrations. In addition to the directly funded support, parent institutions, almost without exception, provided their staff members with a variety of on-campus and other institutional aid. This is gratefully acknowledged at this time on behalf of the entire program.

The success of this program is to a large extent due to the untiring optimism and perseverance of one person, Dr. Jerry Brown of CRREL, who served as director from 1970–1975. His managerial skills allowed many of us to concentrate more effort on scientific problems.

Research in the U.S. Tundra Biome has been an extremely exciting and rewarding experience. To a large extent this has occurred because the program assembled more than 20 diverse projects most of which were committed to the "program," to full cooperation, and to an experimental assessment of plant function and productivity in tundra. It is perhaps this commitment to understand plant function and the basis for productivity which distinguishes our program. The dedication of most projects under the difficult conditions of remote field experimentation is also highly appreciated.

   The program was enriched by the high degree of cooperation and criticism from international colleagues and U.S. participants in other biome programs. The opportunity for international meetings and exchanges as well as U.S. Specialists meetings is greatly appreciated for it helped clarify experimental approaches at regular intervals. May this international spirit of cooperation and camaraderie, especially among the young scientists, be one of the lasting results of the IBP Tundra Biome Program.

L. L. TIESZEN

References

Bliss, L. C. (Ed.) (1977) *Truelove Lowland, Devon Island, Canada: A High Arctic Ecosystem*. Edmonton: University of Alberta Press, 714 p.

Bliss, L. C. and F. E. Wielgolaski (Eds.) (1973) *Primary Production and Production Processes, Tundra Biome*. Proceedings of Conference, Dublin, Ireland, April 1973. International Tundra Biome Steering Committee, Stockholm, 256 p.

Brown, J., P. C. Miller, L. L. Tieszen, F. L. Bunnell, and S. F. MacLean, Jr. (Eds.) (in prep.) *An Arctic Ecosystem: The Coastal Tundra at Barrow, Alaska*. Stroudsburg, Pa.: Dowden, Hutchinson and Ross.

Rosswall, T. and O. W. Heal (Eds.) (1975) *Structure and Function of Tundra Ecosystems*. Swedish Natural Science Research Council, Ecol. Bull. NFR 20, 450 p.

Wielgolaski, F. E. (Ed.) (1975) Fennoscandian Tundra Ecosystems, Parts 1 and 2. New York: Springer–Verlag, 366 and 337 p., respectively.

Contents

Contents   xiii

Contributors

ALEXANDER, V.	Institute of Marine Science, University of Alaska, Fairbanks, Alaska 99701, USA
ALLESSIO, M.	Department of Biology, Rider College, Lawrenceville, New Jersey 08648, USA
BILLINGS, W. D.	Department of Botany, Duke University, Durham, North Carolina 27706, USA
BILLINGTON, M.	Institute of Marine Science, University of Alaska, Fairbanks, Alaska 99701, USA
CALDWELL, M.	Range Science Department, Utah State University, Logan, Utah 84321, USA
CAMERON, R. E.	Argonne National Laboratory, Environmental Statement Project Division, Argonne, Illinois 60439, USA
CHAPIN, F. S., III	Institute of Arctic Biology, University of Alaska, Fairbanks, Alaska 99701, USA
CLEGG, B.	University of California, Lawrence Livermore Laboratory, Livermore, California 94550, USA
COYNE, P. I.	University of California, Lawrence Livermore Laboratory, Livermore, California 94550, USA
DENNIS, J. G.	Office of the Chief Scientist, National Park Service, Washington, D.C. 20240, USA
EHLERINGER, J. R.	Department of Biology, University of Utah, Salt Lake City, Utah 84112, USA
FAREED, M.	10350 Wyton Drive, Los Angeles, California 90024, USA
GERSPER, P. L.	Soils and Plant Nutrition, University of California, Berkeley, California 94720, USA
JOHNSON, D.	Crops Research Laboratory, Utah State University, Logan, Utah 84322, USA
KELLEY, J. J., JR.	Naval Arctic Research Laboratory, Barrow, Alaska 99723, USA
KNOX, A.	Argonne National Laboratory, Environmental Statement Project Division, Argonne, Illinois 60439, USA
KORANDA, J.	University of California, Lawrence Livermore Laboratory, Livermore, California 94550, USA
LAURSEN, G.	Naval Arctic Research Laboratory, Barrow, Alaska 99723, USA
LAWRENCE, B. A.	Biology Department, San Diego State University, San Diego, California 92115, USA
LEWIS, M. C.	Department of Biology, York University, Downsview, Ontario, Canada M3J 1P3
McCOWN, B. H.	Institute for Environmental Studies, University of Wisconsin, Madison, Wisconsin 53706, USA
MITCHELL, G. A.	University of Georgia, Coastal Plain Station, Department of Agronomy, Tifton, Georgia 31794, USA
MILLER, O. K.	Department of Biology, Virginia Polytechnic Institute and State University, Blacksburg, Virginia 24061, USA
McKENDRICK, J. D.	Agricultural Experiment Station, Palmer, Alaska 99645, USA
MORELLI, F. A.	Jet Propulsion Laboratory, California Institute of Technology, Pasadena, California 91107, USA

MURRAY, B. M.   Institute of Arctic Biology and the Museum, University of Alaska, Fairbanks, Alaska 99701, USA

MURRAY, D. F.   Institute of Arctic Biology and the Museum, University of Alaska, Fairbanks, Alaska 99701, USA

OECHEL, W. C.   Biology Department, McGill University, Montreal, PQ, Canada H3A 1B1

OTT, V.   3842 Monterey Avenue, Baldwin, California 91706, USA

PETERSON, K. M.   Department of Botany, Duke University, Durham, North Carolina 27706, USA

PRASHER, D. C.   Department of Biochemistry, Ohio State University, Columbus, Ohio 43210, USA

RASTORFER, J. R.   Department of Biological Sciences, Chicago State University, Chicago, Illinois 60628, USA

RUDOLPH, E. D.   Department of Botany and Institute of Polar Studies, The Ohio State University, Columbus, Ohio 43210, USA

SCHELL, D.   Institute of Marine Science, University of Alaska, Fairbanks, Alaska 99701, USA

SCHOFIELD, E. A.   The Institute of Ecology, Holcomb Research Institute, Butler University, Indianapolis, Indiana 46208, USA

SHAVER, G. R.   Department of Botany, Duke University, Durham, North Carolina 27706, USA

STEERE, W. C.   New York Botanical Garden, Bronx, New York 10460, USA

STUART, M.   University of California, Lawrence Livermore Laboratory, Lawrence, California 94550, USA

SVEINBJORNSSON, B.   Biology Department, McGill University, Montreal, PQ, Canada H3A 1B1
Present address: Abisko Sci. Res. Stn., 980 24 Abisko, Sweden

TIESZEN, L. L.   Biology Department, Augustana College, Sioux Falls, South Dakota 57102, USA
Present address: Division of International Programs, National Science Foundation, Washington, D.C. 20550, USA

ULRICH, A.   Soils and Plant Nutrition, University of California, Berkeley, California 94720, USA

VETTER, M.   Department of Botany, Duke University, Durham, North Carolina 27706, USA

WEBBER, P. J.   Institute of Arctic and Alpine Research, University of Colorado, Boulder, Colorado 80309, USA

WILLIAMS, M. E.   Department of Botany, The Ohio State University, Columbus, Ohio 43210, USA

Section 1
Floristics, Vegetation, and Primary Production

Section 1a: Vascular Plants

The "Floristics, Vegetation, and Primary Production" section establishes the long-term perspective necessary to understand tundra adaptations, relates the distribution of vascular plants, bryophytes, and lichens to environmental factors, and documents the variabilities and magnitudes of primary production and standing crops. Much of this information is new because of the relatively poor research base in the U.S. tundra and is urgently needed as development problems confront us.

The first subsection is concerned with vascular plants and provides, in Chapter 2, a general review and integration of tundra diversity and zonation in the Northern Hemisphere. It compares and clarifies the diverse and confusing nomenclature of tundra zonation and locates the U.S. sites in these classifications. Furthermore, the physiographic provinces of Northern Alaska are described, and the origin of tundra as well as floristic relationships are presented. The Prudhoe Bay system is substantially different from the Barrow system because of a more continental climate, a calcareous alluvium, and more complex vegetation units.

Webber's chapter (3) documents the spatial heterogeneity of the Coastal Plain near Barrow and locates the research sites and species in the broader environmental complex. Eight vegetation noda are defined. The microrelief, primarily because of its relationships to soil moisture, aeration, and phosphate availability, determines the character and performance of these units. The system is dominated by bryophytes and monocotyledons. Aboveground production varies from 18.1 to 118.5 g m^{-2} yr^{-1} and increases along the moisture gradient as species diversity decreases. The spatial patterns of microrelief and vegetation are incorporated in a theoretical successional framework based upon the thaw lake cycle.

The seasonal dynamics of primary production and the effects of some ecosystem experiments and perturbations are reviewed and described in Chapter 4. The intensive site is dominated by three monocotyledons and shows an aboveground production which approaches 102 g m^{-2} yr^{-1} and which displays a leaf area index of over 1.0. The addition of the aboveground increment appears to be quite linear from snowmelt until early August, at which time senescence becomes dominant. Seasonal changes in the belowground compartments are very difficult to assess because of both the high plant density and small individual size and the organic nature of the active layer. Simulated grazing and mulching decreased production whereas increased air temperatures and added nutrients increased production.

This community response to fertilizer was greater for phosphorus than nitrogen and especially enhanced stem base and rhizome biomass.

Section 1b: Nonvascular plants

Steere (Chapter 5) provides a comprehensive floristic overview for bryophytes which parallels the treatment by Murray. Bryophytes were of special interest in our study because of their common occurrence, large cover, and potential effects on nutrient cycling and thermal phenomena. Again, the need for more distributional and descriptive ecological information in Northern Alaska is apparent. On a broad scale the most important factors affecting distributions are water availability, pH, and surface stability. The more localized studies (Chapter 6) again illustrate the interesting differences between Barrow and Prudhoe Bay with no sphagnum and only one polytrichaceous species at the calcareous Prudhoe Bay sites. Although exceedingly common and abundant, the production of bryophytes is low relative to that of vascular plants on the same sites.

Because of the wet nature of our research sites lichens are less abundant than might be expected in a tundra area. Their distribution (Chapter 7) is again closely related to the moisture gradient. Their diversity tends to be greatest on high-centered polygons, including frost boils, but their growth and production are greater at intermediate moisture levels. They may play important roles in nutrient trapping and nitrogen fixation.

Very little information is available on the subjects discussed in the last two chapters (8 and 9) in this section. The wet nature of our terrestrial system suggested that algae might be common components. Several taxonomic groups are widespread at Barrow, especially in the upper 1 or 2 cm. Their ecological significance is yet to be established. Certainly they comprise a food source for other microorganisms and may be important in nitrogen fixation. We are also just beginning to document the occurrence of mycorrhizae among many tundra species. As we learn more about the functional roles of these fungi, wherever they occur, we can expect to develop our understanding of their ecological significance in the nutrient-poor active layer at Barrow.

1. Introduction

Larry L. Tieszen

Background

The U.S. Tundra Biome Program was initiated in the spring of 1970. Its initial development was directed toward two goals. On the one hand the program represented one of five biomes to be included in the U.S. IBP studies and as such its orientation was to the total system. On the other hand, the recent discovery of large oil reserves on the Arctic Slope dramatized the urgent need for a basic understanding of the environment, flora, fauna, and vegetation of the tundra as well as specific information on the response of the system to perturbations and methods of revegetation. Thus, in response to the increasing rate of human activity on the arctic tundra, we supplemented the basic research objectives of the IBP to address applied questions related to the immediate environmental concerns of Northern Alaska. Rather than undertake applied projects, we conducted basic ecological experiments whose results could be useful in solving environmental problems.

Tundra exists in two general areas in the U.S.: latitudinally beyond treeline to the North and altitudinally above timberline in various parts of the country (Figure 1). Since it was impossible to study both alpine and arctic sites in detail, we decided to concentrate our efforts at one arctic site and to coordinate a few projects at other sites thereby allowing some degree of confidence in our extrapolations and generalizations. Arctic tundra was emphasized for a variety of reasons. A group of scientists from several disciplines was already active in the Arctic and provided a central core for the Tundra Biome Program; logistic support and a well-equipped laboratory (NARL) were in existence at Barrow; and national priorities were then on the Arctic Slope of Alaska. Furthermore, an informal agreement with the Canadian Tundra Biome group was established which allowed them to emphasize research on high-latitude, relatively dry tundra and which allowed us to emphasize research on the wet end of the tundra spectrum.

Various potential sites were evaluated in Northern Alaska including Umiat on the Colville River, Atkasook on the Meade River, Cape Thompson, Prudhoe Bay, and Barrow. The substantial data base at Cape Thompson (Wilimovsky and Wolfe, 1966) was outweighed by its remoteness and logistic problems. Similarly, Umiat and Meade River were excluded, even though substantial recent work had

NORTH AMERICAN TUNDRA

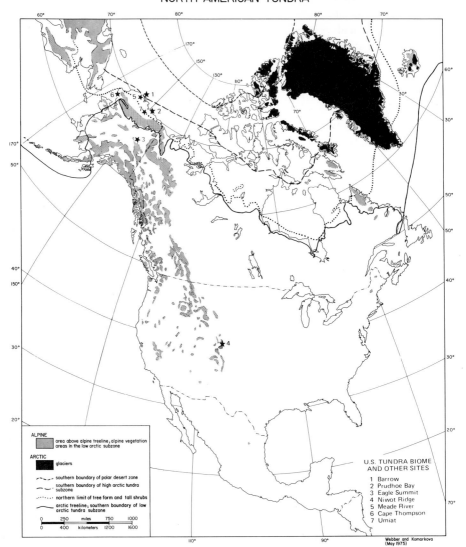

Figure 1.   The distribution of Tundra vegetation in North America (adapted from Komár-ková and Webber, 1978). The Arctic region is subdivided into a Polar Desert zone and a Tundra zone. The latter zone is subdivided into High Arctic Tundra and Low Arctic Tundra (Webber, 1974). These correspond to the arctic and subarctic subzones of Alexandrova (1970). A discussion of Arctic zonation is given in Murray (1978). The location of the U.S. IBP Tundra Biome sites and some others pertinent to this volume are shown.

been conducted (Johnson and Kelley, 1970). Thus, Barrow became our main research site. This site possessed excellent logistic support (NARL), permitted continuous electrical power to be delivered to the field, had been studied intensively by a substantial number of ecologists, and, perhaps most important, was familiar to most potential participants. Perhaps its main disadvantages were that it had undergone significant human disturbance by Barrow residents and field scientists, and that the maritime climatic influence made it less representative of Alaska tundra than inland sites.

Three other sites were included in the U.S. Tundra Biome Program. Niwot Ridge, Colorado, was included to provide a low-latitude high alpine site at which field experiments that could not be undertaken at Barrow were established. Eagle Summit, Alaska, was included as a high-latitude low alpine site. Prudhoe Bay, at the source of the oil pipeline, was included to provide baseline information for a coastal tundra relatively unknown to tundra biologists (Brown, 1975).

Research Design and Site Selections

The design of the plant producer program was guided by several broad objectives. Our foremost objective was to develop a predictive understanding of primary production and its control in tundra, particularly as exemplified by the wet coastal tundra of the Arctic Coastal Plain. A second and intimately related objective was to develop the conceptual understanding underlying production processes and to acquire the necessary data base to model and simulate these production processes in a variety of environmental or ecological situations. The third objective was to relate our understanding to environmental problems on the Arctic Slope, especially those which affected canopy structures and problems associated with revegetation.

Inherent in these objectives was the need to describe both vegetation structure and associated processes. Our structural studies incorporated analyses of species richness, growth form classifications, plant densities, and biomass estimates of various compartments for a number of habitat types and geographical areas. Our detailed process studies, however, required additional structural information which could be obtained for only a few areas. These included detailed canopy studies of both plant architectural features and micrometeorological parameters. Because of time and budget constraints, it was our decision from the beginning to concentrate our detailed structural studies and process studies at one site and on only a few representative species.

The site (site 2) selected for intensive studies (Figures 2 and 3) consists of weakly expressed, low-centered polygons near the midpoint of the complex moisture gradient. The wet meadow vegetation is relatively homogeneous with *Dupontia fisheri, Carex aquatilis,* and *Eriophorum angustifolium* (Tieszen, 1972) dominating. This site, established in 1970, remained the principal site for plant process studies and whole system experiments. An adjacent drier site (site 1) was also established, mainly for experimental studies related to environmental problems associated with the development of oil. In 1971, site 4 was incorporated to provide spatial variation in habitats across microrelief, moisture, vegetation, and soil gradients. These sites are described in more detail by Webber (1978). Replicated plots (6 × 6 m) were established on sites 1 and 2 for control and experimental studies. Random sample points were generated along 1-m grids, and these served as the sites for comparative soil and vegetation analyses. Similar plots were established to assess the response of the system to simulated grazing, mulching, fertilization, heating, vehicular impact, and oil applications. The control plots were used to provide the seasonal progressions and variations in biomass, canopy architecture, plant and soil nutrient status, and other parameters.

The process studies were developed concurrently with the expansion and revision of the primary production model which had already been designed (Miller and Tieszen, 1972). We thus structured our process studies around the canopy as a vegetation unit with physiological parameter values at the species level. The presumed high degree of control of production processes by physical factors of the environments caused us to emphasize these early in the program. Thus, gas exchange processes and climatic interactions were stressed. As our understanding of these processes increased, we began to emphasize the less known and perhaps less tractable controls over production. Thus, concurrently with our development of growth and allocation models, we began to emphasize

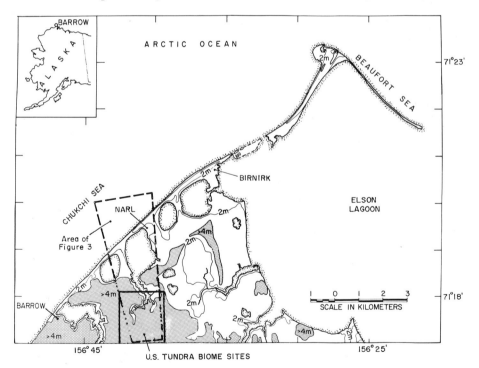

Figure 2.   Index map of Alaska showing the vicinity of the Barrow research sites. Two- and four-meter contour intervals are illustrated. The hatched area is depicted by the photograph in Figure 3.

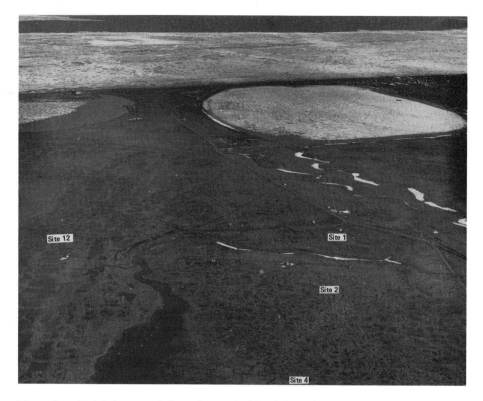

Figure 3. Aerial photograph from the south side of the main research site looking toward the Arctic Ocean in the background. The panorama included in the photograph is delimited in Figure 2. Note the polygonal nature of the area.

soil temperature, moisture, nutrients, and grazing interactions. As these interactions became spatially oriented, we realized the necessity for a population approach, and population dynamics projects were initiated.

Overview of the Barrow Ecological System

This overview provides a general framework for the interpretation of primary producer components. Much of the information has been extracted from Brown et al. (in prep.).

Geological

Barrow, the northernmost point of the U.S., is at the northern extremity of the Arctic Coastal Plain (71°18′ N latitude, 156°40′W longitude). The study sites (Figures 2 and 3) are located 3 km inland from the Chukchi Sea on a complex sequence of drained lakes. This area was not glaciated during the Pleistocene.

Nevertheless, the surfaces are still quite recent since at least some of the near-surface marine sediments (the Gubik Formation) were deposited no earlier than 25,000 to 35,000 yr B.P. by the mid-Wisconsin invasion of the sea. The regression of the sea from this area was followed by the development of tundra (Brown, 1965; Colinvaux, 1964). On the study site marine sediments are overlain in places by a thin layer of peat dated at 12,600 ± 200 yr B.P. (Lewellen, 1972). This peat was buried and/or reworked and buried by sediments from at least one large but shallow lake which covered the area. The lake drained as a natural consequence of the thaw lake cycle (Britton, 1967). The present soil surface is believed to be between 5000 and 10,000 yr old.

Soil profiles commonly start with a relatively recent organic-rich surface horizon underlain by a mineral horizon of silty clay or silty loam and a lower layer representing the older buried peat materials (see Figure 9 in Webber, 1978). The rooting zone, therefore, exploits the recent organic horizon (5 to 10 cm thick), a somewhat variable depth (5 to 25 cm) of mineral soil, and occasionally the lower buried peat. This stratification in combination with slight relief and impermeable permafrost has important consequences. Soils generally remain saturated below a depth of 5 cm. Some soils are saturated up to the moss surface. Water potentials are therefore high (>-1.0 bar) throughout the growing season. Oxygen contents decrease with depth especially in and beneath the zone of mineral soil. Percentage oxygen saturation also decreases as soils become wetter along the moisture gradient. This may have important consequences for species selection along the topographic–moisture gradient and for the depth distributions of roots and root activity.

At site 2 most of the soils are acidic and generally range from a pH of 4.0 to 5.5. This low pH has important consequences for the primary producer component. For one thing, this acidity results in the presence or absence of specific floristic elements among both the vascular plants and the bryophytes (see Murray, 1978; Steere, 1978). This effect is especially striking in the comparison with the Prudhoe Bay system which is more alkaline. In terms of proximate functional effects, this low pH will certainly influence nutrient availability and uptake.

The soil is carbon rich and the accumulation of organic matter itself influences the availability of essential plant nutrients. Nitrogen and phosphorus, two nutrients most commonly considered "limiting" in many tundra areas (McKendrick et al., 1978; Chapin, et al., 1975), are closely associated with the organic carbon component. The total carbon to nitrogen ratio is very constant at Barrow (around 20). In contrast, the ratio of total carbon to phosphorus is more variable and ranges from 116 to 573. These total values are of interest in terms of long-term nutrient cycling, and they suggest rather large pools of nitrogen and phosphorus. However, the large amounts also suggest a slow mineralization and a long-term immobilization. Consequently they may bear only incidental relationships to the nutrient pools available for plant use.

The organic matter provides the bulk of the cation exchange capacity at Barrow. During 1971 the sum of exchangeable cations ranged from 82 to 108 meq 100 g^{-1}, although much of this is exchangeable H$^+$ and aluminum. Significant seasonal variations also occur in the soil solution, especially for N, P, and K. Nitrogen is generally not present in the nitrate form, and NH_4^+-N predominates, especially in the wetter soils. Only low levels of any of these nutrients are available and then only in the confined areas above the permafrost.

One of the major features of any arctic system is the presence of permafrost or perennially frozen ground. This development reflects the low mean annual air temperatures and may extend to depths of 300 to 600 m. Several important features are related to permafrost. First, it presents an impervious layer beneath the active layer which seasonally thaws and freezes. Thus, its presence accounts, in part, for the saturated soils as well as the low temperatures of the active layer. The temperature effect is particularly important since the large water content and the change of state impose a seasonal lag in temperature relative to the above-ground part of the system (Figure 4). Thus, the insulating effect of the late winter snow cover maintains plant temperatures near the mean equilibrium temperature of the permafrost or lower than air temperatures (Tieszen, 1974). The snow pack becomes isothermal as meltwater flows through the snow. This generally occurs shortly (1 or a few days) before the area becomes predominantly snow free.

As the summer season progresses, the active layer thaws to varying depths

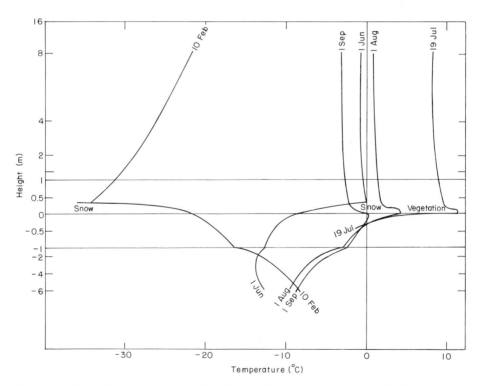

Figure 4. Typical temperature profiles through the air, canopy, and soil. Note the low soil temperatures during the growing season, the significant, although small, canopy effect, and the low soil surface temperatures at 1 June due to the snow cover. (Adapted from Weller and Holmgren, 1974.)

(Nakano and Brown, 1972), generally 25 to 35 cm, by early August. Freezing of the active layer occurs after the aboveground vegetation has senesced or been killed by increasingly lower air temperatures. Thus, root activity may occur in the thawed active layer longer than would be expected on the basis of shoot activity. Of course, these roots will still be in frozen soil after the shoots resume growth next spring.

The second important consequence of the development of permafrost is the patterned ground that results from the freezing and cracking of soil and water in the formation of regularly spaced ice wedge polygons (Figure 5). The V-shaped

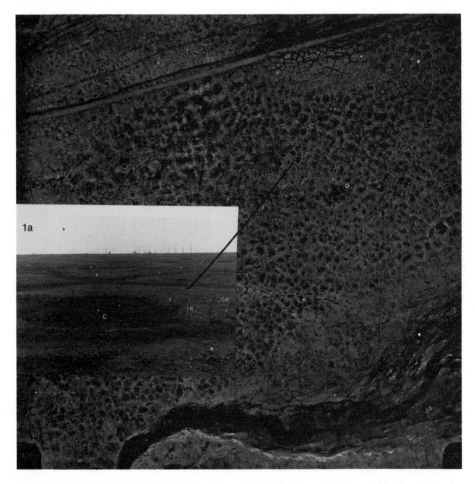

Figure 5. Aerial view of polygonal wet tundra surface near Barrow, Alaska. Note the regularity of the interconnected ice wedges. The localized entrapment of water occurs in the low centers and troughs. Also, note the development of thermokarst in the upper part of the picture where the access road has facilitated drainage. C = center, IR = inner rim, trough not indicated. (Aerial vertical photo, 1971, courtesy of USACRREL. Insert courtesy of K. E. Everett, Ohio State University.)

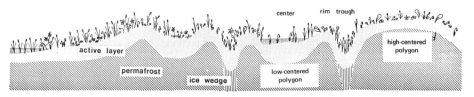

Figure 6. Diagrammatic representation of low-centered and high-centered polygons. Note the position of the ice wedges, rims, and troughs, the active layer, permafrost, and the diagrammatic representation of growth forms. (Provided by P. Bunnell.)

vertical ice wedges form near the surface and grow laterally and downward as the contraction cracks fill with meltwater which freezes and adds to the mass of the wedge. Polygons, which are so obvious from aerial views, are the surficial expression of the interconnected wedges. The polygons are variable in size but in our system were usually 5 to 12 m in diameter.

The polygons impose a significant amount of microrelief in an area which is otherwise extremely flat. As wedges expand, surface layers adjacent and parallel to the wedges are thrust upward to produce rims which surround the centers of the polygons (Figure 6). Troughs, the slight depressions overlying the ice wedges between the rims, develop as rims increase in height and as tops of ice wedges melt. The polygons are either low-centered (centers lower than their associated rims), as at our site 2, or high-centered. The low centers of polygons and troughs frequently contain standing water for a good portion of the summer. These seemingly slight features are of considerable ecologic significance (Wiggins, 1951) since they also affect thermal conditions and soil profiles. Thus different plant species are associated (see Figure 9 in Webber, 1978) with these features, and they support different levels of lemming activity. These geomorphological features are important components of this tundra system and are the direct result of the permafrost and low temperatures characteristic of these high latitudes.

Climatic

Perhaps the most dramatic environmental feature of Barrow is the low temperature both on an annual basis and during the growing season. This is clearly evident from the long-term mean (Table 1) which shows a mean yearly temperature of $-12.6°C$, the product of 9 months (> 278 days) with mean temperatures below freezing. The extremely low mean temperatures ($-26.2°C$) plants experience for the 4 months of December, January, February, and March are perhaps of less biological significance than are the low temperatures during the growing season, because the plants are dormant or quiescent at this time.

As indicated earlier, initiation of plant growth is determined by the time of snowmelt since only then are plants released from the low subnivean temperatures. Thus in June air temperatures become positive (Table 2) and the productive season is initiated. At Barrow these temperatures remain low as is indicated by the mean long-term temperature for July of 3.7°C. In part these very low

Table 1. Long-term (1941 to 1970) Climatic Means for Solar Radiation (kJ cm^{-2}), Air Temperature (°C), Precipitation (mm), and Windspeed (m sec^{-1}) as Recorded at Barrow, Alaska (National Weather Service)

	J	F	M	A	M	J	J	A	S	O	N	D	Year
Solar radiation (kJ cm^{-2})[a]	0	4.4	22.8	48.2	68.0	68.9	57.3	33.5	15.0	5.4	0.5	0	324
Temperature (°C)	−25.9	−28.1	−26.2	−18.3	−7.2	0.6	3.7	3.1	−0.9	−9.3	−18.1	−24.6	−12.6
Precipitation (mm)	5.8	5.1	4.8	5.3	4.3	8.9	22.4	26.4	14.7	14.0	7.6	4.8	124.2
Windspeed (m sec^{-1})	5.0	4.9	5.0	5.2	5.2	5.1	5.2	5.5	5.9	6.0	5.6	5.0	5.3

[a] 14-year mean.

Table 2. Mean Monthly Temperature during the Main Productive Season at Barrow, Alaska[a]

| | Temperature (°C) | | | |
	June	July	August	Season departure
1970	0.5(−0.1)	3.2(−0.5)	1.7(−1.4)	−0.7
1971	1.7(+1.1)	4.7(+1.1)	0.8(−2.3)	−0.1
1972	0.3(−0.3)	6.1(+2.4)	4.8(+1.7)	+1.3
1973	0.7(+0.1)	4.3(+0.6)	4.3(+1.2)	+0.6
1970–1973 $\bar{X}$	0.8(+0.2)	4.6(+0.9)	2.9(−0.2)	
Long-term mean	0.6	3.7	3.1	

[a] Departures from the long-term mean are indicated in parentheses. The mean of the departure for the season is also indicated.

temperatures are caused by the proximity of the ice-covered Arctic Ocean which cools the lower surface of the atmosphere (Barry, personal communication). The low air temperature, high humidity, and on-shore winds also result in a very high incidence of cloud cover and fog, especially in July and August. Further indication of the cold summer climate is shown by the fact that frosts and snowstorms occur at any time; and, in fact, 43 % of the mean hourly temperatures in July are at or below 0°C (Rayner, 1961).

The freezing conditions that are common throughout the season are ameliorated somewhat by canopy effects (Miller and Tieszen, 1972); however, the combination of moderately high winds (Table 1) and narrow graminoid leaves results in a close coupling of leaf and air temperatures. In cushion plants and topographically protected areas, temperatures are likely to be somewhat higher. Observations have shown that when air temperatures drop to nearly −4°C, some plants are damaged. Mature leaf tips of *Carex aquatilis* and *Dupontia fisheri* are killed and all *Dupontia* leaves suffer low temperature damage in the zone of elongating cells. This becomes apparent several days after frost as a leaf with a nonchlorophyllous band (several mm wide) is produced.

There is significant intra- and interseasonal variability in air temperatures as can be seen in a comparison of data from 1970 to 1973 (Table 2). Air temperatures were lower than the long-term mean in 1970 and significantly higher in 1972. The mean monthly temperatures in different years differed by as much as 3 to 4°C, an amount which is likely to be very significant for plant processes occurring near their lower cardinal temperatures.

The low temperature regime of Barrow is perhaps a direct consequence of the radiative balance for this area. Incoming solar radiation is obviously seasonal, since the sun does not drop below the horizon until 2 August. Obviously, it is beneath the horizon for a similar period on each side of the winter solstice. Barry (personal communication) indicates that the total mean annual solar radiation at Barrow is 76 kcal cm^{-2} yr^{-1}. This is approximately one-half that received at Niwot Ridge, Colorado. Both sites receive similar daily amounts in June; however, at Barrow these similar daily amounts are spread over 24 hr and irradiances only rarely exceed 1.0 cal cm^{-2} min^{-1} on a horizontal surface. Thus, instanta-

Table 3.   Mean Values of Visible Radiation (400 to 700 nm) for the 10-Day Periods Ending as Indicated [a]

| Date | Cal cm^{-2} day^{-1} | Cal cm^{-2} hr^{-1} | | Hours sunlight per day |
		Minimum	Maximum	
June 24	205			24
July 4	173	1.14	17.7	24
July 14	176	0.34	19.9	24
July 24	177	0.24	18.7	24
August 3	137	0.062	15.3	22.2
August 13	104	0.001	15.9	19.6
August 23	81	0.00	12.7	17.3
August 26	53	0.00	8.4	16.9

[a]   Daily totals were determined by the summation of the integration of 48 half-hourly scans. Hours sunlight were estimated from Smithsonian Meteorological Tables for 71°20′ N (from Tieszen, 1972).

neous irradiances at Barrow are low even on clear days. Clouds and fog are a common feature and during the snow-free period cloud cover is around 85%. Thus, the irradiance is primarily diffuse rather than direct solar. As is suggested by the sample data for 1970 (Table 3), however, there are marked differences in irradiance at solar noon and solar midnight. Near the summer solstice irradiances at noon were 15 times those at midnight; and this ratio increased substantially as the season progressed, until around 15 August when midnight irradiances approached zero.

Thus, the Barrow system is characterized by limited incoming solar radiation on a yearly basis, reasonably high daily totals early in the growth season, and relatively low irradiances. Of special significance is the fact that plant growth is initiated only near or at the summer solstice and that canopy development and intercepting capability occur in a radiation regime which will soon rapidly diminish.

The extreme nature of the Barrow environment and the seasonal differences are to a large extent dependent upon the balance of radiation or the net radiation regime. For arctic tundra, the annual net radiation is only slightly positive (<15 kcal cm^{-2} yr^{-1}) (Barry and Hare, 1974) and should equilibrate with a mean annual surface temperature of -10 to $-15°C$. This, in fact, is about the mean air temperature for the Barrow area ($-12.6°C$). Thus, the low temperatures are a direct consequence of the low net radiation input for this area.

Weller and Holmgren's (1974) analysis of the energy balance components identified six periods which differed either in net radiation or in the magnitude of the components. These periods for the Barrow area are depicted in Figure 7 and clearly define the major seasonal events which influence or determine the seasonality of plant productivity. The winter period is the longest and, because of the limited incoming solar radiation, is the only period characterized by a negative net radiation balance. To a large extent this is also related to the continuous snow cover which during this period results in an albedo around 85%.

As solar radiation increases in late winter, a premelting period is defined in which net radiation becomes positive as substantial energy is used to heat surface snow and soil. Ultimately (usually early in June) sufficient heat energy is accumulated and the snow surface begins to melt. The albedo now changes rapidly and substantial energy is used in melting surface snow. As the melt-water percolates through the accumulated snow pack, air temperatures in the canopy rapidly increase and approach 0°C (Tieszen, 1974). Prior to this time plants are frozen and at temperatures too low for normal activity.

The immediate postmelt period is a period of high solar irradiance immediately prior to or including the summer solstice. The albedo remains low and rapid thawing of the upper soil surface occurs. Plant growth now resumes in this basically aquatic medium at the top of the thawing active layer. In the summer period, surface water decreases and the canopy develops with a consequent increase in albedo to about 19%. The fall freeze-up period is probably the least predictable period at Barrow. However, it generally is initiated near late August after night temperatures drop sufficiently low to either induce plant senescence or kill exposed vegetation directly. Of course, during this period solar irradiance and photoperiod rapidly decline as well. The transition to the winter condition occurs mainly as a result of the accumulation of winter snow cover which again restores the characteristic high albedo. For some time during this period, however, the active layer may remain near 0°C and belowground biological activity may continue.

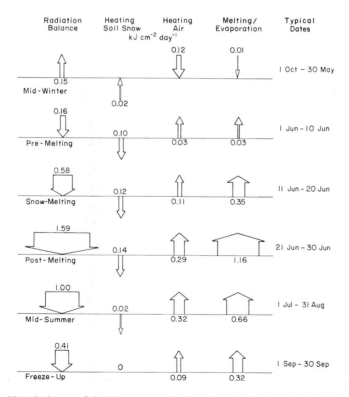

Figure 7.   Heat balance of the Barrow tundra for six periods described and adapted from Weller and Holmgren (1974). Storage and loss components of the balance are indicated by the direction of the arrows. Magnitude of the components is indicated in kJ cm^{-2} day^{-1}.

Biotic

One of the features that made the Barrow system attractive for total ecosystem research was the relatively low biological diversity. This low diversity is apparent in the vascular plants (see Murray, 1978; Webber, 1978) but is more striking among the animal components. Relatively few species of invertebrates or vertebrates are of direct importance to the producer components of the ecosystem. For example, there are very few canopy dwelling invertebrate herbivores at Barrow. This is verified by the fact that in nearly all species of vascular plants, leaves show no signs of insect herbivory. In fact, the leaves are further characterized by the virtual absence of any noticeable fungal infestations. Thus, loss of carbon or nutrients to aboveground microherbivores appears minimal.

The situation belowground is somewhat less certain. However, the intensive recent research of MacLean (in Brown et al., in prep.) has shown a high density and biomass of invertebrates, especially in the upper portions of the soil profile. The production is, however, less than one might expect since the low temperatures result in overlapping generations during any one summer. Only a few of these organisms (e.g., some nematodes) feed on roots directly and appear less important than in other systems. The high correlation of invertebrate biomass with vascular productivity (and therefore low correlation with accumulated organic matter), however, suggests that at least the invertebrates are utilizing the recently produced root and rhizome material. The direct importance of root grazing and the magnitude of root exudation or other losses, however, are unknown.

The relationship with vertebrate components is somewhat clearer. The dominant herbivore is clearly the brown lemming *(Lemmus sibericus)*, although collared lemmings *(Dicrostonyx groenlandicus)* are also common. Other potentially important herbivores, the caribou *(Rangifer tarandus)* and ground squirrel *(Spermophilus parryi)*, are essentially absent from Barrow, presumably partly because of the large amount of human activity. A few birds are also occasionally important herbivores, for example, the Lapland longspur, snow bunting, and red poll are extensive seed eaters at least during certain periods.

It is the lemming, however, which has the greatest impact on both the primary producers and the ecosystem as a whole. To a large extent this impact occurs because of the large fluctuations in population density. During a population low, densities may be on the order of 1 animal ha^{-1} (Pitelka, 1973). In these conditions about 0.1% of the aboveground primary production may be utilized. This contrasts with a population high in which densities may approach 200 animals ha^{-1}. These densities may utilize a large percentage of the primary production. Thus, the potential impact is great and may even be accentuated by the fact that high densities may occur at times when plant biomass is low (winter or spring). Under these circumstances lemmings not only cut down the plant canopy as they feed but must also dig out rhizomes and stem bases as has been reported by Dennis (1968). The overall impact on plant population changes, succession, and thaw phenomena may be very significant.

The invertebrate saprovores and vertebrate herbivores support a rather diverse assemblage of predators which, generally, do not influence the producer component directly. The snowy owls and arctic foxes, however, are important because the denning and feeding habits of these animals produce substantial microrelief in the flat tundra and serve to concentrate nutrients. Thus, specific plants inhabit or become dominant in these areas. These are important spatial features but are not of great significance at our specific research sites.

Historic

The extent to which the system has been influenced by indigenous and present human elements is not clear. Eskimo inhabitants occupied the Birnirk site (Figure 2) until it was inundated by the sea 1200 to 1500 yr B.P. (Brown and Sellmann, 1973). Eskimo hunting in the Barrow region has undoubtedly been great and perhaps accounts for the near absence of caribou and ground squirrels at the present time. In part, the removal of these herbivores from the Barrow system may result in the wider population oscillations reported among the lemmings here compared to other Alaska sites. Although a number of species of plants were used as food sources by these indigenous people, it seems unlikely that this would have had a major impact on the present system.

More recent activities have had more dramatic effects. The use of power-driven vehicles has had a substantial impact upon the tundra system. This is especially obvious where erosion of the nearly flat terrain has occurred at very rapid rates as a result of alterations of the thermal properties of tundra surface and the melting of ground ice. Drainage and erosion of our research sites, for example, have been accelerated because of the thermokarst and thermal erosion caused by a variety of tracked vehicles. Similarly, whole lakes have been drained and impoundments formed by seemingly minor disturbances. This disruption of the normal thaw lake cycle could have serious consequences as further activity and development continue on the Arctic Slope. Succeeding chapters will help lay the groundwork for a better understanding of tundra dynamics and hopefully lead to a more rational approach to land use.

References

Alexandrova, V. D. (1970) The vegetation of the tundra zones in the USSR and data about its productivity. In *Productivity and Conservation in Northern Circumpolar Lands* (W. A. Fuller and P. G. Kevan, Eds.). Int. Union Conserv. Natur., Morges, Switzerland, Publ. 16, N.S., pp. 93–114.

Barry, R. G., and F. K. Hare. (1974) Arctic climate. In *Arctic and Alpine Environments* (J. D. Ives and R. G. Barry, Eds.). London: Methuen, pp. 17–54.

Brown, J. (1965) Radiocarbon dating, Barrow, Alaska. *Arctic,* **18**: 36–48.

Brown, J. (Ed.). (1975) *Ecological Investigations of the Tundra Biome in the Prudhoe Bay Region, Alaska*. Biol. Pap. Univ. Alaska, Spec. Rep. No. 2, 215 pp.

Brown, J., and P. V. Sellmann. (1973) Permafrost and coastal plain history of arctic Alaska. In *Alaskan Arctic Tundra* (M. E. Britton, Ed.). Arct. Inst. North Amer. Tech. Pap. No. 25, pp. 31–37.

Brown, J., P. C. Miller, L. L. Tieszen, F. L. Bunnell, and S. F. MacLean, Jr. (Eds.), (in prep) *An Arctic Ecosystem: The Coastal Tundra at Barrow, Alaska*. Stroudsburg, Pa.: Dowden, Hutchinson and Ross.

Britton, M. E. (1967) Vegetation of the arctic tundra. In *Arctic Biology: 18th Biology Colloquium* (H. P. Hansen, Ed.). Corvallis, Oregon: Oregon State University Press, 2nd ed., pp. 67–130.

Chapin, F. S., III, K. Van Cleve, and L. L. Tieszen. (1975) Seasonal nutrient dynamics of tundra vegetation at Barrow, Alaska. *Arct. Alp. Res.*, **7**: 209–226.

Colinvaux, P. A. (1964) Origin of ice ages; pollen evidence from Arctic Alaska. *Science*, **145**: 707–708.

Dennis, J. G. (1968) Growth of tundra vegetation in relation to arctic microenvironments at Barrow, Alaska. Ph.D. dissertation, Duke University, 289 pp.

Johnson, P. L., and J. J. Kelley, Jr. (1970) Dynamics of carbon dioxide and productivity in an arctic biosphere. *Ecology,* **51**: 73–80.

Komárková, V., and P. J. Webber. (1978) An alpine vegetation map of Niwot Ridge, Colorado. *Arct. Alp. Res.,* **10**: 1–29

Lewellen, R. I. (1972) *Studies on the Fluvial Environment: Arctic Coastal Plain Province Northern Alaska*. Publ. by the author (P.O. Box 1068, Littleton, Colorado 80120), 2 vols., 282 pp. and plates.

McKendrick, J. D., V. J. Ott, and G. A. Mitchell. (1978) Effects of nitrogen and phosphorus fertilization on carbohydrate and nutrient levels in *Dupontia fisheri* and *Arctagrostis latifolia*. In *Vegetation and Production Ecology of an Alaskan Arctic Tundra* (L. L. Tieszen, Ed.). New York: Springer-Verlag, Chap. 22.

Miller, P. C., and L. L. Tieszen. (1972) A preliminary model of processes affecting primary production in the arctic tundra. *Arct. Alp. Res.,* **4**: 1–18.

Murray, D. F. (1978) Vegetation, floristics, and phytogeography of northern Alaska. In *Vegetation and Production Ecology of an Alaskan Arctic Tundra* (L. L. Tieszen, Ed.). New York: Springer-Verlag, Chap. 2.

Nakano, Y., and J. Brown. (1972) Mathematical modeling and validation of the thermal regimes in tundra soils, Barrow, Alaska. *Arct. Alp. Res.,* **4**: 19–38.

Pitelka, F. A. (1973) Cyclic pattern in lemming populations near Barrow, Alaska. In *Alaskan Arctic Tundra* (M. E. Britton, Ed.). Arctic Institute of North America, Tech. Pap. 25, pp. 199–215.

Rayner, J. M. (1961) *Atlas of Surface Temperature Frequencies for North America and Greenland*. McGill University, Arctic Meteorology Research Group, Publ. Meteorol. No. 33.

Steere, W. C. (1978) Floristic, phytogeography, and ecology of arctic Alaskan bryophytes. In *Plant Ecology of an Alaskan Arctic Tundra* (L. L. Tieszen, Ed.). New York: Springer-Verlag.

Tieszen, L. L. (1972) The seasonal course of aboveground production and chlorophyll distribution in a wet arctic tundra at Barrow, Alaska. *Arct. Alp. Res.,* **4**: 307–324.

Tieszen, L. L. (1974) Photosynthetic competence of the subnivean vegetation of an arctic tundra. *Arct. Alp. Res.,* **6**: 253–256.

Webber, P. J. (1978) Spatial and temporal variation of the vegetation and its production, Barrow, Alaska. In *Vegetation and Production Ecology of an Alaskan Arctic Tundra* (L. L. Tieszen, Ed.). New York: Springer-Verlag, Chap. 3.

Weller, G., and B. Holmgren. (1974) The microclimate of the arctic tundra. *J. Appl. Meteorol.,* **13**: 854–862.

Wiggins, I. L. (1951) The distribution of vascular plants on polygonal ground near Point Barrow, Alaska. *Contrib. Dudley Herb.,* **4**: 41–56.

Wilimovsky, N. J., and J. N. Wolfe (Eds.). (1966) *Environment of the Cape Thompson Region, Alaska*. Oak Ridge, Tennessee: United States Atomic Energy Commission, Division of Technical Information, 1250 pp.

2. Vegetation, Floristics, and Phytogeography of Northern Alaska

DAVID F. MURRAY

Tundra Landscapes and Vegetation

Diversity and Zonation

The word *tundra* means very different things to different people; nevertheless, except for its occasional usage in the vernacular for treeless bogs in the subarctic interior, it denotes both the circumpolar treeless region north (and south) of the latitudinal treeline and the less extensive mountain landscapes above altitudinal treeline. Originally the word *tundra* was applied to treeless rolling plains of the Eurasian far north, thus connoting a regional climate, a landscape, and a vegetative cover. Since the history of the landscape and biota is not the same throughout tundra regions, since climates vary, and since floras differ from place to place, tundra, as it might apply to vegetation, is an ambiguous term (Griggs, 1934). Depending upon one's frame of reference, it is possible to develop three basic stereotypes: a steppe-desert, a heathland, and a bog and meadow landscape. All are correct; treelessness is the shared feature.

Arctic tundra covers vast areas over a substantial range of latitude. The vegetation exhibits zonation marked by physiognomic and floristic changes that correlate roughly with increasing northern position and concomitant reductions of summer warmth. Middendorf (1864, cited in Griggs, 1934) first divided tundra into a northern high tundra for the desert type and a southern low tundra of bogs, meadows, and heaths. This generalization persists more or less intact today in the recognition of a northern High Arctic and a more southern Low Arctic. The former is best characterized by discontinuous vegetation of various herbs and prostrate shrubs limited to mesic sites, with bare mineral soil an important, even dominant component of the landscape. The latter zone has continuous mats of vegetation in all but the most exposed and xeric settings, greater floristic diversity, and erect shrubs especially in riparian situations. Polunin (1951), after several years of experience in the Canadian Eastern Arctic, refined the definition of zones and added to the classification his concept of the Middle Arctic. In keeping with its location between the High and Low Arctic, it is a region with intermediate characteristics. There are fewer species than in the Low Arctic; the vegetation is more continuous than in the High Arctic; erect shrubs are absent;

the tussock-forming cottongrass *(Eriophorum vaginatum)*, so abundant in the Low Arctic, has only a minor role; and grass and sedge meadows are very important. Although the concept of the Middle Arctic has not been widely applied, virtually all of the North American literature on tundra has employed subdivisions of High and Low Arctic following Polunin's outlines.

Soviet botanists (Alexandrova, 1970), as well as some North Americans (Porsild, 1951), restrict the use of the term tundra as a zone of vegetation to areas with a generally continuous plant cover on mesic surfaces. They differentiate and exclude from tundra the Polar Desert. Tundra therefore includes all of the area south of Polar Desert and north of latitudinal tree limit. Naturally, this Tundra zone is not uniform and subdivisions have been employed; Soviet botanists often refer to Arctic and Subarctic Tundra subzones of arctic tundra. Some difficulty arises from having subarctic apply to both a subdivision of the Arctic and a separate zone below (south) and adjacent to the Arctic (Barry and Ives, 1974). Webber (1974, 1978) used High and Low Arctic for subdivisions of Tundra that are consistent with current usage here but not identical to the High and Low Arctic zones *sensu* Polunin (Figure 1).

Alpine tundra is often treated as distinct from its arctic counterpart, especially since most of the studies, and therefore also the comparisons, have been made of tundras disjunct by many kilometers and degrees of latitude. In Northern Alaska, where mountains occur in a belt across the arctic zone and mountain (alpine) tundra is confluent with the tundra of the arctic lowland, distinctions between the two are less clear. Physiographic aspect, local landform, and geomorphic pro-

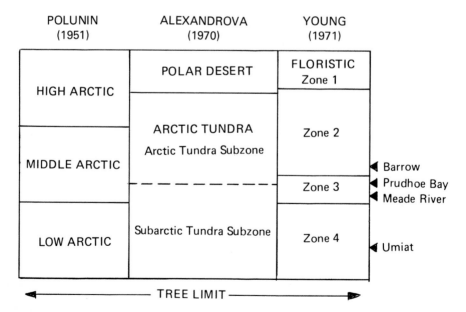

Figure 1.   Relative latitudinal position of major zonation in the North American Arctic from three points of view. The Arctic Tundra Subzone of Alexandrova is synonymous with the High Arctic Subzone of Webber (1974, 1978), and Alexandrova's Subarctic Tundra Subzone is Webber's Low Arctic Subzone.

cesses are at least quantitatively different in each major tundra type, thus there are vegetational and floristic generalizations that can be applied separately to alpine and lowland arctic tundra. In addition, not only is there an alpine setting at high elevations in the Arctic but also alpine conditions of drainage and microclimate typical of mountains occur in areas with only moderate relief (cf. Young, 1975). At sites on the Arctic Coastal Plain that offer alpine habitats (river terraces, active floodplains, dunes, pingos) are several vascular plant species that are "out of place" on the coast but are common in the Brooks Range to the south (discussed further below).

Tundra is a cold climate landscape having a vegetation without trees. The absence of trees is caused by a complex of conditions that is ultimately related to regional climate. This regional aspect distinguishes tundra from treeless bogs and similar local areas without trees due to edaphic extremes in areas that otherwise support a forest cover. The various definitions of tundra gleaned from the literature by Dagon (1966) do not do justice to the diverse vegetation types which outside of the Arctic would be called desert, grassland, heathland, and savanna (Beschel, 1970). Tundra retains meaning when it is applied in a geographic sense. There is such variation of climate, physiography, and vegetative cover in regions accurately classified as tundra that the term without qualification can imply a unity that does not exist. Use of the term in a restrictive way demands amplification, otherwise generalizations made without context can lead to misunderstandings (see the comments by Steere, 1976 and 1978, regarding the conflicting views on the relative importance of bryophytes in tundra vegetation).

In Alaska

The tundra vegetation in Alaska has been described on a number of scales. The early attempts to map and evaluate timber resources for the entire state led, naturally, to the most generalized statements for vegetation; tundra was then simply unspecified in the category of untimbered lands. A refinement in the early period of vegetation study was the differentiation of treeless landscapes as grasslands, arctic tundra, and alpine tundra (cf. Küchler, 1965). Vegetation maps produced in the last 15 yr (at scales of 1:1,000,000 and smaller) have subdivided the arctic tundra of northern Alaska into major subdivisions that correspond roughly with the physiographic regions found there. The Arctic Slope (a name being supplanted by North Slope) extends from the crest of the Brooks Range north to the Arctic Ocean. It has long been viewed as consisting of three physiographic provinces (Payne et al., 1951): mountains (Brooks Range), foothills (Arctic Foothills with northern and southern sections), and coastal plain (Arctic Coastal Plain with Teshekpuk Lake and White Hills sections). Subsequent workers (Wahrhaftig, 1965; Walker, 1973) have followed this scheme with only slight modifications, for the integrity of these divisions is obvious. Northern Alaska is used here in a specific geographic sense to include the Arctic Slope *and* the alpine tundra occurring above 600 m on the south slope of the Brooks Range that can be separated only arbitrarily from tundra on the northern flank of these mountains.

Maps by Spetzman (1963), Küchler (1966), Viereck (1971), and the joint Federal–State Land Use Planning Commission for Alaska (1973) have divided the Arctic Slope vegetation into wet tundra, moist tundra, dry (alpine) tundra, and high brush or their equivalents (Küchler used other names for essentially the same formations). Wet tundra dominated by graminoid taxa characterizes the lowlands of the Arctic Coastal Plain; moist tundra is found on upland surfaces of the coastal plain and throughout the Arctic Foothills; dry or alpine tundra occurs on summits in the foothills and is, of course, the basic unit of the Brooks Range. The high brush type consists of extensive shrub thickets on the floodplains of large arctic rivers and is absent only on the most coastal portion of the coastal plain. There is also a distinct, but narrow and somewhat discontinuous, belt of vegetation peculiar to strands, saline marshes, and estuaries. These coastal wetlands are under the influence of marine systems in which salt water and sea ice are among the factors determining the biological environment. This vegetation type is distinct from the wet tundra unit with which it is usually placed in studies on small map scales.

Such treatments of tundra are only appropriate for mapping of large areas in which synthesis is demanded for practical reasons. Descriptions of vegetation on larger map scales convey more information about the landscape and also quickly become much more complex. More precise physiognomic criteria can be useful for defining major plant assemblages, particularly when the dominant taxa are highly predictable as they are for riparian shrubs, heaths, and tussock tundra. On the other hand, the grass–sedge–forb type is a single physiognomic unit, but one that combines obviously disparate assemblages and consequently is not parallel in floristic homogeneity with the other units. It is the lack of obvious community structure that limits the usefulness of this approach.

The next level of detail has been the definition of plant communities in the traditional North American sense in which boundaries are determined by discontinuities of the dominant, or primary, species. The difficulty in defining, describing, and classifying tundra vegetation has been discussed by Griggs (1934), Raup (1941, 1951), Sigafoos (1951, 1952), Böcher (1954), and Savile (1960). The structure of vegetation in the Arctic is not as obvious as the differentiated, temperate nontundra vegetation viewed on the same scale. The apparent versatility (ecological amplitude) of primary species frequently precludes their use as faithful indicators of specific site conditions and vegetation types (cf. Raup, 1975). The distinct units that do occur often coincide with the equally distinct geomorphic features peculiar to cold climates. These provide the steep environmental gradients that in turn account for abrupt changes in plant cover and well-marked communities. Therefore, the frost-related geomorphic processes and attendant surface instability must be understood in order to clarify plant patterns. Britton (1967) treated this relationship particularly well in his descriptions of Arctic Slope vegetation. Surface conditions change repeatedly over short distances and produce a complex mosaic of plant assemblages leading to a vast array of both well-marked and subtle changes in plant composition.

Descriptions of vegetation on these different scales provide the background for more detailed regional and local plant studies. Whereas it is possible to sample and estimate biomass, carbon, etc. for communities, the experimental studies generally involve individuals of populations perceived as single species. When these taxa tolerate a range of conditions, it might be surmised that there are, within them, differently adapted subsets of the species, or ecotypes. Obviously, the synecological approach yields limited information for this scale of inquiry, and the individual plant–site interaction becomes important to consider. Raup's (1969) discussion of the need to keep the study of site and plant cover on commensurate scales should be read by all students of tundra vegetation.

Over the years of describing the distribution of plant taxa, one develops, at the very least, an intuitive sense for autecology and frequently an accurate predictive ability to characterize plant–site relationships. Local patterns of species are determined by the interplay of physical factors and plant reaction. There are no horizontal plane land surfaces; tundra landscapes that appear level are usually rich in microtopographic relief. The mesotopographic gradient of Billings (1973), when viewed as existing in an interconnected and endlessly repeated convex and concave surface, provides a framework for integrating cold climate physical processes and products with slope aspect. Slope angle (and, therefore, also slope stability), soil texture, and soil moisture vary along the gradient. Convex slopes are characterized by active mass wasting, and the concave slopes by accumulation of the surface material from above. Bedrock exposed along ridge crests is weathered and wasted downslope. The product of disintegration forms a thin, coarse mineral soil of barely altered fragments on the hilltops and a progressively thicker soil of smaller and smaller particles below. The moisture gradient parallels the development of finer and deeper soils (cf. Figure 2, Johnson and Tieszen, 1973).

The basic model presented in the preceding paragraph reveals my own bias developed in mountains where bedrock is close to the surface and frequently exposed. On the coastal plain, since marine sediments mantle the bedrock, roughly comparable mesotopographic gradients are limited to raised beach ridges, river terraces, coastal bluffs, the slopes of drained lake basins, and even such diminutive structures as peat ridges and earth hummocks. Soil moisture varies with the slightest topographic relief and improved drainage. Where peaty soils are saturated (in spite of low annual precipitation) due to conditions determined by permafrost, floristic and vegetational changes can be noticed when the change of relief is on the scale of only centimeters. Similarly, differences of soil texture produce over short linear distances remarkable discontinuities in the vegetation indicative of changes in soil moisture, such as from wet peat to well-drained alluvium of river terraces.

Tundra Origins

The flora of Alaska and neighboring territories (*sensu* Hultén, 1968) is of great interest to the biogeographer who interprets patterns of distribution in the extant and fossil floras and provides reconstructions of geologic events and environmental conditions with estimates of probable ecological and evolutionary responses by the biota. The extensive Pleistocene refugia, the past continental

connections via the Bering Land Bridge, the Asian affinity of the Alaskan tundra
flora, the species restricted to and radiating from Beringia, and the vicariant taxa
of Chukotka and Alaska endemic to their respective continents offer a number of
intriguing problems in phytogeography. It is difficult, however, to organize a
coherent and convincing history of climate, landscape, and biota when the
relevant time period is so long and the data upon which our present record is
based are often distant in space and time (see also the comments by Hopkins,
1972, p. 122). Dated materials and stratigraphic relationships can be questioned,
the application of the uniformitarian principle of geology to the behavior of
organisms does not acknowledge evolutionary change (Packer, 1971), and conclusions are necessarily based on a mixture of data interpretation, inference, and
pure speculation. However, the fascination and challenge of biogeography arise
from the fact that so many of the questions are moot.

Tundra vegetation of northern Alaska was preceded in the Tertiary by forests
that in late Miocene were predominantly coniferous, with a minor broadleaved
deciduous element that is not today a part of the taiga (Wolfe, 1972). Probable
Miocene fossil assemblages from the late Tertiary Beaufort Formation in the
Canadian Arctic Archipelago have suggested to Matthews (1974a, 1976) the
existence of a type of forest-tundra. From this he inferred a Miocene lowland
tundra on the northernmost portions of Axel Heiberg Island, Ellesmere Island,
and Greenland. However, younger remains (Miocene–Pliocene) from the
Seward Peninsula indicate a rich coniferous forest similar to that typical of
southern British Columbia and Washington today (Hopkins et al., 1971). Hopkins et al. interpreted as taiga the pollen remains of similar age found farther
north along the Alaskan coast at Kivalina (68°58′ N). Consequently, lowland
tundra, if present at all during early Pliocene, occurred only on the northern
fringe of Alaska. Yet, it has been inferred and widely accepted that by late
Pliocene, the climate became sufficiently cool to cause the replacement of the
Tertiary temperate flora in the far north with a flora and major zonation much as
it exists today (Arnold, 1959; Dorf, 1959).

Theories for the origin of tundra are derived largely from mid-19th-century
outlines developed by Forbes and Darwin and subsequently restated with botanical emphasis by J. D. Hooker. They postulated as many geographers will today
(Löve and Löve, 1974) a widespread late Pliocene tundra flora. This flora was
forced south from the Arctic (or destroyed there) during glaciations of the
Pleistocene; part became isolated high in the Rocky Mountains and part returned
to the Arctic in the immediate postglacial period accompanied by certain of the
southern alpine taxa (Hooker, 1861; and in Gray and Hooker, 1880).

The notion that arctic vegetation was obliterated by widespread glaciation had
to be modified when it became clear that large regions in Alaska escaped the
direct effects of glaciers, and, therefore, the persistence of plants in periglacial
refugia could be invoked. This helps to explain disjunctions as relicts of once
more continuous populations. Johnson and Packer (1967) stated explicitly that
"the distribution of the extant arctic-alpine flora . . . , when considered along
with modern concepts of rates of migration and evolution in the angiosperms, is

best explained by supporting an essentially circumpolar arctic-alpine flora during the Tertiary upon which Pleistocene events have been superimposed.''

Whether or not tundra vegetation had developed to any extent on arctic lowlands by the end of the Pliocene or developed during the Quaternary, which is also likely, it is generally accepted that the arctic flora has ancient origins. It is derivative of steppe, cold-temperate, and boreal taxa of specialized habitats that preadapted them to arctic conditions (Tolmatchev, 1960; Savile, 1972), taxa capable of the necessary adaptive adjustments *in situ* (Löve, 1959), and Tertiary alpine plants from the mountains of Asia and North America (Holm, 1922; Hultén, 1958; Weber, 1965; Savile, 1972; Yurtsev, 1972; Billings, 1974). It is the role of the Tertiary alpine flora as a precursor of the arctic lowland flora and not the reverse that represents a significant shift of emphasis.

Events of the Quaternary in northern Alaska profoundly affected the types of habitats available to plants. Valley glaciers extended from the Brooks Range beyond the mountain front as much as 50 km during glacial maxima (Hamilton and Porter, 1975), and the coastal plain was repeatedly inundated during marine transgressions of interglacials (Sellmann and Brown, 1973). However, when glaciers reached their maximum extent, sea level was lower and the coastal plain extended out to what is now the Beaufort Sea; when, during the transgressions, the seas and marine sediments reached far inland, glacier termini were presumably back within the mountains. Therefore, throughout the Pleistocene extensive and diverse landscapes were available for the persistence of biota, primarily in the Arctic Foothills.

With the emergence of the Bering Land Bridge during glacial maxima, there appeared a surface for the exchange of fauna and flora between the extensive periglacial areas of Alaska–Yukon and northeastern Asia. At the time of maximum extent, the land bridge was immense and probably experienced climatic zonation in which the latitudinal differences were marked (Yurtsev, 1972, 1973). The arctic ice pack to the north and the Brooks Range to the south greatly reduce the flow of moist air to northernmost Alaska. Given an extensive land bridge and exposed continental shelf, this area must have been decidedly continental in climate and even extremely arid (Hopkins, 1972). Thus, Barrow and environs, on the northeast corner of Beringia, would have received the propagules of Asian plants adapted to a cold, dry continental climate during glacial intervals, with a gradual shift toward more oceanic species favored by the climate of the waning land bridge during marine transgressions (cf. Yurtsev, 1972, 1973).

The question remains as to the extent to which the vegetation became altered during interglacial periods. Tree-line did fluctuate and reached Cape Deceit on the Seward Peninsula during the penultimate interglacial (Yarmouth) and was apparently close to this point during Sangamon (Guthrie and Matthews, 1971). Palynological data show little change in the tundra vegetation of northern Alaska for over 30,000 yr (Colinvaux, 1967); thus, a cold climate is assumed for at least this interval. Certainly there must have been some minor shifts of tundra vegetation zones if tree-line was fluctuating, but the differences are not revealed in the pollen spectrum. Presumably the arctic-alpine flora of the Brooks Range would have had better access to the northern plains during colder periods than when warmer summers resulted in the expansion of a boreal and low arctic–hypoarctic flora between the mountains and the coast (cf. Yurtsev, 1972).

Evidence is accumulating, particularly from interior Alaska, that argues persuasively for the existence of steppe–tundra[1] vegetation during late Pleistocene. This is a departure from a long-held belief, based mainly on a single paper by Chaney and Mason (1936), that the vegetation was then much as it is today. It is now believed their macrofossils were probably of Sangamon, not Wisconsin, age (Guthrie, personal communication, 1976). Guthrie (1966, 1968), Matthews (1970, 1974a, 1974b), and Ager (1975) have provided evidence for shifts in the flora and herbivorous fauna from a probable Wisconsin grassland with a grazing megafauna of bison, horse, and mammoth to a Holocene dwarf birch shrub tundra and eventually a spruce forest with browsers like moose and caribou the important animals. The concept of the steppe is based on the abundance of *Artemisia* and grass pollen, a lesser amount of sedge and willow pollen, and very little or no birch, alder, or spruce pollen.

Péwé's report (1975) from central Alaska of alder and dwarf birch pollen as well as macrofossil remains of alder, spruce, willow, cottonwood, and paper birch in late Wisconsin sediments raises the possibility of gallery forests in certain river valleys. However, given the herbivores that were relatively abundant during late glacial intervals, the vegetation invoked must be consistent with their evident success. Therefore, steppe vegetation on a broad scale is a reasonable assumption. One of these grazers, the saiga, survives today on the Asian steppes. Since there are also saiga fossils from north of the Brooks Range (Guthrie, 1968; Hopkins, 1972, cf. Figure 5), the steppe character of late glacial vegetation could have extended to the Arctic Slope.

The final deposition of marine sediments on the Arctic Coastal Plain occurred during the mid-Wisconsin transgression. Marine recession during late Wisconsin and uplift of the coastal plain left the surface exposed. Species from the Beringian lowlands and those from adjacent highlands in Northern Alaska that escaped the direct effects of both glaciers and inundation moved onto the new surface. Fluvial, lacustrine, and aeolian sediments were added to the marine deposits; and, in the cold, dry environment of the glacial interval, permafrost, ice-wedges, and polygonal ground formed.

Uplift of the plain compensated for rising sea level during postglacial warming. A sharp warming trend began about 10,000 yr ago and reached a peak about 5000 B.P. (Hopkins, 1972), and this peak correlates with the appearance of birch pollen in the Barrow spectrum (Colinvaux, 1967). This would place the final transformation of tundra landscape and biota into the ecosystem we recognize today within mid- to late- Holocene time. Although the flora has great antiquity, it is probable that the composition of the modern vegetation and the plant–site relationships of the lowland tundra have developed during the past 5000 yr for

---

[1] Guthrie (personal communication, 1976) prefers the name mammoth-steppe to link a characteristic mammal with the landscape and also to carry the implication of extinction for the vegetation type as well as the animal.

which the pollen record indicates relative stability (cf. Matthews, 1974c; Péwé, 1975).

Floristics and Phytogeography

The Brooks Range is a massive belt of rugged mountains that reach altitudes over 2700 m in the eastern sector where there are active glaciers. The range is important not only for its alpine aspect but also for its east-west arcuate trend across northern Alaska that presents a barrier to air masses and effectively separates the polar basin climate from that of the continental interior.

Glaciation altered the mountain landscape through the erosion of bedrock and the deposition of till, and slope-forming processes continue to modify the summits and slopes. Bedrock, blockfields, and screes support very little vegetation, but a small flora of taxa adapted to surface instability occupies the relatively protected sites. Solifluction on the vegetated uplands signals instability, and the irregular features produced add another measure of variability to slopes dissected by screes and stream channels. Cushion- and mat-forming plants on higher steep slopes grade into forb–sedge–grass complexes on warm exposures and low shrub (heath) assemblages in cool sites on deeper soils downslope. Tussock tundra forms on gentle slopes with fine-grained soils and impeded drainage. Streambeds occupy the entire floor of narrow V-shaped valleys at high elevations; where broader floodplains form there is the potential for riparian willow thickets and localized wet meadows. Seldom is the terrain so even as to support uniform swards like those on the coastal plain. The frequent abrupt changes of slope exposure, angle, and stability as well as the variations in soil texture, depth, and moisture common to alpine areas result in species aggregations of great diversity.

The southern section of the Arctic Foothills has some of the structural complexity of the Brooks Range, although the rugged features of the mountains are less extensive, and some of the more prominent mountains, hills, and ridges occur isolated from each other. The northern section has greater topographic regularity—rounded summits and gently rolling terrain where bedrock exposures are smaller and infrequent. The summit and upland slope components in the Arctic Foothills have many alpine characteristics, albeit on a smaller scale than in the Brooks Range. For many people, the seemingly endless expanses of tussock tundra define the Foothills Province. Except for the dry summits and ridges and the wetter valleys and interfluves, tussock tundra is the dominant vegetation type. Associated with the major rivers are gravel terraces, bluffs, and a suite of habitats on the active floodplains. At the foot of the terraces and bluffs and in the V-shaped notches cut by tributary streams are sites of snow accumulation. The combination of snowbeds, screes, and coarse- and fine-grained substrates accounts for floristic diversity one might think precluded by the general aspect of foothills landscapes.

The Arctic Coastal Plain is distinctive for the lack of both conspicuous gradient and substantial topographic relief. The plain is composed of Quaternary

marine sediments over Cretaceous bedrock that have been modified since the
final marine transgression of mid-Wisconsin time. The modern emergent surface
is the product of cumulative geomorphic processes from late Wisconsin to the
present that have led to the incorporation of fluvial, lacustrine, and aeolian
sediments, the development of thick peats, and the formation of permafrost and
frost-related surface features (Brown and Sellmann, 1973). Conspicuous are the
numerous oriented thaw-lakes and myriad smaller ponds that interrupt systems
of low- and high-centered ice-wedge polygons. Topographic prominences are
limited to coastal bluffs, beach ridges, river terraces and stream banks, drained-
lake basins, pingos, sand dunes, and bird mounds. Over most of the area, upland
vegetation forms on surfaces 1 to 2 m above base level. The vegetation reflects
predominantly saturated soils maintained by continuous permafrost.

Although we know quite a lot about the vascular flora of the Arctic Slope (cf.
Wiggins and Thomas, 1962; Hultén, 1968), with each new, serious floristic
venture, significant range extensions are discovered. That distribution patterns
are still to some degree influenced by our ignorance demands caution when
generalizing for so large and diverse an area. The Brooks Range is really a
number of mountain ranges dissected by large rivers that flow north to the
Beaufort Sea, west to the Bering Strait, or south to the Yukon River. It is hard to
seek biogeographic threads for the Arctic Slope and not be drawn into a
consideration of the south slope of the range as well. The physical connections of
northern Alaska to Beringia, the uplands of the boreal Interior, and the Rocky
Mountains system account for the Beringian, circumpolar arctic, boreal, and
arctic-alpine floristic elements. The relative importance of each floristic contribution varies in any one of the physiographic provinces, which differ from each
other and from physiographic counterparts in other areas of the north.

Taxa endemic to and widely distributed throughout Northern Alaska are
Erigeron muirii and *Artemisia arctica* ssp. *comata*. *Oxytropis kokrinensis* is
known from only a few localities in the western Brooks Range and the Kokrines
Mountains to the south; *O. koyukukensis* is confined to the central Brooks
Range; and *Festuca ovina* ssp. *alaskensis (F. auriculata)* and *Oxytropis campes-
tris* ssp. *jordalii* are known the full extent of the Brooks Range but also have
widely separated outliers in other mountain ranges. Three narrowly distributed
radiants of the Beringian Flora common at the highest elevations in the mountains are *Novosieversia glacialis (Geum glaciale)*, *Potentilla elegans*, and *Douglasia ochotensis*.

A number of decidedly boreal species common in the taiga on the south slope
of the range, but absent at high elevations of the range crest, reappear again on
the Arctic Slope in the valleys of the major tributaries to the Colville River, along
the Colville itself, and on warm slopes to the north of the mountains in the
Foothills of the eastern sector. Thickets of *Alnus crispa* in the Colville drainage
have such forest species as *Platanthera obtusata*, *Ribes triste*, *Galium boreale*,
and *Boschniakia rossica*. Isolated stands of *Populus balsamifera* are found on
gravel bars of the Killik and Anaktuvuk rivers in the central Brooks Range and
on the Ivishak and Canning rivers farther east, well north of the last white spruce
(Picea glauca). A number of aquatic plants are known from the ponds in the

Arctic Foothills (Umiat region) that are also separated from their main range by the alpine slopes of the range crest: *Equisetum fluviatile, Potamogeton alpinus* ssp. *tenuifolius, P. subsibiricus, Triglochin maritima, Ranunculus trichophyllus, R. reptans,* and *Myriophyllum spicatum* (Hultén, 1968). I must emphasize that we know the least about the foothills flora, and Umiat may be significant more for the extent of the collecting activity there than for any mix of environmental factors unique to that portion of the Arctic Foothills. Nevertheless, the importance of the bryophyte discoveries (Steere, 1976, 1978), the substantial list of boreal taxa, and the sole North American locality for *Potentilla stipularis* (a significant disjunction from its range in Chukotka) may be more than intriguing coincidences. Sand dunes, dry bluffs, and river terraces harbor the northernmost populations of *Juniperis communis, Carex supina* ssp. *spaniocarpa, Pulsatilla patens, Rosa acicularis, Linum lewisii,* and *Artemisia frigida.* The recent discovery of *Geum macrophyllum, Cystopteris montana, Carex praticola, C. atrosquama, Salix pseudomonticola,* and *Moehringia lateriflora* along the Canning River on the lower slopes of the Shublik Mountains (Murray, unpublished data) together with outliers of *Viburnum edule, Silene repens,* and *Arctostaphylos-uva-ursi* already known from this area add a greater floristic complexity to the Arctic Foothills than previously appreciated. Interpretations will require more exploration and analysis, although one is tempted to assign relict status to some of these disjunctions, since they have affinities with more steppe-like conditions that are being postulated for much of late Wisconsin Alaska.

On the Arctic Coastal Plain are three outliers of taxa once thought restricted to the Canadian Arctic or Greenland. *Draba adamsii* occurs on a raised beach ridge at Barrow but is otherwise a plant of the Arctic Archipelago and Greenland. *Mertensia drummondii* is found in Alaska at Meade River and otherwise in Canada from just a few localities on the arctic mainland and Victoria Island. *Thlaspi arcticum* has been found in the foothills along the Sadlerochit River and at Prudhoe Bay in Alaska, and it is known from the nearby Canadian mainland and Herschel Island with populations disjunct in the St. Elias Mountains of southwestern Yukon Territory and British Columbia.

U.S. IBP Tundra Biome research dealt with a number of different tundra sites, but the intensive studies were limited to Barrow and Prudhoe Bay on the Arctic Coastal Plain. The coastal plain varies in width from about 160 km at Barrow in the west to 100 km at Prudhoe Bay and tapers to a mere 30 km at the eastern extreme in Alaska. The immediate coastal segment in which the IBP study areas are located fits very well Polunin's (1951) description of the Middle Arctic. However, the modest increase of elevation, gradient, and continentality of climate with distance from the coast is accompanied by changes in landscape, vegetation, and flora. Clebsch and Shanks (1968) reported an amelioration of climate and an increase of the shrubby component in the vegetation at Meade River, just 100 km southwest of Barrow and only 45 km from the coast. Drainage is better defined, topography more varied, and edaphic contrasts are strong. Erect shrubs occur in sheltered sites, and the appearance of heath and tussock tundra indicates the transition to Low Arctic conditions (cf. vegetation map by Komárková and Webber in Batzli and Brown, 1976). More recent climatic data from the Prudhoe Bay area show how dramatically monthly means and thawing degree-days differ at coastal and inland stations during the growing season (Brown et al., 1975).

Young (1971) devised a zonation of the arctic flora on the basis of climate and taxa present. His measure of accumulated summer warmth is the sum of the mean temperatures of all the months having a mean above 0°C. This is correlated with the northern limits of a number of widely distributed arctic taxa, presumably having ranges determined more by tolerance to environmental features of the region than historical factors. There is a close correspondence between the southern limit of the Arctic Tundra Subzone of Tundra (Alexandrova, 1970) and the southern limit of arctic floristic zone 2 (Young, 1971) (Figure 1). Both approaches to the major zonation place Barrow and Prudhoe in zones apart from each other and from the remainder of the coastal plain. Thus, there is on the northern coast of Alaska on the Beaufort Sea a narrow zone with conditions similar to expanses even further north in the Canadian Arctic and different, at least floristically, from lowland tundra of the Chukchi and Bering Sea regions. Biogeographic transitions on the Arctic Slope occur within only very short distances south from the coast.

Tundra Biome studies were conducted on plots selected from the moist to wet end of the soil moisture gradient, since those settings are so characteristic of Barrow. Research extended to Prudhoe Bay where comparable plots were examined has shown that differences in climate and soils of the two sites are manifested in flora and fauna.

The wet habitats at both localities have a vegetation rich in bryophytes and rhizomatous monocots and poor in species of dicots and lichens. Toward the drier end of the soil moisture spectrum there is a great diversity of species and life forms for all major plant groups, with a lower biomass overall, but a higher one for dicots and lichens. The strand, or littoral zone in the strict sense, has a small, predictable flora of ecological specialists for their respective sites. *Honckenya peploides* and *Mertensia maritima* occur scattered on otherwise barren sand spits disturbed by waves, whereas *Elymus mollis* is an important stabilizer of beaches and dunes shoreward. These three and the common *Stellaria humifusa* are widespread along Alaskan coastal lowlands. Mudflats, silt banks, and sandy beaches have mats of *Puccinellia phryganodes* and *Carex subspathacea*, which are essentially circumpolar species and extend from the Bering Sea north well into the Canadian Arctic Archipelago. Clumps and tufts of *Phippsia algida*, *Puccinellia andersonii*, *Carex ursina*, *Braya purpurascens*, *Cochlearia officinalis*, and *Potentilla pulchella* give beaches on the coastal plain a flora and vegetation typical of such beaches on arctic lowlands. The arctic aspect is heightened floristically by the local occurrence of *Colpodium vahlianum* and *Pleuropogon sabinei* (Halliday, 1977) otherwise very common farther north in the Archipelago.

Fresh water ponds are characteristically outlined by growths of *Arctophila fulva* in deeper water and mixtures of *Eriophorum angustifolium*, *Carex aquatilis*, and *Caltha palustris* in the shallows. Wet meadows with little topographic relief, even lacking features on the scale of ice-wedge polygon troughs and rims, are composed of *Alopecurus alpinus, Dupontia fisheri, Hierochloe pauciflora, Eriophorum angustifolium, E. scheuchzeri, Carex aquatilis,* and *Salix planifolia* ssp. *pulchra;* the first three taxa are essentially arctic in distribution, but the last four are also widespread in boreal regions and exemplify that portion of the boreal element that reaches the arctic and flourishes there.

The mesic to dry-mesic sites on the coastal plain have the greatest floristic diversity and an appreciable biomass of the common fruticose lichens (Williams et al., 1975; Williams et al., 1978), dicot herbs, and prostrate woody species. The number of possible combinations of taxa that form the vegetation make generalizations less trustworthy than they are for wet meadows. Some plant taxa show affinities only to the Arctic *(Salix ovalifolia* var. *arctolitoralis, Papaver hultenii, Braya pilosa,* and *Taraxacum phymatocarpum),* whereas many of the others are found in both arctic and alpine tundra in western North America; *Deschampsia cespitosa, Festuca brachyphylla, Poa arctica, Trisetum spicatum, Kobresia myosuroides, Juncus biglumis, Lloydia serotina, Polygonum viviparum, Oxyria digyna, Minuartia rubella, Saxifraga cernua, S. rivularis, S. hirculus, S. caespitosa, Androsace chamaejasme,* and *Artemisia arctica.*

There are marked differences in flora and vegetation between Barrow and Prudhoe Bay. The vascular flora of Barrow consists of about 125 taxa and that of Prudhoe Bay about 172 (Murray and Murray, this volume, Appendix I), however, only 88 species are common to both study areas. Given the large total range of many of the arctic species and the apparent broad ecological tolerances of quite a number of these, one might expect greater correspondence. Some of the difference may reflect what is implied in Young's (1971) classification of floristic zones. In his scheme, Barrow is placed in floristic zone 2 and Prudhoe Bay in zone 3 (Figure 1). About 20% of the taxa known at Prudhoe Bay but absent at Barrow are those which Young listed as usually having their northern limits in zone 3. But until the more exposed coastal areas near Prudhoe Bay, such as Point McIntyre, can be carefully collected, the climatic influence cannot be fully evaluated.

More obvious is the influence of major features of the landscape. Barrow, of all coastal points, is the farthest from the foothills and mountains. There are few large rivers, and those that do enter the ocean there, such as the Inaru River, have their headwaters on the coastal plain. Wet, acid peaty soils predominate, but stream erosion and frost action have exposed less acid and even slightly basic mineral soils in places. Prudhoe Bay is about 320 km east and experiences a more continental climate (Brown et al., 1975). Snowmelt is earlier and summer temperatures are higher than at Barrow. Prudhoe Bay is closer to the foothills and mountains, but of greater importance in determining local floristic patterns is the presence of two large rivers, the Kuparuk and Sagavanirktok, that head deep within the Brooks Range. Since limestone is a common rock type in the mountains at their headwaters, these rivers bring to the coastal plain calcareous alluvium. A glance at a map of the Prudhoe Bay area is sufficient to show the characteristic expanses of polygonal ground and numerous thaw-lakes consistent with coastal plain topography and not unlike Barrow. But the differences accrue due to microhabitats that lie on the mesic to xeric end of the moisture gradient and to soils so charged with carbonates that white crusts precipitate on the surface of rocks, pebbles, and aquatic mosses. The river systems at Prudhoe Bay provide, in addition to the calcareous soils, a number of distinct landscape units and habitats lacking or very restricted at Barrow (cf. Webber and Walker, 1975; and Webber, 1978).

The river terraces are well-drained substrates of varying textures from coarse gravels to fine silt. Overbank deposits of variable thicknesses mantle the gravels and influence drainage as intricately on the floodplain as does permafrost in adjacent areas that have not been recently reworked by fluvial processes. Deeply incised drainage channels and steep river banks are sites of snow accumulation, and snowbanks can persist well into July, thus accounting for snowbed plant assemblages. The broad floodplain with low terraces at modern river levels and the extensive gravel bars are less well vegetated than the terraces and adjacent meadows, but large stands of *Dryas integrifolia, Deschampsia cespitosa,* and *Carex aquatilis* occur on gravels, sands, and silts respectively. Terraces have a flora composed of alpine species associated with screes and fellfields high in the mountains or on coarse morainal substrates widespread on lower alpine slopes and valley bottoms: *Silene acaulis, Oxytropis borealis, O. deflexa* var. *foliolosa, O. nigrescens, Epilobium latifolium, Pedicularis lanata, Artemisia glomerata,* and *Senecio resedifolius.* Stabilized river banks and the sheltered slopes of pingos have a well-integrated turf of such species as *Carex scirpoidea, Lloydia serotina, Anemone parviflora, A. richardsonii, Parnassia kotzebuei, Oxytropis maydelliana, Pedicularis capitata, Artemisia arctica,* and *Saussurea angustifolia* which are not uncommon on mesic alpine uplands.

Sand dunes at the river mouths are frequently capped with a turf of *Elymus mollis, Trisetum spicatum,* and *Artemisia borealis* but are otherwise wind eroded on the sides. Prudhoe Bay also has a concentration of pingos, which are minor in terms of total area but very important in determining floristic diversity. The large pingos have a well-marked pattern of contrasting summits, slopes, and bases. The steep slopes with a southerly exposure are warm and dry and support such alpine taxa as *Lesquerella arctica, Saxifraga tricuspidata, Potentilla hookeriana, P. uniflora,* and *Oxytropis arctica.* The bright colors of the profusely blooming potentillas and the oxytrope are striking in contrast with the surrounding dull-toned sedge meadow.

The large rivers between the Canning and Colville Rivers and the extensive mudflats of their deltas provide waterfowl habitat uncommon to the west. Pitelka (1974) suggested that, at least in terms of avifauna, the portion of the Arctic Slope with large rivers constitutes a biogeographic sector. Evidence of biological discontinuities from the study of landform, soils, and flora suggests a wider application of this classification is possible. To the west of the Colville delta, rivers that head in the mountains do not flow to the coast but are intercepted by the Colville which in its upper reaches parallels the mountain ranges. East of the Colville, rivers do reach the coastal plain with their calcareous sediments and, I believe, also provide the migration route for alpine taxa out of the mountains through areas of extensive tussock tundra to special microhabitats on the arctic coast. The distance from the mountain slopes to coastal equivalents decreases sharply eastward as the northeastward trend of the Brooks Range brings alpine landscapes within only 25 km of the coast near Demarcation Point. Mountain plants such as *Corydalis pauciflora, Potentilla biflora,* and *Vaccinium uligi-nosum* (Murray, unpublished data) that are not found anywhere else on the coastal plain occur here within 1 to 2 km of the coast. Thus not only are there biogeographic subdivisions along a north–south transect from the coast to the mountains but also western, central, and eastern sectors within each of the subdivisions.

Acknowledgments. This research was supported by the National Science Foundation under Grant GV 29342 to the University of Alaska. It was performed under the joint sponsorship of the International Biological Programme and the Office of Polar Programs and was coordinated by the U.S. Tundra Biome. Field and laboratory activities at Barrow were supported by the Naval Arctic Research Laboratory of the Office of Naval Research.

I am grateful for valuable assistance from Drs. A. W. Johnson, H. M. Raup, W. C. Steere, L. L. Tieszen, and W. A. Weber who carefully read and commented upon drafts of this paper.

References

Ager, T. A. (1975) Late Quaternary environmental history of the Tanana Valley, Alaska. *Ohio State Univ., Inst. Polar Stud. Rep.* No. 54, 117 pp.

Alexandrova, V. D. (1970) The vegetation of the tundra zones in the USSR and data about its productivity. In *Proceedings of the Conference on Productivity and Conservation in Northern Circumpolar Lands* (W. A. Fuller and P. G. Kevan, Eds.). Int. Union Conserv. Natur., Morges, Switzerland, Publ. 16, N.S., pp. 93–114.

Arnold, C. A. (1959) Some paleobotanical aspects of tundra development. *Ecology,* **40**: 146–148.

Barry, R. G., and J. D. Ives. (1974) Introduction. In *Arctic and Alpine Environments* (J. D. Ives and R. G. Barry, Eds.). London: Methuen, pp. 1–13.

Batzli, G. O., and J. Brown. (1976) RATE—The influence of grazing on arctic tundra ecosystems. *Arct. Bull.,* **2**: 153–160.

Beschel, R. E. (1970) The diversity of tundra vegetation. In *Proceedings of the Conference on Productivity and Conservation in Northern Circumpolar Lands* (W. A. Fuller and P. G. Kevan, Eds.). Int. Union Conserv. Natur., Morges, Switzerland, Publ. 16, N.S., pp. 85–92.

Billings, W. D. (1973) Arctic and alpine vegetations: Similarities, differences and susceptibility to disturbance. *Bioscience,* **23**: 697–704.

Billings, W. D. (1974) Adaptations and origins of alpine plants. *Arct. Alp. Res.,* **6**: 129–142.

Böcher, T. W. (1954) Oceanic and continental vegetational complexes in southwest Greenland. *Medd. Grønland,* **149**: 1–336.

Britton, M. E. (1967) Vegetation of the arctic tundra. In *Arctic Biology: 18th Biology Colloquium* (H. P. Hansen, Ed.). Corvallis, Oregon: Oregon State University Press, 2nd ed. pp. 67–130.

Brown, J., R. K. Haugen, and S. Parrish. (1975) Selected climatic and soil thermal characteristics of the Prudhoe Bay region. In *Ecological Investigations of the Tundra Biome in the Prudhoe Bay Region, Alaska* (J. Brown, Ed.). Biol. Pap. Univ. Alaska, Spec. Rep. No. 2, pp. 2–11.

Brown, J., and P. V. Sellmann. (1973) Permafrost and coastal plain history of arctic Alaska. In *Alaskan Arctic Tundra* (M. E. Britton, Ed.). Arct. Inst. North Amer. Tech. Pap. No. 25, pp. 31–37.

Chaney, R. W. and H. L. Mason. (1936) A Pleistocene flora from Fairbanks, Alaska. *Amer. Mus. Novit.,* **8877**: 1–17.

Clebsch, E. E. C., and R. E. Shanks. (1968) Summer climatic gradients and vegetation near Barrow, Alaska. *Arctic,* **21**: 161–171.

Colinvaux, P. A. (1967) Quaternary vegetational history of arctic Alaska. In *The Bering Land Bridge* (D. M. Hopkins, Ed.). Stanford: Stanford University Press, pp. 207–231.

Dagon, R. R. (1966) Tundra—A definition and structural description. *Polar Notes,* **6**: 22–35.

Dorf, E. (1959) Climatic changes of the past and present. *Contrib. Mus. Paleontol.,* University of Michigan, **13**: 181–210.

Gray, A., and J. D. Hooker. (1880) The vegetation of the Rocky Mountain region. *Bull. U.S. Geol. Geogr. Surv. Territ.,* **6**: 1–77.

Griggs, R. F. (1934) The problem of arctic vegetation. *J. Wash. Acad. Sci.,* **24**: 153–175.

Guthrie, R. D. (1966) The extinct wapiti of Alaska and Yukon Territory. *Can. J. Zool.,* **44**: 47–57.

Guthrie, R. D. (1968) Paleoecology of the large-mammal community in interior Alaska during the late Pleistocene. *Amer. Midl. Nat.* **79**: 346–363.

Guthrie, R. D., and J. V. Matthews, Jr. (1971) The Cape Deceit fauna—Early Pleistocene mammalian assemblage from Alaskan arctic. *Quat. Res.,* **1**: 474–510.

Halliday, G. (1977) New and notable finds in the Alaskan vascular flora. *Can. Field-Natur.,* **91**: 319–322.

Hamilton, T. D., and S. C. Porter. (1975) Itkillik glaciation in the Brooks Range, northern Alaska. *Quat. Res.,* **5**: 471–497.

Holm, T. (1922) Contributions to the morphology, synonymy, and geographical distribution of arctic plants. *Reports of the Canadian Arctic Expedition 1913–18,* Vol. 5, part B, pp. 3–139.

Hooker, J. D. (1861) Outlines of the distribution of arctic plants. *Trans. Linnaean Soc. (London),* **23**: 251–348.

Hopkins, D. M. (1972) The paleogeography and climatic history of Beringia during the late Cenozoic time. *Inter-Nord,* **12**: 121–150.

Hopkins, D. M., J. V. Matthews, Jr., J. A. Wolfe, and M. L. Silberman. (1971) A Pliocene flora and insect fauna from the Bering Strait region. *Paleogeogr., Paleoclimatol. and Paleoecol.,* **9**: 211–231.

Hultén, E. (1958) *The Amphi-Atlantic Plants.* Stockholm: Almqvist and Wiksell, 340pp.

Hultén, E. (1968) *Flora of Alaska and Neighboring Territories.* Stanford: Stanford University Press, 1008 pp.

Johnson, A. W., and J. G. Packer. (1967) Distribution, ecology and cytology of the Ogotoruk Creek flora and the history of Beringia. In *The Bering Land Bridge* (D. M. Hopkins, Ed.). Stanford: Stanford University Press, pp. 245–265.

Johnson, P. L. and L. L. Tieszen. (1973) Vegetative research in arctic Alaska. In *Alaskan Arctic Tundra* (M. E. Britton, Ed.). Arct. Inst. North Amer. Tech. Pap. No. 25, pp. 169–198.

Küchler, A. W. (1965) *International Bibliography of Vegetation Maps,* Vol. 1. University of Kansas, Publ., Library Series 21, 453pp.

Küchler, A. W. (1966) Potential natural vegetation of Alaska. *National Atlas (Vegetation),* Sheet 89. U.S. Geological Survey, Washington, D.C.

Löve, Á. (1959) Problems of the Pleistocene and Arctic. *McGill Univ. Museum Publ.,* **1**: 82–95.

Löve, Á., and D. Löve. (1974) Origin and evolution of the arctic and alpine floras. In *Arctic and Alpine Environments* (J. D. Ives and R. G. Barry, Eds.). London: Methuen, pp. 571–603.

Matthews, J. V., Jr. (1970) Quaternary environmental history of interior Alaska: Pollen samples from organic colluvium and peats. *Arct. Alp. Res.,* **2**: 241–251.

Matthews, J. V., Jr. (1974a) A preliminary list of insect fossils from the Beaufort

Formation, Meighen Island, District of Franklin. *Geol. Surv. Can. Pap.*, **74-1A**: 203–206.

Matthews, J. V., Jr. (1974b) Wisconsin environment of interior Alaska: Pollen and macrofossil analysis of a 27-meter core from the Isabella Basin (Fairbanks, Alaska). *Can. J. Earth Sci.*, **11**: 828–841.

Matthews, J. V. Jr. (1974c) Quaternary environments at Cape Deceit (Seward Peninsula, Alaska): Evolution of a tundra ecosystem. *Geol. Soc. Amer. Bull.*, **85**: 1353–1384.

Matthews, J. V., Jr. (1976) Insect fossils from the Beaufort Formation: Geological and biological significance. *Geol. Surv. Can. Pap.*, **76-1B**: 217–227.

Packer, J. G. (1971) Endemism in the flora of western Canada. *Natur. Can.*, **98**: 131–144.

Payne, T. C., *et al.* (1951) *Geology of the Arctic Slope of Alaska.* Oil and Gas Invest. Map OM 126, U.S. Geological Survey, Washington, D.C.

Péwé, T. L. (1975) Quaternary geology of Alaska. *U.S. Geol. Surv. Prof. Pap.*, 835, 139 pp.

Pitelka, F. A. (1974) An avifaunal review for the Barrow region and North Slope of arctic Alaska. *Arct. Alp. Res.*, **6**: 161–184.

Polunin, N. (1951) The real Arctic: Suggestions for its delimitation, subdivision and characterization. *J. Ecol.*, **39**: 308–315.

Porsild, A. E. (1951) Plant life in the Arctic. *Can. Geogr. J.*, **42**: 120–145.

Raup, H. M. (1941) Botanical problems in boreal America. *Bot. Rev.*, 7: 147–248.

Raup, H. M. (1951) Vegetation and cryoplanation. *Ohio J. Sci.*, **51**: 105–116.

Raup, H. M. (1969) The relation of the vascular flora to some factors of site in the Mesters Vig District, Northeast Greenland. *Medd. Grønland*, **176**: 1–80.

Raup, H. M. (1975) Species versatility in shore habitats. *J. Arnold Arboretum*, **56**: 126–163.

Savile, D. B. O. (1960) Limitations of the competitive exclusion principle. *Science,* **132**: 1761.

Savile, D. B. O. (1972) *Arctic Adaptations in Plants.* Can. Dep. Agric. Monogr. 6, 81 pp.

Sellmann, P. V., and J. Brown. (1973) Stratigraphy and diagenesis of perennially frozen sediments in the Barrow, Alaska, region. In *Permafrost: North American Contribution to the Second International Conference.* Washington, D.C.: National Academy of Sciences, pp. 171–181.

Sigafoos, R. S. (1951) Soil instability in tundra vegetation. *Ohio J. Sci.*, **51**: 105–116.

Sigafoos, R. S. (1952) Frost action as a primary physical factor in tundra plant communities. *Ecology,* **33**: 480–487.

Spetzman, L. A. (1963) *Vegetation, Terrain Study of Alaska.* Part V. Military Branch, U.S. Geological Survey, Washington, D.C. (map).

Steere, W. C. (1976) Ecology, phytogeography and floristics of arctic Alaskan bryophytes. *J. Hattori Bot. Lab.*, **41**: 47–72.

Steere, W. C. (1978) Floristics, phytogeography, and ecology of arctic Alaskan bryophytes. In *Vegetation and Production Ecology of an Alaskan Arctic Tundra* (L. L. Tieszen, Ed.). New York: Springer-Verlag, Chap. 5.

Tolmatchev, A. I. (1960) Der Autochthone Grundstock der arktischen Flora und ihre Beziehungen zu den Hochgebirgsfloren Nord- und Zentralasiens. *Bot. Tidsskr.*, **55**: 269–276.

Viereck, L. A. (1971) Vegetation map of Alaska. In *Alaska Trees and Shrubs* (L. A. Viereck and E. L. Little, Jr., Eds.). Agricultural Handbook No. 410, Forest Service, U.S. Department of Agriculture (in pocket).

Wahrhaftig, C. (1965) Physiographic divisions of Alaska. *U.S. Geol. Surv. Prof. Pap.* 482, 52 pp.

Walker, H. F. (1973) Morphology of the North Slope. In *Alaskan Arctic Tundra* (M. E. Britton, Ed.). *Arct. Inst. North Amer. Tech. Pap.* 25, pp. 49–92.

Webber, P. J. (1974) Tundra primary productivity. In *Arctic and Alpine Environments* (J. D. Ives and R. G. Barry, Eds.). London: Methuen, pp. 445–473.

Webber, P. J. (1978) Spatial and temporal variation of the vegetation and its production, Barrow, Alaska. In *Vegetation and Production Ecology of an Alaskan Arctic Tundra* (L. L. Tieszen, Ed.). New York: Springer-Verlag, Chap. 3.

Webber, P. J., and D. A. Walker. (1975) Vegetation and landscape analysis at Prudhoe Bay, Alaska: A vegetation map of the Tundra Biome study area. In *Ecological Investigations of the Tundra Biome in the Prudhoe Bay Region, Alaska* (J. Brown, Ed.). Biol. Pap. Univ. Alaska, Spec. Rep. No. 2, pp. 81–91.

Weber, W. A. (1965) Plant geography in the southern Rocky Mountains. In *The Quaternary of the United States* (H. E. Wright and D. G. Frey, Eds.). Princeton: Princeton University Press, pp. 453–468.

Wiggins, I. L., and J. H. Thomas. (1962) *A Flora of the Alaskan Arctic Slope.* Arctic Institute of North America Spec. Publ. 4. Toronto: University of Toronto Press, 425 pp.

Williams, M. E., E. D. Rudolph, and E. A. Schofield. (1975) Selected data on lichens, mosses and vascular plants on the Prudhoe Bay tundra. In *Ecological Investigations of the Tundra Biome in the Prudhoe Bay Region, Alaska* (J. Brown, Ed.). Biol. Pap. Univ. Alaska, Spec. Rep. No. 2, pp. 213–215.

Williams, M. E., E. D. Rudolph, E. A. Schofield, and D. C. Prasher. (1978) The role of lichens in the structure, productivity and mineral cycling of the wet coastal Alaskan tundra. In *Vegetation and Production Ecology of an Alaskan Arctic Tundra* (L. L. Tieszen, Ed.). New York: Springer-Verlag, Chap. 7.

Wolfe, J. A. (1972) An interpretation of Alaskan Tertiary floras. In *Floristics and Paleofloristics of Asia and Eastern North America* (A. Graham, Ed.). New York: Elsevier Publishing Company, pp. 201–233.

Young, S. B. (1971) The vascular flora of St. Lawrence Island with special reference to floristic zonation in the arctic regions. *Contrib. Gray Herb.,* 201: 11–115.

Young, S. B. (1975) The phytogeography of the Northern American Arctic. Manuscript of paper presented at symposium on Floristic Delimitation and Subdivision of the Arctic. XII International Botanical Congress, Leningrad.

Yurtsev, B. A. (1972) Phytogeography of northeastern Asia and the problems of transberingian floristic interrelations. In *Floristics and Paleofloristics of Asia and Eastern North America* (A. Graham, Ed.). New York: Elsevier Publishing Company, pp. 19–54.

Yurtsev, B. A. (1973) The problems of the late Cenozoic paleogeography of Beringia as elucidated by the phytogeographic data. In *Theses of the Reports of All-Union Symposium on the Bering Land Bridge and Its Role for the History of Holarctic Floras and Faunas in the Late Cenozoic* (R. Ye. Giterman et al., Eds.). Khabarovsk, pp. 1–7.

3. Spatial and Temporal Variation of the Vegetation and Its Production, Barrow, Alaska

PATRICK J. WEBBER

Introduction and Site Overview

This chapter examines the factors which control the distribution of vegetation, growth-forms, species, broad productivity measures, seasonal patterns, and plant succession of the International Biological Programme (IBP) Tundra Biome site at Barrow, Alaska. It provides a framework against which the detailed physiological and modeling experiments of other production studies in this volume can be viewed. The pragmatic decisions of the program focused the experiments on a limited set of plant communities, growth-forms, and species. Since all vegetation and all species could not be modeled or studied to the depth desired, the project was designed to provide an ability to generalize spatially.

Systematic observations on the vegetation of the Arctic Coastal Plain have been made for the last 30 years (Wiggins, 1951; Spetzman, 1951, 1959; Churchill, 1955; Bliss and Cantlon, 1957; Britton, 1966; Koranda, 1954; Johnson et al., 1966). Only four pre-IBP era studies of primary production have been made (Thompson, 1955; Shanks in Bliss, 1962a; Pieper, 1963; Dennis, 1968), and all of these deal with Barrow. Since the start of the IBP program, several papers dealing briefly with vegetation composition and production in the Barrow area have appeared (Tieszen, 1972a and 1972b; Johnson and Tieszen, 1973; Smith, 1974; Neiland and Hok, 1975; Webber and Walker, 1975; Bunnell et al., 1975).

A general description of the environment, soils, geology, and history of the Barrow area was presented in Chapter 1. The IBP study sites (see Figures 2 and 3 in Chapter 1 of Tieszen, 1978) are approximately 110 ha in area and are characterized by shallow ponds, ice-wedge polygons, vertical relief less than a few meters, organic and acidic soils, and an active layer from 30 to 40 cm. The climatic data are summarized in Figure 1.

When compared with other regions of the Arctic Coastal Plain, the flora and number of plant communities of the Barrow area are impoverished (see Chapter 2, Murray, 1978; Chapter 5, Steere, 1978). This impoverishment is related to the combined effects of a severe climate and low habitat diversity. The immediate Barrow region is within the High Arctic Tundra Zone (*sensu* Webber, 1974) where there is an absence of erect shrubs on mesic sites; this is equivalent to the arctic subzone of the Tundra Zone (*sensu* Alexandrova, 1970; see Chapter 2,

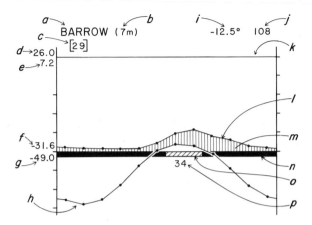

Figure 1.   Climatic diagram for Barrow, Alaska (after Walter and Lieth, 1967) on the
basis of data from the National Weather Service. Abscissa: months (Jan–Dec); ordinate:
one division = 10°C or 20-mm precipitation. a, station name; b, elevation in meters; c,
duration of observation in years; d, highest recorded temperature in °C; e, mean daily
maximum temperature of the warmest month; f, mean daily minimum temperature of the
coldest month; g, lowest recorded temperature; h, curve of mean monthly temperature; i,
mean annual temperature; j, mean annual precipitation in mm; k, line of 100-mm precipitation; l, curve of mean monthly precipitation; m, the relative humid season; n, months with
mean daily minimum temperature below 0°C; o, months with absolute minimum below
0°C; p, mean duration of frost-free period in days (after Walter, 1973).

Murray, 1978, for discussion). Wet, sedge meadows cover well over half of the
land surface (Spetzman, 1959) and about three-quarters of the IBP study area.
These meadows are dominated by one species, *Carex aquatilis,* and often have
only a few secondary species such as *Eriophorum angustifolium, Eriophorum
scheuchzeri,* and *Dupontia fisheri.* They also have a large moss component
consisting of species of *Calliergon* and *Drepanocladus.* Lichens are only a minor
component of these meadows.

The most striking feature of the vegetation of the IBP study site is that it
changes character every few meters in response to changes of microrelief and
drainage. The vegetation changes are not always reflected in the dominant
species, many of which, for example *Carex aquatilis,* are ubiquitous, but in the
subordinate species, especially the mosses and grasses.

Although much of the Barrow region is severely disturbed by construction and
off-road vehicle traffic, the IBP site is relatively undisturbed. The principal
disturbance has been the headward erosion by thermokarsting of Footprint
Creek which flows eastward between sites 1 and 2 (see Figure 3 in Chapter 1).
This erosion was initially triggered in the late 1940s by vehicle disturbance
(Lewellen, 1972). The drainage for this creek has also been affected by the
purposeful draining of Footprint Lake in 1950 which is 2 km to the south of the
IBP site (Lewellen, 1972). The drained Footprint Lake provides a fortuitous
analogue of a naturally drained *thaw lake* (Britton, 1966) basin. The entire IBP
site is a series of overlapping former thaw-lake basins. Britton (1966) has

described the dynamic aspects of the thaw-lake cycle and the development of the ice-wedge polygon landscape. In this cycle polygonization occurs in the meadows of former lake basins, the low-centered polygons coalesce to form ponds which further coalesce to form lakes. Drainage of these lakes allows the cycle to repeat. The time frame of this cycle is on the order of hundreds of years. Thus the vegetation and surfaces in the study are very dynamic. The tundra is not only influenced by geomorphic changes but also by herbivore activities. A spectacular disruption is commonly seen with a frequency of 3 to 5 years when lemming *(Lemmus sibericus)* densities become very high, and their grazing causes widespread destruction of the vegetation (Thompson, 1955; Pitelka, 1957; Pieper, 1963). This heavy grazing may have many subsequent effects on the vegetation resulting from increased nutrient release and increased thaw depth (Schultz, 1964, 1969).

Methods

Vegetation may be viewed either as a continuum or as a series of distinct plant communities (Major, 1961; Whittaker, 1967; Webber, 1971; Shimwell, 1972). The continuum is studied through *gradient analysis* (Whittaker, 1967) and the vegetation units through *classification*. Each set of techniques gives different and useful information about vegetation and they are used to complement one another in this study. Classification, in which plant communities are recognized and named, permits easier communication, mapping, and plot replication and experimentation. Gradient analysis permits a more powerful identification of environmental controls and provides an environmental framework for the description of species distribution and production.

Sampling began during 1972 and continued for four growing seasons. Forty-three permanent plots (1 × 10 m) were established to represent the spectrum of undisturbed vegetation and to allow measurements of seasonal variations of biotic and abiotic factors. This plot size approximates the minimal area of most tundra vegetation and the shape is compatible with most tundra patterns: for example, along polygon rims, troughs, and bog-strings (Webber, 1971). Species composition, cover and frequency were established for each plot and an importance value called species index was derived for each species by summing their relative cover and relative frequency values within the plot (Webber, 1971). Both vascular plants and cryptogams were sampled. Simple abiotic descriptions such as aspect, slope, drainage characteristics, and stability were recorded for each plot at the time of sampling. Soil samples were collected from just below the vegetation mat to a depth of 10 cm. A one-time record was made in mid-August 1973 of the degree to which the smell of hydrogen sulfide was noticeable in soil at spade depth; a four-point subjective scale based on odor was used for this purpose. Chemical and physical analysis of soils was performed in the laboratory using standard techniques. Percentages of sands, silts, and clays, percentage of organic content, water holding capacity, and hydrogen ion concentration (pH) were determined for each soil sample. Soluble soil phosphate was determined for a few selected soils. Gravimetric moisture was determined from two samples at a depth of 10 cm on each plot at intervals of 10 days during two growing seasons. Soil thaw surveys were conducted with a thin, graduated steel rod every 10 days. Snow depth, snow cover, and date of snowmelt surveys were conducted just prior to and at the start of each growing season. Soil thaw and snow depths were determined from 10 measurements in each plot.

Classification was achieved by clustering plots or stands on the basis of their species composition using the average linkage method of Sokal and Sneath (1963). The resulting clusters of stands have similar composition and were found to have a similar environment. These units represent broad vegetation types somewhat above the level of association (*sensu* Braun-Blanquet, 1932). Each cluster is called a *stand nodum* and is named on the basis of its species composition and habitat or physiognomic characteristics (Lambert and Dale, 1964; Webber, 1971). Nomenclature follows Murray and Murray (1978) in Appendix I except that *Salix pulchra* (= *Salix planifolia* ssp. *pulchra*) is retained in this volume to conform to recent use in the literature from the Alaskan arctic.

A map was made by outlining units of uniform landform assemblages. The vegetation map was made by coding these units on the basis of their most abundant nodum. This method solved the problem presented by the fine mosaic of geomorphic and vegetational patterns. A final scale of 1:5000 was found to be convenient.

Gradient analysis was achieved using the indirect stand ordination method of Bray and Curtis (1957). Ordinations based on stand species composition and subsequent correlation of the resulting ordination axes with measured stand environmental variables suggest and rank the major environmental controls of the vegetation. It is important to evaluate the use of the term environmental controls since a cause and effect relationship is not established because a correlation exists. However, in the present exercise those environmental factors which show high correlation with the ordination axes are hypothesized to be controlling. A further point of explanation with regard to ordination is required. The axes of the ordinations are best described as complex environmental gradients (*sensu* Whittaker, 1967). A complex gradient is a complex of interrelated and interdependent factors, some of which may be identified because they correlate with the axis while others may remain unknown because they have not been measured. Thus the factor with the highest correlation is unlikely to be the ultimate cause. An ordination is an expression of the variation within the sampled universe, here the Barrow tundra, and thus indicates the controls on the vegetation within the study site.

Since the early work of Raunkiaer (1934), the importance of plant growth forms as strategic adaptations to the tundra environment has always been emphasized (Bliss, 1956, 1962b; Tikhomirov, 1963; Chabot and Billings, 1972). The growth-forms shown in Figure 2 are based primarily on the nature of the shoot habit. Some categories, however, tend to be more systematic or phylogenetic in character, for example, *bryophytes* and *lichens*. Woody shrubs at Barrow are all of low stature and many are prostrate. Shrubs are subdivided on the basis of being evergreen or deciduous. The genus *Salix* is the principal

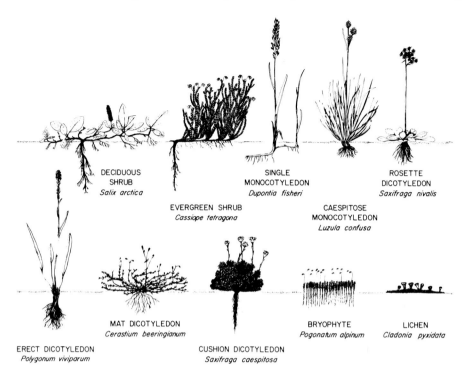

Figure 2.   The 10 plant-growth forms of tundra plants which are recognized in this study (after Wielgolaski and Webber, 1973).

representative of the *deciduous shrubs* while *Cassiope* and *Vaccinium vitis-idaea* represent *evergreen shrubs*. Herbaceous plants with elongated, narrow leaves are represented by the monocotyledonous sedges, grasses, and rushes. These graminoids may be subdivided into those with crowded, bunched shoots, here called *caespitose monocotyledons,* and those with well-spaced individual shoots, here called *single monocotyledons*. *Luzula confusa* and *Eriophorum vaginatum* are examples of caespitose monocotyledons and *Dupontia fisheri* and *Carex aquatilis* represent the single monocotyledons. Four growth-forms are recognized for the broad-leaved herbs or forbs. Acaulescent (essentially stem-less) plants with a rosette of radicle leaves are called *rosette dicotyledons*. *Saxifraga nivalis* and *Pedicularis lanata* are representative of this growth-form. Forbs with erect leaves or leaves supported into the canopy on long petioles or on an erect stem are called *erect dicotyledons*. Representatives of these are *Polygonum viviparum* and *Petasites frigidus*. The remaining forb growth-forms are *mat* and *cushion dicotyledons*. The former have tightly matted, often long, prostrate stems with leaves along the length of the stems. The latter have short crowded stems, often coming from a single tap root, which give a hemispheric shape. Examples of the mat form are *Cerastium jenisejense* and *Stellaria humifusa*. The cushion form is illustrated by *Saxifraga caespitosa* and *Silene acaulis*.

The spatial variation of standing crop and net primary production was established by harvesting two 0.1-m² (20 × 50-cm) quadrats from each plot during the period of peak aboveground vascular biomass. Standing crop refers to living plus dead plant material while biomass refers only to live material. Peak aboveground biomass usually occurs in the first week of August (Tieszen, 1972a). The herbage was clipped at soil level or just beneath any live moss carpet and the contents of each quadrat were kept separate and frozen for later processing. Two soil and root cores (7.6 cm in diameter) were taken to the maximum possible depth (usually 25 cm) in each quadrat. The soil cores were also frozen until they could be washed and sieved in order to retrieve the roots and other belowground organs. The thawed herbage was partitioned into vascular plants, lichens, and bryophytes. The vascular plants were further partitioned into 10 functional fractions. Current year's growth and previous standing material were separated into woody and herbaceous dead and live material. Stem bases and litter and prostrate dead comprised the remaining two fractions. Belowground material was separated into stem bases, rhizomes, live roots, and dead roots; the separation of roots into live and dead was made only on the basis of color and texture.

Estimates of net primary production were made only for the aboveground vascular plant portion of the vegetation. These estimates were based on the weight of the growth of the current season from the clipped material. These are, at best, crude because such small areas were clipped and no allowance was made for grazing, for shedding, for annual stem increments to perennial structures, or for current season die back of graminoid individuals. Turnover rates for various standing crop fractions were calculated by dividing the fraction in question by the estimate of its annual production.

Leaf area index (LAI) was estimated for each plot at the time of clip-harvesting. This includes petioles, stems, and floral structures in addition to blades proper. Sampling was done using a point-quadrat frame with 10 vertical pins spaced at regular intervals along a 1-m linear frame. Ten frames were sampled for each plot. Contacts were recorded for vascular plants, cryptogamic water, stones, bare ground, and lemming feces.

The temporal variation of standing crop and net primary production was only assessed for one vegetation type, a moist *Carex aquatilis*-dominated meadow. The course of standing crop and primary production for the 1971 growing season was compared with the peak season harvests of several years (1970–1974). Collection and analysis of the 1970 and 1971 data were supervised by Dr. John Dennis and results have been reported by Dennis and Tieszen (1972) and Dennis, et al. (1978).

A few measurements were made of the seasonal progression of plant stature of *Carex aquatilis* and the stem elongation of willow *(Salix pulchra)* at site 2. These were made in order to compare the year-to-year variation in plant growth. The plant stature measurements were made on 10 tagged individuals of *Carex aquatilis* at 5-day intervals throughout the growing seasons of 1972 and 1973. The willow measurements were derived from 10 plants harvested in the summer of 1974 and included measurements of stem increments for 1969 to 1973. Phenological studies of a few selected flowering plants were made in 1972 and 1973. These plants were *Carex aquatilis, Dupontia fisheri, Petasites frigidus, Ranunculus nivalis, Pedicularis lanata,* and *Salix pulchra.* These species were chosen to represent a spectrum from early to late bloomers. Each species was recorded in its characteristic nodum; the same individuals were observed in each year. Weekly observations were made from snowmelt to late summer.

Results

Vegetation Units

Identity and Distribution. Eight noda were established (Figure 3) from the 43 plots representing the universe of the sedge meadow complex of the IBP site. Each nodum was given a Roman numeral and a name based on its characteristic species, position on the moisture gradient, and habitat or physiognomy (Table 1). The numbers I to VII are in a general sequence of increasing substrate moisture regime. Nodum VIII represents pioneer vegetation which occurs on recent alluvium. Figures 4–7 illustrate some of the noda. The units of a landform map of the study area were coded for their most abundant nodal vegetation cover (Figure 8). A planimetry estimate of the area occupied by each nodum is given in Table 1. A larger scale, more detailed version of this map is available elsewhere (Walker, 1977). A further picture of the spatial distribution of the noda is given by an idealized profile of the major landforms of the study area (Figure 9). The mean of Species Index, the total number of species in each of the major taxonomic entities, the mean relative cover of each growth-form, and some means of selected substrate variables for each nodum are given in Tables 2, 3, 4, and 5, respectively.

Vegetation dominated by *Carex aquatilis* forms most of the surface cover of the Barrow area. Noda III, IV, and VI, which are dominated by *Carex aquatilis,* occupy 77% of the mapped area. Although the noda seem fairly distinct in the dendrogram (Figure 3), it is difficult to single out clearly diagnostic or character-istic species for each nodum. Few species are unique to a nodum and most occur in several noda. The following descriptions point out the principal characteristics of each nodum.

Nodum I, which is called *Dry, Luzula confusa Heath,* is characteristic of high-centered polygons just to the south of site 1 (Figures 4 and 8). The use of the epithet heath refers to the barren steppe appearance of this vegetation (*sensu* Rübel, 1914) rather than the presence of ericaceous species. Ericaceous species such as *Cassiope tetragona* and *Vaccinium vitis-idaea* are rare at Barrow and are restricted to the flanks of the highest ridges such as raised beaches. This community occupies only 3% of the study area. Its barrenness is due to its dryness and elevation above the water table, low snow cover, and exposure to winds. The characteristic growth-forms are caespitose monocotyledons and

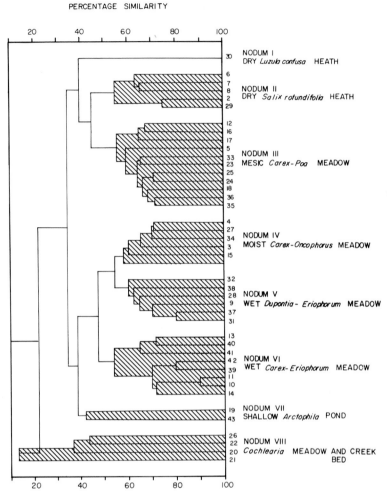

Figure 3. Dendrogram showing the similarity relationships between the 43 sampled vegetation stands. Eight clusters or noda are identified and have been named on the basis of habitat and characteristic species.

lichens (Table 5). In addition to *Luzula confusa* the following are characteristic species: *Potentilla hyparctica, Alectoria nigricans,* and *Pogonatum alpinum.*

Nodum II, *Mesic,* Salix rotundifolia *Heath,* is characteristic of the sloping creek banks between sites 1 and 2 (Figure 8). More abundant than Nodum I, but still small in extent (7%), this vegetation also occurs in the center of some low-centered polygons which drain readily. *Salix rotundifolia,* which is a prostrate deciduous shrub, characterizes this vegetation as do *Arctagrostis latifolia* and *Saxifraga punctata.* This nodum has the highest number of herbaceous dicotyledons (22) and the highest overall number of species (70) of any of the noda of the study area. The well-drained, sandy nature of the soils of this community leads to the deepest active layer of the principal noda.

Table 1. Numbers, Names, Characteristic Species, and Major Microrelief Types of the Eight Vegetation Noda Recognized for the Barrow IBP Site

Nodum		Characteristic species	Major microrelief type(s)
I.	Dry *Luzula confusa* heath	*Luzula confusa, Potentilla hyparctica, Alectoria nigricans, Pogonatum alpinum,* and *Psilopilum cavifolium*	High-centered-polygons
II.	Mesic *Salix rotundifolia* heath	*Salix rotundifolia, Arctagrostis latifolia, Saxifraga punctata, Sphaerophorus globosus,* and *Brachythecium salebrosum*	Low-centered-polygons and sloping creek banks
III.	Mesic *Carex aquatilis–Poa arctica* meadow	*Carex aquatilis, Poa arctica, Luzula arctica, Cetraria richardsonii,* and *Pogonatum alpinum*	Hummocky polygon rims and centers and dry, flat polygonized sites
IV.	Moist *Carex aquatilis–Oncophorus wahlenbergii* meadow	*Carex aquatilis, Oncophorus wahlenbergii, Dupontia fisheri, Peltigera aphthosa,* and *Aulacomnium turgidum*	Moist, flat sites and drained polygon troughs
V.	Wet *Dupontia fisheri–Eriophorum angustifolium* meadow	*Dupontia fisheri, Eriophorum angustifolium, Cerastium jenisejense, Peltigera canina,* and *Campylium stellatum*	Wet, flat sites and polygon troughs
VI.	Wet *Carex aquatilis–Eriophorum russeolum* meadow	*Carex aquatilis, Eriophorum russeolum, Saxifraga foliolosa, Calliergon sarmentosum,* and *Drepanocladus brevifolius*	Low polygon center and pond margins
VII.	*Arctophila fulva* pond margin	*Arctophila fulva, Ranunculus pallasii, Ranunculus gmelini, Eriophorum russeolum,* and *Calliergon giganteum*	Pond and stream margins
VIII.	*Cochlearia officinalis* pioneer meadow	*Cochlearia officinalis, Phippsia algida, Ranunculus pygmaeus, Stellaria humifusa,* and *Saxifraga rivularis*	Snowbeds, creek banks and creek sides

Figure 4. Nodum I—*Dry Luzula confusa heath* situated on the higher center of a tundra polygon at site 1. White crust lichens *(Ochrolechia frigida)* are visible on the sides of the raised center. The trough in the left foreground contains vegetation of Nodum V.

Figure 5. Nodum V—*Wet Dupontia fisheri–Eriophorum angustifolium meadow.* This nodum is often found in wet polygon troughs, but it can form extensive meadows in suitably wet lowlands. Nodum IV occupies the polygon centers of this figure.

Figure 6. An extensive stand representing Nodum V. The white inflorescences of *Eriophorum russeolum* appear at the center of the figure.

Figure 7. Nodum VI—*Wet Carex aquatilis–Eriophorum russeolum meadow* in low center polygon. This is the barren extreme of this nodum and has a characteristic dark appearance.

47

Nodum III, *Mesic, Carex aquatilis–Poa arctica Meadow,* is the most extensive (41%) vegetation type of the study area. Bryophytes, primarily mosses, and lichens have greater relative covers (Table 5) than the more conspicuous graminoids. *Dicranum elongatum* and *Dactylina arctica* are a common moss and lichen, respectively. More characteristic of the nodum are the moss *Pogonatum alpinum* and the lichen *Cetraria richardsonii.* Noda IV, V, and VI all have the same superficial meadow physiognomy. The critical difference which separated Nodum III from them is the abundance of *Poa arctica,* which is very prominant when it is in anthesis, and the highest lichen diversity of any of the noda. Nodum III is best developed in the parts of the polygonized sedge meadow complex where it is fairly dry and where large, flat, and sometimes slightly raised polygon centers with barely discernible troughs and rims occur. It also occurs commonly on hummocky polygon rims in wetter areas with more pronounced polygon development as at site 4 (Figures 8 and 9).

Nodum IV, *Moist Carex aquatilis–Oncophorus wahlenbergii Meadow,* the second most extensive nodum (21%), develops on similar but moister sites than Nodum III and where thaw depth is slightly less. Most surfaces of sites 1 and 2 are classified in this nodum. It occurs on flat polygon centers and in drained, shallow polygon troughs. It appears very much like Nodum III but may be distinguished from it on the basis of less lichen cover and the presence of mosses such as *Calliergon sarmentosum* and *Oncophorus wahlenbergii.* The presence of *Dupontia fisheri, Peltigera aphthosa,* and *Aulacomnium turgidum* also distinguishes Nodum IV from III. Nodum IV has the most species of monocotyledons and it is the most intensively studied vegetation of the Tundra Biome program.

Nodum V, *Wet Dupontia fisheri–Eriophorum angustifolium Meadow,* occupies only 7% of the mapped area. It is characteristic of flat, slowly draining sites and wet polygon troughs. *Eriophorum russeolum* is also common in this nodum (Figure 6), but the low cover of *Carex aquatilis* and the greater abundance of *Dupontia* and *Eriophorum angustifolium* distinguish it from the other meadow noda. Woody dicotyledons are essentially absent but several herbaceous dicotyledons occur and *Cerastium jenisejense, Stellaria edwardsii,* and *Stellaria laeta* may form distinctive mats; *Saxifraga cernua* is a frequent erect forb.

The vegetation which is placed in Nodum VI is variable and is called *Wet Carex aquatilis–Eriophorum russeolum Meadow.* It includes the sparse vegetation of low-centered polygons (Figure 7) which contrasts with that of some pond margins, but there is an underlying floristic commonness between these examples and frequent intergrading stands. It has no woody species and only a few lichens. *Saxifraga foliolosa* is characteristic of the low-centered polygons while *Calliergon sarmentosum* is characteristic of the pond margin meadows. *Drepanocladus brevifolius* is common to all facies of this nodum. A thick organic mat, highly organic soils, and low species diversity also set this nodum apart from the other meadow noda. Spring snow cover on this nodum is moderately deep and late to recede.

Nodum VII is called *Arctophila fulva Pond Margin.* It is also found in slow-flowing streams; it is the most distinct of the principal noda, but it only occupies 2% of the area. Woody plants and lichens are totally absent from this nodum and

VEGETATION

U. S. Tundra Biome Site, Barrow, Alaska

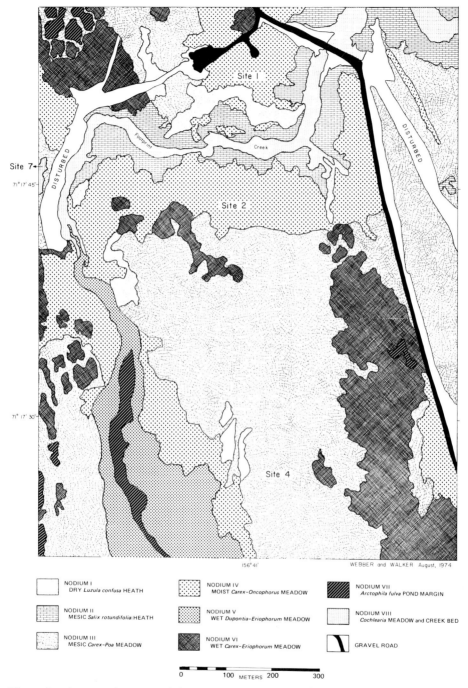

NODIUM I DRY *Luzula confusa* HEATH	NODIUM IV MOIST *Carex-Oncophorus* MEADOW	NODIUM VII *Arctophila fulva* POND MARGIN
NODIUM II MESIC *Salix rotundifolia* HEATH	NODIUM V WET *Dupontia-Eriophorum* MEADOW	NODIUM VIII *Cochlearia* MEADOW and CREEK BED
NODIUM III MESIC *Carex-Poa* MEADOW	NODIUM VI WET *Carex-Eriophorum* MEADOW	GRAVEL ROAD

0   100   METERS 200   300

Figure 8.   A vegetation map of the area surrounding the U.S. Tundra Biome terrestrial sites at Barrow, Alaska. In this map, land-units have been coded according to their most abundant vegetation nodum.

49

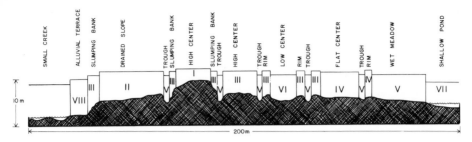

Figure 9.   Idealized profile of the Intensive Study Site showing the principal microrelief features and the distribution of the vegetation noda.

the single monocotyledons, erect herbaceous dicotyledons, and bryophytes have an equal importance, although graminoid growth-form represented by the emergent *Archtophila* is most characteristic. *Ranunculus pallasii* and *Caltha palustris* are occasionally present. The pleurocarpous mosses *Calliergon giganteum* and *Drepanocladus brevifolius* are the principal bryophytes. Late in the growing season the substrate of this nodum may occasionally become noticeably anaerobic and have a characteristic odor of hydrogen sulfide.

Nodum VIII, *Cochlearia officinalis Pioneer Meadow,* is a rudimentary community which develops on recent alluvium in the creek bed which crosses the study area (Figure 8). Frequently these sites are in the shelter of the creek bank where snow may accumulate and be late to melt. Its sandy, moist alluvium thaws very deeply and the active layer may exceed 1 m in late summer. This nodum is floristically rich, but, as Figure 3 illustrates, the sampled plots comprising the nodum are extremely variable. *Cochlearia officinalis, Phippsia algida,* and *Stellaria humifusa* are the principal species.

Standing Crop and Production.   Each nodum has a reasonably distinct standing crop composition at the period of peak aboveground biomass (Table 6). As a generalization the following aboveground pattern is seen for the dry weight (g m^{-2}) of the various fractions for Barrow tundra vegetation at the period of peak season: herbaceous dicotyledon biomass (4) < woody dicotyledon biomass (9) < lichen biomass (17) < monocotyledon biomass (35) < vascular standing dead (40) < bryophyte biomass (50) < total vascular biomass (50) < vascular litter and prostrate dead (80). If belowground standing crops are now considered, the sequence essentially continues: vascular stem base biomass (65) < rhizome biomass (95) < root biomass (560) < total belowground biomass (735) < belowground dead (2565).

Monocotyledon aboveground biomass increases from Nodum II *(mesic heath)* along the moisture gradient to Nodum VII *(pond margin).* Nodum I *(dry heath)* has a higher monocotyledon biomass than Nodum II *(mesic heath)* as a result of caespitose graminoids such as *Luzula confusa* which are abundant on dry sites. In all noda except Noda I and II *(heaths),* which are dry, the standing dead aboveground monocotyledon fraction is less than the live fraction at it is readily incorporated into litter and prostrate dead fractions. The same trend

Table 2. A Complete List of the Species Occurring in the 43 Sampled Stands and Their Nodal Occurrence[a]

Taxon	Growth-form	Vegetation							
		I(1)	II(5)	III(11)	IV(6)	V(6)	VI(8)	VII(2)	VIII(4)
Shrubs									
Cassiope tetragona	ES	—	5.6	—	—	—	—	—	—
Dryas integrifolia	DS	—	1.0	—	—	—	—	—	—
Salix phlebophylla	DS	—	0.1	—	—	—	—	—	—
Salix pulchra	DS	—	0.9	3.0	2.6	—	—	—	—
Salix rotundifolia	DS	10.5	33.2	6.3	2.7	0.6	—	—	0.4
Vaccinium vitis-idaea	ES	—	—	—	—	—	—	—	—
Herbaceous dicotyledons									
Caltha palustris	ED	—	—	—	1.7	4.7	1.3	3.3	1.9
Cardamine pratensis	ED	—	—	—	0.3	4.8	—	0.4	8.7
Cerastium jenisejense	MD	—	—	—	0.4	0.2	—	1.9	0.5
Chrysosplenium tetrandrum	ED	—	—	0.1	0.1	—	—	—	7.0
Cochlearia officinalis	RD	—	—	—	—	—	—	—	—
Draba bellii	RD	—	0.5	—	—	—	—	—	—
Draba oblongata	RD	0.8	2.2	—	0.1	—	—	—	—
Eutrema edwardsii	ED	—	2.2	—	—	—	—	—	—
Papaver hulténii	RD	3.3	1.5	—	—	—	—	—	—
Papaver macounii	RD	—	1.1	—	—	—	—	—	—
Pedicularis lanata	ED	4.6	3.4	0.4	—	—	—	—	—
Pedicularis sudetica	ED	—	—	—	—	—	0.7	—	8.6
Petasites frigidus	ED	—	0.1	1.2	4.6	5.2	—	—	1.3
Polygonum viviparum	ED	—	1.5	—	0.8	—	—	—	0.2
Potentilla hyparctica	ED	16.1	2.0	—	—	—	—	—	—
Ranunculus gmelini	SG	—	—	—	—	—	—	0.4	0.3
Ranunculus nivalis	ED	0.8	0.9	1.6	1.6	2.5	—	—	—
Ranunculus pallasii	ED	—	—	—	—	0.9	—	61.0	—
Ranunculus pygmaeus	ED	—	—	—	—	—	—	—	4.2
Rumex arcticus	ED	—	1.5	—	—	—	—	—	—
Sagina intermedia	RD	—	0.2	—	—	—	—	··	—
Saxifraga caespitosa	CD	—	0.6	—	—	—	—	—	—

Table 2. A Complete List of the Species Occurring in the 43 Sampled Stands and Their Nodal Occurrence[a] (*Continued*)

Taxon	Growth-form	Vegetation							
		I(1)	II(5)	III(11)	IV(6)	V(6)	VI(8)	VII(2)	VIII(4)
Saxifraga cernua	ED	0.8	1.2	5.2	6.7	10.5	5.5	—	9.9
Saxifraga foliolosa	RD	—	0.5	2.6	4.6	2.8	6.3	—	1.5
Saxifraga hieracifolia	MD	—	0.3	0.3	1.2	0.3	—	—	—
Saxifraga hirculus	ED	—	—	—	—	—	—	—	0.3
Saxifraga oppositifolia	MD	—	0.1	0.1	—	—	—	—	—
Saxifraga punctata	ED	—	4.8	—	1.2	0.1	—	—	—
Saxifraga rivularis	ED	—	—	—	—	—	—	—	3.1
Senecio atropurpureus	ED	—	4.4	0.9	—	—	—	—	—
Stellaria edwardsii	MD	5.4	4.1	1.4	3.4	4.3	0.4	—	1.9
Stellaria humifusa	MD	—	—	—	—	—	—	—	12.8
Stellaria laeta	MD	—	0.3	1.0	1.6	3.1	0.7	—	3.1
Monocotyledons									
Alopecurus alpinus	SG	—	1.5	0.5	—	0.5	0.6	—	2.5
Arctagrostis latifolia	SG	—	7.4	2.4	0.1	0.1	—	—	7.1
Arctophila fulva	SG	—	—	—	—	—	0.3	32.0	9.0
Calamagrostis holmii	SG	1.6	1.6	3.2	1.8	4.2	0.6	—	0.5
Carex aquatilis	SG	—	0.1	16.6	23.7	4.2	46.2	10.6	17.4
Dupontia fisheri	SF	—	—	5.2	8.3	18.3	7.7	10.0	—
Eriophorum angustifolium	SG	—	1.2	4.9	4.5	14.1	6.0	10.0	0.5
Eriophorum russeolum	SG	—	.2	4.8	8.0	8.7	17.3	7.3	—
Eriophorum scheuchzeri	SG	—	—	—	1.5	0.9	—	—	—
Festuca brachyphylla	CG	4.0	—	—	—	0.1	—	—	—
Hierochloë pauciflora	SG	—	0.1	—	0.9	—	—	—	—
Juncus biglumis	CG	—	0.4	—	3.4	0.5	0.8	—	—
Luzula arctica	CG	6.2	3.2	3.7	0.8	—	0.4	—	—
Luzula confusa	CG	20.9	5.0	0.6	1.3	—	—	—	0.3
Phippsia algida	CG	—	—	—	—	—	—	—	5.6
Poa arctica	SG	6.4	3.6	11.4	4.0	5.4	1.8	—	4.8

Bryophytes									
Aulacomnium palustre	BR	—	0.2	4.4	2.7	2.1	2.0	—	1.8
Aulacomnium turgidum	BR	—	3.0	3.3	1.4	—	0.2	—	—
Brachythecium cf. *turgidum*	BR	2.0	11.6	0.7	0.7	9.1	10.8	7.3	1.5
Bryum spp.	BR	—	3.6	11.8	14.5	24.5	4.8	28.5	27.4
Calliergon giganteum	BR	—	—	—	4.3	—	34.4	—	—
Calliergon sarmentosum	BR	—	0.2	0.6	12.4	15.7	—	13.6	7.5
Campylium stellatum	BR	—	0.3	—	0.2	4.3	0.2	—	7.4
Conostomum tetragonum	BR	9.2	16.8	0.3	0.5	—	—	—	—
Dicranum elongatum	BR	—	3.5	24.5	—	0.6	0.2	—	4.8
Distichium capillaceum	BR	—	—	0.3	7.4	—	—	—	—
Drepanocladus brevifolius	BR	—	6.1	0.8	2.1	7.2	28.2	13.6	6.5
Drepanocladus uncinatus	BR	—	0.2	2.5	2.7	—	0.2	—	0.5
Hepaticae	BR	—	3.0	4.4	2.3	2.1	2.0	—	1.8
Hylocomium alaskanum	BR	—	0.6	4.5	1.6	—	—	—	—
Myurella julacea	BR	—	1.6	0.7	18.9	—	4.9	—	1.0
Oncophorus wahlenbergii	BR	42.2	3.9	7.5	19.2	7.3	8.4	—	2.1
Pogonatum alpinum	BR	18.9	4.5	13.6	4.6	16.5	0.2	—	—
Polytrichum juniperinum	BR	2.0	—	4.7	—	3.9	0.3	—	—
Psilopilum cavifolium	BR	—	3.9	—	0.5	—	—	—	—
Rhacomitrium spp.	BR	—	—	3.0	0.4	—	—	—	—
Sphagnum spp.	BR	—	0.5	0.3	0.6	—	—	—	—
Tetraplodon mnioides	BR	—	—	—	—	—	—	—	—
Lichens									
Alectoria nigricans	LI	10.9	3.0	1.2	—	—	—	—	1.7
Cetraria andrejevii	LI	—	—	—	—	1.0	—	—	0.3
Cetraria cucullata	LI	7.6	4.5	4.0	0.5	0.1	0.4	—	1.5
Cetraria delisei	LI	0.8	0.1	0.3	0.2	0.1	—	—	0.2
Cetraria islandica	LI	3.7	4.9	6.1	3.3	1.1	1.8	—	2.6
Cetraria nivalis	LI	0.8	—	0.1	0.1	—	—	—	0.2
Cetraria richardsonii	LI	—	0.1	1.5	0.3	0.1	—	—	—
Cetraria simmonsii	LI	—	—	—	—	—	—	—	0.2

Table 2. A Complete List of the Species Occurring in the 43 Sampled Stands and Their Nodal Occurrence[a] (Continued)

Taxon	Growth-form	Vegetation							
		I(1)	II(5)	III(11)	IV(6)	V(6)	VI(8)	VII(2)	VIII(4)
Cladonia amaurocraea	LI	—	—	1.9	—	—	—	—	1.6
Cladonia bellidiflora	LI	—	—	0.3	0.1	—	—	—	—
Cladonia carneola	LI	—	0.1	0.2	—	0.1	—	—	0.2
Cladonia cenotea	LI	—	—	—	—	—	—	—	—
Cladonia chlorophaea	LI	0.8	1.4	1.6	—	—	—	—	1.5
Cladonia cryptochlorophaea	LI	5.5	0.3	—	—	—	—	—	0.2
Cladonia mitis	LI	—	—	0.2	—	—	—	—	—
Cladonia pleurota	LI	0.8	—	0.2	0.3	—	—	—	—
Cladonia pyxidata	LI	—	—	0.1	0.1	—	—	—	0.2
Cladonia uncialis	LI	—	1.7	2.2	—	0.1	—	—	0.3
Cornicularia divergens	LI	—	4.4	2.1	2.1	1.3	2.8	—	3.5
Dactylina arctica	LI	4.7	5.6	8.1	0.1	—	—	—	—
Lobaria linita	LI	—	0.2	0.2	—	—	—	—	—
Peltigera aphthosa	LI	0.8	3.0	1.7	2.9	3.1	1.8	—	—
Peltigera canina	LI	—	0.2	1.4	1.1	3.6	0.2	—	—
Peltigera membranacea	LI	—	01.	—	—	0.1	—	—	—
Pseudoevernia consocians	LI	—	0.3	—	—	—	—	—	7.2
Solorina crocea	LI	—	0.1	—	—	—	—	—	—
Sphaerophorus globosus	LI	—	2.4	1.4	0.1	—	—	—	1.7
Stereocaulon cf. tomentosum	LI	—	1.5	0.1	—	—	—	—	1.0
Thamnolia vermicularis	LI	8.3	4.3	5.0	0.8	—	1.4	—	1.8

[a] The mean nodal values of Species Index are given. The number of stands per nodum is given in parentheses after the nodal number. The species are listed within the major groups and alphabetically arranged within these groups. Symbols are as follows: DS, deciduous shrub; ES, evergreen shrub; SM, single monocotyledon; CC, caespitose monocotyledon; RD, rosette dicotyledon; ED, erect dicotyledon; CD, cushion dicotyledon; MD, mat dicotyledon; BR, bryophyte; LI, lichen.

Table 3.   The Total Number of Sampled Species in Each Growth-Form for the
Vegetation Noda at Barrow

Growth-form	Vegetation noda								Total
	I	II	III	IV	V	VI	VII	VIII	
Woody dicotyledon	1	4	2	2	1	0	0	1	6
Herbaceous dicotyledon	7	22	11	14	12	6	5	16	33
Monocotyledon	5	11	10	12	11	10	5	9	16
Bryophyte	5	17	19	19	10	12	4	10	28
Lichen	11	16	25	15	10	6	0	19	29
Total	29	70	67	62	44	34	14	55	112

exists for the herbaceous dicotyledon standing dead material, and these patterns
are interpreted as the result of slower decay rates on dry sites and the ease with
which standing dead material gets incorporated into the prostrate dead and litter
fraction on wetter sites. Herbaceous dicotyledon production is uniformly low in
all noda except Nodum VIII *(pioneer meadow)*, but even here it does not exceed
the production by monocotyledons. In only one nodum, Nodum II *(mesic
heath)*, does the dicotyledon production exceed that of monocotyledons; here
both herbaceous and woody dicotyledon productions are larger. Bryophyte
biomass is highest in Nodum III *(mesic meadow)*, where many acrocarpous
mosses such as *Dicranum* and *Pogonatum* are abundant; however, bryophyte
biomass is fairly high in all noda. In all but the pioneer vegetation (Nodum VIII),
the dry weight of dead belowground material is several times that of live
belowground material. Presumably, the dead fraction is still accumulating in the
pioneer sediments.

The belowground stem base component is variable from nodum to nodum. It
is low to absent in those noda with little or no organic mat (Nodum II, *mesic
heath;* Nodum VIII, *pioneer meadow*) and also in the pond vegetation (Nodum
VII). In the sequence from Nodum I to Nodum V there is a general trend of
increasing rhizome and total belowground biomass. From Nodum V *(wet
Dupontia meadow)* through Nodum VI *(wet Carex meadow)* to Nodum VII
(pond margin) the rhizome and total belowground biomass decreases.
Belowground biomass is least in the *pioneer meadow* (VIII). The ratio of above-
to belowground biomass does not change in such a clear fashion as that for
belowground biomass alone. The ratio varies from a mean of 1:3.2 in the pioneer
Nodum VIII to a mean value of 1:29.6 in Nodum I. The overall average ratio is
1:16.3 while that based on an average value for aboveground and belowground
biomasses is 1:13.5.

Net aboveground production of vascular plants for the 1972 growing season
ranged from 18.1 g m^{-2} yr^{-1} to 118.5 g m^{-2} yr^{-1} with a mean of 48.1 g m^{-2} yr^{-1}.
Vascular production is positively correlated with site moisture regime and it
essentially increases through the nodal sequence I through VII. *Dry heaths* (I)
have a mean value of 18.1 g m^{-2} yr^{-1} and *pond margins* (VII) have a mean value
of 115.0 g m^{-2} yr^{-1}. Pioneer vegetation is variable from a low of 20.7 g m^{-2} yr^{-1} to
a high of 61.8 g m^{-2} yr^{-1}. Nodum VI, *moist Carex–Eriophorum meadow*, with a

Table 4. The Distribution of the Major Plant Growth-Form Types (Wielgolaski and Webber, 1973) in the Barrow Vegetation Noda[a]

Growth form	Examples	Vegetation noda							
		I	II	III	IV	V	VI	VII	VIII
Deciduous shrub	*Salix rotundifolia*	5.0	*16.5*	3.2	1.4	0.4	—	—	—
Evergreen shrub	*Cassiope tetragona*	—	3.3	1.5	1.3	—	—	—	0.2
Caespitose monocotyledon	*Luzula confusa*	*15.1*	4.4	2.1	2.8	0.4	0.6	—	2.8
Single monocotyledon	*Carex aquatilis*	4.0	8.1	*16.7*	*26.5*	*28.4*	*40.2*	*35.5*	20.8
Cushion dicotyledon	*Saxifraga caespitosa*	—	0.8	—	—	—	—	—	—
Mat-forming dicotyledon	*Cerastium beeringianum*	2.3	2.3	1.4	3.4	6.4	0.5	1.0	*12.9*
Rosette dicotyledon	*Saxifraga nivalis*	2.0	3.0	1.3	2.5	1.5	3.2	—	*4.3*
Erect dicotyledon	*Saxifraga cernua*	11.3	11.1	4.9	8.8	*12.1*	3.8	*32.3*	*15.2*
Bryophyte	*Pogonatum alpinum*	*37.0*	31.7	*45.3*	46.2	45.4	*47.4*	*31.2*	30.2
Lichen	*Cetraria islandica*	23.3	18.8	*23.6*	7.1	5.4	4.3	—	13.6

[a] Values are mean relative cover. For each nodum the value(s) of the characteristic growth-form or combination of growth-forms is (are) in italics.

Table 5. Some Selected Substrate Factors for the Eight Stand or Vegetation Noda
from the Barrow Intensive Site[a]

Substrate factors	Vegetation noda							
	I(1)	II(5)	III(11)	IV(6)	V(6)	VI(8)	VII(2)	VIII(4)
Soil moisture (%) July 29, 1973	77	64	183	287	299	379	500	42
Snow depth (cm) June 6, 1973	0	4	21	29	43	36	28	86
Thaw depth (cm) August 17, 1973	37	54	36	32	32	32	37	76
Soil odor (scaled) August 17, 1973	0	0.4	0.3	2.2	1.7	1.7	2.0	0
Sand (%)	13	34	28	14	9	29	5	75
Soil reaction (pH)	3.9	5.3	4.2	4.1	4.4	4.2	4.5	5.1
Organic matter (%)	24	12	38	61	55	73	53	5
Organic horizon (cm)	4.4	3.7	6.5	11.8	9.5	18.2	2.0	1.2
Soluble phosphate (mg gdw^{-1} soil)[b]	15.5	21.5	18.5	10.1	59.8	4.5	41.2	6.0

[a] Values are means from the plots within each nodum. The number of stands per nodum is given
in parentheses after the nodal number.
[b] Based on only two samples per nodum (except Nodum I with only one sample).

mean value of 43.9 g m^{-2} yr^{-1}, also does not fit the sequence. The variability of
this nodum has already been noted. The low-centered polygons of this nodum
may only produce 9.6 g m^{-2} yr^{-1} of vascular material while the better developed
meadows as much as 106.8 g m^{-2} yr^{-1}.

The map permits an estimate of the overall productivity of the 110 ha of the
study area, if an average value of net productivity for each nodum is used in
conjunction with the estimates of areal extent (Table 7). This provides an average
value of aboveground vascular productivity of 42.4 g m^{-2} yr^{-1} for the whole map
area for the 1972 growing season.

Turnover rates of aboveground vascular biomass (biomass ÷ annual production) are close to 1 yr for the principally herbaceous noda (meadows) where there
is little perennial aboveground living material (Table 6). For the drier noda where
there are woody species and some rosette plants, the turnover rate of aboveground biomass reaches 1.6 yr. For the woody plants themselves this would be
higher; several individuals of Salix pulchra had a minimum age of 20 yr based on
counts of terminal bud scars of their branches. Turnover rates for standing crop
of biomass plus standing dead are slower than for biomass alone (Table 6). The
rates of the former reflect the rate of incorporation of aboveground biomass and
standing dead into litter and prostrate dead. The dry noda (I, II, and III) have a
mean turnover rate of biomass and standing dead of 3.2 yr while that for the
remaining wetter noda is 1.8 yr. Thus rates of incorporation of biomass and
standing dead into surface litter is higher in wet sites than in dry sites. The
turnover rate of litter and prostrate dead has an average value of 2.1 yr while the
average turnover rate of all aboveground vascular material (biomass + standing

Table 6. Some Selected Standing Crop Fractions, Production, and Derived Values (Ratios and Turnover Rates) for the Vegetation Noda[a]

Variables	Vegetation noda								
	I(1)	II(5)	III(11)	IV(6)	V(6)	VI(8)	VII(2)	VIII(4)	Mean
Aboveground biomass									
Herbaceous dicotyledon	4.5	5.2	0.9	3.3	3.4	1.8	40.3	14.7	9.3
Woody dicotyledon	5.4	31.0	31.1	2.6	0.8	0	0	0	8.9
Monocotyledon	9.4	4.1	31.1	41.4	49.1	44.8	78.8	32.9	36.1
Total vascular plant	19.3	40.3	63.1	47.3	53.3	4.8	118.6	47.6	54.3
Lichen	14.8	37.3	55.1	4.6	11.3	0.2	0	13.7	17.1
Bryophyte	15.6	7.5	244.0	40.0	19.1	36.7	18.0	20.0	50.1
Total biomass	49.7	85.1	362.2	91.9	83.7	81.7	136.6	81.3	121.5
Aboveground vascular annual production[b]	18.1	25.3	39.3	45.1	51.1	43.9	115.0	47.0	48.1
Belowground biomass[b]									
Stem bases	112.6	0	64.4	259.6	76.8	111.2	2.2	3.0	66.2
Live rhizomes	59.2	108.1	110.0	91.3	219.8	77.1	43.4	44.2	94.1
Live roots	399.0	363.3	626.2	644.6	1008.3	866.2	466.4	105.9	560.0
Total biomass	570.8	471.4	800.6	995.5	1304.8	1054.5	512.0	153.1	733.5
Ratio[b]									
Above: belowground vascular biomass	29.6	11.7	12.7	21.1	24.5	23.5	4.3	3.2	16.3
Aboveground dead (vascular only)									
Standing dead	57.4	32.7	31.1	35.5	51.4	43.3	35.6	32.3	39.9
Litter + prostrate dead (L&PD)	41.4	126.8	121.4	67.0	75.2	46.9	63.9	84.6	78.4
Belowground dead[b]									
Intact plant organs	2804.9	1399.6	6367.4	3370.1	2105.6	2950.7	1227.5	291.3	2564.6
Turnover rates (years)									
Aboveground vascular biomass	1.1	1.6	1.6	1.1	1.0	1.0	1.0	1.0	1.4
Aboveground biomass + standing dead	4.2	2.9	2.4	1.8	2.1	2.0	1.3	1.7	2.3
Aboveground litter + prostrate dead[c]	2.3	5.0	3.1	1.5	1.5	1.1	0.6	1.8	2.1
Aboveground standing crop (+L&PD)	6.5	7.9	5.5	3.3	3.5	3.1	1.9	3.5	4.4
Belowground biomass	31.5	18.6	20.4	22.1	25.5	24.0	4.5	3.3	19.3

[a] Material was harvested at the period of peak aboveground vascular biomass in 1972. Values are means from the stands within each nodum. The number of stands per nodum is given in parentheses after the nodal number. Except for derived values, all units are g m^{-2}.
[b] Based on three stands per nodum except for Noda I and VII with only one and two stands, respectively, per nodum.
[c] Inverse of vascular decay index (Fig. 18).

Table 7.   Net Aboveground Annual Vascular Primary Production of the Vegetation Noda and Their Percentage Contributions to the Total Aboveground Vascular Productivity of the Entire Mapped Area Surrounding the Biome Sites for the 1972 Growing Season[a]

Nodum	Number of plots[b]	Maximum observed	Mean (g m^{-2} yr^{-1})	Percentage of map	Percentage contribution
I	1	18.1	18.1	3.0	1.3
II	5	28.6	25.3	7.2	4.3
III	11	72.6	39.3	41.0	38.0
IV	6	74.5	45.1	20.9	22.2
V	6	74.9	51.1	6.9	8.3
VI	8	106.8	43.9	14.6	15.1
VII	2	118.5	115.0	2.3	6.3
VIII	4	61.8	47.0	4.1	4.5
Mean	—	—	48.1	—	42.4[b]

[a]   The productivity estimates are based on a peak season harvest.
[b]   The mean net aboveground productivity of the map (Fig. 8).

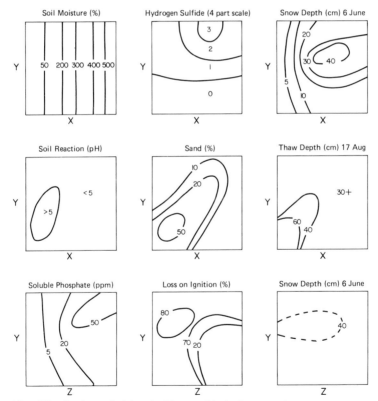

Figure 10.   Distribution of eight significant abiotic factors within the space provided by the ordination. The substrate trends indicated are used to interpret the environmental distribution of various components of the vegetation. All measurements were made in 1973.

dead + litter and prostrate dead) is 4.4 yr. These last two derivatives may be regarded as indicators of surface decay rates.

Gradient Analysis

The Ordination and Its Gradients.   The indirect ordination was constructed from 39 of the 43 sampled stands; four stands belonging to the pioneer vegetation of Nodum VIII were excluded as being too dissimilar from the central sedge meadow complex. The axes of the ordination were correlated with 16 measured substrate variables. Only those variables which had significant correlations are discussed. Figure 10 shows the distribution of these variables within the three-dimensional ordination space. This ordination framework provided by the phyto-sociological axes is the basis for describing the distribution of the vegetation noda, species, growth-forms, and various production and standing crop fractions in terms of the controlling environmental variables. The first axis of the ordination has the highest correlation with soil moisture, the second axis with the presence of hydrogen sulfide in the soil, and the third axis with soluble soil phosphate. These factors are thus used to name the major controlling complex gradients of the study area. In addition to the latter three factors, snow depth, pH of the surface horizon, percentage of sand, late season depth of thaw, and percentage of soil organic matter also show distinct patterns within the ordination space (Figure 10). The factors which correlate with the ordination axes are seldom, if ever, linear or orthogonal, and their interrelationships are complex (Figure 10). The nature of the complex gradients can also be seen when the substrate factors are clustered according to their intercorrelation with one another and with the ordination axes (Table 8 and Figure 11). Four clusters can be identified and each axis is a member of a separate cluster. A complete matrix of values for soluble soil phosphate is not available and in Figure 11 this factor is placed in an approximate position according to its highest correlation, which was with the third axis.

The diagonal trend of increasing concentration of hydrogen sulfide and soil moisture across the space provided by the first two axes reflects an intercorrelation and interdependence which is to be expected for these factors. Similar intercorrelating patterns are seen in Figure 10 between the amount of sand, soil reaction (pH), and depth of thaw. It should be noted that this latter group of variables cluster together (Figure 11).

Snow cover does not emerge from these analyses as a strong major controlling factor at Barrow because the variations in microrelief are insufficient to produce a pronounced snow effect. Where ravines, creek banks, ridges, and snowfences occur, snow cover appears as a controlling factor. Across the sedge meadow complex, shallow ponds, centers of low-centered polygons, and polygon troughs have the greatest snow accumulation and are the last sites to be free of snow, while raised polygon rims and centers of high-centered polygons have only a thin snow cover and are first to be snow free.

Plots from the different microrelief features or habitats have similar locations within the ordination space (Figure 12). Each vegetation nodum has a unique distribution within the ordination (Figure 13). The interpretation of the environment for each nodum from the ordination is compatible with the average value for environmental factors for each nodum (Table 5).

In addition to vegetation characteristics, other ecosystem phenomena such as consumer activities can be described within the ordination framework; for example, lemming activity (clipping, grubbing, and burrowing) and the amount of lemming feces show very distinct patterns within the ordination (Figure 14). Lemmings appear to use the drier sites and concentrate their activities in parts of Noda III and IV. There is a strong correspondence between the distribution of lemming feces and other lemming activities. It is interesting to note the concentration of lemming feces in the upper reaches of the phosphate gradient, but any connection between these is not established.

Distribution of Species and Growth-Forms. The distribution of 33 common species within the ordination is given in Figure 15. Each plot of a species within the ordination framework can be interpreted as a statement concerning the environmental tolerances or controls of that species. For example, *Salix rotundifolia* (top left, Figure 15) is most abundant on dry, well-aerated soils with high levels of soluble phosphate. Many species are wide ranging while others have restricted distributions. For example, *Saxifraga cernua, Carex aquatilis, Eriophorum angustifolium, Eriophorum russeolum, Poa arctica, Pogonatum alpinum,* and *Dactylina arctica* are wide ranging while *Cerastium jenisejense, Pedicularis lanata, Potentilla hyparctica, Eriophorum scheuchzeri, Calliergon*

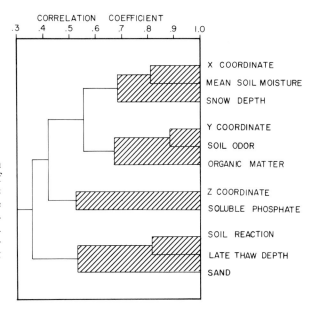

Figure 11. Dendrogram showing the clustering of the eight significant abiotic substrate factors and the three ordination coordinates. The position of soluble phosphate is based only on measurements from 15 plots. For $N = 39$ and $P = 0.001$, $r = 0.52$, and for $P = 0.05$, $r = 0.32$.

Table 8. List of Significant Correlations between Stand Ordination Coordinates and 16 Abiotic Variables[a]

Variable	Variables with significant correlations
Cluster 1	
First ordination coordinate (XAXIS)	LMOIS, MMOIS, YAXIS, SODOR, EMOIS, SNODE
Mean seasonal gravimetric water content (%) (MMOIS)	LMOIS, EMOIS, XAXIS, WABSO, BDENS, ORHORI, YAXIS, SODOR, ORMAT
Late season—August 17—water content (%) (LMOIS)	MMOIS, XAXIS, BDENS, WABSO, OHORI, SODOR, YAXIS, EMOIS
Early season—June 25—water content (%) (EMOIS)	MMOIS, WABSO, OHORI, XAXIS, SODOR, BDENS, LMOIS, YAXIS
Thickness of organic horizon (cm) (OHORI)	MMOIS, WABSO, EMOIS, LMOIS
Cluster 2	
Second ordination coordinate (YAXIS)	SODOR, XAXIS, MMOIS, WABSO, EMOIS, LMOIS
Soil odor on a four-part scale (SODOR)	YAXIS, XAXIS, MMOIS, WABSO, EMOIS, BDENS, LMOIS
Water absorption by dry sample (%) (WABSO)	ORMAT, MMOIS, EMOIS, OHORI, BDENS, LMOIS, SODOR, YAXIS
Loss on ignition (%) (ORMAT)	WABSO, BDENS, MMOIS
Cluster 3	
Third ordination coordinate (ZAXIS)	SPHOS
Soluble soil phosphate (ppm) (SPHOS)	ZAXIS
Early season—June 20—thaw depth (cm) (ETHAW)	SNODE
Early season—June 6—snow depth (cm) (SNODE)	SNOME, ETHAW, XAXIS
Duration of snow melt after June 6 (days) (SNOME)	SNODE
Cluster 4	
Sand in 2-mm soil fraction (%) (PSAND)	LTHAW
Bulk density of oven dry soil (BDENS)	MMOIS, LMOIS, WABSO, ORMAT, EMOIS, LTHAW, MTHAW, SODOR SOREA
Soil reaction of 2-m soil (pH) (SOREA)	MTHAW, LTHAW, BDENS
Late season—August 17—thaw depth (cm) (LTHAW)	BDENS, PSAND
Mean seasonal thaw depth (cm) (MTHAW)	BDENS

[a] Measurements from thirty-nine stands were used. Correlation was established with the Pearson's product–moment correlation coefficient (r) and significance was arbitrarily defined as any r-value with $P < 0.01$. All soil factors listed here were measured on the 0–5 cm soil core segments. Negative correlations are underlined. Correlation variables in each row are listed in sequence of decreasing correlation. All dates are for 1973.

giganteum, and *Alectoria nigricans* are narrow ranging. Some species may have similar distribution along one or two axes, but each is unique when all three axes are considered. For example, *Carex aquatilis* and *Eriophorum angustifolium* and *Dupontia fisheri* and *Eriophorum russeolum* have similar soil moisture and redox tolerances but differ with respect to phosphorus requirements or competition for phosphorus.

The floristic richness, as expressed by a number of species per 10-m² plot, of the vegetation varies from almost pure monospecific stands to rich stands with more than 40 species. There is a trend of decreasing floristic richness along the moisture gradient; dry-mesic sites have the most species, with very dry sites having slightly fewer and wet sites having the fewest species (Figure 16). Floristic richness is also lower in the phosphate-poor and anaerobic segments of the continuum such as low-centered polygons. Each major taxonomic entity has distinct diversity patterns within the continuum. Monocotyledons and bryo-

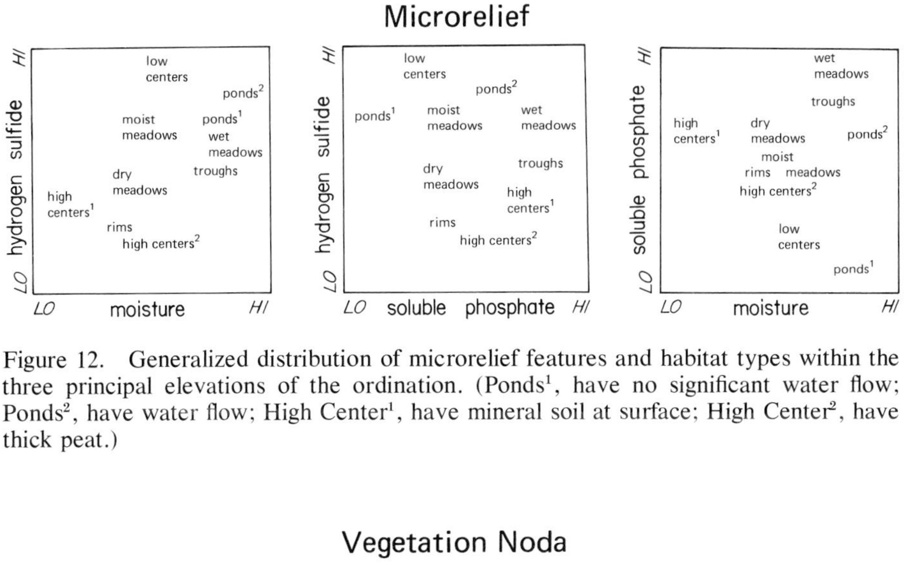

Microrelief

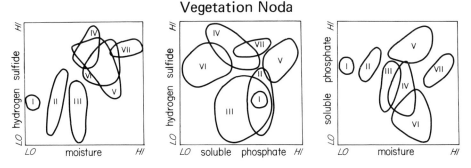

Figure 12.   Generalized distribution of microrelief features and habitat types within the three principal elevations of the ordination. (Ponds¹, have no significant water flow; Ponds², have water flow; High Center¹, have mineral soil at surface; High Center², have thick peat.)

Vegetation Noda

Figure 13.   The distribution of the seven mature noda within the three principal elevations of the ordination. The nodal numbers are those given in Table 1.

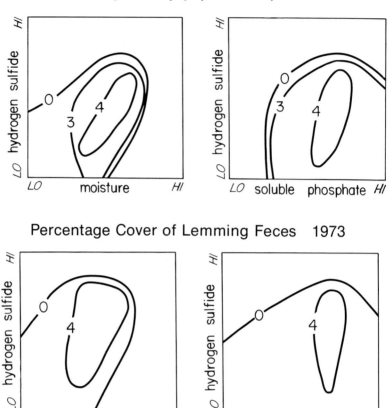

Figure 14.　The distribution of lemming activities within the ordination. Lemming activity was based on evidence of the combined signs of grazing and burrowing. Activity was expressed on a four-part scale from 0 (no signs) to 4 (heavy signs). Percentage cover of feces was derived from the number of contacts of 100 vertical point quadrats per ordination plot.

phytes have most species in mesic sites, herbaceous dicotyledons have most species in dry to mesic sites, and woody dicotyledons have most species on dry, well-aerated sites.

Each growth form also has a distinct distribution within the ordination (Figure 17). Evergreen and deciduous shrubs, caespitose monocotyledons, rosette and cushion dicotyledons, and lichens are more abundant on the drier habitats. Single monocotyledons, erect and rosette dicotyledons, and bryophytes are more abun-

Figure 15.　Distribution of 33 species within the three principal axes of the Barrow ordination. The axis positions correspond to those in the first two frames of Figure 10. Isoline values are Species Index.

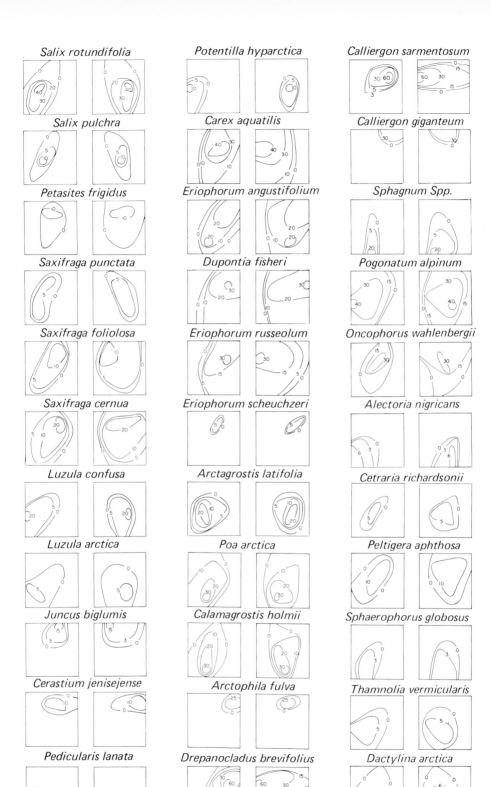

Salix rotundifolia

Salix pulchra

Petasites frigidus

Saxifraga punctata

Saxifraga foliolosa

Saxifraga cernua

Luzula confusa

Luzula arctica

Juncus biglumis

Cerastium jenisejense

Pedicularis lanata

Potentilla hyparctica

Carex aquatilis

Eriophorum angustifolium

Dupontia fisheri

Eriophorum russeolum

Eriophorum scheuchzeri

Arctagrostis latifolia

Poa arctica

Calamagrostis holmii

Arctophila fulva

Drepanocladus brevifolius

Calliergon sarmentosum

Calliergon giganteum

Sphagnum Spp.

Pogonatum alpinum

Oncophorus wahlenbergii

Alectoria nigricans

Cetraria richardsonii

Peltigera aphthosa

Sphaerophorus globosus

Thamnolia vermicularis

Dactylina arctica

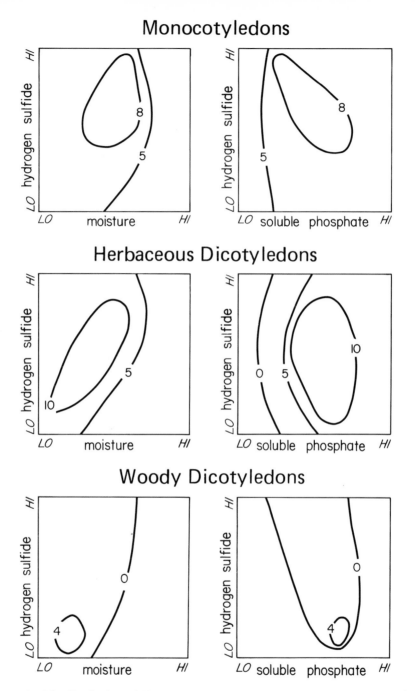

Figure 16.   The distribution within the ordination of the number of species per 10-m² plot of each major taxonomic entity.

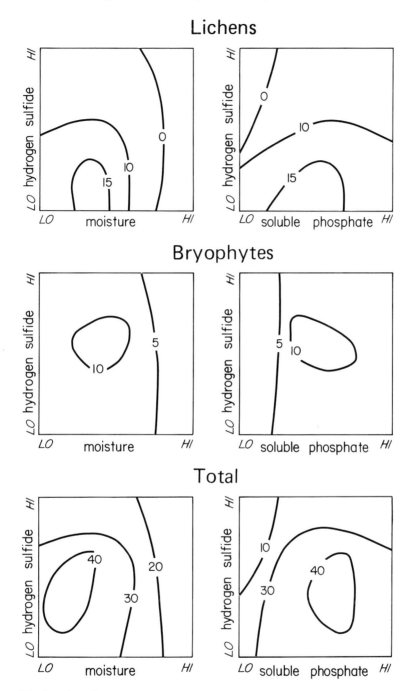

Figure 16 (continued).

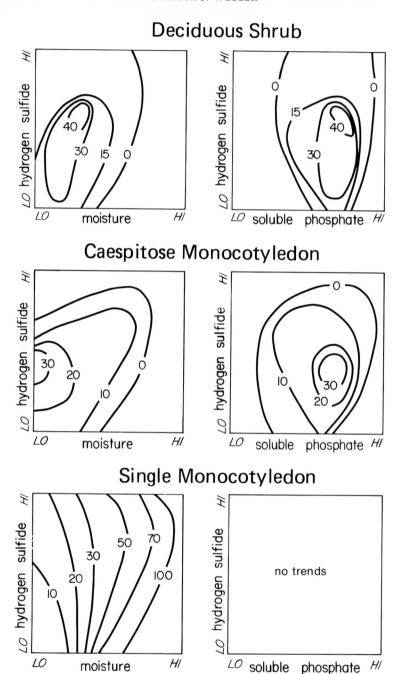

Figure 17. Distribution of 10 growth forms within the two principal elevations of the ordination. Isolines of Species Index have been drawn for each growth form.

Figure 17   (continued).

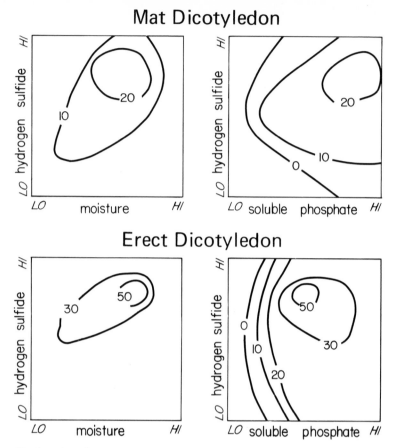

Figure 17   (continued).

dant on the wetter habitats. The cover of caespitose monocotyledons is highest on dry sites with moderate soil aeration and moderate phosphorus. The cover of the single monocotyledons increases with moisture and is independent of soil aeration and soil phosphorus. Evergreen shrubs occur with low moisture, moderately high aeration, and moderate phosphorus. Deciduous shrubs occur with slightly higher soil moisture, slightly lower soil aeration, and higher phosphorus than evergreen shrubs. Mat and erect dicotyledons reach their greatest abundance on moist, relatively anaerobic sites with moderate to high phosphate levels. Rosette dicotyledons are most abundant on dry to moist sites with medium levels of aeration and phosphate. Cushion dicotyledons have a distribution which is restricted to dry, well-aerated, phosphate-rich sites. Lichens occur with moderately low soil moisture, high soil aeration, and moderate phosphorus. Bryophytes, by contrast, occur with moderately high soil moisture, low soil aeration, and low phosphorus.

Bryophyte

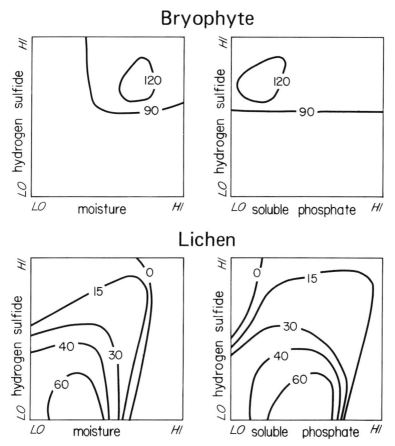

Figure 17   (continued).

The environment with the greatest diversity of a taxonomic entity is not the environment with the greatest abundance of that entity. For example, the number of bryophyte species is highest in mesic segments of each gradient and lowest in wet sites with low phosphate and high hydrogen sulfide levels (Figure 16).

Distribution of Standing Crop and Production.   The quantity of the various harvested herbage fractions and of vascular plant leaf area index and net productivity also form clear interpretable patterns within the ordination (Figure 18). Net aboveground vascular production increases along the moisture gradient to a maximum of just over 100 g m^{-2} yr^{-1} in the wettest sites. These sites have highly reducing soils and intermediate to high phosphorus levels. The patterns of monocotyledon and herbaceous dicotyledon peak-season biomass are very similar to actual production patterns because only small amounts of herbaceous live

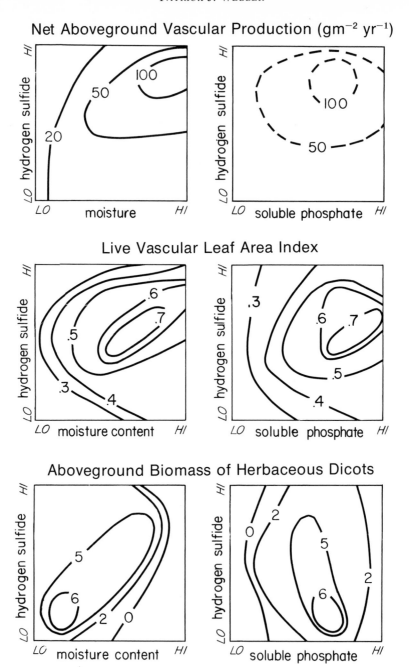

Figure 18. The distribution of several standing crop fractions and productivity related measures within the two principal elevations of the ordination. Measurements were made in 1972.

Standing Crop of Green Bryophyte Biomass (g m⁻²)

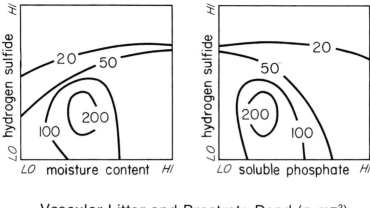

Vascular Litter and Prostrate Dead (g m⁻²)

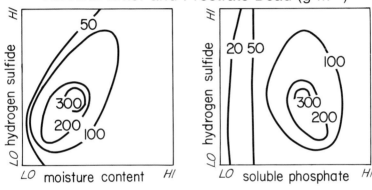

Aboveground Vascular Decay Index

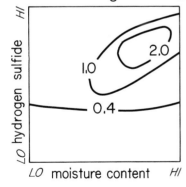
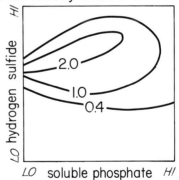

Figure 18 (continued).

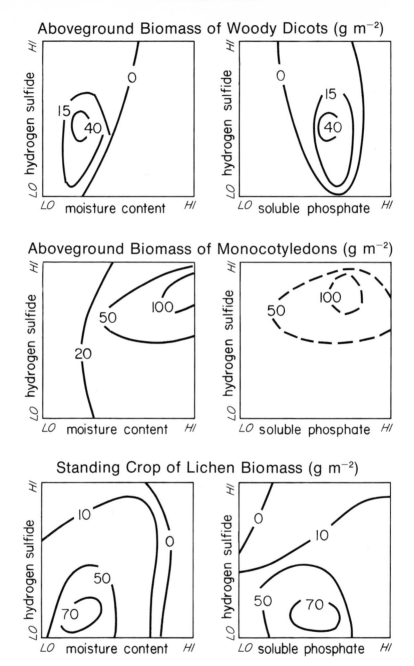

Figure 18   (continued).

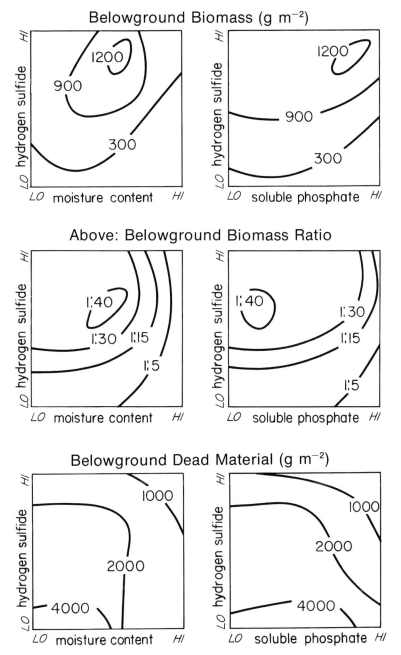

Figure 18 (continued).

material is carried over from previous years and only small amounts of the current year's growth will have been shed or have died or been grazed. Thus the similarity between the vascular production pattern and that of monocotyledon biomass with respect to both magnitude and distribution emphasizes the fact that monocotyledons contribute the majority of the vascular plant production of the IBP site. The production of herbaceous dicotyledons rarely exceeds 6 g m^{-2} yr^{-1} and is highest in dry, oxidizing sites with moderate levels of phosphate. Woody dicotyledon biomass has a similar distribution to that of herbaceous dicotyledons; in optimal sites it just exceeds 40 g m^{-2} and in dry sites the actual production of woody plants occasionally reaches 20 g m^{-2} yr^{-1}, which may equal the herbaceous production in these sites.

The distribution of lichen biomass within the ordination (Figure 18) mirrors that for the total species index of lichens (Figure 17); however, the distributions of bryophyte biomass and species index (Figures 17 and 18) do not correspond. Bryophyte species index, which is a fair measure of total bryophyte cover, is highest in wet, anaerobic, low phosphate sites, whereas biomass is highest on dry-mesic, well-aerated sites with low phosphate.

The distribution of belowground material within the ordination is different from that of aboveground material (Figure 18). Belowground living material is greatest in the most anaerobic sites. The ratio of above- to belowground biomass presents a different picture (Figure 18), with the lowest ratio (1:40) in moist, partly anaerobic sites with low soluble phosphate levels. The ratio of above- to belowground biomass increases along the moisture gradient and from anaerobic to aerobic sites.

Aboveground vascular litter and prostrate dead material and belowground dead material are both most abundant on dry sites and least abundant on wet sites. Litter accumulation is a function of the balance of production and decay. In order to examine this in terms of the principal controlling gradient, an index of aboveground vascular decay was plotted within the ordination (Figure 18). This index is the ratio of the peak season aboveground vascular litter and prostrate dead fraction to the annual aboveground vascular production. This index is the inverse of the litter and prostrate dead turnover rate (Table 6). The index, in the absence of surface transportation or grazing of litter, may be interpreted in terms of surface decay. This suggests that the highest decay rates are found on wet, anaerobic sites with low phosphate and that the lowest rates are found on dry, well-aerated sites with high phosphate. Similar trends for belowground decay are suggested by the distribution of belowground, intact dead matter (Figure 18). The highest accumulations of organic matter, up to 4000 g m^{-2}, occur on dry, well-aerated sites, and the lowest accumulations, about 1000 g m^{-2}, occur on wet, poorly aerated sites.

Seasonal Variation

The Growing Season. The data for the seasonal progression of vascular plant biomass were collected from ordination plots belonging to Nodum IV (*moist meadow*). The growing season starts in the first or second week of June. In both

1970 and 1971, the onset of vascular plant growth corresponds with the point when the mean air temperature exceeds 0°C (Figure 19). Subsequent accumulation of aboveground vascular biomass is linear until it reaches a maximum or peak in the first week of August. The standing crop of aboveground vascular

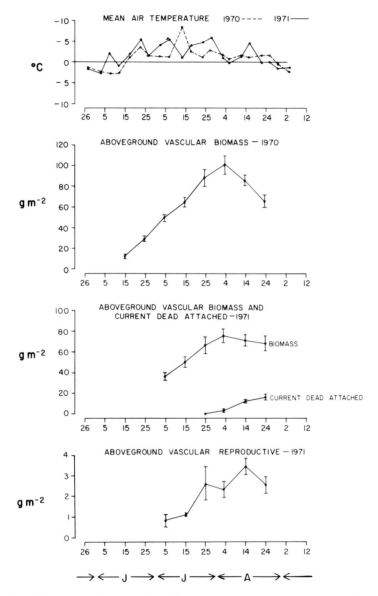

Figure 19.   The seasonal progression of mean air temperature and some aboveground vascular plant standing crop fractions for the 1970 and 1971 growing seasons. Standing crop data are from Dennis and Tieszen (1972) and Dennis *et al.* (1978). The bars at each point represent standard errors of the mean.

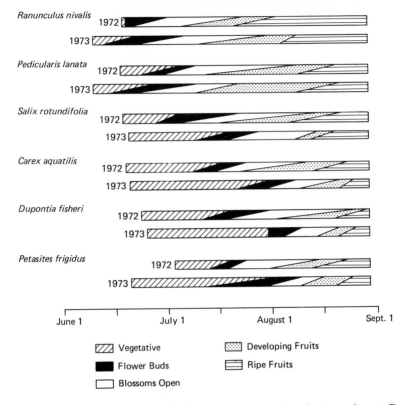

Figure 20. Phenophase diagrams of six common vascular plant species at Barrow, Alaska. The data for two growing seasons, 1972 and 1973, are compared. The species are arranged down the diagram in a sequence from early to late developing.

sexual structures (inflorescences) also reaches its peak in the first week of August.

The onset of vascular plant growth is also closely correlated with the disappearance of snow and the start of soil thaw (see Tieszen, 1974). Those sites which are the first to become free of snow show the first signs of new growth. These sites are the high-centered polygons and raised rims of polygons which belong to Noda I and II *(heaths)* (Figures 10 and 12). The first signs of growth on these sites is that of a little leaf growth followed closely by the flower bud expansion of *Pedicularis lanata* and *Salix rotundifolia*. However, the first prominent flower at the IBP site is that of *Ranunculus nivalis*. It is characteristic of Noda III and IV *(mesic and moist meadows)* which receive moderate snow cover. Nevertheless, the noda can be arranged in phenological sequence. The sequence corresponds to the numerical sequence of the noda, Nodum I *(dry heath)* being the first to begin growth and show mature fruits and Nodum VIII *(pioneer meadow)* being the last. This sequence corresponds to a gradient of increasing duration of snow cover and increasing soil moisture (Table 5). It is probably representative of a complex temperature gradient. Nodum VIII, although well drained, is found in snowbank situations and is the last to be free of snow and hence the last to have blossoms.

A further description of plant responses within the growing season can be obtained from phenophase diagrams for six common species for the years 1972 and 1973 (Figure 20). The first signs of vegetative growth occur toward the end of the first week of June. Although some open blossoms appear toward the end of June, peak blooming for the Barrow tundra is not until mid- or late July. Fruits begin forming in mid-July, but few ripe fruits with well-formed seeds appear before mid-August. Most flowering is complete by the first week in August and the vegetation begins to senesce at this time. Although each species may begin growth and flowering at different times, there is less variation in the timing of the appearance of ripe fruits which is usually complete before the last week of August when freezing mean air temperatures return (Figure 19). *Ranunculus nivalis* produces well-formed seed within 35 days of snow melt. The period between onset of growth for a species and the appearance of the first mature fruits averages 53 days for the studied species. *Carex aquatilis* is one of the last species to produce mature fruits and it took 63 days in 1973 which was not a particularly cold year (Table 9). *Dupontia fisheri* seemed to be close to the last species to complete its fruit development, while *Pedicularis lanata* took the longest (73 days) to pass from onset of growth to final maturing of the last flower.

Variation between Seasons. Willow elongation for the years 1969 to 1973 showed a positive but insignificant correlation ($r = 0.6$) with the mean July temperature (Table 9). In the 1970 growing season which was cooler and later to start than the 1971 season (Figure 19, Table 9) the aboveground vascular production up to peak season was 34% less. This lack of positive correlation between production and growing season temperature is shown further by a comparison of 1972, when the mean July temperature was 2.34°C warmer than the normal (30-yr mean), with 1970, which was −0.50°C cooler than normal. In 1970, the production was 67% higher than in 1972; 1972 was the warmest of the 4 years.

Table 9. Comparison of Plant Growth from Site 2 and Some Climatic Parameters for the Years 1969–1973

Parameter	Season					
	1969	1970	1971	1972	1973	Mean
Willow elongation[a] (cm)	2.32	2.76	4.12	4.13	3.76	3.42
Deviation from 5-yr mean[a]	−1.10	−0.66	+0.70	+0.71	+0.34	—
Aboveground vascular net production (g m^{-2} yr^{-1})	—	101.5	80.1	74.7	57.4	78.43
Mean July temperature (°C)	1.50	3.22	4.67	6.06	4.28	3.94
Deviation of mean July temperature from normal[b] (°C)	−2.22	−0.50	+0.95	+2.34	+0.56	3.72

[a] *Salix pulchra* mean of 10 individual plants.
[b] Based on U.S. Weather Service records for 1941–1970.

In 1973, the onset of growth began earlier than in 1972 (Figure 20). However, the late blooming *Carex aquatilis* and *Dupontia fisheri* flowered later in 1973 and the final stature of *Carex* was less in 1973. Vegetation production was also less in 1973 than in 1972. The thaw season did begin earlier in 1973 than in 1972 but June and July were cooler in 1973 (Table 9).

Plant Succession

A Generalized Scheme.   The scheme shown in Figure 21 is based on field observations and is thus inferential and untested. The vegetation noda are arranged in two connected sequences: one as it occurs on well-drained, fine-grained alluvium such as to be found in a floodplain of a small creek and the other on fine-grained, organic-rich, lacustrine sediments in former lake-basins. The successional changes are controlled primarily by changes in microrelief and thus drainage regimes. Only the most frequent and evident changes are shown; several possible interchanges or other sequences are omitted.

Plant colonization on stable floodplain alluvium is rapid. In a few years, an almost complete cover of species such as *Cochlearia officinalis, Stellaria laeta, Phippsia algida, Alopecurus alpinus, Poa arctica, Saxifraga cernua,* and *Bryum* develops. *Dupontia fisheri, Petasites frigidus,* and many other plants, including lichens, soon follow and stands belonging to Nodum VIII (*pioneer meadow*) result. In sites which are not dominated by snow accumulation, stabilization of sediments and consolidation of the vegetation form stands belonging to Nodum III *(mesic meadow)*. These stabilized sites may become drier either by further alluvial deposition, by aeolian deposition, by raised polygon formation, or by increased drainage as local water courses deepen through thermokarsting. These drier sites will support the vegetation assemblages of Noda II and I *(heaths)*.

The colonization of drained lake sediments is also rapid. The basin of Foot-print Lake has become covered with highly productive vegetation within 25 yr. The present vegetation of the basin varies according to local drainage and substrate composition. It is generally representative of Noda IV, V, VI, and VII *(moist* to *wet meadows* and *pond margins)*. Some vegetation seems to have changed very little since it was first colonized shortly after drainage, and it appears that once a surface has a complete cover of vegetation it seems slow to change to another type. Nevertheless, as the ice-wedge polygons form a variety of microrelief and moisture regime, the vegetation of the Footprint basin will become similar to the current IBP site. However, many decades or longer must elapse to provide the variety to be found at the IBP site. Nodum V, the *wet meadow,* is the most abundant vegetation in the Footprint basin. The annotations on the arrows in Figure 21 suggest the mechanisms necessary for change into other nodal types. The drier Nodum VI *(moist meadow)* can form when polygon troughs drain the polygon centers. Trough formation in Nodum IV can reverse the trend by producing wetter microsites. As polygonization continues, the diversity of surfaces increases. Drier polygon rims or slightly raised centers of

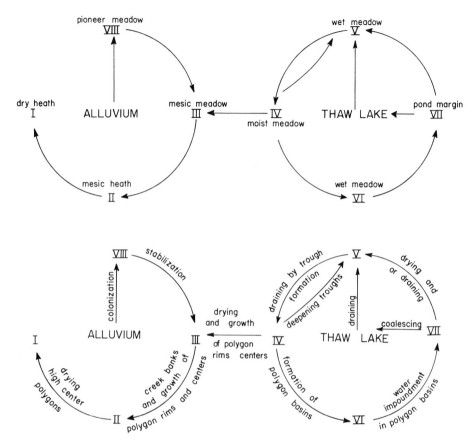

Figure 21. The principal successional trends of the vegetation noda and the principal allogenic geomorphic processes controlling these successional trends.

Nodum IV can transform to Nodum III *(mesic meadow)*. The creation of Nodum III from drained lake sediments forms a link with the alluvial succession. The formation of low polygon centers which may become wet and/or impounded can transform Nodum IV to Nodum VI *(wet Carex meadow)*. Further depression of centers and impounding of water readily produces ponds and the pond margin vegetation of Nodum VII. Some ponds may breach their impounding rims and drain while others may fill with sediments. In this manner Nodum V *(wet Dupontia meadow)* may be reformed. The cycle may also be closed by the coalescence of ponds as their margins or rims are eroded by thermokarsting, and thus large ponds and even thaw lakes may be reformed.

The draining of thaw lakes and the formation of extensive new alluvium in flood plains are relatively rare events in the study area. Both instances in this study are related to man's activities, but both processes can also be of natural occurrence. Several other examples of man's impact on plant succession are available in the study area, for example, vegetation changes triggered by tracked vehicle damage and water impoundments resulting from road construction. These changes will fit into the successional scheme presented here. Catastrophic thermokarsting of any nodum would lead to either the alluvial or thaw lake starting points. The most frequent effect of moderate impact of impoundment in the IBP area is to cause a deepening of troughs through partial ice-wedge melting. In many instances this has caused drier intertrough areas (Nodum V to IV or Nodum IV to III) and wetter troughs (Nodum IV to V or Nodum V to VIII); these changes may occur within a very few years.

Patterns within the Succession. When the presence of various species is plotted on the basic successional frame formed by the nodal sequence, orderly patterns develop (Figure 22). Some species occur in only one nodum or successional stage [*Hierochloë alpinum* (Figure 22b), *Cassiope tetragona* (Figure 22f), *Caltha palustris* (Figure 22t), *Phippsia algida* (Figure 22d), *Saxifraga rivularis* (Figure 22d)]; other species occupy several noda [*Poa arctica* (Figure 22n), *Dupontia fisheri* (Figure 22o), *Saxifraga foliolosa* (Figure 22c), *Carex aquatilis* (Figure 22p), *Eriophorum angustifolium* (Figure 22p)]. The patterns of those species occurring in more than one nodum usually occur sequentially and are not interrupted or haphazard. The single exception to this generalization among the plotted species is *Arctophila fulva* (Figure 22r); it occurs in the wet noda (V and VII) and the pioneer nodum (VIII). Orderly patterns are also formed when the numbers of the various major taxa or growth forms per nodum and various production measures are plotted in the successional sequence (Figure 23). In the sequence from *pioneer meadow* (Nodum VIII) to *raised polygon center* (Nodum I), the number of most taxa increases from Nodum VIII through Nodum III to Nodum II but suddenly decreases from Nodum II to Nodum I. In the thaw-lake cycle from Nodum V through Noda IV, VI, and VII, the number of all taxa first increases from Nodum V to Nodum IV and then decreases from Nodum IV through Nodum VI to Nodum VII. Nodum VII has the lowest overall diversity.

The highest proportion of monocotyledons is found in Nodum VI and the least

---

Figure 22. The distribution of 26 species within the successional sequence diagram. A dot in the position of the nodal number indicates the presence of a species. Roman numerals represent the position of the vegetation noda and the letters A and T represent the starting points of alluvium and thaw lake successions. The six-letter codes for species are as follows: b, HIEALP—*Hierochloë alpina;* c, POTHYP—*Potentilla hyparctica,* PAPHUL—*Papaver hulténii;* d, PHIALG—*Phippsia algida,* SAXRIV—*Saxifraga rivularis;* e, COCOFF—*Cochlearia officinalis;* f, CASTET—*Cassiope tetragona,* SAXOPP—*Saxifraga oppositifolia:* g. SENATR—*Senecio atropurpureus;* h, PEDLAN— *Pedicularis lanata;* i, SPHGLO—*Sphaerophorus globosus;* j, LUZCON—*Luzula confusa;* k, SALPUL—*Salix pulchra,* AULPAL—*Aulacomnium plaustre;* l, SPHAGNUM—*Sphagnum* species; m, CETRIC—*Cetraria richardsonii;* n, POAARC—*Poa arctica;* o, DUPFIS—*Dupontia fisheri,* SAXFOL—*Saxifraga foliolosa;* p, CARAQU— *Carex aquatilis,* ERIANG—*Eriophorum angustifolium;* q, CALGIG—*Calliergon giganteum;* r, ARCFUL—*Arctophila fulva;* s, RANPAL—*Ranunculus pallasii;* t, RANGME—*Ranunculus gmelini,* CALPAL—*Caltha palustris.*

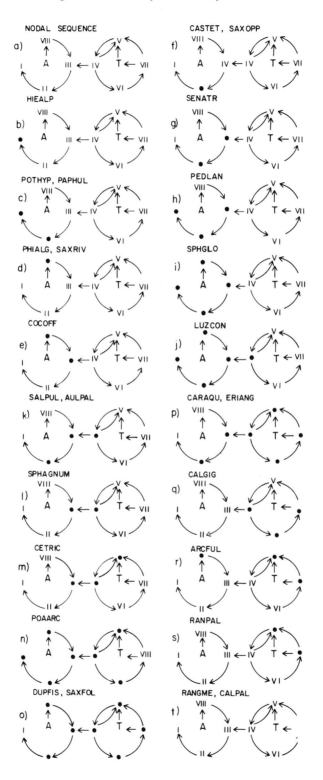

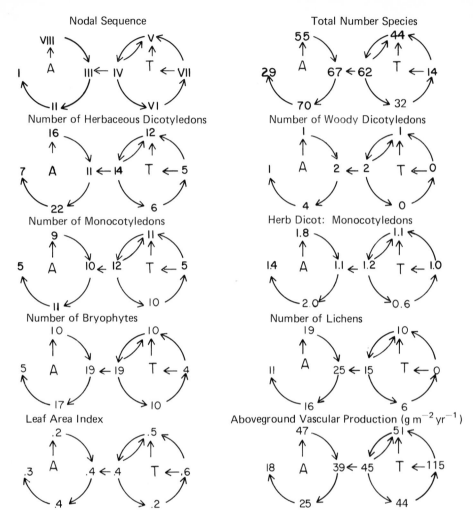

Figure 23. The distribution of the species diversity (number of species per 10 m²) of the major life forms, leaf area index, and annual net aboveground vascular production within the successional sequence diagram. Roman numerals represent the position of the vegetation noda and the letters A and T represent the starting points of alluvium and thaw lake successions. Leaf Area Index only refers to vascular plants.

in Nodum II. The noda found on dry sites, that is those in the left half of Figure 23, have the highest proportion of dicotyledons. Nodum VIII has several typical pioneer species which are dicotyledons, for example, *Cochlearia officinalis, Saxifraga rivularis,* and *Stellaria humifusa.* Leaf area index has a low value of 0.2 in Nodum VIII, and this increases in a sequence to Nodum II and then decreases slightly to Nodum I. It has a high value of 0.6 in Nodum VII and this decreases around the thaw-lake cycle to Nodum IV where it has a value of 0.2.

Annual aboveground vascular production steadily decreases from a value of 47 g m^{-2} yr^{-1} in Nodum VIII to a value of 18 g m^{-2} yr^{-1} in Nodum I, and from Nodum VII through Nodum V to Nodum IV it decreases from a high of 115 g m^{-2} yr^{-1} to a low of 44 g m^{-2} yr^{-1}.

Discussion

Environmental Controls

The microrelief emerges as the principal control of the plant environment within the flat coastal tundra at Barrow. This is shown by the presence of distinct vegetation types, each of which occupies a distinct set of microrelief features. It is also shown by the grouping of points with similar microrelief within the continuum provided by the ordination of the vegetation (Figure 12). The microrelief controls the drainage and other substrate relations which in turn control distribution of the plants. This conclusion supports the earlier studies of the Barrow vegetation which emphasized the controlling influence of microrelief (Wiggins, 1951; Britton, 1966). Matveyeva et al. (1973), who have studied similar vegetation on the Western Taimyr Peninsula, have also stressed the importance of microrelief.

The interrelations of the substrate factors are very complex. The cluster diagram (Figure 10) shows them to be interdependent and the ordination (Figure 10) shows that they are not linearly related. Nevertheless, the cluster diagram shows four relatively independent complexes of substrate factors and the ordination serves to rank them in order of importance: a water complex > an aeration complex > a phosphate complex > a thaw complex (Figure 11). Three other complexes (wind, lemming activity, and soil stability) are also important. The interrelationships between these factors and their controls are summarized in Figure 24. The arrows indicate the direction of control. The complete matrix of arrows is not given and only the strongly substantiated or inferred controls are indicated. The thickness of the arrows indicates their relative importance. Thus the strong control of relief over the water complex and its subsequent control over vegetation and other substrate factors, such as aeration and soil thaw, is shown by heavy arrows. The term relief is used in Figure 24 in order to include the topographic features outside the immediate IBP site. This broadens the application of the diagram just beyond the microrelief of the polygon complex and the creek banks (Figure 9) to include raised beach ridges and lake bluffs.

In contrast to most studies of the controls of tundra vegetation, several factors, for example, wind, snow, active layer thickness, and disturbance (Sørenson, 1941; Gjaerevoll, 1956; Bliss and Cantlon, 1957; Bliss, 1963; Scott and Billings, 1964; Webber, 1971; Webber et al., 1976; Webber and May, 1977), do not emerge as major controls of the vegetation of the IBP site. This is not to say that they are unimportant; it only tells us that these factors may be relatively uniform across the site and do not contribute much to the variation of the vegetation.

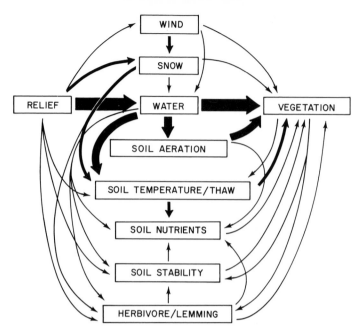

Figure 24. The principal components of the environmental complex as it controls the vegetation at Barrow, Alaska. The major controls are indicated by heavy arrows. The complex matrix of feedback arrows is not given; only substantiated or strongly inferred interactions are shown.

Wind is frequently strong enough to dry the vegetation and redistribute the snow, but these effects are seldom strong because of the moist soil, high relative humidities, and lack of great topographic relief. Nevertheless, the onset of plant growth in spring is strongly geared to the disappearance of snow.

The odor of hydrogen sulfide, used here as a measure of soil aeration, is crude and subjective. The fact that it emerged as a factor with a distinct pattern within the ordination, however, made it difficult to disregard. Gersper and Arkley's observations of soil oxygen levels across different features of microrelief agree broadly with the present observations on soil odor. The balance of oxidizing or reducing environments closely correlates with soil porosity and drainage and may be expected to have a major effect on the root environment and soil microbiology (Dennis, 1968).

Well-drained sites on mineral soils thaw deeply and are the warmest, and wet sites with organic soils have only shallow depths of thaw and are the coldest (Dennis, 1968). The thickness of the active layer and the presence of the permafrost barrier close to the surface must be a strong controller of vegetation type. However, like snowcover, it does not emerge from these studies as a major control.

Phosphate showed moderately distinct patterns with the various other factors and, because of its vital nature, it must have some effect on the vegetation, if it is present in less than adequate supply. Figure 24 shows a general nutrient box which is under some control of thaw, water, and aeration. The deeper thaw provides a greater amount of substrate from which nutrients can be withdrawn, and sites with deeper thaw may have more microbial activity and greater nutrient release.

On the whole, the substrates of the study area are remarkably stable. Cryoturbation, which is active enough to disturb the vegetation, is limited to a few bare spots associated with frost boils, some polygon rims, and the centers of a few high-centered polygons. Surface instability is also a factor on new surfaces in creek bottoms or drained ponds. Where such unstable surfaces occur, the vegetation is very characteristic and hence instability can be a very important factor. It is closely tied to microrelief and topographic control.

It is important to note that in Figure 24 only a few arrows are shown from the vegetation to the substrate factors. Plant canopy development affects thaw depth by changing albedo and insulation regimes (Johnson, 1966; Ng and Miller, 1977). Also, soil stability is affected by the absence or presence of soil-binding vegetation roots. Similarly, lemming activity is dependent upon the availability of forage. Other feedback effects from the vegetation must exist for all factors, for example, wind and snow modification and rates of nutrient return from litter. However, these are relatively minor influences compared to the more direct effect of the substrates on the vegetation. This suggests that the principal forces affecting the vegetation are abiotic or related to animal activity.

Species, Growth Form, and Community Considerations

The wide ecological amplitude of these tundra species has been observed before at Barrow (Wiggins, 1951; Britton, 1966) and elsewhere (Oosting, 1948; Polunin, 1948; Böcher, 1954; Drury, 1962; McVean, 1969; Raup, 1969; Webber, 1971). Some wide-ranging species, for example, *Carex aquatilis, Pogonatum alpinum,* and *Poa arctica,* are dominant, while others, for example, *Saxifraga cernua, Dactylina arctica,* and *Cetraria islandica,* are subordinate (Figure 15, Table 10). It is the combination of ubiquity and dominance which, according to many botanists, has contributed to the difficulty in classifying tundra vegetation (Summerhayes and Elton, 1928; Griggs, 1934; Polunin, 1934, 1934–1935, 1948; Raup, 1941, 1951, 1953; Bruggeman and Calder, 1953; Dansereau, 1954; Savile, 1960, 1961). These botanists have also considered that physical environmental controls dominate tundra vegetation to such an extent that community patterns or clear vegetation associations do not develop and often vegetation is a bewildering array of species with little repetition over the landscape. Griggs (1934) and Savile (1960) argued that unstable surfaces with pioneer or rudimentary vegetation were largely the cause of the difficulty of dealing with tundra vegetation. This should not be the case at Barrow where most vegetation cover is complete and most surfaces are stable. The problem at Barrow originates with the flat terrain and the complexity of the patterns of the ice-wedge polygons (Brown and Johnson, 1965; Dennis, 1968). The presumed dominance of the physical environment over the vegetation composition is primarily the reason why many North American tundra botanists have relied on physical, habitat, and physiognomic descriptors

Table 10.   The Percentage Cover and Percentage Frequency within the Mapped Area of the Most Abundant 40 of the 112 Sampled Taxa[a]

Rank	Taxon	Growth form	Percentage frequency	Percentage cover	Cumulative relative cover
1	*Carex aquatilis*	SG	72.09	18.78	12.91
2	*Pogonatum alpinum*	BR	76.74	14.44	22.84
3	*Bryum* spp.	BR	88.37	13.64	32.22
4	*Dicranum* spp.	BR	55.81	12.59	40.87
5	*Calliergon sarmentosum*	BR	53.49	8.72	46.86
6	*Oncophorus wahlenbergii*	BR	51.16	7.71	52.16
7	*Drepanocladus brevifolius*	BR	51.16	6.42	56.57
8	*Salix rotundifolia*	DS	44.19	6.27	60.88
9	*Poa arctica*	SG	74.42	5.29	64.52
10	*Polytrichum juniperinum*	BR	46.51	3.60	66.99
11	*Dupontia fisheri*	SF	74.42	3.58	69.45
12	*Eriophorum russeolum*	SG	72.09	3.31	71.73
13	*Eriophorum angustifolium*	SG	55.81	3.30	74.00
14	*Hylocomium alaskanum*	BR	20.93	2.80	75.92
15	*Hepaticae*	BR	58.14	2.57	77.69
16	*Saxifraga cernua*	ED	81.40	2.31	79.28
17	*Drepanocladus uncinatus*	BR	18.60	2.17	80.77
18	*Brachythecium* cf. *turgidum*	BR	41.86	2.10	82.21
19	*Salix pulchra*	DS	13.95	2.02	83.60
20	*Aulacomnium turgidum*	BR	37.21	1.85	84.87
21	*Dactylina arctica*	LI	69.77	1.84	86.13
22	*Calliergon giganteum*	BR	9.30	1.77	87.35
23	*Calamagrostis holmii*	SG	30.23	1.60	88.45
24	*Petasites frigidus*	ED	47.60	1.19	89.27
25	*Sphagnum* spp.	BR	13.95	1.11	90.03
26	*Arctagrostis latifolia*	SG	30.23	1.00	90.72
27	*Campylium stellatum*	BR	18.60	0.93	91.36
28	*Cetraria islandica*	LI	69.77	0.87	91.96
29	*Oncophorus* spp.	BR	4.65	0.86	92.55
30	*Myurella julacea*	BR	16.28	0.62	92.98
31	*Luzula confusa*	CG	37.21	0.61	93.40
32	*Ranunculus pallasii*	ED	9.30	0.55	93.78
33	*Stellaria edwardsii*	MD	58.14	0.50	94.12
34	*Peltigera aphthosa*	LI	58.14	0.49	94.46
35	*Arctophila fulva*	SM	11.63	0.47	94.78
36	*Juncus biglumis*	CG	23.26	0.41	95.06
37	*Cetraria cucullata*	LI	60.47	0.38	95.32
38	*Cassiope tetragona*	ES	6.98	0.37	95.57
39	*Distichium capillaceum*	BR	6.98	0.36	95.82
40	*Stellaria laeta*	MD	25.58	0.35	96.06
Totals for all 112 sampled taxa				145.48	100.00

[a]   Percentage frequency is derived from the number of plots of occurrence of a taxon. The cover values are computed from the nodal averages and nodal areas in the vegetation map. Percentage cover values have been converted to relative cover and are accumulated in the last column. The taxa are arranged in order of their cumulative relative cover. Growth-form symbols are as in Table 2.

to name their plant communities rather than species alone (Polunin, 1934, 1934–1935, 1948; Churchill, 1955; Spetzman, 1959; Bliss, 1963; Britton, 1966; Johnson et al., 1966; Potter, 1972).

It is my opinion that tundra vegetation is no more difficult to classify or study than that of more temperate systems. This is in essential agreement with Bliss (1963) who felt that the mosaic of patterns may only appear exaggerated in low stature vegetation where there is an absence of trees. Some level of classification always seems possible after familiarization and detailed analysis. For example, Webber (1971), Barrett (1972), and Smith (1974) have made classifications of high arctic tundra vegetation. Webber (1971) used the same methods as used in this study on Baffin Island; Barrett (1972) used the more rigorous Zürich–Montpelier methods (Braun-Blanquet, 1932) on Devon Island; and Smith, who worked at Barrow on bryophyte communities, used an association analysis based on a χ^2 test (modified from Goodall, 1953).

The vegetation units or noda recognized here are rather broad and consist of several traditional associations or stand types. The *pioneer nodum* (VIII) is the most heterogeneous nodum (Figure 3). The distinct positions of the other noda in the ordination are seen as a reflection of environmental discontinuities and/or an oversampling of certain more frequent habitat types rather than as evidence for support of the association-unit hypothesis (Whittaker, 1962; Daubenmire, 1966). Because the noda are abstractions, it is not always possible to classify each new stand with confidence. However, the mapping experience has shown that the classification system works quite well within the sampled universe, but some difficulties were encountered toward the periphery. The vegetation assemblages of the polygonal ground areas of the Coastal Plain, which correspond to the universe of the IBP site and which have been described by the previous studies (Wiggins, 1951; Spetzman, 1959; Britton, 1966; Potter, 1972), fit into the present classification quite well.

The unique, nonrandom distribution of each species demonstrates that they are not haphazardly arranged, and this in itself is likely to allow for some measure of classification or division of the environmental or vegetational continuum. There is no evidence from these results to support the notion that the competitive exclusion principle is not operating at Barrow as implied for some more severe tundra situations (Savile, 1960). In fact, closely related species which would provide the best evidence for a violation of this test (Cole, 1960) have less similar distributions than unrelated species. For example, the distribution of *Luzula arctica* is more similar to that of *Potentilla hyparctica* than to *Luzula confusa,* and the distribution of *Eriophorum russeolum* is more similar to that of *Carex aquatilis* than to *Eriophorum scheuchzeri*. In a competitive exclusion context, it is unlikely that *Luzula confusa* and *Potentilla hyparctica* would be complete competitors because of their different growth forms. Even those species which do have somewhat similar distributions and aboveground form, for example, *Carex aquatilis, Dupontia fisheri,* and *Eriophorum angustifolium,* have different rooting patterns (Billings et al., 1978). The results on species distribution from this study are in close agreement with others from the same area (Wiggins, 1951; Koranda, 1954; Britton, 1966, Dennis, 1968; Smith,

1974; Rastorfer, 1978; Steere, 1978; Williams et al., 1978). The strength of the present study rests on its quantification of the response of individual species to the principal environmental controls of the IBP site proper.

Tundra is not unique with respect to the dominance of a few common species. Raunkiaer's Law of Frequency suggests that this pattern is generally universal (Kenoyer, 1927), as do other studies of dominance and diversity relations (for example, Patrick et al., 1954; Beschel and Webber, 1963; Whittaker, 1965). Nevertheless, the dominance by a few species seems particularly striking in tundra, especially because the floras are small in number of taxa. On the IBP site, 112 different taxa were sampled and of these 6 taxa make up 50% of the total plant cover, 14 make up 75%, 20 make up 85%, and 40 make up 96% (Table 10). The remaining 4% of the plant cover is provided by 72 species. On Bylot Island, N.W.T., Drury (1962) reported that 34 species comprised 85% of the plant cover, and on central Baffin Island, N.W.T., Webber (1971) found that 38 species comprised 85% of the plant cover. Both Bylot and central Baffin Island have overall floristic diversities similar to those of Barrow. However, the lower figure of 20 taxa comprising 85% of the cover at Barrow is a reflection of the low overall habitat diversity of the present study area compared to the Eastern Arctic sites.

The percentage cover values in Table 10 are means for each species obtained from nodal averages with a correction made for the areal extent of each nodum within the map. Perhaps the most striking feature of Table 10 in which the species are ranked according to their percentage cover is that although *Carex aquatilis* is dominant, the next six species are bryophytes. In fact, of the 10 most abundant species 7 are bryophytes. They comprise 58.1% of the total species cover of 100. Monocotyledons comprise 26.7%, shrubs 6.1%, forbs 5.3%, and lichens 3.8% of the remaining relative cover. In a similar analysis of some central Baffin Island vegetation, Webber (1971) found 5 of the 10 dominants to be bryophytes. Matveyeva et al. (1973) have noted that bryophytes are the dominant plants of their IBP site on the Taimyr Peninsula in the central Siberian Arctic.

The distribution of the major growth forms along the complex moisture gradient in terms of abundance and of number of species per plot in each growth form is very similar to that seen at two other quite different sites, Cape Thompson, Alaska (Johnson et al., 1966), and central Baffin Island, N.W.T. (Webber, 1971). The similarities are so strong that it is a fair assumption to consider these patterns to be universal to tundra. The patterns of growth form distribution corresponding to the redox and phosphate gradients have not been examined in any other studies. Because the growth forms are found in particular segments of the complex gradients, it appears that they provide a meaningful unit for the study of plant–environment responses in the tundra.

The observation that there is not necessarily a correspondence between growth form abundance as measured by the species index, the number of species, and the biomass of a growth form along the various gradients requires examination (Figures 15–17). The pattern for lichens is the same, that is, those segments of the gradients with the greatest number of lichen species also have the greatest species index and biomass. However, the patterns for bryophytes and monocotyledons are not the same when the distribution of these three measures is considered. Several interpretations are possible. The first and most obvious interpretation is that the particular growth form category is invalid and does not comprise species with similar morphological and functional traits. Certainly the bryophyte category is as much phylogenetic as functional A splitting into acrocarpous mosses, pleurocarpous mosses, and leafy liverworts may provide more consistent results (see Rastorfer, 1978).

The plant growth forms recognized here are not unique to tundra (Lewis and Callaghan, 1976). No unique morphological or anatomical structures are known for tundra plants. Even the proportions of the various growth forms are unlikely to be unique although clearly certain combinations are characteristic of tundra as a whole and of specific habitats in particular (Bliss, 1962b; Tikhomirov, 1963). The absence of unique structural features should not be unexpected since it is generally agreed that the present tundra flora is young and is derived largely from preadapted more temperate forms (Savile, 1972; Billings, 1974; Löve and Löve, 1974). Perhaps the absence of specific growth forms such as annuals or trees deserves more attention.

Standing Crop, Production, and Seasonal Growth

There is a substantial number of measurements on standing crops and primary production of tundra (for example, Bliss, 1962b; Alexandrova, 1970; Webber, 1974; Wielgolaski, 1975), but there are only a few estimates of the production of large areas. Too often only the more productive communities have been measured rather than the impoverished communities which make up a large portion of the total arctic landscape (Webber, 1971). There are some measurements of unproductive sites; for example, Warren Wilson (1957) found only 3 g m^{-2} yr^{-1} in an arctic willow barren on Cornwallis Island, and Beschel (1970) estimated that the annual production of lichens on rocks may amount to only a few milligrams per square meter. The Canadian Tundra Biome program has been cognizant of this problem and each production paper in Bliss and Wielgolaski (1973) from Devon Island allowed for the areal extent of the various communities contributing to the total primary production.

The value for aboveground vascular production of 42.4 g m^{-2} yr^{-1} for the 1972 growing season obtained from the mapping estimates is a little less than the nodal mean of 48.1 and the nodal median of 50.2 (Table 7). This is an indication of the relatively few sites with low or high production, but it serves to indicate that significantly large areas of low or high production will significantly affect the productivity of the overall ecosystem. There is a general trend to agreement between measurements of live, vascular leaf area index, aboveground monocotyledon biomass, and aboveground monocotyledon production (Figure 18). Webber et al. (1976) and Dennis et al. (1978) have established a high level of correlation between vascular plant production and leaf area index of tundra. Aboveground biomass and production of vascular plants are also highly correlated in this tundra where so much of the aboveground biomass is an annual phenomenon. Woody plants, lichens, and mosses provide some departure from this generalization because they have a large fraction of aboveground perennial tissue, but productivity for these forms with small or no root system is proportional to the biomass at the beginning of the season (Rastorfer, 1978; Williams et al., 1978). The discrepancy between moss abundance as shown by species index and biomass (Figure 17) may be explained by the greater specific weight of acrocarpous mosses compared to pleurocarpous mosses. Acrocarpous mosses are prevalent on drier sites, are erect, and provide less cover per unit weight than pleurocarpous mosses which dominate wet sites. Nevertheless, overall percentage cover and species index are fair surrogates of actual biomass or productivity. The ordination figures in which species index for various species or growth forms are plotted (Figures 15 and 17) may be interpreted in terms of relative biomass or productivity for their respective taxa.

The dominance of the site by Noda III, IV, and VI leads to the overall dominance of *Carex aquatilis* as previously indicated by Spetzman (1959). The importance of the various major growth forms has already been discussed using cover as a basis (Table 10). An estimate of the importance of these same growth forms with respect to biomass was also obtained from the mapping exercise which allowed the nodal mean biomass values to be corrected for the areal extent of each nodum (Table 11). Again, bryophytes are dominant with a mean value of 117.9 g m^{-2} or 60.8% of the total aboveground biomass. Monocotyledons form only 18.3% of the aboveground biomass with 35.5 g m^{-2}; shrubs form 5.0%, forbs 1.4%, and lichens, which are heavy in relation to their coverage, comprise 14.5%

Table 11.   Percentage Cover, Relative Cover, Peak Season Aboveground Biomass, and Net Aboveground and Total Primary Production of the Principal Growth-Forms of the Terrestrial Vegetation of the IBP Site [a]

Variables	Principal growth-forms					
	Byro-phytes	Mono-cotyledons	Woody shrubs	Herbaceous dicotyledons	Lichens	Totals
Percentage cover	84.5	38.8	9.0	7.7	5.5	145.5
Relative cover	58.1	26.7	6.1	5.3	3.8	100.0
Peak season aboveground biomass (g m $^{-2}$)	117.9	35.5	9.7	2.7	28.1	193.9
Percentage of aboveground peak season biomass	60.8	18.3	5.0	1.4	14.5	100.0
Net production as a percentage of aboveground peak season biomass	21.3[b]	97.2	54.8	96.3	4.0[c]	—
Net aboveground production (g m^{-2} yr^{-1})	25.1	34.5	5.3	2.6	1.1	68.6
Percentage of net aboveground production	36.6	50.3	7.7	3.8	1.6	100.0
Net total production (g m^{-2} yr^{-1})	25.1	69.0	10.6	5.2	1.1	111.0
Percentage of net total production	22.6	62.1	9.6	4.7	1.0	100.0

[a]   Cover and biomass values are derived from a consideration of nodal means and the areal extent of each nodum over the site map. Although it is probably an underestimate, belowground vascular production is assumed equal to that aboveground.
[b]   Estimated from Rastorfer (1978).
[c]   Estimated from Williams *et al.* (1978).

of the biomass. This ranking provides the same sequence as did the nodal means (Table 6), but the biomass values for bryophytes and lichens are higher for the entire map area (Table 11). This is because of the abundance of these growth forms in Nodum III *(mesic meadow)*, which occupies 41% of the map area and is thus the most extensive nodum. By estimating the percentage of peak season biomass, which represents net production of each growth form, an estimate of mean net production for the map area was obtained (Table 11). The productivity estimates for bryophytes were obtained from Rastorfer (1978) and for lichens from Williams et al. (1978). The values for the vascular plants were taken from the peak season clip harvest of 1972 by comparing the start of season biomass with annual net production. Rastorfer (1978) and Williams et al. (1978) derived their values from studies at site 2 in vegetation corresponding to Nodum IV. Using the mean biomass values of the present study, these percentage production values provide a net production value of 25.1 g m^{-2} yr^{-1} for bryophytes and 1.1 g m^{-2} for lichens. The aboveground net production of monocotyledons, shrubs, and forbs was 34.5, 5.3, and 2.6 g m^{-2} yr^{-1}, respectively, which represents an aboveground net production for vascular plants of 42.4 g m^{-2}. This value is the same as derived by slightly different means for the whole map area (Table 7). Therefore, vascular plants have greater production per unit of aboveground biomass. However, on a total end-of-season basis of, say, 750 g m^{-2} (that is, above- plus belowground) and assuming aboveground production equals that of belowground (Johnson and Kelley, 1970), the percentage would only be 11.3. The clip harvesting methods in this study estimated only 3% of the peak season herbaceous vascular material as originating from the previous year (Table 11). This agrees with Dennis and Tieszen (1972) and Tieszen (1972a) who show rather more than this and 8 to 10 g m^{-2} is recorded from site 2 (Figure 19). In the more northerly, but similar, ecosystem on Devon Island, Muc (1973) found from 4.0 to 7.1 g m^{-2} of green material at the onset of growth which represents 8.8 to 15.6% of the annual vascular production estimated to be 45.7 g m^{-2}.

The above discussion points to the importance of bryophytes in the Barrow ecosystem. Bryophytes not only comprise the largest portion of the plant cover, they also are the most productive in terms of a total perennial biomass and contribute to 36.6% of the aboveground net production and 22.6% to the total net production (Table 11). This supports Rastorfer's (1978) contention that bryophytes are an important major component in arctic tundra. Nevertheless, estimates of total production show that vascular plants contribute up to 61.8% of the aboveground or 76.4% of the above- and belowground productivity. The contribution of lichens is very low but may be important for herbivores (White et al., 1975) and nitrogen fixation (Alexander et al., 1974).

The above generalized values for the aboveground vascular production for the tundra of the IBP site are within the ranges reported by others for Barrow tundra (Thompson, 1955; Shanks in Bliss, 1962a; Pieper, 1963; Dennis, 1968; Dennis et al., 1978) and for similar ecosystems elsewhere in the Arctic (Bliss, 1956; Alexandrova, 1970; Wielgolaski, 1972, 1975; Muc, 1973; and Tikhomirov et al., 1975). There are few published values for cryptogams. However, the estimates presented here are very comparable to those for the wet sedge meadows of

Devon Island where Pakarinen and Vitt (1973) estimated a value of 30 g m^{-2} yr^{-1} for bryophytes and Richardson and Finegan (1973) estimated a value between 0.4 and 2.1 g m^{-2} yr^{-1} for lichens.

Our values for standing crop of belowground biomass and its proportion with that found aboveground at the time of peak aboveground vascular biomass are within the several published ranges. All studies of belowground biomass in tundra emphasize the tendency for small aboveground to belowground ratios (A:B) (Alexandrova, 1958, 1970; Bliss, 1962b, 1966, 1970; Mooney and Billings, 1960; Scott and Billings, 1964; Rodin and Basilevich, 1967; Dennis, 1968, 1977; Dennis and Johnson, 1970; Wielgolaski, 1972, 1975; Muc, 1973; Webber, 1974, 1977; Webber and May, 1977). It is appropriate to compare only data which deal with the same or similar vegetation [for example, Dennis (1977) for Barrow and Muc (1973) for Devon Island]. The magnitude of live rhizomes and roots reported by Dennis (1977) match very closely with those in this study. The values reported by Muc (1973) suggest that on Devon Island there is only about half the rhizome biomass but a similar amount of root biomass to that found at Barrow. Vascular aboveground production for wet meadows on Devon Island is about two-thirds that of Barrow. However, Muc (1973) reports a much higher belowground production of around four times that of aboveground. The A:B biomass ratios of Dennis and Johnson (1970), although sometime smaller, are generally similar to the present. The small values were explainable on the basis of unusually low aboveground current production. This points to the need to have either an average value from several years for aboveground production or a value representative of an average year. This need for an average value is not so critical for belowground biomass, which is a perennial system and which is more likely to be a damped system. On Devon Island, the A:B biomass ratios at peak season range from 1:6.1 to 1:13.6 and these seem, on the whole, a little larger than at Barrow. However, Muc (1973) did not include stem bases with the belowground portion. If stem bases are regarded as aboveground material the ratios at Barrow would change considerably from around 1:13.5 to around 1:5.5. This illustration emphasizes the need for further care in the use of A:B ratios and for the standardization of definitions, especially the boundary between above- and belowground in situations of aggrading substrates (Webber, 1977).

The present study did not measure belowground production at Barrow. Dennis (1977), using a regression approach, estimated that belowground production was 143 g m^{-2} yr^{-1} in 1971, which is 1.6 times greater than the aboveground vascular production. Previously, Dennis (1968) had used another indirect method of estimating belowground production, based on an estimate of decay rates. Using a decay rate of 140 g m^{-2} yr^{-1} (Douglas and Tedrow, 1959) and the assumption that the system was in steady state, he calculated on the basis of 70 g m^{-2} yr^{-1} aboveground production rate and an equal rate for belowground production. The current IBP effort has provided new estimates of total site decay. Flanagan et al. (in preparation) consider the average decay loss at Barrow to be 310 g m^{-2} yr^{-1}. If we use Dennis' (1968) method and the present estimate of average aboveground production for vascular plants and cryptogams to be 68.6 g m^{-2} yr^{-1} (Table 12), then belowground production would equal 241.4 g m^{-2} yr^{-1}

Table 12.   Results of the Planimetry of the Three Available Vegetation Maps[a]

Nodum	Percentage cover		
	Map 1 (1:300) Lemming Grid (2.5 ha)	Map 2(1:300) Topographic Grid (4.3 ha)	Map 3 (1:5000) entire IBP site (110 ha)
I	0.8	0.5	3.0
II	13.0	11.2	7.2
III	30.3	22.1	41.0
IV	21.3	24.7	20.9
V	10.1	26.7	6.9
VI	23.8	14.4	14.6
VII	0.7	0.4	2.3
VIII	0	0	4.1
Mean production $(g\ m^{-2}\ yr^{-1})$	41.1	43.0	42.4

[a] The percentage areal extent of each nodum and the mean aboveground vascular production per nodum (Table 7) were used to estimate mean overall aboveground vascular production of the area covered by each map. Map 3 is Figure 10.

or 5.7 times aboveground vascular production. Such an estimate is rather tenuous in the absence of knowledge of whether or not the system is in steady state (see Bunnell et al., 1975) and also because several reports indicate belowground production to range from equal to or half as much again as the aboveground vascular production (Wielgolaski, 1975). In this study, the more conservative estimate of Johnson and Kelley (1970) and Dennis and Johnson (1970) in which belowground production equals that aboveground has been used to estimate total production and to estimate various biomass and standing crop turnover rates.

There are only a few published turnover rates for tundra (Dennis, 1968, 1977; Dennis and Johnson, 1970; Johnson and Kelley, 1970; Bliss and Mark, 1974; Shaver and Billings, 1975; Webber and May, 1977). The previous estimates for Barrow tundra suggest that the aboveground vascular standing crop of live and dead material, including litter, takes between 4 and 5 yr to turn over (Dennis and Johnson, 1970; Johnson and Kelley, 1970). The values for this study range from 1.9 to 7.9 yr along the increasing moisture gradient with a mean of 4.5 yr. For drier alpine tundras, Bliss and Mark (1974) and Webber and May (1977) have estimated rates of turnover of aboveground standing crop to be respectively 8 and 15 yr and from 4.1 to 9.6 yr. These latter values also increased along a site moisture gradient. These above observations demonstrate the positive correlation of site moisture with aboveground turnover rate. Thus the turnover rate of the short-lived aboveground system, which is a measure of the rate of decay of surface material, is water limited. Surface decay rates are highest in the emergent Nodum VII and the pond margin vegetation (Nodum VII), and least in the much drier heaths (Noda I and II).

Dennis (1968) originally estimated turnover rates for belowground biomass to be at least 10 to 20 yr, but after further study (Dennis, 1977) he proposed 10 to 15 yr for rhizomes and 3 to 6 yr for roots. On this basis, the mean turnover time is likely to be less than 10 yr since roots represent the majority (ca. 70%) of the belowground biomass. In Table 6 belowground biomass turnover rates based on aboveground production range from 4.5 to 31.5 yr for the principal noda with a mean of 20.9 yr. If belowground production is substantially higher, then so would be the turnover rate. On the basis of a belowground production of 241.4 g m^{-2} yr^{-1} and a mean belowground biomass of 816.5 g m^{-2} for the principal noda, the turnover time or average residency for belowground structures would be 3.4 yr. Such a turnover rate seems too rapid in view of the expectation that in systems of low productivity one might expect low turnover rates of plant parts (Shaver and Billings, 1975). However, in view of the average ages of tundra plants to be obtained from the literature a mean turnover rate of 20 yr seems too slow. Sørenson (1941) found the shoots of most tundra geophytes, including *Carex aquatilis,* to live from 4 to 9 yr, and Shaver and Billings (1975) have shown that the roots of *Carex aquatilis* may live from 5 to 8 and even 10 yr, and that those of *Dupontia fisheri* may live from 4 to 6 yr, but that the roots of *Eriophorum angustifolium* are annual. Some woody roots may exceed 50 yr of age (Beschel and Webb, 1964), but these appear exceptional. At the present time the above evidence and arguments suggest that belowground production is greater than aboveground vascular production, and that a mean turnover rate of 5 to 10 yr would be reasonable. The same trends of decay can also be deduced from the distribution of belowground dead intact matter (Figure 18). Here, the driest sites, in spite of low production, have the highest concentrations and presumably accumulations of organic matter; the wet sites reverse this pattern.

The extreme values of standing crop fractions and aboveground production shown in Table 6 and Figure 18 do not always match. For example, litter and prostrate dead are higher and A:B biomass ratios are lower in the ordination than the nodal means. These discrepancies are the result of the nodal averaging which masks extremes while the ordination method is more likely to emphasize trends and extremes. Nevertheless, the nodal trends (Table 6) match those of the ordination quite closely, except perhaps for the A:B biomass ratios. In this instance, the within-nodal ranges were very large and have masked completely the pattern shown by the ordination.

Several factors contribute to the small ratios of A:B biomass in tundra. The growth form of tundra plants is the prime cause. These are predominantly hemicryptophytes or cryptophytes which have essentially annual aboveground systems and perennial, long-lived, belowground storing systems. Under experimental conditions and along gradients of increasing environmental severity, such as decreasing nutrients, low temperatures, increased wind, and high moisture levels, the A:B biomass ratio decreases in temperate and tundra plants (Scott and Billings, 1964; Whittington, 1967; Chapin, 1974; Wielgolaski, 1975). Thus the tundra environment may further reduce the size of the ratio. A study such as the present, where measurements have been made along several gradients, provides an opportunity to look for evidence of decreasing A:B ratios corresponding to gradients of environmental severity. The lowest ratios from the IBP site are from a few sites belonging to Nodum VI *(wet Carex meadow).* These sites are not the wettest although they have standing water in the spring. They also have low phosphate and low soil aeration. This supports the hypothesis that small A:B ratios may be a response to harsh environmental conditions and that a large belowground biomass is advantageous in nutrient-poor sites with poor aeration. The patterns of belowground biomass and the A:B biomass ratio in Figure 18 are distinctly different and clearly show the effect of the depressed aboveground peak season biomass in the few sites of Nodum VI which have low production. These sites, however, did not manifest themselves in the ordination framework depicting the distribution of production (Figure 18).

Webber and May (1977) have used identical analyses to examine the distribution of standing crop and productivity fractions on Niwot Ridge, Colorado. They found that A:B biomass ratios ranged from 1:3 to 1:25 and generally decreased along a complex moisture gradient. The smallest ratios (1:25) occurred in a moist tundra nodum and the largest ratios (1:3) in a shrub tundra nodum. The average ratio for this predominantly dry tundra is 1:10. This may be compared with the corresponding value of 1:13.5 for the Barrow tundra. The somewhat lower extremes at Barrow may well be a response to the more severe and especially wetter Barrow environment. Niwot on the average has three to four times greater aboveground net vascular production than Barrow and four to six times greater total vascular biomass. Alexandrova (1970) and Wielgolaski (1975) have noted that A:B biomass ratios tend to decrease along complex moisture gradients. This general trend is seen on Niwot Ridge but is not apparent at Barrow where the ratio first decreases and then increases (Figure 24). At the risk of pushing the results and the value of the A:B ratio too far, I submit that in general in the Arctic Coastal Plain the ratio will be found to decrease from dry, former beach ridges and lake and river bluffs to moist and wet meadows with poorer nutrient and oxygen levels and then increase again in the very wet meadows where nutrient supply is more favorable. Even on Niwot Ridge, the wettest sites show a tendency to have ratios a little larger than the less wet, moist tundra (Webber and May, 1977; Figure 1). However, the significance of the A:B biomass ratio is questionable (Dennis and Johnson, 1970; Dennis, 1977; Webber and May, 1977) and the measure has several serious weaknesses, for example: the difficulty of sampling belowground material; the uncertainties of distinguishing live from dead material; the problems of defining the actual ground level boundary; and finally, the problem associated with this relative measure which tells nothing of the absolute biomass values from which it was derived. Thus caution should be exercised against an overinterpretation of the A:B biomass ratio.

Although Muc (1972) has reported some growth to occur under snow cover on Devon Island, Tieszen (1974) has shown that plant growth does not begin at site 2 (Nodum IV) at Barrow until the snow has melted. The onset of soil thaw also matches very closely with the disappearance of snow (Brown et al., 1974). Thus, although the onset of plant growth as shown by Figure 19 coincides with the advent of nonfreezing air temperatures, the important trigger to growth is the warming of the ground. Nevertheless, over much of the IBP site, snow melt happens almost simultaneously and is closely correlated with air temperatures which are above freezing. The correspondence of growth with thawing soil matches the very careful observations of Sørensen (1941) from Greenland. He found a noticeable lag of 3 to 4 weeks of positive mean air temperatures behind those of soil and found that plant growth was keyed to soil temperature rather than air temperature, although he acknowledged the causal effect of air temperature in the first instance. Sørensen (1941) also noted that soil thaw did not start until the snow had disappeared and he emphasized the great control of snow cover over plant life and development in the tundra. Holway and Ward (1965) have also noted the control of snow cover over plant phenology and the onset of growth.

Several studies at Barrow have noted the timing of the period of peak biomass to within the first week of August (Pieper, 1963; Dennis and Tieszen, 1972; Tieszen, 1972a). Tieszen (1972b) has shown this is also the time when there is an increase of translocation of products of photosynthesis to belowground organs and when photosynthesis is no longer positive throughout the day. Some sort of extrinsic or intrinsic timing mechanism comes into play but its nature is not known (Bunnell et al., 1975). The onset of plant growth varies from year to year (Figure 20), but the later phenophases such as the development and ripening of fruits are more synchronous. Experimental studies using snow-fences to increase snow cover and delay snow melt by up to 14 days in the Niwot Ridge alpine have also shown that the later phenophases are in close synchrony with the control plots which had no snow fences (Webber et al., 1976). Wielgolaski's (1974) data also show that the fruiting phenophases have a greater degree of synchroneity between sites than do the early phenophases. Again, some sort of timing mechanism which controls plant development and senescence is suggested.

The Barrow species seem remarkably adapted to the short growing season, as the first fruits are mature before the return of mean freezing air temperatures. The mean length of the thaw season with mean air temperatures above freezing is 87 days and the longest time needed for development was 73 days for *Pedicularis lanata* which occurs on those sites which are first to become free of snow. Pieper (1963) and Dennis (1968) made phenological observations on many of these same species in 1962, 1964, and 1965, respectively. They ranked the species in much the same phenologic sequence as the present one (Figure 22). The only remarkable difference was the very early development and flowering of *Petasites* in 1964 (Dennis, 1968). In spite of the ability of these plants to complete their sexual cycle within a short season an early frost often prevents seed set and will reduce germinability (Webber et al., 1976), and this may be quite frequent at Barrow where the frost-free period averages only 34 days (Figure 1). Late flowering species such as *Carex aquatilis* and *Dupontia fisheri* will thus be particularly vulnerable to early frost. In fact, Dennis (1968) stressed that these plants rely much more heavily on vegetative reproduction than some of the early developing species such as *Ranunculus nivalis*.

Temperature has a strong effect on plant growth and production. The onset and end of growth, the rate of phenological progression, the stature of plants, and the amount of willow stem elongation are all positively correlated with air temperatures. However, production as measured by the clip-harvesting showed no clear correlation with the temperature of the current year or even, as Sørensen (1941) has noted, the previous year.

Plant Succession

Plant colonization of new surfaces or of disturbed surfaces and subsequent plant succession is an important topic in the Arctic, especially with the prospect of increasing disturbance and development of arctic resources. Unfortunately, the problem is complex and really no better understood at the present time than when Churchill and Hanson (1958) wrote their comprehensive review of tundra succession. Too often, as in this present report, successional patterns are interpreted by inference without either an adequate set of observations through time or dated surfaces (Drury and Nisbet, 1973; for tundra examples, see Oosting, 1956; Polunin, 1935; Billings and Mooney, 1959; Spetzman, 1959; Marr, 1969; Johnson and Tieszen, 1973). The scheme which has been adapted from Britton's (1966) observations to suit the present classification is oversimplified and may apply only to the immediate study area and no doubt will be revised.

The present observations on the rapid colonization of Footprint Lake concur with others at Barrow. For example, Carson (1968) observed drained lake strands to revegetate in 10 to 20 yr, and Britton (1966) has suggested that only a few decades are necessary to produce patterned ground and sorting of populations to create vegetation similar to surrounding, much older, former lake basins.

The entire successional scheme presented here emphasizes the dominance of the microrelief and geomorphic processes on the control of vegetation composition. The observation that the population of drained lake surfaces varies by species according to local microrelief drainage conditions and substrate composition and the fact that some of the resulting vegetation may be slow to change to another type suggests that Muller's (1952) concept of *Autosuccession* may apply in some situations. This concept, which was developed from tundra observations, allows for succession in which there are no intermediate stages so that the final stable self-perpetuating plant assemblage has a similar composition to that of the pioneer stand. However, it may be too soon to judge the stability of the Footprint Lake sediments since the lifespan of many of the individuals of a species may be close to 10 yr (Sørensen, 1941; Shaver and Billings, 1975). Autosuccession implies that competition between species and ensuing integration of the plant community is minimal and that environmental controls are dominant. In such a succession, *autogenic* (plant) controls are secondary to *allogenic* (environmental) controls (see Tansley, 1935). Dansereau (1954) has commented that in the Arctic, allogenic controls override autogenic controls in the determination of vegetation composition. Further evidence that environmental controls are predominant over plant controls is indicated by the rapid response of vegetation to even subtle changes in substrate conditions. For example, natural or man-induced nondestructive thermokarsting and road impoundments create changes in vegetation composition in a few seasons. However, as Tansley (1935) emphasized, both allogenic and autogenic controls can act simultaneously and are not mutually exclusive. The closed nature of the vegetation at Barrow and the accumulation of organic material containing nutrients attest to at least a measure of control of the plants themselves over the environment. From time to time, *biogenic* (principally animal) controls (Tansley, 1935; Dansereau, 1957, 1974) are also striking; lemmings provide a documented example of such a control (Thompson, 1955; Dennis, 1968). Catastrophic changes are not regarded as part of succession. They are simply means of providing a new surface for succession (Tansley, 1935). Massive thermokarsting would be in this latter category.

The basic frame of the successional scheme presented here can be viewed as an ordination in which the various pathways are complex gradients. Time is one of the components of these gradients, but it is not a constant from step to step. The rates of change are extremely variable. On one hand, there are the rapid processes such as colonization of new, stable surfaces and responses to subtle changes of the substrate and also to fluctuating lemming pressures. These changes may all be clearly visible from between one to several years. On the other hand, however, there is an apparent steady state over most of the landscape and apart from normal, small, year-to-year fluctuations, the vegetation seems to have been unchanged for decades and even longer. Even many of the cryoturbating surfaces, such as frost boils, appear to have been almost indefinitely in their present state. It is generally thought that it has been several thousand years since the site was last occupied by a thaw lake (Brown, 1965) but the age of the site does not equate to surface stability, especially in the Arctic. Nevertheless, the ice-wedge polygons might be from 500 to 600 yr in age (Leffingwell, 1919; Lachenbruch, 1962), and the large accumulations of organic matter, which are frequently 100 to 500 times that of a liberal estimate of belowground annual production, suggest that several generations of tundra individuals have occurred. Therefore, as a working hypothesis, most of the Barrow vegetation is considered to be in a steady state. There is, however, little strong evidence for this steady state concept and the testing of this hypothesis must wait until the present permanent plots can be reexamined.

The present successional scheme, which is cyclic, precludes the identification of one climax type for the area because any one of the stable mesic stages might qualify as a climax. Both the *polyclimax* (Tansley, 1935) and the *climax pattern* (Whittaker, 1953) concepts would fit this scheme. The latter conforms best from a theoretical point of view because of the continuous nature of the vegetation along the various gradients. However, from a pragmatic viewpoint, the *polyclimax* concept is best. This allows for a practical application of the present nodal concept and accommodates the distinct pattern of habitats provided by the icewedge polygon complex. Thus any mature nodum is a topoedaphic climax (*sensu* Marr, 1961).

The orderly progression of species, number of species per plot, and various productivity and standing crop measures within the successional scheme give it a sense of credibility. However, the scheme is more a moisture gradient sequence than a time-controlled sequence and may remain so after testing. Therefore, at this time it is sensible to refrain from trying to interpret species diversity and productivity trends in terms of successional sequences. The literature for temperate vegetation conflicts sufficiently (Whittaker, 1965; Pielou, 1966; Loucks, 1970) without attempting to test these data for generalizations such as those of Margalef (1968) and Odum (1969).

Representativeness

The earlier studies of the program were heavily process oriented and had to be based on the study of a few species (*Carex aquatilis, Dupontia fisheri, Eriophorum angustifolium, Pogonatum alpinum, Dicranum fuscescens,* and *Peltigera aphthosa*) at sites 1 and 2 (Miller and Tieszen, 1972; Tieszen, 1972a, 1972b; Collins and Oechel, 1973; Allessio and Tieszen, 1974; Oechel and Collins, 1974; Coyne and Kelley, 1975; Stoner and Miller, 1975; Billings et al., 1977). Later attention was turned to a greater diversity of habitats and some additional species (Alexander and Schell, 1973; MacLean, 1974; Miller and Laursen, 1974; Rastorfer, 1978; Williams et al., 1978). My assessment of the representativeness shows that indeed the experimental plots and species which underwent intensive examination were representative of the IBP site at Barrow.

The map records the distribution of landform assemblages and their most abundant nodal vegetation type. This can be misleading because the abundant vegetation type may occupy as little as 20% of the landform unit and five other noda might also be present in more or less equal amounts. In other words, is the map useful as a generalization of the distribution of the noda and as a means of estimating the primary production of the entire site? In an attempt to assess this problem, a comparative analysis was made of the nodal composition of some large scale maps which covered small areas and which mapped individual polygons and polygon parts. Two maps (1:300) were made of areas of site 4, which is the most complex and therefore problematical area to map at a smaller scale. One map, the Lemming Grid, was 2.5 ha (158 × 158 m) in area. The other map, the Topographic Grid, because it was the site of detailed vegetation, microrelief, and soil studies (Walker, 1977), was 4.3 ha (390 × 110 m) in area. The Lemming Grid is typical of the part of site 4 which is coded as Nodum III in the 1:5000 map, while the Topographic Grid ranges across several landform assemblages each dominated in turn by the major noda. The planimetric analyses of all three vegetation maps are provided in Table 12. The prevalance of Nodum III seen in the Lemming Grid and the more or less equal overall presence of Noda III, IV, and V in the Topographic Grid match with the information for the 1:5000 map. Apart from the absence of Nodum VIII in the two large scale maps, the relative representation of each nodal type is similar in all three maps. Also the mean aboveground production of each map calculated from the mean nodal production and the map area dominated by a nodum are very similar. The slightly higher mean productivity of the Topographic Grid is explainable on the basis of a higher occurrence of Nodum V in the Topographic Grids than in the other areas. Nodum V has a higher productivity than Noda III and IV (Tables 6 and 7). This mapping exercise indicates that the information provided for the overall IBP site by the 1:5000 map (Fig. 10) is both usable and representative with regard to the distribution of noda.

The 43 plots used for the ordination, production, and standing crop studies were selected to represent the overall range of undisturbed vegetation of the IBP site. These plots contained 112 taxa of which 55 were vascular plants, 28 were bryophytes, and 29 were lichens. In the entire Barrow region, the flora has been estimated at 295 taxa of which 125 are vascular plants, 95 are bryophytes, and 75 are lichens (Murray and Murray, 1978). Thus, of the total available flora, only 38% of the taxa were encountered and this comprised 44% of the available vascular plants, 29% of the available bryophytes, and 39% of the available lichens.

Such a low representation of the regional taxa by the sampled plots is primarily a result of the low diversity of the IBP sites compared to the whole Barrow region which contains habitats such as beaches, salt marshes, and extensive well-drained ridges which do not occur within the IBP area. Two other factors contribute to the low representation. These are the small number and size of the plots and my level of taxonomic competence. Undoubtedly, more and larger plots would have encountered additional species, but the dominance diversity curve plotted from the data in Table 10 and for the remaining 72 species makes a strong lognormal curve with a long asymptotic upper part, a good indication of a thorough sampling of an available universe (Patrick et al., 1954; Webber, 1971). I am experienced with the arctic vascular flora although less so with the cryptograms. Only those bryophytes and lichens which I felt we could recognize with certainty were documented. Several lichens and many mosses which occurred in the sampled quadrats were ignored. No attempt was made to collect exhaustively every unrecognized scrap or strand for later determination. Smith (1974) and Rastorfer (1978) have shown how many species of bryophytes can be added within the IBP site through meticulous testing of small sods of bryophytes. In conclusion, the 43 sampled plots do represent the IBP site with regard to the vascular species, but they were undersampled for cryptogams. The studies of Smith (1974) and Rastorfer (1978) make up for some of this deficiency.

The selection process of sites, species, and growth forms by the scientists has a natural tendency to ensure that the area is represented. This is because only those units which are sufficiently extensive or abundant to allow for adequate replication and control are selected. It is these abundant and extensive units which not only characterize the area but also provide the best clues to the understanding of the adaptation and functions of the regional ecosystem.

The majority of primary production process studies were made at sites 1 and 2. Sites 1 and 2 are located on landscape units dominated by Nodum IV (moist Carex meadow) which occupies an estimated 21% of the region (Table 7). Although Nodum IV is not as abundant as Nodum III, it offers larger, more uniform areas than Nodum III (mesic meadow), which is often restricted to hummocky polygon rims and centers (Table 1). Nodum IV also offers a much more average aspect of the total vegetation in terms of species composition (Table 2, Figure 23), position on the major gradients (Table 5, Figure 13), and productivity (Table 6). Thus, sites 1 and 2, especially site 2, which is farther from vehicle and other disturbances, are well suited as locations for making detailed physiological experiments. Site 4 was selected to provide a greater diversity of habitats and plant life than that provided by sites 1 and 2. The Topographic Grid covers much of site 4 and it contains all the noda except the *pioneer meadow* (Nodum VIII). Site 4 is a complex of ice-wedge polygons of all denominations ranging from incipient through flat center and low center to high-center types. Thus, site 4 provides a high degree of diversity and its study ensured representation of the area. The lack of Nodum VIII *(pioneer meadow)* is not considered critical.

The most intensively studied species were the monocotyledons *Carex aquatilis, Dupontia fisheri,* and *Eriophorum angustifolium* (for example, Tieszen, 1972a and 1972b; Allessio and Tieszen, 1974; Chapin and Bloom, 1976; Billings et al., 1978; Stoner et al., 1978). This decision seems fair since monocotyledons are the most important vascular plants of the region and *Carex aquatilis* is the most abundant species (Table 10). *Poa arctica,* which is the second most important monocotyledon, has not been studied so intensively as the other monocotyledons. If abundance is an important criterion for study, *Poa* should have been given preference over *Dupontia* as a tundra graminoid, especially as it is much more wide ranging. However, *Dupontia* is a uniquely arctic genus and this alone would be sufficient justification for its inclusion. The second most abundant vascular plant, according to Table 10, is *Salix rotundifolia,* which has been largely neglected. The principal woody plant to have been studied in physiological terms is the less abundant *Salix pulchra* (Stoner and Miller, 1975). The more massive stems and leaves of *Salix pulchra* compared to those of *Salix rotundifolia* may have been the reason for its selection. Herbaceous dicotyledons were somewhat neglected as experimental organisms. Some work was done with *Petasites frigidus* and *Potentilla hyparctica.* While *Petasites* is fairly frequent throughout the study area, *Potentilla* is restricted to the dry, slightly unstable surfaces of Nodum I *(dry Luzula heath).* This latter point in itself may be reason enough for study. However, a better herbaceous dicotyledon for intensive study, at least on the basis of high frequency and abundance, would have been *Saxifraga cernua.* The principal bryophytes which were studied were *Pogonatum alpinum, Dicranum elongatum, Dicranum fuscescens,* and *Calliergon sarmentosum* (Collins and Oechel, 1973; Oechel and Collins, 1974). All four species are excellent choices as they together comprise 25% of the total relative cover of all taxa sampled by the 43 extensive plots (Table 10) and also they range across the complex moisture gradient with *Pogonatum* prevalent on dry sites, the *Dicranum* spp. on mesic sites, and *Calliergon* on wet sites. The principal lichens to be examined were *Dactylina arctica, Cetraria richardsonii* (Williams et al., 1978), and *Peltigera aphthosa* (Alexander and Schell, 1973). *Dactylina* is the most ubiquitous and abundant lichen on the IBP site. *Cetraria richardsonii* and *Peltigera aphthosa,* although occasional, are never abundant. *Cetraria* is an important arctic genus, but it would have been better represented by *Cetraria islandica,* which has a more typical lichen behavior. Both *Cetraria richardsonii* and *Peltigera aphthosa* occur in quite wet sites which is at odds with the distribution of the majority of tundra lichens (Table 2). *Cetraria richardsonii* does, however, have a large, strong thallus and *Peltigera* is able to fix nitrogen, and these characteristics may have played a part in their selection.

Summary and Conclusions

1. The site is characteristic of the seaward areas of the Coastal Plain physiographic province of the Alaskan Arctic Slope. It has a cool, moist climate and it is flat, poorly drained, and underlain by permafrost. The vegetation changes character every few meters in concert with the microrelief of ice-wedge polygon complexes.

2. Eight broad plant communities or noda were recognized for the study area on the basis of species composition: Nodum I, *dry Luzula heath;* Nodum II, *mesic Salix heath;* Nodum III, *mesic Carex–Poa meadow;* Nodum IV, *moist Carex–Oncophorus meadow;* Nodum V, *wet Dupontia meadow;* Nodum VI, *wet Carex–Eriophorum meadow;* Nodum VII, *Arctophila pond margin;* and Nodum VIII, *Cochlearia pioneer meadow.* The noda permit the classification of experimental plots and also the construction of a vegetation map of the IBP area. The map permitted an areal estimate to be made of the extent of each nodum and an estimate of the overall aboveground production.

3. The axes of a three-dimensional indirect stand ordination correlated with substrate complexes relating to moisture, aeration, and phosphate. The environmental framework provided by the axes of the ordination can be used to describe the distribution of species, growth forms, noda, various standing crop fractions and ratios, and production in terms of these gradients. Other environmental controls were identified; among these were wind, snow cover, temperature, depth of active layer, substrate stability, and the effects of lemming grazing. The interrelation of the environmental variables is complex but they are all initially controlled by the microrelief. Thus, within the climatic and historic setting of the region, it is the microrelief which determines the character and performance of the vegetation and its components.

4. The undisturbed vegetation cover of the site is complete and dominated by bryophytes and monocotyledons. Bryophytes form 58% of the cover, monocotyledons 27%, shrubs 6%, forbs 5%, and lichens 4%. The most abundant species is *Carex aquatilis,* which forms 13% of the cover. Six species form 50% of the total cover, 40 species form 96%, and the remaining 70 of the 112 species recorded form only 4% of the cover. The dominant species are wide ranging along the principal environmental gradients and occur in nearly all noda. Nevertheless, each species has a unique distribution and characteristic location within the ordination. Ten plant growth forms were recognized and each of these has a characteristic distribution.

5. The nodal means of aboveground net production of vascular plants ranged from 18.1 to 118.5 g m^{-2} yr^{-1}. On the basis of overall decay rates and longevity of belowground structures, the belowground production may be up to five to seven times higher than that aboveground. The mean nodal standing crop of belowground biomass ranges from 153 to 1305 g m^{-2} with a mean value of 734 g m^{-2}. Aboveground vascular production increases along the complex moisture gradient, but the ratio of aboveground to belowground biomass does not show such a simple pattern. The latter is lowest in wet anaerobic sites with low levels of soluble phosphate. It is higher in dry, well-aerated sites, and wet, phosphate high sites. Various turnover rates of above- and belowground standing crop

fractions show that decay rates increase with increasing site and soil moisture. The diversity or number of species per vegetation plot is highest on dry to mesic sites and low on very dry sites and on very wet sites. The first signs of plant growth occur soon after snowmelt. Plants can complete the majority of the production in 60 days and can produce seeds within 75 days. Although the mean length of the thaw season is 87 days, the frost-free season only averages 34 days. This emphasizes 'the importance of vegetative propagation and the perennial habit in tundra plants.

6.   Colonization of new or disturbed surfaces takes place rapidly and a complete cover of vegetation can be produced in 20 yr. Physical rather than plant controls dominate the Barrow vegetation, and the principal noda appear to be able to persist in the landscape for long periods until changed by some physical process.

7. Several topics have emerged from this study as prime targets for future research on the producer component of the Barrow ecosystem. The principal topics in no special sequence are: plant succession including competition, assessment and study of tundra growth forms, production of bryophytes and lichens, production of the belowground system, and reproductive aspects of phenology.

Acknowledgments. This work was made possible by a grant from the National Science Foundation, GV-29350, to the University of Colorado. I wish to thank my colleagues and graduate students and the staff of the Institute of Arctic and Alpine Research, University of Colorado, for many hours of laboratory support. In particular, I wish to thank Dr. J. D. Ives for his unflagging efforts to provide all that I thought was necessary; Dr. Vera Komárková, Dr. J. C. Emerick, Dr. Diane May, Dr. T. May, Mr. D. A. Walker, Mr. J. Batty, Mr. K. Sutherland, and Ms. Susan Vetter Clark were among the many willing field assistants who gathered and helped analyze the raw samples and data; Mr. Rolf Kihl made, with great enthusiasm, the soil analyses; Mrs. Margaret Eccles gave graciously of her computer expertise; Parry Donnelly, Vicki Dow, and Karen Sproule drafted the figures; and Karen Seabert and Laura Kohn typed the manuscript. Dr. L. C. Bliss read and made helpful comments on an early version of the manuscript. Finally, I wish to thank my scientific colleagues of the U.S. Tundra Biome program. In particular, I wish to thank Dr. J. Brown and Dr. L. L. Tieszen for their support and patience.

References

Alexandrova, V. D. (1958) An attempt to measure the overground and underground productivity of plant communities in the arctic tundra. *Bot. Zh.,* **43**: 1748–1762 (in Russian).

Alexandrova, V. D. (1970) The vegetation of the tundra zones in the USSR and data about its productivity. In *Productivity and Conservation in Northern Circumpolar Lands* (W. A. Fuller and P. G. Kevan, Eds.). Internat. Union Conserv. Natur., Morges, Switzerland, Publ. **16**: pp. 93–114.

Alexander, V., and D. M. Schell. (1973) Seasonal and spatial variation of nitrogen fixation in the Barrow, Alaska, tundra. *Arct. Alp. Res.,* **5**: 77–88.

Alexander, V., M. Billington and D. M. Schell (1974) The influence of abiotic factors on nitrogen fixation rates in the Barrow, Alaska, arctic tundra. Report Kevo Subarctic Research Station **11**: 3–11.

Allessio, M. L., and L. L. Tieszen. (1974) Leaf age on translocation and distribution of C14 photoassimilate in *Dupontia fisheri* at Barrow, Alaska. *Arct. Alp. Res.,* **7**: 3–12.

Barrett, P. E. (1972) Phytogeocoenoses of a coastal lowland ecosystem, Devon Island, N.W.T. PhD. dissertation, University of British Columbia, Vancouver, 292 pp.

Beschel, R. E. (1970) The diversity of tundra vegetation. In *Productivity and Conservation in Northern Circumpolar Lands* (W. A. Fuller and P. G. Kevan, Eds.). Internat. Union Conserv. Natur., Morges, Switzerland, Publ. **16**: pp. 85–92.

Beschel, R. E., and D. Webb. (1964) Growth ring studies on arctic willows. In *Axel Heiberg Island, Preliminary Report* (F. Müller, Ed.). McGill University, Montreal, pp. 189–198.

Beschel, R. E., and P. J. Webber. (1963) Bermerkungen zur log-normalen Struktur der Vegetation. Berichte des Naturwissenschaftlich Medizinischen Vereins in Innsbruck. *Festschrift Helmut Gams,* **53**(1959–1963): 9–22.

Billings, W. D. (1974) Adaptations and origins of alpine plants. *Arct. Alp. Res.,* **6**: 129–142.

Billings, W. D., and H. A. Mooney. (1959) An apparent frost-sorted polygon cycle in the alpine tundra of Wyoming. *Ecology,* **40**: 16–20.

Billings, W. D., K. M. Peterson, and G. R. Shaver. (1978) Growth, turnover, and respiration rates of roots and tillers of tundra graminoids. In *Vegetation and Production Ecology of an Alaskan Arctic Tundra* (L. L. Tieszen, Ed.). New York: Springer-Verlag, Chap. 18.

Bliss, L. C. (1956) A comparison of plant development in microenvironments of arctic and alpine tundras. *Ecol. Monogr.,* **26**: 303–337.

Bliss, L. C. (1962a) Net primary production of tundra ecosystems. In *Die Stoffproduction der Pflazendecke* (H. Lieth, Ed.). Stuttgart: Fischer, pp. 35–46.

Bliss, L. C. (1962b) Adaptations of arctic and alpine plants to environmental conditions. *Arctic,* **15**: 117–144.

Bliss, L. C. (1963) Alpine plant communities of the Presidential Range, New Hampshire. *Ecology,* **44**: 678–697.

Bliss, L. C. (1966) Plant productivity in alpine microenvironments on Mt. Washington, New Hampshire. *Ecol. Monogr.,* **36**: 125–155.

Bliss, L. C. (1970) Primary production within arctic tundra ecosystems. In *Productivity and Conservation in Northern Circumpolar Lands* (W. A. Fuller and P. G. Kevan, Eds.). Internat. Union Conserv. Natur., Morges, Switzerland, Publ. **16**: pp. 17–85.

Bliss, L. C., and J. E. Cantlon. (1957) Succession on river alluvium in northern Alaska. *Amer. Midl. Nat.,* **58**: 452–469.

Bliss, L. C., and A. F. Mark. (1974) High-alpine environments and primary production on the Rock and Pillar Range, Central Otago, New Zealand. *N. Z. J. Bot.,* **12**: 445–483.

Bliss, L. C., and F. E. Wielgolaski. (1973) *Primary Production and Production Processes, Tundra Biome*. Tundra Biome Steering Committee, Edmonton-Oslo, 256 pp.

Böcher, T. W. (1954) Oceanic and continental vegetation complexes in Southwest Greenland. *Medd. Grøn.,* **148**: 1–336.

Braun-Blanquet, J. (1932) *Plant Sociology, the Study of Plant Communities*. (Transl. from German by G. D. F. Fuller and H. S. Conrad.) New York: McGraw–Hill, 439 pp.

Bray, J. R., and J. T. Curtis. (1957) An ordination of the upland forest communities of southern Wisconsin. *Ecol. Monogr.,* **27**: 325–349.

Britton, M. E. (1966) Vegetation of the arctic tundra. In *Arctic Biology* (H. P. Hansen, Ed.) Oregon State University Press, Corvallis. pp. 67–113.

Brown, J. (1965) Radiocarbon dating, Barrow, Alaska. *Arctic,* **18**: 36–48.

Brown, J., and P. L. Johnson. (1965) *Pedo-ecological Investigations, Barrow, Alaska.* CRREL Tech. Rep. 159, 38 pp.

Brown, J., P. J. Webber, S. F. MacLean, Jr., P. L. Gersper, and P. S. Flint. (1974) Barrow thaw data 1970–1974. *U.S. Tundra Biome Data Rep.* 74–18, 7 pp.

Bruggemann, P. F., and J. A. Calder. (1953) Botanical investigations in northeastern Ellesmere Island, 1951. *Can. Field-Natur.,* **67**: 157–174.

Bunnell, F. L., S. F. MacLean, Jr., and J. Brown. (1975) Structure and function of tundra ecosystems. In *Ecological Bulletin* (T. Rosswall and O. W. Heal, Eds.) Swedish Natural Science Research Council, Stockholm. pp. 73–124.

Carson, C. E. (1968) Radiocarbon dating of lacustrine strands in arctic Alaska. *Arctic,* **21** 12–26.

Chabot, B. F., and W. D. Billings. (1972) Origins and ecology of the Sierran alpine flora and vegetation. *Ecol. Monogr.*, **42**: 163–199.

Chapin, F. S., III. (1974) Morphological and physiological mechanisms of temperature compensation in phosphate absorption along a latitudinal gradient. *Ecology*, **55**: 1180–1198.

Chapin, F. S. III, and A. Bloom. (1976) Phosphate absorption: Adaptations of tundra graminoids to a low temperature, low phosphorus environment. *Oikos*, **27**: 111–121.

Churchill, E. D. (1955) Phytosociological and environmental characteristics of some plant communities in the Umiat Region of Alaska. *Ecology*, **36**: 606–627.

Churchill, E. D., and H. C. Hanson. (1958) The concept of climax in arctic and alpine vegetation. *Bot. Rev.*, **24**: 127–191.

Cole, L. C. (1960) Competitive exclusion. *Science*, **132**: 348–349.

Collins, M. J., and W. C. Oechel. (1973) The pattern of growth and translocation of photosynthate in a tundra moss, *Polytrichum alpinum*. *Can. J. Bot.*, **52**: 355–363.

Coyne, P. I., and J. J. Kelley. (1975) Carbon dioxide exchange over the Alaskan arctic tundra: Meteorological assessment by an aerodynamic method. *J. Appl. Ecol.*, **12**: 587–611.

Dansereau, P. (1954) Studies on central Baffin vegetation. I. Bray Island. *Vegetatio*, **5, 6**: 329–339.

Dansereau, P. (1957) *Biogeography: An Ecological Perspective*. New York: Ronald Press, 394 pp.

Dansereau, P. (1974) Types of succession. In *Vegetation Dynamics* (R. Knapp, Ed.). The Hague: Junk, pp. 125–135.

Daubenmire, R. F. (1966) Vegetation: Identification of typal communities. *Science*, **151**: 291–298.

Dennis, J. G. (1968) Growth of tundra vegetation in relation to arctic microenvironments at Barrow, Alaska. Ph.D. dissertation, Duke University, 289 pp.

Dennis, J. G. (1977) Distribution patterns of belowground standing crop in arctic tundra at Barrow, Alaska. *Arct. Alp. Res.*, **9**: 111–125.

Dennis, J. G., and P. L. Johnson. (1970) Shoot and rhizome-root standing crops of tundra vegetation at Barrow, Alaska. *Arct. Alp. Res.*, **2**: 253–266.

Dennis, J. G., and L. L. Tieszen. (1972) Seasonal course of dry matter and chlorophyll by species at Barrow, Alaska. *Proceedings 1972 Tundra Biome Symposium*, pp. 16–21.

Dennis, J. G., L. L. Tieszen, and M. A. Vetter. (1978) Seasonal dynamics of above- and belowground production of vascular plants at Barrow, Alaska. In *Vegetation and Production Ecology of an Alaskan Arctic Tundra* (L. L. Tieszen, Ed.). New York: Springer-Verlag, Chap. 4.

Douglas, L. A., and J. C. F. Tedrow. (1959) Organic matter decomposition rates in arctic soils. *Soil Sci.*, **88**: 305–312.

Drury, W. H. (1962) *Patterned Ground and Vegetation on Southern Bylot Island, Northwest Territories, Canada*. The Gray Herb. Of Harvard Univ., Cambridge, Mass. Publ. No. 190.

Drury, W. H., and I. C. T. Nisbet. (1973) Succession. *J. Arnold Arboretum*, **54**(3): 331–368.

Gjaerevoll, O. (1956) The plant communities of the Scandinavian alpine snow-beds. *Det. Kgl. Norske Vidensk. Selskabs Skrifter*, 1956(1). 405 pp.

Goodall, D. W. (1953) Objective methods for the classification of vegetation. I. The use of positive interspecific correlation. *Aust. J. Bot.*, **1**: 39–63.

Griggs, R. F. (1934) The problem of arctic vegetation *J. Wash. Acad. Sci.*, **24**: 153–175.

Holway, J. G., and R. T. Ward. (1965) Phenology of alpine plants in an area of the Colorado alpine. *Ecology*, **46**: 73–83.

Johnson, A. W. (1966) *Plant Ecology of Permafrost Areas*. Proceedings of the First Internat. Permafrost Conference. NAS-NRC Publ. No. 1287, Washington, D.C.

Johnson, A. W., L. A. Viereck, R. E. Johnson, and H. Melchior. (1966) Vegetation and the flora of the Cape Thompson Region, Alaska. In *Environment of the Cape Thomp-*

son Region, Alaska (N. J. Wilimovsky and J. N. Wolfe, Eds.). Oak Ridge, Tennessee: U. S. Atomic Energy Commission, pp. 277–354.

Johnson, P. L., and J. J. Kelley, Jr. (1970) Dynamics of carbon dioxide and productivity in an arctic biosphere. *Ecology,* **51**: 73–80.

Johnson, P. L., and L. L. Tieszen. (1973) Vegetative research in arctic Alaska. In *Alaska Arctic Tundra* (M. E. Britton, Ed.). Arctic Inst. North Amer. Tech. Pap., **25**: pp. 169– 198.

Kenoyer, L. A. (1927) A study of Raunkiaer's law of frequency. *Ecology,* **8**: 341–349.

Koranda, J. J. (1954) A phytosociological study of an uplifted marine beach ridge near Point Barrow, Alaska. M.S. thesis, Michigan State College.

Lachenbruch, A. H. (1962) Mechanics of thermal contraction cracks and ice-wedge polygons in permafrost. *Geol. Soc. Amer. Spec. Pap.* 70. 69 pp.

Lambert, J. M., and M. B. Dale. (1964) The use of statistics in phytosociology. *Adv. Ecol. Res.,* **2**: 59–99.

Leffingwell, E. de K. (1919) *The Canning River Region.* U.S. Geological Survey Prof. Pap. 109, pp. 1–251.

Lewellen, R. I. (1972) Studies on the fluvial environment, Arctic Coastal Plain Province, northern Alaska. Published by the author, Littleton, Colorado, Vol. 1, 282 pp.

Lewis, M. C., and T. V. Callaghan. (1976) Tundra. In *Vegetation and the Atmosphere* (J. L. Monteith, Ed.). London: Academic Press, Vol. 2, pp. 399–433.

Loucks, O. L. (1970) Evolution of diversity, efficiency and community stability. *Amer. Zool.,* **10**: 17–25.

Löve, Á., and D. Löve. (1974) Origin and evolution of the arctic and alpine floras. In *Arctic and Alpine Environments* (J. D. Ives and R. G. Barry, Eds.). London: Methuen, pp. 572–603.

MacLean, S. F. (1974) Production, decomposition, and the activity of soil invertebrates in tundra ecosystems: a hypothesis. In *Soil Organisms and Decomposition in Tundra. Proceedings of a Working Group Meeting in Fairbanks, Alaska, August, 1973* (A. J. Holding, O. W. Heal, S. F. MacLean, and P. W. Flanagan, Eds.). Tundra Biome Steering Committee, Stockholm, Sweden, pp. 197–206.

McVean, D. N. (1969) Alpine vegetation of the Central Snowy Mountains of New South Wales. *J. Ecol.,* **57**: 67–86.

Major, J. (1961) On two trends in phytosociology. (A translation of a paper by V. M. Ponyatovskaya.) *Vegetatio,* **10**: 373–385.

Margalef, D. R. (1968) *Perspective in Ecological Theory.* Chicago: University of Chicago Press. 111 pp.

Marr, J. W. (1961) Ecosystems of the east slope of the Front Range in Colorado. *Univ. of Colorado Stud. Ser. in Biology,* No. 8. 138 pp.

Marr, J. W. (1969) Cyclic change in a patterned-ground ecosystem, Thule, Greenland. In *The Periglacial Environment* (T. L. Péwé, Ed.). McGill–Queen's University Press, Montreal, pp. 177–201.

Matveyeva, N. V., T. G. Polozova, L. S. Blagodatskykh, and E. V. Dorogostaiskaya. (1973) A brief essay on the vegetation in the vicinity of the Taimyr Biogeocoenological Station (Kratkii ockerk restitel' nosti okrestnostei Taimyrskogo biogeotsenologisheskogo statsionara). In *Biogeocenoses of Taimyr Tundra and Their Productivity* (Biogeotsenozy Taimyrskoi tundry i ikh productnost'), Vol. 2, pp. 7–49. Leningrad: Nauka. (International Tundra Biome Translation 13, January 1975; Translator: Doris Löve, 51 pp.)

Miller, O. K., Jr., and G. A. Laursen. (1974) Belowground fungal biomass on U.S. Tundra Biome sites at Barrow, Alaska. In *Soil Organisms and Decomposition in Tundra* (A. J. Holding et al., Eds.). IBP Tundra Biome Steering Committee, Stockholm, pp. 151–158.

Miller, P. C., and L. L. Tieszen. (1972) A preliminary model of processes affecting primary production in the arctic tundra. *Arct. Alp. Res.,* **4**: 1–18.

Mooney, H. A., and W. D. Billings. (1960) The annual carbohydrate cycle of alpine plants as related to growth. *Amer. J. Bot.,* **47**: 594–598.

Muc, M. (1972) Vascular plant production in the sedge meadows of Truelove Lowland. In *Devon Island IBP Project, High Arctic Ecosystem, Project Report 1970 and 1971* (L. C. Bliss, Ed.). Dept. of Botany, University of Alberta, Edmonton, pp. 113–145.

Muc, M. (1973) Primary production of plant communities of the Truelove Lowland, Devon Island, Canada—Sedge meadows. In *Primary Production and Production Processes, Tundra Biome* (L. C. Bliss and F. E. Wielgolaski, Eds.). Tundra Biome Steering Committee, Edmonton-Oslo, pp. 3–14.

Muller, C. H. (1952) Plant succession in arctic heath and tundra in northern Scandinavia. *Bull. Torr. Bot. Club,* **79**: 296–309.

Murray, D. F. (1978) Ecology, floristics, and phytogeography of northern Alaska. In *Vegetation and Production Ecology of an Alaskan Arctic Tundra* (L. L. Tieszen, Ed.). New York: Springer-Verlag, Chap. 2.

Murray, B. M., and D. F. Murray. (1978) Checklists of vascular plants, bryophytes, and lichens for the Alaskan U.S. IBP Tundra Biome study areas—Barrow, Prudhoe Bay, Eagle Summit. In *Vegetation and Production Ecology of an Alaskan Arctic Tundra* (L. L. Tieszen, Ed.). New York: Springer-Verlag, Appendix I.

Neiland, B. J., and J. R. Hok. (1975) Vegetation survey of the Prudhoe Bay region. In *Ecological Investigations of the Tundra Biome in the Prudhoe Bay Region, Alaska* (J. Brown, Ed.). Biological Pap. of the Univ. of Alaska, Spec. Rep. 2, pp. 73–78.

Ng, E., and P. C. Miller. (1977) Validation of a model of the effect of tundra vegetation on soil temperature. *Arct. Alp. Res., 9*: 89–104.

Odum, E. P. (1969) The strategy of ecosystem development. *Science, 164*: 262–270.

Oechel, W. C., and N. J. Collins. (1974) Seasonal patterns of CO_2 exchange in bryophytes at Barrow, Alaska. In *Primary Production and Production Processes, Tundra Biome* (L. C. Bliss and F. E. Wielgolaski, Eds.). IBP Tundra Biome Steering Committee, Stockholm, pp. 197–203.

Oosting, H. J. (1948) Ecological notes on the flora of East Greenland and Jan Mayen. In *The Coast of Northeast Greenland* (L. A. Boyd, Ed.). Amer. Geogr. Soc. Spec. Publ., **30**: pp. 225–269.

Oosting, H. J. (1956) *The Study of Plant Communities.* San Francisco: Freeman and Co., 2nd ed., 440 pp.

Pakarinen, P., and D. H. Vitt. (1973) Primary production of plant communities of Truelove Lowland, Devon Island, Canada—Moss communities. In *Primary Production and Production Processes, Tundra Biome* (L. C. Bliss and F. E. Wielgolaski, Eds.). Tundra Biome Steering Committee, Edmonton-Oslo, pp. 37–46.

Parrick, R., M. H. Hohn, and J. H. Wallace. (1954) A new method of determining the pattern of a diatom flora. *Notulae Naturae Acad. Nat. Sci. Philadephia, 259*: 1–12.

Pielou, E. C. (1966) Species-diversity and pattern-diversity in the study of ecological succession. *J. Theoret. Biol., 10*: 370–383.

Pieper, R. D. (1963) Production and chemical composition of arctic tundra vegetation and their relation to the lemming cycle. Ph.D. dissertation, University of California, Berkeley, 95 pp.

Pitelka, F. A. (1957) Some characteristics of microtine cycles in the Arctic. In *Arctic Biology* (H. P. Hansen, Ed.). Oregon State University Press. Corvallis, pp. 153–184.

Polunin, N. (1934) The flora of Akpatok Island, Hudson Strait. *J. Bot., 72*: 197–204.

Polunin, N. (1934–35) The vegetation of Akpatok Island. I-II. *J. Ecol., 22*: 337–395; **23**: 161–209.

Polunin, N. (1948) *Botany of the Canadian Eastern Arctic, Part III.* Veg. and Ecol. Nat. Mus. Can. Bull. No. 104, Can. Dept. Mines and Resources, Ottawa. 304 pp.

Potter, L. D. (1972) Plant ecology of the Walakpa Bay area, Alaska. *Arctic, 25*: 115–130.

Rastorfer, J. R. (1978) Composition and bryomass of the moss layers of two wet-meadow tundra communities near Barrow, Alaska. In *Vegetation and Production Ecology of an Alaskan Arctic Tundra* (L. L. Tieszen, Ed.). New York: Springer-Verlag, Chap. 6.

Raunkiaer, C. (1934) *The Life Forms of Plants and Statistical Plant Geography,* Oxford: Clarendon Press. 632 pp.

Raup, H. M. (1941) Botanical problems in boreal America. *Bot. Rev.*, **7**: 147–248.
Raup, H. M. (1951) Vegetation and cryoplanation. *Ohio. J. Sci.*, **51**(3): 105–116.
Raup, H. M. (1953) Some botanical problems of the arctic and subarctic regions. *Arctic*, 6:68–74.
Raup, H. M. (1969) The relation of the vascular flora to some factors of site in the Mesters Vig district of Northeast Greenland. *Medd. Grøn.*, **176**(5): 1–80.
Richardson, D. H. S., and E. J. Finegan. (1973) Primary production of plant communities of the Truelove Lowland, Devon Island, Canada—Lichen communities. In *Primary Production and Production Processes, Tundra Biome* (L. C. Bliss and F. E. Wielgolaski, Eds.). Tundra Biome Steering Committee, Edmonton-Oslo, pp. 47–55.
Rodin, L. E., and N. I. Bazilevich. (1967) *Production and Mineral Cycling in Terrestrial Vegetation.* Edinburgh: Oliver and Boyd, 288 pp.
Rübel, E. A. (1914) Heath and Steppe, Macchia and Garigue. *J. Ecol.*, **2**: 232–237.
Savile, D. B. O. (1960) Limitations of the competitive exclusion principle. *Science*, **132**: 1761.
Savile, D. B. O. (1961) The botany of the northwestern Queen Elizabeth Islands. *Can. J. Bot.*, **39**: 909–942.
Savile, D. B. O. (1972) *Arctic Adaptations in Plants.* Can. Dept. Agric. Monogr. 6, 81 pp.
Schultz, A. M. (1964) The nutrient recovery hypothesis for arctic microtine cycles. II. Ecosystem variables in relation to arctic microtine cycles. In *Grazing in Terrestrial and Marine Environments* (D. J. Crisp, Ed.). Oxford: Blackwell Scientific Publ., pp. 57–68.
Schultz, A. M. (1969) A study of an ecosystem: The arctic tundra. In *The Ecosystem Concept in Natural Resource Management* (G. M. Van Dyne, Ed.). Academic Press, New York, pp. 77–93.
Scott, D., and W. D. Billings. (1964) Effects of environmental factors on standing crop and productivity of an alpine tundra. *Ecol. Monogr.*, **34**: 243–270.
Shaver, G. R., and W. D. Billings. (1975) Root production and root turnover in a wet tundra ecosystem, Barrow, Alaska. *Ecology*, **56**: 401–409.
Shimwell, D. W. (1972) *The Description and Classification of Vegetation.* Seattle: University of Washington Press, 322 pp.
Smith, D. K. (1974) Floristic, ecologic, and phytogeographic studies of the bryophytes in the tundra around Barrow, Alaska. Ph.D. dissertation, University of Tennessee. 191 pp.
Sokal, R. R., and P. H. A. Sneath. (1963) *Principles of Numerical Taxonomy.* San Francisco: W. H. Freeman, 359 pp.
Sørensen, T. (1941) Temperature relations and phenology of the Northeast Greenland flowering plants. *Medd. Grøn.*, **125**: 1–305.
Spetzman, L. A. (1951) Plant geology and ecology of the Arctic Slope of Alaska. M.S. thesis, University of Minnesota.
Spetzman, L. A. (1959) *Vegetation of the Arctic Slope of Alaska.* U.S. Geol. Surv. Prof. Pap. No. 302, pp. 19–58.
Steere, W. C. (1978) Floristics, phytogeography, and ecology of arctic alaskan bryophytes. In *Vegetation and Production Ecology of an Alaskan Arctic Tundra* (L. L. Tieszen, Ed.). New York: Springer-Verlag, Chap. 5.
Stoner, W. A., and P. C. Miller. (1975) Water relations of plant species in the wet coastal tundra at Barrow, Alaska. *Arct. Alp. Res.*, 7:109–124.
Stoner, W. A., P. C. Miller, and L. L. Tieszen. (1978) A model of plant growth and phosphorus allocation for *Dupontia fisheri* in coastal, wet meadow tundra. In *Vegetation and Production Ecology of an Alaskan Arctic Tundra* (L. L. Tieszen, Ed.). New York: Springer-Verlag, Chap. 24.
Summerhayes, V. A., and C. S. Elton. (1928) Further contributions to the ecology of Spitzbergen. *J. Ecol.*, **16**: 13–267.
Tansley, A. G. (1935) The use and abuse of vegetational concepts and terms. *Ecology*, **16**: 284–307.
Thompson, D. Q. (1955) *The Role of Food and Cover in Population Fluctuations of the*

Brown Lemming at Point Barrow, Alaska. Transactions 20th North Amer. Wildlife Conf., pp. 166–176.

Tieszen, L. L. (1972a) The seasonal course of aboveground production and chlorophyll distribution in wet arctic tundra at Barrow, Alaska. *Arct. Alp. Res.,* **4**: 307–324.

Tieszen, L. L. (1972b) Photosynthesis in relation to primary production. In *Proceedings IV International Meeting on the Biological Productivity of Tundra, Leningrad, October 1971* (F. E. Wielgolaski and Th. Rosswall, Eds.). Tundra Biome Steering Committee, Stockholm, pp. 52–62.

Tieszen, L. L. (1974) Photosynthetic competence of the subnivean vegetation of an arctic tundra. *Arct. Alp. Res.,* **6**: 253–256.

Tieszen, L. L. (1978) Photosynthesis in the principal Barrow, Alaska, species: A summary of field and laboratory responses. In *Vegetation and Production Ecology of an Alaskan Arctic Tundra* (L. L. Tieszen, Ed.). New York: Springer-Verlag, Chap. 10.

Tikhomirov, B. A. (1963) *Contribution to the Biology of Arctic Plants.* (Text in Russian.) Dokl. Akad. Nauk, USSR. 154 pp.

Tikhomirov, B. A., V. F. Shamurin, and V. D. Alexandrova. (1975) Primary production of tundra communities. Unpublished manuscript.

Walker, D. A. (1977) Analysis of effectiveness of a television scanning densitometer for indicating geobotanical features in an ice-wedge polygon complex at Barrow, Alaska. M.A. thesis, University of Colorado. 132 pp.

Walter, H. (1973). *Vegetation of the Earth.* New York: Springer-Verlag. 237 pp.

Walter, H. and H. Lieth. (1967) *Klimadiagramm-Weltatlas.* Gustav Fischer, Jena.

Warren Wilson, J. (1957) Arctic plant growth. *Adv. Sci.,* **13**: 383–388.

Warren Wilson, J. (1959) Analysis of the distribution of the foliage area in grassland. In *The Measurement of Grassland Productivity* (J. D. Ivins, Ed.). London: Butterworth. pp. 51–61.

Webber, P. J. (1971) Gradient analysis of vegetation around the Lewis Valley, north-central Baffin Island, Northwest Territories, Canada. Ph.D. dissertation, Queen's University, Kingston. 366 pp.

Webber, P. J. (1974) Tundra primary productivity. In *Arctic and Alpine Environments* (J. D. Ives and R. G. Barry, Eds.). London: Methuen, pp. 445–473.

Webber, P. J. (1977) Belowground tundra research: A commentary. *Arct. Alp. Res.,* **9**: 105–111.

Webber, P. J., J. C. Emerick, D. C. E. May, and V. Komárková. (1976) The impact of increased snowfall on alpine vegetation. In *Ecological Impacts of Snowpack Augmentation in the San Juan Mountains, Colorado* (H. W. Steinhoff and J. D. Ives, Eds.). Final Report of the San Juan Ecology Project. Colorado State University, CSU-FNR-7052-1, pp. 201–264.

Webber, P. J., and D. C. E. May. (1977) The distribution and magnitude of belowground plant structures in the alpine tundra of Niwot Ridge, Colorado. *Arct. Alp. Res.,* **9**: 155–166.

Webber, P. J., and D. A. Walker. (1975) Vegetation and landscape analysis at Prudhoe Bay, Alaska: A vegetation map of the Tundra Biome study site. In *Ecological Investigations of the Tundra Biome in the Prudhoe Bay Region, Alaska* (J. Brown, Ed.). Biological Papers of the University of Alaska, Special Report 2, pp. 81–91.

White, R. G., B. R. Thompson, T. Skogland, S. J. Person, D. E. Russell, D. F. Holleman, and J. R. Luick (1975) Ecology of caribou at Prudhoe Bay, Alaska. In *Ecological Investigations of the Tundra Biome in the Prudhoe Bay Region, Alaska* (J. Brown, Ed.). Biological Papers of the University of Alaska, Special Report 2, pp. 151–187.

Whittaker, R. H. (1953) A consideration of climax theory: The climax as a population and pattern. *Ecol. Monogr.,* **23**: 41–78.

Whittaker, R. H. (1962) Classification of natural communities. *Bot. Rev.,* **28**: 1–239.

Whittaker, R. H. (1965) Dominance and diversity in land plant communities. *Science,* **147**: 250–260.

Whittaker, R. H. (1967) Gradient analysis of vegetation. *Biol. Rev.,* **42**: 207–264.

Whittington, W. J. (1967) *Root Growth*. London: Butterworth. 450 pp.

Wielgolaski, F. E. (1972) Vegetation types and plant biomass in tundra. *Arct. Alp. Res.*, **4**: 291–306.

Wielgolaski, F. E. (1974) Phenology studies in tundra. In *Phenology and Seasonal Modelling* (H. Lieth, Ed.). New York: Springer, pp. 209–214.

Wielgolaski, F. E. (1975) Productivity of Tundra Ecosystems. In *Proceedings of a Symposium on Productivity of World Ecosystems*. National Academy of Sciences, Washington, D.C., pp. 1–12.

Wielgolaski, F. E., and P. J. Webber. (1973) Classification and ordination of circumpolar arctic and alpine vegetation. *International Tundra Biome Newsletter*, **9**: 24–31.

Wiggins, I. L. (1951) The distribution of vascular plants on polygonal ground near Point Barrow, Alaska. *Dudl. Herb. Cont.*, **4**: 41–52.

Williams, M. E., E. D. Rudolph, E. A. Schofield, and D. C. Prasher. (1978) The role of lichens in the structure, productivity, and mineral cycling of the wet coastal Alaskan tundra. In *Vegetation and Production Ecology of an Alaskan Arctic Tundra* (L. L. Tieszen, Ed.). New York: Springer-Verlag. Chap. 7.

4. Seasonal Dynamics of Above- and Belowground Production of Vascular Plants at Barrow, Alaska

J. G. Dennis, L. L. Tieszen, and M. A. Vetter

Introduction

An analysis of the functional relationships of any ecosystem requires an under-standing of the mechanisms and controls of primary production—the photosyn-thetic conversion of solar energy to energy contained in plant biomass. Prior to the support of the IBP, analyses of primary production in U.S. arctic tundra ecosystems were limited primarily to measurements of aboveground standing crop. Dennis and Johnson (1970) summarized the results of many of these analyses and concluded that at Barrow the range of aboveground primary production was 30 to 224 g m^{-2} when herbivores were excluded. Aboveground standing crop alone, however, may grossly underestimate the magnitude of total standing crop and production since belowground standing crops are large. Scott and Billings (1964) found that a maximum of over 80% of the total standing crop was located belowground in a mesic alpine tundra site. Alexandrova (1958) clearly has documented maximum belowground standing crops of over 800 g m^{-2} in U.S.S.R. arctic tundra sites. Dennis and Johnson (1970) reported belowground standing crops at Barrow of up to 1435 g m^{-2}, and Khodachek (1969) measured belowground standing crops as high as 4700 g m^{-2} in the western Taimyr region. The large size of the belowground standing crops suggests that the annual increment of belowground material is an important component in total primary production. Our IBP-supported analysis of the course and magnitude of primary production at Barrow, Alaska, was undertaken to provide a more detailed description of seasonal events and to increase our knowledge of the control of this production by such factors as climatological conditions, soil and nutrient limitations, and the effect of consumers.

The arctic tundra is characterized by low mean daily summer temperatures combined with a short growing season (Billings, 1973). At Barrow, Alaska, for example, the combination of a persistent snow cover, which prevents significant rates of photosynthesis (Tieszen, 1974), and frozen soil, which impedes plant uptake of water and minerals, greatly limits the length of the growing season. Plant growth starts approximately 15 June when the tundra first becomes snow-free and mean ambient temperatures rise above 0°C. The termination of the productive season is less distinct but occurs sometime near early September when mean ambient temperatures fall below 0°C and the soil begins to refreeze from the surface downward. Subzero air and soil temperatures throughout most of the year thus limit the period available for aboveground growth to 54 to 111 days, with the mean thaw season length for Barrow being 87 days (Lewellen, 1972). At the onset of the growing season, when the tundra first becomes essentially free of snow, the sun has almost achieved the summer solstice. Incoming daily solar radiation in mid-June occurs at the highest rate of the year at 600 to 800 cal cm^{-2} 24-hr day^{-1} (300 to 3000 nm). It declines to highs of around 300 to 500 cal cm^{-2} day^{-1} by late August. Although radiation is received for 24 hr day^{-1} until 2 August, there is a marked diurnal variation in visible (400 to 700 nm) radiation even in early July. Mean ambient temperatures reach the seasonal high of near 5°C in mid-July and, as daily radiation begins to decrease, mean ambient temperatures also decrease, reaching values near 0°C by the end of August.

Periodic, acute grazing pressure is a characteristic feature of the tundra ecosystem at Barrow, and the effect of the lemming, the principal herbivore, has been studied intensively. In the past, lemming populations have undergone large fluctuations in density and have attained high numbers every three to five years (Pitelka, 1972). During a lemming high the tundra ecosystem is subject to acute grazing pressure. Melchior (1972), for example, has calculated that a juvenile brown lemming can consume 13.6 g dry wt day^{-1}. This rate of consumption in combination with lemming densities that may reach 125 to 250 animals ha^{-1} (Bliss et al., 1973) results in the removal of whole areas of aboveground vegetation. The high proportion of belowground relative to aboveground plant biomass may be an adaptation allowing the tundra system to recover from such periodic and intense grazing pressure.

Soils at Barrow range from the well-drained arctic brown to the poorly drained meadow tundra and are underlain by permafrost. The low maximum soil temperature at the bottom of the active layer appears to limit plant root growth (Dennis and Johnson, 1970; Billings et al., 1978), and it is possible that the varying thaw depths in different soils may partly determine species distributions on the basis of different rooting patterns (Dennis and Johnson, 1970). The IBP site 2 is a polygonized wet meadow area having predominantly meadow tundra soil. This soil is nearly saturated to saturated most of the time and has a thick surface organic mat underlain by a wet layer of mineral soil. Permafrost under this soil starts approximately 25 to 35 cm beneath the surface.

Our analysis of terrestrial vascular production by arctic tundra at Barrow was conducted to achieve the following principal goals:

(1) To document the seasonal course and magnitude of aboveground production and to determine the principal contributing species. (2) To estimate the allocation to each plant compartment of biomass and ultimately nutrients. Allocation was determined by measuring the relative proportions of standing crop and leaf area index present in each compartment. In addition, the caloric content and mineral (see Chapin, 1978) and organic nutrient (see McCown, 1978) status of each compartment were analyzed. (3) To document the seasonal progression

of efficiency of energy interception and utilization and to provide information necessary for construction of a simulation model that would predict canopy photosynthesis and canopy microclimate. The structure of the Barrow canopy was characterized in terms of leaf area index and leaf inclinations. Microenvironmental information was collected to determine possible physiological significance of the observed variations of leaf area index and leaf inclination. (4) In addition to the characterization of the canopy structure in the intensive site, the canopies in plots in several additional areas were analyzed to relate canopy development and canopy structure to an array of microenvironmental features characteristic of Barrow.

Portions of the information in this chapter have been published in Tieszen (1972), Caldwell et al. (1974), Tieszen (1974), Chapin et al. (1975), and Dennis (1977).

Methods and Materials

This study was conducted at the site 2 intensive study site of the U.S.–IBP Tundra Biome at Barrow, Alaska (described by Tieszen, 1972; see Webber, 1978). Methods have been described by Tieszen (1972) and Caldwell et al. (1974). Study plots, each 6×6 m centered on a flat polygon center, were chosen subjectively in an area of relatively flat polygons, in early June, 1970. Sampling of these plots for both production and LAI was scheduled at approximately 10-day intervals from mid-June to the end of August in both 1970 and 1971. Leaf area index was estimated with inclined point quadrats, and leaf inclination was estimated with a protractor. In addition, one to several quadrats from plots scattered in other sites that represented wet meadow, polygon tops and troughs, low polygons, and pure species stands were sampled for LAI at the peak season of both years. Aboveground standing crop was sampled by the harvest method. Two, previously unclipped, circular quadrats of approximately 0.1 m^2 area were clipped from each of five control plots, except for 5 June, 15 June, 24 August, and 4 September 1971, when eight 0.0045-m^2 quadrats from one plot, five 0.1-m^2 quadrats from one plot, six 0.1-m^2 quadrats from four plots, and four 0.01-m^2 quadrats from two plots, respectively, were clipped. Two subsamples, one for analysis of chlorophyll content and the other for analysis of either carbohydrate or caloric content, were removed. Caloric contents were determined with a Parr oxygen bomb calorimeter. When sufficient material was available for a parallel sample, ash content was determined by combusting at 500°C for 3 hr.

Belowground standing crop was sampled in 1971 by means of soil cores of 0.0045-m^2 surface area from each clipped quadrat. Plant material was separated manually into component fractions including live stem base, dead stem base, live rhizome, dead rhizome, live tan root, live white root, and dead root. The stem base consisted of vertically oriented belowground stem material extending from the air–moss interface down to the rhizome. Live material was distinguished from dead material on the basis of light rather than dark color, shiny rather than dull surface, high rather than low tensile strength, and dense rather than vacuous appearance. Depth of soil thaw was determined concurrently from the soil cores (Dennis and Johnson, 1970).

Results

Aboveground Standing Crop

Of the approximately 120 species found in the Barrow area (Dennis and Johnson, 1970), only 22 were encountered in site 2 in 1970 and 1971. The majority of these species were graminoids and included the sedges *Carex aquatilis*, *Eriophorum angustifolium*, *E. scheuchzeri*, and *E. russeolum*, and the grasses *Calamagrostis holmii*, *Dupontia fisheri*, and *Poa arctica*. Only two forbs, *Saxifraga cernua* and *Petasites frigidus*, were present in substantial amounts. The graminoid dominance of the community is shown clearly by the data. Three graminoid species, *Carex aquatilis*, *Dupontia fisheri*, and *Eriophorum angustifolium* accounted for 73% of the aboveground standing crop in both years. Forbs and shrubs accounted for only 7% of the aboveground standing crop, despite 10 of the 22 species present being dicotyledons.

The general trend of the seasonal progression of current year aboveground standing crop has been described in detail by Tieszen (1972) and is summarized for 1970 and 1971 in Figure 1. The early season pattern of current year aboveground standing crop was similar in both years, with a peak occurring in early August. The magnitude of this estimate of peak season, current year standing crop varied between 1970 and 1971, being 101.5 g m^{-2} in 1970 and only 86.0 g m^{-2} at a later time in 1971. The 1970 data indicate that when the area became snow free around 15 June, nearly 12 g m^{-2} of live material already was present. Similarly, the 1971 data show that in other nearby areas 7 g m^{-2} of green standing crop was present on 15 June, only a day or two after snow disappearance. As much as 23 g of material that appeared green was present in selected nearby areas on 5 June beneath 40 to 50 cm snow. Assuming a 55-day period from the initiation of growth until the attainment of peak current year biomass (10 June to 4 August for 1970), and assuming that all of this biomass in fact was produced in the current year, the daily increase in aboveground dry matter was 1.85 g m^{-2} in 1970 and 1.46 g m^{-2} in 1971. In both years, this increase occurred at nearly a linear rate throughout the first 55 days of the growing period. In 1971, however, production of foliage continued beyond 4 August and attained a maximum of 86.6 g m^{-2}. Senescence then occurred very rapidly, and by 4 September there was only 49.6 g m^{-2} present, nearly half of which was dead. This difference between 1970 and 1971 is also suggested by the progression of leaf area index (see Figure 7). Of the material making up this new standing crop in each year, 70 to 90% was blade material, with the blade fraction increasing slightly in relative proportion through the season (Figure 2).

The standing crop of chlorophyll in 1970 increased more rapidly early in the season then did that of aboveground dry matter, and the chlorophyll peak (459

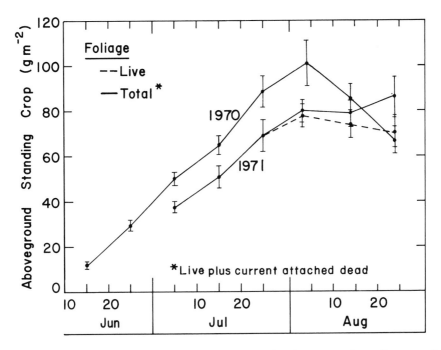

Figure 1.   Seasonal progression of total and live vascular aboveground standing crops for 1970 and 1971, site 2.

mg m^{-2}) was reached on 25 July, approximately 10 days before the peak in dry matter. Chlorophyll standing crop then decreased fairly rapidly through the remainder of the season (Figure 3). The chlorophyll concentration on a dry weight basis started at 3 mg g dry wt^{-1} in the beginning of the season and increased to a maximum of near 6 mg g dry wt^{-1} by 15 July. Concentration of chlorophyll then declined steadily (Tieszen, 1972) to an end of season level of 1.5 mg g dry wt^{-1}. This decrease in chlorophyll concentration was paralleled by an early initiation of leaf senescence. The leaf area index measurements for the same year showed 10.3% of all new leaves to be dead by 3 August and 39.8% to be dead by 22 August. The harvest data for the following year showed 1.9% of the current year's production to be dead on 3 August and 18.5% to be dead on 24 August. Species differences in the seasonal progressions were not apparent within, but were noticeable between, the groupings monocotyledons and dicotyledons. In contrast to the trend for an initial increase of chlorophyll concentration shown by the monocotyledons, chlorophyll concentration in the dicotyledons peaked early in the season and declined continuously thereafter (Tieszen, 1972). Chlorophyll contents generally were higher for the dicots than for the monocots. In all forms, however, senescence was well developed as early as 1 August.

Standing dead material and loose litter are an important part of this community. During 1970 we measured a standing crop of around 128 g m^{-2} throughout the season. In 1971, the measurements of dead standing crop varied between 110

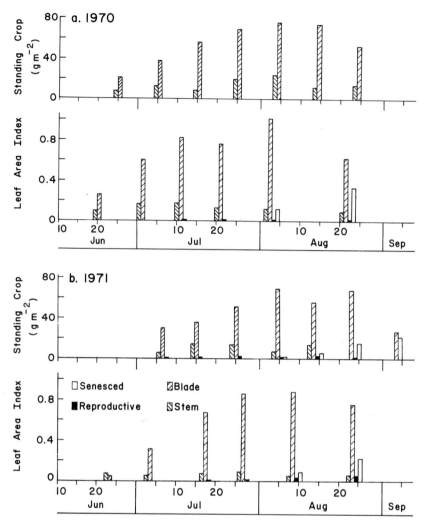

Figure 2. Seasonal progression of relative contributions of plant compartments to total aboveground standing crop in 1970 and 1971, site 2.

and 181 g m⁻². This dead material from previous year's production in 1970 possessed a leaf area index of 1.23, a value which was sufficiently large to have an impact on light penetration to the green components (Miller and Tieszen, 1972). This material was appressed close to the soil surface, with 81% in the first 5 cm, 18% between 5 and 10 cm, and only 1% higher than 10 cm above the soil surface.

A number of environmental manipulations were conducted to determine major effects on canopy microclimates and plant response. Harvesting of standing crop at peak season was used to measure plant response (Table 1). Plots which at the beginning of the season were clipped to simulate lemming grazing

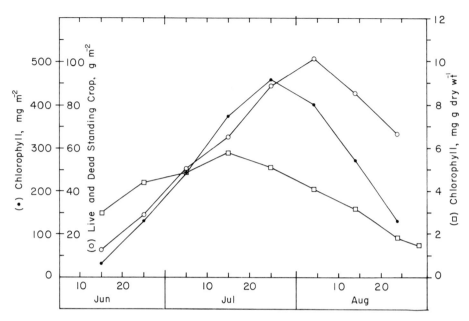

Figure 3. Seasonal progression of the mean standing crops of dry weight and chlorophyll and of chlorophyll concentration on a dry weight basis for total vascular foliage, site 2, 1970. (Adapted from Tieszen, 1972.)

and then cleared of the resulting litter produced less biomass than the controls. Frost damage to upright leaf blades in these plots was apparent after 1 August. Plots which received an added mulch layer of magnitude equal to the standing crop of dead material normally present also produced less material. Chlorophyll concentrations were reduced in these plots, but in no others of the treated plots, perhaps as a result of the enhanced shading and the observable etiolation in some leaves. The fertilized plots showed increased production, and plants from these

Table 1. Comparison of Aboveground Biomass (g m^{-2}) in Control and Experimental Plots on 4 August 1970[a]

Plot	g dry wt m^{-2}	mg Chl g dry wt^{-1}
Control	101.5 ± 9.7	4.2
Clipped and cleared	57 to 69	4.2
Mulched	66 to 79	3.7
Fertilized (20–10–10)[b]	101 to 119	4.0
Fertilized (8–32–16)[b]	141 to 150	4.2
Outside greenhouse	111 to 135	4.3
Outside greenhouse (on 19 August)	(103)	(4.2)
Schultz fertilized plot[c]	102	
Schultz control plot[c]	68	

[a] N for control = 10. For descriptions see text.
[b] Applied after snowmelt at 450 kg ha^{-1}.
[c] Plots fertilized with N, P, K, and Ca between 1960 and 1964 by Schultz (1964).

plots had statistically significant increases in plant phosphorus and potassium but not nitrogen (Chapin et al., 1975). The long term effect of fertilizer additions is readily apparent from the experiments established by Schultz (1964). Ten years after application was initiated, the leaf area index on treated plots was nearly two times greater than on the controls (Table 1). Furthermore, the frequency of flowering (16.5%) on treated plots was three times greater than on control plots (5%). Shoot density averaged 4918 shoots m^{-2} and live and dead production averaged 102 g m^{-2} on the treated areas compared to 1897 shoots m^{-2} and 68 g m^{-2} on the control plots. These control plots vegetationally were very similar to the control plots at site 2.

One treatment plot (outside greenhouse) was partially enclosed in clear plastic to elevate air temperature several degrees. The treatment did not increase plant standing crop substantially during the first year, although height growth was obviously greater and growth continued longer into the fall as indicated both by the high biomass values on 19 August and especially by the high chlorophyll concentrations (4.2 mg g dry wt^{-1}) at that time (Table 1). In contrast, chlorophyll concentrations on the control plots were around 2.7 mg g dry wt^{-1} at that late date. The small increase in air temperature produced by the treatment thus lengthened the growing season in terms of prolonging higher chlorophyll concentrations. In 1971, the LAI on the treated plot (Table 2) attained a value of 1.87, or nearly double that of the control. There was no difference, however, in the percentage of plants (6%) which had flowered.

Caloric Contents

Caloric contents are known to vary among species (Golley, 1961) and depend mainly upon the proportion of lipid materials. A survey of caloric contents of the principal species of each Tundra Biome site was undertaken to test suggestions (Golley, 1961; Bliss, 1962) that caloric contents increase with latitude and altitude. This survey (Table 3) showed little variation in caloric content across the Biome. Caloric contents were not significantly different between fertilized and control plots. Reproductive structures had caloric contents that were only 4 to 10% higher than leaves. These data do not support the hypothesis being tested. On the other hand, all species tested from a phytometer experiment with peas, ryegrass, and timothy showed an increased caloric content with increased severity of the tundra environment. Such data do support the hypothesis and may reflect proportionately higher lipid levels.

The seasonal progression of caloric content in the three main Barrow monocotyledons is given in Figure 4. The caloric content of the total shoot material remained quite stable through the season, with maximum values occurring on 4 August in all three species. The sedges *Carex aquatilis* and *Eriophorum angustifolium* had higher maximum caloric values (4500 and 4536 cal g dry wt^{-1}, respectively) than the grass *Dupontia fisheri* (4397 cal g dry wt^{-1}). In *Carex* the caloric content of the blade component was higher through the season than the caloric content of the whole shoot except at the last sampling date, a trend that did not occur with either *Dupontia* or *Eriophorum,* in which the caloric content

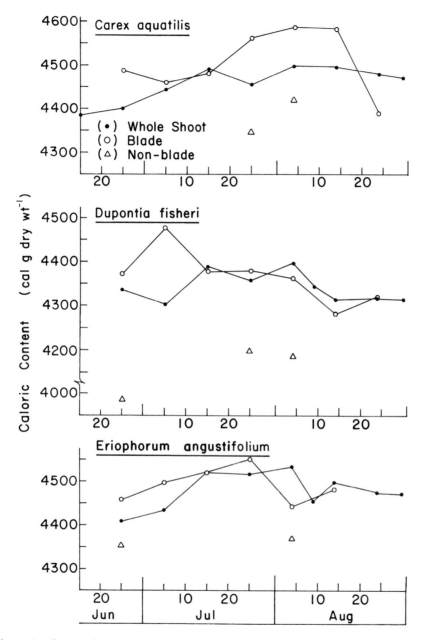

Figure 4. Seasonal progression of caloric content (ash-free) in the three main Barrow monocotyledons, site 2, 1970.

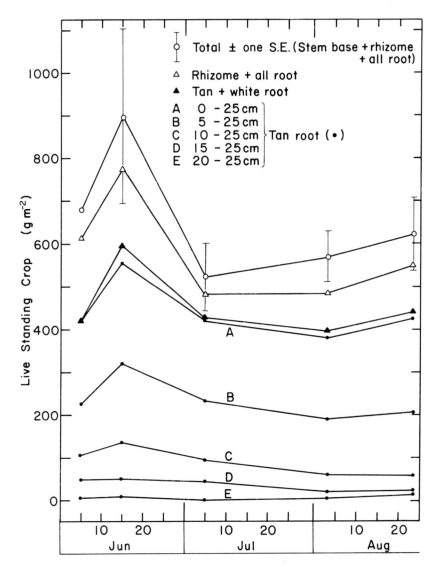

Figure 5. Seasonal progression of live belowground standing crop for site 2, 1971. Vertical bars represent standard errors for $N = 6$ to 10.

of the blade component more closely followed that of the whole shoot throughout the season. In general, the caloric contents of all three species were quite similar and constant throughout the season.

Belowground Standing Crop

The 1971 seasonal progressions of belowground live and dead standing crops of stem bases, roots, and rhizomes are illustrated in Figures 5 and 6 and in Tables 4

Table 2. Live Leaf Area Index in a Variety of Natural and Experimental Plots[a]

Nodum	Plot	Canopy height (cm)						Total
		20–25	15–20	10–15	5–10	2.5–5	0–2.5	
				3 August 1970				
Wet meadow V	401			0.058	0.293	0.874	0.438	1.66
Polygon trough V	402			0.112	0.315	0.600	0.692	1.72
Mesic tundra III	403				0.141	0.363	0.511	1.02
Mesic tundra low polygon IV	404				0.057	0.561	0.623	1.24
Polygon top III	405					0.089	0.475	0.56
Polygon top III	406				0.176	0.322	0.235	0.73
				18 August 1971				
Wet meadow V	407					0.330	0.357	0.69
Low polygon VI	409					0.132	0.110	0.24
Polygon trough VI	410				0.306	0.183	0.216	0.71
Low polygon sunken basin VI	411					0.022	0.066	0.09
Low polygon sunken basin VI	412					0.020	0.061	0.08
Arctophila slough VII	413		0.044	0.067	0.404	0.180	0.112	0.81
Polygon top III	405							

Table 2. Live Leaf Area Index in a Variety of Natural and Experimental Plots[a] (*Continued*)

	Canopy height (cm)									
	30+	25–30	20–25	15–20	10–15	5–10	2.5–5	0–2.5	Total
Pure stands of main				2 August 1970					
Barrow species									
Dupontia fisheri	0.052	0.280	0.351	0.709	1.343	1.402	0.582	0.527	5.25
Arctophila fulva	0.055	0.605	1.375	1.870	1.980	1.815	0.467	0.303	8.48
Experimental plots				18 August 1971					
Outside									
greenhouse			0.022	0.024	0.169	0.384	0.813	0.458	1.87
				2 August 1971					
Schultz control[b]				0.011	0.000	0.154	0.099	0.132	0.40
Schultz fertilized[b]				0.011	0.033	0.176	0.440	0.077	0.74

[a] For description of Noda, see Webber (1978).
[b] See Table 1.

Table 3. Mean Caloric Values for Arctic and Alpine Tundra Species[a]

Species	Caloric content Mean	SE	Species	Caloric content Mean	SE
		Barrow			
Alopecurus alpinus	4559	±45	*Eriophorum scheuchzeri*	4642	66
Arctagrostis latifolia	4601	22	*Luzula arctica*	4470	23
Arctophila fulva	4770	54	*Poa arctica*	4363	12
Calamagrostis holmii	4451	38	*Petasites frigidus*	4333	40
Carex aquatilis	4640	13	*Saxifraga cernua*	4152	17
Dupontia fisheri	4576	17	*Stellaria laeta*	4499	92
Eriophorum angustifolium	4629	13	*Salix pulchra*	4305	70
		Eagle Summit			
Carex aquatilis	4908	17	*Poa alpina*	4594	20
Dupontia fisheri	4639	19	*Dryas octopetala*	5149	64
Eriophorum angustifolium	4665	50	*Salix arctica*	4946	70
		Beaufort Lagoon			
Eriophorum angustifolium	4620	77	*Carex aquatilis*	4631	26
Eriophorum scheuchzeri	4494	51			
		Niwot Ridge			
Calamagrostis purpurascens	4591	61	*Castillejia occidentalis*	4549	80
Carex rupestris	4714	20	*Erigeron simplex*	4694	43
Carex scopulorum	4587	55	*Geum rossii*	4468	25
Deschampsia cespitosa	4667	43	*Hymenoxis acaulis*	4583	109
Luzula spicata	4591	61	*Lloydia serotina*	4756	
Kobresia myosuroides	4787	21	*Pedicularis groenlandica*	4595	80
Artemisia scopulorum	4743	47	*Polygonum bistortoides*	4709	80
Caltha leptosepala	4447	103	*Potentilla consimilis*	4654	20
			Trifolium dasyphyllum	4654	75

[a] Values are in calories g^{-1} ash free weight. More original data available from Data Report 74-3. N varied from 2 to 56 and was commonly 12.

and 5. The maximum total live standing crop of 899 ±202 g m^{-2} was measured in mid-June, immediately before the onset of spring growth. The minimum total live standing crop of 522 $\pm$ 79 g m^{-2} was measured in early July, approximately 25 days after the designated onset of aboveground growth on 10 June. The 5 June sample came from frozen soil under 40 to 50 cm of snow, the 15 June sample came from frozen soil in an area from which the snow had disappeared no more than a day or two earlier, and the 24 August sample came from soil that had thawed to a depth of 25 cm. The live stem base standing crop varied between 42 ±15 and 133 ±58 g m^{-2}, while the dead stem base component varied between 21 ±13 and 115 ±29 g m^{-2}. The range of live rhizome standing crop was 52 ±15 to 192 ±30 g m^{-2}, while the dead rhizome values ranged between 44 ±11 and 88 ±28 g m^{-2}.

Live roots varied between 397 ±44 and 596 ±128 g m^{-2}, while dead roots ranged between 556 ±44 and 1125 ±137 g m^{-2}. The root component of the total

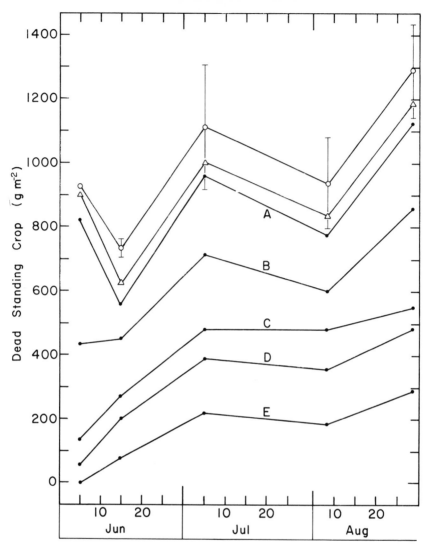

Figure 6.   Seasonal progression of dead belowground standing crop for site 2, 1971. Vertical bars represent standard errors for $N = 6$ to 10. (See Figure 5.)

live standing crop reached its minimum value later in the season, 3 August, than did the rhizome plus stem base component (Figure 5). Stem base and rhizome material occurred almost entirely within the top 10 cm of soil, with the majority found in the top 5 cm (Table 5). Root material was distributed throughout the active layer, although the major part of the live root standing crop again occurred in the top 10 cm of soil. The live root to live rhizome plus stem base ratio increased from less than 2:1 in early season to greater than 4:1 on 5 July, then

Table 4. Distribution by Category of Average Belowground Plant Standing Crop on the IBP Tundra Biome Site 2 Control Plots at Barrow, Alaska, in 1971[a]

| | Frozen soil | | | | | | Thawed soil | | | | | | | | | Average |
| | 5 June | | | 15 (16) June[b] | | | 5 (7) July | | | 3 (4) August | | | 24 (28)August | | | |
Category	$\bar{X}$[c]	SE[c]	%	$\bar{X}$	SE	%	$\bar{X}$	SE	%	$\bar{X}$	SE	%	$\bar{X}$	SE	%	
Live																
Rhizome	192	30	28	170	42	19	52	15	10	88	16	15	111	17	18	123
Stem base	66	27	10	133	58	15	52	15	8	85	18	15	74	14	12	80
Total shoot	258	32	38	303	88	34	94	20	18	173	27	30	185	26	30	203
White root	0	0	0	36	13	4	8	2	1	18	4	3	14	5	2	15
Tan root	420	—	62	560	116	62	421	63	81	379	41	67	424	75	68	441
Total root	420	—	62	596	128	66	428	64	82	397	44	70	438	77	70	456
Total live	679	—	100	899	202	100	522	79	100	570	58	100	623	87	100	659
Dead																
Rhizome	88	28	10	62	18	8	44	11	4	62	11	7	56	16	5	62
Stem base	21	13	2	115	29	16	107	30	10	102	17	11	108	11	8	91
Total shoot	109	31	12	177	38	24	151	35	14	164	19	18	164	25	13	153
Root	816	—	88	556	44	76	960	189	86	772	148	82	1125	137	87	846
Total dead	925	—	100	733	25	100	1111	195	100	936	142	100	1289	145	100	999
Live plus dead																
Total	1604	—		1632	221	—	1633	232		1506	172		1912	189	—	1658
Percentage live			42			55			32			38			33	40
Percentage dead			58			45			68			62			67	60

[a] From Dennis (1977).

[b] First date shows when the aboveground sample was harvested; date in parentheses indicates when the belowground sample was harvested.

[c] $\bar{X}$ = mean standing crop; SE = standard error of the mean (both g m^{-2}).

Table 5.   Average Belowground Standing Crop (g m^{-2}) by Plant Part and Depth, Site 2, Control Plots, 1971[a]

Depth (cm)	5 June $\overline{X}$	15 (16) June[b] $\overline{X}$	5 (7) July $\overline{X}$	3 (4) August $\overline{X}$	24 (28) August $\overline{X}$
Live stem base					
0–5	66	133	42	85	74
5–10	0	0	0	0	0
10–15	0	0	0	0	0
15–20	0	0	0	0	0
20–25	0	0	0	0	0
Total	66	133	42	85	74
Live rhizome					
0–5	191	166	47	85	107
5–10	1	4	4	2	4
10–15	0	0	T	0	0
15–20	0	0	1	0	0
20–25	0	0	0	0	0
Total	192	170	52	87	111
Current live root					
0–5	0	5	4	8	4
5–10	0	10	1	3	4
10–15	0	10	2	5	5
15–20	0	5	1	2	1
20–25	0	6	0	T	1
Total	0	36	8	18	14
Prior live root					
0–5	196	238	191	190	218
5–10	119	186	135	125	142
10–15	54	84	49	41	40
15–20	49	43	43	17	9
20–25	2	9	3	5	15
Total	420	560	421	379	424
Dead stem base					
0–5	21	112	97	101	98
5–10	0	3	10	1	10
10–15	0	0	0	0	0
15–20	0	0	0	0	0
20–25	0	0	0	0	0
Total	21	115	107	102	108
Dead rhizome					
0–5	86	55	30	54	50
5–10	2	7	12	8	6
10–15	0	0	2	0	0
15–20	0	0	T	0	0
20–25	0	0	0	0	0
Total	88	62	44	62	56
Dead root					
0–5	384	111	247	170	268
5–10	294	177	236	124	311
10–15	83	65	87	118	70
15–20	51	125	172	179	183
20–25	4	78	217	181	293
Total	816	556	959	772	1125

[a] $T = < 0.5$.
[b] Dates in parentheses indicate the dates soil cores were obtained when the belowground was sampled on different days than the aboveground.

decreased again to about 2:1 by the end of the growing season (Table 6). The corresponding dead root to dead rhizome plus stem base ratio fluctuated with no pattern between 3.1:1 and 7.5:1.

The live belowground to live aboveground ratio was 128:1 at the onset of spring growth in mid-June, declined to 7:1 by the early August peak of aboveground live standing crop, and then increased to 9:1 during the end of August period of aboveground senescence. Belowground production measured between 5 July and 24 August included 101 g m^{-2} of live material and 179 g m^{-2} of dead material. Modeled belowground production for the entire season included a loss of 227 g m^{-2} of live material and an increase of 369 g m^{-2} of dead material, for a net gain of 143 g m^{-2} (Dennis, 1977).

The belowground system in the long-term fertilizer experiment had a response in magnitude of standing crop similar to that of the aboveground system (Dennis, 1977). Total live standing crop was 867 g m^{-2} on the treated plots and 457 g m^{-2} on the controls (Table 7). The stem base and rhizome portions constituted greater percentages of the total on the fertilized than on the control plots, being 8 and 17%, respectively, compared to 3 and 9%, respectively. Similarly, white roots were more abundant in the fertilized than in the control standing crops, comprising 5 and 3% of the totals, respectively. On the fertilized plot there was 5.0 times as much live stem base, 3.5 times as much live rhizome, 3.5 times as much live white root, and 1.5 times as much live tan root as on the control plot. These differences between treatments did not occur in the dead standing crop. Of the total dead standing crop of 717 g m^{-2} on the control plot and 697 g m^{-2} on the fertilized plot, 17 and 17%, respectively, was stem base, 7 and 11%, rhizome, and

Table 6. Belowground Plant Standing Crops during 1971

Plant component	5 June	15 June	5 July	3 August	24 August
Live standing crop (% of total)					
Rhizome + stem base	38.0	34.0	18.0	30	30
Total root	62	66	82	70	70
Root:rhizome + stem base ratio	1.6	1.9	4.6	2.3	2.3
Dead standing crop (% of total)					
Rhizome + stem base	12	24	14	18	13
Root	88	76	86	82	87
Root:rhizome + stem base ratio	7.5	3.1	6.4	4.7	6.9
Total standing crop (g m^{-2})					
Live aboveground	7	23	37	79	71
Live belowground	679	899	522	570	623
Below:aboveground ratio	29.5	128.4	14.1	7.2	8.8
Live + current dead aboveground	7	23	37	80	87
Below:aboveground	29.5	128.4	14.1	7.1	7.2

Table 7. Aboveground and Belowground Plant Standing Crops on Fertilized and Adjacent Unfertilized Areas of the Schultz Plots, Barrow, Alaska[a]

| | Treatment | | | | |
| | Unfertilized | | Fertilized | | |
Sample	Mean	As percentage of root + shoot	Mean	As percentage of root + shoot	Fertilized ÷ unfertilized
Live belowground standing crop (g m^{-2})					
Stem base	13	3	67	8	5.2
Rhizome	42	9	152	17	3.6
Total shoot	55	12	219	25	4.0
White root	12	3	41	5	3.4
Tan root	390	85	607	70	1.6
Total root	402	88	648	75	1.6
Total root + shoot	457	100	867	100	1.9
Dead belowground standing crop (g m^{-2})					
Stem base	121	17	118	17	1.0
Rhizome	54	7	77	11	1.4
Total shoot	75	24	195	28	1.1
White root					
Tan root					
Total root	542	76	502	72	0.9
Total root + shoot	717	100	697	100	1.0
Aboveground standing crop (g m^{-2})					
Current live + dead	68		102		1.5
Prior dead	103		142		1.4
Soil characteristics (cm)					
Thaw	24		20		
Sample depth	34		30		
Core length	22		22		
Core compression	12		8		

[a] Samples collected 2 August 1971 (from Dennis, 1977).

76 and 72%, root. For the dead standing crops, the ratios of fertilized to control ranged between 0.9 (roots) and 1.4 (rhizomes).

Leaf Area Index

The seasonal progression of live leaf area index (Caldwell et al., 1974, 1978) closely followed that of live standing crop with the peak occurring in late July and early August and with higher values in 1970 than 1971 (Figure 7). Canopy development began earlier, and, for higher canopy increments, continued longer in 1970 than in 1971 (Figure 8). Although the canopy was developed to a maximum height of 20 cm, more than half the total leaf area index occurred in the lowest 5 cm. The plant material that extended into the 15- to 20-cm height increment consisted predominantly of reproductive structures. As with the standing crop data, the LAI data show that the blade component increased in relative importance through the season from 70 to 90% of total LAI. Approximately 10% of the peak season LAI was accounted for by "stem" material, most of which was concentrated in the 0- to 5-cm canopy height interval. In agreement with the standing crop data, the majority of the peak season leaf area index, 72 to 76 per cent, consisted of the three main graminoid species. The dicotyledon species accounted for less than 16% of the peak season LAI. Comparison of total foliage leaf area index data with total standing crop data yielded correlation coefficients of 0.92 for 1970 and 0.96 for 1971 with the correlation coefficients for *Carex, Dupontia* and *Eriophorum* being similarly high.

The distribution of leaf inclinations in each 30° interval (Table 8) illustrates the primarily erectophilic nature of the vascular plant canopy of wet tundra vegetation. Seventy-two percent of all leaves were inclined at 30° or greater and 27% of all leaves were inclined 80 to 89°. Leaves of graminoid species tended to be inclined more toward the vertical than leaves of dicotyledon species. For example, 79% of graminoid leaves were inclined at 30° or greater whereas only 39% of

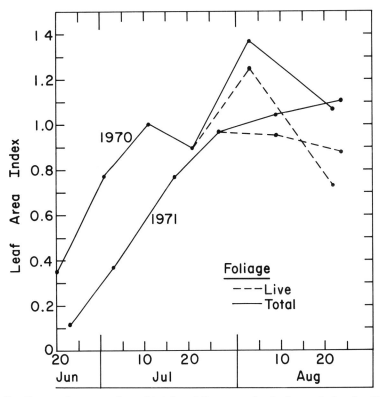

Figure 7. Seasonal progression of total and live vascular leaf area index for 1970 and 1971, site 2.

dicotyledon leaves were inclined 30° or greater (Table 8). Of the three main Barrow monocotyledons, 77% of the leaves of *Carex aquatilis* were inclined 60° or more, 57% of *Dupontia fisheri* leaves fell within that range, but only 18% of the leaves of *Eriophorum angustifolium* were inclined 60° or more. Of the two most abundant dicots, on the other hand, *Petasites frigidus* had only 7% of its leaves inclined 60° or more and *Salix pulchra* had 16% of its leaves in that range.

In addition to the seasonal sampling in site 2, other plots comparable to some sampled by Webber (1978) were sampled for leaf area index in August 1970 and 1971 (Table 2). A large range of LAI values, from 0.08 to 1.72, occurred in these

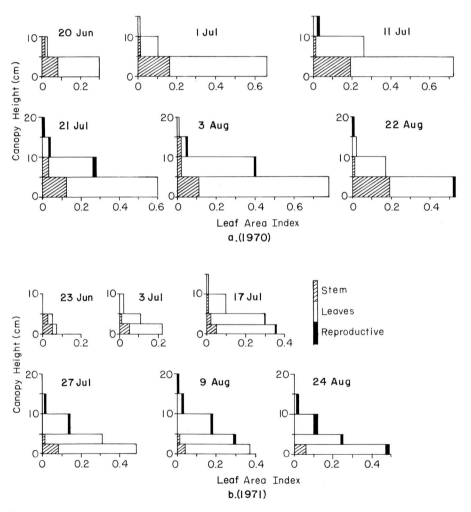

Figure 8.   Seasonal progression of canopy development in height increments with relative contributions of stem, leaves, and reproductive fractions. Note that in 1971 the 0- to 5-cm increment was subdivided into 0 to 2.5 and 2.5 to 5.0 cm.

Table 8.   Site 2 Distribution of Leaf Inclination

Species	Percentage leaves each 30° interval		
	0–29	30–59	60–89
Monocotyledons			
Calamagrostis holmii	100	0	0
Carex aquatilis	9	14	77
Dupontia fisheri	9	33	57
Eriophorum angustifolium	38	42	18
Eriophorum scheuchzeri	14	48	38
Poa arctica	47	29	24
Total	21	32	47
Dicotyledons			
Cardamine pratensis	100	0	0
Petasites frigidus	79	14	7
Ranunculus nivalis	100	0	0
Salix pulchra	37	47	16
Saxifraga cernua	67	33	0
Stellaria laeta	100	0	0
Total	61	30	9
Total all species	28	31	41

plots and characterizes the mosaic of the Barrow area. In keeping with the trend observed at site 2 plots, LAI values were higher in all comparable plots in 1970 than in 1971. Among the plots sampled in 1970, the lowest LAI values were found on two polygon tops, whereas the highest values occurred in a polygon trough and a wet meadow. In 1971, the highest values were found in an *Arctophila* slough and again in polygon trough and wet meadow plots, although in the latter plots the absolute values were only about 40% of those from comparable plots in 1970. The polygon top plot in 1971 had an LAI that was only 70% that of the same plot in 1970. The lowest values in 1971 were found in the low polygon, sunken basin plots, which were not measured in 1970.

As the survey of this diversity of natural communities was being carried out, it became obvious that a number of small disturbed sites possessed exceptionally high leaf area indices and aboveground productivities. These small areas generally occurred in impoundments which were caused by vehicular activity and which retained standing water throughout the growing season. The vegetation usually consisted of monospecific stands of either *Arctophila fulva* or *Dupontia fisheri*, although *Ranunculus gmelinii* was present as a subordinate in part of the stand. Two of these pure species swards were sampled in 1970 and 1971. One, an *Arctophila* sward, had an LAI in 1970 of 8.47, the highest LAI recorded at Barrow. The other, a pure *Dupontia* sward, had an LAI of 5.25 in 1970. In both cases, the canopy was developed to a height of 30 cm. The greatest portion of the leaf area fell between 10 and 15 cm. Leaves on these plants also were steeply inclined; *Arctophila* possessed a mean inclination of 76°, *Dupontia* 81°.

Discussion

The seasonal maxima of aboveground, current year standing crops for IBP site 2 of 102 g m^{-2} in 1970 and 87 g m^{-2} in 1971 are in agreement with and near the middle of the range of those reported by Bliss (1962) in his review of production data for several tundra communities at Point Barrow. The plots in site 2 occupy centers of flat-centered polygons, generally are free of standing water during the growing season (Tieszen, 1972), and occur on meadow tundra soils. The community is dominated by graminoid species, of which only three, *Carex aquatilis, Dupontia fisheri,* and *Eriophorum angustifolium,* account for over 70% of both aboveground standing crop and LAI. The belowground standing crop in this graminoid vegetation at peak season is seven times larger than the standing crop of the aboveground portion.

Our below-to-aboveground standing crop ratios are in agreement with those compiled by Bliss (1970) for regions of cold temperatures. Dennis (1968, 1977) and Shaver and Billings (1975) have suggested that annual belowground turnover in at least some arctic species is low and that the large belowground biomass is the result of many years accumulative production. Chapin (1974) and Billings et al. (1977) have shown that cold soil temperatures (near 0°C) do not impede substantial rates of root growth in *Eriophorum angustifolium;* and Chapin (1974) discussed the hypothesis that high root:shoot ratios represent a plant compensation to enhance nutrient uptake in a cold and nutrient-poor environment. On the other hand, Allessio and Tieszen (1975), using radiotracer studies, have shown not only that allocation of photosynthate to belowground components is high, but also that it can occur in early spring before soil thaw is advanced. The apparent seasonal fluctuation in live belowground biomass at Barrow of an early season maximum followed by the seasonal low reached approximately 25 days after the onset of aboveground growth is in keeping with seasonal changes of carbohydrate levels in belowground structures of alpine plants shown by Mooney and Billings (1960).

Such observations support the hypothesis that belowground biomass in part consists of reserve materials that are utilized for an early season growth of aboveground tissue when ambient temperatures are still low but solar radiation receipt is at its seasonal high. These reserves are replaced later in the season when photosynthetic tissues still are active but net increase in aboveground standing crop has ceased. The timing of peak solar radiation at the very beginning of a short growing season creates an environment in which it is highly adaptive for a plant to exsert as much photosynthetic material as early after snowmelt as possible. Allessio and Tieszen (1975) have shown that the stem bases and rhizomes can serve as storage reservoirs of photosynthate, and, moreover, that materials stored in stem bases and rhizomes of a tiller with completely senesced aboveground portions can become available to other tillers within the tiller system. The belowground standing crop data show that the apparent decrease of belowground standing crop in the first 25 days of the season occurs primarily in the rhizome–stem base component, not the root component. Although the standard errors are very large as a result of differences among plots, and interpretations must be made with this error in mind, the proportionately large early season drop in stem base–rhizome live standing crop further supports the hypothesis that it is the stem base–rhizome fraction which serves as the storage compartment to fund early season belowground and aboveground growth. The apparent end of season increase in the stem base–rhizome compartment also is consistent with the hypothesis that stored material is used in the spring and replaced in the fall.

McCown (1978) found that, although there was not a large seasonal fluctuation in carbohydrate level, there was a seasonal variation in the sugar pool of stem bases and rhizomes. He interpreted this finding to indicate that these structures may act as storage reservoirs, supplying material to active growth areas in the early season and becoming replenished by the end of the growing season. In addition, he showed that the nature of the nonstructural photosynthate levels in plant tissue varies such that during the nongrowing season there is an apparent storage of tissue precursors which presumably can be drawn upon for the rapid growth of early spring. Our observations on the reduction and subsequent increase in standing crop of the stem base and rhizome belowground compartments suggest a general base level stability in magnitude of belowground standing crop due to structural elements accompanied by a seasonal or perhaps one- to several-year (Dennis, 1977) variability due to storage elements.

The maximum depth of belowground plant penetration is determined by the maximum depth of seasonal thaw that occurs in the permafrost-underlain terrain characteristic of Barrow. In 1971, the soil thawed to a depth of 20 cm by the end of July and 25 cm by the end of August; yet more than 80% of the total, and all of the stem base and rhizome, live standing crop occurred within the top 10 cm of soil. Although environmental factors influencing this vertical distribution of plant standing crop have not been established clearly, it is reasonable to hypothesize that the majority of the standing crop occurs near the surface because of warmer temperature, earlier thaw, and hence a longer growing season and probable better aeration. The ideas of Chapin (1974) and Schultz (1964), respectively, suggest that the downward growth of the root system that does occur despite the colder temperature, slower rate of thaw, and probable poor aeration that exist at depth may occur in response to a temporary, seasonal, growth-induced depletion of available soil nutrients near the surface and to a gradual release during thawing of nutrients held in the seasonally still-frozen portions of the active layer. The growth restrictive environment that occurs at the bottom of the active layer also makes it reasonable to hypothesize that the comparatively large standing crop of dead roots that we observed in the 20- to 25-cm depth increment may represent the gradual accumulation of a small annual production and senescence of roots in an anaerobic environment probably characterized by a very slow rate of decay.

Use of data from Table 5 for the 20- to 25-cm depth increment to derive average values for live root standing crop of 7 g m^{-2} and dead root standing crop of 155 g m^{-2} suggests that the turnover time of dead root material at that depth may be 20 yr or more. In contrast to this apparent long turnover time at depth, turnover times of dead standing crop computed in the same manner for the top 10 cm of soil are much shorter—1.1 for the stem base fraction, 0.5 for rhizomes, and

1.3 for roots. These estimated turnover rates are too rapid since the live standing crops are composed of perennial, rather than annual, structures. The differences in dead turnover times among the three structures may reflect relative differences in their functioning as storage reservoirs, with rhizomes providing the most storage capacity and roots the least. The difficulties of dealing with frozen ground, of distinguishing live from recently dead material, and of separating underground plant structures from the particulate organic matter so abundant in the high organic content meadow tundra soil, have contributed to the uncertainties apparent in our assessment of the meaning of the seasonal and spatial variabilities found in the belowground standing crop. The large sampling errors of our data reflect these difficulties and limit the certainty of our interpretations, but the trends do appear to follow the dynamics hypothesized for wet tundra belowground vegetation.

Net primary production rates during the 1970 and 1971 seasons were comparable to those reported by Bliss (1962) for a *Carex* meadow community at Point Barrow. Assuming that 9.7×10^7 cal m^{-2} of photosynthetically available radiation (400 to 700 nm) is received during a 55-day growing season at Barrow (Tieszen, 1972), and that the mean caloric value of the Barrow aboveground vegetation is 4500 cal g^{-1}, the efficiency of aboveground production was around 0.46% for 1970 and somewhat lower for 1971 (assuming the same radiation receipt for both years). As we do not have caloric values for belowground plant components, and given the difficulties in assessing net belowground production discussed earlier, belowground efficiencies are not estimated here, and, therefore, efficiency of total solar conversion is underestimated. The 0.46% value is, however, in good agreement with the range of 0.20 to 0.45% for aboveground, growing season efficiencies for Barrow and Devon Island reported in the review of Bliss et al. (1973). Tieszen's data (1978) suggest that net CO_2 uptake by photosynthesis is substantially greater.

The data presented by Tieszen (1972) show a seasonally varying relationship between chlorophyll standing crop and aboveground vegetation standing crop at Barrow. The peak chlorophyll standing crop is attained 10 days earlier than the peak of aboveground dry matter. The higher chlorophyll contents of the dicots compared to the monocots probably are reflective of less structural tissue and suggestive of a more highly nutritious, although less abundant, resource in the dicots than in the monocots. Differences in the seasonal progression in dicots vs monocots are related to differing life strategies. The dicots develop peak chlorophyll standing crops early in the season, not only when the high sun angle is more favorable to the receipt of radiation by their more horizontal leaves, but also before the graminoid canopy has grown into the higher canopy levels. Later in the season the sun angle is reduced, at which time more nearly vertically inclined leaves at a high LAI can better utilize the incoming radiation (Miller and Tieszen, 1972). The graminoid component of the community then attains its peak standing crop of chlorophyll. These differences in seasonal kinetics also reflect the more sequential leaf growth pattern of the monocots, which are characterized by a continuous turnover of leaves. Strikingly similar relationships are seen in the seasonal nutrient data (Chapin et al. 1975). The peak in community standing crop of chlorophyll of 459 g m^{-2} is lower than the 756 g m^{-2} reported by Tieszen and Johnson (1968) for an arctic community at Meade River, Alaska, but is in close agreement with the 460 g m^{-2} reported by Bliss (1966) for alpine tundra on Mt. Washington, New Hampshire (Tieszen, 1972).

The seasonal progression of leaf area index was very similar to that of aboveground standing crop. The increase early in the season was linear, and by the time of the summer solstice the canopy had developed an LAI of 0.35 in 1970 and 0.10 in 1971. This development of a sizable canopy shortly after the vegetation becomes essentially snow free supports the hypothesis that belowground reserves are utilized for early season aboveground growth. The maximum LAI for communities of mixed-species composition of around 1.10 to 1.37 was consistent among plots in site 2. Interestingly, the LAI for a *Kobresia* meadow community on Niwot Ridge, Colorado, was higher at a maximum of 3.6 (Caldwell et al., 1974) than that of the Barrow community. The lower LAI of the Barrow community is as efficient in utilizing the Barrow regime of solar energy received at a lower solar angle for a 24-hr period as is the higher LAI *Kobresia* community in utilizing the Niwot Ridge regime of solar energy characterized by a greater sun angle but lasting for only a 14-hr period (Caldwell et al., 1978).

Although LAI was consistent among the polygon center plots of site 2, LAI and leaf inclination varied greatly among the different plots outside of site 2. The highest LAIs were found in stands of pure-species composition. The model of net primary production in the Arctic (Miller and Tieszen, 1972) predicts that maximum production would occur in an area of high leaf area index and steeply inclined leaves with little standing dead when water is not limiting. From this it would be predicted that the *Arctophila fulva* and *Dupontia fisheri* pure stands would be the most productive communities with the highest LAIs and most steeply inclined leaves. The graminoids, especially, tended to have steeply inclined leaves as shown by the high percentage of graminoid leaves inclined 30° or greater. Of the three main Barrow species, *Carex aquatilis* and *Dupontia fisheri* had more steeply inclined leaves than *Eriophorum angustifolium*. Dicotyledon species had, on the whole, more horizontally inclined leaves. The relationship between leaf inclination and LAI is not seen clearly with our data since leaf inclination was not recorded for all plots. However, in 1971 when leaf inclinations were recorded, the plots with the greatest majority of steeply inclined leaves, e.g., polygon trough plot 410, possessed the highest LAIs. These observations indicate that the angle of leaf inclination is one of many factors, biotic and abiotic, that work together to limit leaf area production.

The model of production at Barrow (Miller and Tieszen, 1972) has indicated the sensitivity of the system to canopy structure, climatic factors, and grazing by herbivores. The importance of rapid, early season canopy development to permit maximum absorption of the higher early season solar radiation shows the adaptive significance of utilization of belowground reserves to fund early season growth when low temperatures and minimum leaf area prevent efficient photosynthetic use of incoming radiation. In addition, the overwintering within the protective sheath of green material of a leaf that only partially developed the preceding year (Tieszen and Johnson, 1975) represents an adaptation for the rapid exsertion of green material. The large proportion of blade relative to stem material, and the abundance of chlorophyll in what stem material is present (Tieszen, 1972), both indicate that in the short growing season characteristic of Barrow, it is perhaps too costly for a plant to expend energy in the development of nonphotosynthetic aboveground material. This conclusion perhaps explains the loss of shrubs that occurs at higher latitudes. Although the seasonal aboveground production is low, daily rates of production are comparable to those of some temperate ecosystems (Billings and Mooney, 1968). The dwarf nature of tundra vegetation in general has been described (Bliss, 1962; Billings and Mooney, 1968), and the short nature of the canopy has been suggested as an adaptation by which the plant invests its resources in the production of a small amount of highly efficient material (Chapin, 1974) rather than a large amount of less efficient plant material. It is the adaptations of early season growth, the investment of resources in belowground biomass for a reservoir of biosynthetic materials and as a possible resistance to grazing stress, the developmental senescence of aboveground material with the conservation of nutrients (McCown, 1978), and the compact canopy structure that enable tundra plants to survive in an environment which characterizes the cold extreme of climatic conditions.

Acknowledgments. This research was supported by the National Science Foundation under Grant 29343 to Augustana College. It was performed under the joint sponsorship of the International Biological Programme and the Office of Polar Programs and was coordinated by the U.S. Tundra Biome. Field and laboratory activities at Barrow were supported by the Naval Arctic Research Laboratory of the Office of Naval Research.

We are especially grateful to the many assistants who participated in the field and laboratory work.

References

Alexandrova, V. D. (1958) An attempt to determine the aboveground productivity of plant communities in the arctic tundra. *Bot. Zh.,* **43**: 1748–1761.

Allessio, M. L., and L. L. Tieszen. (1975) Patterns of carbon allocation in an arctic tundra grass, *Dupontia fischeri* (Gramineae), at Barrow, Alaska. *Amer. J. Bot.,* **62**: 797–807.

Billings, W. D. (1973) Arctic and Alpine vegetations: Similarities, differences and susceptibility to disturbance. *Bioscience,* **23**: 697–704.

Billings, W. D., and H. A. Mooney. (1968) The ecology of arctic and alpine plants. *Biol. Rev.,* **43**: 481–529.

Billings, W. D., K. M. Peterson, G. R. Shaver, and A. W. Trent. (1977) Root growth, respiration, and carbon dioxide evolution in an arctic tundra soil. *Arct. Alp. Res.,* **9**: 127–135.

Billings, W. D., K. M. Peterson, and G. R. Shaver. (1978) Growth, turnover, and respiration rates of roots and tillers in tundra graminoids. In *Vegetation and Production Ecology of an Alaskan Arctic Tundra* (L. L. Tieszen, Ed.). New York: Springer-Verlag, Chap. 18.

Bliss, L. C. (1962) Adaptations of arctic and alpine plants to environmental conditions. *Arctic,* **15**: 117–144.

Bliss, L. C. (1966) Plant productivity in alpine microenvironments on Mt. Washington, New Hampshire. *Ecol. Monogr.,* **36**: 125–155.

Bliss, L. C. (1970) Primary production within arctic tundra ecosystems. In *Proceedings of the Conference on Productivity and Conservation in Northern Circumpolar Lands, Edmonton, Alberta, 15 to 17 October 1969* (W. A. Fuller and P. G. Kevan, Eds.). Int. Union Conserv. Natur., Morges, Switzerland, Publ. 16, N.S., pp. 77–85.

Bliss, L. C., G. M. Courtin, D. L. Pattie, R. R. Riewe, D. W. A. Whitfield, and P. Widden. (1973) Arctic tundra ecosystems. *Ann. Rev. Ecol. Syst.,* **4**: 359–399.

Brown, J., H. Coulombe, and F. Pitelka. (1970) Structure and functions of the tundra ecosystem at Barrow, Alaska. In *Proceedings of Conference on Productivity and Conservation in Northern Circumpolar Lands, Edmonton, Alberta, 15 to 17 October 1969* (W. A. Fuller and P. G. Kevan, Eds.). Int. Union Conserv. Natur., Morges, Switzerland, Publ. 16, *N. S.,* pp. 41–71.

Caldwell, M. M., L. L. Tieszen, and M. Fareed. (1974) The canopy structure of tundra plant communities at Barrow, Alaska, and Niwot Ridge, Colorado. *Arct. Alp. Res.,* **6**: 151–159.

Caldwell, M. M.,.D. A. Johnson, and M. Fareed. (1978) Constraints on tundra productivity: Photosynthetic capacity in relation to solar radiation utilization and water stress in arctic and alpine tundras. In *Vegetation and Production Ecology of an Alaskan Arctic Tundra* (L. L. Tieszen, Ed.). New York: Springer-Verlag.

Chapin, F. S., III (1974) Morphological and physiological mechanisms of temperature compensation in phosphate absorption along a latitudinal gradient. *Ecology,* **55**: 1180–1198.

Chapin, F. S. III (1978) Phosphate uptake and nutrient utilization by Barrow tundra vegetation. In *Vegetation and Production Ecology of an Alaskan Arctic Tundra* (L. L. Tieszen, Ed.). New York: Springer-Verlag, Chap. 21.

Chapin, F. S., III, K. Van Cleve, and L. L. Tieszen. (1975) Seasonal nutrient dynamics of tundra vegetation at Barrow, Alaska. *Arct. Alp. Res.,* **7**: 209–226.

Dennis, J. G. (1968) Growth of tundra vegetation in relation to arctic microenvironments at Barrow, Alaska. Ph.D. dissertation, Duke University, 289 pp.

Dennis, J. G. (1977) Distribution patterns of belowground standing crop in arctic tundra at Barrow, Alaska. *Arct. Alp. Res.,* **9**: 111–125.

Dennis, J. G., and P. L. Johnson. (1970) Shoot and rhizome-root standing crops of tundra vegetation at Barrow, Alaska. *Arct. Alp. Res.,* **2**: 253–266.

Golley, F. B. (1961) Energy values of ecological material. *Ecology,* **42**: 581–583.

Hultén, E. (1968) *Flora of Alaska and Neighboring Territories.* Stanford: Stanford University Press, 1008 pp.

Johnson, P. L., and L. L. Tieszen. (1973) Vegetative research in arctic Alaska. In *Alaskan Arctic Tundra* (M. E. Britton, Ed.). Arctic Institute of North America Tech. Pap. No. 25, pp. 169–198.

Khodachek, E. A. (1969) Vegetal matter of tundra phytocoenoses in the western part of Taimyr Peninsula. *Bot. Zh.,* **54**(7): 1059–1073. (International Tundra Biome Translation 5, December 1971; Translator: G. Belkov, 12 pp.)

Lewellen, R. I. (1972) *Studies on the Fluvial Environment: Arctic Coastal Plain Province, Northern Alaska.* Publ. by the author (P.O. Box 1068, Littleton, Colorado 80120), 2 Vols., 282 pp. and plates.

McCown, B. H. (1978) The interaction of organic nutrients, soil nitrogen, and soil temperature on plant growth and survival in the arctic environment. In *Vegetation and Production Ecology of an Alaskan Arctic Tundra* (L. L. Tieszen, Ed.). New York: Springer-Verlag, Chap. 29.

Melchior, H. R. (1972) Summer herbivory by the brown lemming in Barrow, Alaska. In *Proceedings 1972 Tundra Biome Symposium, Lake Wilderness Center, University of Washington* (S. Bowen, Ed.). U.S. Tundra Biome, U.S. International Biological Program and U.S. Arctic Research Program, pp. 136–138.

Miller, P. C., and L. L. Tieszen. (1972) A preliminary model of processes affecting primary production in the arctic tundra. *Arct. Alp. Res.,* **4**: 1–18.

Mooney, H. A., and W. D. Billings. (1960) The annual carbohydrate cycle of alpine plants as related to growth. *Amer. J. Bot.,* **47**: 594–598.

Pitelka, F. A. (1972) Cycle pattern in lemming populations near Barrow, Alaska. In *Proceedings 1972 Tundra Biome Symposium, Lake Wilderness Center, University of Washington* (S. Bowen, Ed.). U.S. Tundra Biome, U.S. International Biological Program, U.S. Arctic Research Program, pp. 132–135.

Schultz, A. M. (1964) The nutrient recovery hypothesis for arctic microtine cycles. II. Ecosystem variables in relation to arctic microtine cycles. In *Grazing in Terrestrial and Marine Environments* (D. J. Crisp, Ed.). Oxford: Blackwell Scientific Publications, pp. 57–68.

Scott, D., and W. D. Billings. (1964) Effects of environmental factors on standing crop and productivity of an alpine tundra. *Ecol. Monogr.,* **34**: 243–270.

Shaver, G. R., and W. D. Billings. (1975) Root production and root turnover in a wet tundra ecosystem, Barrow, Alaska. *Ecology,* **56**: 401–409.

Tieszen, L. L. (1972) The seasonal course of aboveground production and chlorophyll distribution in a wet arctic tundra at Barrow, Alaska. *Arct. Alp. Res.,* **4**: 307–324.

Tieszen, L. L. (1974) Photosynthetic competence of the subnivean vegetation of an arctic tundra. *Arct. Alp. Res.,* **6**: 253–256.

Tieszen, L. L. (1978) Photosynthesis in the principal Barrow, Alaska, species: A summary of field and laboratory responses. In *Vegetation and Production Ecology of an Alaskan Arctic Tundra* (L. L. Tieszen, Ed.). New York: Springer-Verlag, Chap. 10.

Tieszen, L. L., and P. L. Johnson. (1968) Pigment structure of some arctic tundra communities. *Ecology,* **49**: 370–373.

Tieszen, L. L., and D. A. Johnson. (1975) Seasonal pattern of photosynthesis in individual grass leaves and other plant parts in arctic Alaska with a portable $^{14}CO_2$ system. *Bot. Gaz.,* **136**: 99–105.

Webber, P. J. (1978) Spatial and temporal variation of the vegetation and its production, Barrow, Alaska. In *Vegetation and Production Ecology of an Alaskan Arctic Tundra* (L. L. Tieszen, Ed.). New York: Springer-Verlag, Chap. 3.

5. Floristics, Phytogeography, and Ecology of Arctic Alaskan Bryophytes

WILLIAM CAMPBELL STEERE

The Bryophyte Flora and Its Floristic Elements

The Flora

Our knowledge of the bryophytes of Arctic Alaska has been much enhanced in recent years by the appearance of a series of useful catalogues and checklists compiled from the literature or based, at least in part, on actual field exploration and original investigation (Murray, 1974; Murray and Murray, 1974, 1978; Rastorfer, 1972; Rastorfer et al., 1973; Smith, 1974; Worley, 1970; Worley and Iwatsuki, 1970, 1971; Steere, 1978). The bryophyte flora of Arctic Alaska is surprisingly large, even after only a relatively short period of study, and with remarkably few bryologists contributing to our knowledge of it. At present 420 species of mosses and 135 species of hepatics (Steere and Inoue, 1978) are known for a total number of known species of bryophytes of 555 (Table 1), many of which have not yet been reported. Within a few years, if bryological exploration continues, this number will likely reach 600. This prediction is an educated guess in that (1) several as yet undescribed species are in hand, (2) some species of uncertain identity, and still not identified, are omitted from the count, (3) new and capable bryologists are entering the field, (4) the enormous, difficult, and untidy genus *Bryum* has not yet received the careful treatment that will certainly double the number of known species in Arctic Alaska, and (5) the hepatics, likewise, have been insufficiently collected and studied by professional hepaticologists. An interesting comparison is that Crum (1972, 1976) reports only 245 species of mosses from the Douglas Lake region of the northern part of the Lower Peninsula of Michigan, bryologically one of the best known areas of the middle west. The whole state of Michigan has some 345 species of mosses, but it has a much more favorable climate and a much more diversified vegetation than Arctic Alaska—yet the latter area has more species.

As another reference point, Wiggins and Thomas (1962), in their treatment of the vascular plants, including ferns, of the Arctic Slope of Alaska, cover 435 species, not including subspecific taxa. Because of the field investigations conducted by Dr. I. L. Wiggins and his students, as well as by Dr. Eric Hultén, the chances of major increases in the vascular flora are relatively low, in strong

Table 1. Summary of the Bryophytes of Arctic Alaska

Mosses		
Boreal species	350	
Arctic species	55	
Asiatic disjuncts	5	
Cosmopolitan	5	Subtotal 420
Hepatics		
Boreal species	97	
Arctic species	30	
Asiatic disjuncts	7	
Cosmopolitan	1	Subtotal 135
		Total 555 species (subject to
		continued revision)

contrast to the situation in bryophytes, in which substantial discoveries still remain to be made. Although intensive investigations on bryophytes of restricted localities have been carried on at Barrow and Prudhoe Bay during the 3 yr of activity of the Tundra Biome program, 1971 through 1973 (Murray, 1974; Murray and Murray, 1974, 1978; Rastorfer, 1972; Rastorfer et al., 1973; Smith, 1974), the Foothills Province and the Brooks Range have been neglected, except at isolated points, as Umiat and Peters Lake. Continued field work in Arctic Alaska is therefore urgently necessary and will be abundantly justified by the resulting contributions to science. For example, in my own field work, no field season has passed without the discovery of several species variously new to the Arctic, to Alaska, to North America, or to science.

Floristic and Bryogeographic Elements

The bryophyte flora of Arctic Alaska contains several well-marked floristic and bryogeographic elements, some of which have been difficult to understand until the past few decades, during which it became increasingly clear that much of the Arctic Slope was unglaciated by continental ice sheets during the Pleistocene Epoch (Coulter et al., 1965). Thus it was available as a refugium for Tertiary plants throughout the Pleistocene, a situation that has great phytogeographic significance. Schofield (1972), Schofield and Crum (1972), and Schuster (1966–1974) have made especially substantial contributions to our understanding of disjunct geographic distributions in northern bryophytes.

1. The Boreal and Subarctic Element. By far the largest, commonest, and best known element, as well as a complex element that will eventually be subdivided as its history comes to be better known, consists of 400 species, or approximately 80% of the known bryophytes of Arctic Alaska, and is an extension northward of those species and associations equally common and well known in forests, swamps, mountains, marshes, bogs, and fens in boreal and subarctic areas farther south. Most of them are included in standard manuals by Grout (1928–1940) for North American and by Nyholm (1954–1969) for Fennoscandia, and will not be listed here. Some measure of the length of time needed for the migration of boreal mosses is given in a very thoughtful paper by Crum (1972), who points out that 245 species of mosses are now present in the northern part of the Lower Peninsula of Michigan. However, this region was completely covered by continental glaciers until 10,000 yr ago, after which it was inundated by great freshwater glacial seas, Lakes Algonquin and Nipissing, as late as 2500 yr ago.

2. Temperate Disjunct Species.   This element represents a group of temperate species that were evidently stranded in Arctic Alaska during the Pleistocene Epoch, and that have been able to survive there, even under a much more rigorous climatic regime now. Some of them are separated from the closest known populations to the south by great distances (Steere, 1965). In fact, many of the species are characteristic of the upper Mississippi River, and the midwestern and eastern states at the same latitude (Crum, 1976; Redfearn, 1972), although some of them have western affinities. Several of these species grow on soil or humus, but the more critical ones occur on calcareous sandstone cliffs or at the protected base of large boulders that have separated from their original rock outcrops through frost action. As such rock habitats do not occur in the Coastal Plain Province, but only in the Foothills Province and in the Brooks Range and seem to be especially favorable at Umiat, in terms of the numbers of species present there, I have called this anomalous relict element the "Umiat Syndrome" (Steere, 1965). That this is a relict element seems to be clear from the normal geographic range of the species involved, i.e., more temperate areas of North America. In fact, from the previously known distribution of these species, they would never have been predicted by bryologists to occur in the Arctic before their discovery there.

Prominent species in the disjunct "Umiat Syndrome" element:

Desmatodon obtusifolius	*Radula complanata*
Grimmia plagiopodia	*Seligeria calcarea*
Gyroweisia tenuis	*S. campylopoda*
Frullania eboracensis	*S. diversifolia*
F. tamarisci subsp. *nisquallensis*	*S. pusilla*
Molendoa sendtneri	*Tetraphis pellucida*
M. tenerrima	*Tetrodontium repandum*
Neckera pennata	

3. Circumpolar Arctic Element.   Perhaps the most striking and thought provoking element in arctic bryophytes is the large group of circumpolar species. Being restricted to high latitudes, these bryophates are not often seen in herbaria or in the field and are therefore novelties to most bryologists. This relatively coherent floristic element (Steere, 1953, 1965) consists of some 75 species or approximately 20% of the Arctic Alaskan bryophyte flora. Some species occur only within the Arctic Circle whereas others, although more frequent in arctic regions, extend southward where high altitude or climatic factors provide appropriate environmental conditions (Gams, 1955; Holmen, 1955, 1957, 1960; Steere, 1953, 1965). However, several species that had previously seemed to be restricted to the Arctic have now turned up somewhat unexpectedly in the northern Rocky Mountains in areas that appear to have escaped Pleistocene glaciation, or at least some phases of it (Packer and Vitt, 1974), for example, *Didymodon johansenii* (Crum, 1965, 1969), *Bryobrittonia pellucida* (Vitt, 1974), *Oreas martiana* (Weber, 1960, 1973; Schofield, 1972), and *Hypnum procerrimum* (Bird, 1968; Schofield, 1972). In evaluating the southward extension of otherwise arctic species of bryophytes into the Rocky Mountains and Canadian Rockies, it should be kept in mind that the Brooks Range is actually the northwesternmost extension of the Rocky Mountain System in North America. Further search will undoubtedly yield information on the occurrence of other members of the arctic element considerably farther south in North America than we would originally have anticipated, just as some of the otherwise arctic species in Europe persist in the Alps, and as many otherwise arctic hepatics occur in the "tundra zone" in the immediate vicinity of Lake Superior (Schuster, 1966–1974) and Great Bear Lake, N.W.T. (Steere, 1977).

The geographic evolution of the circumpolar arctic species is most enigmatic. None of them is closely related to species that occur in more temperate climates to the south and many of them belong to monotypic genera or even families; in other words, they are not simply reduced or impoverished forms of temperate species. Quite to the contrary, they are often larger than the other species of the same genus, as *Aulacomnium acuminatum, Didymodon asperifolius, Scapania simmonsii, Tortella arctica,* and *Trichostonum cuspidatissimum,* and are so distinctly set off from their congeners that some of them have been given supraspecific rank in their own section or subgenus. The unusual vigor of these species, perhaps not to be expected in plants growing under exceptionally rigorous climatic conditions, seems to be no artifact or accident of observation in the field. My colleague, Dr. D. Basile, to whom I sent living arctic hepatics, told me that the arctic species were growing more rapidly and more luxuriantly in his cultures than collections from more temperate regions.

Several of the arctic hepatics appear to be related to tropical or subtropical species of the Southern Hemisphere (Steere, 1953), which would suggest a very ancient relationship and a very long period of isolation of Tertiary species in north polar regions. These anomalous relationships have been commented upon by Schuster (1974) in establishing a new monotypic subsection for *Scapania simmonsii,* as follows, "Only the type species, a very isolated taxon with a unique aspect, belongs here. It seems to be a remnant of the Early Tertiary or pre-Tertiary, essentially nonarctic, flora that has persisted in some loci (chiefly not glaciated) in high arctic regions." He then goes on to add *Hygrolejeunea alaskana, Ascidiota blepharophylla,* and *Pseudolepicolea fryei* (as *Lophochaete*) with the remark that "All of these taxa occur on the unglaciated north slope of the Brooks Range of Alaska, and all show nonarctic floristic affinities." Persson (1952) made the same point with reference to the moss, *Trichostomum cuspidatissimum,* as follows, "It is rather surprising to find such a southern species in the Arctic. It is not the only example. Of the same type are *Radula prolifera* Arn. (Persson, 1947), and for all *Entosthodon polaris* Bryhn. The latter species is said to be most nearly related to a species from the Mediterranean region of Europe. It is quite possible that these species are the last of a flora that existed there under conditions of a temperate–subtropical climate.'' As we know that much of the Arctic Slope of Alaska was not affected by continental glaciation during the Pleistocene (Coulter et al., 1965), it is clear that Tertiary species, which were much more widely distributed geographically in earlier times characterized by widespread warm and temperate climates, survived the Ice Ages in large refugia in arctic North America and Asia, north of the primary continental glaciers and their scouring action.

The selection of the species for a list of members of the circumboreal arctic floristic element has been both more difficult and more arbitrary than I had anticipated, with the exception of those unquestionable species which are obviously more at home in arctic climates than south of them, such as *Aplodon wormskjoldii, Aulacomnium acuminatum, Bryum wrightii, B. cryophilum, Cinclidium arcticum, C. latifolium, Fissidens arcticus, Mesoptychia sahlbergii, Radula prolifera, Scapania simmonsii, Tortella arctica, Trichostomum cuspidatissimum, Voitia hyperborea* and many more. The species much more difficult to appraise are those that have disjunct populations in mountains far to the south, and may eventually be discovered to comprise a new or different floristic element. Schuster (1966–1974) has noted the difficulty in interpreting whether some species of hepatics have originated in the Arctic and migrated southward or vice versa, and I have omitted from the following list many of the species that he terms ''arctic–alpine,'' or even ''arctic,'' because of disjunct populations, especially in Central Europe, that occur in mountains far south of the Arctic Circle. In any case, I have probably erred as much by including some species in this list as I have by omitting others; therefore, I may, quite by accident, have achieved some kind of balance.

Members of the Circumpolar Arctic Bryophyte Element

Andreaea obovata
Aplodon wormskjoldii
Arctoa anderssonii
Arnellia fennica
Aulacomnium acuminatum
Bryobrittonia pellucida
Bryum arcticum
B. cryophilum
B. wrightii
Cephaloziella arctica
Ceratodon heterophyllus
Chandonanthus setiformis
Cinclidium arcticum
C. latifolium
Coscinodon latifolius
Cratoneuron arcticum
Cynodontium glaucescens
Cyrtomnium hymenophylloides

C. hymenophyllum
Desmatodon leucostomus
Didymodon asperifolius
D. johansenii
Distichium hagenii
Desmatodon heimii var. *arcticus*
Drepanocladus badius
D. brevifolius
D. lycopodioides
Fissidens arcticus
Fossombronia alaskana
Funaria polaris
Grimmia tenera
Gymnomitrium corallioides
Hygrohypnum polare
Hygrolejeunea alaskana
Kiaeria glacialis
Lophozia binsteadii

L. cavifolia
L. latifolia
L. quadriloba
Marchantia alpestris
Marsupella arctica
Mesoptychia sahlbergii
Mnium blyttii
Oligotrichum falcatum
Peltolepis grandis
Philonotis tomentella
Philocrya aspera
Plagiochila arctica
Pohlia crudoides
Polytrichum hyperboreum
Pseudolepicolea fryei
Psilopilum cavifolium

P. laevigatum
Pterygoneurum arcticum
Radula prolifera
Scapania crassiretis
S. degenii
S. polaris
S. simmonsii
S. spitsbergensis
S. uliginosa
Schistidium holmenianum
Seligeria polaris
Solenostoma polaris
Timmia comata
Tortella arctica
Trichostomum cuspidatissimum
Voitia hyperborea

4. Disjunct Bryophytes on Calcareous Silt. A highly specialized habitat for bryophytes in arctic Alaska is the fine-grained calcareous silt that is extruded in the form of "frost boils" under the stresses of expansion as soil layers, thawed during the brief summer, refreeze. When wet, this silt is soft and sticky, and contains much water, so much so that if agitated vigorously, it will lose its firmness and become semifluid. However, when it dries, the silt either develops a cement-like hardness or becomes a fine white powder, depending on its original consistency and particle size. Exposed to full sun and wind, and without other plants for protection, the frost boils which are generally bare except for bryophytes must be xeric habitats, indeed. The species of bryophytes found on them are variously (1) members of the arctic circumpolar element, as *Bryum wrightii* (Steere and Murray, 1974), *Desmatodon leucostoma, Fissidens arcticus* (Steere and Brassard, 1974), *Fossombronia alaskana* (Steere and Inoue, 1974), and *Pterygoneurum arcticum* (Steere, 1959); (2) disjunct steppe mosses such as *Aloina brevirostris* and *Stegonia latifolia* (Gams, 1934); or (3) disjunct desert mosses such as *Pterygoneurum subsessile* (Steere and Iwatsuki, 1974). All these bryophytes, regardless of their phytogeographic relationships and affinities, are largely restricted in their habitat to calcareous silt, a curious and undoubtedly ancient specialization that may have developed in the warm and dry climate of steppes and subdeserts, and perhaps not in the cold and dry one of the Arctic.

5. Disjunct Asiatic Bryophytes in Arctic Alaska. This topic has been discussed in detail (Steere, 1965, 1969) and so will be reviewed briefly. *Habrodon leucotrichus,* a common corticolous moss in the mountains of Japan, was first reported from Alaska by Persson (1946). Since then, it has turned out to have a relatively wide distribution in Alaska, but has been reported from only one locality north of the Arctic Circle, namely, from Walker Lake on the South Slope of the Brooks Range, from a collection made by Louis Jordal in 1950 (Sherrard,

1955). Steere and Schofield (1956) reported *Myuroclada maximowiczii* from coastal Alaska, collected by Schofield at King Salmon, as new to North America. It was first discovered north of the Arctic Circle in Alaska at Ogotoruk Creek in 1963 (Steere, 1969); subsequent collections have also been made there. During the summer of 1974, moreover, a most unexpected collection was made at Cascade Lake in the central Brooks Range, which implies a much wider distribution in Arctic Alaska than I would have predicted. This species is widely distributed in boreal and Arctic Asia (Lindberg and Arnell, 1890) and is now known in North America from several other nonarctic Alaskan collections, most of them unreported. A most puzzling phytogeographic question was raised by my discovery in 1951 of the hepatic, *Ascidiota blepharophylla,* at Driftwood Creek in Arctic Alaska. This very distinctive species is known only from the type specimen collected in Central China, from the abundant collections made at Driftwood Creek (Steere and Schuster, 1960), and from a small collection made at Ogotoruk Creek by Hultén (Persson, 1962); this distribution has been mapped by Schofield and Crum (1972, Fig. 13). The Driftwood Creek locality, near the western end of the Brooks Range, is extraordinarily rich in bryophytes of unusual phytogeographic interest. It is the only known Alaskan station for another hepatic, *Frullania jackii,* even though just a single collection was made, unfortunately, because of its close resemblance in the field to the relatively common *F. tamarisci* subsp. *nisquallensis* (Hattori, 1972), which is abundant in the Brooks Range. This collection agrees better with the Asiatic race of *F. jackii* than it does with European populations from the Alps (M. Fulford and S. Hattori, personal communication). A third hepatic first collected at Driftwood Creek was described as a new species, *Hygrolejeunea alaskana* (Schuster and Steere, 1958), whose closest relatives appear to be Asiatic. This is a most improbable genus for the Arctic, yet it has been found in several other localities in the Brooks Range, and as far east as Peters Lake, where it occurs abundantly. Mosses of special phytogeographical interest at Driftwood Creek are *Gollania turgens (G. densepinnata), Herzogiella adscendens* (Iwatsuki and Schofield, 1973), and *Claopodium pellucinerve* (Noguchi, 1964); these are all Asiatic species with disjunct stations in Arctic Alaska. *Gollania turgens* has a geographic distribution essentially parallel to that of *Ascidiota blepharophylla,* as, in addition to four Brooks Range localities, it is otherwise known only from Shensi and western Kansu provinces in central China (Ando, 1966; Ando et al., 1957; Persson and Gjaerevoll, 1961; Schofield and Crum, 1972, Fig. 12). Finally, two further hepatics, *Macrodiplophyllum microdontum* and *M. plicatum,* are Asiatic species that also occur in the Brooks Range of Arctic Alaska (Abramova, 1956; Persson, 1949), although not previously reported.

6. Oceanic Element.   This floristic element is not particularly homogeneous and its members have also been included in other elements. However, its peculiar habitat requirements or preferences certainly have some relevance to past climates and phytogeographic questions of disjunction. The so-called oceanic species of hepatics that compose this element are usually found together on steep, wet, north-facing ridges that rise abruptly above surrounding tundra and consequently catch and precipitate low-hanging fog, or on which fog hangs regularly in bad weather, so that they receive far more moisture than the annual level of precipitation would normally provide them; moreover, they are protected during the winter by snow-banks, which also provide an abundant supply of melt-water in early summer. *Hygrolejeunea alaskana* is so far an endemic species of the Brooks Range, although it is almost certain to turn up in Arctic Asia. *Radula prolifera,* like *Mesoptychia sahlbergii* (Steere and Inoue, 1975), is not restricted to this habitat, in which it produces large and pure growths, but it also occurs in sloping tundras in wet flushes, so that it receives extra moisture from drainage channels. *Frullania jackii* is a disjunct species of the Alps, the Himalayas, China, Japan, and with only one known station in the Brooks Range. *Frullania tamarisci* subsp. *nisquallensis,* which occurs in abundance in many localities throughout the Brooks Range fryei, was first described from fog-shrouded St. Lawrence Island in the Bering Sea, and is now known from many localities in the Brooks Range, in the Foothills Province, and as far north in the Coastal Plain as Barrow and Cape Simpson, where it is common and abundant. In the wet acid tundra it usually occurs associated with *Sphagnum,* again, where much water is available. A closely related species, *P. trollei,* occurs in Japan (Ando, 1963). *Odontoschisma macounii* is a member of the wide-ranging circumpolar arctic element without other phytogeographic significance, which also flourishes in other very wet places, yet is not an aquatic. *Herberta sakuraii* is a disjunct species with its type locality in Japan and a wide distribution in Arctic Alaska, from one end of the Brooks Range to the other, although the exact identity of the Alaskan race still seems to be open to some question (Schuster, 1957; Steere, 1969). The members of genera that elsewhere in their geographic range are normally corticolous, especially *Hygrolejeunea, Frullania,* and *Radula,* grow in Arctic Alaska in open tundra, usually wet and steep, as characterized above. I have suggested elsewhere (Steere, 1953) that, since the disappearance of trees in northernmost Alaska with the deterioration of the climate, these hepatics have adapted themselves to living among and on tundra tussocks which, like the bark of trees, are relatively low in minerals and nutritive materials and generally receive water from fog, in addition to normal precipitation. A further possibility is that before the geologically recent uplift of the subterranean mountain range that now forms the Aleutian chain of islands, the Japanese Current undoubtedly flowed into the present Bering Sea and perhaps even into the Arctic Ocean, which would have produced a climatic regime wholly different from the present one.

7. The Cosmopolitan Element.   In most areas, we think of those weedy bryophytes that are truly cosmopolitan as being carried about by man, and as being weeds in artificially disturbed habitats. However, in many parts of Arctic Alaska, *Marchantia polymorpha, Bryum argenteum, Ceratodon purpureus, Funaria hygrometrica,* and *Leptobryum pyriforme,* the most frequent of all weedy bryophytes, are both common and abundant in areas that have never been disturbed by man's activities, but where the disturbance is caused by other animals and by natural events. For example, wind-driven ice rafts along lake shores cause the upheaval of soil and the formation of ice ramparts, storm erosion exposes new soil, and the burrowing of ground squirrels and other rodents brings new soil to the surface. Moose and caribou trails along lakeshores and elsewhere also bring rather considerable disturbance to the habitat. Newly exposed soil is soon colonized by weedy species, without any help from humans. The annual flood-scouring of river banks at the time of ice and snow melt removes old soil and exposes new, often in the form of vertical silt banks that soon support, in addition to cosmopolitan weedy bryophytes, some very interesting and relatively rare species as *Fissidens arcticus, Mniobryum atropurpureum, Pohlia annotina, P. proligera,* and *Desmatodon heimii,* among many others that are also characteristic invaders of naturally disturbed soil. The annual migration of vast numbers of caribou through silty or sandy areas likewise exposes new surfaces for colonization by weedy and interesting species alike and the caribou droppings are soon colonized by members of the *Splachnaceae. Bryum argenteum* and *Tortula ruralis* are frequent inhabitants of enriched areas around owl perches and bird cliffs. *Andreaea rupestris* is cosmopolitan in the sense that it is widely distributed on all continents, but can hardly be considered as weedy because of its restriction to acidic rock and its preference for cool or cold climates, at high altitudes in warmer regions. Species of *Grimmia* and *Schistidium* that are cosmopolitan in terms of the extremely wide geographic distribution also fall into much the same category as *Andreaea* because of their narrow habitat tolerance.

Ecology and Physical Factors of the Environment

The ecology of the bryophytes of the Barrow area has been investigated and discussed by Smith (1974) and has been reviewed briefly for the Prudhoe Bay area by Rastorfer et al. (1973). The lack of careful and controlled ecological observations in most other areas of Arctic Alaska, and the resulting lack of publications dealing with this topic, constitutes a serious problem for bryologists who wish to know more about it; but to remedy the lack immediately would strain both the manpower and financial resources of the field. However, now that the floristic aspects of arctic bryology are becoming reasonably well elucidated, a serious, comprehensive, and highly quantified study of bryological ecology should be established as the next order of priority. In the paragraphs that follow, I have drawn freely from my own field experience and field notebooks that have resulted from many years of observation in Arctic Alaska and elsewhere in arctic areas, including Canada, Greenland, and Lappland, as well as from the botanical literature.

The Physical Factors of the Environment

1. Water.   There can be no doubt that the single most important factor of the arctic environment in controlling the rapidity and luxuriance of growth of bryophytes is the amount of water available to them. Along a decreasing gradient of water availability we find hydrophytes, hygrophytes, mesophytes, hemixerophytes, and xerophytes, with a decreasing luxuriance of growth paralleling the diminishing water supply. The water gradient may be gradual, extending over a kilometer or more of gradually sloping tundra, with a stream or lake at the lower end of the slope and a relatively dry to very dry area at the upper level, or it may be very steep, running the whole gamut from wet to dry in a matter of a few decimeters or even centimeters. For example, in the generally wet tundra in the vicinity of Barrow, both high centered and low centered ice-wedge polygons are abundant (Britton, 1967; Smith, 1974). In an area of high-centered polygons, which vary in height up to a meter or somewhat more, there is normally a wet trough between them, in which hygrophytic bryophytes flourish. On the sides of the polygons, mesophytic species occur, whereas at the top are those species adapted for survival in dry habitats, or in habitats that are subjected periodically to desiccation. In low-centered polygons, on the other hand, the hygrophytic species occur in the center of the polygon, which may be a pool, especially early in the season, and the mesophytic species occur at the edge of the polygon and on the usually low ridge that separates one low-centered polygon from another. The relation of each species to its position on the water-availability gradient is subtle and difficult to interpret because (1) the amount of available water fluctuates greatly during the frost-free season, (2) of the occurrence of intense frost churning of both soil and vegetation, and (3) of the great mixture of species that coexist, seemingly without competition, in any one microniche. It is clear that most mesophytic species and even some xerophytes tolerate complete or partial submergence during the early part of the growing season, when most low-centered polygons are filled with water, whereas later in the season they may have to resist desiccation.

Relative to the remarkable ability of bryophytes to withstand and to recover from desiccation, several examples of interesting, illuminating, and relevant research can be cited. Malta (1921) found that different species of mosses demonstrate extremely divergent degrees of resistance to drought, and that there exists a strong correlation between the degree of resistance and the amount of water in their normal habitat. For example, *Fontinalis squamosa,* an aquatic moss restricted to flowing water, did not survive a week of air-drying, and *Philonotis fontana,* a moss of wet to very wet places, was able to survive for 15 to 20 weeks. However, a rock-inhabiting xerophytic moss, *Grimmia pulvinata,* survived for 60 weeks in a desiccator at 20°C.

Although many adaptations to a xerophytic habitat have evolved in bryophytes, as shown by the pioneer study of Watson (1913), none of the morphological adaptations serves to keep the gametophytic vegetative plant from drying out to some degree, or even from total desiccation. Rather than being morphological, the adaptations that seem to function best for survival of drought are physiological and biochemical. For example, species of *Grimmia* growing even on the driest and hottest rocks will still produce sporophytes and develop mature and viable spores, even though the vegetative plant may appear to be completely desiccated. The internal cells of the capsule, however, especially the sporocytes, are filled with lipids and complex mucilaginous carbohydrates that absorb and hold water until the vital process of meiosis has been completed and the extremely drought-resistant spores are liberated (Steere, 1954).

Von Noerr (1974a) studied the course of drought resistance of several common mosses during the whole growing season and found that drought resistance increases from a spring depression to a summer maximum and then decreases during the autumn, although the curves for the different species are not strictly parallel.

Bewley and Thorpe (1974) found no perceptible respiration in dried plants of *Tortula ruralis,* but that their recovery from desiccation was accompanied by an immediate and rapid increase in respiration (measured as O_2 uptake) either at 25.5 or 3.5°C. The respiratory burst was greater on rehydration of plants that had been rapidly desiccated over silica gel than those that had been more slowly dried in atmospheres of high relative humidity. Unlike many higher plants under stress, this moss did not exhibit any changes in its starch and sugar content during or following desiccation, nor were there any changes in free proline levels. An increased activity of the Embden–Meyerhof–Parnas pathway of glucose oxidation on rehydration of *Tortula* was observed.

Tucker et al. (1975) studied the ultrastructure of the same moss, *Tortula ruralis,* on recovery from desiccation. They found that dehydration causes condensation of the protoplasm at the distal and proximal cell ends, a rounding and compacting of the chloroplasts, and vesicle disappearance. Immediately upon rehydration, the chloroplasts and mitochondria swell and their internal membrane structures are rearranged. During a 24-hr rehydrating period the cell structure returns to that of the normal state. Gẃódzdź and Bewley (1975) continued the study of *T. ruralis* with special reference to its protein synthesis. When dried slowly, the moss retained fewer polyribosomes on desiccation, but more ribosomes than when rapidly dried. Even in the completely desiccated moss the polyribosomes or ribosomes, or both if present, retained their synthetic capacities. On rehydration, the slowly dried moss resumed protein synthesis more quickly than when desiccated rapidly. Extracted polyribosomes can withstand desiccation to a significant extent, which suggests that they may not need protection by the cytoplasm. These workers also investigated the aquatic moss, *Hygrohypnum luridum,* and found that it is able to retain polyribosomal and ribosomal activity during desiccation, but that this activity decreases greatly on rehydration.

2. Hydrogen-Ion Concentration.   Like water, hydrogen-ion concentration (pH) is a physical factor of the environment that clearly influences the distribution of bryophytes, simply because most mosses and hepatics fall into one of two classes, those that prefer calcareous habitats and those that prefer acidic habitats. However, in spite of these fundamental and characteristic preferences, many of the species are surprisingly tolerant of conditions that might seem to be unfavorable, and other species have a broad spectrum of tolerance for the presence or absence of calcium. For example, according to Tallis (1958) *Rhacomitrium lanuginosum* occurs in Great Britain in both calcareous and acid habitats and does best where there is high humidity and little competition from higher plants.

In the Barrow area and for some distance along the coast, both east and west, the boggy superficial soils, derived from peats, are predominantly acid, so that the bryophyte flora of the northernmost region of Alaska is largely that characteristic of acid areas, with much *Sphagnum, Lophozia,* and many members of the Polytrichaceae and Dicranaceae. Only on frost boils where the underlying calcareous silt has been extruded from the Gubik Formation, on the same silt exposed in ravines and stream cuts, and on the tops of ridges of calcareous silt, do calciphilous species of bryophytes appear in any quantity.

At some distance southeast of Barrow, where northward-flowing streams such as the Meade, Chipp, and Ikpikpuk rivers meander through the tundra, much sand has been brought downstream through the millennia, so that dunes and sand flats are commonplace. In general, these sands are somewhat to strongly calcareous, depending on the place of origin of the river and of its tributaries. Still farther to the east, several major rivers, as the Colville, Sagavanirktok, and Canning, that arise in the Brooks Range and whose tributaries come from limestone mountains, have churned up their flood plains by means of their changes of course over thousands of years, and have filled their valleys with deposits of calcareous silts and sands. Except for the rare to occasional occurrence of acid peats at the surface produced by the vegetation cover, especially *Eriophorum,* all habitats influenced by these rivers tend to be calcareous. At Prudhoe Bay the whole region is so uniformly calcareous that no species of *Sphagnum* has yet been found, and only one species of the *Polytrichaceae* has been reported (Murray and Murray, 1974, 1978; Rastorfer et al., 1973); if others are eventually discovered, it is safe to predict that they will occur in small quantity and very locally. In the Foothills Province and in the Brooks Range, the occurrence of bryophytes of various genera on rocks is likewise determined by the presence or absence of calcium. *Andreaea rupestris, Chandonanthus setiformis,* and *Ulota curvifolia,* for example, are intolerant of much calcium ion, which, on the contrary, seems to be a requirement for the occurrence of most species of *Grimmia, Amphidium, Seligeria, Orthotrichum,* and many other genera. In addition to calcium ion, copper and other metallic ions have been reported to influence the distribution of a small group of species that seem to prefer mineralized rocks of some complexity (Loeske, 1930; Persson, 1948; Persson and Gjaerevoll, 1961). Two of these so-called "copper mosses," *Mielichhoferia mielichhoferi* and *Coscinodon cribrosus,* occur on reddish metamorphic rock on a low mountain at the western end of the peninsula that separates Peters Lake from Schrader Lake. Schatz (1955), however, has suggested that sulfur ion may be the important factor in this curious distribution pattern instead of metallic cations. Also, in general, away from the calcareous river flood plains, it appears that the drier the habitat is, the more apt it is to be calcareous, because of the lack of leaching, as is so evident in the polar deserts and their bryophyte floras.

3. Temperature. Factors of the physical environment other than water availability and the hydrogen-ion concentration of the substratum have a much less dramatic effect in determining the presence or absence of specific bryophytes, and, as a consequence, are much more difficult to evaluate, and, especially, to measure. In the Coastal Plain Province surface relief is generally so slight that the effect of temperature gradients caused by differential insolation on the vegetation of slopes is not particularly apparent, if it exists at all. In the more rugged topography characteristic of the Foothills Province and the Brooks Range, however, one may easily recognize well-developed temperature effects. A remarkably clear example is provided by the first high ridge approximately 1 mile north of the airstrip at Umiat, where the south-facing slope, which receives maximum solar energy for its latitude, supports an assemblage of bryophytes that are astonishingly temperate in their affiliations, and many of the species do not occur otherwise for long distances to the south (Steere, 1965). However, on the north-facing side of the same ridge, which receives much less solar heat because of the lower angle of the midnight sun (at 69° N. latitude), one finds a preponderance of truly arctic bryophytes, such as *Mesoptychia sahlbergii, Arnellia fennica, Radula prolifera, Scapania simmonsii, Mnium andrewsianum, M. hymenophyllum, Dicranum angustum, D. elongatum, Aplodon wormskjoldii, Orthothecium chryseum, Philonotis tomentella,* etc., that could not possibly survive on the dry south-facing slope. Whether the temperature, by itself, and not just as a drying factor, is limiting to the distribution of bryophytes is still a moot question and needs much experimental work.

The only pronounced examples of temperature effect on bryophytes in the Barrow area are rather narrowly local. The first is the flourishing of normally mesophytic species in water-filled low-centered polygons early in the season. The reason for this anomalous behavior is apparently the warming effect that the long summer days, with 24 hr of light, have on the shallow water which in turn enables the bryophytes to grow more luxuriantly than they could if exposed to the cold, drying winds of early summer. Moreover, as these pools normally dry up during the summer or at least lose their standing water, the hygrophytic species do not compete too severely with the mesophytic ones, although the line between them is very delicate and finely drawn. The second example is a much more quantitative one and was derived from temperature data from Point Barrow Base recorded by Dr. Max E. Britton, at 2 mm and 2 cm above the surface of the soil (Steere, 1954, Table 2). These levels represent the warmest air temperatures recorded and are of special and urgent importance to the life processes of bryophytes. The 2-mm level, at the surface of the soil, represents the point of activity of the growing green gametophyte, which carries on photosynthesis and eventually produces *antheridia* and *archegonia,* either on the same plant or on different ones. At or near this same level, then, fertilization takes place and the young moss sporophyte begins its upward growth. The 2-cm level, on the other hand, apparently determines the distance above the surface of the soil at which the apex of the seta swells out and differentiates into the capsule. Also, at this warmer level, the sporocytes in the capsule are able to carry on meiosis, even when the official temperature, as recorded by the U.S. Weather Bureau at Barrow, was below freezing. This localized warmer belt of air undoubtedly explains why mosses of the colder tundras, as at Barrow, have a relatively shorter seta, which does not elongate into the colder air farther above the ground, especially in early summer. Stevenson et al. (1972) have reviewed the several environmental factors, especially temperature, that influence seta elongation in mosses. With regard to seta elongation and capsule development, another adaptation of mosses to low temperature should be mentioned. Anyone who studies the sporocytes of arctic mosses at the time of meiosis will discover, when the "squash" technique is used on living material, that these cells are so filled with lipids and other fatty materials which undoubtedly serve to prevent them from freezing, that cytological observations are impeded.

No real correlation has yet been shown between the presence or depth of permafrost and the occurrence of any special species or groups of bryophytes, so that they are not particularly useful as indicators of permafrost conditions; also, in winter, when the active layer freezes, they become an integral part of the permafrost. However, it is obvious that the bryophyte cover, when unbroken, is closely related to the formation of permafrost, as a differential heat filter. In the summer, the bryophyte cover, especially *Sphagnum,* becomes relatively dry on the surface and acts as an insulating blanket over the surface of the ground. In the winter, however, when the moss cover is wet, the escape of heat from the underlying soil is facilitated, thus lowering the temperature of the subsurface levels. Deep bryophyte mats in wet tundra thaw rather shallowly and permafrost can be demonstrated in them by probing or digging sometimes only a few inches from the surface, throughout the summer. An especially clear demonstration of the relationship between bryophyte cover and the persistence of permafrost is to be found in high-centered polygons, where the tops, usually with sparse or no moss cover, thaw to a considerable depth. The wet moss-filled troughs between them, however, thaw shallowly.

Probably no species found elsewhere in Arctic Alaska is unable to tolerate the low temperature, per se, of the Barrow area; the limiting factors are water availability, hydrogen-ion concentration, the frost-churning of the tundra surface, and the lack of suitable habitats. My evidence for this conclusion is the fact that many species of bryophytes extend to the ultimate northern margins of land, to beyond 83° N. latitude in northern Ellesmere Island (Brassard, 1971a, 1971b; Schuster et al., 1959), and in northernmost Greenland (Holmen, 1955, 1960). As both of these areas are arid, and properly classified as polar desert, the physical factor that limits the number of species that occur in them is clearly the amount of water available in the environment, and not the temperature regime, as rigorous as it may be. In fact, the probability that bryophytes are resistant to even lower temperatures is implicit in our conclusion that bryophytes survived north of the Brooks Range through at least some if not all phases of the Pleistocene epoch, with its undoubtedly colder and presumably wetter climate. Evidence for this conclusion was derived from two diverse sources. First, it has been shown recently (Coulter et al., 1965) that the Arctic Slope escaped continental glaciation, apparently because the level of precipitation prevalent during the Pleistocene was insufficient to produce massive ice sheets outflowing from the Brooks Range. Second, the persistence of a large element of what are apparently Tertiary species of bryophytes on the Arctic Slope and in other circumpolar unglaciated areas furnished the original evidence for this hypothesis (Steere, 1953, 1965), which already postulated a lack of glaciation at least in local areas which could serve as refugia and, eventually, as points of dispersal.

Some experimental work has been carried out on the effect of freezing and subfreezing temperatures on bryophytes. Irmscher (1912) investigated the effect of low temperature on a broad spectrum of mosses and found that the majority of species survived freezing in the normal hydrated condition at −10°C, that some could survive temperatures of −15 or −20°C, but that all were killed at −30°C. In a parallel investigation on Japanese mosses, Ochi (1952) found that 18 species withstood temperatures as low as −20°C, and that seven of them survived −27°C. In a much more sophisticated series of experiments, Dilks and Proctor (1975) studied the responses of bryophytes to low temperatures, with special reference to assimilation, respiration, and freezing damage. All species investigated tolerated temperatures down to −30°C without freezing damage as long as they were in a dry condition, but two species (one hepatic and one moss) were killed by the rigorous desiccation procedure used. Of some 35 species which were frozen in the turgid condition (their Table 2), they report that only three species, all thalloid hepatics, were unequivocally killed at −5°C. Several other species were damaged and failed to survive, even though they still showed substantial net assimilation rates on thawing. The majority of species survived freezing at −5°C without damage, and two species withstood freezing at −10°C without evidence of damage. However, the species of *Andreaea* appeared to be resistant to the very lowest temperatures used in the experiment and showed so little change that it was difficult for the authors to determine whether the plants were living or dead. The authors suggest that the rapid rate of cooling used in the experiment was directly related to the surprisingly high mortality of bryophytes at freezing temperatures approaching −10°C, and that they would undoubtedly be much more resistant at the rates of cooling that occur naturally under field conditions. In any event, it is obvious that bryophytes in Arctic Alaska must be able to withstand winter temperatures of −40°C or lower without appreciable damage, and are then able to renew their growth when temperatures again become more favorable.

The foregoing experimental work would suggest that careful and quantitative investigations should be made of the actual conditions under which arctic bryophytes become frozen, as it would seem that the critical temperature for polar bryophytes must be the freezing point or slightly below it, when the protoplasm freezes and metabolism comes to a halt. Although many bryophytes are snow-covered during the winter, those in exposed places, as at the top of a high-centered polygon and on the vertical sides of ravines, are subject to the full rigors of the climate, the strong winds as well as the low temperatures. Those species that cannot tolerate freezing, for reasons still unknown, cannot survive in northern climates, so that temperature can obviously become a limiting factor for tropical bryophytes. For the arctic species, however, once they are frozen the level to which the ambient temperature falls would seem to be immaterial; the protoplasm appears to be insensitive to further loss of heat, once frozen, at least in nature. Some insight is given into the cold-tolerance of bryophytes by the experiment of Morrill (1950), who placed living mosses in liquid air and found that, when they were removed and thawed, they continued growth in a normal fashion, even the most delicate structures, as the protonemata. In a much more recent investigation, Bewley and Thorpe (1974) found that dried plants of *Tortula ruralis* placed in liquid N_2 resumed respiration on rewarming and rehydration, but plants that had been frozen in the hydrated condition respired to a lesser extent and showed signs of freeze damage.

In addition to their phenomenal resistance to low temperatures, especially in dry condition, bryophytes can also tolerate surprisingly high temperatures. Mosses growing on hot dry rocks in any climate rapidly dehydrate and become so brittle that they crumble to dust when subjected to pressure. Lange (1955) ran a series of experiments to determine the heat resistance of more than 50 species of bryophytes in the dry state. Four species tolerated a temperature of 100°C without damage, two a temperature of 105°C, and one a temperature of 110°C. Each of these species showed damage at an increase of 5°C above the resistance indicated, and were killed by an increase of 10°C. It is curious that some species, that we think of as being unusually xerophytic and drought resistant, are less tolerant of high temperatures. For example, in Lange's experiments *Polytrichum piliferum* suffered damage at 95°C, and *Rhacomitrium lanuginosum, Rhytidium rugosum,* and *Orthotrichum anomalum* were damaged by 100°C. Hepatics seem to be much less resistant to heat than mosses; Lange found that two species were killed at 75°C, and three at 90°C; no hepatic tested survived higher temperatures. The primary conclusion that can be drawn from Lange's data is that bryophytes growing directly on rocks are more heat resistant than those normally from moister habitats. In a more recent investigation Von Noerr (1974b) found that the lethal limit of dried mosses (eight species) is between 85 and 110°C and of normally hydrated mosses is between 42 and 52°C.

4. Light.   The distribution of tufts or mats of bryophytes on a vertical rock face seems to reflect more the presence of water drainage channels than a response to the intensity of light. In Arctic Alaska, a few species of bryophytes are generally restricted in their habitat to deep rock crevices. However, this preference appears to indicate a requirement for the higher humidity that occurs out of the wind and direct sun and not a reaction to light intensity, per se; some examples are *Radula complanata, Neckera pennata, Leskeella nervosa, Pseudoleskeella* spp., *Amphidium* spp., and *Seligeria* spp. My evidence for this conclusion is that in more temperate climates with normally higher atmospheric humidity, these very same species may grow in the open on the trunks of trees and on the sides of rocks, not in full sunlight, but certainly not restricted to crevices for protection against drying winds. To sum up, tundra bryophytes, with very few exceptions, are wholly tolerant of insolation, and many of them develop a "sun-red" pigmentation as the result of full insolation, as *Mesoptychia sahlbergii* and *Cinclidium arcticum,* which are often associated and yet which may lack the red pigment when protected from the direct sun. In fact, most tundra bryophytes tend to be considerably more pigmented in full sun than when growing in shade.

Bryophytes that normally occur under the canopy of coniferous forests tend to be somewhat smaller and less well developed when growing in open tundra exposed to full light. A conspicuous and classic case has long been the stunted arctic form of *Hylocomium splendens,* which has been treated variously as an independent species, *H. alaskanum,* as a variety, as a form, and as a simple synonym. Other typical forest species, such as *Pleurozium scheberi, Ptilium crista-castrensis,* and *Rhytidiadelphus triquetrus,* although not as reduced as *Hylocomium splendens* under tundra conditions, are still less luxuriant than when they are growing in the more humid and less insolated environment under the forest canopy. The reduction in luxuriance seems to be a direct effect of the more intense irradiation and perhaps the reduced humidity of the tundra atmosphere. In addition to these factors, Tamm (1950) has made the following suggestion with regard to *H. splendens:* "The decrease in amount of moss growth with increasing distance from the area covered by a tree canopy may well be due to variations in the supply of one or more nutrients provided either from the litter shed by the trees or from rain washings off the crowns." Species of open bogs, marshes, fens, and pools, however, grow as luxuriantly at high latitudes as they do at lower ones, and sometimes more so, probably because of reduced competition and shading by vascular plants in the far North. As has been emphasized earlier, the only absolutely basic requirement for abundant growth of bryophytes seems to be an adequate water supply. Likewise, montane and alpine species that normally occur on rocks do grow as well under arctic conditions as in boreal or subarctic ones, insofar as individual plants are concerned, although they may be less abundant because of reduced humidity, and their tufts and mats may be correspondingly smaller.

Bryophyte Associations and Communities

I. Associations.   Without roots, bryophytes reflect very accurately the nature of their superficial environment and substratum, in terms of the physical factors that impinge upon them. As a consequence, species with similar or identical habitat preferences very naturally tend to occur together, and a few species have a fairly high degree of fidelity in their association. For example, *Entodon concinnus* nearly always grows with *Abietinella abietina* and *Rhytidium rugosum, Mesoptychia sahlbergii* with *Cinclidium arcticum,* and *Anastrophyllum minutum* with *Dicranum elongatum;* there is up to a 50% probability that a herbarium packet containing one species will also contain the other. However, most bryophyte associations are more complex in the numbers of species involved in them, as I have pointed out in a discussion of associations represented in the mosses of Cornwallis Island in the Canadian Arctic Archipelago (Steere, 1951). Since many collections of each species were available, I made a simple tally of the association present in each collection, and, on a purely arithmetic basis, I concluded that "Although there are few firmly fixed associations, obvious tendencies for species to occur together can be illustrated" by the material studied. In this same connection, a much larger data base of the same sort is easily available and should respond well to a computerized analysis. Bryhn (1906–1907), in his report on the bryophytes of the Second Expedition of the "Fram" collected over a 4-yr period by Simmons largely in central and southern Ellesmere Island, tabulated in a separate appendix all the species that occurred together in each one of hundreds of individual collections, collection by collection, which adds up to a remarkable but totally undigested information source. In this day of easy and rapid calculation of the degree of correlation between numerous separate and unrelated items by means of computer techniques, it would be extremely rewarding to determine the degree of fidelity in the association of the various species from Bryhn's data. Although Bryhn made numerous errors of identification, he did so with reasonable consistency (for example, he consistently identified specimens of *Aloina brevirostris* as *A. rigida*), so that correction for this factor is not difficult, and, in any event, it could not affect statistically the whole picture to be gained—as long as the extensive list of "new" species of *Bryum* were to be omitted.

As the physical factors of each microhabitat and microniche limit and control what species of bryophytes are able to exist there, by the physiological and ecological reaction of the bryophytes to these factors, it seems more appropriate here to identify the more prominent tundra associations in terms of their special habitat than to attempt to give percentages of the species found to be associated in each habitat. The following examples of special habitats are intended to be informative illustrations and not a comprehensive catalogue, either of all species or all microniches. It should be noted that some species have so broad a spectrum of tolerance to a variety of environmental conditions that they are able to occur in two or even more habitats.

1. Rather dry and calcareous gravel, sand, or soil, usually sloping:

Abietinella abietina	*Hylocomium splendens*
Aulacomnium acuminatum	*Hypnum revolutum*
Drepanocladus uncinatus	*Rhytidium rugosum*
Entodon concinnus	

2. Moist alkaline soil:

Aongstroemia longipes	*Hypnum procerrimum*
Barbula icmadophila	*Peltolepis grandis*
Bryoerythrophyllum recurvirostrum	*Plagiobryum demissum*
Bryobrittonia pellucida	*Preissia quadrata*
Clevea hyalina	*Timmia* spp.
Desmatodon spp.	*Tortella* spp.
Didymodon asperifolius	*Trematodon brevicollis*
Encalypta spp.	*Trichodon cylindricus*

3. Relatively "bare" mineral soil, acid to neutral:

Conostomum tetragonum	*Pohlia proligera*
Cynodontium spp.	*Polytrichastrum alpinum*
Dicranella spp.	*Polytrichum* spp.
Dicranum spp.	*Psilopilum cavifolium*
Pogonatum dentatum	*Ptilidium ciliare*
Pohlia annotina	

4. Moist calcareous silt, usually in frost boils:

Aloina brevirostris   *Nardia* spp.
Bryum wrightii   *Pterygoneurum* spp.
Desmatodon spp.   *Stegonia latifolia*
Fissidens arcticus   +var. *pilifera*
Fossombronia alaskana   *Tortula mucronifolia*

5. On soil disturbed by man, rodents or caribou:

Bryum argenteum   *Leptobryum pyriforme*
Ceratodon purpureus   *Marchantia polymorpha*
Funaria hygrometrica

6. On steep, moist slopes on calcareous soil:

Bartramia pomiformis   *Mnium orthorrhynchum*
Bryum cryophilum   *Myurella* spp.
Cinclidium arcticum   *Orthothecium* spp.
Cyrtomnium spp.   *Plagiobryum demissum*
Dicranum spp.   *Plagiopus oederiana*
Encalypta spp.   *Philocrya aspera*
Gollania turgens   *Platydictya jungermannioides*
Isopterygium pulchellum   *Ptilium crista-castrensis*
Mesoptychia sahlbergii   *Rhizomnium andrewsianum*
Mnium blyttii

7. In very wet acid tundra:

Anastrophyllum minutum   *Lophozia* spp. (many)
Calliergon stramineum   *Pseudolepicolea fryei*
Dicranum elongatum   *Sphagnum* spp. (many)

8. In sloping calcareous fens, in seepage channels:

Catoscopium nigritum   *Meesea uliginosa*
Claopodium pellucinerve   *Myurella* sp.
Cratoneuron filicinum   *Radula prolifera*
Ctenidium molluscum   *Scapania simmonsii*
Didymodon asperifolius   *Tortella arctica*
Encalypta procera   *T. fragilis*
Geheebia gigantea   *Trichostomum cuspidatissimum*

9. In pools of varying depth, frankly calcareous:

Calliergon trifarium   *Scorpidium scorpioides*
Drepanocladus lycopodioides   *S. turgescens*
D. brevifolius

10. Very wet, somewhat acid to neutral or mildly calcareous low-center polygons:

Aulacomnium palustre   *Calliergon orbicularicordatum*
A. turgidum   *C. richardsonii*

C. sarmentosum

Campylium stellatum

Chiloscyphus pallescens

Cinclidium spp.

Dicranum angustum

Drepanocladus revolvens

Lophozia rutheana

Meesia triquetra

Oncophorus wahlenbergii

Paludella squarrosa

Plagiothecium cavifolium

Pohlia nutans

Riccardia pinguis

Splachnum vasculosum

11. On neutral soil, high-center polygons, beach ridges, etc.:

Bartramia ithyphylla

Bryum spp. (many)

Cirriphyllum cirrosum

Eurhynchium pulchellum

Hylocomium splendens

Pohlia cruda

12. On north-facing steep slopes frequently moistened by fog:

Ascidota blepharophylla

Frullania jackii

F. tamarisci ssp. nisquallensis

Herberta sakuraii

Hygrolejeunea alaskana

Odontoschisma macounii

Pseudolepicolea fryei

Radula prolifera

13. On enriched substrata:

Owl perches:

Bryum argenteum

Dicranum elongatum

Orthotrichum speciosum

Tortula ruralis

Lemming runways and burrows:

Bryum wrightii

Desmatodon leucostoma

Funaria polaris

Enriched soil and weathered
    caribou or moose dung:

Aplodon wormskjoldii

Splachnum luteum

S. sphaericum

Tayloria ssp.

Tetraplodon mnioides

T. paradoxus

Voitia hyperborea

14. On rocks in streams:

Dichodontium pellucidum

Grimmia platyphylla

Grimmia rivularis

Hygrohypnum spp.

15. On dry acid rocks:

Anastrophyllum saxicola

Andreaea rupestris

Chandonanthus setiformia

Grimmia ovalis

Ulota curivolia

16. On wet acid shale:

Andreaea obovata

Oligotrichum falcatum

17. On somewhat to strongly calcareous rock:

Amphidium ssp.   *Hypnum* ssp.
Didymodon johansenii   *Orthotrichum* ssp.
Grimmia ssp.   *Pterigynandrum filiforme*

18. On mineralized metamorphic rock:

Coscinodon cribrosus   *Mielichhoferia mielichhoferi*

19. In sheltered rock crevices:

Brachythecium trachypodium   *Neckera pennata*
Campylophyllum halleri   *Pseudoleskeella* ssp.
Leskeella nervosa   *Seligeria* ssp.

20. In anomalous habitats (several normally saxicolous species occur in the Alaska arctic tundra on wet soil, on humus, or mixed with other bryophytes not on rock):

Andreaea rupestris (on acid sand)
Orthotrichum speciosum (on moist sand)
Rhacomitrium canescens (on wet sand)
R. lanuginosum (on humus ridges in tundra)
Schistidium apocarpum (on moist soil)
S. gracile (on moist soil)
Seligeria polaris (on shale chips in frost boils)

II. Communities.   In the recently published proceedings of a symposium on polar deserts (Smiley and Zumberge, 1974), one finds that the definition of "polar desert" is quite as elusive and enigmatic as "arctic" (Britton, 1967), and that, in the same way, the professional background of an individual will very naturally influence his definition. Thus, in his chapter on geomorphic processes, Troy Péwé, a geologist, includes the Arctic Slope of Alaska within his definition of polar desert in his map, which is that zone north of the mean July isotherm of 10°C, and the line representing mean annual precipitation of 25 cm, whichever is farther north. M. Bovis and R. Barry, in a climatological analysis, used the Turc evaporation formula as a guide, and indicate Arctic Alaska as being in the semidesert zone, with the area of true polar desert being very restricted and existing largely in northern Ellesmere Island and northernmost Greenland. J. C. F. Tedrow, in his chapter on soils of the High Arctic and, of course, using soils as a criterion, has indicated three major soil zones on his map, the polar desert, the subpolar desert, and the tundra, in the latter of which he places Arctic Alaska in his map. In spite of the confusion that may well arise from this juxtaposition of three approaches that reach such different conclusions, I believe that the bryological data will support Tedrow's opinion. Bryologically, the relatively well-watered tundra of northernmost Alaska comes off very favorably in comparison with the polar deserts of northern Ellesmere Island and, especially, northernmost Greenland, in numbers of species, complexity of wet tundra bryological communities, and importance in vegetation cover.

Two major bryological investigations of the communities of North American polar deserts (in the strict sense) have been made. Holmen (1955, 1960) worked at length in Peary Land in northernmost Greenland, including a full winter of observation there, which gave him valuable information on the winter snow-cover of the several communities he identified. In his evaluation of the bryophyte vegetation (Holmen, 1955), he recognizes 15 bryological communities in the High Arctic Desert District and six communities in the Fell-Field District. He characterized these communities clearly and indicated which ones may intergrade. His remark that "the delimitation of these as one community will often be difficult and they will very often form an even transition to other communities" highlights a major problem immediately encountered by anyone who attempts to identify or classify bryophyte communities in any arctic area, especially where adequate water is available, as in northernmost Alaska. Brassard (1971a, pp. 241–248, Table 1) has identified and described 20 moss communities for the area of northern Ellesmere Island that he studied in depth for several summer field seasons. His communities coincide in part but not entirely with those proposed by Holmen (1955), largely because, although also classified under any definition as polar desert, northern Ellesmere Island obviously has more water available to bryophytes than Peary Land, as can be seen from Brassard's photographs as well as from the much larger number of species of mosses that he records (Brassard, 1971b).

Because of the subtle but inexorable effect of gradients of water availability in determining the distribution or even the existence of bryophytes in the tundra, as well as the infinite variability of other environmental factors, it is really not possible to establish formal bryophyte communities that a majority of bryologists or ecologists would accept, even for a relatively small area such as Barrow or Prudhoe Bay. Communities that appear to be distinct and coherent at the center merge imperceptibly into several other neighboring communities at the edges, so that very arbitrary and artificial lines of demarcation would have to be established in order to maintain the integrity of any one community. For this reason, I have resisted the temptation to create a detailed classification of the bryophyte communities of the North Slope.

III. The Relative Importance of Bryophytes in Vegetation Cover. Schuster (in Schuster et al., 1959, p. 11) has the following to say about the vegetation cover of northernmost Ellesmere Island in the vicinity of Alert: "Most impressive was the extremely restricted development of the bryophyte vegetation. The Arctic is often considered as an area where moss and lichens abound and where the bryophyte cover exceeds by far that of the vascular plants. This concept is largely erroneous." With reference to northernmost Greenland, in Peary Land, Holmen (1955) says, "All the species of Hepatics and the majority of the Musci were of no real importance in the formation of vegetation, many were found in only a few places." However, in evaluating these remarks just quoted, one must keep in mind that both Schuster and Holmen were referring to the true polar desert, where summer precipitation is slight and where soil moisture is derived largely from the melting of snow accumulation during the 10-month winter; it is especially significant, in this light, that neither author (nor Brassard) found *Sphagnum*, a critical indicator of both standing water and low pH. In his important summing-up paper on arctic bryology, Schofield (1972) also appears, from his remarks, to be out of sympathy with the concept of continuous bryophyte cover in arctic areas, a point of view that is wholly understandable when one considers that his experience in the Arctic was gained largely on Cornwallis Island (Steere, 1951), which, again, is a part of the polar desert of the northern part of the Canadian Arctic Archipelago.

In great contrast to the polar desert, the tundra or subpolar desert zone of Arctic Alaska is provided with adequate and sometimes superfluous water, especially in the Coastal Plain Province, in which the landscape consists of an almost unbroken expanse of bog, marshy meadows, and fens, usually in the form of numerous low-centered ice-wedge polygons interrupted by innumerable lakes, to such an extent that in some areas the water surface is greater than the land surface. In such an area bryophytes—and especially *Sphagnum*—may become the dominant and continuous vegetation over large areas, usually associated with grasses, carices, and *Eriophorum*. It is safe to say that in such areas bryophytes form the major component of the total biomass. Nonetheless it is easily predictable, on the basis of the positive correlation between the supply of available water and the luxuriance of the bryophyte vegetation already elucidated, that where the habitat does not have an adequate water supply during the whole growing season, the growth of bryophytes will be less luxuriant and they will not form continuous vegetation cover. And thus it actually is. Even in the Coastal Plain of Arctic Alaska, so much wetter than any part of the true polar desert, where there is well-drained gravel, dry sand, broken rock, or other habitats that do not retain water, bryophytes exist on a catch-as-catch-can basis, or are lacking; the continuous vegetation cover of bryophytes, either by themselves or in association with vascular plants, can exist in the Arctic only in low wet tundra. As a consequence, any definition or concept of vegetation cover in the Arctic, especially with relation to bryophytes, must first take into consideration the amount of water available for their growth. Where water abounds, mosses grow luxuriantly; where water is inadequate, bryophytes become a much less important factor in the vegetation. In north-central Banks Island in the Canadian Arctic Archipelago, for example, the rough silty uplands are vegetated with bryophytes only in the crevices between frost-heaved clods and soil masses which are uniformly calcareous, whereas in swales where melt-water accumulates and stands for much of the short summer, a solid moss cover develops (Steere, unpublished field data).

Acknowledgments. This research was supported by the National Science Foundation. Field and laboratory activities at Barrow were supported by the Naval Arctic Research Laboratory of the Office of Naval Research.

An abbreviated version of this article was published by the Hattori Botanical Laboratory (Steere, 1976).

References

Abramova, A. L. (1956) Mkhi severo-vostoka Azii. (The mosses of southeastern Asia.) *Trudy Bot. Inst. Komarov* Akad. Nauk SSSR, Ser. II, **10**: 490–511.

Ando, H. (1963) *Pseudolepicolea* found in the middle Honshu of Japan. *Hikobia*, **3**: 177–183.

Ando, H. (1966) A revision of the Chinese Cupressinae described by C. Mueller. *Bot. Mag. Tokyo*, **79**: 759–769.

Ando, H., H. Persson and E. M. Sherrard. (1957) The first record of *Gollania* in North America. *Bryologist*, **60**: 326–335.

Bewley, J. D., and T. A. Thorpe. (1974) On the metabolism of *Tortula ruralis* following desiccation and freezing: Respiration and carbohydrate oxidation. *Physiol. Plant*, **32**: 147–153.

Bird, C. D. (1968) New or otherwise interesting mosses from Alberta. *Bryologist*, **71**: 358–361.

Brassard, G. R. (1971a) The mosses of northern Ellesmere Island, Arctic Canada. I. Ecology and phytogeography, with an analysis for the Queen Elizabeth Islands. *Bryologist*, **74**: 233–281.

Brassard, G. R. (1971b) The mosses of northern Ellesmere Island, Arctic Canada. II. Annotated list of the taxa. *Bryologist*, **74**: 282–311.

Britton, M. E. (1967) Vegetation of the arctic tundra. In *Arctic Biology: 18th Biology Colloquium* (H. P. Hansen, Ed.). Corvallis, Oregon: Oregon State University Press, 2nd ed., pp. 67–130.

Bryhn, N. (1906–1907) Bryophyta in itinere polari norvagorum secundo collecta. *Report 2nd Norwegian Arctic Expedition "Fram"*, Vol. 11, pp. 1–260.

Coulter, H. W., D. M. Hopkins, T. N. V. Karlstrom, T. L. Péwé, C. Wahrhaftig, and J. R. Williams. (1965) *Map Showing Extent of Glaciations in Alaska*. Misc. Geol. Investigations Map I-415, U.S. Geological Survey, Washington, D.C.

Crum, H. A. (1965) *Barbula johansenii*, an arctic disjunct in the Canadian Rocky Mountains. *Bryologist*, **68**: 344–345.

Crum, H. A. (1969) A reconsideration of the relationship of *Barbula johansenii* (Musci.). *Can. Field Natur.*, **83**: 156–157.

Crum, H. A. (1972) The geographic origins of the mosses of North America's eastern deciduous forest. *J. Hattori Bot. Lab.*, **35**: 269–298.

Crum, H. A. (1976) Mosses of the Great Lakes Forest. *Contrib. Univ. Mich. Herb.*, Vol. 10, rev. ed., pp. 1–404.

Dilks, T. J. K., and M. C. F. Proctor. (1975) Comparative experiments on temperature responses of bryophytes: Assimilation, respiration and freezing damage. *J. Bryol.*, **8**: 317–336.

Gams, H. (1934) Beitrage zur Kenntnis der Steppenmoose. *Ann. Bryol.*, **7**: 37–56.

Gams, H. (1955) Zur Arealgeschichte der arktischen und arktisch-oreophytischen Moose. *Feddes Repert.*, **58**: 80–92.

Grout, A. J. (1928–1940) *Moss Flora of North America North of Mexico*. 3 vols., Newfane, Vermont. (Reissued, partly in facsimile, by Hafner Publishing Co., New York, 1972.)

Gwóźdź, E. A., and J. D. Bewley. (1975) Plant desiccation and protein synthesis: An *in vitro* system from dry and hydrated mosses using endogenous and synthetic messenger ribonucleic acid. *Plant Physiol.*, **55**: 340–345.

Hattori, S. (1972) *Frullania tamarisci*-complex and the species concept. *J. Hattori Bot. Lab.*, **35**: 202–251.

Holmen, K. (1955) Notes on the bryophyte vegetation of Peary Land, North Greenland. *Mitteil. Thuringisch. Bot. Ges. (Theodore-Herzog-Festschrift)*, **1**: 96–106.

Holmen, K. (1957) Three west arctic moss species in Greenland. On the occurrence of

Cinclidium latifolium, Aulacomnium acuminatum and *Trichostomun cuspidatissimum. Medd. Grøn.,* **156**: 1–16.

Holmen, K. (1960) The mosses of Peary Land, North Greenland. *Medd. Grøn.,* **163**: 1–98.

Irmscher, E. (1912) Über die Resistenz der Laubmoose gegen Austrocknung und Kalte. *Jahrb. Wiss. Bot.,* **50**: 387–449.

Iwatsuki, Z., and W. B. Schofield. (1973) The taxonomic position of *Campylium adscendens. J. Hattori Bot. Lab.,* **37**: 609–615.

Lange, O. L. (1955) Untersuchungen über die Hitzeresistenz der Moose in Beziehung zu ihrer Verbreitung. I. Der Resistenz stark ausgetrockneter Moose. *Flora,* **142**: 381–399.

Lindberg, S. O., and H. W. Arnell. (1890) Musci Asiae Boreales. Beschreibung der von den schwedischen Expeditionen nach Sibirien in den Jahren 1875 and 1876 gesammelten Moose mit Berücksichtigung aller fruheren bryologischen Angaben für das russische Nord-Asien. II. Laubmoose. *Kgl. Svenska Vetensk.-Akag. Handl.,* **23**: 1–163.

Loeske, L. (1930) Monographie der europaischen Grimmiaceen. *Biblioth. Bot.,* **101**(i–x): 1–236.

Malta, N. (1921) Versuch über die Wiederstandsfahigkeit der Moose gegen Austrocknung. *Acta Univ. Latviensis,* **1**: 125–129.

Morrill, J. B. (1950) Mosses in liquid air. *Bryologist,* **53**: 163–164.

Murray, B. (1974) *Catalog of Bryophytes and Lichens of the Central Brooks Range, Alaska: A Literature Review.* University of Alaska Museum, 46 pp.

Murray, B., and D. F. Murray. (1974) *Provisional Checklist to the Vascular Bryophytes and Lichen Flora of Prudhoe Bay.* University of Alaska Museum, 21 pp.

Murray, B. M., and D. F. Murray. (1978) Checklists of vascular plants, bryophytes, and lichens for the Alaska U.S. IBP Tundra Biome study areas—Barrow, Prudhoe Bay, Eagle Summit. In *Vegetation and Production Ecology of an Alaskan Arctic Tundra* (L. L. Tieszen, Ed.). New York: Springer-Verlag, Appendix I.

Noerr, M. von (1974a) Trockenresistenz bei Moosen. *Flora,* **163**: 371–387.

Noerr, M. von (1974b) Hitzeresistenz bei Moosen. *Flora,* **163**: 388–397.

Noguchi, A. (1964) A revision of the genus *Claopodium. J. Hattori Bot. Lab.,* **27**: 20–46.

Nyholm, E. (1954–1969) *Illustrated Moss Flora of Fennoscandia* II. *Musci.* Nat. Sci. Res. Council, Stockholm, and CWK Gleerup, Lund. 799 pp.

Ochi, H. (1952) The preliminary report on the osmotic value, permeability, drought and cold resistance of mosses. *Bot. Mag. Tokyo,* **65**: 10–12.

Packer, J. G., and D. H. Vitt. (1974) Mountain Park: A plant refugium in the Canadian Rocky Mountains. *Can. J. Bot.,* **52**: 1393–1409.

Persson, H. (1946) The genus *Habrodon* discovered in North America. *Svensk. Bot. Tidskr.,* **40**: 317–324.

Persson, H. (1947) Further notes on Alaskan–Yukon bryophytes. *Bryologist,* **50**: 279–310.

Persson, H. (1948) On the discovery of *Merceya ligulata* in the Azores with a discussion of the so-called "copper mosses." *Rev. Bryol. Lichénol. N.S.,* **17**: 75–78.

Persson, H. (1949) Studies in the bryophyte flora of Alaska—Yukon. *Svensk. Bot. Tidskr.,* **43**: 491–533.

Persson, H. (1952) Critical or otherwise interesting bryophytes from Alaska–Yukon. *Bryologist,* **55**: 1–25, 88–116.

Persson, H. (1962) Bryophytes from Alaska collected by E. Hultén and others. *Svensk. Bot. Tidskr.,* **56**: 1–35.

Persson, H., and O. Gjaerevoll. (1961) New records of Alaskan bryophytes. *Kgl. Norske Vidensk. Selsk. Skrift.,* **1961**: 1–26.

Rastorfer, J. R. (1972) Bryophyte taxa lists of the high Alaskan arctic. Institute of Polar Studies, Ohio State University, Columbus. 54 pp.

Rastorfer, J. R., H. J. Webster, and D. K. Smith. (1973) Floristic and ecologic studies of bryophytes of selected habitats at Prudhoe Bay, Alaska. *Inst. Polar Stud., Ohio State Univ., Rep.* 49, 20 pp.

Redfearn, P. L. (1972) Mosses of the interior highlands of North America. *Ann. Missouri Bot. Gard.,* **59**: 1–103.

Schatz, A. (1955) Speculations on the ecology and photosynthesis of the "copper mosses." *Bryologist,* **58**: 113–120.

Schofield, W. B. (1972) Bryology in arctic and boreal North America and Greenland. *Can. J. Bot.,* **50**: 1111–1133.

Schofield, W. B., and H. A. Crum. (1972) Disjunction in bryophytes. *Ann. Missouri Bot. Gard.,* **59**: 174–202.

Schuster, R. M. (1957) Notes on Nearctic Hepaticae, XV. *Herberta. Rev. Bryol. Lichénol.,* **26**: 123–145.

Schuster, R. M. (1966–1974) *The Hepaticae and Anthocerotae of North America, East of the Hundredth Meridian.* New York: Columbia University Press, 3 vols. and continuing.

Schuster, R. M., and W. C. Steere. (1958) *Hygrolejeunea alaskana* sp. n., a critical endemic of northern Alaska. *Bull. Torrey Bot. Club,* **85**: 188–196.

Schuster, R. M., W. C. Steere, and J. W. Thomson. (1959) The terrestrial cryptogams of northern Ellesmere Island. *Nat. Mus. Can. Bull.,* **164**: 1–132.

Sherrard, E. M. (1955) Bryophytes of Alaska. II. Some mosses from the southern slopes of the Brooks Range. *Bryologist,* **58**: 225–236.

Smiley, T. L., and J. H. Zumberge. (Eds.) (1974) *Polar Deserts and Modern Man.* Tucson: University of Arizona Press, 173 pp.

Smith, D. K. (1974) Floristic, ecologic and phytogeographic studies of the bryophytes in the tundra around Barrow, Alaska. Ph.D. dissertation, University of Tennessee, Knoxville, 191 pp.

Steere, W. C. (1951) Bryophyta of Arctic America. IV. The mosses of Cornwallis Island. *Bryologist,* **54**: 181–202.

Steere, W. C. (1953) On the geographical distribution of arctic bryophytes. *Stanford Univ. Publ. Biol. Sci.,* **11**: 30–47.

Steere, W. C. (1954) Chromosome number and behavior in arctic mosses. *Bot. Gaz.,* **116**: 93–133.

Steere, W. C. (1959) *Pterygoneurum arcticum,* a new species from northern Alaska. *Bryologist,* **62**: 215–221.

Steere, W. C. (1965) The boreal bryophyte flora as affected by Quaternary glaciation. In *The Quaternary of the United States* (H. E. Wright, Jr., and D. G. Frey, Eds.). Princeton: Princeton University Press, pp. 485–495.

Steere, W. C. (1969) Asiatic elements in the bryophyte flora of western North America. *Bryologist,* **72**: 507–512.

Steere, W. C. (1976) Ecology, phytogeography and floristics of arctic Alaskan bryophytes. *J. Hattori Bot. Lab.,* **41**: 47–72.

Steere, W. C. (1977) Bryophytes from Great Bear Lake and Coppermine, Northwest Territories, Canada. *J. Hattori Bot. Lab.,* **42**: 425–465.

Steere, W. C. (1978) *The Mosses of Arctic Alaska.* 508 pages; 23 figs; 48 maps. Lehre, Germany: J. Cramer.

Steere, W. C., and G. R. Brassard. (1974) The systematic position and geographical distribution of *Fissidens arcticus. Bryologist,* **77**: 195–202.

Steere, W. C., and H. Inoue. (1974) *Fossombronia alaskana,* a new hepatic from arctic Alaska. *Bryologist,* **77**: 63–71.

Steere, W. C., and H. Inoue (1975) Contributions to our knowledge of *Mesoptychia sahlbergii. Bull. Nat. Sci. Mus. (Tokyo),* Ser. B (Botany), 1(2): 59–72.

Steere, W. C. and H. Inoue. (1978) The Hepaticae of Arctic Alaska. *J. Hattori Bot. Lab.*

Steere, W. C., and Z. Iwatsuki. (1974) The discovery of *Pterygoneurum subsessile* (Brid.) Jur. in arctic Alaska. *J. Hattori Bot. Lab.,* **38**: 463–473.

Steere, W. C., and B. M. Murray. (1974) The geographical distribution of *Bryum wrightii* in arctic and boreal North America. *Bryologist,* **77**: 172–178.

Steere, W. C., and W. B. Schofield. (1956) *Myuroclada,* a genus new to North America. *Bryologist,* **59**: 1–5.

Steere, W. C., and R. M. Schuster. (1960) The hepatic genus *Ascidiota* Massalongo new to North America. *Bull Torrey Bot. Club,* **87**: 209–215.

Stevenson, D. W., J. R. Rastorfer, and R. E. Showman. (1972) Effects of temperature on seta elongation in *Atrichum undulatum. Ohio J. Sci.,* **72**: 146–152.

Tallis, J. H. (1958) Studies in the biology and ecology of *Rhacomitrium lanuginosum* Brid. I. Distribution and ecology. *J. Ecol.,* **46**: 271–288.

Tamm, C. O. (1950) Growth and plant nutrient concentration in *Hylocomium proliferum* (L.) Lindb. in relation to tree canopy. *Oikos,* **2**: 60–64.

Tucker, E. B., J. W. Costerton, and J. D. Bewley (1975) The ultrastructure of the moss *Tortula ruralis* on recovery from desiccation. *Can. J. Bot.,* **53**: 94–101.

Vitt, D. H. (1974) The distribution of *Bryobrittonia pellucida* Williams (Musci). *Arctic,* **27**: 237–241.

Watson, W. (1913) Xerophytic adaptations of bryophytes in relation to habitat. *New Phytol.,* **13**: 149–169, 181–190.

Weber, W. A. (1960) A second American record for *Oreas martiana,* from Colorado. *Bryologist,* **63**: 241–244.

Weber, W. A. (1973) Guide to the mosses of Colorado. Keys and ecological notes based on field and herbarium studies. *Inst. Arct. Alp. Res., Univ. Colorado Occas. Pap.* 6, 48 pp.

Wiggins, I. L., and J. H. Thomas. (1962) *A Flora of the Alaskan Arctic Slope.* Arctic Institute of North America Spec. Publ. 4. Toronto: University of Toronto Press, 425 pp.

Worley, I. A. (1970) A checklist of the Hepaticae of Alaska. *Bryologist,* **73**: 32–38.

Worley, I. A., and Z. Iwatsuki. (1970) A checklist of the mosses of Alaska. *Bryologist,* **73**: 59–71.

Worley, I. A., and Z. Iwatsuki. (1971) Additions to the checklist of Alaskan mosses. *Bryologist,* **74**: 376.

6. Composition and Bryomass of the Moss Layers of Two Wet-Tundra-Meadow Communities Near Barrow, Alaska

JAMES R. RASTORFER

Introduction

Bryophytes are major components of arctic tundra vegetation (Steere, 1978) and in some plant communities bryophytes often exceed vascular plants in biomass (Kil'dyushevskii, 1964; Britton, 1967; Khodachek, 1969; Pavlova, 1969; Clarke et al., 1971). Therefore, the status and role of bryophytes with respect to colonization, species composition, primary production, and mineral nutrient cycling must be elucidated before the structure and function of an arctic tundra ecosystem can be understood.

The principal objectives of the present investigation were to determine the general species composition and bryomass of the moss and liverwort layers of selected arctic tundra habitats near Barrow, Alaska. The landscape in the vicinity of Barrow is characterized by ice-wedge polygons which vary considerably with respect to microrelief, edaphic, biotic, and community factors (Drew and Tedrow, 1962; Britton, 1967; Smith, 1974). Although several types of ice-wedge polygons occur in the Barrow region, the present study focused on two areas distinguished by low-centered polygons which support wet-tundra-meadow communities. Both study areas represented portions of two of the U.S. Tundra Biome Program intensive study sites which were designated IBP sites 2 and 4.

Materials and Methods

Description of Study Areas

The stand of vegetation sampled in site 2 was located in the eastern sector of the site (see site map, Introduction; Webber, 1978). This section of the site was considered a wet-tundra-meadow characterized by poorly defined low-centered polygons. Physiognomically, the vegetation appeared homogeneous consisting of herbaceous (mostly graminoid) vascular plants. However, bryophytes were abundant but were generally inconspicuous because of the dense vascular plant overstory. Exceptions to the inconspicuous occurrence of mosses and liverworts were some relatively pure swards of *Campylium stellatum* which filled slight depressions, and species of *Calliergon, Drepanocladus,* and *Campylium* which formed mats in small shallow pools. Although considered discrete habitats, the slight depressions and shallow pools represented such a small percentage of the total site area that they were eliminated from the present study.

The stands of vegetation sampled in site 4 were located in the western sector of the site (see site map, Introduction; Webber, 1978). This section was also a wet-tundra-meadow community but was characterized by well-defined low-centered polygons. These polygons had broad flat centers with relatively narrow, but distinct, interpolygonal troughs. Generally, the vegetation consisted of a compact graminoid overstory with a dense bryophyte understory. However, in interpolygonal troughs mosses formed the dominant layer with only a few scattered vascular plants. Therefore, in this site the polygonal centers and the interpolygonal troughs were sampled separately.

Sampling

Sampling was done in the field with a stainless-steel corer—46 cm long with a cross-sectional area of 9.8 cm^2. Sample cores were taken by briskly pushing the corer down through the vegetation into mineral soil or down to frozen soil. Each sample core was removed from the corer by using a plunger to push the sample core from the bottom out through the top of the corer. This precaution prevented compression and damage to the moss turf. Cores usually consisted of a layer of vegetation, a layer of fibrous peat, a thin layer of highly organic material, and then mineral soil. Only the upper portions, including 2 to 3 cm of mineral soil, were retained for laboratory use. Subsequently, the sample cores were taken into the laboratory and oven-dried for 24 hr at 70°C.

In site 2, three study plots (Moss Plots 201, 202, and 203) were established in a stand that appeared physiognomically uniform and free from human disturbance. The plots were 1 × 20 m arranged parallel to each other approximately 10 m apart. Sample cores were taken within a 20 × 50-cm frame at 1-m intervals along the northern edge of each plot. The inner area of the frame was divided into 10 equal quadrats. These numbered quadrats were used to randomly locate different sample positions within the frame. Twenty sample cores were collected from each plot on each of the following dates during the summer of 1972: 22 June, 24 July, and 24 August.

In site 4, sample cores were collected from polygonal centers and interpolygonal troughs at 1-m intervals along temporary line transects. Only the central portions of the polygonal centers were sampled to avoid transition zones at the polygonal margins. Likewise, only the central axes of the interpolygonal troughs were sampled. Eleven sample cores were collected from each transect on 28 July

1973. Four polygonal centers (Transect Numbers 405, 406, 409, and 410) and two interpolygonal troughs (Transect Numbers 407 and 408) were sampled.

Analyses of Sample Cores

After each sample core was soaked in water for several minutes, the mosses and liverworts were removed and then placed in open petri dishes to ascertain the species composition. All plant fragments that could be identified as a moss or liverwort were retained for bryomass determinations (gram dry weight per square meter). Prior to dry weight measurements each sample was cleaned of extraneous materials, oven-dried for 24 hr at 70°C, and then cooled over $CaCl_2$ in a desiccator.

In reference to site 2, the mosses and liverworts were segregated into different morphological groups prior to weighing for each sample core collected on 22 June (1972). On the other hand, all of the bryophytes were weighed collectively for each sample core collected 24 July (1972) and 24 August (1972). Not all cores could be processed during the season in which they were collected. Those collected in Moss Plots 201 and 202 on 22 June (1972) were analyzed during that summer. The remaining sample cores were analyzed during the summer of 1973.

In reference to site 4, each bryophyte taxon of each sample core was assigned a relative abundance value based on a scale of one to five (from very rare to very abundant; Oosting, 1956). All of the sample cores were analyzed during the summer (1973) in which they were collected. Lichens that occurred in these samples were analyzed also, but the amounts were negligible.

Taxonomy

The Latin names of the taxa and the order in which they are presented in the tables of the results follow the system used by Rastorfer (1972). The authors of the species are given in the checklist of plants prepared for this volume by Murray and Murray (Appendix I). Exceptions to this concern several recent nomenclatural changes which Murray and Murray have included in their check-list. These changes include: *Orthocaulis quadrilobus* = *Lophozia quadrilobus; Riccardia pinquis* = *Aneura pinquis; Mnium andrewsianum* = *Rhizomnium andrewsianum; Mnium rugicum* = *Plagiomnium ellipticum; Polytrichum com-mune* var. *jensenii* = *P. commune* var. *maximoviczii;* and *Hylocomium splen-dens* var. *alaskanum* = *H. splendens* var. *obtusifolium.*

Several taxonomic references were used during the investigation, especially those of Grout (1928–40), Frye and Clark (1937–47), Nyholm (1954–69), Arnell (1956), Abramova et al. (1961), Lawton (1971), and others mentioned in Steere (1971) and Rastorfer (1972). The present report does not deal with the bryoflora of the Barrow (Alaska) area from the phytogeographic point of view. This aspect of the bryoflora has been treated in the recent works of Brassard (1971), Steere (1971, 1978), and Smith (1974).

Results

Site Two

Bryophytes (Table 1) and graminoid vascular plants were the principal components of the stand of vegetation investigated. The bryophyte synusia typically formed an understory to the vascular plants and consisted of a heterogeneous assemblage of approximately 30 mosses and liverworts. The morphological groups (growth forms) of bryophytes included leafy liverworts and acrocarpous

Table 1. List of Taxa and Frequencies (%) of Bryophytes of a Wet-Tundra-Meadow Community in IBP Site 2[a]

Taxon	Sampling date (1972) and plot number								
	22 June			24 July			24 August		
	201	202	203	201	202	203	201	202	203
Leafy liverworts									
Blepharostoma trichophyllum	45	30	—	30	15	—	30	15	—
Cephaloziella arctica	50	30	—	5	10	15	20	10	5
Chiloscyphus pallescens	40	35	5	10	5	25	—	5	10
Lophozia cf. ventricosa	20	15	10	50	15	15	25	20	20
Orthocaulis quadrilobus	15	—	—	—	—	—	5	—	—
Ptilidium ciliare	10	5	—	—	—	—	5	—	—
Scapania sp.	—	—	10	—	5	20	—	—	—
Tritomaria quinquedentata	25	—	5	20	5	5	15	—	5
Thalloid liverworts	—	—	—	—	—	—	—	—	—
Sphagnum mosses	—	—	—	—	5	—	—	5	—
Acrocarpous mosses									
Bryaceae–Mniaceae[b]	100	85	100	100	100	100	100	100	100
Ceratodon heterophyllus	—	5	—	—	—	—	—	—	—
Dicranum elongatum	15	—	—	5	—	—	15	10	—
Bryoerythrophyllum recurvirostrum	15	—	—	—	—	—	—	—	—
Distichium capillaceum	55	10	—	40	5	15	40	10	10
Oncophorus wahlenbergii	40	15	30	40	10	—	20	15	15
Pogonatum alpinum[c]	5	20	15	15	20	—	—	—	5
Tortella fragilis	—	5	—	—	—	—	—	—	—
Polytrichum commune[d]	—	—	20	—	5	15	5	—	30
Pleurocarpous mosses									
Brachythecium cf. turgidum	40	10	5	55	10	5	35	15	5
Calliergon giganteum	5	—	5	5	—	5	15	10	—
Calliergon sarmentosum	100	95	85	80	85	75	100	95	95
Campylium stellatum	95	95	90	95	100	95	95	100	95
Drepanocladus cf. aduncus	15	—	—	—	5	5	15	10	15
Drepanocladus uncinatus	45	70	65	65	30	55	45	45	75
Tomenthypnum nitens	15	10	5	30	—	10	20	5	—
Hylocomium splendens[e]	30	25	—	25	25	5	20	15	5
Myurella julacea	20	5	—	20	—	—	15	—	5
Orthothecium chryseum	40	5	—	5	—	—	5	—	—
Plagiothecium denticulatum	15	50	—	—	25	15	35	35	10

[a] Each percentage value was based on 20 samples.
[b] The following taxa were identified in this group: Bryum spp. Cinclidium sp., Cyrtomnium sp., Mnium rugicum, Pohlia sp., and Mnium andrewsianum.
[c] P. alpinum var. septentrionale.
[d] P. commune var. jensenii.
[e] H. splendens var. alaskanum.

and pleurocarpous mosses. No thalloid liverworts were found in any of the samples collected during the summer of 1972. Specimens of *Sphagnum* were found in only two samples of one of the plots and the amounts were very small.

Based on frequency percentages (Table 1), the most common bryophytes were three pleurocarpous mosses, namely, *Campylium stellatum*, *Calliergon sarmentosum*, and *Drepanocladus uncinatus*. Among the acrocarpous mosses, *Distichium capillaceum*, *Oncophorus wahlenbergii*, and the Bryaceae–Mniaceae group were the most common. A few leafy liverworts were widely distributed throughout the stand, particularly *Lophozia* cf. *ventricosa*, *Cephaloziella arctica*, and *Chiloscyphus pallescens*.

Even though the plots appeared to be in a homogeneous stand, there was considerable diversity among the plots with respect to species composition. On an average for the three collecting dates, there were 21, 18, and 15 species (excluding the Bryaceae and Mniaceae) for plots 201, 202, and 203, respectively. Particularly noteworthy were the higher frequencies of *Distichium capillaceum*, *Brachythecium* cf. *turgidum*, and *Myurella julacea* in plot 201 than in the other two plots (Table 1).

Bryomass varied considerably in the different morphological groups (Table 2). Approximately 77% of the total bryomass was composed of pleurocarpous mosses, whereas only 21 and 3% of the remaining bryomass consisted of acrocarpous mosses and leafy liverworts, respectively. Surprisingly, 69% of the bryomass consisted of only two pleurocarpous-moss species, *Campylium stellatum* (54%) and *Calliergon sarmentosum* (15%). Most of the bryomass of the acrocarpous mosses consisted of the Bryaceae–Mniaceae group.

In addition to the above bryomass data determined from sample cores collected on 22 June (1972), bryomass was ascertained for the bryophytes collectively from sample cores collected on 24 July and 24 August (1972) (Table 3). The average bryomass for the three collection dates of each plot differed, and amounted to 107, 128, and 157 g m^{-2} for plots 201, 202, and 203, respectively. Seasonal changes in bryomass were consistent in the three plots and net increases in bryomass were obtained between the initial measurement for 22 June (1972) and the final measurement for 24 August (1972). These net increases were

Table 2.   Percentage Bryomass for the Major Bryophyte Morphological Groups of a Wet-Tundra-Meadow Community in IBP Site 2

Morphological component	Plot number and bryomass (%)[a]			
	201	202	203	Average[b]
Leafy liverworts	5	2	1	2.7
Thalloid liverworts	—	—	—	—
Sphagnum mosses	—	—	—	—
Acrocarpous mosses	23	21	18	20.7
Pleurocarpous mosses	72	77	81	76.6

[a] Bryomass = grams dry weight of the bryophyte standing crop per square meter. The percentage values were determined from samples collected on 22 June 1972.
[b] Mean of the three percentage values for each morphological group.

Table 3.   Total Bryomass Values for the Bryophyte Synusia of a Wet-
Tundra-Meadow Community in IBP Site 2

Date	Plot number and bryomass[a]			
	201	202	203	Average[b]
22 June 1972	98	121	126	115
24 July 1972	112	121	169	134
24 August 1972	111	141	175	142

[a]   Bryomass = grams dry weight of bryophyte standing crop per square meter.
Each bryomass value is a mean of 20 sample measurements.

13, 20, and 49 g m^{-2} or 13, 17, and 39% for plots 201, 202, and 203, respectively.
The mean net change from 22 June (1972) to 24 August (1972) was 27 g m^{-2} or a
net increase of 23%.

Site Four

Generally, the bryophytes (Table 4) formed a dense understory to the graminoid
vascular plants of this wet-tundra-meadow community. The principal components were leafy liverworts and acrocarpous and pleurocarpous mosses in both
the polygonal centers and the interpolygonal troughs. Although a thalloid liverwort and a sphagnum moss were found, they were rare and occurred in very
small amounts. The most common species, based on frequency percentages and
relative abundance values, were *Calliergon sarmentosum, Drepanocladus revolvens, Chiloscyphus pallescens, Calliergon giganteum,* and *Campylium stellatum*
(Table 4). In addition, species of *Bryum* and *Mnium* were also considered
common.

Similarities and differences between polygonal centers and troughs with
respect to species composition were quite apparent. The number of taxa varied
from 23 to 29 in the polygonal centers, whereas only about 17 taxa occurred in
the interpolygonal troughs (Table 5). In addition, different distribution patterns
were recognized at the generic level. Genera that occurred in both polygonal
centers and interpolygonal troughs were *Chiloscyphus, Bryum, Cinclidium,
Pohlia, Mnium, Calliergon, Campylium,* and *Drepanocladus.* In contrast, other
genera appeared to be more restricted and were found only in the polygonal
centers. These genera included *Blepharostoma, Lophozia, Tritomaria, Oncophorus, Distichium, Pogonatum, Polytrichum, Cirriphyllum,* and *Plagiothecium. Oncophorus wahlenbergii* and *Plagiothecium denticulatum* were particularly abundant. Apparently, there were no genera that occurred exclusively in
interpolygonal troughs (Table 4).

Important distribution patterns were also noted at the species level. *Calliergon sarmentosum* was uniformly distributed, whereas *Calliergon richardsonii*
and especially *Calliergon giganteum* were more abundant in interpolygonal
troughs than in polygonal centers. *Drepanocladus revolvens* was apparently the
most abundant *Drepanocladus* species; however, in one polygonal center (Tran-

Table 4. List of Taxa, Frequencies, and Relative Abundances of Bryophytes for Polygonal Centers and Interpolygonal Troughs of a Wet-Tundra-Meadow Community in IBP Site 4

| | Frequency (%)[a] | | | | | | Relative abundance[b] | | | | | |
| | Center | | | | Trough | | Center | | | | Trough | |
Taxon	405	406	409	410	407	408	405	406	409	410	407	408
Leafy liverworts												
Anastrophyllum minutum	9	—	—	—	—	—	0.1	—	—	—	—	—
Blepharostoma trichophyllum	—	—	55	54	—	—	—	—	1.0	0.8	—	—
Chiloscyphus pallescens	82	64	64	73	82	54	1.4	1.0	0.9	0.8	1.5	0.7
Diplophyllum taxifolium	—	—	9	—	—	—	—	—	0.1	—	—	—
Gymnocolea inflata	91	36	91	82	9	18	1.5	0.7	2.6	1.9	0.1	0.2
Lophozia spp.	—	54	63	45	—	—	—	1.0	1.0	0.4	—	—
Orthocaulis spp.	9	18	18	18	—	—	0.4	0.3	0.5	0.2	—	—
Scapania irrigua	64	—	36	18	18	18	1.0	—	0.4	0.2	0.2	0.2
Tritomaria spp.	9	9	45	9	—	—	0.2	0.1	0.9	0.1	—	—
Thalloid liverworts												
Riccardia cf. pinguis	9	—	—	9	—	—	0.2	—	—	0.1	—	—
Sphagnum mosses												
Sphagnum sp.	—	—	27	—	—	—	—	—	0.3	—	—	—
Acrocarpous mosses												
Bryum spp.	64	91	91	100	55	73	1.3	2.2	2.2	2.7	1.2	1.3
Cinclidium subrotundum	27	—	54	18	45	—	0.8	—	0.8	0.4	1.3	—
Cinclidium sp.	45	36	—	9	45	18	1.3	0.7	—	0.2	0.8	0.3
Dicranum sp.	—	—	99	—	—	—	—	—	0.1	—	—	—
Bryoerythrophyllum recurvirostrum	—	45	—	9	—	—	—	0.6	—	0.1	—	—

Table 4. List of Taxa, Frequencies, and Relative Abundances of Bryophytes for Polygonal Centers and Interpolygonal Troughs of a Wet-Tundra-Meadow Community in IBP Site 4 (*Continued*)

Taxon	Frequency[a]						Relative abundance[b]					
Distichium capillaceum	9	9	—	82	—	—	0.2	0.2	—	2.5	—	—
Mnium rugicum	45	27	9	9	27	27	1.1	0.8	0.2	0.4	0.5	0.5
Mnium sp.	36	64	45	82	45	36	0.7	1.2	0.9	1.7	1.0	0.7
Oncorphorus wahlenbergii	54	27	82	82	—	—	1.2	0.3	2.0	1.9	—	—
Pogonatum alpinum[c]	—	45	63	72	9	9	—	1.2	1.1	1.0	0.1	0.1
Pohlia spp.	36	18	36	27	45	—	0.7	0.2	0.5	0.6	1.0	—
Polytrichum commune[d]	54	—	9	9	—	—	0.8	—	0.1	0.1	—	—
Polytrichum juniperinum	9	—	—	—	—	—	0.1	—	—	—	—	—
Pleurocarpous mosses												
Brachythecium cf. *turgidum*	54	45	—	—	64	27	1.5	0.6	—	—	1.2	0.3
Calliergon giganteum	9	18	18	9	45	100	0.1	0.3	0.2	0.1	1.0	3.0
Calliergon richardsonii	9	—	—	9	18	18	0.1	—	—	0.3	0.4	0.4
Calliergon sarmentosum	91	82	91	100	91	100	2.5	1.9	2.2	2.0	3.8	4.0
Campylium stellatum	—	18	63	72	18	64	—	0.4	0.8	1.1	0.3	1.1
Cirriphyllum cirrosum	18	18	45	—	—	—	0.2	0.4	0.4	—	—	—
Drepanocladus revolvens	73	9	82	27	91	100	2.1	0.2	2.1	0.5	2.4	2.5
Drepanocladus uncinatus	27	91	27	9	9	27	0.4	2.6	0.7	0.1	0.2	0.5
Drepanocladus sp.	9	—	9	18	9	9	0.1	—	0.1	0.2	0.2	0.1
Tomenthypnum nitens	—	9	18	27	—	18	—	0.2	0.2	0.3	—	0.2
Hylocomium splendens[e]	—	—	18	9	—	—	—	—	0.2	0.1	—	—
Plagiothecium denticulatum	27	54	64	64	—	—	0.3	0.6	1.0	1.1	—	—

[a] Percentages based on 11 samples.
[b] Mean of 11 samples.
[c] *P. alpinum* var. *septentrionale*.
[d] *P. commune* var. *jensenii*.
[e] *H. splendens* var. *alaskanum*.

Table 5. Bryomass Values and Numbers of Bryophyte Taxa for Polygonal Centers and Interpolygonal Troughs of a Wet-Tundra-Meadow Community in IBP Site 4

Parameter	Center				Trough	
	405	406	409	410	407	408
Bryomass (g m^{-2})[a]	165	156	153	122	170	410
Number of taxa	26	23	28	29	18	17

[a] Mean of 11 samples.

sect Number 406) *Drepanocladus uncinatus* was much more frequent and abundant than *Drepanocladus revolvens* (Table 4). In addition to these observations, *Blepharostoma trichophyllum* was found in only two of the polygonal centers (Transect Numbers 409 and 410). But surprisingly, *Brachythecium* cf. *turgidum* was found in all of the polygonal features sampled except for the polygonal centers in which *Blepharostoma trichophyllum* was found. Similarly, *Pogonatum alpinum* was apparently absent from only one of the polygonal centers (Transect Number 405), whereas *Polytrichum commune* occurred quite frequently and abundantly in the polygonal center in which *Pogonatum alpinum* was seemingly absent. Although both species had occurred together in two of the polygonal centers (Transect Numbers 409 and 410), *Pogonatum alpinum* was the most frequent and abundant (Table 4).

Total bryomass values for each polygonal feature sampled are shown in Table 5. The bryomass values were actually quite close for the four polygonal centers and ranged from 122 to 165 g m^{-2}. In contrast, the bryomass values tended to be higher for the two interpolygonal troughs and ranged from 170 to 410 g m^{-2}.

Discussion

Taxonomic Composition

The two areas investigated were considered wet-tundra-meadows; however, they differed with respect to the dominant bryophyte species. The dominant species of site 2 was *Campylium stellatum* with *Calliergon sarmentosum* as an associate species. In contrast, the dominant species of site 4 was *Calliergon sarmentosum* with *Campylium stellatum* as an associate species (in both polygonal centers and interpolygonal troughs). Other common taxa found in the two sites included *Brachythecium* cf. *turgidum*, *Chiloscyphus pallescens*, *Drepanocladus uncinatus*, and species of *Bryum* and *Mnium* (Tables 1 and 4).

A number of species were fairly common in both site 2 and the polygonal centers of site 4 but apparently were absent from the interpolygonal troughs of site 4. These species included *Blepharostoma trichophyllum*, *Distichium capillaceum*, *Lophozia* cf. *ventricosa*, *Oncophorus wahlenbergii*, and *Plagiothecium denticulatum*.

Other than *Cephaloziella arctica* and *Myurella julacea*, there appeared to be no taxa of appreciable abundance that were unique to site 2. On the other hand,

Calliergon richardsonii, Drepanocladus revolvens, and *Gymnocolea inflata*
were found only in site 4 where they were considered common. Of these species,
Gymnocolea inflata was more abundant in polygonal centers than in interpolygonal troughs. In contrast, *Calliergon richardsonii* and *Drepanocladus revolvens*
were more abundant in interpolygonal troughs than in polygonal centers. In
addition to the latter two species, *Brachythecium* cf. *turgidum, Calliergon
giganteum, Calliergon sarmentosum, Campylium stellatum,* and species of
Bryum and *Mnium* were considered important components of interpolygonal
troughs.

The occurrence of *Calliergon giganteum, Calliergon richardsonii,* and *Drepanocladus revolvens* along with *Calliergon sarmentosum* and *Campylium stellatum* is probably indicative of generally wetter habitats in site 4, especially in the
interpolygonal troughs, than in site 2. This appears to be in agreement with the
findings of Smith (1974) for another locality in the Barrow area. On the other
hand, associations of *Campylium stellatum, Calliergon sarmentosum,* and *Drepanocladus uncinatus* with species of *Blepharostoma, Distichium, Oncophorus, Pogonatum,* and *Plagiothecium* indicate somewhat drier habitats (at least
periodically) in the polygonal centers (and in site 2) than in the interpolygonal
troughs of site 4.

Various habitats approximating mesic conditions in the Barrow area are likely
to have *Dicranum elongatum, Blepharostoma trichophyllum, Hylocomium
splendens, Tomenthypnum nitens, Ptilidium ciliare, Distichium capillaceum,* and
Drepanocladus uncinatus among the dominant taxa, whereas habitats approximating xeric conditions are likely to have *Pohlia nutans, Cephaloziella arctica,
Blepharostoma trichophyllum, Cephalozia pleniceps, Dicranum elongatum,
Anastrophyllum minutum,* and members of the Polytrichaceae among the dominant taxa (Smith, 1974).

A comparison of the known bryofloras for the Barrow area and Prudhoe Bay
area tundras reveals differences which are indicative of dissimilarities in edaphic
factors. Particularly noticeable is the occurrence of more acidophilous mosses in
the Barrow area tundra than in the Prudhoe Bay area tundra. Bryophyte taxa
especially associated with acid soils include species of *Lophozia, Sphagnum,*
and members of the Polytrichaceae. At the present time only one polytrichaceous species (*Pogonatum alpinum* var. *septentrionale*) has been reported for
Prudhoe Bay, and as yet no species of *Sphagnum* has been found (Rastorfer et
al., 1973). Bryophyte taxa which are generally considered calciphiles (e.g.,
Catoscopium nigritum) are apparently more common in the Prudhoe Bay area
tundra than in the Barrow area tundra (Rastorfer et al., 1973; Steere, 1978).

Bryomass

Bryomass varied within habitats and among different habitats. In site 2, the
bryomass ranged from 111 to 175 g m^{-2} (Table 3). In site 4, the bryomass ranged
from 122 to 165 g m^{-2} for the polygonal centers (Table 5). However, the average
values of bryomass for these two habitats were essentially the same and
amounted to 142 and 149 g m^{-2}, respectively.

In contrast, the bryomass ranged from 170 to 410 g m^{-2} and averaged 290 g m^{-2} for the interpolygonal troughs of site 4 (Table 5). Similarly, two relatively pure stands of *Campylium stellatum* located in wet depressions of site 2 had bryomass values of 560 and 437 g m^{-2} (D. K. Smith, unpublished data, samples collected on 5 July 1972; J. R. Rastorfer, unpublished data, samples collected on 21 July 1973, Moss Plot 203E).

High values of bryomass were not necessarily found only in wet interpolygonal troughs and depressions of low-center polygonal ground. For example, a bryomass of 582 g m^{-2} was found in a mesic habitat in which the bryophyte layer consisted mostly of *Tomenthypnum nitens* (J. R. Rastorfer, unpublished data, samples collected on 11 July 1973, IBP site Three, Moss Plot 302).

Estimates of bryomass have been reported for other arctic tundra localities, and except for xeric habitats they have ranged from approximately 90 to 1600 g m^{-2} (Pavlova, 1969; Khodachek, 1969; Pakarinen and Vitt, 1972). Lower bryomass values ranging from approximately 7 to 70 g m^{-2} were found for various habitats in a subarctic zone of Finnish Lapland (Kärenlampi, 1973). On the other hand, Collins (1973) reports bryomass values in excess of 2000 g m^{-2} for a *Drepanocladus uncinatus* carpet on Signy Island, Antarctica. Somewhat lower values of bryomass were found for turfs of *Pohlia* and mixed turfs of *Pohlia–Philonotis* occurring on the sub-antarctic island of South Georgia (Clarke et al., 1971).

Net Production

An estimate of net production of the bryophyte synusia in site 2 was obtained from the difference between the average bryomass of 115 g m^{-2} for 22 June (1972) and the average bryomass of 142 g m^{-2} for 24 August (1972), which equals 27 g m^{-2} for that 63-day period (Table 3). This value of net production may seem low; however, it represents an increase of 23% in bryomass, and it is in agreement with other estimates of net production. For example, the relatively pure stand of *Campylium stellatum* mentioned earlier (D. K. Smith, unpublished data) had a bryomass of 560 g m^{-2} of which 398 g m^{-2} was considered past growth and 162 g m^{-2} or 29% was considered new growth. Likewise, measurements of the green and nongreen portions of *Campylium stellatum* were 53 and 174 g m^{-2}, respectively, and those of *Tomenthypnum nitens* were 123 and 269 g m^{-2}, respectively (J. R. Rastorfer, unpublished data). These values amounted to 23 and 33% of the bryomass being attributable to new growth for these two species.

The value of 27 g m^{-2} for the bryophyte synusia of site 2 represents an approximation of the annual net production; however, the actual value is likely to be 10 to 20% larger for a growing season extending over an 80- to 90-day period. This estimate of annual net production for the bryophytes appears to be sound because it is in good agreement with values reported by Oechel and Sveinbjörnsson (1978). Even though they used different methods than those employed by the author of the present report, their estimates of annual net production in site 2 (western sector) ranged from 22 g m^{-2} yr^{-1} for the bryophyte synusia under the vascular plant canopy to 39 g m^{-2} yr^{-1} for the bryophyte synusia exposed to full sunlight. Of course, yearly differences would be expected in net production depending on prevailing weather conditions.

Studies in the high arctic on Devon Island revealed net bryophyte productions of 33 g m^{-2} yr^{-1} for a mesic meadow community, 60 g m^{-2} yr^{-1} for a hydric meadow community, and 350 g m^{-2} yr^{-1} for a stream-side community (Pakarinen and Vitt, 1972). In an alpine tundra environment, annual net production (green bryophyte tissues) ranged from 37 to 97 g m^{-2} yr^{-1} (Wielgolaski and Kjelvik, 1973). In contrast, large annual net productions of bryomass were reported for several bryophyte species occurring in south polar regions that ranged from approximately 200 to 1000 g m^{-2} yr^{-1} (Longton, 1970; Collins, 1973).

Conclusion

Bryophytes and graminoid vascular plants were the principal components of the stands of vegetation investigated. Although lichens, especially *Peltigera* spp., did occur in the wet-tundra-meadow communities, they were minor components with respect to the total vegetational biomass. However, lichens were better represented in other communities in the Barrow area (Williams et al., 1978).

The bryophyte synusia, in the wet-tundra-meadow communities, typically formed an understory to the vascular plants and consisted of a heterogeneous assemblage of mosses and liverworts. Similar observations were made by Smith (1974) for other study areas in the Barrow area and by Rastorfer et al. (1973) for the Prudhoe Bay area. The most abundant bryophytes were usually pleurocarpous mosses with two or three species comprising the bulk of the bryomass; and thus, the structural features of the bryophyte synusia were characteristic of the pleurocarpous growth habit. The associate mosses and liverworts generally grew interspersed among the shoots of the dominant species; however, some clumping did occur especially among liverworts.

An assessment of the ecological status of the bryophytes in comparison to the vascular plants of the arctic tundra is not an easy task because the aerial shoots of the two groups are morphologically very different, and because of the massive underground rhizome and root systems of the vascular plants (Dennis and Johnson 1970; Billings et al., 1973). However, comparisons of the standing crops (aboveground) between bryophytes and vascular plants are useful in a general way because these components are functionally similar with respect to primary production.

In reference to the aboveground vegetation in the wet-tundra-meadow of site 2, Dennis et al. (1978) found the maximum living aboveground standing crop for the vascular plants to be 102 and 87 g m^{-2} in 1970 and 1971, respectively. In contrast, the largest bryophyte standing crop (bryomass) determined in 1972 was 142 g m^{-2} (Table 3). In addition, the species compositions differed also. Dennis et al. found 22 species of vascular plants, whereas over 30 species of mosses were found (Table 1). Based on these comparisons (and until further details are available on comparative productivities), the bryophytes and vascular plants should be considered codominants of the vegetation in the wet-tundra-meadow communities of site 2 and the polygonal centers of site 4.

On the other hand, bryophytes can be considered the dominant vegetational components in some arctic tundra communities. Elsewhere in the Barrow area, Dennis and Tieszen found that the maximum seasonal (1972) aboveground standing crop for the vascular plants and lichens was approximately 80 g m^{-2} of which about 80% consisted of monocots. In contrast, an estimate of bryomass amounted to about 150 g m^{-2}. In addition, Matveyeva et al. (1973) considered mosses the dominant vegetational components apparently for most of the tundra communities in the vicinity of Taimyr (U.S.S.R.).

Acknowledgments. This research was supported by the National Science Foundation under Grant GV 29342 to the University of Alaska. It was performed under the joint sponsorship of the International Biological Programme and the Office of Polar Programs and was coordinated by the U.S. Tundra Biome. Field and laboratory activities at Barrow were supported by the Naval Arctic Research Laboratory of the Office of Naval Research.

I acknowledge with sincere gratitude the contributions made to the project by co-workers Dr. David K. Smith and Dr. Harold J. Webster during the 1972 boreal summer and by co-workers Mr. James R. Haeberlin and Dr. Allen C. Skorepa during the 1973 boreal summer. Working closely, often inching across plots on hands and knees, with these fellow scientists was an honor and a pleasure. I am grateful to Dr. Hiroshi Inoue, Dr. Zennoske Iwatsuki and Dr. William C. Steere for their willingness to help identify several difficult species, and for the joint field trips taken in the vicinity of Barrow. Special thanks are due to Dr. David K. Smith and Mrs. Judith B. Rastorfer for reviewing the manuscript.

References

Abramova, A. L., L. I. Savich-Lyubitskaya, and N. Z. Smirnova. (1961) *Manual of Leafy Mosses of Arctic U.S.S.R.* V.L. Komarov Botanical Institute, Academy of Sciences, U.S.S.R., Leningrad (in Russian).

Arnell, S. (1956) *Illustrated Moss Flora of Fennoscandia.* I. *Hepaticae.* Stockholm: Nat. Sci. Res. Counc., and Lund: CWK Gleerup, 315 pp.

Billings, W. D., G. R. Shaver, and A. W. Trent. (1973) Temperature effects on growth and respiration of roots and rhizomes in tundra graminoids. In *Proceedings of the Conference on Primary Production and Production Processes, Tundra Biome, Dublin, Ireland, April 1973* (L. C. Bliss and F. E. Wielgolaski, Eds.). Stockholm: International Biological Programme, Tundra Biome Steering Committee, pp. 57–63.

Brassard, G. R. (1971) The mosses of northern Ellesmere Island, arctic Canada. I. Ecology and phytogeography, with an analysis for the Queen Elizabeth Islands. *Bryologist, 74*: 233–281.

Britton, M. E. (1967) Vegetation of the arctic tundra. In *Arctic Biology: 18th Biology Colloquium* (H. P. Hansen, Ed.). Corvallis, Oregon: Oregon State University Press, 2nd ed., pp. 67–130.

Clarke, G. C. S., S. W. Greene, and D. M. Greene. (1971) Productivity of bryophytes in polar regions. *Ann. Bot., 35*: 99–108.

Collins, N. J. (1973) Productivity of selected bryophyte communities in the Maritime Antarctic. In *Proceedings of the Conference on Primary Production and Production Processes, Tundra Biome, Dublin, Ireland, April 1973* (L. C. Bliss and F. E. Wielgolaski, Eds.). Stockholm: International Biological Programme, Tundra Biome Steering Committee, pp. 177–183.

Dennis, J. G., and P. L. Johnson. (1970) Shoot and rhizome-root standing crops of tundra vegetation at Barrow, Alaska. *Arct. Alp. Res., 2*: 253–266.

Dennis, J. G., L. L. Tieszen, and M. A. Vetter. (1978) Seasonal dynamics of above- and belowground production of vascular plants at Barrow, Alaska. In *Vegetation and Production Ecology of an Alaskan Arctic Tundra* (L. L. Tieszen, Ed.). New York: Springer-Verlag, Chap. 4.

Drew, J. V., and J. C. F. Tedrow. (1962) Arctic soil classification and patterned ground. *Arctic,* **15**: 109–116.

Frye, T. C., and L. Clark. (1937–47) *Hepaticae of North America.* University of Washington Publications, Biology, Vol. 6, 1022 pp.

Grout, A. J. (1928–1940) *Moss flora of North America north of Mexico.* 3 vols., Newfane, Vermont (Reissued, partly in facsimile, by Hafner Publishing Co., New York, 1972.).

Kärenlampi, L. (1973) Biomass and estimated yearly net production of the ground vegetation at Kevo. In *Proceedings of the Conference on Primary Production and Production Processes, Tundra Biome, Dublin, Ireland, April 1973* (L. C. Bliss and F. E. Wielgolaski, Eds.). Stockholm: International Biological Programme Tundra Biome Steering Committee, pp. 111–114.

Khodachek, E. A. (1969) Vegetal matter of tundra phytocoenoses in the western part of Taimyr Peninsula. *Bot. Zh.,* **54**(7): 1059–1073 (International Tundra Biome Translation 5, December 1971; Translator: G. Belkov, 12 pp.).

Kil'dyushevskii, I. D. (1964) On the ecology of mosses—dominants of the plant cover in the north. *Problemy Severa,* **8**: 83–87 (National Research Council of Canada Translation: Problems of the North, Vol. 8, pp. 85–89, 1965.).

Lawton, E. (1971) *Moss Flora of the Pacific Northwest.* Nichinan, Miyazaki: The Hattori Botanical Laboratory, 362 pp. and 195 plates.

Longton, R. E. (1970) Growth and productivity of the moss *Polytrichum alpestre* Hoppe in antarctic regions. In *Antarctic Ecology* (M. W. Holdgate, Ed.). New York: Academic Press, pp. 818–837.

Matveyeva, N. V., T. G. Polozova, L. S. Blagodatskykh, and E. V. Dorogostaiskaya. (1973) A brief essay on the vegetation in the vicinity of the Taimyr Biogeocoenological Station (Kratkii ocherk rastitel'nosti okrestnostei taimyrskogo biogeotsenologisheskogo statsionara). In *Biogeocenoses of Taimyr Tundra and Their Productivity, (Biogeotsenozy Taimyrskoi tundry i ikh produktivnost')* Vol. 2, pp. 7–49. Leningrad: Nauka (International Tundra Biome Translation 13, January 1975: Translator: Doris Löve, 51 pp.).

Nyholm, E. (1954–1969) *Illustrated Moss Flora of Fennoscandia II. Musci.* Nat. Sci. Res. Council. Stockholm, and CWK Gleerup, Lund, 799 pp.

Oechel, W. C., and B. Sveinbjörnsson. (1978) Primary production processes in arctic bryophytes at Barrow, Alaska. In *Vegetation and Production Ecology of an Alaskan Arctic Tundra* (L. L. Tieszen, Ed.). New York: Springer-Verlag, Chap. 11.

Oosting, H. J. (1956) *The Study of Plant Communities.* San Francisco: W. H. Freeman and Company, 2nd ed., 440 pp.

Pakarinen, P., and D. H. Vitt. (1972) The ecology of bryophytes in Truelove Lowland, Devon Island. In *Devon Island IBP Project, High Arctic Ecosystem, Project Report 1970 and 1971* (L. C. Bliss, Ed.). Dep. of Botany, University of Alberta, pp. 185–196.

Pavlova, E. B. (1969) Vegetal mass of the tundras of western Taimyr (O rastitel' noi masse tundr Zapadnoi Taimyra). *Vestnik Moskovskogo Universiteta,* No. 5, pp. 62–67 (International Tundra Biome Translation 3, April 1971; Translator: G. Belkov, 5 pp.).

Rastorfer, J. R. (1972) Bryophyte taxa lists of the high Alaskan arctic. Institute of Polar Studies, Ohio State University, Columbus, 54 pp.

Rastorfer, J. R., H. J. Webster, and D. K. Smith. (1973) Floristic and ecologic studies of bryophytes of selected habitats at Prudhoe Bay, Alaska. *Inst. Polar Stud., Ohio State Univ., Rep.* 49, 20 pp.

Smith, D. K. (1974) Floristic, ecologic and phytogeographic studies of the bryophytes in the tundra around Barrow, Alaska. Ph.D. dissertation, University of Tennessee, Knoxville, 191 pp.

Steere, W. C. (1971) A review of arctic bryology. *Bryologist,* **74**: 428–441.

Steere, W. C. (1978) Floristics, phytogeography, and ecology of arctic Alaskan bryophytes. In *Vegetation and Production Ecology of an Alaskan Arctic Tundra* (L. L.
Tieszen, Ed.). New York: Springer-Verlag, Chap. 5.

Webber, P. J. (1978) Spatial and temporal variation of the vegetation and its production,
Barrow, Alaska. In *Vegetation and Production Ecology of an Alaskan Arctic Tundra*
(L. L. Tieszen, Ed.). New York: Springer-Verlag, Chap. 3.

Wielgolaski, F. E., and S. Kjelvik. (1973) Production of plants (vascular plants and
cryptogams) in alpine tundra, Hardangervidda. In *Primary Production and Production
Processes* (L. C. Bliss and F. E. Wielgolaski, Eds.). Tundra Biome Steering Committee, Stockholm, pp. 75–86.

Williams, M. E., E. D. Rudolph, E. A. Schofield, and D. C. Prasher. (1978) The role of
lichens in the structure, productivity, and mineral cycling of the wet coastal Alaskan
tundra. In *Vegetation and Production Ecology of an Alaskan Arctic Tundra* (L. L.
Tieszen, Ed.). New York: Springer-Verlag, Cahp. 7.

7. The Role of Lichens in the Structure, Productivity, and Mineral Cycling of the Wet Coastal Alaskan Tundra[1]

M. E. WILLIAMS, E. D. RUDOLPH, E. A. SCHOFIELD, AND D. C. PRASHER

Introduction

Lichens are a constant, though variable, component of arctic tundra ecosystems (Wielgolaski, 1972, 1975). In the Barrow tundra lichens have been reported as a minor part of the vegetation (Britton, 1967; Spetzman, 1959; Wielgolaski, 1972). We have examined the lichens in the vicinity of Barrow and have attempted to quantify their occurrence, biomass, and productivity, as well as to discover the relationships of lichens to their microenvironments. We measured the elemental contents of selected lichens and their substrates in an attempt to discover if significant mineral accumulation was occurring in the lichens.

A major function of plants in any ecosystem is providing the base of most food chains. Caribou, the principal herbivores reported to eat lichens heavily in winter, are present in many Alaskan Arctic Slope areas, but seldom migrate into the Barrow area, where lemmings are the primary grazers. The analysis of gut contents by members of the Tundra Biome study group has shown lichens to be essentially absent from the diet of these rodents. The remainder of the fauna at Barrow are unlikely grazers on lichens. This is supported by the lack of evidence of grazing in the numerous lichen samples collected by us. It appears that lichens are not substantial parts of the food chains in the Barrow tundra.

Environmental Setting

The wet coastal tundra in the vicinity of Barrow, Alaska, appears to contradict the common impression that lichens and mosses are the hallmark of tundra ecosystems. Indeed, lichens contribute a relatively small proportion of the standing vegetation in many areas of this grass- and sedge-dominated tundra. Upon closer inspection, however, the prominence of both lichens and bryo-

---

[1] Contribution No. 356 of the Institute of Polar Studies, The Ohio State University, Columbus, Ohio.

phytes (Rastorfer, 1978) in the interstem spaces of all but the wettest terrestrial habitats is readily observed.

Climate is, of course, a major controlling factor in determining the plant structure of the ecosystem. Though the normal precipitation is low, it exceeds evapotranspiration, making the tundra very wet. In addition, for about 3 weeks after snow melt in early- to mid-June, temporary thaw ponds in the basins of low-centered polygons and the numerous permanent lakes provide a water cover over much of the area. Such conditions are unfavorable for the survival and growth of most lichens. Freezing temperatures may occur at any time during the short growing season, and extremes of +25.5 and −49°C have been recorded. Lichens, evergreen perennial plants, must be able to withstand these extreme and fluctuating temperatures.

A most striking feature of the topography in the Barrow area is the near absence of relief. The highest points, less than 6 m above sea level, are on former beach ridges (Clebsch and Shanks, 1968). In spite of the flatness of the terrain, a surprising number of discrete microhabitats for lichens occur on the two basic types of ice-wedge polygons, the low-centered and high-centered polygons. Individual polygons may be 20 or more m across, and are delimited by troughs. Many polygons intermediate to the two basic types are also found, as well as areas in which they are poorly developed, giving these level-appearing areas a meadow-like aspect. The lichen study sites included here were mostly free from direct marine effects such as salt spray or inundation with saltwater.

Lichen Occurrence

The sunken centers of low-centered polygons normally are very wet, with standing water for at least the first 2 to 3 weeks of the June to August growing season. *Siphula ceratites* and *Cetraria delisei* are often present near the margin (Figure 1), as are the thalli of other species which are washed or blown in from the margin. Moss cover is very sparse, consisting primarily of *Pogonatum alpinum*. Vascular plants are equally sparse, with *Carex aquatilis* and *Saxifraga foliolosa* the most common.

The raised margins of low-centered polygons typically are covered with dense moss tufts and several species of grasses, sedges, willows, and other vascular plants. Small frost boils may form in some areas, with the same general characteristics as those described below. A highly diverse lichen flora occurs on these margins, often in the depressions between moss tufts, the most common being *Alectoria nigricans, Cornicularia divergens, Cetraria cucullata, C. islandica, Cladonia amaurocraea, C. uncialis, C. rangiferina, Dactylina arctica, Ochrolechia frigida, Sphaerophorus globosus,* and *Thamnolia subuliformis.* The substrate may remain moist throughout the season or may become quite dry, depending on the climatic conditions of the year in question.

The trough bottoms and trough slopes of low-centered polygons support an interesting vegetative spectrum as one moves from the often water-saturated bottoms up to the relatively dry margin tops. Lichens are usually absent from

W = ICE WEDGE

T = TROUGH

M = MARGIN

C = CENTER

S = TROUGH SLOPE

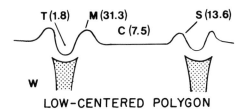

LOW-CENTERED POLYGON

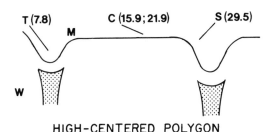

HIGH-CENTERED POLYGON

Figure 1. Cross-sectional profiles of basic polygon types at Barrow, Alaska. Figures represent lichen percentages of total standing crop (see Table 1). The stippled areas represent ice wedges.

trough bottoms in which there is standing water for longer than the first few weeks of the growing season. Those that drain more rapidly often support the growth of *Peltigera aphthosa*, *P. canina*, and *Cetraria delisei*. These species are also common on the lower trough slopes, where other taxa such as *Cetraria islandica*, *Dactylina arctica*, *Cladonia uncialis*, *C. rangiferina*, *C. amaurocraea*, *Sphaerophorus globosus*, and *Thamnolia subuliformis* appear and increase in density as one approaches the margin top. The moss cover is often very dense, while vascular plants such as *Dupontia fisheri*, *Carex aquatilis*, *Eriophorum angustifolium*, *Arctagrostis latifolia*, *Petasites frigidus*, *Salix rotundifolia*, *Saxifraga cernua*, and *Stellaria laeta* are common but usually make up a smaller proportion of the aboveground biomass than do the mosses. *Cardamine bellidiflora* and colonies of the blue–green alga *Nostoc* sp. may often be found in the trough bottoms.

Because of their xeric nature, high-centered polygons differ from low-centered in both lichen diversity and abundance. The trough bottoms and slopes are of the same general composition as those associated with low-centered polygons, though the moss density is somewhat lower and lichens are more prevalent.

Cetraria cucullata, C. delisei, Peltigera aphthosa, and *P. canina* are especially common on the trough bottoms and lower slopes, while the upper slopes and margins may contain virtually any macrolichen species found on the polygon center.

High-centered polygon tops are of two basic types. One variety is the "grassy top" on which grasses, sedges, and *Saxifraga cernua* predominate, the moss cover is sparse, and lichens are well represented. *Dactylina arctica, Cetraria islandica, C. cucullata, Alectoria nigricans, Thamnolia subuliformis,* and several *Cladonia* species are especially common. Often a considerable amount of vascular plant litter collects there. The other type is the frost boil, on which the soil has been heaved by intensive frost action near the surface. Mosses and such vascular plants as *Salix rotundifolia* and some grasses and sedges are scattered; the soil is sandy and with an almost nonexistent humus layer, very dry, and exposed to much wind blasting. These sites thus present the most severe growth conditions in terms of substrate moisture and temperature, especially when these polygons occur on ridges (Billings and Mooney, 1968). A large proportion of the surface is covered by crustose lichens such as *Caloplaca cinnamonea, Psoroma hypnorum, Rinodina* spp., *Ochrolechia frigida,* and several sterile taxa. *Alectoria nigricans* and *Cornicularia divergens* are often found in intertwined tufts, and *Cetraria* and *Cladonia* species are extremely abundant on these boils in addition to *Dactylina arctica, Thamnolia subuliformis,* and *Sphaerophorus globosus.*

Many sectors of tundra in the Barrow area have no well-defined polygonal structure, but appear as meadows containing numerous 5- to 20-cm-high tussocks. Grasses, sedges, and willows make up much of the vascular flora. The moss cover is both dense and extremely diverse, frequently having more than 20 species in a 10-cm^2 area (Rastorfer et al., 1973). Wet meadows usually have poorly defined tussocks, and the lichens *Peltigera aphthosa, P. canina,* and *Lobaria linita* are common. These species are also found in the depressions between tussocks in better drained meadows, as well as *Cetraria richardsonii, C. delisei,* and *C. cucullata.* The blue–green alga *Nostoc* commonly colonizes the depressions when soil moisture is adequate. The same lichen species are also present on the tussocks, in addition to *Dactylina arctica, Thamnolia subuliformis,* and several *Cladonia* species. *Psoroma hypnorum, Rinodina* spp., *Alectoria nigricans, Ochrolechia frigida* f. *thelephoroides,* and *Cornicularia divergens* are also common.

Methods

Standing Crop

In order to quantify the relative standing crop of lichens, random sample cores of 100 or 41 cm^2 were collected in sites 4 and 12. The aboveground plant material was separated into moss, vascular plant, and lichen components, the latter being separated by species. Vascular plant litter and crustose lichens were not included. Each component was washed to remove any extraneous material and

Table 1. Relative Aboveground Plant Standing Crop and Soil pH at Barrow, Alaska. Live and Attached Dead Were Not Separated

Microhabitat[a]	Mean pH	Mean g m^{-2} (dry weight)				Mean percentage total standing crop			Number of lichen species	Number of samples
		Lichens	Mosses	Vasculars[b]	Total	Lichens	Mosses	Vasculars		
LCP-center	5.0	2.4	9.9	19.9	32.2	7.5	33.2	59.1	4	5
LCP-margin	4.6	180.1	333.2	70.8	584.1	31.3	56.4	12.3	15	10
LCP-trough slope	4.7	40.7	320.1	70.0	430.8	13.6	62.2	24.2	14	10
LCP-trough bottom	5.0	4.8	194.9	101.8	301.5	1.8	65.3	33.0	1	8
HCP-frost boil	4.0	60.1	149.2	67.5	276.8	21.9	48.6	28.7	17	12
HCP-grassy top	4.3	51.8	258.9	93.8	404.5	15.9	58.8	25.4	14	9
HCP-trough slope	4.3	101.6	96.4	125.5	323.5	29.5	28.0	42.5	10	6
HCP-trough bottom	4.6	20.5	127.1	82.6	230.2	7.8	50.8	41.3	5	6
M-tussock	5.4	100.8	161.8	243.6	506.2	20.1	31.7	48.2	14	3
M-depression[c]	5.4	35.9	99.0	143.4	278.3	10.0	25.3	54.6	8	3

a LCP, low-centered polygon; HCP, high-centered polygon; M, meadow.
b Vascular plant litter not included.
c Also contained 19.5 g m^{-2} Nostoc (10.1%).

Table 2. Percentage Soil Moisture in Various Lichen Microhabitats at Barrow, Alaska

Microhabitat[a]	Date (1973)						Mean
	6/26	7/2	7/9	7/16	7/24	7/30	
LCP-trough slope	159.0	151.9	150.3	142.0	105.0	143.1	141.9
LCP-margin	120.5	150.6	154.9	121.4	102.9	154.7	134.2
HCP-trough bottom	28.1	69.3	90.2	46.1	72.6	53.9	60.0
HCP-trough slope	19.9	41.1	29.1	47.8	35.4	40.5	35.6
HCP-top	25.1	46.7	17.5	31.2	34.6	38.6	32.3
M-tussock	25.3	21.1	20.4	20.9	22.6	24.9	22.5
M-depression	42.8	28.2	43.8	61.4	42.2	119.3	56.3

[a] LCP, low-centered polygon; HCP, high-centered polygon; M, meadow.

then dried for 24 hr at 70 to 80°C. Both the g dry wt m^{-2} and the percentage total biomass were calculated (Table 1) from an average of 5 to 10 samples.

Lichens were identified by thallus morphology and spot tests with potassium hydroxide, calcium hypochlorite, iodine, and p-phenylenediamine. The presence or absence of characteristic lichen acids was confirmed, when necessary, by thin-layer chromatography (Santesson, 1967). Taxonomic work was complicated by the fact that the relatively small amount of pertinent literature is widely scattered, and is of little help in identifying the often reduced and nonfruiting forms encountered in the Barrow area.

Soil Moisture

Soil moisture data were obtained during the 1973 field season by taking random cores to a depth of 10 cm in the various microhabitats at weekly intervals. The surface vegetation and any loose humus was cut away and the samples were placed in covered drying cans. They were then weighed, dried at 110°C for 24 hr, and reweighed (Table 2). The thickness of the thaw layer was measured at the same time the soil moisture samples were collected, as was the thickness of the humus layer (Table 3). The humus layer was defined as the layer between the mineral soil and the base of intact, growing moss or vascular plants. Data were not obtained for low-centered polygon centers and trough bottoms due to the presence of standing water during the entire study period. Values for high-centered polygon grassy tops and frost boils were pooled. All samples were collected in the IBP site 12 moisture gradient.

Productivity: Direct Measurements

Seasonal growth was determined for *Dactylina arctica* (Hook.) Nyl. because of the ease with which this species could be accurately subjected to morphological analysis. The thalli were divided subjectively into three components: "new growth" (the greenish-yellow, actively growing tips and buds of the thallus,

Table 3. Humus Thickness and Depth to Permafrost in Various Lichen Microhabitats at Barrow, Alaska

Microhabitat[a]	Humus layer (cm)	Thaw depth (cm)—Date (1973)[b]				
		6/26	7/2	7/16	7/24	7/30
LCP-trough slope	4.7	7.3	8.7	16.5	23.3	27.7
LCP-margin	4.1	9.5	14.0	24.0	19.6	31.3
HCP-trough bottom	3.1	6.5	15.9	22.0	23.3	27.8
HCP-trough slope	2.2	15.0	15.4	17.0	19.5	24.2
HCP-top	0.8	28.5	>30	>30	21.3	39.7
M-tussock	1.4	12.5	20.2	32.0	32.0	38.3
M-depression	3.5	10.0	13.7	19.7	21.3	26.3

[a] LCP, low-centered polygon; HCP, high-centered polygon; M, meadow.
[b] No measurements are available for July 9th.

containing most of the viable algae); "standing old growth" (the yellow, intact portion of the thallus consisting primarily of fungal hyphae); and "decomposing material" (the dark brown, lower portion of the thallus, typically pitted and partially fragmented).

A direct measurement of the growth of *Cetraria richardsonii* Hook. was made during the 1973 field season by a method analogous to that of Kärenlampi (1971), but without using boxes. Samples of the lichen were collected, dried over calcium hydroxide in a desiccator for 24 hr at room temperature, and then weighed. They were later attached to threads, labeled, and returned to the collection site, where the threads were fastened to a stake. As this species normally is found lying free among vascular plant stems in mesic sites, its removal and subsequent replacement among the plants represented a minimum of disturbance in growth conditions. After 22 days, the samples were returned to the laboratory, dried as before, and reweighed.

Productivity: Photosynthesis and Respiration

In order to test the metabolic activity over a range of temperatures, and to allow comparison of lichen metabolism with that of mosses and vascular plants, respiration and photosynthesis rates were determined for nine of the most abundant lichen species by Gilson differential respirometry. Samples, still attached to their substrates, were collected on 3 August 1973 from mesic meadow sites, placed in plastic bags, packed in ice, and shipped to Columbus, Ohio, where they were stored at 2°C until tested. Testing was begun 4 days after collection, using a manometric method similar to that of Rastorfer and Higinbotham (1968). Small samples to be tested were cleaned of extraneous material, rinsed twice in sterile distilled water, and placed upper surface down in the reaction vessels. One milliliter of sterile distilled water was added to the side arm and 0.1 ml was placed in the bottom of the reaction vessel to keep the thalli near 100% saturation. A constant CO_2 pressure of 1.0% CO_2 in air (v/v) was maintained by adding 1 ml of Pardee buffer (diethanolamine) to the center well (Umbreit et al., 1957). A 4-cm^2 piece of filter paper was folded accordion-style and placed in the center well of the reaction vessel to increase the surface area for CO_2 exchange. After all the reaction vessels were attached and the water bath temperature had reached the desired level, the entire system was flushed with 1% CO_2 in air (v/v) for 3 to 5 min to establish equilibrium before initiating the measurement period. Manometer readings indicating the amount of O_2 utilized or evolved were made after 4 to 8 hr. Following each experiment the lichens were removed from the reaction vessels, placed in aluminum tins, dried at 80°C for 24 hr, and weighed to determine dry weight.

Respiration measurements were made in the dark, while photosynthesis readings were taken with the flasks illuminated from beneath the water bath by 10 incandescent reflector lamps (110 W). The light intensity at the level of the reaction vessels was 1.4×10^5 erg cm^{-2} sec^{-1}. Since lichens generally, and arctic and alpine lichens in particular, have low light compensation points (Bliss and Hadley, 1964), it is reasonable to assume that the light intensity employed was at least as high as the compensation points for the species studied, possibly may have been too high for the expression of maximum photosynthetic rates. Showman (1972) found the same illumination to be photosaturating for isolated phycobionts of temperate lichens.

Mineral Accumulation

Lichen samples were collected for the species listed in Tables 4 and 5, separated from any extraneous matter, and lightly rinsed with distilled water. After gentle blotting the specimens were air dried at room temperature and placed in polyethylene bags until analyzed. The two morphologically indistinguishable but chemically different species of *Thamnolia* were separated by their fluorescence in ultraviolet light (Sato, 1963). *T. subuliformis* (Ehrh.) W. Culb. fluoresces (bright white) under uv due to the presence of squamatic acid, while *T. vermicularis* (Sw.) Adr. ex Schaer. lacks this substance and therefore appears lavender–brown under illumination. The mineral determinations were made at the Forest Soils Laboratory of the University of Alaska, using the methods of Jackson (L. Oliver, personal communication). Nitrogen was determined using the micro-Kjeldahl method. Phosphorus and cations were analyzed on aliquots of nitric-perchloric acid digest. Phosphorus was determined by the molybdenum blue method using an amino-naphthol-sulfonic acid reducing agent. Cations were quantified on an aliquot of the digest using an atomic-absorption spectrophotometer. Strontium chloride was added to the Ca and Mg aliquots to suppress anion interference. Total elemental concentrations were expressed as percentage dry weight of the sample (Tables 4 and 5).

Soil samples were collected from the various microhabitats described above. The sample cores, 7.2 cm in diameter and usually about 10 cm in depth, were freed of surface vegetation and loose humus, air dried, and stored in polyethylene bags prior to analysis by the Ohio State University Soil Testing Laboratory. Soil pH was determined by adding 5 g of distilled water to 5 g of soil which had been passed through a 10-mesh stainless-steel screen.

Phosphorus concentration was determined by adding 10 ml of HCl–ammonium fluoride (pH 2.6) to 1 g of soil, shaking 5 min, and pouring onto moist folded filter paper in a filter vial. Then, 10.2 ml of ammonium molybdate–P-free sulfuric acid [9.6 g NH_2SO_4 with 2.5 g of $(NH_4)_6Mo_7O_{24}\cdot 4H_2O$/100 ml of H_2SO_4 saturated with borate] were mixed with 5 ml of filtrate. After this, 0.2 ml of stannous chloride (50 g of $SnCl_2\cdot 2H_2O$ in 200 ml conc. HCl, diluted to 2:1) was added, the resulting solution was shaken, and it was read, after 10 min, in a colorimeter at 680 nm. The K concentration was obtained by adding 5 g of soil to 15 ml of 1 N ammonium acetate. After shaking 5 min and filtering, a 2-ml aliquot was added to 6 ml of the ammonium acetate extracting solution and the solution was read on an atomic absorption spectrophotometer. Ca and Mg were determined by shaking 1.5 g of soil and 15 ml of 1 N ammonium acetate for 10 min, filtering, and diluting a 0.5-ml aliquot to 25 ml with a 1% lanthanum solution. Ca and then Mg were read as for P and checked against standards. NO_3–N was determined by adding 1 g of soil to 5 ml of distilled H_2O and taking millivolt readings, after standing 10

Table 4.　Mineral Analyses of Lichens of Various Microhabitats, Barrow, Alaska

Species and collection site[a]	Collection date (1973)	(% of dry weight)							
		Ca	Fe	K	Mg	Mn	P	Zn	N
Cetraria cucullata	7/1	0.83	0.01	0.22	0.12	0.010	0.04	0.002	0.62
(HCP trough slope)	7/13	0.75	0.00	0.24	0.14	0.003	0.05	0.002	0.52
Cetraria delisei	7/13	0.03	0.12	0.26	0.05	0.000	0.08	0.005	1.69
(LCP trough bottom)	7/29	0.06	0.12	0.68	0.06	0.000	0.10	0.005	1.69
Cetraria islandica	6/22	0.62	0.02	0.18	0.08	0.000	0.01	0.003	0.51
(LCP margin top)	7/13	1.01	0.04	0.18	0.11	0.004	0.04	0.002	0.40
Cetraria richardsonii	7/1	0.50	0.01	0.13	0.06	0.008	0.03	0.001	0.52
(Meadow depression)	7/13	0.59	0.01	0.13	0.06	0.002	0.04	0.001	0.43
Cladonia amaurocraea	7/1	0.06	0.02	0.15	0.07	0.003	0.07	0.001	0.47
(LCP margin top)	7/13	0.04	0.02	0.17	0.08	0.004	0.05	0.002	0.59
Dactylina arctica	6/22	0.23	0.01	0.10	0.07	0.002	0.03	0.001	0.53
(LCP margin top)	7/13	0.18	0.04	0.17	0.14	0.005	0.06	0.001	0.40
Peltigera aphthosa	6/27	0.14	0.09	0.63	0.12	0.008	0.08	0.002	2.52
(Meadow tussock)	7/13	0.19	0.03	0.57	0.13	0.006	0.11	0.002	2.38
Peltigera canina	7/1	0.16	0.10	0.99	0.15	0.019	0.15	0.004	4.17
(Meadow tussock)	7/13	0.51	0.02	0.56	0.20	0.008	0.12	0.005	4.19
Peltigera scabrosa	7/13	0.11	0.11	0.72	0.15	0.005	0.13	0.002	3.92
(Meadow tussock)	7/29	0.13	0.20	0.57	0.17	0.010	0.12	0.003	2.61
Thamnolia subuliformis	6/22	1.04	0.00	0.25	0.05	0.002	0.09	0.002	0.72
(LCP margin top)	6/22	1.85	0.02	0.28	0.06	0.004	0.09	0.002	0.57
Thamnolia vermicularis	7/25	0.50	0.02	0.34	0.05	0.004	0.08	0.003	0.83
(LCP margin top)	7/25	0.46	0.01	0.28	0.04	0.003	0.08	0.003	0.67

[a] HCP, high-centered polygon; LCP, low-centered polygon.

Table 5. Soil Sample Mineral Analyses, Barrow, Alaska[a]

Microhabitat[b]	(ppm)							
	Ca	Fe	K	Mg	Mn	P	Zn	NO_3–N
LCP-margin	1366	474	112	520	8.3	2.0	3.0	5.8
LCP-trough slope	1116	1997	40	419	3.8	4.0	21.9	8.5
HCP-trough	1105	1856	78	407	0.2	4.5	3.9	2.5
HCP-trough slope	1000	427	54	344	0.2	1.8	0.7	1.3
HCP-top	800	332	118	341	6.7	1.3	0.6	1.8
M-tussock	1185	2080	97	405	33.1	3.8	12.9	1.0
M-depression	1512	176	88	458	6.4	2.3	0.9	2.3

[a] At least three samples averaged.
[b] LCP, low-centered polygon; HCP, high-centered polygon; M, meadow.

min, with a nitrate-ion electrode, using a calomel reference electrode. The Mn, Zn, and Fe values were determined by adding 10 g of soil to 20 ml of extracting solution (0.005 M diethylenetriamine pentaacetic acid, 0.1 M triethanolamine, and 0.01 M $CaCl_2$, pH 7.3). The mixture was shaken 2 hr, centrifuged, and filtered prior to reading on the mass spectrometer.

Results and Discussion

Lichen Distribution Related to Microhabitats

The quantitative and qualitative data (Figure 2, Tables 1 and 2) demonstrate substrate moisture to be one of the most prominent factors related to lichen distribution. The poikilohydric nature of lichens makes them directly responsive to environmental moisture, or the lack of it. Liquid water enters the void spaces of the thallus by capillary action at a rate such that many foliose and fruticose species reach full saturation from a dry state in only 2 to 4 min (Blum, 1973). This condition, which would be a common state for lichens occurring in wet sites at Barrow, would be clearly unfavorable for the growth of most lichens due to the low diffusion of CO_2 to the phycobiont (Kappen, 1973). However, the *Peltigera* species are able to tolerate wet habitats due to a loose thallus structure, which permits CO_2 exchange in saturated state (Kappen, 1973).

Unlike higher plants, lichens are capable of directly using atmospheric moisture, a factor of considerable value to lichens colonizing dry habitats, especially the tops of high-centered polygons, and it may in fact be the principal reason that

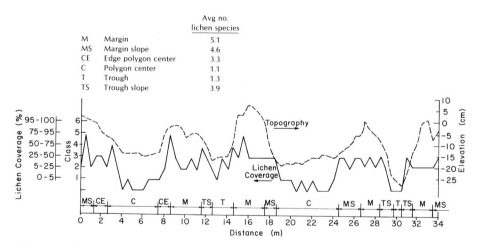

Figure 2.   Transect across area of low-centered polygons at IBP Tundra Biome site 4, Barrow. Coverage estimates are given in percentage classes, topography is indicated by the dashed line and by letters indicating microhabitat, and average lichen species numbers for each microhabitat type are provided.

lichens are able to dominate such sites. The presence of a high relative humidity during the growing season in the Barrow area is well documented.

The ability of lichens to rapidly take up liquid water and to use atmospheric moisture is an advantage over vascular plants under certain conditions. Lichens are further able to initiate respiratory and photosynthetic processes very rapidly in response to the occurrence of adequate moisture, provided the temperature and light conditions are sufficient. Lichens are well known for their adaptation to long periods of desiccation, both in summer and in winter. These periods of dormancy can last for considerable lengths of time, obviating the need to produce new biomass each year (Billings and Mooney, 1968). As this capacity is also inherent to lichen vegetative propagules, it is possible for these disseminules to remain viable for several years, initiating growth when favorable conditions arise. Such advantages are more pronounced in environments more limiting to plant growth, as occurs in the Antarctic (Rudolph, 1971).

Even though lichens are capable of taking up moisture rapidly, they have no (or poorly developed) structures to prevent the rapid loss of that water. This problem is less pronounced in *Peltigera* (and a few other genera) due to the retention of water in the gelatinous sheath of their blue–green algal symbiont (Blum, 1973), permitting relatively longer periods of metabolic activity as compared to lichens containing only nonsheathed green algae. Dehydration, however, is of importance to the long-term survival of the thallus due to its "hardening" effect; this should be especially true as the water loss prevents the formation of intercellular ice crystals during periods of freezing temperatures.

The close correlation between lichens and certain microhabitats was clearly demonstrated by a study of the vegetation of a 34-m transect of low-centered polygons, a study performed in conjunction with J. Rastorfer (bryophytes) and P. Webber (vascular plants). Estimates of lichen coverage were made within a decimeter-square frame placed at 0.5-m intervals along the transect, 25 cm from the stake line. A scale of 1 to 6 was used to grade the percentage lichen cover (see legend, Figure 2). Lichen species were also noted: *Cetraria islandica* (L.) Ach., *Dactylina arctica* (Hook.) Nyl., and, to a lesser extent, several *Cladonia* species were abundant on the margins, while *Peltigera canina* (L.) Willd. and *P. spuria* (Ach.) DC. were the only lichen taxa regularly encountered in the polygon centers and troughs. The results of the coverage estimates are presented in Figure 2, along with a cross-section profile of the transect (Rastorfer, personal communication). The average number of macrolichen species in each decimeter-square sampling of the individual microhabitats are listed in the legend to Figure 2. It should be noted that these values may be somewhat low, as it was not possible to locate and identify small isolated thalli among the vegetation.

Standing Crops

A summary of the results of the random sample cores (each figure representing an average of 3 to 12 samples) is given in Table 1. The highest standing crops (> 400 g m^{-2}) for all plants are found on the margins and slopes of the low-centered polygons, the grassy tops of the high-centered polygons, and the meadow tussocks. The lowest standing crop, by far, is found in the centers of the low-centered polygons. If only the lichens are considered, the greatest standing crops are found on the margins of the low-centered polygons where also a relatively large number of species is represented. The trough slopes of high-centered polygons and meadow tussocks also have high standing crops and species numbers. By far the lowest standing crops of lichens are found in the centers of low-centered polygons and the trough bottoms of high-centered polygons where conditions are very wet.

It is interesting that the highest number of lichen species was found on frost boils in high-centered polygons where the lichen, as well as the total, standing crop is of intermediate value. This could be the result of the more unstable condition of the surface which results in lower over-all plant cover. The cover of lichens (Figure 2) on margins of low-centered polygons is very high, as it is on the slopes of the margins. The percentage of lichens on the frost boil high-centered polygons is slightly greater than on grassy topped ones. This might indicate that lichens are more tolerant of surface disturbance than are mosses.

Low-centered polygons are consistently wetter (Table 2) than the high-centered polygons and meadow areas. It thus appears that lichens grow best in intermediate moisture conditions as represented by the drier parts of the low-centered polygons. However, they can tolerate low soil moisture conditions and have great species diversity in such places (Table 1). Lichens are almost absent from the centers and trough bottoms where standing water is frequently present. Such absence may be the direct result of the wetness or the indirect effect of anaerobic conditions.

Lichen Productivity: Direct Measurements

Using the direct method of dry weight measurements of the different morphologically distinct parts of the *Dactylina arctica* thalli and examining the chlorophyll content of these portions indicates that the new growth during the season accounts for an increase of 8% only. The percentage of new growth increased from 11% of the total dry weight on 22 June 1972 to 20% by 2 August, standing old growth remained constant at 35%, while the amount of decomposing material decreased from 54 to 45% for the same period. A chlorophyll analysis of a 27 June sample of the thallus divisions confirmed that most of the algae were in the tips and buds, though surprisingly 20% of the chlorophyll was in the decomposing material. This condition has obvious value in the propagation of the species, as the fragmenting decomposing material would be a source from which new buds could be dispersed. Indeed, by the latter part of the growing season many new thalli were observed to be budding from such decomposing fragments. Thallus fragments are probably of primary importance for lichen reproduction and dispersal in view of the fact that most lichens in the Barrow tundra lack sexual fruiting structures and are poorly developed or lacking in vegetative propagules, such as soredia.

The samples of *Cetraria richardsonii* attached to threads over a 22-day period generally increased in weight a very small amount. Eight of ten samples increased in weight. The other two decreased due to decomposition and/or loss of small fragments. On the average, each thallus increased in dry weight by 1.1%. Unfortunately, this method and that described for *Dactylina arctica* are not applicable to most other species due to different modes of growth (e.g., closely appressed to the substrate) or the lack of clear divisions between old and new growth. In any case, the results indicate that seasonal growth accounts for less than 8% of the weight of the plants.

Lichen Productivity: Respiration and Photosynthesis

The respiration rates increased in a linear fashion for all species tested, except *Cetraria cucullata* (Figure 3). The temperature response agrees quite well with the conclusions of Wager (1941) and James (1953) (cited by Kappen, 1973) that lichens as well as vascular plants from cold regions show higher respiration rates with increasing temperatures than those from moderate and warm climates. In contrast to the respiration response to temperature, photosynthetic rates were

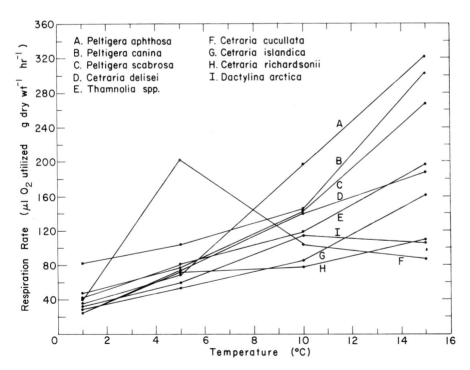

Figure 3.   Respiration rates for selected lichens of the Barrow area at different temperatures from 1 to 15°C (pooled readings), in the dark.

found to be highest between 5 and 10°C for *Cetraria cucullata, C. delisei,* and the three *Peltigera* species (Figure 4).

Scholander et al. (1952) found little evidence of temperature adaptation in a comparison of the respiration rates of arctic (Barrow) and tropical lichens. The only exceptions to this were species of *Peltigera* and *Sticta,* in which the arctic specimens had as high respiration rates at 10°C as did the tropical specimens at 20°C. The relatively high *Peltigera* photosynthetic levels observed in the present study appear to support the above conclusion that these species are metaboli-

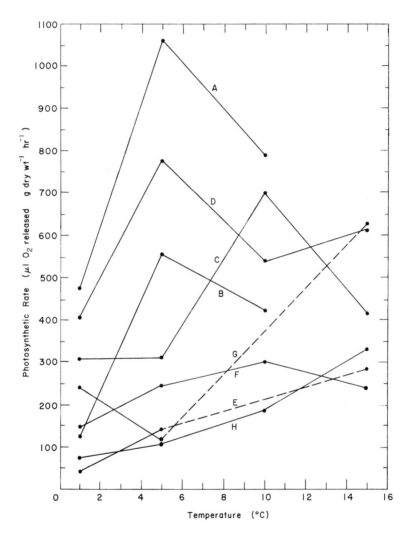

Figure 4. Photosynthetic rates for selected lichens of the Barrow area at different temperatures from 1 to 15°C (pooled readings). Light was maintained at 1.4×10^5 erg cm^{-2} sec^{-2}.

cally adapted to cold climates. The phenomenon may, however, be a result of the loose thallus structure of *Peltigera*, which permits CO_2 exchange to occur even when the thalli are fully saturated with water (Kappen, 1973).

The relative respiration rates of *Dactylina arctica* and the *Cetraria* and *Thamnolia* species correlate well with the values presented by Scholander et al. (1952) for the same species at Barrow and seem to support their conclusion that these species are not temperature adapted. *Cetraria cucullata*, which shows peak respiratory activity at 5°C, is the only apparent exception. Bliss and Hadley (1964) also found no evidence for temperature adaptation in a study of the photosynthetic and respiratory rates of alpine specimens of *Cetraria islandica*, *C. nivalis*, and *Cladina rangiferina*, but conclude that the three lichens are well adapted to a wet alpine environment with low light levels, conditions not unlike those of the Arctic. The peak photosynthetic activity of *Cetraria cucullata* at 5°C and of *C. delisei* at 10°C observed in the present study, however, supports the conclusion that these taxa as well as the *Peltigera* species are cold-temperature adapted.

Though the relative metabolic activities and the temperature responses of the lichens tested in the present study appear to agree with the results of previous work (Scholander et al., 1952; Bliss and Hadley, 1964; Schofield and Atlas, unpublished data), the absolute values may be questioned for several reasons. The samples may have been damaged somewhat during collection and transit, and it is possible that they may have been less active metabolically than at earlier times during the growing season. A considerable variation between replicates of some samples was also encountered, which may have been attributable to differences in the ratio of viable to inactive portions of the thalli included in each run. Thallus water content was maintained during the test period within the 150 to 250% of dry weight regime cited as optimal by Kallio and Heinonen (1971).

Though for the most part the above studies lack absolute evidence for cold-temperature adaptation by arctic lichens, the general observation of many workers is that the metabolism of most lichens is better adapted to cool or temperate rather than very warm climates (Kappen, 1973). This is dramatically supported by the work of Lange (1963, 1965) and Schulze and Lange (1968), in which lichens from various habitats in the world, including tropical rain forest, are able to photosynthesize at temperatures below 0°C. Kallio and Heinonen (1971) demonstrated assimilation of CO_2 at $-20°C$ by *Cetraria nivalis*, while *Nephroma arcticum*, *Cladina alpestris*, *Parmelia olivacea*, *Solorina crocea*, *Hypogymnia physodes*, and *Xanthoria parietina* showed net assimilation at -5 to $-15°C$. These authors also found rapid recovery of respiration in nearly all species tested after chilling to -20 to $-30°C$. *Nephroma arcticum*, *Solorina crocea*, and *Umbilicaria vellea* exhibited a peak of respiration between -10 and 0°C in the ascendant temperature phase after chilling. *Hypogymnia physodes* has been shown to synthesize chlorophylls and β-carotenes at $-7°C$ (Godnev et al., 1966, cited by Kappen, 1973). Respiration generally survives low temperatures better than photosynthesis (Lange, 1965), as is demonstrated by the detectable levels of respiration of *Dactylina arctica* and *Thamnolia vermicularis* at $-26°C$ (Scholander et al., 1952). Thus, the ability of lichens to resist freezing and to survive long periods of inactivity in a frozen state (Kappen, 1973), as well as their ability to photosynthesize at subzero temperatures and to recover photosynthetic capacities immediately after experiencing very low temperatures (Kallio and Heinonen, 1971), make them able to survive in an arctic environment.

The role of lichens in contributing to the seasonal CO_2 flux at Barrow can be considered minimal. For example, in site 2, lichens form only a small percentage of the total phytomass and even under optimal conditions have much lower respiration and photosynthesis rates than do the mosses and vascular plants. Due to the suboptimal thallus moisture experienced by lichens in the dry sites, in which they form a more significant portion of the phytomass, CO_2 flux by lichens is still slight because of their metabolic inactivity during most of the growing season.

Mineral Accumulation

Our analysis of lichens and their substrates (Tables 4 and 5) indicates considerable variation in amounts of mineral elements in both lichens and substrates. Certain elements, for example K, N, Mn, and Mg, accumulate to a significant amount in *Peltigera* spp. Others, for example Ca and Zn, accumulate more in other lichens. This may be a reflection of the high moisture in the *Peltigera* localities as compared to others. Compared to lichen data of other workers (Scotter, 1972; Scotter and Miltimore, 1973) from farther south in the Canadian Reindeer Preserve, Northwest Territories, our lichen contents appear relatively high. This suggests that lichens may be more important accumulators and trappers of some minerals in wet northern tundra areas.

Syers and Iskandar (1973) note that lichens are capable of accumulating minerals to surprisingly high levels. These authors cite as especially important the ability of lichens to accumulate P and to transform Ca into forms useful to subsequent plant forms. Such mineral-trapping effects, with release upon decomposition in forms useful to surrounding plants, would be of considerable import in nutrient-poor systems such as the tundra. N has been shown to be especially limiting in the tundra soils (Bliss, 1971; McKendrick et al., 1978). Lange and Ziegler (1963) found Fe and Ca levels in lichens to be above the tolerable levels for isolated fungi and algae. The reason for lichen tolerance is apparently because most of the mineral is deposited in an insoluble form on the thallus surface. It is possible that much of the mineral is bound in chelated form by the so-called "lichen acids" of the thallus (Williams and Rudolph, 1974). In this case, then, one can envisage an effective removal, or trapping, of certain minerals from the environment via the uptake of soluble ions in water entering the thallus. The removal of these ions by conversion to an insoluble form would in effect set up a diffusion gradient analogous to that observed across a cell membrane, where the rapid utilization of transported material results in continuous transport with no energy expenditure, i.e., by "facilitated transport." This system would cease uptake only when equilibrium is reached, or could switch to energy-dependent transport processes to further concentrate the ion or substrate. Launomaa (see Tuominen and Jaakkola, 1973) noted a concentration of Fe in dead basal segments of reindeer lichens that was two to three times higher than that found in the living tops. The Fe apparently came from the substrate. The translocation of nutrients within thalli from the bases to the tops of *Cladina rangiferina* was demonstrated by Barashkova (Tuominen and Jaakkola, 1973).

Lichens are also capable of concentrating minerals present in atmospheric moisture and dust. This has considerable significance as a result of the atmospheric testing of nuclear weapons over polar regions in recent years. Radionuclides, such as ^{137}Cs and ^{90}Sr, have been found to remain in lichens in concentrated form for some time. As this occurs in reindeer lichens (as well as other species) which are the base of the lichen–reindeer/caribou–man food chain, the nuclides have been found to be present in significantly high levels in Eskimos and Lapps, where they concentrate with time in the bone marrow. As exposure to such radiation is known to be a cause of various neoplastic diseases, such as leukemia, this problem had the potential to become one of considerable public health importance prior to the banning of atmospheric nuclear testing. An excellent discussion of the subject is provided by Tuominen and Jaakkola (1973).

Perhaps the most important role of lichens in mineral accumulation in the wet coastal tundra is the nitrogen fixation by the blue–green phycobionts in cephalodia of *Peltigera* species (Schell and Alexander, 1973). Indeed, these species have higher nitrogen contents than most vascular plants or mosses (T. Chapin, personal communication) (Table 4). Milbank and Kershaw (1969) report that virtually all of the nitrogen fixed by the algae is secreted into the remainder of the lichen thallus. It is reasonable to assume that much of this nitrogen is leached or otherwise lost to the underlying substrate. This is supported by the relatively high soil levels of nitrogen in sites colonized by the *Peltigera* (T. Chapin, personal communication), though free-living blue–green algae would also contribute fixed nitrogen in these areas (Schell and Alexander, 1973). The relatively high nitrogen level found in *Cetraria delisei* (Table 4), a nonfixing lichen common in many wet sites along with *Peltigera,* is a reflection of the high substrate nitrogen.

Milbank and Kershaw (1973) cite some 16 species of lichens as being N-fixers. Of these 16 species, 5 are *Collema,* 6 are *Peltigera,* and an unspecified number are *Stereocaulon* species. The latter 2 genera are of importance in Barrow tundra, while the first is common at Prudhoe Bay.

The N values of 3.0 and 3.3% of thallus dry weight reported by these authors for *Peltigera aphthosa* and *P. canina* agree with those of the present report. Though Milbank and Kershaw (1973) list a thallus moisture of at least 80 to 90% as the most important initial factor, a temperature greater than 0°C is required for significant nitrogenase activity and there is a distinct light dependency. As was the case for respiration and photosynthesis, N-fixation began rapidly with favorable conditions even after long periods of desiccation and dormancy. The nitrogenase activity of *Peltigera* species *in situ* at Barrow, then, may be considered high during the growing season due to adequate moisture in *Peltigera* habitats, the 24-hr daylight, and temperatures averaging about 3°C.

Conclusions

Lichen composition and biomass in the Alaskan coastal tundra is reasonably well correlated with substrate moisture. Lichens are a significant component of the tundra at Barrow, where under certain favorable conditions as much as 30% of the plant biomass may be lichens. They are usually absent from microhabitats where standing water is present for more than the first few weeks of the growing season. Few species, particularly those of the genus *Peltigera,* and low lichen standing crop, less than 2% of the plant material, are characteristic of the wetter sites. Many species, particularly in the genera *Alectoria, Cetraria, Cladonia, Dactylina,* and *Thamnolia,* and high lichen standing crops, up to 30% of the plant material, are found in drier sites. In laboratory tests, respiration rates of nine tundra lichen species increased linearly with increasing temperatures between 1 and 15°C, except for one *Cetraria* species whose respiration peaked at 5°C. Six of the nine species had peak photosynthetic rates at 10°C or lower, indicating high productivity at low temperatures. Some lichens, particularly those in the less well-drained sites, contain significant amounts of mineral nutrients—for example potassium, manganese, and zinc—and may serve as important nutrient traps. In many tundra areas lichens are a major vegetational component; however, in the Alaskan coastal tundra lichens are less dominant and productive than other plants and apparently are not major links in the food chain.

In the wet coastal tundra where nitrogen and other nutrients are limiting, lichens may play a significant role in fixing nitrogen and in trapping certain other elements which otherwise may be lost to the system. Subsequent decay, very slow for lichens, could release these essential nutrients over long periods rather than having them lost by run-off during the snow-melting period.

Acknowledgments. This research was supported by the National Science Foundation under Grant GV 33851 to Ohio State University. It was performed under the joint sponsorship of the International Biological Programme and the Office of Polar Programs and was coordinated by the U.S. Tundra Biome. Field and laboratory activities at Barrow were supported by the Naval Arctic Research Laboratory of the Office of Naval Research.

The Department of Botany and The Institute of Polar Studies, Ohio State University, provided support for the part of the work done in Columbus. We appreciated the helpful cooperation of other Tundra Biome Program investigators.

References

Billings, W. D., and H. A. Mooney. (1968) The ecology of arctic and alpine plants. *Biol. Rev.,* **43**: 481–529.

Bliss, L. C. (1971) Arctic and alpine plant life cycles. *Ann. Rev. Ecol. Syst.,* **2**: 405–438.

Bliss, L. C., and E. B. Hadley. (1964) Photosynthesis and respiration of alpine lichens. *Amer. J. Bot.,* **51**: 870–874.

Blum, O. B. (1973) Water relations. In *The Lichens* (V. Ahmadjian and M. E. Hale, Eds.). New York: Academic Press, pp. 381–400.

Britton, M. E. (1967) Vegetation of the arctic tundra. In *Arctic Biology: 18th Biology Colloquium* (H. P. Hansen, Ed.). Corvallis, Oregon: Oregon State University Press, 2nd ed., pp. 67–130.

Clebsch, E. E. C., and R. E. Shanks. (1968) Summer climatic gradients and vegetation near Barrow, Alaska. *Arctic,* **21**: 161–171.

Kallio, P., and S. Heinonen. (1971) Influence of short-term low temperature on net photosynthesis in some subarctic lichens. *Rep. Kevo Subarct. Res. Sta.,* **8**: 63–72.

Kappen, L. (1973) Response of lichens to extreme environments. In *The Lichens* (V. Ahmadjian and M. E. Hale, Eds.). New York: Academic Press, pp. 310–330.

Kärenlampi, L. (1971) Studies on the relative growth rate of some fruticose lichens. *Rep. Kevo Subarct. Res. Sta.,* **7**: 33–39.

Lange, O. L. (1963) Die Photosynthese der Flechten bei tiefen Temperaturen und nach Frostperioden. *Ber. Dtsch. Bot. Ges.,* **75**: 351–352.

Lange, O. L. (1965) Der CO_2 Gaswechsel von Flechten bei tiefen Temperaturen. *Planta,* **64**: 1–49.

Lange, O. L., and H. Ziegler. (1963) Der Schwermetallgehalt von Flechten aus dem *Acarosporetum sinopicae* auf Erzchlackenholden des Harzes. *Mitt. Florist.-soziol. Arbeitsgem.,* **10**: 156.

McKendrick, J. D., V. J. Ott, and G. A. Mitchell. (1978) Effects of nitrogen and phosphorus fertilization on the carbohydrate and nutrient levels in *Dupontia fisheri* and *Arctagrostis latifolia.* In *Vegetation and Production Ecology of an Alaskan Arctic Tundra* (L. L. Tieszen, Ed.). New York: Springer-Verlag, Chap. 22.

Milbank, J. W., and K. A. Kershaw. (1973) Nitrogen metabolism. In *The Lichens* (V. Ahmadjian and M. E. Hale, Eds.). New York: Academic Press, pp. 289–309.

Rastorfer, J. R., and N. Higinbotham. (1968) Rates of photosynthesis and respiration of the moss *Bryum sandbergiias* influenced by light intensity and temperature. *Amer. J. Bot.,* **55**: 1225–1229.

Rastorfer, J. R., H. J. Webster, and D. K. Smith. (1973) Floristic and ecologic studies of bryophytes of selected habitats at Prudhoe Bay, Alaska. *Inst. Polar Stud., Ohio State Univ., Rep.* 49, 20 pp.

Rastorfer, J. R. (1978) Composition and bryomass of the moss layers of two wet-meadow tundra communities near Barrow, Alaska. In *Vegetation and Production Ecology of an Alaskan Arctic Tundra* (L. L. Tieszen, Ed.). New York: Springer-Verlag, Chap. 6.

Rudolph E. D. (1971) Ecology of land plants in Antarctica. In *Research in the Antarctic* (L. O. Quam, Ed.). Washington, D.C.: Amer. Assoc. Adv. Sci. Publ. 93, pp. 191–222.

Santesson, J. (1967) Chemical studies on lichens. 4. Thin layer chromatography of lichen substances. *Acta Chem. Scand.,* **21**: 1162–1172.

Sato, M. (1963) Mixture ratio of the lichen genus *Thamnolia. Nova Hedwigia,* **5**: 149–155.

Schell, D. M., and V. Alexander. (1973) Nitrogen fixation in arctic coastal tundra in relation to vegetation and micro-relief. *Arctic,* **26**: 130–137.

Scholander, P. F., W. Flagg, V. Walters, and L. Irving. (1952) Respiration in some arctic and tropical lichens in relation to temperature. *Amer. J. Bot.,* **39**: 707–713.

Schulze, E. D., and O. L. Lange. (1968) CO_2-Gaswechsel der Flechte *Hypogymnia physodes* bei tiefen Temperaturen in Freiland. *Flora,* **158**: 180–184.

Scotter, G. W. (1972) Chemical composition of forage plants from the Reindeer Preserve, Northwest Territories. *Arctic,* **25**: 21–27.

Scotter, G. W., and J. E. Miltimore. (1973) Mineral content of forage plants from the Reindeer Preserve, Northwest Territories. *Can. J. Plant Sci.,* **53**: 263–268.

Showman, R. E. (1972) Residual effects of sulfur dioxide on the net photosynthetic and respiratory rates of lichen thalli and cultured lichen symbionts. *Bryologist,* **75**: 335–341.

Spetzman, L. A. (1959) Vegetation of the Arctic Slope of Alaska. Exploration of Naval Petroleum Reserve No. 4 and adjacent areas, northern Alaska, 1944–1953. Part 2, Regional studies. *U.S. Geol. Surv. Prof. Pap.,* **302-B**: 19–58.

Syers, J. K., and I. K. Iskandar. (1973) Pedogenic significance of lichens. In *The Lichens* (V. Ahmadjian and M. E. Hale, Eds.). New York: Academic Press, pp. 225–248.

Tuominen, Y., and T. Jaakkola. (1973) Absorption and accumulation of mineral elements and radioactive nuclides. In *The Lichens* (V. Ahmadjian and M. E. Hale, Eds.). New York: Academic Press, pp. 185–224.

Umbreit, W. W., R. H. Burris, and J. F. Stauffer. (1957) *Manometric Techniques.*
    Minneapolis: Burgess, 305 pp.
Wielgolaski, F. E. (1972) Vegetation types and plant biomass in tundra. *Arct. Alp. Res., 4*:
    291–305.
Wielgolaski, F. E. (1975) Primary productivity of alpine meadow communities. In *Fenno-
    scandian Tundra Ecosystems. Part I. Plants and Microorganisms* (F. E. Wielgolaski,
    Ed.). New York: Springer-Verlag, pp. 120–128.
Williams, M. E., and E. D. Rudolph. (1974) The role of lichens and associated fungi in the
    chemical weathering of rock. *Mycologia, 66*: 648–660.

8. The Role of Algae in Tundra Soils

R. E. CAMERON, A. D. KNOX, AND F. A. MORELLI

Introduction

Algal investigations of the Arctic and Subarctic have been undertaken primarily for aquatic environments, including lakes and shallow ponds (Croasdale, 1965; Foged, 1971; Hilliard and Tash, 1966; Kalff, 1967, 1970; Kling, 1972; Maruyama, 1967; Prescott, 1953, 1954, 1963a,b,c; Prescott and Vinyard, 1965). Although taxonomic synonymity must be considered (Prescott, 1964), between 3000 and 4000 species have been identified for the North American Arctic and Subarctic (Prescott, 1959, 1963a). They are a substantial component of the soil vegetal mass (Tikhomirov, 1971), and their importance at Barrow is recognized.

In the Soviet Arctic about 240 shallow pond algae have been recorded even though this represents a collection for only one summer in the vicinity of the Taimyr Station (Ermolaev et al., 1971). Most of these were diatoms and green algae, followed by blue–greens. Similar relative abundances also have been shown for a Canadian Subarctic lake (Sheath and Munawar, 1974). More variable results have been obtained for Alaskan Arctic environments, especially in terms of biomass contributions by the various phytoplankton components. In other lakes, diatoms may comprise much of the taxa (Hilliard, 1966; Hilliard and Tash, 1966). Most of the species found in Arctic lake phytoplankton can also be found during the winter in north temperate zone lowland and mountain lakes (Kalff, 1970). Furthermore, it has been stated that both the terrestrial and aquatic algal flora of the Arctic show little distinctive variety in different geographical areas, apart from local climatic effects, and the species are seldom exclusive to the Arctic (except possibly for some desmids) (Taylor, 1956).

In an earlier study of Arctic lake ecosystems (Prescott, 1959), it was assumed that water chemistry was highly important in determining the species distribution and the most important factor in delimiting the blue–green algae. Later studies have subsequently shown that both the species diversity and the morphologically dominating form are probably determined mainly by the physical and chemical state of lakes at all latitudes (Sheath and Munawar, 1974). Arctic lakes are well known to be low in nutrients, such as phosphates and nitrates (Prescott and Vinyard, 1965), and responses have been shown following the addition of phosphate.

Intermediate in habitat between aquatic and soil habitats are the algae of ice and snow, such as those found on Alaska's Columbia Glacier (Kol, 1942), and the cryoconite of Greenland. Within this frigid habitat, algae of various colors absorb heat from the sun to form a film of water from melted ice and snow. The cryoconite usually consists of fine-grained mineral matter bound together by the mucilagenous filaments of blue–green algae (Gerdel and Drouet, 1960), and the resulting community is composed not only of blue–greens, but also greens and diatoms, rotifers, and fungi (Prescott, 1972), which, in species composition and diversity (including bacteria), approach that of the tundra soil habitat. Certain of the green algae of these cryophyte habitats are recognized in cultures of peat and sandy soils not only in the Arctic, but also in the Antarctic and alpine areas (Akiyama, 1970). Some of the habitats are increased in organic matter content through the addition of pollen, insects, and even small dead birds, to make them even more similar to a tundra soil habitat (Charlesworth, 1957).

The Arctic soil algae have been much less studied than the aquatic algae in the same general areas. Reviews of earlier work, both freshwater and soil for various Arctic regions, have been given earlier (Taylor, 1956; Ross, 1956). Among the "subaerial" (soil surface) algae, *Prasiola* spp., *Gloeocapsa* spp., and *Nostoc* spp. were considered the most abundant, with the oscillatorioid *Phormidium* spp. and *Schizothrix* spp. also considered prominent. In Greenland, large masses of *Nostoc commune* were common (Taylor, 1956). In Iceland, a large variety of blue–greens and diatoms have been reported, but many were collected in the vicinity of populated areas (Boye Peterson, 1923, 1928).

Prior to the IBP, tundra soils were examined in the Soviet Arctic in the vicinity of the port of Tiksi, Bol'shoi Lyskhovskii Island, and Chukotsk Peninsula (Dorogostaiskaya, 1959). Fifty-eight species of algae, predominantly unicellular and colonial greens, were identified which were similar to others occurring widely throughout Eurasia, with the exception of blue–greens found around hot springs in the tundra (Dorogostaiskaya, 1959). Approximately 150 to 160 species of soil algae have been recorded for the various habitats in the tundra of Taimyr Station District and the limits of the Western Taimyr including blue–greens, greens and yellow–greens, and diatoms (Dorogostaiskaya and Sdobnitkova, 1973; Tikhomirov, 1974). Mainly yellow–greens and unicellular and colonial greens have developed as soil crusts on the spotted and polygonal tundra of the Yamal and Gudan Peninsulas, but *Nostoc commune* is widespread in the spotted tundra of Taimyr (Novichkova-Ivanova, 1972). *N. commune* is most certainly widespread in the arctic as well as the subarctic regions (Taylor, 1956). On the basis of previous work, it can be concluded that the primary algal flora of the soil include greens, blue–greens, yellow–greens, and diatoms in a variety of tundra habitats of Alaska, Canada, Greenland, Iceland, Swedish Lappland, Jan Mayen, Spitsbergen, the Faeroes, Novaya Zemyla, Franz Josef Land, and other areas of the Soviet arctic (Akiyama, 1970; Dorogostaiskaya and Novichkova-Ivanova, 1967; Gollerbach and Shtina, 1969; Kobayasi, 1969; Stutz and Cook, personal communication; Taylor, 1956).

In preparation for the present study of tundra soil algae, samples of soil were collected near the present Barrow sites in August 1968, as well as from the Putu Dunes on the Colville River, and from IBP sites 2 and 4 in August 1971. A total of

11 samples were collected from the surface 2 and 15 cm and these were analyzed in addition to a sample obtained form the Tuktoyaktuk Peninsula, Canada.[1] Partial results of these analyses, including not only algal abundance, but also bacteria, yeasts, molds, and soil analyses, have been published previously (Cameron, 1972a, 1974). Algae were found in all of the samples and they varied in numbers from 10^4 to 10^8 g^{-1} of wet weight of soil in the surface 2 cm to as few as 10 g^{-1} of wet weight in the subsurface. The algae, as obtained by culture methods, were predominantly greens, filamentous blue–greens, and diatoms. *Chlorococcum, Chlamydomonas,* and *Chlorella* spp. were prominent among the green algae. The small filamentous oscillatorioid blue–green, *Schizothrix calcicola,* which is widely distributed in many habitats throughout the world, was prominent in some of the samples, and *N. commune* also was present.

For the present study, the major objectives were:

1. To determine the composition of the algal community and to enumerate the species insofar as possible at selected sites and plots.

2. To determine seasonal changes in algal composition (groups) and biomass.

3. To determine algal abundance and distribution within the surface (moss) and subsurface (soil) layers.

4. In a related study, to determine the presence of kinds of algal associates, especially the protozoa.

Methods

Algal Abundance, Distribution, and Diversity

Soil samples were collected for a single season during the period of 8 June to 6 September 1973 in strata from the surface few centimeters to depths approaching 15 cm. The 26 plots included polygon rims, basins, troughs, and meadows. Samples were periodically collected in the field by aseptic technique using a hand trowel sterilized with a propane torch. Samples were then placed in presterilized "Whirlpak" plastic bags for processing. The samples were stored in a refrigerator at 4°C until prepared for incubation within 24 hr. Ten-gram (wet weight) samples of unsieved materials were then placed in 90 ml of sterile water in 99-ml milk dilution bottles to give a first dilution of 1:10. Subsequent dilutions were made by transferring 1 ml of diluent to successive 1:10 dilutions to obtain a terminal dilution of 10^{-6}.

The culture medium, originally formulated for soil bacteria (Allen, 1957), consisted of Thornton's "basal salts" of the following composition per liter of distilled water: KNO_3, 0.5 g; K_2HPO_4, 1.0 g; $MgSO_4 \cdot 7H_2O$, 0.2 g; $CaCl_2$, 0.1 g; NaCl, 0.1 g; $FeCl_3$, 0.002 g; and EDTA, 0.001 g. All of the samples were incubated under fluorescent lights at intensities of 25 to 50 ftc intensity and up to 210 days incubation before final enumerations were made.

After tubes or bottles showed positive indications of algal growth, aliquots of each dilution were subsequently examined with a light microscope. Positives for

---

[1] Canadian sample obtained January 1972, from Mr. Randy Gossen, University of Calgary.

growth were recorded at the highest dilution. Algae were identified by reference to standard text or works recognized for determining the various taxa (Collins, 1909; Drouet, 1959, 1968, 1973; Drouet and Daily, 1956; Hazen, 1902; Patrick, 1959; Patrick and Reimer, 1966; Prescott, 1931, 1951, 1954; Starr, 1955; Thompson, 1959; Tiffany and Britton, 1952; Whitford and Schumacher, 1969). In addition to algal abundance determined for each plot, identified algae were listed for each plot by depth. Small aliquots of each algal sample also were air-dried on plastic cover slips for future reference and study.

Algal Biomass Determinations

Soil samples for the estimation of algal biomass were collected at 5-day intervals during the early part of the season and at 10-day intervals during the latter part of the season. The samples were taken from representative habitats in five plots at depths of 0- to 1-cm "surface moss layer" and the 1- to 2-cm layer of "soil," although the actual degree of biotic and abiotic substrate varied with the habitat. Cores taken from the cooperative sites had a 5-cm diameter, but those taken from the wet meadow plots were taken with a smaller corer. Cores of all materials were placed in "Whirlpak" bags and returned to the Laboratory at NARL.

Plots sampled in cooperation with other investigators included:
Site 4, plot 440—polygon *trough,* aquatic moss-sedge.
Site 4, plot 441—polygon *rim,* dry with lichens.
Site 4, plot 442—low-centered polygon *basin,* saxifrage-sedge.
Site 2, plot 200—polygon *flat top,* low flat wet meadow, grass-sedge.
Site 12, plot 100—very wet *aquatic meadow,* mosses, including sphagnum, and sedges.

A 2.5-g aliquot of each wet sample was taken from the 0- to 1-cm and 1-cm segment of each core and subsequently blended in 20 ml of H_2O for 30 sec and then mixed with 20 ml of 1.2 percent agar solution. Samples were then pipetted into Neubauer haemacytometer chambers, and the algae were counted using a Whipple disk grid. Twenty-five fields on four slides were counted for a total of 100 fields. Algae were recorded according to three size categories within three taxa: (1) blue–greens, (2) "greens," but also consisting of yellow–greens and euglenoids, and (3) diatoms. Only live cells were counted, with the exception of diatoms, when all whole cells, including frustules, were enumerated.

Results and Discussion

Algal Abundance, Distribution and Diversity

Since the algae were collected essentially from plots varying in physical factors of topography, orientation, exposure, drainage, soil, and vegetation, these sites would be expected to vary in decomposition rates, primary production levels, organic matter accumulation, and nitrogen fixation capacity. There were also some variations in the abundance, distributions, and kinds of algae at these plots.

Major groups of algae enumerated either by the culture method or by direct microscopic observation included the greens (Chlorophyta), blue–greens (Cyanophyta), diatoms (Bacillariophyta), and yellow–greens (Xanthophyta). Of all the habitats sampled, there were at least 59 species present (Table 1). These included the following numbers per algal group: 14 small semiaquatic and epiphytic greens, 12 filamentous greens, 9 coccoid greens, 7 blue–greens, 12 diatoms, 4 euglenoids, and 1 yellow–green. Among the above taxa, the diatoms appeared to have the greatest abundance as determined by direct microscopic

Table 1. Algae Observed in Tundra Soils, Arranged by Groups

Greens
  Coccoids
    Carteria spp.
    Chlamydomonas spp.
    Chlorella ellipsoidea
    Chlorella vulgaris
    Chlorococcum minutum
    Gloeocystis spp.
    Planktosphaeria gelatinosa
    Protoderma viridae
    Spongiochloris spongiosa
  Filamentous
    Cylindrocapsa geminella
    Hormidium klebsii
    Hormidium subtile
    Microspora tumidula
    Microthammion strictissimum
    Raphidionema sempervirens
    Stichococcus bacillaris
    Stichococcus subtilus
    Stigeoclonium nannum
    Ulothrix cylindricum
    Ulothrix subtilissima
    Ulothrix tenuissima
  Small semiaquatics and epiphytes
    Ankistrodesmus braunii
    Ankistrodesmus falcatus
    Ankistrodesmus fractus
    Bumilleriopsis brevis
    Characiopsis longipes
    Characiopsis polycholoris
    Characium acuminatum
    Chlorocloster pirenigera
    Closteriopsis longissima
    Elakatothrix gelatinosa
    Elakatothrix nannum
    Elakatothrix viridis
    Quadrigula lacustris
    Scenedesmus quadricauda

Blue–greens
    Anacystis marina
    Anacystis montana
    Microcoleus lyngbyaceous
    Microcoleus vaginatus
    Nostoc commune
    Schizothrix arenaria
    Schizothrix calcicola
Diatoms
    Amphora spp.
    Diatomella spp.
    Eunotia spp.
    Fragilaria spp.
    Frustulia spp.
    Hantzschia spp.
    Hantzschia amphioxys
    Navicula spp.
    Nitzschia spp.
    Pinnularia spp.
    Rhopalodia spp.
    Synedra spp.
Euglenoids
    Astasia klebsii
    Euglena acus
    Euglena acuta
    Euglena elastica
Yellow–greens
    Botrydium granulatum

observation, whereas the greens were more abundant by the dilution culture method.

The green algae included at least 35 identifiable taxa, many determinable to the species level. In order of frequency observed in culture, the predominant green algae were *Chlamydomonas* spp., *Chlorella vulgaris,* and *Stichococcus bacillaris,* closely followed by *Chlorococcum minutum* and *Ulothrix subtilissima.* However, since the gametes of *Ulothrix* spp. resemble cells of *Chlamydomonas* spp., and the mitospores of *Chlamydomonas* spp. resemble the quadriflagellate *Carteria* spp., the *Ulothrix* spp. and related species may be more prevalent than realized (Hazen, 1902). *Prasiola* spp. were not collected at any of the sites or habitats, but it was noticed to form an observable cover in association with owl droppings (E. A. Schofield, personal communication). In the Prudhoe area, mats of *Prasiola crispa* were observed to form extensive green mats over caribou carcasses (Atlas et al., 1976). Species of this algae are irregularly distributed throughout the Arctic, as well as in some Antarctic habitats (Kobayasi, 1967, 1969).

Other algal groups were less frequently encountered. Blue–greens were found at 20 of the 26 habitats sampled, and these were predominately the filamentous oscillatorioid forms: *Schizothrix calcicola* (or its small-diameter ecophenes, e.g., *Plectonema nostocorum, Lyngyba diguetii, L. Lagerheimii*) (Drouet, 1968), and *Microcoleus vaginatus. Nostoc commune,* occurring as discrete macroplants on the soil surface, particularly in low areas, was not usually found in culture. Euglenoids and yellow–green algae were even less frequently encountered upon examination of cultures, but *Bumilleria* and *Tribonema* spp. were more commonly observed in the direct microscopic examination of samples. *Navicula* sp. was a fairly frequently observed diatom.

In a comparison of habitats, the polygon rims showed the greatest number of species, 37, representing 63% of all possible species (Table 2). As shown in Figure 1, percentage ranking of the species by habitat were rim, 37% > flat top, 33% > trough, 30% > basin = aquatic meadow, 25%. Ranking the habitats by diversity index ($h = \Sigma - \pi \log_2 \pi$) these were flat top, 4.68 > rim, 4.61 > trough, 4.56 > aquatic meadow, 4.36 > basin, 4.30; and ranking the habitats by proportions equitability ($E = h/\log_2 N$ for given number of species, n), aquatic meadow, 0.94 > flat top, 0.93 > trough = basin, 0.93 > rim, 0.89. These measures indicate that the species diversity was primarily a function of the number of species, rather than the distribution of abundance (as reflected by π = probability of finding the ith species). However, it must be considered that the estimate of π was rather crude, since it must be obtained by the number of replicas of the particular environment wherein the species appeared, and these results are therefore not surprising. The number of species present in a habitat is probably the best single integrative measure of diversity.

The distribution of species groups between habitats also is shown in Figure 1. The blue–greens have the greatest number of species in the rims, where they also have the greatest biomass based upon species composition. Their smallest species count and biomass were both in the basin habitat. Combining these estimates, it appears that the suitability of the habitats for the blue–green algae as

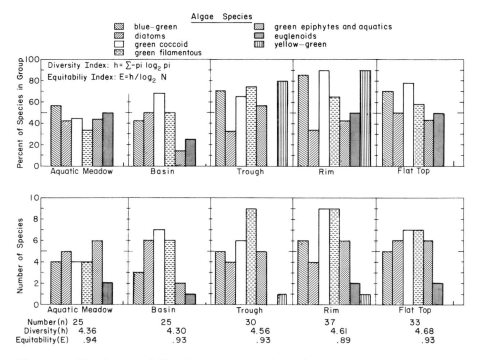

Figure 1. Abundance and diversity of algae species at five habitats.

measured by both species diversity plus seasonal average biomass was ordered as follows: rim > trough = aquatic meadow > basin.

For the green algae, species ranks were rim > trough > basin > aquatic meadow, whereas, according to biomass, it was trough > rim > aquatic meadow > basin. The greens were divided into three groups—the coccoids, filamentous, and semiaquatics plus epiphytes. The coccoid species were most numerous in rims and troughs and least in the aquatic meadows, whereas the semiaquatics plus epiphytes were equally numerous in rims and aquatic meadows plus flat tops, and least in the basins.

For the remaining algal groups, the euglenoids were represented by two species on the rims, aquatic meadows, plus flat tops, one in the basin, and none at all in the troughs, where light would be most limiting. The single species of culturable yellow–green algae was found only in troughs plus rims, although, as indicated previously, yellow–greens were more prevalent upon direct microscopic examination.

In terms of species diversity, the coccoid greens generally showed the largest percentage (25%), followed by filamentous greens (20%), with blue–greens and diatoms each accounting for about 15%, followed by semiaquatic plus epiphytic greens with about 12%, and the euglenoids with only 2 to 3% (Figure 2). The percentage distributions appeared to be consistent from habitat to habitat, with the following exceptions: (1) blue–greens dominated in the troughs (25%) and

Review references: Fourcans and Jain (1974) and Lenaz (1977).

Table 2.   Diversity of Algae Occurring in Tundra Sphagnum Plots

Plot 100—very wet aquatic meadow, including true mosses and sedges

Surface	Subsurface
Characium acuminatum	Not determined
Chlorella vulgaris	
Chlorococcum minutum	
Elakatothrix gelatinosa	
Microspora tumidula	
Navicula sp.	
Nostoc commune	
Schizothrix calcicola	
Stichococcus bacillaris	
Stigeoclonium nannum	

Plot 200—polygon flat top, low flat wet meadow, grass-sedge

Surface	Subsurface
Carteria sp.	*Carteria* sp.
Chlamydomonas sp.	*Characiopsis longipes*
Chlorella vulgaris	*Chlamydomonas* sp.
Chlorococcum minutum	*Chlorococcum minutum*
Cylindrocapsa geminella	*Cylindrocapsa geminella*
Elakatothrix gelatinosa	*Elakatothrix gelatinosa*
Elakatothrix viridis	*Elakatothrix viridis*
Fragilaria sp.	*Fragilaria* sp.
Gloeocystis sp.	*Gloeocystis* sp.
Microcoleus vaginatus	*Microcoleus vaginatus*
Navicula sp.	*Navicula* sp.
Nostoc commune	*Nostoc commune*
Planktosphaeria gelatinosa	*Planktosphaeria gelatinosa*
Schizothrix arenaria	*Schizothrix calcicola*
Schizothrix calcicola	*Stichococcus bacillaris*
Stichococcus bacillaris	
Ulothrix subtilissima	

Plot 440—polygon trough, aquatic moss-sedge

Surface	Subsurface
Characiopsis longipes	*Chlamydomonas* sp.
Characiopsis polycholoris	*Chlorella vulgaris*
Chlamydomonas sp.	*Chlorococcum minutum*
Chlorella vulgaris	*Elakatothrix gelatinosa*
Chlorococcum minutum	*Fragilaria* sp.
Diatomella sp.	*Microcoleus vaginatus*
Elakatothrix gelatinosa	*Microthamnion strictissimum*
Fragilaria sp.	*Navicula* sp.
Hormidium klebsii	*Nostoc commune*
Microcoleus vaginatus	*Planktosphaeria gelatinosa*
Microspora tumidula	*Schizothrix arenaria*
Navicula sp.	*Schizothrix calcicola*
Nostoc commune	*Stichococcus bacillaris*
Protoderma viridae	*Ulothrix subtilissima*
Schizothrix arenaria	
Schizothrix calcicola	
Stichococcus bacillaris	
Stichococcus subtilus	
Ulothrix subtilissima	

Table 2. Diversity of Algae Occurring in Tundra Sphagnum Plots

Plot 441—polygon rim, dry with lichens

Surface	Subsurface
Anacystis marina	*Carteria* sp.
Bumilleriopsis brevis	*Chlamydomonas* sp.
Carteria sp.	*Chlorella ellipsoidea*
Chlamydomonas sp.	*Chlorococcum minutum*
Chlorococcum minutum	*Cylindrocapsa geminella*
Cylindrocapsa geminella	*Elakatothrix gelatinosa*
Elakatothrix gelatinosa	*Gloeocystis* sp.
Fragilaria sp.	*Microcoleus vaginatus*
Gloeocystis sp.	*Navicula* sp.
Microcoleus vaginatus	*Schizothrix arenaria*
Navicula sp.	*Schizothrix calcicola*
Schizothrix calcicola	*Stichococcus bacillaris*
Stichococcus bacillaris	*Ulothrix subtilissima*
Ulothrix subtilissima	
Ulothrix tenuissima	

Plot 442—polygon low-centered basin, saxifrage-sedge

Surface	Subsurface
Amphora sp.	*Carteria* sp.
Astasia klebsii	*Chlamydomonas* sp.
Carteria sp.	*Chlorococcum minutum*
Chlamydomonas sp.	*Elakatothrix gelatinosa*
Chlorococcum minutum	*Elakatothrix viridis*
Cylindrocapsa geminella	*Fragilaria* sp.
Elakatothrix gelatinosa	*Navicula* sp.
Euglena acus	*Schizothrix calcicola*
Fragilaria sp.	*Stichococcus bacillaris*
Microcoleus vaginatus	*Ulothrix subtilissima*
Microspora tumidula	
Navicula sp.	
Schizothrix calcicola	
Stichococcus bacillaris	
Stigeoclonium nannum	
Ulothrix subtilissima	

were very low in the rims (10%); (2) diatoms were particularly dominant (in terms of number of species) in the rims (35%); and (3) euglenoids were totally undetected in the troughs (Figure 2).

Analysis of species diversity by depth showed that there were more species in the top 0 to 1 cm than appear in the 1 to 2 cm depth (or lower), depending on the habitat (Figure 2). Also, there are a few species which do not appear in the top 1 to 2 cm; the number of these varies from two to five, again depending on the habitat. However, most species are found at both depths. The percentage overlap by depth, including both 0- to 1- and 1- to 2-cm layers, was as follows: rim, 59%; basin, 60%; trough, 67%; aquatic meadow, 52%; and flat top, 70%.

Algae were found at all sites but not at all depths during the period of sampling, especially in the subsurface "soil" layer. Abundance ranged from 10 to 10^6 g^{-1} of wet material. There appeared to be an increase in abundance of algae at some of the habitats at the "end" of the season. Whether or not this was a real

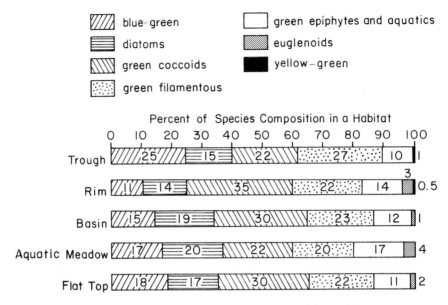

Figure 2.   Percentage composition by algal group in each of five habitats.

phenomenon was not readily apparent because the increased incubation period may have allowed for an increase in the total numbers of algae as well as an increase in species diversity as we have found for the accompanying protozoa.

Algal Biomass

Algal biomass was determined for representative habitats based upon five plots with samples taken at 0- to 1- and 1- to 2-cm depths at intervals of about 5 days. Both the algal biomass and constituent groups of organisms were determined by a direct microscopic counting technique. The organisms were grouped within three taxa. In addition to differences between sites, there were significant changes through the season with groups generally showing synchronous patterns.

Both the blue–green and "green" algae produced the highest mean seasonal biomasses (0.116 and 0.061 g m^{-2}, respectively) calculated on a dry weight basis for the very wet aquatic meadow. The lowest mean seasonal biomasses (0.015 and 0.017 g m^{-2}, respectively) were produced in the polygon basin. For other habitats biomasses of both taxa decreased from polygon trough to polygon rim to flat top. In consideration of the blue–greens, their mean seasonal cell number was greatest (406 $\times$ 10^6 m^{-2}) in the wet meadow. For the "greens" the mean seasonal cell number was highest for the rim habitat (290 $\times$ 10^6 m^{-2}). Both the blue–greens and "greens" had the lowest mean numbers in the basin (117 and 145 $\times$ 10^6 m^{-2}, respectively). For both taxa over 60% of the mean seasonal biomass occurred in the 0- to 1-cm depth, except for the rim habitat, where the reverse situation was noted for the blue–greens.

Another pattern emerged when only the diatoms were considered. Their highest mean seasonal biomass (0.600 g m^{-2}) occurred in the basin, and the lowest biomass (0.120 g m^{-2}) occurred in the low, very wet aquatic meadow habitat of frequent standing water, and it decreased from rim to trough to flat top. In all the habitats, 63 to 68% of the diatom biomass occurred at the 1- to 2-cm level. When mean seasonal cell numbers are considered (Figure 3), diatoms occurred in decreasing order in the basin (500 × 10^6 m^{-2}), rim (493 × 10^6 m^{-2}), aquatic meadow (225 × 10^6 m^{-2}), flat top (154 × 10^6 m^{-2}), and trough (118 × 10^6 m^{-2}) habitats, reflecting the differences in habitat and environmental factors of temperature, moisture, and concurrent phenomena. The diatoms, as well as the yellow–greens, are more susceptible to drought, higher temperatures, and salt (Lund, 1962). When all taxa are grouped together (Fig. 3), mean seasonal biomass decreases from basin (0.600 g m^{-2}) to rim (0.280 g m^{-2}) to trough (0.205 g m^{-2}) to flat top (0.172 g m^{-2}) to the even moister aquatic meadow (0.120 g m^{-2}), paralleling the diatom pattern. Mean seasonal cell numbers decreased from rim (995 × 10^6 m^{-2}) to flat top (804 × 10^6 m^{-2}) to basin (785 × 10^6 m^{-2}) to trough (559 × 10^6 m^{-2}) to the aquatic meadow (487 × 10^6 m^{-2}) habitat.

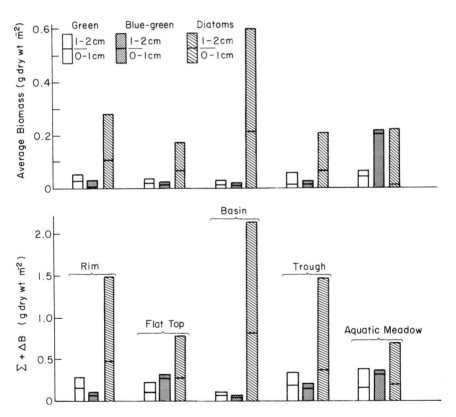

Figure 3. Mean seasonal biomass by algal group for each of five habitats and the summation of positive changes in biomass.

At various times during the study of algal biomass at the five habitats, three different methods were used to determine the numbers of algal cells. At mid-season, samples from all the representative habitats were collected, and the algae were enumerated using all three methods. For these samples, the direct counts were higher than any of the MPN (Most Probable Number) counts in the serial dilutions, probably indicating the influence of total numbers of diatoms, both living and dead, in the direct microscopic counts. Direct counts on a gram dry weight basis ranged from a maximum of 2.7×10^6 cells at the moistest habitat to a minimum of no observable cells in a wet meadow sample. The culture method, using 1:10 serial dilutions in Thornton's salt medium, showed a minimum of 10 g^{-1} of sample to a maximum of 10^8 g^{-1} of sample.

Algal-Abiotic Relationships

The algae are involved in cycling of tundra soil nutrients, especially nitrogen and phosphorus. These are the main elements known to stimulate the growth of algae (Lund, 1962), although they are not as important for the soil diatoms (Lund, 1945). There were variations in ammonia, nitrate, and phosphorus in the tundra, and these nutrients, being available to the algae, would in turn be reflected in fluctuations in algal abundances. In a polygonal basin, for example, both ammonia and nitrate showed mid-season increases, but for the rim, these increased in late season, while in a trough, ammonia was high early in the season and lower at mid- and late season. Nitrate showed peaks opposite to temperature optima. A high phosphorus level may be found in the trough in the organic layer to give a high fertility level with a correspondingly lusher vegetation, but with a much higher concentration in the trough. This was not reflected in the macrovegetation, but in the algae. However, increases in phosphorus concentration did not correspond to temperature optima or to elevations in algal biomass in either the trough or basins during the season.

Organic nutrients may have a greater effect on the algae in terms of fertilization than inorganic nutrients (Ketchum, 1951). Eutrophication may delimit the diversity of species (Sheath and Munawar, 1974). The presence of euglenoids in some of the samples (although not in the troughs) might indicate a response to organics, such as from lemming droppings. Also, the predilection of some yellow–greens for fecal materials, as well as their preference for shade and subsequent abundance a few centimeters below the surface (Lund, 1962), may explain their absence form a number of the cultured samples since they were not organically enriched. *Chlorella* spp. and *Chlamydomonas* spp., two of the most frequently observed and abundant green algae in our cultures, also have a predilection for organic materials and, along with *Scenedesmus* spp., are common algae of sewage treatment plants (Doyle, 1964). Aerobic bacterial breakdown products of sewage are also primarily photosynthetic requirements for these algae (Doyle, 1964), and the same conditions can be applied to the tundra, and to include the small green algae, *Elakatothrix* spp.

The blue–greens, especially the oscillatorioid species, *Schizothrix* spp. and *Microcoleus* spp. (or their ecophenes), also can thrive in fecally contaminated habitats, and they were the most prevalent blue–green forms in the Barrow soil ecosystem. Whether or not there was an increase in blue–greens and greens during mid-summer, with a reduction in diatoms, as shown for a subarctic lake when it changes from oligotrophic to eutrophic status, was not known for the tundra soil ecosystem.

The algae not only derive their nutrients from the soil, but they return nutrients to the ecosystem upon their death and subsequent decay or by release of metabolites from their cells. Both the greens and blue–greens can release polysaccharides or polypeptides into solution and produce growth substances and antibiotics. The nitrogen-fixing *N. commune* undoubtedly release fixed nitrogen into the soil solution. Some of the algae are known to benefit from the CO_2 supplied by live as well as decaying roots (De and Sulaiman, 1950), and there was a significant amount of CO_2 production in the wet meadow which also may be attributed to microbial respiration and metabolism. Production of CO_2 very possibly may be one of the factors contributing to the more frequent occurrence of discrete macroplants of *N. commune* in low, wet areas in the tundra, and in damp interpolygonal troughs, where abundant mats of *Nostoc commune* actively fixed nitrogen (Alexander et al., 1978). Although not specifically mentioned in a tundra soil vehicular perturbation study (Challinor and Gersper, 1975), the effects of tundra depression, with consequent changes in microtopography, temperature, nutrient status, and drainage, favor the growth of mats of *N. commune*.

With regard to soil pH there were no obvious correlations with the soil algae, and, although acid, the presence of blue–greens in all plots showed that it was not detrimental, even though the blue–greens would tend to be near or at the surface when the pH was not too low. Nearly neutral or alkaline soils have the most species, but not necessarily the largest crop; most blue–greens prefer neutral or alkaline soils, but the green algae have a wider range of pH tolerance and may be found in highly acid soils where the blue–greens and diatoms may be absent (Lund, 1962). As the algae utilize the nitrates, phosphates, and CO_2, the pH of the soil solution should increase. In this same environment the number of algae may increase while those of the bacteria decrease. The pH was negatively correlated with bulk density and positively correlated with inorganic phosphorus in solution (Barél and Barsdate, 1978), and there may therefore be an indirect correlation with the abundance, distribution, or kinds of algae. For the three polygon habitats sampled for algal biomass, the bulk density was highest for the rim, next for the basin, and least for the trough (Miller et al., 1978). As indicated previously, the percentage distribution of blue–greens was highest in the troughs and lowest in the rims, whereas the diatoms were particularly dominant in the rims.

Grazing was evident by protozoa and rotifers observed in algal cultures and a number of nematodes also were observed by direct microscopic observation of samples collected for biomass determinations. Grazing may have some effect on the periodicity of the algae, with ingestion of smaller genera, such as *Stichococcus, Elakatothrix,* and *Navicula* spp. Some protozoa may be selective for algal species as for the bacteria, and the blue–greens may not be as subject to grazing as the other taxa (Round, 1965), although fragments and whole cells of blue–greens were observed in some protozoa, e.g., *Amoeba* spp. Fungal activity and diseases also will have an effect and those of Chytridiaceous and Phycomycetous fungi may particularly attack desmids (Round, 1965). No desmids were encountered in our samples, although based on observations in other organic soils in polar regions, they should be expected, and it has been stated that they constitute about 75% of the freshwater Arctic flora (Prescott, 1959).

In the aquatic ecosystem at Barrow, the desmids were a significant proportion of the green algal species (Alexander and Coulon, personal communication). Different filamentous greens were found in the aquatic as compared to the terrestrial habitat, and, surprisingly, "planktonic" forms, *Quadrigula lacustris* and *Scenedesmus quadricauda,* were cultured from the wet meadow plots, but not listed for the aquatic habitats. *Ankistrodesmus* spp., *Chlamydomonas* spp., and *Elakatothrix lacustris* were found in both aquatic and terrestrial habitats. Different filamentous greens were found in the aquatic as compared to the terrestrial habitats. Coccoid greens, both *Chlorella* and *Chlorococcum* spp., while found in nearly all cultures of terrestrial samples, were noticeably absent from the aquatic environment. For the diatoms, *Navicula* spp. were the most prevalent in both aquatic and terrestrial ecosystems, and *Fragilaria* spp. also were abundant in both ecosystems.

A greater diversity of blue–greens was observed in the aquatic habitat, and these were predominately aquatic, coccoid forms, whereas the oscillatorioid forms were more predominant in the terrestrial ecosystem. *Nostoc* spp. were not reported for the aquatic ecosystem although *Anabaena* sp. was present, and *N. commune* has a planktonic Anabaena-like ecophene in aquatic environments of ponds, lakes, and streams. In terrestrial environments, *N. commune* was found in both drier and moister habitats, although forming larger plants in the moister habitats, and sometimes forming microscopic plants, such as the ecophenes of *N. microscopicum* in the drier habitats. The more typical macroplants were formed in the wet meadows and troughs where shallow standing water was available for a sufficient time for formation of discrete spreading mats. As in other harsh environmental areas, *N. commune,* because it is less mobile in mat form, tends to become parasitized and lichenized, and, as a "climax" form, it stabilizes as the more desiccative-resistant *Peltigera* and *Sterocaulon* spp. *N. commune* is world-wide in distribution, forming mat-like growth in the tropics as well as in the Antarctic (Cameron, 1972b; Cameron and Blank, 1966).

Temperature and moisture are two obviously important factors for the algae, and especially when considered together as the "solclime" to give favorable conditions of warm temperatures and available water. In the basin, 10-day mean temperatures were only slightly higher than for the trough and showed an elevated pattern at the end of the season. The moisture level was much lower in the basin than in the trough, decreasing in mid-season and showing similar levels at the beginning and the end of the season with an indication of increasing moisture in late season. The algal groups did not show any definite concurrent responses to slight changes in the moisture status. When moisture was sufficient early in the season, there was an excellent response to temperature by all algal groups. There was a slight response by blue-greens at the 0- to 1-cm depth and "greens" at the 1- to 2-cm depth in response to three 10-day mean peaks of temperature, and the diatoms noticeably responded to the mid-season temperature peak. At the end of the season, diatoms at both levels appeared to be responding to a slight increase in moisture.

For the rim habitat, a general pattern of algal group response was not particularly evident for moisture or temperature. Moisture levels were generally low except at the very end of the season, when algal abundance was also low and the immediate response was not apparent. There was an initial low response by "greens" and blue–greens. Although the abundance level of the "greens" at the 1- to 2-cm depth was less than for diatoms and blue–greens, there were low responses by the "greens" to 10-day running mean temperature peaks. The diatoms showed the closest correlation with daily mean temperatures in the polygon trough in early to mid-season when moisture also was available. Later in the season, when moisture was a limiting factor, variations in temperature were not reflected in changes in numbers of diatoms. The "greens," and also the blue–greens, generally responded slower in both time and numbers than the diatoms, but they showed more of a positive response to temperature later in the season than the diatoms.

For the wet meadow habitats, where moisture was not limiting, the responses by all algal groups to temperature peaks were more evident than for the sites on the polygon rim, basins, and trough. At the 0- to 1-cm depth the abundances were blue–greens followed by "greens" and diatoms. At the 1- to 2-cm depth, it was blue–greens followed by diatoms and "greens." The blue–greens showed a pronounced decrease in abundance at the beginning of the season, but the other two groups also showed a less pronounced decrease.

Comparison with Other Biomes

In the Antarctic, where the environmental conditions approach much harsher conditions than for the Arctic, soils are generally much less developed, dry, and either barren or irregularly spotted by a few cryptogams; there is a negligible quantity of organic matter, some salt accumulations, higher pH values, and, consequently, a lower diversity of species and lower abundance of algae (Cameron, 1972b). In culture the algae may be undetectable or approach abundance levels of 10^7 g^{-1} of dry weight of soil. The algae are commonly coccoid and oscillatorioid blue–greens in association with chlorococcoid greens (Cameron, 1972b), although, in lower latitudes of the Antarctic, a chlorophycean dominancy may be evident (Akiyama, 1970). In lower Antarctic latitudes, the numbers of blue–greens also may rival those of the yellow–greens (Akiyama, 1970). In the moister and "warmer" areas the filamentous green alga *Prasiola* sp. may be found on sand and rocks (Holm-Hansen, 1964; Kobayasi, 1969). *Prasiola crispa* has been noted to form an abundant ground cover on wet surfaces in the Antarctic (Rudolph, 1966), and it seems to do well in areas contaminated by penguin guano or that of other sea birds. In other geographical areas some species of *Prasiola* grow only in habitats rich in nitrogen compounds, or in cold, swiftly flowing streams (Smith, 1950).

In temperate grassland regions the blue–green and coccoid green algae are the predominant forms (Durrell, 1959), and they can number between 10^3 to 10^4 g^{-1} of soil (Thayer, 1972). In drier regions of the temperate zone, blue–greens and coccoid greens also are the predominant forms, and, although abundance levels may approach more than 10^7 g^{-1} of soil, other soil samples may not show any recoverable algae (Cameron and Blank, 1966). Algal biomass has been shown to vary from 96 to 239 kg/ha in a sagebrush desert area (Lynn and Cameron, 1973). Among the various taxa, *Chlorococcum* spp. may be the most widely distributed algal species in the world (Prescott, 1954), but it has been stated that the oscillatorioid blue–greens are the most common terrestrial forms in the Antarctic (Drouet, 1962). *Schizothrix calcicola* or its ecophenes is the most frequently observed algal species, it appears to be the most common blue–green in the Arctic tundra soil ecosystem, and it also furnishes the largest algal biomass in temperate desert soils (Cameron, 1972b) as well as in the grassland soils.

In the Antarctic *N. commune* is generally a "hydrophilic" species, occurring in low, moist areas as in the tundra, and it is recognized to be a nitrogen-fixing microorganism in that locality (Holm-Hansen, 1963). As a phycobiont of the lichens *Stereocaulon* and *Peltigera* spp., it has shown a significant rate of nitrogen-fixation in tundra soils, especially with an increase in temperature (Schell and Alexander, 1973). In nonpolar areas *N. commune* also is important in nitrogen-fixation, occurring in diverse habitats such as paddy fields, in partially shaded areas between grasses and shrubs, and on dry, barren desert soils (Shields and Durrell, 1964).

Conclusions

Green, blue–green, yellow–green, and diatom algal floras are widely distributed throughout the surface of the tundra soil ecosystem, both in the 0- to 1-cm "moss" layer and in the 1- to 2 (or more)-cm "soil" layer. They exist in a community of microorganisms including bacteria, protozoa, rotifers, nematodes, yeasts, molds, and other microflora and microfauna in addition to the macro biota.

Algae were recovered from all of the various 26 plots and habitats throughout the single season that the study was undertaken, but not for each time of collection, especially in the subsurface layers. By means of the culture method utilizing Thornton's basal salt solution, an abundance as high as 10^8 g^{-1} and frequently 10^6 g^{-1} of wet sample was obtained, with the higher values generally produced at the "end" of the season in early September. In general, the total number of algae was as high, or a log unit or two lower, in the subsurface "soil" layer as compared to the surface "moss" layer.

The terrestrial Arctic algae included at least 59 identifiable taxa, including 14 small semiaquatic and epiphytic greens, 12 filamentous greens, 9 coccoid greens, 7 blue–greens, 12 diatoms, 4 euglenoids, and 1 yellow–green. Among the above taxa, the diatoms appeared to be in largest abundance as determined by direct microscopic examination, and the greens as obtained by dilution culture. Prominent species among the greens included *Chlamydomonas* spp., *Chlorella vulgaris, Stichococcus bacillaris, Chlorococcum minutum,* and *Ulothrix subtilissima.* A small *Navicula* sp. was most frequently observed among the diatoms, and the small oscillatorioid blue–green, *Schizothrix calcicola,* was commonly cultured form many of the habitats. *N. commune,* a well-known nitrogen-fixing blue–green, was observed in discrete mat forms in the field, but it was not easily cultured in the laboratory. A number of the same species encountered in tundra soils are found in many other biomes, including the Antarctic.

  Calculated on a dry weight basis, both blue–green and "green" algae produced the highest mean seasonal biomasses (0.116 and 0.061 g m^{-2}) in the very wet aquatic meadow—a low area of standing water and sphagnum moss. The lowest mean seasonal biomasses (0.015 and 0.017 g m^{-2}, respectively) were obtained for the polygon basin. For the remaining habitats, the biomasses of both taxa decreased from the polygon trough to rim to wet meadow.

The diatoms produced the highest mean seasonal biomass of all the groups (0.600 g m^{-2}), and this occurred in the basin, lowest in the very wet meadow, and usually at the 1- to 2-cm level. The mean seasonal cell number for blue–greens was greatest in the wet meadow, greatest for the "greens" in the rim habitat, and greatest for the diatoms in the basin, possibly indicating a "niche preference" by the various groups.

The algal groups did not show any definite concurrent responses to slight changes in the moisture status, but, when moisture was sufficient early in the season, there was an excellent response to temperature early in the season by all the groups, and, at the end of the season, diatoms at both depths appeared to be responding to a slight increase in moisture. For the wet meadow, where moisture was not limiting, the responses by all algal groups to temperature peaks were more evident than for the habitats on the polygon rim, basin, and trough.

Acknowledgments. This project was performed under the joint NSF sponsorship of the International Biological Programme and was directed under the auspices of the U.S. Tundra Biome. Field and laboratory activities at Barrow were supported by the Naval Arctic Research Laboratory of the Office of Naval Research.

The research was supported primarily by the National Science Foundation via NASA Contract NAS7-100 under NASA codes 382-04-00-01-55 and 382-10-00-01-55 to the Jet Propulsion Laboratory, California Institute of Technology, Pasadena, California. It also was supported under NSF Grant GV 29342 to the University of Alaska, through a subcontract to Virginia Polytechnic and State University, and through purchase order No. 82693 from the University of Alaska. Additional funds and the principal investigator's time were supplied by the Darwin Research Institute, to complete the laboratory and necessary final report.

Special "thanks" are extended to Dr. Robert E. Benoit, Dr. Jerry Brown, Dr. Gary Laursen, Dr. Fred Bunnell, Pille Bunnell, Dr. George Llano, Dr. Orson Miller, Dr. Edmund A. Schofield, and Don Sanders for their direct or indirect assistance to our project.

References

Akiyama, M. (1970) Some soil algae from the arctic Alaska, Canada, and Greenland. *Proc. Shimane Univ. Natur. Sci., Mem. Fac. Educ.,* **4**: 53–75.

Alexander, V., M. Billington, and D. M. Schell. (1978) Nitrogen fixation in arctic and alpine tundra. In *Vegetation and Production Ecology of an Alaskan Arctic Tundra* (L. L. Tieszen, Ed.). New York: Springer-Verlag, Chap. 23.

Allen, O. N. (1957) *Experiments in Soil Bacteriology.* Minneapolis: Rev. Burgess Publishing Company, 3rd ed., 117 pp.

Atlas, R. M., E. A. Schofield, F. A. Morelli, and R. E. Cameron. (1976) Effects of petroleum pollutants on arctic microbial populations. *Environ. Pollut.,* **10**: 35–43.

Barèl, D., and R. J. Barsdate. (1978) Phosphorus dynamics of wet coastal tundra soils near Barrow, Alaska. In *Environmental Chemistry and Cycling Processes* (D. C. Adriano and I. L. Brisbin, Eds.). Dept. of Energy. Symposium Series. CONF.-760-429 (in press).

Boye Peterson, J. (1923) The fresh-water cyanophyceae of Iceland. In *The Botany of Iceland* (L. K. Rosenvinge and E. Warming, Eds.). Copenhagen: J. Frimodt, Vol. 2, part 2, pp. 250–324.

Boye Peterson, J. (1928) The aerial algae of Iceland. In *The Botany of Iceland* (L. K. Rosenvinge and E. Warming, Eds.). Copenhagen: J. Frimodt, Vol. 2, part 2, pp. 329–447.

Cameron, R. E. (1972a) A comparison of soil microbial ecosystems in hot, cold and polar desert regions. In *Ecophysiological Foundation of Ecosystems Productivity in Arid Zones* (L. E. Rodin, Ed.). Leningrad: Nauka, pp. 185–192.

Cameron, R. E. (1972b) Ecology of blue–green algae in antarctic soils. In *Symposium on Taxonomy and Ecology of Blue–Green Algae* (T. V. Desikachary, Ed.). Madras: Bangalore Press, pp. 253–386.

Cameron, R. E. (1974) Application of low-latitude microbial ecology to high-latitude deserts. In *Polar Deserts and Modern Man* (T. L. Smiley and J. H. Zumberge, Eds.). Tucson: University of Arizona Press, pp. 71–90.

Cameron, R. E., and G. B. Blank. (1966) Desert algae: Soil crusts and diaphanous substrata as algal habitats. *Jet Propulsion Lab. Tech. Rep.* 32-971, 15 July, Pasadena, 41 pp.

Challinor, J. L., and P. L. Gersper. (1975) Vehicle perturbation effects upon a tundra soil–plant system. II. Effects on the chemical regime. *Soil Sci. Soc. Amer. Proc.,* **39**: 689–695.

Charlesworth, J. K. (1957) *The Quaternary Era.* London: Edward Arnold, Vol. 1, pp. 1–60.

Collins, F. S. (1909) The green algae of North America. *Tufts Coll. Stud.,* **2**: 1–480.

Croasdale, H. (1965) Desmids of Devon Island, N.W.T., Canada. *Trans. Amer. Microsc. Soc.,* **84**: 301–335.

De, P. K. and M. Sulaiman. (1950) Fixation of nitrogen in rice soils by algae as influenced by crop, CO_2, and inorganic substances. *Soil Sci.,* **70**: 137–151.

Dorogostaiskaya, E. V. (1959) On the problem of soil algae in the far north. *Bot. Zh.,* **44**: 312–321 (in Russian).

Dorogostaiskaya, E. V. and L. N. Novichkova-Ivanova. (1967) On the changes in the algal flora of tundra soils resulting from their reclamation. *Bot. Zh.,* **52**: 461–468 (in Russian).

Dorogostaiskaya, E. V., and N. V. Sdobnitkova. (1973) Soil algae of the western Taimyr tundras (Pochvennye vodorosti tundr zapadnogo Taimyra). In *Biogeotsenozy Taimyrskoi Tundry i ikh Produktivnost' (Biogeocenoses of Taimyr tundra and their productivity).* Leningrad: Nauka, Vol. 2, pp. 128–138 (in Russian with English abstract).

Doyle, W. T. (1964) *Nonvascular Plants: Form and Function.* Belmont: Wadsworth Publishing Company, Inc., pp. 61–62.

Drouet, F. (1959) Myxophyceae. In *Freshwater Biology* (W. T. Edmondson, Ed.). New York: John Wiley and Sons, 2nd ed. pp. 95–114.

Drouet, F. (1962) The oscillatoriaceae and distribution in Antarctica. *Polar Rec.*, **11**: 320–321.

Drouet, F. (1968) *Revision of the Classification of the Oscillatoriaceae*. Monograph 15, The Academy of Natural Sciences, Philadelphia. Lancaster, Pa.: Fulton Press, 370 pp.

Drouet, F. (1973) *Revision of the Nostocaceae with Cylindrical Trichomes*. New York: Hafner Press, 292 pp.

Drouet, F., and W. A. Daily. (1956) Revision of the coccoid myxophyceae. *Butler Univ. Bot. Stud.*, **12**: 1–218.

Durrell, L. W. (1959) Algae in Colorado soils. *Amer. Midl. Natur.*, **61**; 322–328.

Ermolaev, V. I., G. D. Levadnaya, and T. A. Safonova. (1971) The algal production of some small lakes and pools in the region of the Taimyr Station. In *Biogeotsenozy Taimyrskoi Tundry i ikh Produktivnost' (Biogeocenoses of Taimyr tundra and their productivity)*. Leningrad: Nauka, pp. 116–129 (in Russian with English summary).

Foged, N. (1971) Diatoms found in a bottom sediment sample from a small deep lake on the Northern Slope, Alaska. *Nova Hedwigia*, **21**: 923–1035.

Gerdel, R. W., and F. Drouet. (1960) The cryoconite of the Thule area, Greenland. *Trans. Amer. Microsc. Soc.*, **79**: 256–272.

Gollerbach, M. M., and E. A. Shtina. (1969) *Soil Algae*. Leningrad: Nauka, 228 pp. (in Russian).

Hazen, T. E. (1902) The Ulothricaceae and Chaetophoraceae. *Mem. Torrey Bot. Club*, **11**: 1–250.

Hilliard, D. K. (1966) Studies on chrysophyceae from some ponds and lakes in Alaska. Five notes on the taxonomy and phytoplankton in an Alaskan pond. *Hydrobiologia*, **28**: 553–576.

Hilliard, D. K., and J. C. Tash. (1966) Freshwater algae and zooplankton. In *Environment of the Cape Thompson Region, Alaska* (N. J. Wilimovsky and J. N. Wolfe, Eds.). U.S. Atomic Energy Commission, Oak Ridge, Tennessee, pp. 363–413.

Holm-Hansen, O. (1963) Algae: Nitrogen fixation by antarctic species. *Science*, **139**: 1059–1060.

Holm-Hansen, O. (1964) Isolation and culture of terrestrial and freshwater algae of Antarctica. *Phycologia*, **4**: 43–51.

Kalff, J. (1967) Phytoplankton abundance and primary production rates in two arctic ponds. *Ecology*, **48**: 558–565.

Kalff, J. (1970) Arctic lake ecosystems. In *Antarctic Ecology* (M. W. Holdgate, Ed.). London: Academic Press, Vol. 2, pp. 651–663.

Ketchum, B. H. (1951) Plankton algae and their biological significance. In *Manual of Phycology* (G. M. Smith, Ed.). Waltham, Mass.: Chronica Botanica Co., pp. 335–346.

Kling, H. (1972) Species distribution and abundance in Char and Meretta Lakes, Cornwallis Island. *Char Lake Annual Rep.*, Canadian International Biological Program, 1971, pp. 23–26.

Kobayasi, Y. (1967) *Prasiola crispa* and its allies in the Alaskan arctic and Antarctica. In *Phycological Report of the Japanese Microbiological Expedition to the Alaskan Arctic* (Y. Kobayasi, Ed.). *Bull. Natur. Sci. Mus.*, **10**: 211–220.

Kobayasi, Y. (1969) *Polar regions. Nature with Special Consideration on Plant Life.* Tokyo: Seibundo-Shinosha Publishing Company, Ltd., 208 pp. (In Japanese with English subtitles.)

Kol, E. (1942) The snow and ice algae of Alaska. *Smithsonian Misc. Coll.*, **101**(16): 1–33.

Lund, J. W. G. (1945) Observations on soil algae. *New Phytol.*, **44**: 196–219.

Lund, J. W. G. (1962) Soil algae. In *Physiology and Biochemistry of Algae* (R. A. Lewin, Ed.). New York: Academic Press, pp. 759–770.

Lynn, R. I., and R. E. Cameron. (1973) The role of algae in crust formation and nitrogen cycling in desert soils. *IBP Desert Biome Reports of 1972 Progress*, Vol. 3, *Terrestrial Process Studies, Res. Memo.*, RM 73-40, 26 pp.

Maruyama, K. (1967) Blue–green algae in the Alaskan arctic. In *Phycological Report of the Japanese Microbiological Expedition to the Alaskan Arctic* (Y. Kobayasi, Ed.). *Bull. Natur. Sci. Mus. (Tokyo)*, **10**: 221–239.

Miller, O. K., Jr., and G. A. Laursen. (1978) Ecto- and endomycorrhizae of arctic plants at Barrow, Alaska. In *Vegetation and Production Ecology of an Alaskan Arctic Tundra* (L. L. Tieszen, Ed.). New York: Springer-Verlag, Chap. 9.

Novichkova-Ivanova, L. N. (1972) Soil and aerial algae of polar deserts and arctic tundra. In *Proceedings IV International Meeting on the Biological Productivity of Tundra, Leningrad, USSR* (F. E. Wielgolaski and Th. Rosswall, Eds.). Stockholm: International Biological Programme Tundra Biome Steering Committee, pp. 261–265.

Patrick, R. (1959) Bacillariophyceae. In *Freshwater Biology* (W. T. Edmondson, Ed.). New York: John Wiley and Sons, Inc., pp. 171–189.

Patrick, R., and C. W. Reimer. (1966) *The Diatoms of the United States*. Monograph 13, The Academy of Natural Sciences, Philadelphia. Lancaster: Fulton Press, Inc., 688 pp.

Prescott, G. W. (1931) Iowa algae. *Univ. Iowa Stud. Natur. Hist.*, **13**: 1–235.

Prescott, G. W. (1951) *Algae of the Western Great Lakes Area*. Cranbrook Institute of Science Bulletin no. 31. Bloomington Hills: The Cranbrook Press, 946 pp.

Prescott, G. W. (1953) Preliminary notes on the ecology of freshwater algae of the Arctic Slope, Alaska, with descriptions of some species. *Amer. Midl. Natur.*, **50**: 463–470.

Prescott, G. W. (1954) *How to Know the Freshwater Algae*. Dubuque, Iowa: Wm. C. Brown, Company, 211 pp.

Prescott, G. W. (1959) Ecology of freshwater algae in the arctic. *Int. Bot. Congr.*, **6**: 201–207.

Prescott, G. W. (1963a) Ecology of Alaskan freshwater algae. 2. Introduction: General considerations. *Trans. Amer. Microsc. Soc.*, **82**: 93–98.

Prescott, G. W. (1963b) Ecology of Alaskan freshwater algae. 3. Introduction: General features (additional notes). 4. Additional notes of *Pseudoclonium* and a transfer. *Trans. Amer. Microsc. Soc.*, **82**: 137–143.

Prescott, G. W. (1963c) Algae collecting at the Pole. *Ward's Bull.*, **3**: 1–7.

Prescott, G. W. (1964) Contributions of current research to algal systematics. In *Algae and Man* (D. F. Jackson, Ed.). New York: Plenum Press, pp. 1–30.

Prescott, G. W. (1972) Algae: Hot and cold. *Ward's Bull.*, **11**: 1, 2, 4, and 7.

Prescott, G. W., and W. C. Vinyard. (1965) Ecology of Alaskan freshwater algae. 5. Limnology and flora of Malipuk Lake. *Trans. Amer. Microsc. Soc.*, **84**: 427–478.

Ross, R. (1956) The cryptogamic flora of the arctic. III. Algae: Planktonic. *Bot. Rev.*, **20**: 400–416.

Round, F. E. (1965) *The Biology of the Algae*. New York: St. Martin's Press, 269 pp.

Rudolph, E. (1966) Terrestrial vegetation of Antarctica: Past and present studies. In *Antarctic Soils and Soil Forming Processes* (J. C. F. Tedrow, Ed.). Antarctic Research Series, Vol. 8, Washington, D.C.: American Geophysical Union, pp. 109–124.

Schell, D. M., and V. Alexander. (1973) Nitrogen fixation in arctic coastal tundra in relation to vegetation and micro-relief. *Arctic*, **26**: 130–137.

Sheath, R., and M. Munawar. (1974) Phytoplankton composition of a small subarctic lake in the Northwest Territories, Canada. *Phycologia*, **13**: 149–161.

Shields, L. M., and L. W. Durrell. (1964) Algae in relation to soil fertility. *Bot. Rev.*, **30**: 93–128.

Smith, G. M. (1950) *The Freshwater Algae of the United States*. New York: McGraw–Hill Book Company, 716 pp.

Starr, R. C. (1955) *A Comparative Study of Chlorococcum Meneghini and Other Spherical, Zoospore-producing Genera of the Chlorococcales*. Bloomington: Indiana University Press, 111 pp.

Stutz, R. C. and F. D. Cook. A method of enumerating soil algae. personal communication.

Taylor, W. R. (1956) The cryptogamic flora of the arctic. II. Algae: Nonplanktonic. *Bot. Rev.,* **20**: 363–399.

Thayer, D. W. (1972) Microbiological studies at the Pantex site, 1971. *IBP Grassland Biome, Tech. Rep.* 184, 29 pp.

Thompson, R. H. (1959) Algae. In *Freshwater Biology* (W. T. Edmondson, Ed.). New York: John Wiley and Sons, Inc., 1248 pp.

Tiffany, L. H., and M. E. Britton. (1952) *The Algae of Illinois.* Chicago: University of Chicago Press, 407 pp.

Tikhomirov, B. A. (1971) Summary of U.S.S.R. Tundra Biome studies. In *Proceedings of the Tundra Biome Working Meeting on Analysis of Ecosystems, Kevo, Finland* (O. W. Heal, Ed.). Stockholm: International Biological Programme Tundra Biome Steering Committee, pp. 165–170.

Tikhomirov, B. A. (1974) Peculiarities of the biosphere in the extreme north. *Priroda,* Vol. 11, pp. 30–42. (International Tundra Biome Translation 9, 1974, Translator: Multilingual Services Division, Canada Bureau of Transport, 20 pp.)

Whitford, L. A., and G. J. Schumacher. (1969) A manual of the freshwater algae in North Carolina. *The North Carolina Agric. Exp. Sta. Tech. Bull.* 188, 313 pp.

9. Ecto- and Endomycorrhizae of Arctic Plants at Barrow, Alaska

O. K. MILLER, JR., AND G. A. LAURSEN

Introduction

The increased growth and dependence of many green plants on the presence of mycorrhizae has long been known and appreciated by biologists (Marks and Kozlowski, 1973). The early research dealt with ectomycorrhizal fungi and a series of experiments clearly indicated that greater uptake of nitrogen and phosphorus from infertile soils occurred when mycorrhizae were present. Nutritional studies of selected mycorrhizal fungi by Palmer and Hacskaylo (1970) indicated that most of these fungi utilized simple sugars most efficiently. In fact, they were unable to degrade lignin, cellulose, pectin, and other complex carbohydrates. This nutritional mode contrasted sharply with the typical fungal decomposers which possess this ability along with a faster growth rate and the ability to grow on a wide variety of natural and artificial substrates. In 1965, Lewis and Harley first described the one-way movement of carbohydrates from host roots to the mycorrhizal fungus. In this process sugar is converted by the fungus to trehelose and mannitol and eventually to the storage polysaccharide, glycogen. The host plant cannot utilize these sugars. Bevege, et al. (1975) confirmed this phenomenon with *Pinus radiata* and *Rhizopogon luteolus*. They reported the same process in the endomycorrhizal hyphae of *Endogone* but only a minute amount of sugar was incorporated by the fungus. The difference seems to lie in the ability of the ectomycorrhizal fungus to act as a nutrient sink. Nevertheless, both fungal–root associations have the overall effect of increasing plant growth dramatically. The experiments carried out by Hattingh (1975) leave no doubt that phosphorus uptake is carried out by the endomycorrhizal fungus and that phosphorus must be translocated through the fungal hyphae to the host root. Other recent studies have indicated that sulfur (Gray and Gerdemann, 1973) and zinc (Gilmore, 1971) uptake is increased by means of endomycorrhizal fungi. Bowen (1973) has reported on the ability of the ectomycorrhizal fungus to supply both phosphorus as well as nitrate–N or ammonium–N to host plant roots. It would appear then that both the ecto- and endomycorrhizal fungi are usually unable to enzymatically degrade cell walls or other complex carbohydrates but rely, in fact, on their host for sufficient simple sugars for their needs. In return, they are able to take up necessary metabolites such as phosphorus, nitrogen, sulfur, zinc, and perhaps others (Powell, 1975) and translocate these metabolites to the higher plant. The inability of the higher plant to obtain phosphorus and other required nutrients from poor soils is certainly one of the key factors in the unique role of the mycorrhizal partner. There seems to be no excreted phosphatases to exploit organic soil phosphates (Bevege et al., 1975), and several hypotheses have been advanced to explain the mechanism of phosphorus absorption. In a functional way the ecto- and endomycorrhizal association appears to provide its host with the ability to obtain the same metabolites and greatly extend the functional root system of the host.

Some conspicuous differences seem to exist, however, between the ecto- and endomycorrhizal association. The ectomycorrhizal association involves a well-developed tissue called the sheath (Figures 1b and c) from which hyphae radiate

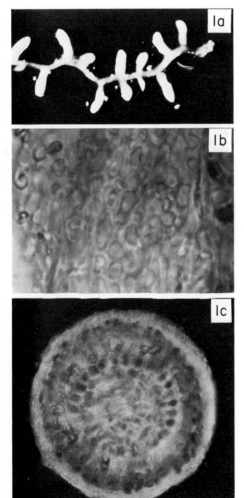

Figure 1. Ectomycorrhizal rootlets of *Salix rotundifolia*. (a) Each rootlet is surrounded by a white mycorrhizal sheath (×5). (b) Cross-section of the sheath showing the compact fungal cells and dark-colored cortical cells on the right (×3500). (c) Cross-section shows the white sheath in phase contrast surrounding the darker cortical cells. The stele located in the center is not invaded by the fungus (×70).

into the soil. A cohesive tissue of this sort has a comparatively large mass and is able to absorb and hold phosphate in a nonmetabolic pool, most likely as orthophosphate (Bowen, 1973), along with carbohydrates. During periods of low phosphate availability the stored phosphate is slowly translocated to the plant and readily available carbohydrates are utilized by the fungus to sustain its active metabolism. Jennings (1964) has reported that ammonium stimulates phosphorus uptake and that phosphorus is removed from this pool by a series of synthetic reactions. In addition, the sheath is often capable of living more than one season, which ensures an immediate functional mycorrhizal relationship as soon as nutrients enter the nutrient pool in the soil. Many ectomycorrhizal fungi also produce rather large fruiting bodies which have elevated levels of phosphorus and nitrogen (Laursen, 1975) over the vegetative hyphae and the host. These nutrients are either back translocated from the fruiting body to the fungal

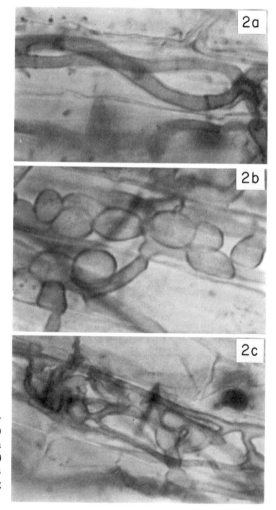

Figure 2.   Endomycorrhizae in the cortical cells of *Ranunculus pygmaeus*. (a)
Filamentous hyphae with frequent septa
(×2000). (b) Vesicular–arbuscular (VA)
mycorrhizae (×2000). (c) Filamentous
and VA endomycorrhizae proliferating
through the cortical cells (×2000).

hyphae, and hence to the host, or may become available to other soil microbes through decomposition of the fruiting bodies. Endomycorrhizal fungi, on the other hand, accumulate few carbohydrates and other nutrients. Their mass is rather small (Figure 2), and they often invade their host seasonally occupying a limited number of cells. They have a profound influence on the increased uptake of phosphorus, but do not appear to act as a sink like ectomycorrhizal fungi. Endomycorrhizae have minute fruiting bodies (0.1 to 10.0 mm) and would differ sharply in this respect from the ectomycorrhizae which often have large, fleshy mushrooms for their sexual stage.

In arctic tundra, the semiaquatic habitats such as low-centered polygon troughs, basins, and inundated wet meadows have few mycorrhizal plant species and those that are there are endomycorrhizal. The ecto- and additional endomy-corrhizal species are in the dryer meadows, polygon rims, high-centered polygon tops, and old beach ridges. It would appear from our soil sampling (Miller and Laursen, 1974; Laursen, 1975) that Basidiomycetes are greatly reduced in high moisture content soils with low available oxygen. Both the distribution of higher plants and the way in which nutrients are taken up and held are altered consider-ably by the habitat type and the subsequent distribution of higher plants in the Arctic Coastal Plain. A more complete discussion of these associations is devel-oped below.

Distribution of Ectomycorrhizae

The Basidiomycetes and Ascomycetes (including Discomycetes) are distributed among 17 families including 30 genera with fewer than 110 species (Miller et al., 1973; Miller and Laursen, 1974; Kobayasi et al., 1967; Laursen et al., 1976). However, on the Barrow IBP Tundra Biome sites, only 7 families and 11 genera, represented by 22 species, appear to be ectomycorrhizal associates of arctic plants. Only 7 species are found frequently each season and often thousands of fruiting bodies of these few species are observed. Of course, we have no way of relating the frequency of mycorrhizal sheaths to the frequency of fruiting bodies. Species never found fruiting at Barrow, such as *Leccinum scabrum,* (Fr.) S. F. Gray could possibly be widely distributed as ectomycorrhizal but never fruit in these habitats. Inland areas such as Meade River (110 km SSW) with extensive stands of *Betula nana* L. have large seasonal fruitings of *L. scabrum.* Numerous observations of this sort with other mycorrhizal fungi lend credence to the relationship observed over time between the frequency of fruiting and the presence of both mycorrhizae and the appropriate host plant.

Inland tundras show dramatic increases in the number of plant species over the coastal areas. The numbers of species of higher fungi, especially mycorrhizal fungi, also substantially increase. At best, however, it is still a small number compared with the deciduous and coniferous biomes where one finds at least 50 families and more than 250 genera of Basidiomycetes and Discomycetes (Miller and Farr, 1975). From the checklists which have been assembled by us in Montana and by others (Thiers, 1975, personal communication), the total species in a given ecosystem would range from 400 to 1200. The numbers of fungi are dependent upon the diversity of the higher plant flora. At Barrow these mycorrhizal fungi are distributed among four Basidiomycete families in the order Agaricales, including the Cortinariaceae, Russulaceae, Tricholomataceae, and the Hygrophoraceae. Two families in the Aphyllophorales are represented by the Clavariaceae and Thelephoraceae. Representatives of the puffball order Lycoperdales are found only in one family, the Lycoperdaceae.

Ectomycorrhizae have been observed on all species of the Salicaceae (Figure 1a) at Barrow. The typical sheath (Figures 1b and c) is found abundantly on the small rootlets. In many instances we have been able to trace the typical clamped hyphae found only in the Basidiomycetes directly to the outer fungal cells of the sheath. All Barrow species in the Ericaceae are also associated with ectomycorrhizae and include *Cassiope tetragona* (L.) Don, *Vaccinium vitis-idaea* L., and *Ledum palustre* L., *Pedicularis kanei*, *Dryas integrifolia M.* Vahl, and *Cochlearia officinalis* L. also appear to have ectomycorrhizal fungus associates.

Distribution of Endomycorrhizae

The endomycorrhizae observed so far in the roots of plants in the Barrow area are largely of the second subgroup described by Gerdemann (1971). They are sparingly septate and produce hyaline swollen hyphal segments (Figure 2) within the cortical cells of the host. They are called vesicular–arbuscular (VA) mycorrhizae. The sporocarps observed by us in soil are numerous, rather small (200 to 400 μm), and contain azygospores or zygospores. They belong in the family Endogonaceae (Gerdemann and Trappe, 1974), but additional study will be necessary to ascertain to which genus they should be referred and whether or not they are described taxa. The Endogonaceae is a family in the Mucorales in the class Zygomycetes (Ainsworth et al., 1973). Mosse (1953, 1956), Gerdemann (1964), and Daft and Nicholson (1966) first clearly showed the strong positive influence of VA endomycorrhizae on growth of species of Monocotyledons, especially Graminae. However, VA mycorrhizae have a much wider host range (Gerdemann, 1971) and are found in many dicotyledonous plant families as well as a variety of other plants.

At Barrow, we have observed VA endomycorrhizae in species of *Ranunculus* (Figure 2) including *R. nivalis* L., *R. palasii* Schlecht, and *R. pygmaeus* Wahlenb, and also in *Caltha palustris* L. in the Ranunculaceae. *Saxifraga punctata* L. and *S. oppositifolia L.* also have VA mycorrhizae or at least endomycorrhizae. We have also observed sporocarps typically of the Endogonaceae and hyphae closely associated with the roots of several genera of grasses including *Poa* and *Alopecurus*. We have also observed similar close associations with the sedge *Eriophorum* and with bryophyte rhizoids of the genus *Pogonatum*. The nature of these root–fungus relationships is being intensively investigated in our laboratories and these mycorrhizae appear to be more widely distributed in the Graminae (A. E. Linkins and R. A. Antibus, personal communication) at Barrow than previously thought.

Distribution of Mycorrhizal Fungi

Mycorrhiza formers are least abundant in the extremely wet sloughs, basins, and polygon troughs (Figure 3). Many species in these habitats, such as *Dupontia fisheri* R. Br. have intricately proliferating rootlets which attenuate to diameters of 10 to 30 μm. The increased surface area of the root system combined with the additional space occupied by fine roots in the soil probably compensates, in part, for the lack of mycorrhizal fungi. Highly productive wet meadows and low-centered polygon rims (Figure 3) have the highest number of mycorrhiza formers consisting of both endomycorrhizae and ectomycorrhizae.

High-centered polygons have a reduced number of mycorrhizal plants with a general tendency towards ectomycorrhizal fungi (Figure 3). A similar pattern involving many of the same species was reported by Stutz (1972) for the Truelove Lowland at Devon Island in the Canadian Arctic. In only two cases where 15 higher plants were examined from both Devon and Barrow does the form of mycorrhizal relationship differ. These differences can be explained by the time of sampling. It is more than likely that the presence of mycorrhizal fungi provides a selective advantage for the associated plant in at least two ways. First, the fungi are able to convert the predominant amino form of nitrogen to ammonia, which can also be converted by bacteria to nitrates or other readily available nitrogen compounds for use by the higher plant. Evidence to support this hypothesis is the sporadic occurrence of nitrate in soil on polygon tops and other elevated areas where mycorrhizal plants are most abundant. Second, any available phosphorus which occurs at low levels in the soil solution can be taken up quickly by the thousands of meters of hyphal cells proliferating through the soil, but associated directly with the plant roots via the mycorrhizal sheath, or the proliferating intracellular hyphae of endomycorrhizae.

We must also keep in mind that many fungi cannot exist in the sloughs and wet areas because of the extremely high moisture content. Similarly, they are restricted from invading the high beach ridges and the north faces of polygons or polygon rims because of their low moisture content. Total fungal biomass is low at soil moisture contents over 500 and under 250 % (Laursen and Miller, 1977). Many arctic plants, particularly the willows, cannot exist in their natural habitat without mycorrhizae. Biotic and abiotic factors which limit the fungus can and do restrict the distribution of the higher plant which must have mycorrhizae.

Carbon, Nitrogen, and Phosphorus

The sporocarps of mycorrhizal fungi have 43 to 48% of their dry weight in carbon (Laursen, 1975), while decomposers and basidiolichens average somewhat less (40 to 44%). The hyphae of these mycorrhizal species that are grown in pure culture under laboratory conditions maintain 48.8 $\pm$1.1% carbon, which is comparable to nonmycorrhizal fungi, 47.9%. The nitrogen content of sporocarps of mycorrhizal fungi is 4.23 $\pm$0.41% compared to 6.39 $\pm$0.21% in nonmycorrhizal sporocarps. These high nitrogen levels are not maintained in the hyphae which have 2.02 $\pm$0.22% of dry weight in mycorrhizal fungi, 3.35 $\pm$0.77% in nonmycor-

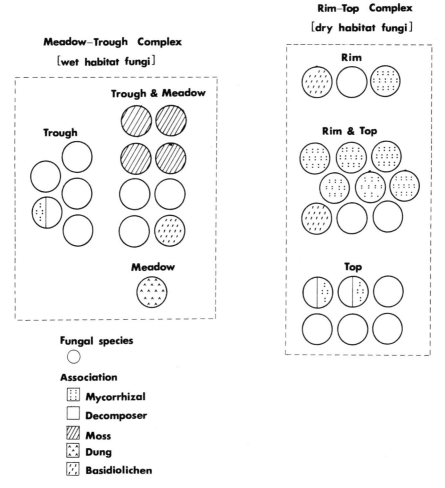

Figure 3.   Functional groups of Basidiomycete species distributed by habitat, based on the appearance of the fruiting body over a 4-year period (1971 to 1974), Barrow, Alaska, Tundra Biome IBP sites.

rhizal fungi, or a combined 2.75 ±0.43% for all species. It is obvious that the sporocarps of Basidiomycetes, while maintaining about the same levels of carbon content, have markedly elevated levels of nitrogen over the vegetative hyphae. Phosphorus content of sporocarps is about 9.3% of dry weight, with nonmycorrhizal fungi averaging 11.07 ±1.42% and mycorrhizal fungi 5.74 ±0.87% (Laursen, 1975).

In mid- to late summer in the arctic, the large fruitings of Basidiomycetes result in the accumulation of a large amount of nitrogen and phosphorus. These nutrients soon become available to other organisms as rapid decomposition of the fruiting bodies follows the completion of spore discharge. The caloric content of these fruiting bodies is 4290 cal g^{-1} and ranges up to 4800 cal g^{-1}, which is comparable to other plant tissue (Laursen, 1975).

Conclusions and Discussion

The growing body of knowledge concerning the role of mycorrhizal fungi suggests several important distinctions between them and the typical fungal saprophyte or parasite. The studies of Bjorkman (1944), Palmer and Hacskaylo (1970), and Hacskaylo (1973), among others, lead to the general conclusion that mycorrhizal fungi are capable of utilizing only soluble carbohydrates effectively. Their value to the plant then is not in the role of a decomposer of the so-called "hard" material such as lignins and cellulose. They rely, in fact, on the ready availability of simple carbohydrates provided by the host roots, or elsewhere, and are able to supply nitrate–N or ammonium–N in combination with phosphate, from a variety of organic sources to the host plant (Bowen, 1973).

Acknowledgments. This research was supported by the National Science Foundation under Grant GV 29354 to the Virginia Polytechnic Institute and State University. It was performed under the joint sponsorship of the International Biological Programme and the Office of Polar Programs and was coordinated by the U.S. Tundra Biome. Field and laboratory activities at Barrow were supported by the Naval Arctic Research Laboratory of the Office of Naval Research.

References

Ainsworth, G. C., F. K. Sparrow, and A. S. Sussman. (1973) *The Fungi, An Advanced Treatise*. New York: Academic Press, Vol. IVB, 505 pp.

Bevege, D. I., G. D. Bowen, and M. F. Skinner. (1975) Comparative carbohydrate physiology of ecto- and endomycorrhizas. In *Endomycorrhizas* (F. E. Sanders, B. Mosse, and P. B. Tinker, Eds.). London: Academic Press, 626 pp.

Bjorkman, E. (1944) The effect of strangulation on the formation of mycorrhiza in pine. *Sven. Bot. Tidskr.,* **38:** 1–14.

Bowen, G. D. (1973) Mineral nutrition of Ectomycorrhizae. In *Ectomycorrhizae, Their Ecology and Physiology* (G. C. Marks and T. T. Kozlowski, Eds.). New York: Academic Press, 444 pp.

Chapin, F. S., III, K. Van Cleve, and L. L. Tieszen. (1975) Seasonal nutrient dynamics of tundra vegetation at Barrow, Alaska. *Arct. Alp. Res.,* **7:** 209–226.

Daft, M. J., and H. T. Nicholson. (1966) Effect of *Endogone* mycorrhiza on plant growth. *New Phytol.,* **65:** 343–350.

Gerdemann, J. W. (1964) The effect of mycorrhiza on the growth of maize. *Mycologia,* **56:** 342–349.

Gerdemann, J. W. (1971) Fungi that form the vesicular–arbuscular type of endomycorrhiza. In *Mycorrhizae* (E. Hacskaylo, Ed.). U.S. Dep. Agric. Misc. Publ. 1189, 255 pp.

Gerdemann, J. W., and J. M. Trappe. (1974) The Endogonaceae in the Pacific Northwest *Mycologia Mem.,* No. 5, 76 pp.

Gilmore, A. E. (1971) The influence of endotropic mycorrhizae on the growth of peach seedlings. *J. Amer. Soc. Hort. Sci.,* **96:** 35–38.

Gray, L. E., and J. W. Gerdemann. (1973) Uptake of sulphur-35 by vesicular–arbuscular mycorrhizae. *Plant Soil,* **39:** 687–689.

Hacskaylo, E. (1973) Carbohydrate physiology of ectomycorrhizae. In *Ectomycorrhizae, Their Ecology and Physiology* (G. C. Marks and T. T. Kozlowski, Eds.). New York: Academic Press, 444 pp.

Hattingh, M. J. (1975) Uptake of ^{32}P-labeled phosphate by endomycorrhizal roots in soil chambers. In *Endomycorrhizas* (F. E. Sanders, B. Mosse, and P. B. Tinker, Eds.). London: Academic Press, 626 pp.

Jennings, D. H. (1964) Changes in the size of orthophosphate pools in mycorrhizal roots of beech with reference to absorption of the ion from the external medium. *New Phytol.*, **63**: 181.

Kobayasi, Y., N. Hiratusuka, K. Aoshima, R. P. Korf, M. Soneda, K. Tubaki, and J. Sugiyama. (1967) Mycological studies of the Alaskan Arctic. *Ann. Rep. Inst. Ferment., Osaka,* **3**, 138 pp.

Laursen, G. A. (1975) Higher fungi in soils of coastal arctic tundra plant communities. Ph.D. dissertation, Virginia Polytechnic Institute and State University, 390 pp.

Laursen, G. A., and O. K. Miller, Jr. (1977) The distribution of fungal hyphae in arctic soil on the International Biological Programme Tundra Biome Site, Barrow, Alaska. *Arct. Alp. Res.,* **9**: 149–156.

Laursen, G. A., O. K. Miller, Jr., and H. E. Bigelow (1976) A new *Clitocybe* from the Alaskan arctic. *Can. J. Bot.,* **54**: 976–980.

Marks, G. C., and T. T. Kozlowski. (Eds.) (1973) *Ectomycorrhizae, Their Ecology and Physiology.* New York: Academic Press, 444 pp.

Miller, O. K. Jr., and D. E. Farr. (1975) Index of the common fungi of North America (synonymy and common names). *Biblio. Mycol.,* **44**: 260.

Miller, O. K. Jr., and G. A. Laursen. (1974) Belowground fungal biomass on U.S. Tundra Biome sites at Barrow, Alaska. In *Soil Organisms and Decomposition in Tundra: Proceedings of the Microbiology, Decomposition and Invertebrate Working Groups Meeting, Fairbanks, Alaska, August 1973* (A. J. Holding *et al.,* Eds.). Stockholm: International Biological Programme Tundra Biome Steering Committee, pp. 151–158.

Miller, O. K., Jr., G. A. Laursen, and B. M. Murray. (1973) Arctic and alpine agarics from Alaska and Canada. *Can. J. Bot.,* **51**: 43–49.

Mosse, B. (1953) Fructifications associated with mycorrhizal strawberry roots. *Nature,* **171**: 974.

Mosse, B. (1956) Fructifications of an *Endogone* species causing endotrophic mycorrhiza on fruit plants. *Ann. Bot.,* **20**: 349–362.

Palmer, J. G., and E. Hacskaylo. (1970) Ectomycorrhizal fungi in pure culture. I. Growth on single carbon sources. *Physiol. Plant.,* **23**: 1187–1197.

Powell, C. L. (1975) Potassium uptake by endotropic mycorrhizas. In *Endomycorrhizas* (F. E. Sanders, B. Mosse, and P. B. Tinker, Eds.). London: Academic Press, 626 pp.

Stutz, R. C. (1972) Survey of mycorrhizal plants. In *Devon Island IBP Project: High Arctic Ecosystem, Project Report 1970 and 1971* (L. C. Bliss, Ed.). Dept. of Botany, University of Alberta, pp. 214–216.

Woods, F. W., and K. Brock. (1964) Interspecific transfer of Ca^{45} and P^{32} by root systems. *Ecology,* **45**: 886–889.

Section II
Photosynthesis, Respiration, and Water Relations

Section IIa: Plant and Community CO$_2$ Exchange

"Photosynthesis, Respiration, and Water Relations" summarizes a portion of the experimental approach to the study of our tundra system. We initially concentrated on those factors (e.g., canopy interactions with irradiance and temperature) we understood best, but developed research capabilities where previous studies had not been attempted. These included studies of moss gas exchange, root growth and activity, and an aerodynamic assessment of system CO$_2$ exchange. Tieszen (Chapter 10) reviews the intensive photosynthesis studies which concentrated on the dominant graminoids and were directly coordinated with the concurrent studies on the mosses (Chapter 11) and the entire system (Chapter 12). These studies were designed to be directly comparable but also to be largely independent. We hoped that the independent estimates of CO$_2$ exchange would both provide a check on our abilities to extrapolate from cuvette studies and in a sense help validate the canopy photosynthesis model which was then extant.

Photosynthetic rates vary among the various growth forms with rapidly growing graminoids possessing high rates. Both vascular plants (10) and mosses (11) photosynthesize well at the low temperatures characteristic of Barrow and approach rates comparable to those of similar growth forms in more temperate zones. The two groups differ in their responses to moisture and irradiance. Both maintain positive rates throughout most of the periods of continuous irradiance. Mosses are much more opportunistic by virtue of their "evergreen" habit and rapid recovery from drought or low temperature imposed inactivity. Similarly, they seem to show clearer patterns of micro-habitat adaptation. Neither group, however, illustrates temperature acclimatization in the field, possibly because of the rather narrow temperature range they experience.

The estimates of seasonal CO$_2$ uptake by mosses agree very closely with those reported by Rastorfer (Section I, Chapter 6). The estimates derived from vascular plants suggest that nearly one-third of the annual carbon dioxide incorporation occurs after the canopy has completed expansion. This seasonal total of 602 g m^{-2} exceeds the estimate derived from harvest studies (Section I, Chapter 3) but is in remarkable agreement with the simulation estimates and the independent estimate derived from the aerodynamic approach. All estimates of CO$_2$ uptake indicate a high dependence upon available leaf area suggesting that the system's ability to incorporate energy may be very sensitive to the date of meltoff, grazing, or other phenomena which impact on canopy development.

Section IIb: Photosynthesis and Water Relations

Chapter 13 assesses the interception characteristics of high latitude (Barrow) and high altitude (Niwot Ridge, Colorado) canopies. The alpine studies were initiated to provide a low latitude contrast and to initiate detailed studies of the effect of water stress on CO_2 exchange. More detailed and comparative studies of species spanning a portion of the moisture gradient at each site were undertaken under controlled laboratory conditions. The results help explain the segregation of species along moisture gradients proposed by Webber (Section I, Chapter 3). Wider ranging species were characterized by greater abilities to maintain high rates of CO_2 uptake as water stress increased.

Miller et al. (Chapter 14) use field data and simulations to compare the responses of Barrow and Niwot Ridge species along a moisture gradient. Again, a mechanistic basis for the distributions is apparent. Although all species are occasionally water stressed to the extent that stomatal closure is initiated, it appears that annual production in existing species–environment complexes is not limited except in very dry years. The tracer studies of Koranda et al. (Chapter 15) reinforce other meteorological and physiological data which suggest that transpirational losses from this system account for only a small percentage of the total evapotranspirational flux.

The final chapter (16) in this section incorporates many of the moss and vascular plant responses in an attempt to analyze competitive relationships between moss and graminoid growth forms in existing canopies. Although competition for inorganic nutrients is not included, the model does incorporate thermal phenomena and water relations. The effect of the vascular canopy appears significant but, interestingly, depends upon the amount of available water.

10. Photosynthesis in the Principal Barrow, Alaska, Species: A Summary of Field and Laboratory Responses

LARRY L. TIESZEN

Introduction

A number of factors suggest that the photosynthetic systems of tundra plants function effectively under arctic environments. For example, the Barrow wet meadow system is capable of accumulating 100 g of dry matter aboveground in a 55-day period (Tieszen, 1972). This production results in a minimal efficiency of primary production of 0.46%. This comparatively high efficiency suggests that the process of photosynthesis is adapted to the low ambient temperatures, a suggestion which differs from the review and theoretical consideration provided by Warren Wilson (1966). One of the main limitations to primary production in this system may be simply the short duration of the growing season.

Recent field photosynthesis studies of the principal Barrow species (Tieszen, 1973, 1975) and reports from other tundras (Shvetsova and Voznessenskii, 1970; Gerasimenko and Zalenskii, 1973) have established directly that photosynthetic rates are comparable to those of temperate zone plants. In combination with relatively high activity at low temperatures, these species were shown to be capable of high instantaneous and daily rates. Comparatively, however, specific adaptations to tundras and even the quantitative coupling of CO_2 uptake to environmental factors are largely unknown. Earlier data derived largely from laboratory studies indicated that both genetic and acclimation adaptations (Mooney and Billings, 1961; Mooney and Johnson, 1965; Tieszen and Helgager, 1968) were important for arctic and alpine species. The data of Billings et al. (1971) suggest that this acclimation potential is greatest in alpine areas. Neither the extent nor the basis for similar responses is known for the arctic.

This chapter summarizes the underlying data for our understanding of vascular photosynthesis at Barrow, Alaska. We have attempted to relate, quantitatively, CO_2 uptake to the principal environmental factors. Furthermore, sufficient field data were obtained to estimate total CO_2 uptake for at least the Barrow wet meadow system. We will only summarize the photosynthesis data in this chapter since this information has been incorporated in a mechanistic model (Miller and Tieszen, 1972) which has been further refined (Miller et al., 1976).

Methods

The field studies were conducted at the intensive site described by Tieszen (1972, 1973) and by Webber (1978). Occasionally plants were selected from other communities for comparative studies but these were always located within about 1 km of the intensive site. Leaf areas were measured directly or with a point frame system as described by Tieszen (1972) and Caldwell et al. (1974). Photosynthesis determinations were undertaken in the field by three methods as described by Tieszen (1973) for leaf studies, Tieszen et al. (1974) for the portable $^{14}CO_2$ system, and Tieszen (1975) for the whole shoot studies. The determinations of resistance components required some additional controls. Air from a 5-m-high port was mixed in a 25-liter container and conditioned by passage through either a drying or humidifying system. Water vapor density was maintained near ambient levels when possible, although frequently it had to be reduced to prevent condensation in the photosynthesis chambers. All tubing and sensors were enclosed in insulation and heated well above the circulating dew point. Air was diverted from the mixing manifold to one of three cuvette systems. Dew point sensors were located adjacent to each remote cuvette providing continuous monitoring of incoming and outgoing water vapor density. Returning air streams were selected by a solenoid switching system which directed the sample through drying tubes and the IRGA.

The entire system was monitored by a VIDAR data acquisition system under the control of a PDP-8 computer. The inputs included averages of leaf temperatures, chamber air temperature, associated mean canopy temperature, chamber light intensity, outputs from incoming and outgoing dew point hygrometers, and the output from the IRGA. In combination with information on leaf area, flow rate, and CO_2 compensation concentrations, the mean photosynthesis, transpiration, and leaf and mesophyll resistances were calculated and printed.

In the study relating leaf growth to photosynthesis and carboxylation activity, matched tillers of *Dupontia fisheri*, *Carex aquatilis*, and *Eriophorum angustifolium* were selected. These tiller systems have been characterized by Mattheis et al. (1976). Leaves were identified, marked, and measured to the tip and to the end of the green portion at intervals of 10 days. Senescence was visually estimated by the loss of green. Assays of carboxylation activity were made on leaves comparable to those used in the growth determinations by the methods previously established (Tieszen and Sigurdson, 1973). Leaf samples approximately 20 mm long were removed from about 10 mm above the ligule. This sometimes necessitated pooling more than 20 individual tillers for each extraction.

Results and Discussion

Maximum Rates

The rates of photosynthesis reported for fully expanded and intact leaves ranged from 7 to nearly 31 mg of CO_2 dm^{-2} hr^{-1} for Barrow grasses (Tieszen, 1973). The lowest value was obtained for *Hierochloe alpina,* a grass characteristic of the exposed ridges of high-centered polygons at Barrow. In contrast, the highest value was obtained for *Elymus arenarius,* a grass restricted to very sandy sites where its growth is very rapid. Photosynthetic rates for the three dominant grasses (*Dupontia fisheri,* 18.1 ± 3.1; *Carex aquatilis,* 18.5 ± 3.1; and *Eriophorum angustifolium,* 20.9 ± 1.6 mg of CO_2 dm^{-2} hr^{-1}, means ± SE, $N = 10$) did not differ significantly. But *Salix pulchra* had a substantially higher rate (28 mg of CO_2 dm^{-2} hr^{-1}). The forbs for which data are available had rates comparable to the dominant graminoids: *Petasites frigidus,* 13.4 and *Ranunculus nivalis* 18 mg of CO_2 dm^{-2} hr^{-1}. These rates were nearly the same for entire intact plants (Tieszen, 1975) as for individual leaves (Tieszen, 1973).

One of the vegetation features associated with increasing latitude is the reduction in aboveground supporting (nonphotosynthetic) tissue. One of the results of this trend is the reduced abundance of some growth forms (Webber, 1978), e.g., shrubs and even deciduous dwarf shrubs, at high latitudes. Also, Tieszen (1970) suggested on the basis of pigment analyses that even in leaf blades there was a reduction in nonchlorophyllous cells at high latitudes, especially in comparison to alpine areas. At the Barrow site most tissues possess some chlorophyll (Tieszen 1972) and are presumably photosynthetic. The bulk of aboveground productivity, however, is accounted for by leaf blades which generally were high in chlorophyll. Most species possessed around two-thirds of their dry weight in blades, but *Eriophorum angustifolium* and *E. scheuchzeri* are substantially higher with 89% of all material in blades. In all cases the concentration of chlorophyll (mg g dry wt^{-1}) was two to three times greater in leaf blade material than in the remaining plant parts. Thus the ratio of standing crops of chlorophyll in blades to nonblades was substantially higher than that for dry weight.

The species differed in the relative photosynthetic rates of their constituent parts (Tieszen and Johnson, 1975). On an area basis the sheaths, exposed stems (culms), and inflorescences of *D. fisheri* had photosynthetic rates that were 50% those of the leaves. The stem portion enclosed by the sheath was least active, although its rate was still 19% that of the leaves. *Alopecurus alpinus* possessed equally active sheaths, although other parts were relatively less efficient. In addition to these differences, potential rates varied markedly through the season as a function of leaf age or developmental stage as also was shown for the Alpine (Johnson and Caldwell, 1974; Caldwell et al. 1978).

Tieszen's data (1975) indicate that intact plants commonly possessed maximum rates around 10 to 25 mg of CO_2 dm^{-2} hr^{-1}. Near the end of August, however, maximum rates were around 5 mg of CO_2 dm^{-2} hr^{-1}, even though earlier in the season periods with similar irradiances and temperature would have resulted in substantially higher rates.

Light Responses

Intact and fully expanded leaves were used to characterize the photosynthetic responses of the principal Barrow species (Tieszen, 1973). Irradiances required for compensation of respiration were very low, ranging from 0.008 to 0.010 cal

cm^{-2} min^{-1} (400 to 700 nm) at 15°C. This suggests that at ambient temperatures leaf respiration is compensated by photosynthesis throughout most days until the sun drops below the horizon. Similarly, this low requirement suggests that other chlorophyllous structures, e.g., stems and inflorescences, would commonly be able to support themselves. Light saturation of photosynthesis was approached in most species at 0.4 cal cm^{-2} min^{-1} (Tieszen, 1973). This irradiance is received only on clear days near noon and suggests that the rates of randomly arranged leaves would be light dependent under most natural canopy conditions. These data apparently can be generalized for all leaves and throughout the season since Tieszen and Johnson (1975) report similar relationships with the portable system when the data from all leaves and species are integrated for the entire season. On the basis of these leaf data we would expect the system to be closely coupled to irradiance and to be above compensation throughout most days.

Temperature Responses

The temperature responses of all species were again similar and not substantially different from most temperate zone grasses. The optimum increased slightly as irradiances increased and at 0.2 cal cm^{-2} min^{-1} was around 15°C. Of greater significance was the broad range over which CO_2 uptake was positive. This occurred in all species tested at 0°C and in *Dupontia* to as low as −5°C. In agreement with data from the U.S.S.R. (Zalenskii et al., 1972; Shvetsova and Voznessenskii, 1970), all species were positve at temperatures as high as 35°C or at least 15°C higher than the ambient temperatures normally experienced. Thus, the ability to extend the range of positive CO_2 uptake about 20°C higher and lower than the optimum provides a high degree of temperature adaptability. In combination with leaf and canopy temperatures that are slightly higher than air temperatures, the photosynthetic system appears to be slightly inhibited by low temperatures but not nearly as closely coupled to temperature as to irradiance. This is corroborated by Tieszen's field data (1975) for intact plants. High rates were relatively independent of temperature. Even at 3°C, *Dupontia* took up 14 mg of CO_2 dm^{-2} hr^{-1} and *Salix* as much as 23 mg of CO_2 dm^{-2} hr^{-1}.

The basis for this broad range and the ability to photosynthesize below 0°C is not known; however, it is likely related to a general adaptation that allows many metabolic processes (including respiration, translocation, uptake, etc.) to occur at high rates and in an integrated manner. This may be related more to alterations in general lipid components than to changes in a large number of independent enzymes. Certainly, our early work on ribulose-1,5-diphosphate carboxylase (Tieszen and Sigurdson, 1973) shows no differences at the extracted enzyme level in thermal stability or activity ($E_\alpha = 1.46 \times 10^4$ cal $mole^{-1}$) between these tundra species and other C_3 forms. Studies of enzyme-substrate affinity at low temperatures and at low (normal) substrate concentrations need to be undertaken.

Field Diurnal and Seasonal Patterns

The short growing season in the Arctic places a severe restriction on net seasonal CO_2 uptake and primary production. This is especially important in the wet meadow system, since this system becomes snow free only shortly before the summer solstice; and with a canopy which is not evergreen the conversion of incident light to $(CH_2O)_n$ must be low early in the season. Suggestions that arctic plants resume growth beneath the snow in a manner comparable to that reported elsewhere (Mooney and Billings, 1960; Fonda and Bliss, 1966) are not supported by data from Barrow (Tieszen, 1974) as suggested in Figure 1. Temperature and light conditions beneath the snow are not conducive to photosynthesis; and the species are not competent at this time. Photosynthesis increases as the canopy becomes snow free, but prior to the time that meltwater percolates through the

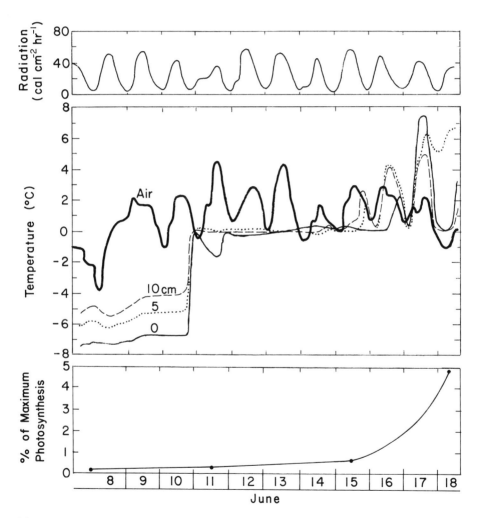

Figure 1. Pattern of temperature changes and the development of photosynthesis in *Dupontia fisheri* during snowmelt in June. Snow temperatures were taken at ground level (0 cm) and 5 and 10 cm above ground level. Meltwater percolated through the snow on 10 June and the sensors became generally snowfree on 15 June. (Adapted from Tieszen, 1974, with permission from *Arctic and Alpine Research*.)

snow temperatures are below $-5°C$. Bryophytes are potentially somewhat more productive prior to snowmelt. For example, on 7 June rates of *Polytrichum* sp. removed from beneath the snow were as high as 0.56 mg of CO_2 g dry wt^{-1}, a value not much lower than its summer maximum (Oechel and Sveinbjörnsson, 1978). Thus mosses are in a position to photosynthesize at significant rates immediately after meltoff since their relative photosynthetic competence is high and since a substantial amount of evergreen tissue is present. The vascular canopy, in contrast, has a low competency and must produce its entire canopy anew. As a consequence interception is low early in the season as shown by Caldwell et al. (1974) for Barrow and generalized for wet arctic tundra by Lewis and Callaghan (1975).

The photosynthetic rates were monitored quantitatively between meltoff and fall senescence. Although meltoff in 1971 occurred around 15 June, the combination of short plants and an inundated wet meadow precluded synoptic whole plant measurements before 5 July. Rates obtained from these early measurements were already high (Tieszen, 1975) and followed solar radiation quite closely. Photosynthesis for the whole shoots was positive throughout the day, and at solar midnight *Carex aquatilis* remained above 3.3 mg of CO_2 dm^{-2} hr^{-1}. These rates in all four species (*Carex aquatilis, Dupontia fisheri, Eriophorum angustifolium,* and *Salix pulchra*) tested were similar and remained high until early August. The diurnal curves shown by Tieszen (1975) indicate that when temperatures were low (e.g., 1 and 2 August) the rate of photosynthesis was markedly sinusoidal due to the daily course of irradiance, but when the air temperature exceeded about $15°C$ a midday depression was sometimes apparent. After mid-August the photosynthetic rates (maxima and daily totals) dropped markedly due apparently to a developing senescence as well as a deteriorating irradiance and thermal environment.

These seasonally varying diurnal patterns are also evident in the seasonal progression of daily maximum and minimum values for CO_2 uptake (Tieszen, 1975). The highest values for CO_2 uptake on a leaf area basis occurred prior to 15 August, and all species showed a consistent decrease in the latter part of the season. The minimum values for CO_2 uptake were generally positive during the early part of the season, were negative on heavily overcast nights, and remained negative after the beginning of August. Near the end of August (Table 1) CO_2 uptake was negative for at least 9 hr each day. However, this dark CO_2 output from plant respiration was only 10 to 20% of the net CO_2 incorporation. Thus, net aboveground CO_2 balance was still definitely positive.

Respiration rates at the low temperatures were generally low and varied among species with *Salix pulchra* possessing substantially higher rates than the monocotyledons. A temperature dependence resulted in higher respiratory rates at higher temperatures, although the narrow temperature range and low rates of CO_2 exchange did not allow us to establish the quantitative relationship in the field. Compensation points were estimated during August since in this period radiation was sufficiently low to routinely cause the plants to develop a pattern of negative CO_2 uptake. The compensation points were low for all species (0.013 to 0.024 cal cm^{-2} min^{-1}) although *Dupontia fisheri* and *Salix pulchra* required

Table 1.   Hours per Day of Negative CO_2 Exchange for Intact Shoots under Ambient Conditions

Species	Date July							August							
	5	15	16	17	18	22	23	1	8	9	14	15	24	25	28
Carex aquatilis	0	0	—	—	—	0	0	3	2	6	5	7	8	9	8
Dupontia fisheri	—	—	2	3	—	2	0	5	5	8	5	7	9	9	—
Eriophorum angustifolium	—	0	—	—	—	0	0	2	3	6	4	5	—	9	8
Salix pulchra	—	—	—	—	—	—	—	4	6	7	6	6	6	9	6

slightly more light to compensate for respiration than the other two species (Tieszen, 1975).

An integration of the hourly values for a 24-hr period provides an estimate of daily net CO_2 uptake by the exposed plants. These daily totals were generally high (to more than 200 mg of CO_2 dm^{-2} day^{-1}) for all species beginning with the earliest samples which were taken. The absolute amounts were greatest on days which had a high receipt of solar radiation. For all species the linear correlations between total CO_2 incorporation were higher with daily totals of radiation than with temperature (Figure 2; see Tieszen, 1975). An extrapolation of the linear equation relating CO_2 uptake to radiation suggests that for the three monocotyledons slightly less than 100 cal cm^{-2} day^{-1} are required to compensate for daily respiratory CO_2 losses. This value is somewhat greater than the compensation points actually measured during ''night'' runs and may suggest a higher respiration rate during ''daytime'' than at ''night.''

Late in the season the combination of shorter photoperiods, reduced irradiance, and a developing senescence resulted in a decrease in the daily incorporation of CO_2. Thus by 25 August, daily totals for the monocots were well below 50 mg of CO_2 dm^{-2} day^{-1}. *Salix pulchra* possessed higher values than the monocotyledons throughout the period in which it was sampled, even though its dark respiration rate was higher.

In an attempt to utilize the data from all 62 days sampled to predict seasonal uptake independent of the physiological data used in the mechanistic model (Miller and Tieszen, 1972; Miller et al., 1976), we undertook a multiple linear regression analysis. Mean hourly values for CO_2 uptake were regressed against mean hourly temperature, radiation, time of day (24 hourly periods), species, and date entered as discrete 10-day periods.

The regression models developed for each species explained a large amount of the variance (76 to 83%) and all variables were significant. The regression analysis suggests that in all species there was a highly significant seasonal change in photosynthesis which was independent of the seasonal changes in radiation and temperature. The variation explained by the inclusion of the sampling date was less for *Salix pulchra* than for the other species, perhaps because of a later initiation of the sampling regime. In all cases photosynthesis decreased as the season progressed as was also indicated by some low maximal photosynthesis

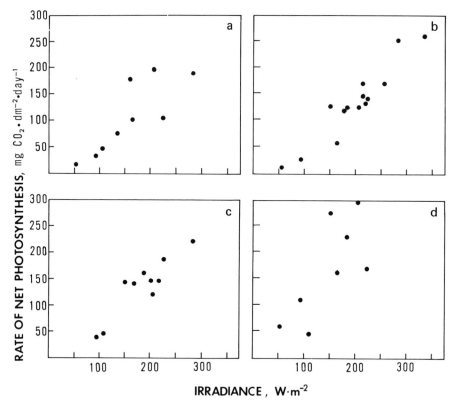

Figure 2.   Net daily uptake of CO_2 in four species and its relationship to mean irradiance.

values late in the season at times when radiation was still quite high. Presumably this indicates a loss of photosynthetic efficiency for the entire plant which could be caused by an increase in the proportion of supporting (nonchlorophyllous) tissues, a developing senescence, or other phenomena. These studies did not include the first 15 to 20 days of the season, a time when photosynthetic competence may have been increasing. Thus the forced linearity of the seasonal change may be artificial.

Abiotic data from 1971 were used with the regression to summarize the predicted daily curves as seen in Figure 3. These simulations of the mean daily course of photosynthesis suggest distinct daily trends, substantial seasonal changes, and species differences. All species possessed similar daily trends with evidence of a depression in CO_2 uptake around solar noon, which is especially evident in *Dupontia fisheri* and *Salix pulchra*. *S. pulchra* and *D. fisheri* also possessed a greater seasonal decrease in the daily course of CO_2 uptake than *Carex aquatilis* and *Eriophorum angustifolium*. The general decrease in all species is a result of decreasing radiation as well as a decrease in photosynthetic activity as is indicated by the negative regression coefficients associated with the period of the season. An integration of the net CO_2 uptake for the mean day of each period provides an estimate of mean daily CO_2 uptake for each species in each period (Table 2). These daily totals (Table 2) indicate the highly productive system with up to 400 mg of CO_2 incorporated by a square decimeter of leaf in one day. The data also illustrate the substantial seasonal decrease per unit leaf area and the major species difference with *Salix pulchra* substantially higher than any of the monocotyledons. Even at the end of August, however, shoots possessed a significantly positive daily CO_2 budget.

In an attempt to determine the relationships between some of the environmental factors and the control of the rate of photosynthesis, subsequent field experiments were established in which the resistances to CO_2 uptake could be calculated. The system monitored three plants simultaneously and with the exception of minor down time was in continuous operation between 1 July and 20 August.

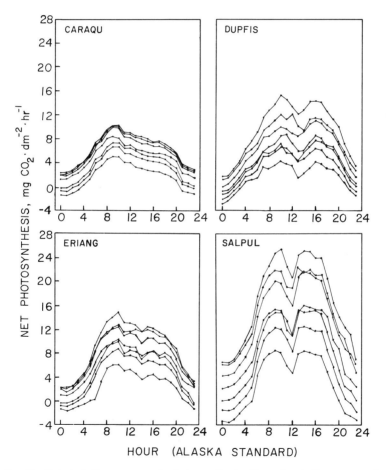

Figure 3. Predicted daily course of photosynthesis throughout the season for four species. These regression estimates were calculated as in Tieszen (1975) and estimate CO_2 uptake for 10-day periods beginning on 25 June (top curve) and ending 3 September (bottom curve).

Table 2.   Regression Estimated Mean Daily Totals of CO_2 Uptake (mg of CO_2 dm^{-2} day^{-1}) by Each Species in Each 10-day Period through the Growing Season

Period	Carex aquatilis	Dupontia fisheri	Eriophorum angustifolium	Salix pulchra
2 24 June–4 July	148.8	235.1	214.5	400.6
3 5 July–14 July	141.6	189.9	186.3	361.2
4 15 July–24 July	148.5	162.9	195.8	318.0
5 25 July–3 Aug	115.8	104.7	142.8	234.1
6 4 Aug–13 Aug	89.1	112.7	125.1	203.9
7 14 Aug–23 Aug	73.1	79.7	103.4	145.5
8 23 Aug–3 Sept	43.1	41.3	61.9	71.9

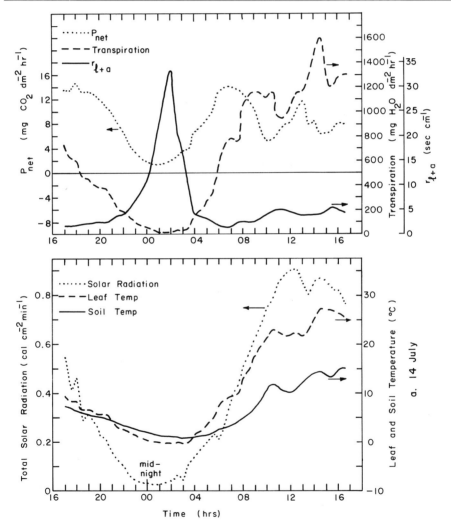

Figure 4.   Representative daily courses of photosynthesis, transpiration, and leaf resistance in *Dupontia fisheri*. (a) 14 July. (b) 7 August.

The photosynthesis responses were similar to those recorded the previous summer, with *Salix pulchra* possessing the highest rates (more than 21 mg of CO_2 dm^{-2} hr^{-1}), *Petasites frigidus* the lowest maximum rates (14 mg of CO_2 dm^{-2} hr^{-1}), and the three monocots again being comparable (*Carex aquatilis* = 16, *Dupontia fisheri* = 18, *Eriophorum angustifolium* = 16 mg of CO_2 dm^{-2} hr^{-1}). The incidence of clear days was greater than the previous summer and consequently canopy temperatures were higher and the tundra surface was subjectively drier, although soil water potential still remained high.

Figure 4 illustrates the daily courses of photosynthesis, transpiration, and leaf resistances from two select days in *Dupontia fisheri*. Leaf resistances in all species became low whenever irradiances exceeded 0.1 to 0.2 cal cm^{-2} min^{-1} and remained low (1 to 5 sec cm^{-1}) throughout the day except on exceptionally clear and warm days. Under these conditions of high heat load (leaf temperatures exceeding 18°C), stomata began to close, as is seen by the rise in r_l, and accounted for most of the decrease in net photosynthesis which was observed. Thus the midday depressions in CO_2 uptake, which were also detected but uncommon in 1971, are accounted for by stomatal closure (see Miller et al., 1978) in contrast to the suggestion of high respiration rates proposed by Shvetsova and

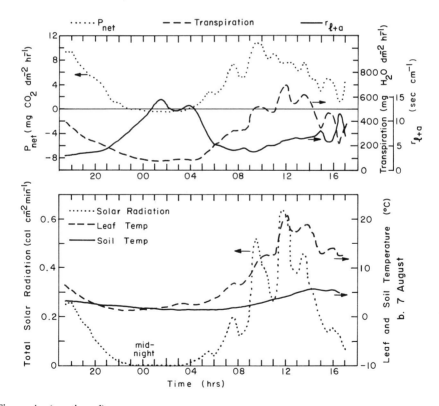

Figure 4   (continued).

Voznessenskii (1970) for similar species in the U.S.S.R. Arctic. It should be pointed out, however, that temperatures in the chamber were a few degrees higher than in the open canopy, and the vapor density gradient was somewhat higher. Thus, water stress was greater in the chamber than in the free canopy, and under natural conditions water stress in these wet meadow species would be uncommon.

Although minimum leaf resistances for all species were low, transpiration rates were not high. The high atmospheric humidities, low temperatures, and close leaf and air temperatures resulted in little water loss. Values as high as 1600 mg of H_2O dm^{-2} hr^{-1} were occasionally transpired; however, frequently the highest daily rates were lower than 1000 mg of H_2O dm^{-2} hr^{-1}. Solar midnight values were thus very low and were commonly less than 100 to 200 mg of H_2O dm^{-2} hr^{-1}. As the season progressed irradiance fell to zero and mean maximum resistances of 21, 16, and 19 sec cm^{-1} were recorded for *Carex, Dupontia,* and *Eriophorum*. These maximum leaf resistances are low but agree with those reported independently by Miller et al. (1978).

The light response for stomatal movement (Figure 5) indicates that the stomates are open at irradiances exceeding 0.2 cal cm^{-2} min^{-1} (solar total). Interestingly, at very high irradiances (> 0.7), there is occasionally an increase in leaf resistances. This stomatal closure is likely a result of water stress under these high irradiances and leaf temperatures (see Miller et al., 1978). The responsiveness of r_l to water stress suggests that this component of the resistance series is

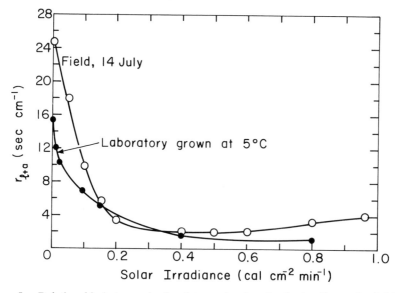

Figure 5.   Relationship between leaf resistance (r_{l+a}) and solar irradiance for field-grown plants and laboratory plants grown at 5°C. Field data are for entire plants and laboratory data are for intact leaves.

Table 3.   Values of Mean Minimum Mesophyll Resistances and Maximum Leaf
Resistances for Intact Plants Enclosed in Cuvettes ($N > 6$)

Species	r_m (Season mean) minimum	r_m (Late Aug.) minimum	r_l (Season mean) maximum
Carex aquatilis	18 ± 3	29 ± 10	21 ± 8
Dupontia fisheri	15 ± 5	37 ± 12	16 ± 5
Eriophorum angustifolium	13 ± 4	22 ± 3	19 ± 5

most important in modulating hourly rates. The higher values for mesophyll
resistance (Table 3), however, suggest that some component of this resistance is
more important in determining the maximum photosynthetic rate. The minimum
mesophyll resistances are substantially higher than the minimum leaf resistances,
although they are comparable to those of similar C_3 grasses in temperate regions.
These whole plant data support the positive correlation shown by Tieszen (1973)
between maximum rates of CO_2 uptake and carboxylation activity for most
tundra grasses. The regression analysis of field photosynthesis responses suggested a seasonal decline which is attributed to leaf senescence. This increased
senescence should be apparent as an increase in mesophyll resistance as photosynthetic competence is lost. Although sufficient comparable data of r_m are not
available for a seasonal course to be established, Table 3 indicates that the late
sample periods were higher than the season mean. Presumably, we would expect
the seasonal mean to resemble that of chlorophyll concentration already established (Tieszen, 1972).

Estimated Seasonal Course of CO_2 Uptake

The field photosynthesis data were used to estimate total incorporation for the
growing season (see Tieszen, 1975). The mean daily CO_2 uptake for each species
from the regression (Table 2) was used to calculate uptake on a land area basis:

$$\text{g of } CO_2 \text{ m}^{-2} \text{ land day}^{-1} = \text{mg of } CO_2 \text{ dm}^{-2} \text{ day}^{-1} \times \text{LAI}$$
$$\times 10^2 \text{ dm}^{-2} \text{ m}^{-2} \times 10^{-3} \text{ g mg}^{-1}.$$

In an attempt to develop an estimate for total CO_2 uptake by vascular plants, we
have assumed that all monocots not accounted for by the three used in the
regression (*Carex, Dupontia,* and *Eriophorum*) possessed daily totals on a leaf
area basis equal to the mean of the other three. Similarly, we have assumed that
all dicots possessed the same rates of CO_2 uptake as *Salix pulchra* and have used
these values to calculate community totals. This procedure will overestimate the
"true" community CO_2 uptake because of the mutual shading which occurs in
the undisturbed canopy. Since the total LAI (live and dead) can exceed 2.0, self-
shading is high, and our estimates may be substantially higher than actually exist.
These estimates (Figure 6) indicate that incorporation of CO_2 started at a low
level (< 1 g of CO_2 m^{-2} day^{-1} at the summer solstice) due mainly to lack of

available photosynthetic tissue. There was a progressive increase in CO_2 uptake until the 4th, 5th, and 6th 10-day periods when maximal incorporation occurred. *Dupontia fisheri* attained its maximum uptake in the 6th period or somewhat later than the other species because its leaf area was still expanding at this time. Maximum uptake for the community of vascular plants occurred in periods 4 or 5 or just preceding the onset of darkness. Although the contribution by the dicots was low early in the season, it increased rapidly, attained levels as high as 4.7 g m^{-2} day^{-1}, and exceeded any of the other groups during a number of the sampling periods.

The summation of estimated mean daily values for each period provides an estimate of the net seasonal incorporation of CO_2. *Eriophorum angustifolium* possessed the greatest net seasonal incorporation (215 g m^{-2}) which is a reflection of its high photosynthetic rates and its large leaf area index. The estimate for all dicotyledons indicates that they account for nearly 30% of incorporated CO_2. Total net incorporation for the vascular community is estimated with this method at 760 g of CO_2, which presumably represents the upper estimate for the vascular components of this community in 1971. The greatest uptake (Figure 6) of CO_2 occurred during periods 4, 5, and 6 or after the period of maximum radiation and highest temperatures. During these periods, however, the leaf area index (Figure 7) approached its maximum value. These data also indicate that the rate of CO_2 accumulation dropped after mid-August and apparently dropped rapidly late in

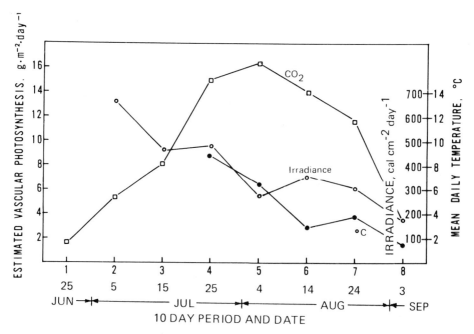

Figure 6.   Seasonal estimate of CO_2 incorporation for all vascular plants and associated seasonal changes in irradiance and temperature. (Adapted, with permission from *Photosynthetica*, from Tieszen, 1975.)

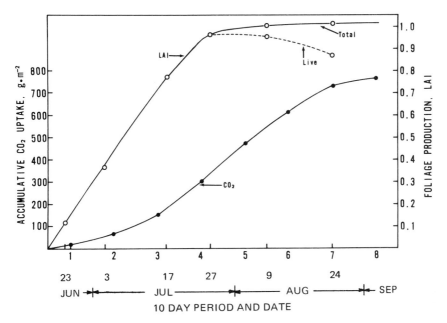

Figure 7. Accumulative CO_2 uptake by all vascular plants estimated from regression analysis and uncorrected for self-shading as related to live and total leaf area index. (Adapted, with permission from *Photosynthetica,* from Tieszen, 1975.)

the season even though substantial live leaf area was still present. This reduction of CO_2 uptake is probably most closely related to an increased leaf senescence and a decrease in radiation. Substantial CO_2 incorporation occurred after above-ground standing crop attained its maximal value. Approximately 262 g of CO_2 m^{-2}, over one-third of the total, was incorporated after peak production, even though leaf efficiency at this time was low due to senescence and a reduction in available radiation.

In an attempt to determine the magnitude of the canopy effect on CO_2 uptake, we used the stand photosynthesis model (Miller and Tieszen, 1972) to calculate the rate of photosynthesis throughout the season for a shaded (natural) and "unshaded" canopy. The "unshaded" canopy consisted of a very small leaf area index of the photosynthesizing species and no other live or dead canopy material. The comparison of the shaded and unshaded rates, expressed on equivalent leaf areas, suggested the magnitude of the canopy effect. This effect, which was substantial and varied somewhat through the season, averaged 21%. Thus it resulted in an estimate of 602 g of CO_2 m^{-2} for the season rather than the regression calculated 760 g.

Relationship between Photosynthesis and Leaf Growth

Previous results have shown a low photosynthetic competence at melt-off and a seasonal decrease in whole plant photosynthesis which we ascribed to a developing senescence (Tieszen, 1975). In an attempt to understand the intratiller dynamics, detailed studies of leaf growth, photosynthesis, and carboxylation were undertaken. These activities were neglible prior to snowmelt (Tieszen, 1974); however, all species possessed some chlorophyll. In *Dupontia,* this was contained mainly in the exposed bases of leaf blades which were to resume growth this season. Often about 1 cm of exposed green tissue was present. The remaining chlorophyll was present in partially developed and unexpanded leaves which were still protected by older and dead but persistent leaf sheaths. These sheaths extended for a height of about 31 mm, thus elevating this tissue above the moss surface. Quiescent tissues of *Carex* and *Eriophorum* were positioned 19 and 9 mm respectively above the moss surface.

The seasonal progressions of leaf elongation for each species are presented in Figure 8. Three leaves were growing on 19 June and only the third leaf had not been exserted the previous season. Thus, the first two leaves possessed dead apical portions. Leaf 1 resumed growth and attained its maximum development

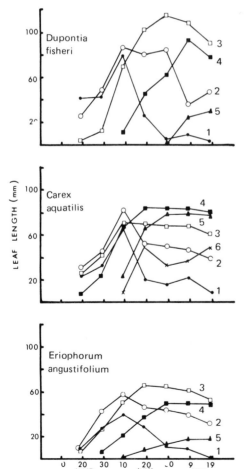

Figure 8. Seasonal progression of live leaf length in *Carex aquatilis, Eriophorum angustifolium,* and *Dupontia fisheri.* Leaf position is indicated by solid circles (oldest = 1), open circles (2), open squares (3), solid squares (4), triangles (5), and × (6 = youngest).

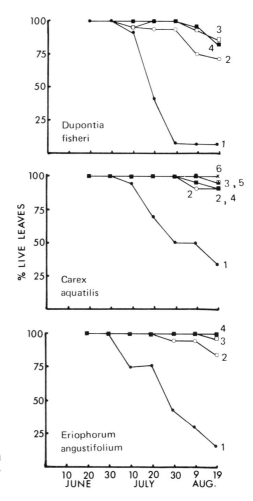

Figure 9. Retention of live leaves in each species as a function of leaf position. Leaf position as in Figure 8.

by 10 July after which it began to senesce rapidly, and by the beginning of August it was dead nearly to the base of the sheath. Leaf 2 became exserted slightly more than the first leaf and remained green in the fully elongated state for about 20 days; it then senesced (end of July) rapidly to the base of the blade, retaining, however, a green sheath. The third and fourth leaves were the only ones which completed or nearly completed expansion during one season. Neither leaf, however, had a long maturation phase as is seen by the decline in green length after 29 July (leaf 3) and 8 August (leaf 4). Most tillers initiated a fifth leaf; however, this leaf began senescing very shortly after exsertion. The net effect was that leaves 1 and 2 had completed growth by 19 August and had died back to the base of the sheaths or blades (Figure 9). Of the 18 plants used in the growth study, 14 produced a fourth leaf but only 4 had exserted leaf 5 by 19 August.

The pattern of growth in *Carex aquatilis* was generally similar to that of *Dupontia*. However, *Carex* exserted more leaves early in the season (some tillers possessed five leaves with green tissue by 19 June). The first two leaves had grown the previous year and attained their maximum length by 9 July after which senescence rapidly developed, especially for leaf 1. Subsequent leaves (e.g., 3, 4, and 5) possessed longer maturation phases and, in fact, were only developing strong senescence by 18 August (Figure 9). Eleven of nineteen tillers produced a sixth leaf and two produced part of a seventh leaf. In all cases, the first three leaves attained complete exsertion during the season. The pattern of growth in *Eriophorum* was more comparable to *Carex* than *Dupontia*. The first two leaves again showed an early senescence with subsequent leaves possessing a more protracted maturation phase. Again, the first three leaves completed expansion. Fifty percent of the tillers produced a fifth leaf and two initiated a sixth leaf by 18 August. There is an indication of a greater degree of synchrony in *Carex* and *Eriophorum* than in *Dupontia* which appears sequential.

Growth rates were generally highest during the earliest growth periods for each leaf. In *Dupontia* the first leaf to be exserted wholly in this season (i.e., leaf 3) possessed the highest mean growth rate (17.8 mm day^{-1}) at the beginning of the season and then decreased throughout the season approaching zero by period 4. Mean maximum growth rates were generally between 2 and 6 mm day^{-1} for all species although there were exceptions. Rates at the last measurement period were either all negative, as in *Dupontia* and *Eriophorum,* or at least greatly reduced in *Carex*. This pattern derived from 10-day intervals suggests that the leaf growth rate decreases very shortly after the leaf is exserted—a response which is not seen in growth chamber experiments. Under these artificial chamber conditions, even at 5°C, growth (increase in length) seems to be maintained for an extended period.

The lack of close synchrony among all tillers precludes an accurate assessment of the durations of the main leaf growth stages: expansion, maturation, and senescence. Therefore, growth rates were calculated for each leaf of each tiller throughout the season, and visual inspection of the seasonal plots was used to estimate the durations of the elongation, maturation (fully elongated) and senescing phases of the leaf life cycle (Table 4). As the data for *Dupontia* indicate, the growth and senescence stages are longer than the mature stages for each leaf. The first leaf apparently grows somewhat during the season and then initiates a senescence which occupies most of the current growth season. Leaves 2 and 3 appear to be the only two which maintain a mature phase, and the duration of this period is about 8 to 10 days. Successive leaves to be produced apparently possess short mature stages because of an environmentally induced senescence. This table also suggests that leaves likely experience more than 53 growing season days, although for most leaves this occurred only with two growing seasons.

The pattern in *Carex aquatilis* is again similar, although it appears that more leaves have a mature phase which may last as many as 27 days. This may be associated with a tendency to also retain leaves somewhat longer (60 days) as is the case with *Eriophorum*. In all species the last leaves to be initiated experience only about half their lifetimes in this season and may begin senescence while still elongating. All species would be expected to have rapidly changing photosynthetic rates and nutrient requirements.

Table 4.   Duration of Three Growth Phases in *Dupontia, Carex,* and *Eriophorum* as Estimated from Seasonal Curves of Leaf Growth Rate for Each Tiller[a]

Species	Leaf number	Mean length of interval (days)			
		Growth	Maturation	Senescent	Leaf life
Dupontia fisheri	1	> 9.8	2.0	32.6	44.4
	2	⩾18.1	9.9	⩾25.3	⩾53.3
	3	20.3	8.0	⩾13.3	⩾41.7
	4	22.7	1.7	⩾ 4.1	⩾28.6
	5	17.5	2.5	⩾ 0.0	⩾20.0
Carex aquatilis	1	> 9.7	11.9	⩾31.6	⩾52.6
	2	>16.8	25.8	⩾17.3	⩾60.0
	3	>23.4	26.9	⩾ 8.2	⩾57.9
	4	27.9	16.1	⩾ 4.4	⩾48.4
	5	27.7	⩾ 6.3	⩾ 2.3	⩾36.3
	6	25.4	⩾ 1.4	⩾ 0.5	⩾28.2
Eriophorum angustifolium	1	>10.8	7.9	31.3	⩾50.0
	2	>20.3	15.2	⩾23.9	⩾59.5
	3	>26.7	10.4	⩾16.1	⩾52.1
	4	25.5	9.1	⩾ 3.1	⩾37.9
	5	20.0	> 0	⩾ 4.4	⩾26.2

[a]   Growth phase = days from visible exsertion to cessation of elongation; mature phase = interval between cessation of elongation and initiation of senescence; senescent phase = interval between start of senescence as judged by loss of green and complete senescence. >, growth began before measurements started in early summer or the previous season; ⩾, senescence was not complete by the end of August.

The seasonal courses of carboxylation activity (Figure 10) illustrate strong seasonal and leaf dependencies. All species possessed activity in green leaves as well as in unexpanded leaf tissues beneath the snow; however, these activities were much lower than rates to be attained in several weeks. Maximum activity in *Dupontia's* first leaf was coincident with melt-off and decreased thereafter. All other leaves showed a marked increase in activity until mid-July. By 19 August levels had decreased to less than half the maximum activity in leaves 2, 3, and 4, although leaf 5 was still increasing in activity. *Eriophorum* and especially *Carex* tended to maintain high levels of carboxylation activity for longer periods than *Dupontia,* and both tended to attain the maximum activities near the beginning of August rather than mid-July. Levels were very low in *Eriophorum* by the end of August. Leaves differed both in maximal rates and in the seasonal progressions.

The photosynthesis data described in Figure 11 were taken under existing field temperature and light conditions, but temperatures were always above 5°C and intensities usually approached saturating values as reported by Tieszen (1973). On 11 August, however, after a protracted period of low irradiances, analyses were made when intensities were as low as 0.31 cal cm^{-2} min^{-1}. These rates approach those reported by Tieszen (1973, 1975) for the same and other species. In all species the first leaf to possess chlorophyll increased its rate of CO_2 uptake as the season progressed but only slightly. In *Dupontia* it approached 0.5 mg of CO_2 dm^{-2} hr^{-1}, whereas *Carex* approached 3 and *Eriophorum* 3.5 mg of CO_2

dm^{-2} hr^{-1}. These leaves lost activity rapidly in *Dupontia* and *Carex*. Subsequent leaves showed a trend of low rates of fixation shortly after exsertion which approached maximal values in late July or early August and then declined. In all three species even the second leaf failed to attain rates as high as subsequent leaves. Thus leaves which had initiated growth the previous season failed to attain maxima as high as newly formed leaves, although they were present and functional early in the season. Newly formed leaves illustrated a sequential development of photosynthetic capacity, although the expression of this was masked by the late summer environmental suppression of photosynthesis.

In combination with leaf length and width data through the season, it is possible for us to construct whole leaf and whole tiller budgets through the season. The seasonal courses of leaf area illustrated in Figure 12 show a

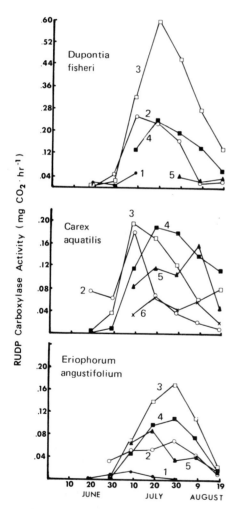

Figure 10. Seasonal course of carboxylation activity in three species at Barrow. Leaf position as in Figure 8.

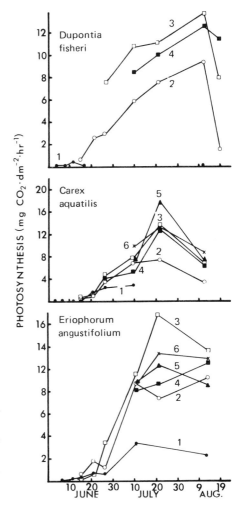

Figure 11. Seasonal progression of photosynthesis in each leaf of three species. CO_2 uptake was determined under ambient temperatures and irradiances approaching saturation with the $^{14}CO_2$ field system. Leaf position as in Figure 8.

sequential development of area which to a large extent is a reflection of leaf length. In all species, the first leaf attains its maximum functional area early in the season (10 July) and then shows a rapid decrease, a pattern which is also similar for leaf number two. Again the third leaf, or in *Carex* the third and fourth leaves, develops the most area and furthermore maintains that area for the greatest period, suggesting that this leaf may contribute most substantially to the entire tiller. The net effect of the leaf area development is that these tillers attain maximal areas relatively early in the season, i.e., between 10 July for *Carex* and *Eriophorum* and 30 July for *Dupontia*. In combination with the slow but continuous production of tillers this supports the data of Tieszen (1972) which showed maximum aboveground standing crop of 102 g m^{-2} by 4 August as well as the data of Caldwell et al. (1974) which showed that maximum live LAI was attained

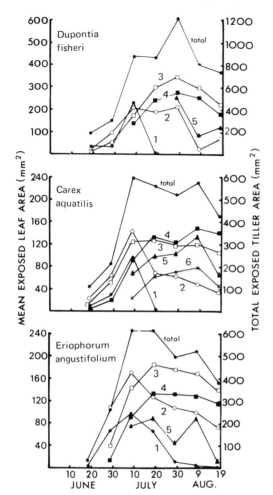

Figure 12. Sequential development of live leaf area for each leaf and species on a tiller basis. Leaf position as in Figure 8.

around the beginning of August. Maximum leaf area per tiller is therefore produced within about a 30-day period and before photoperiods begin to shorten on 2 August.

We can now calculate an estimate of mean leaf and total tiller carboxylation potential and relate it to photosynthesis (Figure 13). In *Dupontia*, leaf 3 is potentially the most active as is the case in *Eriophorum*. In *Carex*, however, leaves 3 and 4 are more similar although sequential in activity. Overall activity is always attained before the end of July and is substantially greater in *Dupontia* than in the other species. In all cases activity is low by the end of August, indicating a well-developed senescence. Thus the details of leaf photosynthesis reveal marked seasonal changes and corroborate the whole plant and cuvette studies previously described.

The seasonal changes in photosynthesis which have been described for whole plants and individual leaves are highly correlated with changes in carboxylation

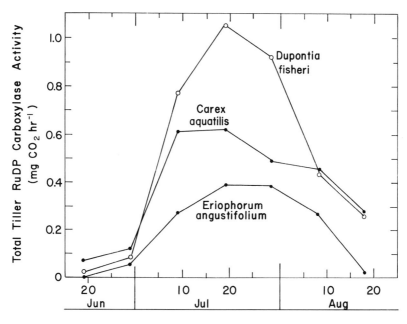

Figure 13.   Seasonal development of total carboxylase activity in three species.

activity (Table 5). The correlations throughout the season for all leaves were highly significant and ranged from $r = 0.74$ for *Dupontia* to 0.81 for *Carex* and 0.75 for *Eriophorum*. Although early season (21 June) values were low and not highly significant, they increased and were generally high for the rest of the season. At this early date both photosynthesis and carboxylation rates were very low and somewhat variable. These data substantiate the effect of senescence inferred from whole shoot responses (Tieszen, 1975). Further, the high correlation (0.76) between maximum photosynthetic rates of all species and carboxylase activity (Tieszen, 1973) supports the causal relationship. Our equally good correlation (0.83) with leaf density thickness (mg dm^{-2}) suggests that there may

Table 5.   Carboxylation Activity (mg of CO_2 g fr wt^{-1} hr^{-1}) of Three Species on the Fertilized and Control Plots of Schultz (See 1964, 1969) ($N = 6$)

Species	Treatment	PEP[a]	RuDP[b]
Carex aquatilis	Fertilized	1.9 ± 0.3	13.8 ± 2.1
	Control	2.9 ± 0.7	15.9 ± 4.1
Dupontia fisheri	Fertilized	0.9 ± 0.3	9.2 ± 1.3
	Control	0.8 ± 0.1	9.4 ± 0.4
Eriophorum angustifolium	Fertilized	2.3 ± 0.2	6.7 ± 0.2
	Control	2.0 ± 0.3	6.4 ± 0.3

[a]   PEP, phosphoenolpyruvate carboxylase.
[b]   RuDP, ribulose-1,5-diphosphate carboxylase.

Table 6. Correlations (r) of Photosynthesis and Carboxylation Activity of All Leaves at Various Dates through the Season and of Each Leaf for the Entire Season

Date	Dupontia		Carex		Eriophorum	
	r	P	r	P	r	P
June 21	0.18	0.75	0.02	—	−0.28	0.80
June 26	—	—	0.82	0.95	0.29	0.75
July 10–11	0.92	0.96	0.87	0.97	0.46	0.84
July 21–22	0.95	0.97	0.90	0.99	0.42	0.76
July 30	0.95	0.97	—	—	—	—
August 2	—	—	0.54	0.84	0.30	0.75
August 11–12	0.96	0.97	0.39	0.75	0.42	0.78
August 21	0.73	0.95	—	—	—	—
Leaf number						
1	0.87	0.97	−0.51	0.91	—	—
2	0.63	0.92	0.83	0.98	0.45	0.78
3	0.77	0.95	0.87	0.99	0.75	0.97
4	—	—	0.82	0.98	0.93	0.99
5	—	—	0.46	0.73	0.86	0.98
6	—	—	0.58	0.79	0.37	0.75

be a relationship with mesophyll thickness rather than enzyme concentration per chloroplast per se.

Nutrients

Numerous studies have shown that the tundra responds to increases in nutrients with an increase in production. This has been shown most dramatically by the large scale fertilizer treatments established by Schultz (1964, 1969). Our initial study of nutrient effects (Table 6) indicated that although plant composition,

Table 7. Range of Nutrients over which There Was No Significant ($P = 0.95$) Correlation with Photosynthesis. Soil Phosphorus Varied by a Factor of Four.

Species	Percentage concentration	
	Minimum	Maximum
Dupontia fisheri		
Leaf K	0.64	1.59
Leaf P	0.07	0.24
Leaf N	1.83	3.28
Carex aquatilis		
Leaf K	0.55	1.55
Leaf P	0.07	0.40
Leaf N	2.71	3.28
Eriophorum angustifolium		
Leaf K	0.45	1.14
Leaf P	0.15	0.31
Leaf N	1.50	3.21

growth, and production varied, the carboxylation activities were not significantly different in control and fertilized plots, suggesting that the major effect of nutrients might be on allocation rather than the photosynthetic process per se. In an attempt to examine the effect of natural ranges in nutrients, a study was conducted in 11 plots along a wide productivity range. Aboveground production varied from 18 g m^{-2} in a low-center polygon to 205 g m^{-2} in a distrubed and wet track. Plant density and stature showed similar variations. A direct comparison of maximum photosynthetic rates of plants from the most and least productive sites showed no significant differences. Also, there were no significant ($P = 0.95$) correlations among photosynthesis and soil or leaf K, N, or P (Table 7). This insensitivity of photosynthesis under field conditions suggests that the major effect of nutrient limitations is on the allocation pattern.

Conclusions

The field data indicate clearly that the maximum photosynthetic rates in the field at Barrow, Alaska, are comparable to those of C_3 counterparts from temperate zones. Thus, the potential for high rates of daily photosynthesis exists. To a large extent this daily potential is achieved, and net rates of CO_2 uptake were commonly greater than 150 mg of CO_2 dm^{-2} day^{-1}. The deciduous shrub, *Salix pulchra,* possessed daily rates even higher than 300 mg of CO_2 dm^{-2}. These high rates reflect the generally high rates for these life forms in tundra.

To a large extent these high daily rates reflect the positive rates of CO_2 uptake throughout the 24-hr day until early August. Under these conditions, however, the leaves are normally not light saturated even though the leaf area index is usually less than 1.0. Photosynthesis in the field is operating at temperatures below the 15°C maximum characteristic of most Barrow species (Tieszen, 1973) as well as other tundra species (Grace and Woolhouse, 1970; Billings et al., 1971). However, these species do not appear to be substantially inhibited at the low ambient temperatures. Perhaps this results from the fact that leaves are normally operating on the light-dependent portion and not the light-saturated portions of the light response curve. Recent simulations (Miller et al., 1976) suggest that these species have a sufficiently broad temperature response curve that with an optimum of 15°C they can effectively exploit the Barrow environment with a mean ambient temperature in the growing season of less than 5°C.

The net effects are high photosynthetic rates on a leaf area basis but photosynthetic rates on a land area basis which are largely limited by the available foliage. Thus, early in the growing season, CO_2 uptake on a land area basis occurs at a relatively low level even though solar irradiances and temperatures are more conducive to uptake than they are later in the season. Similarly, the peak of uptake coincides with the development of significant leaf senescence as estimated by the leaf area index determinations. It appears, therefore, that one of the most significant factors which may influence seasonal uptake is the date of the onset of canopy growth. Growth prior to the summer solstice should substantially increase the seasonal uptake of photosynthesis and, hence, primary production.

Even though the rate of CO_2 uptake decreases after the beginning of August and the canopy ceases development at this time, nearly one-third of the total carbon incorporated occurs after this period. This suggests that substantial carbon must be translocated to belowground reserves late in the season. Autoradiographic evidence to support this large transfer, however, has not been obtained (Allessio and Tieszen, 1975), although allocation to stem bases does increase.

The late season decrease in the rate of CO_2 fixation is due in part to the deteriorating light and temperature regimes. However, our detailed leaf and tiller studies reviewed here indicate that a substantial senescence develops after the 1st of August. Thus, although leaf area is large, the potential contribution as measured by direct $^{14}CO_2$ fixation and by assays of carboxylation activity becomes progressively and rapidly reduced.

Our estimate of net CO_2 uptake of 602 g of CO_2 m^{-2} is substantially greater than is accounted for by the harvest method of determining aboveground primary production. This estimate, however, is in close agreement with the simulated estimates of seasonal photosynthesis (Miller et al., 1976) and the estimate derived from the aerodynamic exchange of CO_2 for this system (Coyne and Kelly, 1978). It therefore appears that substantial amounts of carbon are translocated to belowground storage areas or are used for plant maintenance.

Acknowledgments. This research was supported by the National Science Foundation under Grant GV 29343 to Augustana College. It was performed under the joint sponsorship of the International Biological Programme and the Office of Polar Programs and was coorindated by the U.S. Tundra Biome. Field and laboratory activities at Barrow were supported by the Naval Arctic Research Laboratory of the Office of Naval Research.

Special thanks are given to the many students and assistants who supported this work including: S. Tieszen, D. Johnson, R. Mandsager, B. Oksol, R. Nelson, C. Mikkelson, and C. Sigurdson.

References

Allessio, M. L., and L. L. Tieszen. (1975) Patterns of carbon allocation in an arctic tundra grass, *Dupontia fischeri* (Gramineae), at Barrow, Alaska. *Amer. J. Bot.,* **62**: 797–807.

Billings, W. D., P. J. Godfrey, B. F. Chabot, and D. P. Bourque. (1971) Metabolic acclimation to temperature in arctic and alpine ecotypes of *Oxyria digyna. Arct. Alp. Res.,* **3**: 277–289.

Caldwell, M. M., L. L. Tieszen, and M. Fareed. (1974) The canopy structure of tundra plant communities at Barrow, Alaska, and Niwot Ridge, Colorado. *Arct. Alp. Res.,* **6**: 151–159.

Caldwell, M. M., D. A. Johnson, and M. Fareed. (1978) Constraints on tundra productivity: Photosynthetic capacity in relation to solar radiation utilization and water stress in arctic and alpine tundras. In *Vegetation and Production Ecology of an Alaskan Arctic Tundra* (L. L. Tieszen, Ed.). New York: Springer-Verlag, Chap. 13.

Coyne, P. I., and J. J. Kelley. (1975) CO_2 exchange over the Alaskan arctic tundra: Meteorological assessment by an aerodynamic method. *J. Appl. Ecol.,* **12**: 587–611.

Coyne, P. I., and J. J. Kelley. (1978) Meteorological assessment of CO_2 exchange over an Alaskan arctic tundra. In *Vegetation and Production Ecology of an Alaskan Arctic Tundra* (L. L. Tieszen, Ed.). New York: Springer-Verlag, Chap. 12.

Fonda, R. W., and L. C. Bliss. (1966) Annual carbohydrate cycle of alpine plants on Mt. Washington, New Hampshire. *Bull. Torrey Bot. Club,* **93**: 268–277.

Gerasimenko, T. V., and O. V. Zalenskii. (1973) Diurnal and seasonal dynamics of photosynthesis in plants of Wrangel Island (Sutochnaia i sezonnaia dinamika fotosinteza u rastenii ostrova Vrangelia). *Bot. Zh.,* **58**: 1655–1666.

Grace, J., and H. W. Woolhouse. (1970) A physiological and mathematical study of the growth and productivity of a *Calluna-Sphagnum* community. I. Net photosynthesis of *Calluna vulgaris* (L.) Hull. *J. Appl. Ecol.,* **7**: 363–381.

Johnson, D. A., and M. M. Caldwell. (1974) Field measurements of photosynthesis and leaf growth rate on three alpine plant species. *Arct. Alp. Res.,* **6**: 245–251.

Lewis, M. C., and T. V. Callaghan. (1975) Ecological efficiency of tundra vegetation. In *Vegetation and the Atmosphere* (J. L. Montieth, Ed.). New York: Academic Press, pp. 399–433.

Mattheis, P. J., L. L. Tieszen, and M. C. Lewis. (1976) Responses of *Dupontia fischeri* to simulated lemming grazing in an Alaskan arctic tundra. *Ann. Bot.,* **40**: 179–197.

Miller, P. C., and L. L. Tieszen. (1972) A preliminary model of processes affecting primary production in the arctic tundra. *Arct. Alp. Res.,* **4**: 1–18.

Miller, P. C., W. A. Stoner, and L. L. Tieszen. (1976) A model of stand photosynthesis for the wet meadow tundra at Barrow, Alaska. *Ecology,* **57**: 411–430.

Miller, P. C., W. A. Stoner, and J. R. Ehleringer. (1978) Some aspects of water relations of arctic and alpine regions. In *Vegetation and Production Ecology of an Alaskan Arctic Tundra* (L. L. Tieszen, Ed.). New York: Springer-Verlag, Chap. 14.

Mooney, H. A., and W. D. Billings. (1960) The annual carbohydrate cycle of alpine plants as related to growth. *Amer. J. Bot.,* **47**: 594–598.

Mooney, H. A., and W. D. Billings. (1961) Comparative physiological ecology of arctic and alpine populations of *Oxyria digyna*. *Ecol. Monogr.,* **31**: 1–29.

Mooney, H. A., and A. W. Johnson. (1965) Comparative physiological ecology of an arctic and an alpine population of *Thalictrum alpinum* L. *Ecology,* **4**: 721–727.

Oechel, W. C., and B. Sveinbjörnsson. (1978) Primary production processes in arctic bryophytes at Barrow, Alaska. In *Vegetation and Production Ecology of an Alaskan Arctic Tundra* (L. L. Tieszen, Ed.). New York: Springer-Verlag, Chap. 11.

Schultz, A. M. (1964) The nutrient recovery hypothesis for arctic microtine cycles. II. Ecosystem variables in relation to arctic microtine cycles. In *Grazing in Terrestrial and Marine Environments* (D. J. Crisp, Ed.). Oxford: Blackwell Scientific Publications, pp. 57–68.

Schultz, A. M. (1969) A study of an ecosystem: The arctic tundra. In *The Ecosystem Concept in Natural Resource Management* (G. M. Van Dyne, Ed.). New York: Academic Press, pp. 77–93.

Shvetsova, V. M., and V. L. Voznesenskii. (1970) Diurnal and seasonal changes of intensity of photosynthesis in some plants of the Western Taimyr. *Bot. Zh.,* **55**: 66–76. (International Tundra Biome Translation 2, April 1971; Translator: G. Belkov, 11 pp.)

Stoner, W. A., P. C. Miller, and L. L. Tieszen. (1978) A model of plant growth and phosphorus allocation for *Dupontia fischeri* in coastal, wet-meadow tundra. In *Vegetation and Production Ecology of an Alaskan Arctic Tundra* (L. L. Tieszen, Ed.). New York: Springer-Verlag, Chap. 24.

Tieszen, L. L. (1970) Comparisons of chlorophyll content and leaf structure in arctic and alpine grasses. *Amer. Midl. Natur.,* **83**: 238–253.

Tieszen, L. L. (1972) The seasonal course of aboveground production and chlorophyll distribution in a wet arctic tundra at Barrow, Alaska. *Arct. Alp. Res.,* **4**: 307–324.

Tieszen, L. L. (1973) Photosynthesis and respiration in arctic tundra grasses: Field light intensity and temperature responses. *Arct. Alp. Res.,* **5**: 239–251.

Tieszen, L. L. (1974) Photosynthetic ompetence of the subnivean vegetation of an arctic tundra. *Arct. Alp. Res.*, **6**: 253–256.

Tieszen, L. L. (1975) CO_2 exchange in the Alaskan Arctic Tundra: Seasonal changes in the rate of photosynthesis of four species. *Photosynthetica*, **9**: 376–390.

Tieszen, L. L., and J. A. Helgager. (1968) Genetic and physiological adaptation in the hill reaction of *Deschampsia caespitosa. Nature*, **219**: 1066–1067.

Tieszen, L. L., and D. A. Johnson. (1975) Seasonal pattern of photosynthesis in individual grass leaves and other plant parts in arctic Alaska with a portable $^{14}CO_2$ system. *Bot. Gaz.*, **136**: 99–105.

Tieszen, L. L., and D. C. Sigurdson. (1973) Effect of temperature on carboxylase activity and stability in some Calvin cycle grasses from the arctic. *Arct. Alp. Res.*, **5**: 59–66.

Tieszen, L. L., D. A. Johnson, and M. M. Caldwell. (1974) A portable system for the measurement of photosynthesis using $^{14}CO_2$. *Photosynthetica*, **8**: 151–160.

Treharne, K. J., and J. P. Cooper. (1969) Effect of temperature on the activity of carboxylases in tropical and temperate Gramineae. *J. Exp. Bot.*, **20**: 170–175.

Warren Wilson, J. (1966) An analysis of plant growth and its control in arctic environments. *Ann. Bot.*, **39**: 383–402.

Webber, P. J. (1978) Spatial and temporal variation of the vegetation and its production, Barrow, Alaska. In *Vegetation and Production Ecology of an Alaskan Arctic Tundra* (L. L. Tieszen, Ed.). New York: Springer-Verlag, Chap. 3.

Zalenskii, O. V., V. M. Shvetsova, and V. L. Voznessenskii. (1972) Photosynthesis in some plants of Western Taimyr. In *Proceedings IV International Meeting on the Biological Productivity of Tundra* (F. E. Wielgolaski and Th. Rosswall, Eds.). Stockholm: International Biological Programme Tundra Biome Steering Committee, pp. 182–186.

11. Primary Production Processes in Arctic Bryophytes at Barrow, Alaska

W. C. OECHEL AND B. SVEINBJÖRNSSON

Introduction

Mosses represent an important growth form in arctic areas where they contribute significantly to the production, biomass, and cover (Wielgolaski, 1972; Webber, 1974, 1978; Rastorfer, 1978). For example, bryophytes comprise 91% of the aboveground biomass in a sedge-moss meadow in Western Taimyr, and an average of 38% of the aboveground biomass for all IBP tundra sites reported by Wielgolaski (1972). In drier tundra areas, however, moss cover drops substantially (Wielgolaski and Webber, 1973). Despite their high levels of cover and biomass, aboveground productivity in arctic bryophytes is generally considerably lower than in vascular plants. Wielgolaski (1972) reports a range of aboveground production for bryophytes from 1 to 50 g dry wt m^{-2} yr $^{-1}$ for numerous IBP tundra sites. This compares to values of 9 to 382 g dry wt m^{-2} yr^{-1} for vascular plants. In select antarctic areas, however, production of bryophytes may reach 1000 g dry wt m^{-2} yr^{-1} (Clarke et al., 1971).

In wet tundra areas, as are found in the Barrow area, bryophytes appear to be of particular importance. In the sites at Barrow mapped by Webber (1974, 1978), six of the seven species with the greatest cover were bryophytes. The relative cover of all bryophytes was 56% compared to 26% for monocotyledons.

Aboveground production of bryophytes and vascular plants may be similar at Barrow. In wet meadow areas, Webber (1974) estimated the aboveground production of vascular plants to be about 52 g dry wt m^{-2} yr^{-1}. The value for bryophytes is less than 50 g dry wt m^{-2} yr^{-1}. However, much of the vascular plant production occurs belowground, and total rates of productivity and photosynthesis are considerably higher in vascular species. The lower level of production in bryophytes may be the result of lower intrinsic rates of photosynthesis and greater environmental limitations. In addition, the winters, which are long and nonproductive, are a time when considerable plant material may be lost due to winter damage and lemming grazing activity.

Mosses at Barrow are distributed throughout the vegetation units, but their biomass and productivity are skewed toward the wet end of the moisture gradient suggesting the importance of moisture relations (Webber, 1974, 1978).

Furthermore, their ability to absorb and to tenaciously retain nutrients makes an understanding of their photosynthesis, growth, production, and decomposition processes of fundamental importance. Their relative independence of nonproductive supporting and absorptive tissues provides an extreme contrast to the terrestrial *Dupontia*-based graminoid system. This chapter will attempt to describe the photosynthetic response patterns and growth strategies of mosses at Barrow.

Field Research Site

The research site was established on U.S. Tundra Biome intensive site 2 and was situated in a transitional region between a moderately polygonized area to the southwest and a wet meadow area showing little polygon development to the east. In the polygonized area, three distinct habitats are found: (1) interpolygonal troughs of various depths and widths, (2) low polygon centers, and (3) rims of low-centered polygons and elevated portions of high-centered polygons. The wet meadow is considered a single habitat although diversity can be found in moisture levels and other factors.

Calliergon sarmentosum is a principal moss species which occurs in interpolygonal troughs with sparse cover of *Dupontia fisheri* and occasionally in ponds with *Arctophila fulva. Pogonatum (Polytrichum) alpinum* is common in all but the wettest part of the wet meadow, and it occurs in low polygon centers, sometimes in almost pure stands. In both habitats it is found in association with *Carex aquatilis* and *Dupontia fisheri,* although the latter is rare in low polygon centers. *Polytrichum commune* is especially abundant on polygon ridges and on raised center polygons where the sparse cover of vascular plants is comprised of *Luzula confusa* and several other species. In these habitats small hummocks are formed, principally as a result of the growth of *Dicranum elongatum.* In drier parts of the wet meadow, *Dicranum angustum* clumps are frequently found growing in the understory of *Salix rotundifolia.* In the wettest part of the wet meadow and the deepest polygon centers where standing water is common throughout the growing season, *Drepanocladus revolvens* is frequently found. *Sphagnum* mats can be found in the wet meadow in an intermediate moisture regime.

The distribution mosaic of species, especially those with cushion growth form, is a reflection of microtopographic differences. Figure 1 illustrates the early season distribution of bryophyte green biomass and represents the mean biomass of six transects through low-centered polygons. Bryophyte biomass is highest on the sides of polygon rims where moisture, wind, and soil O_2 levels would be expected to be intermediate between the troughs and the rim tops. Bryophytes are also abundant in the wet meadow areas. Lichens are most prevalent in drier areas (see Webber, 1978), such as on the polygon rims and drier meadow areas. Early in the season there is little living, aboveground vascular plant material. Total bryophyte biomass ranged from about 7 g dry wt m^{-2} in the polygon low

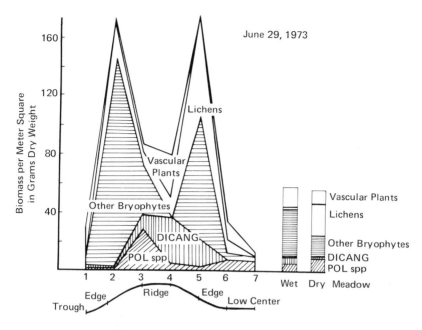

Figure 1.   Distribution of aboveground biomass for lichens, mosses (green tissue), and vascular plants (living tissue). Numbers represent sampling stations of transects across the polygon system. POL spp. = *Pogonatum alpinum* and *Polytrichum commune,* DICANG = *Dicranum angustum.*

centers to 150 g dry wt m^{-2} on the polygon rim edge. Biomass of bryophytes in the wet meadow was 45 g dry wt m^{-2}.

Methods

Determination of Resistances to Water Loss

Values of leaf resistances (r_1) and air resistance (r_a) to water loss were determined through substitution of appropriate values in Eq. (1) (Slatyer, 1957; Kramer, 1969):

$$\text{Transpiration} = \frac{VD_1 - VD_a}{r_1 + r_a} \qquad (1)$$

where VD = water vapor density of the leaf (l) or air (a).

Fully moistened moss samples were mounted in their normal orientation in aluminum weighing trays and placed in Plexiglas cuvettes with an internally mounted fan. Leaf and air temperatures were monitored with thermocouples, wind speed was measured at the moss surface, dew points were monitored with a

Cambridge model 880 dew point hygrometer, and flow rates into the chamber were measured with calibrated flow meters. The shoots were kept fully moistened to determine values of r_a. Effects from r_l were thereby eliminated, and all resistance to water loss was due to r_a. The r_l values as a function of water content were similarly calculated. All results were expressed on a unit area basis (ground cover) so that resistances are in the standard units (sec cm^{-1}) (see Hicklenton, 1975).

During diurnal measurements of photosynthesis, water content was simultaneously determined for the species measured according to Eq (2). Percentage water content (% WC) was calculated as follows:

$$\% \text{ WC} = \frac{\text{fresh wt} - \text{dry wt}}{\text{dry wt}} \times 100. \tag{2}$$

Determination of Net CO_2 Exchange

Intensive characterizations of the net photosynthetic patterns were undertaken on five common species: *Pogonatum alpinum, Polytrichum commune, Calliergon sarmentosum, Dicranum elongatum,* and *D. angustum.* Habitats studied with representative genotypes included: wet meadow, high polygon rim tops, polygon rim sides, polygon low centers, and deep polygon troughs. Some species were studied in more than one habitat (e.g., *Pogonatum alpinum*), and two genera each spanned several habitats. In 1972, the net photosynthetic responses of *C. sarmentosum* and *P. alpinum* to temperature, light and moisture were determined. *In situ* measurements of net photosynthesis for these and other species were also made. In 1973, emphasis was placed on the elucidation of diurnal photosynthetic patterns through continuous monitoring of the species of interest. An intensive sampling pattern was necessary to adequately observe the populations and to sample developmental, environmental, and ecotypic or phenotypic effects on net photosynthesis during the short growing season. In 1974, attention was again turned to field experimentation in an attempt to thoroughly examine acclimatization and senescence effects and interspecific differences in response patterns.

Laboratory and Field Determinations

The temperature response patterns of net photosynthesis were determined under saturating water contents and near optimal irradiance levels of 0.147 to 0.181 cal cm^{-2} min^{-1} PhAR (650 to 800 μE m^{-2} sec^{-1}, ca. 3500 ftc) between temperatures of 0 and 30°C. Dark respiration was measured alternately with net photosynthesis or at the end of the experiment by darkening the chambers and reducing temperature.

Responses to light were determined using moisture-saturated tissue held at near optimal temperatures of 10 to 15°C. Samples were placed in darkened chambers, the rate of respiration was obtained, and the light was gradually increased to 0.430 cal cm^{-2} min^{-1} PhAR (1900 μE m^{-2} sec^{-1}) (Oechel and

Collins, 1976) over a period of 4 to 5 hr. Alternatively, measurements were made at an intermediate light intensity, and then subsequent measurements were made at alternately higher and lower values.

The effect of tissue water content on photosynthetic rates under optimal conditions [10°C, 0.181 cal cm^{-2} min^{-1} (800 μE m^{-2} sec^{-1})] was determined. Moss tissue was fully saturated, and photosynthetic rates and tissue water contents were measured as the tissue desiccated.

For field studies moss samples were placed in cuvettes and maintained fully hydrated under ambient temperature conditions. Samples received either normal incident sunlight or one-half incident illumination (to simulate growth in the open or under a vascular plant overstory). Moss tissue temperatures in the cuvettes were controlled to track ambient tissue levels. Temperature was measured continuously in 1973 in each of the habitats of interest (Oechel and Sveinbjörnsson, 1974). Special emphasis was placed on the monitoring of tissue temperatures in each of the habitats being studied intensively. Light intensity was measured continuously at two levels above vascular plant canopy using a Kipp and Zonen solarimeter and a Lambda Instrument Quantum sensor and below canopy with a Lambda Quantum sensor. Net photosynthetic rates were measured using an "open" infrared gas analysis system employing Beckman 215A or MSA differential gas (CO$_2$) analyzers (Hicklenton and Oechel, 1976, 1977; Vowinckel et al., 1974; Oechel, 1976; Oechel and Collins, 1976).

Results and Discussion

Water Relations

Mosses are poikilohydric and show little control over tissue moisture status. Since they lack roots or functional equivalents, they are largely independent of the stabilizing supply of soil moisture. The uptake of liquid water occurs primarily as a result of capillarity on the outside of the plant (Bowen, 1931, 1933; Magdefrau, 1935). Furthermore, internal conduction of water in mosses and the role of rhizoids in water uptake appear to be minimal in many species (Anderson and Bourdeau, 1955), although dense rhizoid tomentum may increase the capillary uptake of moisture from the substrate. The saturated nature of the Barrow soils and the high incidence of rain and fog suggest that both capillary movement of water from the soil surface and the application of liquid water directly to the leaves and stems are important in supplying moisture to the moss.

The colony growth form increases the rates of water uptake over that achieved by individual shoots. *Pogonatum alpinum* and *P. commune* represent the less dense turf growth form and water relations are dictated primarily by the response of single shoots. *Calliergon sarmentosum*, on the other hand, develops a carpet growth form with high shoot density. In *C. sarmentosum*, water is held externally, and the colony form is much more important in controlling water uptake and loss than in the case of *P. alpinum* (Gimingham and Smith, 1971). However, it appears that appreciable water can be taken up via the stem bases of both species. These rates of water uptake vary considerably. *P. alpinum,* e.g., has a slow rate of uptake as shown by the length of time to recover 50% of the total water content (WC_{50}). The rate of water uptake (from air dried status to WC_{50}) was 0.24 g of H_2O g dry wt^{-1} min^{-1} when the bases were immersed in water to a depth of 2 mm and 0.01 g dry wt^{-1} min^{-1} when the apices were immersed to 2 mm. *Calliergon sarmentosum* took up water much more rapidly and to a larger extent. Water was taken up via the apices at a rate of 4.05 g of H_2O g dry wt^{-1} min^{-1} and WC_{50} was reached in 1.2 min. Water uptake via the bases was faster at a rate of 12.9 g of H_2O g dry wt^{-1} min^{-1} with WC_{50} only 0.3 min (calculated from the authors' data, Gimingham and Smith, 1971). Therefore, the rates of water uptake for these two Barrow species differ widely and approximate the range of extremes found.

In the wet meadow area, Ψ_{soil} is usually high, thus providing a positive gradient for water movement. Water input to the mosses also occurs from above in the form of dew, rain, and drip from the vascular plant overstory. Dew fall and fog deposit water at an average rate of about 20 g m^{-2} day^{-1} (Oechel and Sveinbjörnsson, 1974). If the input of water exceeds the water holding capacity of the moss colony, then the precipitation will percolate through the moss carpet. Water-holding capacities of the moss surface vary by species. For *Polytrichum juniperinum* (from the subarctic) it is about 13 g of H_2O g dry wt^{-1} (520 g of H_2O m^{-2}). *Dicranum angustum* will retain about 18 g of H_2O g dry wt^{-1} (720 g of H_2O m^{-2}) and *Calliergon sarmentosum* will retain the largest amount, 93 g of H_2O g dry wt^{-1} (3720 g of H_2O m^{-2}) (Oechel and Hicklenton, unpublished).

Transpiration rates in mosses are potentially very high due to small resistances to water loss. *Calliergon sarmentosum* is a mesic to hydric species and shows little resistance to desiccation (at a water content of about 400%, water loss is 0.230 g g dry wt^{-1} min^{-1} at 24°C, 20% rh, and wind speed of 1.7 m sec^{-1}). *Pogonatum alpinum* occurs in xeric to mesic areas at Barrow and shows numerous xerophytic adaptations including cuticularization and the ability to fold the leaves against the stem. At saturating water contents of 400% and the same conditions as above, *P. alpinum* loses only about 0.032 g of H_2O g dry wt^{-1} min^{-1}.

Interpretation of r_a and r_l values in mosses is difficult because of differences in morphology and growth habit between mosses and vascular plants. Due to difficulties in measuring wind movement and vapor density at the leaf level within the moss canopy, it was decided that the primary area unit should be ground cover area rather than leaf area. In this way the moss canopy is considered as a rough, horizontal surface. r_a values were similar for both *D. elongatum* and *C. sarmentosum*. At wind speeds between 0.3 and 3 m sec^{-1}, r_a values ranged from 2.2 to 1.0 sec cm^{-1} (Figure 2a). Higher wind speeds had little effect on decreasing r_a. These species, however, showed possible differences with respect to r_l. Over all measured ranges of water content, r_l values were higher in *D. elongatum* than in *C. sarmentosum*. At water contents of about 175%, for example, r_l in *D. elongatum* is about 6.7 sec cm^{-1} as opposed to about 5.8 in *C. sarmentosum* (Figure 2b). Both species show little control of water loss, even under severe desiccation. These responses are in contrast to those of

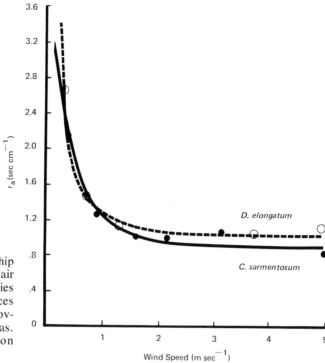

Figure 2a. The relationship between wind speed and air resistance in two moss species from Barrow. Air resistances are based on ground area covered rather than on leaf areas. Lines fitted by regression analysis.

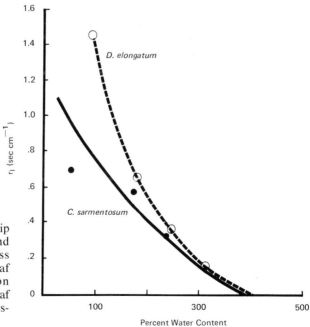

Figure 2b. The relationship between water content and leaf resistance in two moss species from Barrow. Leaf resistances are based on ground area rather than on leaf areas. Lines fitted by regression analysis.

subarctic *P. juniperinum* which shows higher levels of r_l and an ability to increase r_l abruptly as desiccation proceeds (Oechel and Hicklenton, unpublished data). This results in part from a rolling of the leaf edges and the movement of the leaves to a position against the stems. If it can be assumed that *P. alpinum* and *P. commune* contain similarly high values for r_l with moisture availability, then these relationships may present a partial explanation for *Polytrichum* spp. success in drier sites in the Barrow tundra. Increases in r_l in *D. elongatum* and *C. sarmentosum* may result from several factors including desiccation of the most exposed shoot tips and reorientation of the shoots to increase the average path length for water vapor diffusion.

Low r_a and r_l values in bryophytes result in potential rates of transpiration much higher than those in vascular plants. They are, therefore, extremely subject to desiccation and the factor controlling water loss in mosses rests in most cases on the vapor density gradient between the plant tissue and the surrounding air. Bryophytes are, therefore, extremely dependent on their immediate microclimate to maintain a favorable water balance. Despite their susceptibility to desiccation, the Barrow area provides an ideal climate to maintain favorable moisture regimes. The low radiation levels, low temperatures, and low evapotranspiration levels all act to help maintain turgescent moss tissue; and mosses in the Barrow area usually show favorable levels of water content throughout the summer (Figure 3). *P. alpinum* shows moisture levels throughout the summer of 140 to 200% water content, sufficiently high to maintain photosynthesis at near optimal levels (see later sections and Oechel and Collins, 1976).

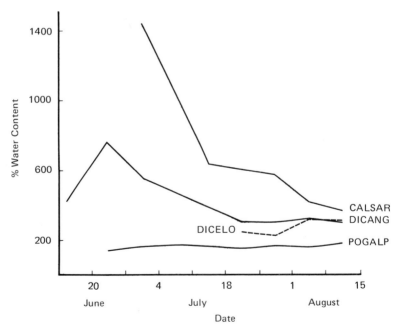

Figure 3.   The seasonal course of the tissue water content in four moss species on IBP site 2. CALSAR = *C. sarmentosum*, DICELO = *D. elongatum*, DICANG = *Dicranum angustum*, and POGALP = *Pogonatum alpinum*.

Calliergon sarmentosum occurs primarily in troughs, ponds, and, to a lesser extent, as an understory element in wet meadow areas. Early in the year, it is often submerged or inundated with water, resulting in water contents exceeding 1400% (internal and external water). As water levels drop and the season progresses, water contents decline to 400%. Even this level, however, maintains photosynthesis near maximum levels, and thus moss water relations at Barrow seem near optimal. Locally, however, tissue moisture relations affect species distribution and survival. Certainly moisture regimes of the polygon rims would not generally allow the growth of *C. sarmentosum* in that area, and the wet nature of many *C. sarmentosum* habitats would depress photosynthesis in *P. alpinum* (Oechel and Collins, 1976).

There are local populations which become desiccated at various times during the summer which are not represented in Figure 3. This is especially common in *C. sarmentosum*. As the water level recedes, populations at higher levels will drain free of water early in the summer and become desiccated. This is especially prevalent for individuals growing at the margin of troughs which are water filled early in the season. Also periods of drought will have a major influence on bryophyte growth and survival. An exceptionally warm dry period of several weeks in 1972 resulted in the death of numerous individuals of *P. alpinum* and *C. sarmentosum* and other species. *C. sarmentosum* mats recovered only through the initiation of new shoots from existing material (Oechel and Collins, 1976).

Photosynthesis and Productivity

Vascular plants in tundra regions occupy habitats which are apparently near the environmental extremes that will support them. However, mosses occurring in the same areas may exist in microenvironments to which they are well adapted. They may be less restricted by the arctic conditions of low solar radiation, low temperatures, anaerobic soil conditions, and short growing season than are vascular species. Presumably, net CO_2 incorporation is a valid indicator of success in arctic mosses. Certainly, a moss in a negative carbon balance cannot maintain itself. Increased CO_2 incorporation may allow greater rates of sexual reproduction and biomass accumulation resulting in a larger photosynthetically active compartment, thereby permitting still greater rates of biomass production. The following sections will describe intrinsic and extrinsic controls on photosynthesis and production in arctic mosses.

Cellular photosynthetic pathways in arctic mosses are assumed to be similar to those described for vascular plants, although resistances to CO_2 uptake and overall photosynthetic rates and response patterns in mosses do diverge from their vascular plant counterparts. As in the case of the vascular plants, photosynthesis can be viewed as a response to a CO_2 concentration gradient and a series of resistances to flow along this gradient. However, as a result of the morphology of mosses, the resistances to flow are somewhat different.

The range of resistances for r_1 of mosses is limited because of the lack of stomates which would regulate water loss. Variation in r_1 is possible, however, by changing leaf orientation and cell wall conductances. r_y is the residual resistance to CO_2 flux when the other resistances are known and is not synonymous with r_m used with higher plants due to the simpler anatomy of mosses. In mosses there is no mesophyll and, in leaves of a single cell layer, there are no cell wall resistances incorporated in r_y. The magnitudes of r_a and r_l to H_2O vapor flux and their response to extrinsic factors, such as wind speed and tissue water content, were discussed earlier.

Intrinsic Factors Affecting CO_2 Exchange

P_{max}, the maximum rate of CO_2 incorporation, varies widely by species and they are similar to those found in temperate, subarctic, and antarctic mosses (Hosokawa et al., 1964; Stålfelt, 1937; Rastorfer, 1972; Kallio and Kärenlampi, 1975). It is of interest to note that the average maximum net photosynthesis values for mosses in the Barrow area, shown in Table 1, are about 50% larger (1.5 vs 2.1 mg of CO_2 g dry wt^{-1} hr^{-1}) than the average of all the values reported by Kallio and Kärenlampi. The highest value reported from Barrow (4.4 mg of CO_2 g dry wt^{-1} hr^{-1}) is about 25% larger than the highest value reported by Kallio and Kärenlampi for temperate and subarctic mosses. Reported rates of photosynthesis in mosses, however, range as high as 16 mg of CO_2 g dry wt^{-1} hr^{-1} for *Polytrichum juniperinum* (Bazzaz et al., 1970, calculated from the authors' data). It appears, therefore, that average maximal rates of photosynthesis are reduced little, if any, in arctic mosses when compared to mosses from temperate and boreal biomes.

These maximal rates of photosynthesis are much lower than in the vascular plants and average about 8 to 9% of the responses in graminoids and nongraminoid herbaceous plants presented by Tieszen (1978). However, aboveground production values for these growth forms are more similar than the photosynthetic values would indicate, due to the high level of belowground translocation in vascular plants for construction and maintenance of belowground organs. The majority of the bryophyte production remains in aboveground structures and is potentially productive.

Although little seasonal effect or senescence has been detected, the age class of the tissue does affect the photosynthetic response. First-year segments of *Polytrichum alpinum* show the maximum photosynthetic response. In second-year segments the rate decreases to 75% of the maximum rate and in the third year the rate is 42% of the first-year rate (Collins and Oechel, 1974). Shoots with 2-yr-old tissue comprise about 15% of the number of 1-yr shoots. Shoots with 3-yr-old tissue are much less common and vary from 0 to about 4% (Collins and Oechel, 1974).

This perennial and evergreen nature of certain arctic moss species and the retention of photosynthetically active leaf material for more than 1 yr allows a higher return on materials invested in photosynthetic tissue than would otherwise be possible. This contrasts with a total leaf life of less than 60 days in graminoids (Tieszen, 1978). The perennial evergreen nature of *Polytrichum* and *Dicranum* species is made possible in part by the very "resistant" physiology of mosses (Kallio and Heinonen, 1973).

The evergreen growth habit and the apparent lack of programmed senescence means that mosses are able to take full advantage of environmental conditions

Table 1. Maximum Photosynthetic Rates (Commonly Observed Maximum Values) of Tundra Mosses

Species	Location	Habitat	P_{max} (mg of CO_2 g dry wt^{-1} hr^{-1})	Notes	Reference
Pogonatum alpinum	Barrow	Wet meadow	4.4	Field grown plants at optimum light and temperature	Oechel and Collins, 1976
Calliergon sarmentosum	Barrow	Polygon troughs	2.7	Field grown plants at optimum light and temperature	Oechel and Collins, 1976
Polytrichum commune	Barrow	Polygon rims	2.9	Under ambient light and temperature	Oechel and Sveinbjörnsson, unpublished
Dicranum angustum	Barrow	Wet meadow	1.0	Under ambient light and temperature	Oechel and Sveinbjörnsson, unpublished
Dicranum elongatum	Barrow	Polygon rims	1.3	Under ambient light and temperature	Oechel and Sveinbjörnsson, unpublished
Dicranum angustum	Barrow	Wet meadow	1.0	Growth chamber grown plants at optimal light and temperature	Hicklenton *et al.*, unpublished
Dicranum fuscescens	Schefferville, Quebec 55° N lat.	Subarctic shrub-tundra under birch and spruce	1.5	Growth chamber grown plants at optimal light and temperature	Hicklenton and Oechel, 1976

conducive to positive photosynthesis as they occur. There is no major withdrawal of organic and inorganic nutrients (Hicklenton, 1975; Oechel and Sveinbjörnsson, unpublished data) in the fall, and photosynthesis can continue as
long as sufficiently high temperature and light conditions persist (see later in this
chapter). As a result, high photosynthetic competencies persist much longer than
in graminoid or other deciduous vegetation (Tieszen, 1978). Tieszen (personal
communication) showed that the photosynthetic rate for *Polytrichum* sp. under
the snow was between 0.34 and 0.61 mg of CO_2 g dry wt^{-1} hr^{-1}. These rates
represent between 8 and 14% of the maximum rate observed for *P. alpinum*
under optimal conditions. In contrast, *Dupontia fisheri,* during the same period,
photosynthesized at values of from 0.4 to 2.4% of its typical maximum, and
Carex aquatilis achieved only 0.3 to 0.5% of its maximum rate (Tieszen, 1974,
1978). The large number of shoots which overwinter until the second spring are
important in early season photosynthesis, since during this period, continuous
light, large solar angles, a poorly developed vascular plant overstory, and high
moisture and nutrient levels allow potentially high photosynthetic rates.

Extrinsic Factors Affecting CO_2 Exchange

Light.   Light intensity, in part, sets the upper limit possible for photosynthetic rates. However, statements that arctic tundras are light-limited ecosystems
may better apply to vascular plants. Arctic mosses tend to reach light saturation
at lower radiation values than do their vascular plant counterparts and in the field
saturate at about 0.14 cal cm^{-2} min^{-1} (PhAR) ($\sim$620 μE m^{-2} sec^{-1} PhAR) (Oechel
and Collins, 1976). This is about 30% of the radiation required for
saturation of vascular plants in the same area. The tendency for light intensities
above saturation to reduce the rate of photosynthesis in some instances may be a
result of photo-inhibition or photo-oxidation of the photosynthetic apparatus
(Oechel and Collins, 1976). This is in contrast to the graminoids which increase in
photosynthesis to radiation levels approaching full sunlight (Tieszen, 1975, 1978),
and represents a major response difference between the two growth forms. As a
result, moss species are likely to be light saturated much of the time and possibly
even light inhibited. At low light intensities, responses are similar to graminoids,
and nearly equivalent irradiances are required for compensation (0.008 to 0.016
cal cm^{-2} min^{-1}). Thus, as with vascular plants, even low levels of light will yield
positive photosynthesis.

Laboratory studies showed negligible seasonal shifts in photosynthetic
responses to light intensity in nonpolytrichous species (Oechel and Sveinbjörnsson, unpublished data). *Dicranum elongatum, D. angustum,* and *Calliergon
sarmentosum* showed no discernible tendency for compensation points, saturation values, or maximum photosynthesis rates to vary significantly with the
season. Thus, effects of acclimatization of these features or senescence appear to
be minimal in these species. In all cases, photosynthesis rates increase with
increasing light intensity to 0.430 cal cm^{-2} min^{-1}, although the increase was small
above 0.068 to 0.181 cal cm^{-2} min^{-1}. *Pogonatum alpinum* and *Polytrichum
commune,* in contrast, showed seasonal shifts in maximum rates of photosynthesis and in the light intensity required for photosynthetic compensation. In the

early summer (late June) light requirement for compensation was found to be about 0.034 to 0.045 cal cm^{-2} min^{-1}. This dropped to 0.007 to 0.014 cal cm^{-2} min^{-1} in late summer (Aug–Sept). Samples of *P. alpinum* growing shaded under the vascular plant canopy in the wet meadow had the lowest late season light requirement for compensation. They also showed marked inhibition of photosynthesis at light intensities higher than 0.181 cal cm^{-2} min^{-1} (800 μE m^{-2} sec^{-1}), which is slightly higher than *in situ* maximal radiation levels. These features are probably phenotypic as artificial low light growing conditions [0.015 cal cm^{-2} min^{-1} (65 μE m^{-2} sec^{-1})] induced similar compensation light requirements in *P. commune,* and some indication of high light intensity inhibition, albeit small, was found (Figure 4).

The light intensity required for net CO_2 exchange, compensation, and saturation is dependent on temperature. In all five species, light requirements for CO_2 compensation increased with increasing temperatures from 0 to 30°C. For example, in *D. angustum,* compensation was reached at intensities of 0.005 cal cm^{-2} min^{-1} at 0°C and 0.070 cal cm^{-2} min^{-1} at 30°C. The light intensity at which saturation occurs also increased with temperature. *D. angustum,* which is light saturated by 0.027 cal cm^{-2} min^{-1} at 0°C, light saturates at about 0.136 cal cm^{-2} min^{-1} at 30°C. *C. sarmentosum* saturates at 0.091 cal cm^{-2} min^{-1} at 0°C and at 30°C is not light saturated at 0.430 cal m^{-2} sec^{-1}.

Total daily radiation as it is normally distributed has only a small effect on

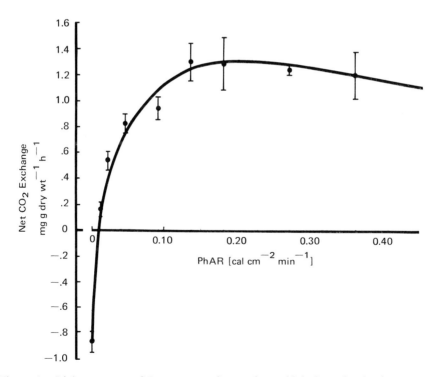

Figure 4.   Light response of *P. commune* from polygon high rims after having grown at low light intensity for 5 weeks. Bars represent one standard deviation.

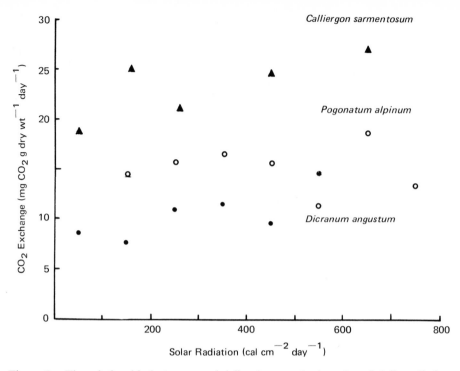

Figure 5. The relationship between total daily photosynthesis and total daily radiation. Each point represents the average daily photosynthetic rate for each daily radiation interval (in 100 calorie intervals, PhAR).

total daily photosynthesis (Figure 5). Correlation coefficients between the two variables were insignificant. Presumably this is because the plants are usually light saturated for much of the day, and increasing radiation during these periods would have little effect on photosynthesis. This situation is in contrast to the situation in vascular plants where increases in daily radiation strongly increase the photosynthetic total as a result of the unsaturated nature of the canopy below 500 cal cm^{-2} day^{-1} (Tieszen, 1978).

    The mosses are never light saturated from 2300 to 0200 hr (solar time) (Figure 6a). However, between 1200 and 1300 hr, the mosses are saturated 81% of the

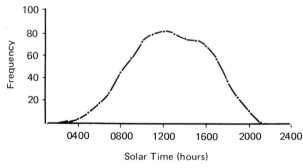

Figure 6a. Percentage frequency of time when radiation is above light saturation for photosynthesis in mosses (about 0.15 cal cm^{-2} min^{-1}, PhAR) from 15 June to 11 August 1973. Data represent hourly means of light intensities recorded at 5-min intervals.

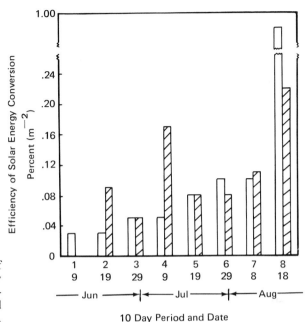

Figure 6b. The efficiency of solar energy conversion by bryophytes growing in full sunlight (open bars) and reduced sunlight (50%, hatched bars).

time. Throughout the 24-hr period, light saturation occurred 34% of the time. If the mosses are growing under a vascular plant overstory, which reduces the light by 50%, the period when no light saturation occurs is increased to from 2100 to 0400 hr, and midday light saturation is found only about 40% of the period between 1100 and 1200 hr.

The efficiency of solar energy conversion was calculated throughout the summer for shaded and nonshaded moss canopies (Figure 6b). The calculated efficiencies were generally similar for shaded and nonshaded mosses. This is the result of the interaction of lower incoming radiation, lower photosynthesis rates, and lower biomass in the shaded areas. Efficiency tended to increase as the season progressed. This is due to the increasing biomass, which resulted in higher efficiencies of energy conversion, and also to lower radiation levels. The extremely high efficiencies in the last period, especially in the unshaded mosses, resulted from a few days with low radiation and high photosynthesis rates. The seasonal efficiency of energy conversion per square meter in full sunlight was 0.18%, and in reduced sunlight (50% of full sunlight) it was 0.11%. The mosses generally are much lower in overall efficiencies than are vascular plants, except under periods of low radiation (Figure 6b).

Temperature. Low arctic temperatures are another factor imposing a potential limitation on carbon accumulation in mosses in the Arctic. However, relatively high photosynthetic rates at low temperatures allow appreciable CO_2 accumulation at arctic temperatures. Mid-season temperature optima are 10 to 15°C, and rates are only slightly decreased at 5°C (Figure 7). *P. alpinum* photosynthesizes at 55% of the maximum rate at 0°C. Upper temperature

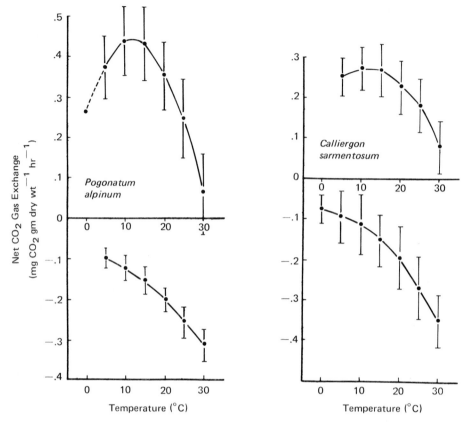

Figure 7.   Midseason response of photosynthesis and respiration to temperature in two moss species (from Oechel and Collins, 1976). Bars represent two standard deviations.

compensation is about 30°C for both species (Oechel and Collins, 1976). While the high rates of photosynthesis at low temperatures are of obvious adaptive significance, temperature optima of from 10 to 15°C may at first seem inordinately high. However, upon closer examination, the high rates at temperatures exceeding 10°C are important to the carbon economy of arctic mosses. Tissue temperatures frequently drop to between −5 and 0°C during the growing season (Figure 8). This places a selective advantage on maintaining positive photosynthesis at temperatures at and below 0°C. Continuous sunlight results in positive net photosynthesis during these periods. However, during periods of high radiation, tissue temperatures exceed air temperature. In 1973, midday temperatures were above 5°C 87% of the time and above 10°C 44% of the time. In 1972, which was warmer and drier, temperatures were above 10°C at midday 73% of the time and reached 30 to 35°C. The broad temperature responses of arctic bryophytes, allowing high rates of photosynthesis from near 0°C to temperatures above 20°C, confer adaptation to the wide range of tissue temperatures encountered in the

Arctic. However, despite these factors, the temperature optimum observed for
D. elongatum decreases carbon uptake by about 25% below the maximum
theoretically possible under another, lower, temperature optimum (Oechel et al.,
1975).

There seems to be considerable uncertainty concerning the acclimation potential of photosynthesis in mosses to temperature. Because of the reduced seasonality during the summer of tissue temperatures, little temperature acclimatization
might be expected at Barrow. Moss surface temperature data for 1974 show little
progression in temperatures, with 5-day mean values running between about 4
and 9°C (Billings et al., unpublished data). Long-term monthly averages of air
temperature support these general trends. Possibly in response to this relatively
constant thermal regime, little field acclimatization of net photosynthesis to
temperature has been observed in nonpolytrichous species (Figure 9). Vascular
plants have also been shown to undergo little field acclimatization to temperature
(Tieszen, 1975). Generally there was little change in absolute rates of net
photosynthesis after the early season (June) sampling periods in *Dicranum
elongatum* (Figure 10). Dark respiration rates, as well, were generally uniform
through the summer. However, *Dicranum* tended to show aberrant early season
response patterns in June. This may indicate an early season lag time during
which the plants are not operating at full metabolic potential.

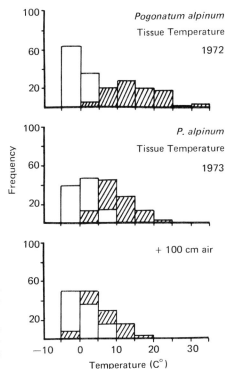

Figure 8. Frequency of temperature in each
temperature class for air (1973) and plant tissue for the period 11 June to 19 August 1973
and 24 June to 10 August 1972. Hatched bars
represent 1000 to 1359 hr in 1972 and 1200 to
1300 hr in 1973. Clear bars represent 2000 to
0159 hr in 1972 and 2300 to 2400 hr in 1973.

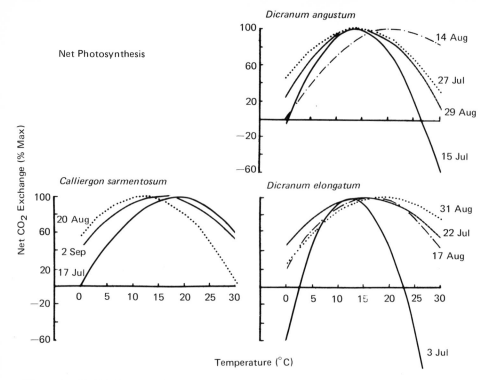

Figure 9.   Relative seasonal response of net photosynthesis to temperature in three moss species during the summer, 1974 (from Oechel, 1976).

The lack of strong acclimatization of photosynthesis to temperature appears to be a response to the Barrow climate, and not a genetic limitation. For example, Hicklenton and Oechel (1976) have shown marked acclimatization in *Dicranum fuscesens* in Schefferville, Quebec, where the environment is much more continental and seasonal. Temperature optima ranged from 5°C in the spring and broadened to a range of from 5 to 25°C in midsummer in a lowland population. An upland population showed a temperature optimum range of 0 to 10°C in June and an optimum of 10 to 20°C in late August (Hicklenton and Oechel, 1976).

Pogonatum alpinum and *Polytrichum commune* showed larger shifts in temperature optima and in absolute rates of photosynthesis. This may reflect differences in growth form, habitat, or development between polytrichous mosses and the others measured. Also, early season temperature optima for photosynthesis tended to be much lower in *Polytrichum* and *Pogonatum* than in *Calliergon* and *Dicranum* spp. *P. alpinum* from wet meadow had the lowest optimum (about 3°C), whereas the same species from low-centered polygons and *P. commune* had optima about 10°C (Figure 11). By early September these optima had shifted upward to about 18°C except in *P. alpinum* from low-centered polygons where the change was minimal. Maximum rates were approximately double that of the low early season values. Absolute rates were similar at 0°C, but the relative

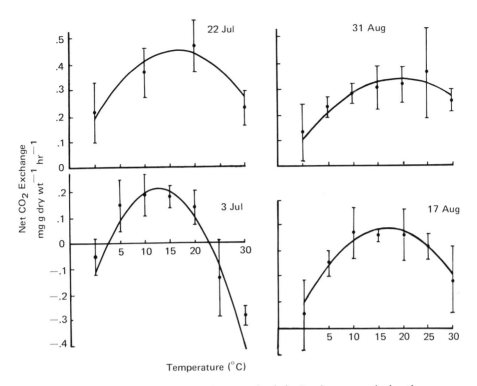

Figure 10.   Seasonal response of net photosynthesis in *D. elongatum* during the summer of 1974.

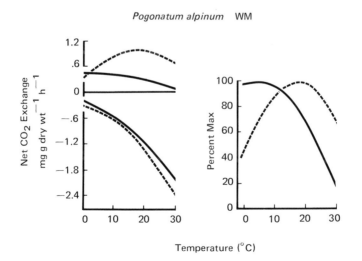

Figure 11.   Seasonal response of net photosynthesis and dark respiration of shoots of *P. alpinum* from wet meadow during the summer of 1974 (solid line, early season; broken line, late season). Curves fit by regression analysis.

percentage had obviously decreased markedly. These marked changes in maximum photosynthetic rates and temperature optima are obviously not caused by dark respiration but probably involve the photosynthetic process itself either through changes in CO_2 diffusion resistance or enzymatic activity.

Moisture. The water relations of the wet meadow tundra are particularly conducive to bryophyte growth. Cool temperatures, low radiation, and greater precipitation than evapotranspiration all act to maintain hydrated tissue. This, in turn, yields high photosynthetic rates and avoids damage due to desiccation. The photosynthetic responses of mosses to water status, however, reflect the water relations of their habitats of origin. *Pogonatum alpinum,* which occurs in drier tundra areas including wet meadows and polygon centers, reaches photosynthetic compensation at 60% water content, and optimal rates are observed at 200 to 350% WC (Figure 12). Higher water contents may actually decrease the rate of photosynthesis by increasing the resistance to CO_2 movement through the accumulation of water on leaf surfaces and in leaf bases (Oechel and Collins, 1976). *Calliergon sarmentosum* occurs in wetter areas, including polygon troughs and the wet meadow, and requires higher moisture contents (75% WC) to reach compensation and maximum photosynthesis (750% WC) than does *P. alpinum.* During the summer of 1973, these species generally remained hydrated allowing photosynthesis to proceed at near maximal rates (Figure 3).

Nutrients. Mosses are apparently quite effective in intercepting and absorbing available nutrients. As a result, moss production was not demonstrably nutrient limited. Fertilization with a dilute nutrient solution showed no increase in either growth or photosynthesis. These results are similar to those found for lichens and feather mosses growing in the subarctic (Carstairs and Oechel, 1978;

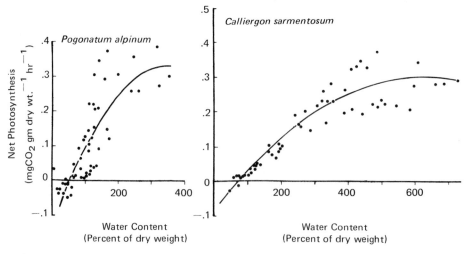

Figure 12.   The response of photosynthesis to moisture status in two moss species (from Oechel and Collins, 1976). Curves fit by regression analysis.

Oechel, unpublished data). In comparison, graminoids growing in the same area are nutrient limited for growth (Chapin, 1978, McKendrick et al., 1978), but not for photosynthesis (Tieszen, 1978).

Diurnal and Seasonal Patterns of CO_2 Exchange

The photosynthetic response patterns can be related to three periods of the growing season. During early summer (mid-June) radiation was near its maximum, there was continuous sunlight, snow was present in many areas, and the moss had recently emerged from snow cover. There was no current season's growth, and the photosynthetically active tissue had overwintered. The second period represents mid-season (mid-July). During this period there was still continuous radiation, but evening radiation values were falling, new tissue was present, and conditions were favorable for growth. There was no snow cover and temperatures were mild. The last period represents the end of season (early August). Snowfall was frequent, there was a dark period at night, and temperature and radiation during this period were decreasing. These periods are presented in Figure 13 and represent the general trend in photosynthesis of mosses through the summer season.

Figure 13a illustrates the early-season photosynthetic response of *Pogonatum alpinum* growing in full sunlight in the site 2 wet meadow. Photosynthesis was positive throughout the day, and at solar midnight was 20% of the daily maximum. Minimum hourly average radiation occurred at 2400 hr and was 0.015 cal cm^{-2} min^{-1}. The maximum rate of photosynthesis of 0.80 mg of CO_2 g dry wt^{-1} hr^{-1} was lower than later in the season. This is presumably a result of the fact that all of the photosynthetic tissue had overwintered at least one season and

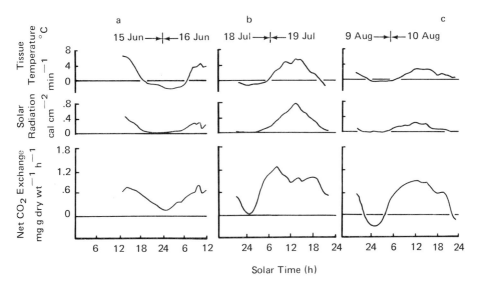

Figure 13.   Diurnal pattern of net CO_2 exchange in *Pogonatum alpinum* in early summer (a), mid-summer (b), and late summer (c), solar radiation, and leaf temperature.

therefore had an intrinsically lower P_{max}. It appears that photosynthesis was limited simultaneously during most of this period by suboptimal light and temperature levels and by the intrinsic rate of P_{max}. Net photosynthesis for the 24-hr period was 11.6 mg of CO_2 g dry wt^{-1} day^{-1}.

The second period is illustrated in Figure 13b. By this time radiation had decreased near solar midnight causing photosynthesis to drop below light compensation. At 0100 hr a light intensity of 0.01 cal cm^{-2} min^{-1} resulted in a loss of 0.01 mg of CO_2 g dry wt^{-1} hr^{-1}. Light saturation was reached at 0.17 cal cm^{-2} min^{-1} at 0800 hr. Further increases in temperature and light resulted in a decrease in the photosynthetic rate. Since temperatures remained below the optimum, the observed decreases in photosynthesis may have been the result of a photoinhibitory process. Total net photosynthesis for the 24-hr period was 19.2 mg of CO_2 g dry wt^{-1} day^{-1}.

Photosynthesis during the third period was both light and temperature limited (Figure 13c). Light intensities for the day approached saturation at 0.13 cal cm^{-2} min^{-1} resulting in photosynthetic rates of 0.90 mg of CO_2 g dry wt^{-1} hr^{-1} at 2.8°C. Photosynthesis was closely coupled with changes in light intensity during these measurements, and maximum rates did not approach those of the previous period due to lower temperature and radiation levels. Radiation was less than 0.005 cal cm^{-2} min^{-1} for 6 hr, and net photosynthesis was below zero for the same period. The net CO_2 accumulation for this period was 10.88 mg g dry wt^{-1} day^{-1}. It is significant that, during this period, daytime rates remained high despite low radiation values, low temperatures, and the lateness of the season. This dramatizes the level of low light and temperature regimes to which mosses are adapted.

By 16 August, the period of negative CO_2 accumulation had increased to 7 hr, which represented a loss of 2.6 mg of CO_2. However, low radiation and near optimal tissue temperatures (5 to 11.5°C) resulted in high rates of CO_2 incorporation. A total light incorporation of 19.3 mg of CO_2 was realized resulting in a net CO_2 uptake of 16.7 mg g dry wt^{-1} day^{-1} at this late period of the year.

The preceding section illustrates the interaction of one genotype, *P. alpinum*, with environmental conditions from one habitat, the wet meadow, receiving full sunlight. However, in the tundra, there are many interactions between genotypes and habitats over short distances. Figure 14 illustrates such combinations. Shown are the responses of photosynthesis in *P. alpinum* from the wet meadow and *P. commune* from the polygon rims. *P. commune* usually occurs without a heavy vascular plant overstory and therefore receives full sunlight. *P. alpinum* from the wet meadow, on the other hand, often occurs under a vascular plant overstory and is shaded. *P. alpinum* in full sunlight was discussed previously (Fig. 13b). The photosynthetic response dropped slightly below compensation at night, and a midday light inhibition reduced the photosynthetic rate. When conditions under the canopy were approximated by reducing the light intensity 50%, higher midday photosynthesis values are found. Reduction in the light intensity at 1200 hr from 0.35 to 0.175 cal cm^{-2} min^{-1} resulted in an increase in photosynthesis of 57% from 0.86 to 1.50 mg of CO_2 g dry wt^{-1} hr^{-1}. Temperatures were similar in both cases, and the increase in photosynthesis seems due

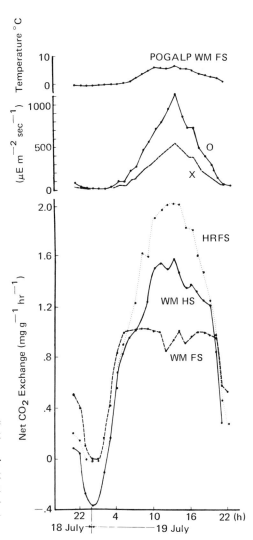

Figure 14. Diurnal patterns of response for *Pogonatum alpinum* from the wet meadow (WM) in full sunlight (FS), *P. alpinum* (WM) under simulated vascular plant overstory receiving half sunlight (HS), and *P. commune* from the polygon rim (HR) receiving full sunlight. Light levels represent full sunlight (○) and half sunlight (×).

entirely to the reduced radiation regime. The daily total net CO_2 incorporation was slightly reduced, however, in the case of the shaded *P. alpinum* as a result of "nighttime" dark respiration. The unshaded *P. alpinum* dropped only slightly below compensation. However, the reduction in "nighttime" light intensity by the simulated canopy to 0.025 cal cm^{-2} min^{-1} (PhAR) resulted in considerable dark respiration amounting to 1.08 mg of CO_2 g dry wt^{-1}. The net increase in daily CO_2 incorporation as a result of shading was 1.66 mg of CO_2 g dry wt^{-1} day^{-1}. As the season progressed, however, and daily light intensities diminished, the shaded populations were adversely affected.

    P. commune from the polygon rims is normally exposed to full sunlight. As a result, this species showed sun-adapted characteristics. It had the highest daily level of CO_2 incorporation, 26.6 mg g dry wt^{-1} day^{-1}, and the highest instanta-

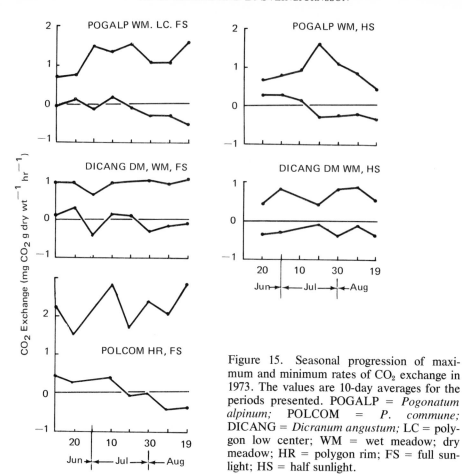

Figure 15. Seasonal progression of maximum and minimum rates of CO_2 exchange in 1973. The values are 10-day averages for the periods presented. POGALP = *Pogonatum alpinum;* POLCOM = *P. commune;* DICANG = *Dicranum angustum;* LC = polygon low center; WM = wet meadow; dry meadow; HR = polygon rim; FS = full sunlight; HS = half sunlight.

neous rate of photosynthesis (2.1 mg g dry wt^{-1} hr^{-1} vs 1.59 mg g dry wt^{-1} hr^{-1} for *P. alpinum*). Also, it did not light saturate over the range observed (0 to 0.4 cal cm^{-2} min^{-1}). These results indicate that there are various adaptive strategies within arctic *Polytrichum* species with respect to light, and the observed responses occur within populations separated by only a few meters. However, it is not clear at this point if the sun and shade responses observed represent different genetic adaptations or if they are simply phenotypic expressions of plastic physiologies. We do know that the response displayed by *P. alpinum* may be displayed in part by *P. commune* when grown under low light intensities.

Figure 15 shows the seasonal progression of average maximum and minimum rates of CO_2 exchange during the summer in 1973. The values presented are averages for 10-day periods and are a result of the interaction of numerous intrinsic and extrinsic factors. Some species were measured under full and one-half full sunlight, to simulate net CO_2 gas exchange in the open and under a vascular plant overstory. While differences exist between responses shown by different species and treatments, some generalizations are apparent. There is a tendency for minimum values of net photosynthesis to be positive early in the season and negative later. The exception is *D. angustum*. *D. angustum* from habitats receiving full solar radiation showed minimal values of net CO_2 exchange of less than CO_2 compensation when measured under one-half normal radiation levels. The same populations receiving full sunlight showed minimal values greater than compensation during one-half of the periods measured. Another striking pattern is the lack of seasonality in the maximum rates of net photosynthesis. This is due in part to the constant temperature conditions which prevail, the high incidence of days in which saturating light intensities occur at midday, the relatively small importance of seasonal changes in P_{max}, and the lack of effects due to acclimatization. Differences in the shape of the diurnal responses of photosynthesis appear to be much more important in determining seasonal patterns of CO_2 uptake than do seasonal changes in maximum and minimum values.

Populations receiving one-half normal solar radiation appear to be especially at a disadvantage at the end of the season when solar radiation levels begin to decrease. At these times, the radiation levels under the canopy may be insufficient to saturate the photosynthetic system, even at midday, and maximum rates of net photosynthesis begin to drop.

Simulations of seasonal patterns of diurnal photosynthesis in four species show that maximal photosynthetic contribution is reached in early to mid season. At this time, the greatest photosynthetic rates and the longest periods above compensation are experienced. As the season progresses, there is a gradual decrease in both parameters (Miller et al., 1978).

Patterns of CO_2 Accumulation

In order to obtain a seasonal estimate of net CO_2 incorporation, bryophytes were divided into two categories based on the light intensity of their habitat, since this is the most important environmental variable controlling production. The first group comprised species and populations normally occurring in full sunlight, and their rates of CO_2 incorporation were measured under this regime. The second group consisted of species and populations found in the wet meadow under reduced light intensity, and these were measured under ambient conditions but with reduced (50% incident) radiation. Photosynthesis rates were averaged over 10-day intervals, and the average rate of CO_2 incorporation per 10-day period for each treatment is presented in Figure 16.

Early in the season, while light intensity was high and often above saturation, the two treatments show similar rates of incorporation. During this period, higher rates at midday under reduced sunlight and protection from photoinhibition offset the effects of reduced levels of CO_2 incorporation in the evenings. However, as the light intensity decreases, especially during the period around solar midnight, and dark respiration increases, the shaded samples drop off in relative rates. When midday radiation values are high, advantage is conferred through

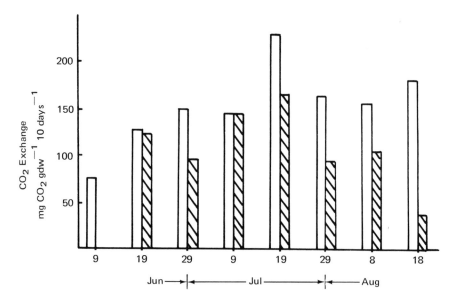

Figure 16. Net CO_2 accumulation for 10-day periods in bryophytes receiving full sunlight (open bars) and 50% sunlight (hatched bars).

shading, but if midday radiation values are below saturation, there is a lowering of photosynthetic rates in response to shading. In addition, of course, evening photosynthesis is also lowered by shading, often below compensation, and the period of dark respiration is increased as a result.

Before the period of 20 to 30 July, there was little effect of shading on daily photosynthesis totals. After this period, however, shading markedly decreased total CO_2 accumulation. The seasonal difference from 20 July to 29 August is 34%. In full sunlight, 1.15 g of CO_2 g dry wt^{-1} were incorporated over the 70-day period. In reduced sunlight, only 0.76 g of CO_2 g dry wt^{-1} were incorporated for the same period.

These rates of CO_2 uptake have been translated into biomass accumulation. For this analysis, it was assumed that there is little death, consumption, or movement from green to nongreen compartments during the growing season. This is generally correct in the absence of droughts and intensive lemming activity. Most of the death of photosynthetic tissue seems to be the result of catastrophic events usually associated with or as a result of winter, and most of the damage due to lemming grazing occurs during winter. Up to 40% of the winter diet of lemmings may be composed of mosses (Batzli, 1975) Also, in the process of grazing and routing, lemmings destroy a large amount of moss tissue which is not ingested. Much of the green material frozen at the end of summer does not recover the following spring and represents another catastrophic loss of tissue. This is especially true for species such as *Calliergon sarmentosum*.

In the estimation of biomass accumulation, the rate of production for 1 g of tissue during the first 10-day period was calculated. This was converted to dry matter, assuming:

$$1 \text{ g of } CO_2 = 0.614 \text{ g dry wt moss tissue.}$$

The amount of tissue produced was then added to the pool of photosynthetically active tissue, and the rate of CO_2 incorporation was calculated for the second 10-day period. Therefore the biomass at the $n + 1$ period was:

$$B_{n+1} = B_n P_{s_n} 0.614 + B_n$$

where B_n = green biomass at 10-day period n, P_{s_n} = photosynthetic rate for the nth photosynthetic period (10-day total), and B_{n+1} = biomass at the beginning of the next 10-day period. In this manner it was possible to estimate the biomass produced in full sunlight and half sunlight for each gram of green biomass present at the beginning of the season.

Over the 70-day period, it is estimated that 1 g of moss biomass growing in an area receiving full sunlight would produce an additional 0.97 g dry wt by the end of the 70-day period. For mosses occurring in the wet meadow area and receiving 50% of the incident sunlight, the total would be 0.56 g dry wt for each initial gram (Figure 17).

The average green biomass for the site 2 area wet meadow in late June is about 40 g dry wt m^{-2} of green material, which yields an average annual production of from 22 g m^{-2} in the vascular plant understory to 39 g m^{-2} in full sunlight. The reduction due to shading is about 44% in total production as a result of the compound effects of biomass accumulation.

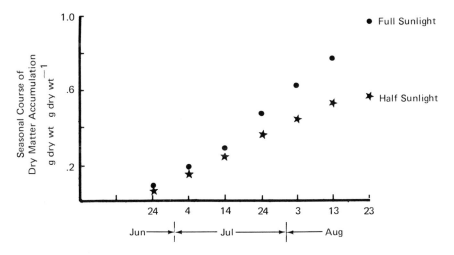

Figure 17.   Seasonal course of biomass accumulation per gram of green moss tissue at the beginning of the season. Estimated from CO_2 gas exchange measurements.

Conclusions

Mosses are a major component of this tundra ecosystem. Not only are they important in terms of cover and biomass, but also in their contribution to net primary production. Estimates of dry matter production using CO_2 gas exchange techniques place annual net primary production of mosses at about 22 g dry wt^{-1} m^{-2} to 40 g m^{-2} depending on the radiation regime within the IBP site 2 area. This places the aboveground production of mosses at about 50 to 95% that of vascular plants growing in the wet meadow as estimated by Webber (1974). The photosynthetically active biomass of mosses, however, varies enormously over short distances in the site 2 and site 4 areas (see also Webber, 1978; Rastorfer, 1978). This results in large spatial heterogeneity in production. The production patterns reported here may not be applicable to other arctic areas, since arctic conditions vary widely, and the environmental factors in the Barrow area appear to be particularly appropriate for moss growth and production.

Temperature optima for photosynthesis are above mean tissue temperatures for the area. However, high relative rates of photosynthesis near 0°C and above result in appreciable rates of CO_2 uptake under the thermal environment normally experienced. A resistance to frost and to freezing greatly extends the period of photosynthetic activity and aids in eliminating the need for late season senescence, thereby extending the period of activity further. Higher rates of production, however, would be achieved by increasing P_{max}, decreasing T_{opt}, or by increasing the relative photosynthetic rate between 0°C and T_{opt} (Oechel et al., 1975).

Low light compensation points and low saturation levels for net photosynthesis help maximize photosynthesis under the arctic environment. The photosynthetic system is saturated most days (83%) at midday for unshaded mosses. Radiation is usually above compensation during the "midnight" periods early in the season. As a result of these factors, changes in daily radiation have little effect on daily productivity.

Solar radiation and temperature are probably near optimal for mosses growing in many arctic areas. The moisture regime in many areas, however, may be limiting. At Barrow, the low evapotranspiration, the high soil moisture content, and the colony growth form of mosses promote tissue moisture contents near optimal levels most of the time in all of the species studied. Tissue desiccation, which may limit production during much of the season in some areas (Hicklenton and Oechel, 1976) or eliminate most moss species altogether, was seldom observed at Barrow. The species present assorted themselves according to the moisture present, their physiological moisture requirements, and their morphological and anatomical structures used to obtain and retain water.

Acknowledgments. This research was supported by the National Science Foundation under Grant GV 29342 to the University of Alaska. It was performed under the joint sponsorship of the International Biological Programme and the Office of Polar Programs and was coordinated by the U.S. Tundra Biome. Field and laboratory activities at Barrow were supported by the Naval Arctic Research Laboratory of the Office of Naval Research.

References

Anderson, L., and P. Bourdeau. (1955) Water relations of two species of terrestrial mosses. *Ecology,* **36**: 206–212.

Batzli, G. (1975) The role of small mammals in arctic ecosystems. In *Small Mammals: Their Production and Population Dynamics* (F. B. Golley, K. Petrusewicz, and L. Ryszkowski, Eds.) Cambridge Univ. Press. pp. 243–268.

Bazzaz, F. A., D. J. Paolillo, and R. H. Jagels. (1970) Photosynthesis and respiration of forest and alpine populations of *Polytrichum juniperinum. Bryologist,* **73**: 579–585.

Bowen, E. J. (1931) Water conduction in *Polytrichum commune. Ann. Bot.,* **45**: 175–200.

Bowen, E. J. (1933) The mechanism of water conduction in the Musci considered in relation to habitat. *Ann. Bot.,* **47**: 889–912.

Carstairs, A. G. and W. C. Oechel. (1978) Effects of several microclimatic factors and nutrients on net carbon dioxide exchange in *Cladonia alpestris* (L.) Rabh. in the subarctic. *Arct. Alp. Res.,* **10**: 81–94.

Chapin, F. S. (1978) Nutrient uptake and utilization by tundra vegetation. *Vegetation and Production Ecology of an Alaskan Arctic Tundra* (L. L. Tieszen, Ed.). New York: Springer Verlag, Chap. 21.

Clarke, G. C. S., S. W. Greene, and D. M. Greene. (1971) Productivity of bryophytes in polar regions. *Ann. Bot.,* **35**: 99–108.

Collins, N. J., and W. C. Oechel. (1974) The pattern of growth and translocation of photosynthate in a tundra moss, *Polytrichum alpinum. Can. J. Bot.,* **52**: 355–363.

Gimingham, C., and R. Smith. (1971) Growth form and water relations of mosses in the maritime Antarctic. *Brit. Antarct. Surv. Bull.,* **25**: 1–21.

Hicklenton, P. (1975) The physiological ecology of *Dicranum fuscescens* Turn. in the subarctic. M.Sc. thesis, McGill University, Montreal.

Hicklenton, P., and W. C. Oechel. (1976) Physiological aspects of the ecology of *Dicranum fuscescens* in the subarctic. I. Acclimatization and acclimation potential of CO_2 exchange in relation to habitat, light and temperature. *Can. J. Bot.,* **54**: 1104–1119.

Hicklenton, P., and W. C. Oechel. (1977) The influence of light intensity and temperature on the field carbon dioxide exchange of *Dicranum fuscescens* in the subarctic. *Arct. Alp. Res.,* **9**: 407–419.

Hosokawa, T., N. Odani, and H. Tagawa. (1964) Causality of the distribution of corticolous species in forests with special reference to the physioecological approach. *Bryologist,* **67**: 396–411.

Kallio, P., and S. Heinonen. (1973) Ecology of *Rhacomitrium lanuginosum* (Hedw.) Brid. *Rep. Kevo Subarct. Res. Sta.,* **10**: 43–54.

Kallio, P., and L. Kärenlampi. (1975) Photosynthetic activity of multicellular lower plants (mosses and lichens). In *IBP Synthesis Meeting on the Functioning of Photosynthetic Systems in Different Environments, Aberysthwyth, 1973: Photosynthesis and Productivity in Different Environments* (J. P. Cooper, Ed.). London: Cambridge University Press, pp. 393–423.

Kramer, P. J. (1969) *Plant and Soil Water Relationships. A Modern Synthesis.* New York: McGraw–Hill, 2nd ed. 482 pp.

Magdefrau, K. (1935) Untersuchungen über die Wasserversorgung des Gametophyten and Sporophyten der Laubmoose. *Bot. Zeit. (Berlin,* **29**: 337–375.

McKendrick, J. D., V. Ott and G. A. Mitchell. (1978) Effects of nitrogen and phosphorus fertilization on the carbohydrate reserve and seasonal nitrogen and phosphorus levels in *Dupontia fisheri* and *Arctogrostis latifolia* shoots and rhizomes near Barrow, Alaska. *Vegetation and Production Ecology of an Alaskan Arctic Tundra* (L. L. Tieszen, Ed.). New York: Springer Verlag, Chap. 22.

Miller, P. C., W. C. Oechel, W. A. Stoner and B. Sveinbjörnsson. (1978) Simulation of

CO$_2$ uptake and water relations of four arctic bryophytes at Point Barrow, Alaska. *Photosynthetica* **12**: 7–20.

Oechel, W. C. (1976) Seasonal patterns of CO$_2$ flux and acclimation in arctic mosses growing *in situ*, *Photosynthetica*, **10**: 447–456.

Oechel, W. C., and N. J. Collins. (1976) Comparative CO$_2$ exchange patterns in mosses from two tundra habitats at Barrow, Alaska. *Can. J. Bot.*, **54**: 1355–1369.

Oechel, W. C., P. Hicklenton, B., Sveinbjörnsson, P. C. Miller, and W. Stoner. (1975) Temperature acclimation of photosynthesis in *D. fuscescens* growing *in situ* in the arctic and subarctic. *Proceedings of Circumpolar Conference on Northern Ecology, Ottawa*. Ottawa: National Research Council of Canada, pp. 1131–1144.

Oechel, W. C., and B. Sveinbjörnsson. (1974) The microenvironment of arctic bryophytes at Barrow, Alaska. *U.S. Tundra Biome Data Rep.* 74-25, 204 pp.

Rastorfer, J. R. (1972) Comparative physiology of four West Antarctic mosses. In *Antarctic Terrestrial Biology* (G. A. Llano, Ed.). Antarctic Research Series Vol. 20. Washington, D.C.: American Geophysical Union, pp. 143–161.

Rastorfer, J. R. (1978) Composition and bryomass of the moss layers of two wet-meadow tundra communities near Barrow, Alaska. In *Vegetation and Production Ecology of an Alaskan Arctic Tundra* (L. L. Tieszen, Ed.). New York: Springer-Verlag, Chap. 6.

Slatyer, R. O. (1957) The influence of progressive increases in total soil moisture stress on transpiration, growth and internal water relationships of plants. *Aust. J. Biol. Sci.*, **10**: 320–336.

Stålfelt, M. G. (1937) Der Gasaustausch der Moose. *Planta*, **27**: 30–60.

Tieszen, L. L. (1974) Photosynthetic competence of the subnivean vegetation of the arctic. *Arct. Alp. Res.*, **6**: 253–256.

Tieszen, L. L. (1975) CO$_2$ exchange in the Alaskan Arctic Tundra: Seasonal changes in the rate of photosynthesis of four species. *Photosynthetica*, **9**: 376–390.

Tieszen, L. L. (1978) Photosynthesis in the principal Barrow, Alaska, species: A summary of field and laboratory responses. In *Vegetation and Production Ecology of an Alaskan Arctic Tundra* (L. L. Tieszen, Ed.). New York: Springer-Verlag, Chap. 10.

Vowinckel, T., W. C. Oechel, and W. Boll. (1974) The effect of climate on the photosynthesis of *Picea mariana* at the subarctic tree line. I. Field measurements. *Can. J. Bot.*, **53**: 604–620.

Webber, P. J. (1974) Tundra primary productivity. In *Arctic and Alpine Environments* (J. D. Ives and R. G. Barry, Eds.). London: Methuen, pp. 445–473.

Webber, P. J. (1978) Spatial and temporal variation of the vegetation and its production, Barrow, Alaska. In *Vegetation and Production Ecology of an Alaskan Arctic Tundra* (L. L. Tieszen, Ed.). New York: Springer-Verlag, Chap. 3.

Tieszen, L. L., D. A. Johnson, and M. M. Caldwell. (1974) A portable system for the measurement of photosynthesis using ^{14}CO$_2$. *Photosynthetica*, **8**: 151–160.

Vowinckel, T., W. C. Oechel, and W. Boll. (1974) The effect of climate on the photosynthesis of *Picea mariana* at the subarctic tree line. I. Field measurements. *Can. J. Bot.*, **53**: 604–620.

Webber, P. J. (1974) Tundra primary productivity. In *Arctic and Alpine Environments* (J. D. Ives and R. G. Barry, Eds.). London: Methuen, pp. 445–473.

Webber, P. J. (1978) Spatial and temporal variation of the vegetation and its production, Barrow, Alaska. In *Vegetation and Production Ecology of an Alaskan Arctic Tundra* (L. L. Tieszen, Ed.). New York: Springer-Verlag, Chap. 3.

Wielgolaski, F. E. (1972) Vegetation types and primary production in tundra. In *Proceedings IV International Meeting on the Biological Productivity of Tundra* (F. E. Wielgolaski and Th. Rosswall, Eds.). Stockholm: International Biological Programme Tundra Biome Steering Committee, pp. 9–34.

Wielgolaski, F. E., and P. J. Webber. (1973) Classification and ordination of circumpolar arctic and alpine vegetation. *Int. Tundra Biome Newsletter*, **9**: 24–31.

12. Meteorological Assessment of CO_2 Exchange over an Alaskan Arctic Tundra

P. I. COYNE AND J. J. KELLEY

Introduction

Kinetic characterization of primary productivity is paramount to understanding energy partition and flow within an ecological system. The central role of primary production has led to considerable research to characterize photosynthesis and processes related to primary production at Barrow. Independent and complementary approaches to document the seasonal incorporation of CO_2 as a function of environmental parameters are desired. Our approach utilized an aerodynamic method to measure CO_2 fluxes above a tundra surface. In comparison with other methods for assessing primary productivity, the aerodynamic method has some desirable attributes. It is nondestructive and is sensitive to immediate environmental conditions, but, equally or more importantly, it responds to CO_2 exchange at the community level of integration so that extrapolation from individual plant data is not required to obtain estimates for the entire plant community. In effect, this approach integrates primary producer and decomposer components of an ecosystem.

Theory

Photosynthesis and respiration are gas exchange processes which effect an interaction between the atmosphere and the biosphere through depletion or repletion of CO_2 in the ambient environment. The net imbalance of these processes at any given time is reflected in CO_2 concentration gradients in the lower atmosphere which give rise to CO_2 fluxes directed toward or away from the biological surface. Meteorological methods provide data on the variability of key parameters with height (gradients) from which vertical fluxes of mass and energy above or within plant stands can be calculated. Two types of methods (Lemon, 1970) have commonly been employed. The energy balance method requires the partition of net absorbed radiation between plant and soil surfaces. The aerodynamic or momentum balance method used in this study depends upon the exchange of mementum between plant and soil surfaces and the wind. Both methods have been developed in detail by numerous authors (Baumgartner,

1969; Denmead, 1970; Denmead and McIlroy, 1971; Inoue, 1968; Lemon, 1960, 1965, 1967, 1969; Monteith, 1962, 1968, 1973; Priestly, 1959; Uchijima, 1970; Webb, 1965), and the assumptions on which the methods are based and their limitations are discussed.

The equation essential to flux calculations resulting from the theoretical derivation is

$$C = \frac{k^2(u_2 - u_1)(c_2 - c_1)}{\left(\ln \dfrac{z_2 - d}{z_1 - d}\right)^2}. \tag{1}$$

Where C = vertical flux of CO_2 (g cm^{-2} sec^{-1}),
k = dimensionless von Karman's constant (0.41; Hicks, 1972; Monteith, 1973),
u = wind speed (cm sec^{-1}),
c = CO_2 concentration (g cm^{-3}), and
z_1 and z_2 = two heights (cm) above the vegetation canopy.

Presence of vegetation extends the surface above the ground to some height d (zero plane displacement) which is the effective vegetation height where turbulent exchange begins. If z is measured from the ground, it is necessary to subtract from z the height d into the canopy where the zero wind reference surface is displaced (Tanner, 1963). Input for the aerodynamic method then requires careful simultaneous measurement of wind speed and CO_2 profiles. Following the convention of Lemon (1967) and Wright and Lemon (1966), negative fluxes are considered upward and positive downward.

For non-neutral atmospheric conditions, Eq. (1) was multiplied by an empirical correction term of the form $(1 - bR_i)^a$ for unstable or lapse conditions and $(1 + bR_i)^{-a}$ for stable or inversion conditions (Lettau, 1962) where R_i is the gradient Richardson number. R_i, a stability parameter, was calculated according to Rose (1966) and is a measure of the importance of buoyancy forces (thermal convection) in producing turbulence as compared to frictional forces (Tanner, 1963). Values assigned to the parameters a and b were 0.25 and 18, respectively (Ellison, 1957; Panofsky et al., 1960; Sellers, 1965).

Experimental Procedure

The study site was located near Barrow, Alaska, at the Tundra Biome intensive site 2. The vegetation (see Webber, 1978; Dennis et al., 1978) consists of a relatively homogeneous wet meadow tundra community and has been described in detail with respect to production and pigment structure by Tieszen (1972). The sampling period was from 27 June to 26 August 1971.

Wind velocity and air temperature data were provided by the IBP micrometeorology project at the study site (Weller et al., 1972). Instrumentation consisted of Climet sensors on a 16-m tower adjacent to the fetch area used for CO_2 measurements. Wind data were integrated half-hourly values from six levels (0.25, 1.25, 2.24, 4.51, 9.06, and 16.61 m). Temperature values were half-hourly instantaneous readings from ventilated thermocouples at the 0.25-, 0.50-, 1.24-, 2.24-, and 4.51-m levels. Ambient air was sampled at five heights (10, 20, 40, 80, and 160 cm) above the tundra canopy for CO_2 analysis. Air was aspirated through approximately 20 m of 1.3-cm diameter stainless-steel tubing from the intake ports to exhaust ports on an automatic switching system which controlled air flow to the analyzer. The lines were continually flushed at a flow rate of about 2 liters min^{-1}. Each of the five air streams and a reference gas of accurately known CO_2 concentration automatically flowed through the CO_2 analyzer in sequence for 5 min every 0.5 hr on a continuous basis throughout the sampling period except for periods of maintenance and calibration.

   CO_2 concentrations were determined by nondispersive infrared gas analysis using a Mine Safety Appliance Model 200 LIRA instrument. The analog output was recorded by a Hewlett–Packard Model 7100B potentiometric, millivolt, stripchart recorder. Other details concerning calibration and this specific study are available from Johnson and Kelley (1970) and Coyne and Kelley (1975).

   The physical size of a relatively homogeneous fetch area required to attain almost constant flux values is dependent upon the height of the profile sensors and the surface roughness (Taylor, 1962). Homogeneity in community floristic composition which affects CO_2 profiles was considered more limiting at the study site than surface roughness homogeneity which affects wind profiles. The height of the CO_2 sensors was therefore considered in the evaluation of fetch/height ratios. Fetch/height ratios ranging from 10 (Panofsky and Townsend, 1964) to greater than 140 (Dyer, 1963) were reported in the literature as reviewed by Sellers (1965). According to Denmead (1970), fully developed profiles can be expected within fetch/height ratios between 50 and 200. In this study, the minimum ratio was about 50/1 relative to the highest CO_2 sensor of 1.6 m. Assuming the effects on CO_2 profiles of a small drainage tributary 80 m east of the sensors were insignificant, the ratio increased to 125/1 in the prevailing wind direction (easterly).

Data Analysis

A least squares fit of temperature (T vs z) and wind speed (u vs $\ln (z - d)$) was computed for each half-hourly interval. The resulting prediction equations were used to compute wind speed and temperature at heights corresponding to the CO_2 sampling ports. The zero plane displacement, d, averaged about 5 cm. Carbon dioxide data were converted from chart ordinates to concentrations in ppm (cm^3 10^{-6} cm^{-3}), adjusted to standard pressure, and multiplied by the density of CO_2 to obtain g cm^{-3}. Since the CO_2 data from each level were not recorded simultaneously, half-hourly values were interpolated using the closest data on either side of the half-hour. Interpolated values were then subjected to least squares analysis (CO_2 vs ln $(z - d)$). Predicted values of wind, CO_2, and temperature at heights z_1 (10 cm) and z_2 (160 cm) were used in calculating CO_2 flux [Eq. (1)].

Estimates of seasonal CO_2 incorporated by the plant community were derived by fitting a regression line which incorporated a CO_2 source (soil and plant respiration) and a CO_2 sink (photosynthesis) to the data. The algebraic sum of the source and sink represents the dependent variable or CO_2 flux. The form of the equation is our much simplified concept of how the ecosystem physically responds to the variables of irradiance, temperature, and chlorophyll. The basic form of the equation is

$$\hat{C} = A + (B - A)\exp(-a_6 R) \tag{2}$$

where $\hat{C}$ is the predicted CO_2 flux (g m^{-2} hr^{-1}), A is the asymptote, B is the intercept, a_6 is the irradiance sensitivity coefficient, and R is the irradiance (cal cm^{-2} 0.5 hr^{-1}) between 400 and 700 nm.

The asymptote of the equation, A, was defined as a function of chlorophyll and soil surface temperature and represents the photosynthetic potential of the ecosystem at a given temperature and chlorophyll standing crop. Soil surface temperature rather than air temperature was used due to the low stature of the vegetation canopy and due to the greater influence of soil surface temperature on soil respiration. The expression for A is

$$A = a_1 \text{Chl} \exp(-a_2(T - a_3)^2). \tag{3}$$

Chlorophyll (mg m^{-2}) data were obtained at approximately 10-day intervals and a linear interpolation was used to calculate values to correspond to the more frequently sampled variables. Chlorophyll was included in the regression equation as an index of the potential of the plant community to respond to radiation. Saturation light intensities of grasses at Barrow have been estimated at about 0.4 cal cm^{-2} min^{-1} (400 to 700 nm) by Tieszen (1973) suggesting that few leaves would be light saturated under field conditions. According to Tieszen (1972), standing crop of chlorophyll is not always a direct indicator of productivity but can be an indicator of an immediate photosynthetic potential which may or may not be realized. The availability of chlorophyll data and the generally subsaturation field light intensities led to the selection of chlorophyll as the plant variable in the regression equation.

Under constant temperature conditions, the asymptote was assumed to increase linearly with chlorophyll standing crop. The slope of the response curve for the asymptote expressed as a function of chlorophyll was further assumed to increase with temperature up to the optimum soil surface temperature (a_3) for atmospheric CO_2 flux. At T greater than a_3 the slope was assumed to decrease. The parameter a_3 is significant in that the proportion of photosynthetically fixed CO_2 derived from the atmosphere decreases at temperatures higher than a_3.

Parameter a_1 represents the maximum atmospheric CO_2 flux per unit chlorophyll at $T = a_3$ and parameter a_2 is the temperature sensitivity coefficient.

The CO_2 source or respiration portion of the regression equation is simply a Q_{10} relationship. The expression is the intercept B of Eq. (2) and represents the dark CO_2 flux from all respiration processes to the atmosphere. B is calculated by the equation

$$B = -a_4^{0.1(T-a_5)}. \tag{4}$$

The negative sign indicates net CO_2 flux to the atmosphere. In Eq. (4), a_4 is the Q_{10} and a_5 represents the soil surface temperature at which CO_2 flux at zero irradiance is -1.0 g m^{-2} hr^{-1}.

Results

Ambient CO_2 Changes and Patterns of Flux

Average CO_2 concentrations above the vegetative canopy between 0.1 and 1.6 m followed a seasonal trend similar to those published by Kelley (1968). Solar noon values at the start of sampling (27 June 1971) were approximately 326 ppm and declined gradually during the growing season to about 312 ppm (26 August) (Figure 1). Solar midnight concentrations were similar in trend but were up to 8 ppm higher. The period of decline was in phase with the annual growth cycle. The hours per day of CO_2 flux in the downward direction (Figure 2) were greatest at the start of the sampling period near the summer solstice and gradually declined with season and daylength. Negative or upward CO_2 flux occupied a larger part of each day as the season progressed. Fluctuations in the seasonal trends reflect local conditions in irradiance and temperature.

Net daily atmospheric CO_2 flux ranged from -1 g m^{-2} day^{-1} in late June when chlorophyll standing crop was relatively low to 8 g m^{-2} day^{-1} during peak season to -1.5 g m^{-2} day^{-1} near the end of the season when plants were approaching quiescence. Net daily fluxes less than -8 g m^{-2} day were observed around 20 August as a result of dense cloud cover with further attenuation of light by 10 cm of snow. Differences in CO_2 concentrations between 1.6 and 0.1 m ranged from about -5.5 to 4.1 ppm from 27 June to 26 August 1971. Gradients for solar noon and midnight are graphed for the season in Figure 3. It should be noted that the observed extremes in CO_2 gradient did not necessarily occur at noon and midnight.

Parameter Estimates

The diurnal course of profiles of CO_2, wind speed, temperature, and eddy diffusivity are illustrated with the data for 20 July 1971 (Figure 4). CO_2 concentra-

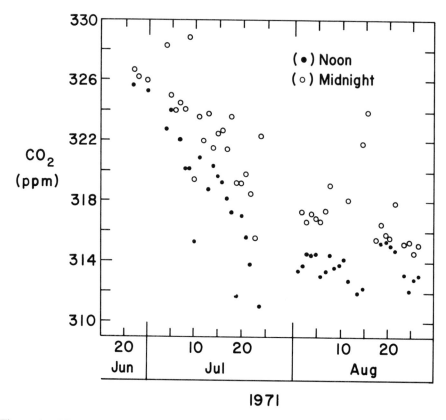

Figure 1.   Mean CO_2 concentrations between 0.1 and 1.6 m at solar noon and solar midnight.

tions decreased with height during the early morning hours but increased with height throughout most of the day. Temperature gradients were greatest near midday becoming nearly isothermal around midnight. Measured CO_2 flux from the atmosphere was 8.0 g m^{-2} day^{-1} for this date. The mean momentum eddy diffusivity (K_m) for the sampling period at five heights is given in Table 1. The magnitude of K_m is a function of atmospheric turbulence (Denmead, 1970) and is typically on the order 10^3 cm^2 sec^{-1} at a height of 1 to 2 m. Values of K_m around 2500 cm^2 sec^{-1} (Priestly, 1959) have been found to be typical at 1.5 m over grassland in winds of about 4 m sec^{-1} and slight to moderate instability.

The nonlinear regression equation was successful in explaining nearly 80% (R^2) of the variation associated with CO_2 flux of the atmosphere on a seasonal basis. Four separate runs of the least squares program with each successive run incorporating more data are summarized in Table 2. Run A was for a single day only which explains the higher multiple R^2. In runs B through D, the curve-fitting program was attempted initially using data from the middle of the growing season when presumably CO_2 flux was most sensitive to changes in irradiance and temperature. The additional runs were made incorporating more data at the

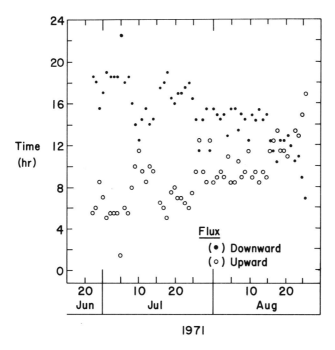

Figure 2.   Seasonal course of the number of hours per day of downward and upward CO$_2$ flux predicted by regression analyses.

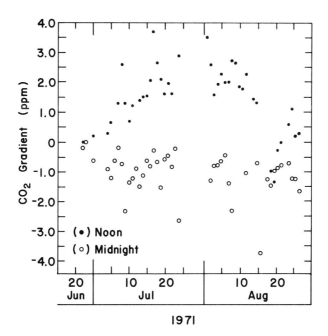

Figure 3.   Seasonal course of the CO$_2$ gradient between 0.1 and 1.6 m. A positive gradient represents increasing CO$_2$ with height.

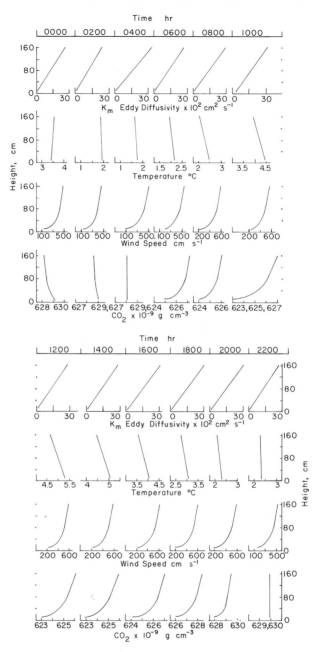

Figure 4.   Diurnal course of CO_2 concentration, wind speed, soil surface temperature, and eddy diffusivity for 20 July 1971. Time = Alaska Standard. (From Coyne and Kelley, 1975; used with permission of Blackwell Scientific Publications LTD.)

Table 1.   Mean K_m for the Sampling Period at Five Heights

Statistic	K_m cm^2 sec^{-1}				
	10 cm	20 cm	40 cm	80 cm	160 cm
Mean for 1786 half-hours	59	176	412	882	1823
Standard deviation	20	59	137	293	605

Used with permission of Blackwell Scientific Publications LTD.

extremes of the sampling period. The decrease in the multiple R^2 was only 1% when data from throughout the sampling period were used.

Parameter values of special interest from the regression are a_3, a_4, and a_5, the optimum temperature for atmospheric CO_2 flux, the Q_{10} for ecosystem respiration, and the temperature at which ecosystem respiration is 1 g m^{-2} hr^{-1}, respectively. Early in the study decline in CO_2 flux from the atmosphere often led the daily maximum irradiance, a response that appeared to be a temperature interaction with light. The estimated optimum temperature of 6.5°C for the sampling period as a whole suggests that on the average, above a soil surface temperature of 6.5°C, the balance of the two CO_2 sources supporting photosynthesis (atmosphere, respiration) shifts in the direction of respiration thereby resulting in a decrease in atmospheric CO_2 flux. The value of a_3 changed (Table 2) as more data were included in the regression indicating that a_3 is not a true constant but rather depends on the photosynthesis potential of the community and light intensity. In general, as a_3 increases, the photosynthetic potential per unit chlorophyll [A Chl^{-1}, Eq. (3)] increases. Thus lower values of a_3 would be expected when the chlorophyll standing crop is low as a result of growth limitations early in the season and as a result of chlorophyll degradation during senescence late in the season.

The temperature × light interaction is illustrated by the diurnal course of CO_2 flux, irradiance, and temperature for 21 July 1971 (Figure 5). The curve of predicted CO_2 flux is based on run A in Table 2. There was good agreement between predicted and observed net daily CO_2 fluxes (see Table 3 for net fluxes) which was expected of regressions computed using data from a single day. The higher optimum temperature a_3 in run A compared to runs B through D reflects a relatively high chlorophyll standing crop as peak season was near and a higher level of irradiance (266 cal cm^{-2} day^{-1}) compared to the seasonal average of 180 cal cm^{-2} day^{-1}. Thus, the photosynthetic potential was higher for 21 July than for the season as a whole, and the soil surface temperature required to increase respiration to the point that atmospheric CO_2 source strength declined was nearly twice as high as for the seasonal average.

Although a_4 or Q_{10} was approximately the same in run A as in runs B through D, a_5 had a high standard error and was not statistically significant. A high standard error does not imply that a_5 should be deleted from the prediction equation but rather a wide range of values was possible. The values of the remaining parameters hold only for a_5 equal to 26. Fixing a_5 at any other value

Table 2. Parameter Estimates and Standard Errors for Eq. (2) from Four Separate Regression Runs

Run[b]	Statistic	Parameters[a]					
		a_1	a_2	a_3	a_4	a_5	a_6
A	Value	0.00190**	0.00192	12.35926**	1.73855*	26.17085	0.01565**
	Standard error	0.00014	0.00110	2.63350	0.81161	19.82300	0.00399
	Multiple R^2	93.6%					
	Std. err. est.	0.081			44 half-hours represented		
B	Value	0.00204**	0.00130**	7.64416**	1.72212**	23.28500**	0.01603**
	Standard error	0.00008	0.00041	2.39320	0.30646	7.33650	0.00147
	Multiple R^2	78.3%					
	Std. err. est.	0.154			893 half-hours represented		
C	Value	0.00205**	0.00141**	8.02017**	1.75172**	22.86668**	0.01570**
	Standard error	0.00008	0.00041	2.08930	0.29982	6.72830	0.00142
	Multiple R^2	78.1%					
	Std. err. est.	0.153			974 half-hours represented		
D[c]	Value	0.00206**	0.00136**	6.49539**	1.77890**	22.73480**	0.01463**
	Standard error	0.00009	0.00038	2.33870	0.25073	5.37780	0.00121
	Multiple R^2	77.1%					
	Std. err. est.	0.146			1240 half-hours represented		

A. July 21.
B. July 8, 10, 11, 13–22, 24; August 1–9, 11, 12, 15, 16.
C. June 27, 30; July 8, 10, 11, 13–22, 24; August 1–9, 11, 12, 15, 16, 26.
D. June 27, 30; July 3–14, 16–22, 24; August 1–12, 14–16, 24–26.
[a] a_1 = maximum atmospheric CO_2 flux per unit chlorophyll at temperature = a_3. a_2 = temperature sensitivity coefficient. a_3 = optimum soil surface temperature. a_4 = Q_{10} for ecosystem respiration. a_5 = soil surface temperature at which CO_2 flux at zero irradiance is -1 g m^{-2} hr^{-1}. a_6 = irradiance sensitivity coefficient.
[b]   Regression based upon data from following 1971 dates:
[c]   Predicted CO_2 fluxes in this paper except for Figure 2 and Table 3 were calculated with Eqs. 2–4 using parameter estimates from this run. Figure 2 predictions used run A and Table 3 predictions were from separate regression equations based on data from the indicated date.
*   Significant at 0.05 probability level.
**   Significant at 0.01 probability level.
Used with permission of Blackwell Scientific Publications LTD.

Table 3. Variation in Regression Parameters among Six Separate Days

Date, 1973	Irradiance (400–700 nm) (Wm^{-2})	Mean soil surface temp. (°C)	Mean wind speed at 25 cm (cm sec^{-1})	Parameter a_3 (°C)	a_4 (°C)	a_5 (°C)	R^2 (%)	Standard error of estimate (g m^{-2} day^{-1})	Atm. CO$_2$ flux (g m^{-2} day^{-1}) Obs.	Pred.
8 July	107	11.5	143	6.4	1.8	26.1	75	0.08	1.75	1.87
14 July	57	3.3	209	10.5	1.7	23.4	95	0.06	1.33	1.25
20 July	95	6.6	321	15.6	1.7	24.1	90	0.11	7.95	7.97
21 July	129	10.2	234	12.4	1.7	26.2	94	0.08	6.19	6.14
2 August	72	4.3	233	11.6	1.8	24.8	89	0.14	7.84	7.76
3 August	52	2.7	265	4.1	1.7	25.7	93	0.11	6.74	6.51

Used with permission of Blackwell Scientific Publications LTD.

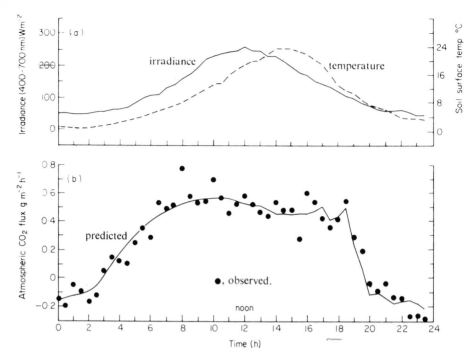

Figure 5.   Diurnal course of atmospheric CO_2 flux, irradiance (400 to 700 nm) and soil surface temperature for 21 July 1971. Time = Alaska Standard. (From Coyne and Kelley, 1975; used with permission of Blackwell Scientific Publications LTD.)

suspected as being more realistic and rerunning the regression changes the remaining parameters but gives essentially the same fit. One must decide whether the other parameters have reasonable values. Since a_4 was estimated at 1.7 as in the other runs and because 12°C (a_3) is the approximate temperature at which CO_2 flux began to decline on 21 July (Figure 5), the value for a_5 is probably a realistic estimate.

Observed and Predicted Fluxes

The standard error of estimate was about 0.15 g m^{-2} hr^{-1} in runs B through D (Table 2). However, standard errors of estimates for individual days may be as high as 0.6 using parameter estimates of run D. Diurnal courses of predicted (run D) and observed CO_2 fluxes were compared for 11 different days during the sampling period (Figure 6). Some of the days have discontinuities in observed and/or predicted fluxes due to either missing observed data or missing irradiance data required as input to the prediction equation. The prediction equation overestimated daily flux on some days and underestimated flux on others. Over the entire sampling period, these differences will partially cancel each other decreasing the discrepancy between predicted and observed seasonal estimates. Predicted atmospheric CO_2 fluxes were in general too low in comparison with observed fluxes during midday. This discrepancy may have been due primarily to a_3 not being a true constant (Table 3). Data in Tables 2 and 3 suggest that all other parameters of the regression equation were fairly constant and that much of the unexplained variability in atmospheric CO_2 flux probably resulted from the inconsistency of a_3.

Daily means of observed and predicted CO_2 fluxes and the independent variables of irradiance, temperature, and chlorophyll (vascular plants only) on which the predictions are based are shown in Figure 7. The 10-day sampling schedule evidently bracketed peak chlorophyll standing crop as evidenced from the center flat portion of the top curve in Figure 7. Tieszen (1972) reported a peak standing crop of chlorophyll in vascular plants at the study site of 460 mg m^{-2} for 1970. In a separate study, analyses of four tundra communities in northern Alaska sampled in late July of 1966 showed that chlorophyll standing crops in vascular plants plus mosses ranged from 320 to 770 mg m^{-2} (Tieszen and Johnson, 1968). The curve in Fig. 7 extrapolates to a peak of about 390 mg m^{-2}.

Seasonal CO_2 Fluxes

An important objective of the analysis was to derive an estimate of seasonal values for atmospheric CO_2 flux. An additional benefit was that the intercept expression of the regression equation [Eq. (4)] was itself an estimate of CO_2 flux from biological respiration. By summing these two fluxes, a crude estimate of gross community CO_2 uptake was derived. The lowermost portion of Fig. 7 shows the seasonal course of CO_2 fluxes. The bottom curve was predicted from Eq. (2) using parameter estimates of run D in Table 2. Data for 20 days during the study period had sufficient sampling points throughout the 24-hr period to allow calculation of net daily atmospheric CO_2 fluxes. These data are plotted as observed values. Sampling was most intense from 4 to 24 July and from 1 to 26 August. Agreement between observed and predicted daily fluxes were generally good except for 20 July and 2 and 3 August.

The middle CO_2 flux curve is an estimate of ecosystem respiration calculated from Eq. (4). As this estimate is the intercept of the plot of atmospheric CO_2 flux versus irradiance, it represents dark respiration extrapolated to existing temperature conditions. The algebraic sum of the two lower curves results in our best estimate of gross community CO_2 uptake. The integrated area provides an estimate of seasonal fluxes. Atmospheric CO_2 flux for the sampling period was estimated at 146 g m^{-2}. Seasonal ecosystem respiration was predicted at 626 g m^{-2}. Therefore, gross community CO_2 uptake would be 772 g m^{-2}. Multiplying by 0.614 (Woodwell and Botkin, 1970) gives CO_2 incorporation in terms of dry matter, 474 g dry matter m^{-2} season^{-1}.

The simple correlation coefficient matrix and the coefficients of multiple determination (R^2) were obtained by running a multiple linear regression in a

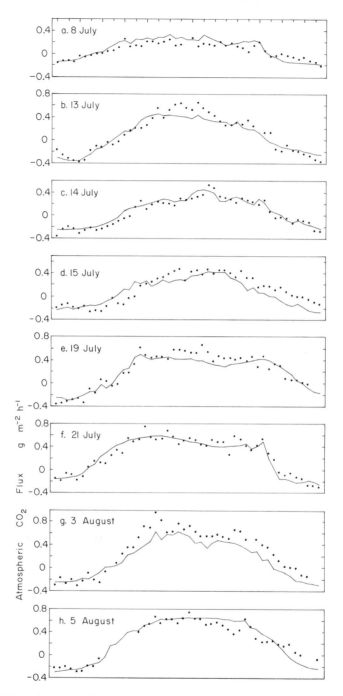

Figure 6. Diurnal course of observed and predicted atmospheric CO_2 flux for 11 days during the sampling period in 1971. Time = Alaska Standard. (From Coyne and Kelley, 1975; used with permission of Blackwell Scientific Publications, LTD.)

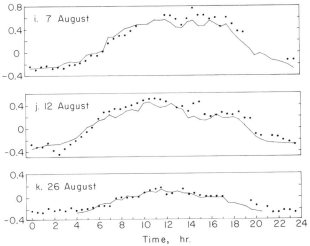

Figure 6 (continued).

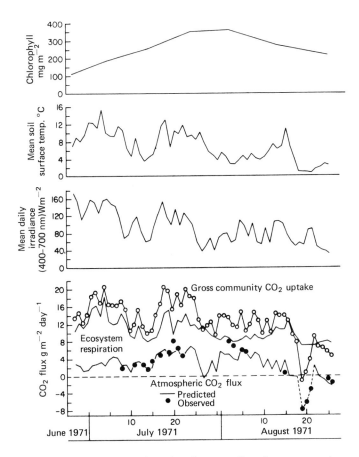

Figure 7. Seasonal courses of CO_2 flux, irradiance, soil surface temperature, and chlorophyll standing crop. (From Coyne and Kelley, 1975; used with permission of Blackwell Scientific Publications LTD.)

Table 4.   Correlation and Stepwise Multiple Linear Regression Analysis[a]

	Simple correlation coefficient matrix					
	CO$_2$	Period	Time	Temp.	IRR.	Chl.
CO$_2$	1	0.12	0.18	0.48	0.66	0.36
Period		1	0.05	−0.21	−0.18	0.45
Time			1	0.18	0.08	0.03
Temp.				1	0.86	−0.07
IRR.					1	−0.04
Chl.						1

Percentage of the variation explained by the regressions,[b] R^2					
All variables	Minus period	Minus time	Minus temp.	Minus IRR.	Minus Chl.
77	76	62	74	74	76

[a] CO$_2$ flux densities were regressed on period during the season (7 each 10 days long), time of day (day divided into 24 1-hr periods), soil surface temperature, irradiance (400–700 nm), and chlorophyll standing crop.
[b] The ratio of variation explained by the regression to the unexplained variation (F test) was significant at the 0.01 probability level in each case.
Used with permission of Blackwell Scientific Publications LTD.

stepwise manner (Table 4). Data used in the regression correspond to those of Run D (Table 2). Deleting time from the regression had the greatest effect on R^2 reducing it from 77 to 62%. Deletion of the remaining variables had little effect. Evidently, most of the curvilinear response of CO$_2$ flux results from diurnal variation in irradiance and temperature rather than seasonal variation. CO$_2$ flux was most highly correlated with irradiance ($r = 0.66$) of the variables tested. The correlation matrix also shows the expected high degree of covariance between irradiance and temperature ($r = 0.86$).

Discussion

Tieszen (1973, 1975, 1978) has found that the Barrow grasses are efficient photosynthetically at low light intensities. Light compensation intensities for most species were between 0.008 and 0.010 cal cm^{-2} min^{-1}. Although the optimum temperature for CO$_2$ fixation in these species is around 15°C, uptake at 5°C was still quite high and uptake even at 0°C was always positive. Light intensities equal to or greater than the compensation points reported by Tieszen are possible at solar midnight from the beginning of the growing season to late July which suggests the possibility of positive CO$_2$ exchange for individual plants throughout a diurnal cycle. No days were observed at Barrow for which atmospheric CO$_2$ flux was downward during the entire day. At the lower light intensities in the late evening and early morning, soil and plant respiration exceeded CO$_2$ uptake resulting in a net upward transfer of CO$_2$ to the atmosphere.

The early July peak in CO_2 fluxes (Figure 7) seems inflated so early in the season when presumably photosynthetic potential is low. The growing season begins during or immediately after snowmelt and, according to Tieszen (1972, 1978), available photosynthetic tissue must limit total community CO_2 uptake. However, throughout most of the growing season, low light intensity and low temperature are also limiting. Tieszen also points out that the life strategy of arctic communities such as that under study includes early development of photosynthetic tissue. Further, partially developed leaves of monocots, the most common of the Barrow species, have the ability to enter dormancy in this condition surrounded by dead leaf sheaths. Upon snowmelt when the basal meristem resumes growth, these protected leaves are exserted and rapidly become chlorophyllous. High irradiance during the period would tend to partially compensate for limiting chlorophyllous tissue. Conversely, the soil surface should be more exposed to the direct solar beam which would elevate surface temperatures. This in turn might inflate respiration estimates since the depth of thaw at this time may not be sufficient to allow soil surface CO_2 flux to attain its potential for a given temperature. This appears doubtful in light of the results of Billings et al. (1977). They found most of the living plant material at the study site to be within the top 10 cm of soil. Further, 60 to 80% of the total soil respiration originated in the top 5 cm of soil and 80 to 90% originated in the upper 10 cm of soil. Thaw depth was at least 10 cm by early July.

Respiration response to temperature once the surface layer of soil thaws (thaw depth is about 12 cm by 25 June, Nakano and Brown, 1972) should be fairly constant under conditions of adequate moisture. Soil moisture at Barrow in general is not considered limiting to photosynthesis (Tieszen, 1972, 1978; Miller et al., 1978). Billings et al. (1977) found that total soil respiration closely followed air temperature during the season but lagged it by about 3 hr. In other ecosystems where moisture may at times become limiting, it is generally of lesser importance in its effect on soil respiration than is temperature (Witkamp, 1966; Van Cleve and Sprague, 1971; Garrett and Cox, 1973; Lemon, 1967).

The estimate of ecosystem respiration is biased toward the August data because the sun at Barrow remains above the horizon throughout July thus limiting the influence of July conditions on the regression line intercept. However, another possibility to explain the peak, if it is real, is that moss productivity may be higher relative to vascular plant productivity earlier in the season because of the more open vascular canopy in June and early July (Oechel and Sveinbjörnsson, 1978). The best independent estimate of dark CO_2 flux we have for comparison is the "very rough estimate" of Peterson and Billings (1975) for the period 9 June to 5 August 1973. They calculated total dark CO_2 flux from the tundra meadow system at Barrow between 220 and 240 g of CO_2 m^{-2}. While the aerodynamic estimate is much greater, the value of Peterson and Billings is low since a significant portion of the thaw period and hence respiration activity occurs after 5 August.

Environmental conditions which cause partial stomatal closure would reduce CO_2 uptake and also reduce a_3. A commonly observed midday drop in net photosynthesis has been attributed to partial stomatal closure in response to leaf drying at higher midday temperatures (Zelitch, 1971). This phenomena has been observed in Barrow species (Tieszen, 1975, 1978). The higher radiation load and temperature on 21 July compared to 20 July suggests that leaf temperatures would have been higher on 21 July and that stomatal closure could have been more pronounced than on the previous day. This might explain part of the discrepancy for these two days (Table 3). It is doubtful that the higher temperatures on 21 July would have reduced net photosynthesis by inducing higher respiration rates since temperatures remained below the optimum temperature of 15°C reported by Tieszen (1973) during much of the day.

Reported annual aboveground vascular dry matter production of arctic and alpine tundras generally ranges from 73 to 219 g m^{-2} (Bliss, 1961, 1962). Tieszen (1972) reported vascular aboveground production at the study site to be 102 g of dry matter m^{-2} yr^{-1} in 1970. Dennis and Johnson (1970) found the average accretion of aboveground dry matter was 97 g m^{-2} yr^{-1} during the 1964 and 1965 growing season at Barrow. Tieszen (1978) estimates that 602 g of CO_2 m^{-2} are incorporated by vascular plant photosynthesis. Our seasonal estimate for atmospheric CO_2 flux was only 146 g of CO_2 m^{-2}. We cannot partition this among mosses and vascular plants but the point to be made is that the largest proportion of the CO_2 incorporated by photosynthesis was derived from respiration sources rather than the atmosphere. This is further seen by comparing atmospheric CO_2 flux with our estimate of CO_2 flux due to ecosystem respiration (626 g m^{-2}) which indicates about 80% of the CO_2 came from soil and plant respiration including aboveground plant material. The net seasonal flux of CO_2 from the atmosphere (146 g m^{-2}) also provides an estimate of the CO_2 which must be returned to the atmosphere by decomposers either before and/or after the production season if the ecosystem is in a steady state.

Uchijima (1970) reported the upward flux of CO_2 from the soil during daylight hours was 10 to 20% of total CO_2 fixation by a corn (*Aea mays* L.) crop. Results reported for other crops are similar (Monteith, 1962; Monteith and Szeicz, 1960). Monteith (1968) found that during the growing season, the CO_2 respired by an old grassland soil supplied about 20% of the total carbon assimilated by field crops. This agrees well with the results of Denmead (1968) and those of Meyer and Koepf and Russell as cited by Monteith (1968). Monteith et al. (1964) indicated that the soil CO_2 contribution can equal 20 to 30% of the total CO_2 requirement. One reason our estimate of the relative importance of respiration-derived CO_2 to total CO_2 uptake is higher than literature values is that ours includes aboveground respiration also. However, the contribution to respiration CO_2 flux from aboveground plant parts should be minimized compared to root and microbial respiration. Living belowground biomass in the arctic tundra is several-fold greater than living aboveground biomass (Dennis and Johnson, 1970; Dennis et al., 1978).

The high proportion of gross CO_2 uptake by plants coming from soil and plant respiration should probably be expected at Barrow. The study area is relatively undisturbed and assumed to be in an approximate steady state. Furthermore, the period of time for which conditions are favorable to appreciable decomposition activity is probably not as long in relation to the period of primary productivity as in temperate latitudes. Thus a relatively higher proportion of the previous year's production is being decomposed concurrently with the period of the current year's production. The proportion of the total upward CO_2 flux emanating from root respiration rather than decomposition may be well over 50%. Billings et al. (1977) estimated that 50 to 90% of total soil respiration at the study site could be accounted for by root and rhizome respiration. For other systems, Lundegärdh (1927) found that root respiration of oats (*Avena sativa* L.) accounted for about 30% of total soil respiration. Kucera and Kirkham (1971) partitioned soil respiration in a Missouri tallgrass prairie into microbial processes (60%) and root metabolism (40%). Wiant (1967b,c) estimated root respiration of white pine (*Pinus strobus* L.) and eastern hemlock (*Tsuga canadensis* (L.) Carr.) as high as 50% of total CO_2 output in a July field experiment. Witkamp and Frank (1969) concluded that root respiration in a 40-yr-old second growth stand of *Pinus echinata* Mill. was not greater than 50% of total CO_2 evolution from the forest floor. Root respiration made up half of total soil resipiration on a particular day in a bean (*Vicia vulgaris*) crop (Monteith, 1962). Reiners (1968) measured CO_2 evolution from the forest floor in Minnesota and found that total CO_2 evolution was three times higher than expected from an equivalent amount of carbon release from annual litter fall. Tree root respiration was thought to be the major cause of this disparity. Most of the CO_2 evolved from the floor of an oak–hickory forest was due to root and associated microorganism respiration rather than litter decomposition (Garrett and Cox, 1973). It is possible therefore that the apparently high proportion of CO_2 originating *in situ* in relation to CO_2 uptake may be due at least in part to a high root standing crop.

Johnson and Kelley (1970) estimated that root respiration was five times larger than top respiration for an arctic tundra ecosystem at Meade River, Alaska, about 110 km SSW of Barrow. Dennis and Johnson (1970) evaluated root:shoot ratios for a variety of arctic tundra sites and found that the live subsurface standing crops ranged from 3 to 48 times greater than the live aboveground standing crop. Thus, in view of the above, it seems logical to hypothesize that the bulk of CO_2 flux from respiration originates from soil microorganisms and plant roots with the latter probably contributing most at the Barrow site and that a much smaller proportion arises from aboveground plant respiration.

No particular adaptive value appears evident for the relatively high rate of CO_2 production from respiration during the growing season. The possibility of CO_2 enrichment in the canopy and a resulting enhancement of photosynthesis is unlikely in the short graminoid canopy and generally turbulent atmospheric conditions of Barrow. More likely the relatively short period in which the soil surface layer is thawed demands a high rate of decomposition to prevent large accumulations of organic matter. In addition, a high rate of decomposition may be required in a nutrient-limited system to provide rapid turnover of the available supply of soil nutrients to support the demands of current production.

A Q_{10} of about 1.7 or 1.8 was indicated for combined respiration processes by the regression analysis. Kucera and Kirkham (1971) reported Q_{10} values from 1.6

to 2.3 for soil respiration in a Missouri tallgrass prairie at soil temperatures between 10 and 30°C. Witkamp (1969) found considerable variation in Q_{10} values for a mixed hardwood forest depending on the substrate. Respiration in litter bags had a Q_{10} of approximately 2.5 and the litter layer had a Q_{10} of 3 while the entire profile had a Q_{10} of 1.5. The Q_{10} for soil respiration in soft and hardwood forest stands ranged between 1.7 and 2 between 20 and 40°C (Wiant, 1967a). Wiant also reported results of earlier studies (Lundegårdh, Koepf) which found the Q_{10} was approximately 2 between 10 and 40°C. Uchijima (1970) used a Q_{10} of 3.5 to approximate soil CO_2 flux from crop land. Tieszen (personal communication) measured Q_{10} for dark respiration of aboveground parts of some Barrow species. He found Q_{10} varied between 1.8 and 2.1 over the temperature range of 0 to 35°C. Apparently, the estimate of Q_{10} derived by the aerodynamic method is within the limits measured for other ecosystems. In contrast, Peterson and Billings (1975) observed a linear rather than a Q_{10} response of respiration to temperature at the Barrow site in 1973. Since they considered most of the CO_2 produced to result from respiration, they attributed the linearity to the effect of summing the various Q_{10} responses of the different organisms present.

The regression equation intercept [Eq. (4)] estimated total respiration for the 61-day sampling period at 626 g of CO_2 m^{-2}, an average of 0.43 g m^{-2} hr^{-1}. During the period, estimates ranged from 0.2 to 1.6 g m^{-2} hr^{-1}. CO_2 evolution from a tall grass prairie (Kucera and Kirkham, 1971) was zero during the winter rising to 0.45 g m^{-2} hr^{-1} in summer. Lundegårdh's (1927) investigations demonstrated rates of CO_2 evolution between 0.13 and 1.36 m^{-2} hr^{-1}. The maximum summer rate of CO_2 evolution from the floor of an oak–hickory forest was 1.2 g m^{-2} hr^{-1} while the maximum winter rate was 0.18 g m^{-2} hr^{-1} (Garrett and Cox, 1973). Total soil respiration measured by Billings et al. (1977) at the study site ranged from 0.08 to 0.13 g m^{-2} hr^{-1} between 16 and 19 July and from 0.15 to 0.30 g m^{-2} hr^{-1} during the final week in July. Similar values were obtained in 1973 by Peterson and Billings (1975). The regression estimate appears too high in light of the data of Billings et al., possibly as a result of overestimation early in the season as pointed out in the results section (Figure 7).

In summary, the aerodynamic method and our approach to data analysis using dark CO_2 fluxes to estimate ecosystem respiration appear to be useful techniques for integrating community productivity in the arctic tundra. Comparisons with independent methods of primary production assessment (Tieszen, 1978) show that the aerodynamic method estimated seasonal CO_2 incorporation within a few percent of the *in situ* cuvette estimate and that both these estimates approximated the value predicted by the mechanistic model of Miller and Tieszen (1972).

Acknowledgments. This research was supported by the National Science Foundation under the joint sponsorship of the International Biological Programme and the Office of Polar Programs and was coordinated by the U.S. Tundra Biome. Field and laboratory activities at Barrow were supported by the Naval Arctic Research Laboratory of the Office of Naval Research.

This research was also supported by the U.S. Army Cold Regions Research and Engineering Laboratory (DA Task 4A061101A91D03, In-House Laboratory Independent Research, Work Unit 147) in cooperation with the University of Alaska's Institute of

Marine Sciences. Supporting data for this study were provided by Drs. J. Brown, J. Dennis, W. Klein, L. Tieszen, and G. Weller. We extend a special thanks to Mrs. Mary Ann Coyne for data reduction assistance and to Drs. Kenneth Burnham, Dennis Hein, and Ivan Frohne for data analysis assistance.

These results were originally published in *J. Appl. Ecol.* **12**: 587–611, 1975.

References

Baumgartner, A. (1969) Meteorological approach to the exchange of CO_2 between the atmosphere and vegetation, particularly forest stands. *Photosynthetica,* **3**: 127–149.

Billings, W. D., and H. A. Mooney. (1968) The ecology of arctic and alpine plants. *Biol. Rev.,* **43**: 481–529.

Billings, W. D., K. M. Peterson, G. R. Shaver, and A. W. Trent. (1977) Root growth, respiration, and carbon dioxide evolution in an arctic tundra soil. *Arct. Alp. Res.,* **9**: 127–135.

Bliss, L. C. (1961) Adaptations of arctic and alpine plants to environmental conditions. *Arctic,* **15**: 117–144.

Bliss, L. C. (1962) Net primary production of tundra ecosystems. In *Die Stoffproduktion der Pflanzendecke* (H. Leith, Ed.). Stuttgart: Fischer, pp. 35–46.

Coyne, P. I., and J. J. Kelley. (1975) CO_2 exchange over the Alaskan arctic tundra: Metererological assessment by an aerodynamic method. *J. Appl. Ecol.,* **12**: 587–611.

Denmead, O. T. (1968) Carbon dioxide exchange in the field: Its measurement and interpretation. In *Proceedings WMO Seminar: Agricultural Meteorology, Melbourne, 1966.* Bureau of Meteorology, Melbourne, Australia, pp. 445–482.

Denmead, O. T. (1970) Transfer processes between vegetation and air: Measurement, interpretation and modelling. In *Prediction and Measurement of Photosynthetic Productivity: Proceedings, IBP/PP Technical Meeting, Trebon, 2–4 September 1969.* Wageningen: Center for Agricultural, Publishing and Documentation, pp. 149–164.

Denmead, O. T., and I. C. McIlroy. (1971) Measurement of carbon dioxide exchange in the field. In *Plant Photosynthetic Production: Manual of Methods* (Z. Šesták, J. Čatský, and P. G. Jarvis, Eds.). The Hague: Dr. W. Junk, pp. 467–516.

Dennis, J. G., and P. L. Johnson. (1970) Shoot and rhizome-root standing crops of tundra vegetation at Barrow, Alaska. *Arct. Alp. Res.,* **2**: 253–266.

Dennis, J. G., L. L. Tieszen, and M. Vetter. (1978) Seasonal dynamics of above- and belowground production of vascular plants at Barrow, Alaska. In *Vegetation and Production Ecology of an Alaskan Arctic Tundra* (L. L. Tieszen, Ed.). New York: Springer-Verlag, Chap. 4.

Dyer, A. J. (1963) The adjustment of profiles and eddy fluxes. *Quart. J. Roy. Meteorol. Soc.,* **89**: 276–280.

Ellison, T. H. (1957) Turbulent transport of heat and momentum from an infinite rough plane. *J. Fluid Mechan.,* **2**: 456–466.

Garrett, H. E., and G. S. Cox. (1973) Carbon dioxide evolution from the floor of an oak–hickory forest. *Soil Sci. Soc. Amer. Proc.,* **37**: 641–644.

Hicks, B. B. (1972) Some evaluations of drag and bulk transfer coefficients over water bodies of different sizes. *Bound. Layer Meteorol.,* **3**: 201–213.

Inoue, E. (1968) The CO_2-concentration profile within crop canopies and its significance for the productivity of plant communities. In *Functioning of Terrestrial Ecosystems at the Primary Production Level* (F. E. Eckardt, Ed.). Paris: Unesco, pp. 359–366.

Johnson, P. L., and J. J. Kelley, Jr. (1970) Dynamics of carbon dioxide and productivity in an arctic biosphere. *Ecology,* **51**: 73–80.

Kelley, J. J., Jr. (1968) Carbon dioxide and ozone in the arctic atmosphere. In *Arctic Drifting Stations* (J. E. Sater, Coordinator). Washington, D.C.: Arctic Institute of North America, pp. 155–166.

Kucera, C. L., and D. R. Kirkham. (1971) Soil respiration studies in tall grass prairie in Missouri. *Ecology,* **52**: 912–915.

Lemon, E. R. (1960) Photosynthesis under field conditions. II. An aerodynamic method for determining the turbulent carbon dioxide exchange between the atmosphere and a corn field. *Agron. J.,* **52**: 697–703.

Lemon, E. R. (1965) Micrometeorology and the physiology of plants in their natural environment. In *Plant Physiology* (F. C. Steward, Ed.). New York: Academic Press, Vol. IV-A, pp. 203–207.

Lemon, E. R. (1967) Aerodynamic studies of CO_2 exchange between the atmosphere and the plant. In *Harvesting the Sun* (A. SanPietro, F. A. Greer, and T. J. Army, Eds.). New York: Academic Press, pp. 263–290.

Lemon, E. R. (1969) Gaseous exchange in crop stands. In *Physiological Aspects of Crop Yield* (J. D. Eastin, F. A. Haskins, C. Y. Sullivan, and C. H. M. Van Baval, Eds.). Madison: American Society of Agronomy, pp. 117–137.

Lemon, E. R. (1970) Mass and energy exchange between plant stands and environment. In *Prediction and Measurement of Photosynthetic Productivity: Proceedings, IBP/PP Technical Meeting, Trebon, 2–4 September 1969.* Wageningen: Center for Agricultural, Publishing and Documentation, pp. 149–164.

Lettau, H. H. (1962) Notes on theoretical models of profile structure in the diabatic surface layer. In *Studies of the Three-Dimensional Structure of the Planetary Boundary Layer.* University of Wisconsin, Madison, pp. 195–226.

Lundegärdh, H. (1927) Carbon dioxide evolution of soil and crop growth. *Soil Sci.,* **23**: 417–453.

Miller, P. C., and L. L. Tieszen. (1972) A preliminary model of processes affecting primary production in the arctic tundra. *Arct. Alp. Res.,* **4**: 1–18.

Miller, P. C., W. A. Stoner, and J. R. Ehleringer. (1978) Some aspects of water relations of arctic and alpine regions. In *Vegetation and Production Ecology of an Alaskan Arctic Tundra* (L. L. Tieszen, Ed.). New York: Springer-Verlag, Chap. 14.

Monteith, J. L. (1962) Measurement and interpretation of carbon dioxide fluxes in the field. *Neth. J. Agric. Sci.,* **10**: 334–346.

Monteith, J. L. (1968) Analysis of the photosynthesis and respiration of field crops from vertical fluxes of carbon dioxide. In *Functioning of Terrestrial Ecosystems at the Primary Production Level* (F. E. Eckardt, Ed.). Paris: Unesco, pp. 349–358.

Monteith, J. L. (1973) *Principles of Environmental Physics.* London: Edward Arnold, 241 pp.

Monteith, J. L., and Szeicz, G. (1960) The carbon dioxide flux over a field of sugar beet. *Quart. J. Roy. Meteorol. Soc.,* **86**: 205–214.

Monteith, J. L., G. Szeicz, and K. Yabuki. (1964) Crop photosynthesis and the flux of carbon dioxide below the canopy. *J. Appl. Ecol.,* **1**: 321–337.

Nakano, Y., and J. Brown. (1972) Mathematical modeling and validation of the thermal regimes in tundra soils, Barrow, Alaska. *Arct. Alp. Res.,* **4**: 19–38.

Oechel, W. C., and B. Sveinbjörnsson. (1978) Primary production processes in arctic bryophytes at Barrow, Alaska. In *Vegetation and Production Ecology of an Alaskan Arctic Tundra* (L. L. Tieszen, Ed.). New York: Springer-Verlag, Chap. 11.

Panofsky, H. A., A. K. Blackadar, and G. E. McVehil. (1960) The diabatic wind profile. *Quart. J. Roy. Meteorol. Soc.,* **86**: 390–898.

Panofsky, H. A., and A. A. Townsend. (1964) Change of terrain roughness and the wind profile. *Quart. J. Roy. Meteorol. Soc.,* **90**: 147–155.

Peterson, K. M., and W. D. Billings. (1975) Carbon dioxide flux from tundra soils and vegetation as related to temperature at Barrow, Alaska. *Amer. Midl. Natur.,* **94**: 88–98.

Priestly, C. H. B. (1959) *Turbulent Transfer in the Lower Atmosphere.* Chicago: University of Chicago Press, 130 pp.

Reiners, W. A. (1968) Carbon dioxide evolution from the floor of three Minnesota forests. *Ecology,* **49**: 471–483.

Rose, C. W. (1966) *Agricultural Physics*. Oxford: Pergamon Press, 226 pp.

Sellers, W. D. (1965) *Physical Climatology*. Chicago: University of Chicago Press, 272 pp.

Tanner, C. B. (1963) Energy relations in plant communities. In *Environmental Control of Plant Growth* (L. T. Evans, Ed.). New York: Academic Press, pp. 141–148.

Taylor, R. J. (1962) Small-scale advection and the neutral wind profile. *J. Fluid Mech.*, **13**: 529–539.

Tieszen, L. L. (1972) The seasonal course of aboveground production and chlorophyll distribution in a wet arctic tundra at Barrow, Alaska. *Arct. Alp. Res.*, **4**: 307–324.

Tieszen, L. L. (1973) Photosynthesis and respiration in arctic tundra grasses: Field light intensity and temperature responses. *Arct. Alp. Res.*, **5**: 239–251.

Tieszen, L. L. (1975) CO_2 exchange in the Alaskan Arctic Tundra: Seasonal changes in the rate of photosynthesis of four species. *Photosynthetica*, **9**: 376–390.

Tieszen, L. L. (1978) Photosynthesis in the principal Barrow, Alaska, species: A summary of field and laboratory responses. In *Vegetation and Production Ecology of an Alaskan Arctic Tundra* (L. L. Tieszen, Ed.). New York: Springer-Verlag, Chap. 10.

Tieszen, L. L., and P. L. Johnson (1968) Pigment structure of some arctic tundra communities. *Ecology*, **49**: 370–373.

Uchijima, Z. (1970) Carbon dioxide environment and flux within a corn crop canopy. In *Prediction and Measurement of Photosynthetic Productivity: Proceedings IBP/PP Technical Meeting, Trebon, 14–21 September 1969*. Wageningen: Center for Agricultural Publishing and Documentation, pp. 179–198.

Van Cleve, K., and D. Sprague. (1971) Respiration rates in the forest floor of birch and aspen stands in interior Alaska. *Arct. Alp. Res.*, **3**: 17–26.

Webb, E. K. (1965) Aerial microclimate. In *Agricultural Meteorology. Meteorol. Monogr.*, **6**: 27–58.

Webber, P. J. (1978) Spatial and temporal variation of the vegetation and its production, Barrow, Alaska. In *Vegetation and Production Ecology of an Alaskan Arctic Tundra* (L. L. Tieszen, Ed.). New York: Springer-Verlag, Chap. 3.

Weller, G., S. Cubley, S. Parker, D. Trabant, and C. Benson. (1972) The tundra microclimate during snow-melt at Barrow, Alaska. *Arctic*, **25**: 291–300.

Wiant, H. V. (1967a) Influence of temperature on rate of soil respiration. *J. Forest.*, **65**: 489–490.

Wiant, H. V. (1967b) Contribution of roots to forest "soil respiration." *Advan. Frontiers Plant Sci.*, **18**: 136–138.

Wiant, H. V. (1967c) Has the contribution of litter decay to forest "soil respiration" been overestimated? *J. Forest.*, **65**: 408–409.

Witkamp, M. (1966) Decomposition of leaf litter in relation to environment, microflora, and microbial respiration. *Ecology*, **47**: 194–201.

Witkamp, M. (1969) Cycles of temperature and carbon dioxide evolution from litter and soil. *Ecology*, **50**: 922–924.

Witkamp, M., and M. L. Frank. (1969) Evolution of CO_2 from litter, humus, and subsoil of a pine stand. *Pedobiologia*, **9**: 358–365.

Woodwell, G. M., and D. B. Botkin. (1970) Metabolism of terrestrial ecosystems by gas exchange techniques: The Brookhaven approach. In *Analysis of Temperate Forest Ecosystems* (D. E. Reichle, Ed.). New York: Springer-Verlag, pp. 73–85.

Wright, J. L., and E. R. Lemon. (1966) Photosynthesis under field conditions. IX. Vertical distribution of photosynthesis within a corn crop. *Agron. J.*, **58**: 265–268.

Zelitch, I. (1971) *Photosynthesis, Photorespiration, and Plant Productivity*. New York: Academic Press, 347 pp.

13.   Constraints on Tundra Productivity: Photosynthetic Capacity in Relation to Solar Radiation Utilization and Water Stress in Arctic and Alpine Tundras

M. M. CALDWELL, D. A. JOHNSON, AND M. FAREED

Introduction

Short growing seasons and low temperatures are the primary constraints on productivity in both arctic and alpine tundra environments. This is reflected in the similar life form and the taxonomic affinities of arctic and alpine tundra floras despite the differences in solar radiation intensity, photoperiod, wind regimes, and edaphic factors that often vary substantially between arctic and alpine habitats (Bliss, 1956, 1971).

Maximizing carbon fixation during the short growing season must involve both a rapid development of photosynthetically active tissue and a capacity for photosynthesis at low temperatures (Billings, 1975). Because of high solar radiation intensities in many alpine regions, the long photoperiods in arctic areas, and the relatively low stature of tundra vegetation, constraints on productivity due to limiting solar radiation or competition for light within the tundra canopy are not normally considered to be significant in tundra areas. Several notable exceptions have been described, such as in the nivale zone of the Austrian Alps, where Moser (1973) reported detailed measurements at the canopy level indicating radiation often limited photosynthesis during the growing season. This was due primarily to cloud cover and periodic, though ephemeral, coverage of the vegetation by snow from summer storms. Similarly, in the Presidential Range of New Hampshire, Hadley and Bliss (1964) suggested productivity was limited by low light intensities due to frequent cloud cover.

Water stress has been suggested as another constraint on productivity of tundra species. Scott and Billings (1964) and Billings and Bliss (1959) reported significant correlations of site productivity and water availability in alpine tundra sites of southern Wyoming. Segregation of species in both arctic and alpine tundra environments along microtopographical gradients may in part be determined by the differential sensitivity of various tundra species to water stress. Although wet meadow sites in the alpine tundra on Niwot Ridge, Colorado, typically exhibited soil water potentials, Ψ_s, above -5 bars for most of the 1973 growing season, windswept fellfield habitats in the same area reached Ψ_s less than -10 bars (Johnson and Caldwell, unpublished data). Similarly, in the Arctic, low-centered polygons and polygon troughs often have standing water during most of the growing season. However, high-center polygons and beach ridges can exhibit Ψ_s values well below -10 bars (Teeri, 1973). This same microsite variability with respect to soil water was also described in some of the first comparisons of arctic and alpine microenvironments by Bliss (1956).

Atmospheric water stress also can be relatively severe in tundra environments. Salisbury and Spomer (1964) have measured leaf-to-air temperature gradients of more than 20°C in alpine tundra areas of Colorado. Though much less frequent, similar gradients may be reached in some arctic tundra sites. For example, Mayo et al. (1973) reported that leaf temperatures of *Dryas integrifolia* may be 30°C above air temperature. With such large temperature gradients between leaves and the ambient air, water vapor difference (WVD) between saturated leaf vapor pressure and ambient air vapor pressure can exceed 20 mbars in some instances. Such atmospheric moisture stress when exacerbated by low soil moisture may play a large role in limiting productivity and competitive ability of wet site species in drier areas—particularly in years of very limited growing season precipitation.

The Tundra Biome program provided the opportunity to conduct coordinated studies in arctic and alpine sites including the dynamics of leaf growth, photosynthetic capacity of leaves of different age and position, and the phenology of canopy development. In addition, the potential for water stress to limit the productivity of tundra species and to play an important role in the spatial segregation of both arctic and alpine species was investigated.

Methods

All of the field studies were conducted at the U.S. IBP Tundra Biome intensive sites on Niwot Ridge, Colorado (40°4′ N, 105°36′ W, 3476 m elev.) and at the wet coastal tundra site at Barrow, Alaska (71°18′ N, 156°46′ W, 3 m elev.). On Niwot Ridge, our studies centered on two principal communities: (1) a *Kobresia* community which is a well-developed mesic meadow dominated by *Kobresia myosuroides* with subdominants *Geum rossii* and *Carex rupestris*, and (2) a hairgrass meadow dominated by *Deschampsia caespitosa, Geum rossii,* and *Artemisia scopulorum*. In this community snowmelt occurs later in the season, and soil water is generally more plentiful longer into the growing season. In the Arctic, a community dominated by *Eriophorum angustifolium, Dupontia fisheri,* and *Carex aquatilis* was used for our field studies. Plant nomenclature generally follows that of Weber (1967) for the alpine sites and Hultén (1968) for the arctic species.

Modified inclined point quadrats (Warren Wilson, 1960) were used to estimate leaf area index (foliage area per unit ground surface, F = LAI) and to quantitatively follow phenological progression in these sites throughout the 1971 growing season (Caldwell et al., 1974; Fareed and Caldwell, 1975).

In 1972 and 1973, leaf photosynthetic capacity measurements were determined using a portable $^{14}CO_2$-labeling system (Tieszen et al., 1974). These determinations were taken under a constant chamber temperature of $10 \pm 0.5°C$, which is representative of mean daytime growing season temperatures at the study area. Leaf temperatures were within $\pm 1°C$ of chamber air temperatures. Constant irradiance of 2700 ± 20 μE m^{-2} sec^{-1} in the 400- to 700-nm waveband was used for these photosynthetic capacity determinations. This approximates maximum midday solar irradiance. Leaf growth rates were also taken in conjunction with the photosynthetic measurements (Johnson and Caldwell, 1974). These studies were in many respects conducted in a manner very similar to that of Tieszen and Johnson (1975) in the arctic site.

During 1974, selected arctic and alpine species were grown in a greenhouse and subjected to varying soil water stress levels. Laboratory gas exchange measurements were carried out using a Siemens gas exchange system (Koch et al., 1971). Carbon dioxide exchange was determined by infrared gas analysis, and water vapor exchange employed a Cambridge dew point hygrometer. Leaf area (one side) was measured with an electronic planimeter (Lambda Corp.). Temperature, humidity, irradiation, and soil water status were controlled in each experiment. By varying soil water and air humidity independently of other environmental factors, it was possible to evaluate the relationship between plant gas exchange, soil water stress, and atmospheric water stress (Johnson and Caldwell, 1975). In addition, the water potential components of leaf discs from these greenhouse-grown species were determined under a range of soil water conditions (Johnson and Caldwell, 1976) using a psychrometric technique described by Brown (1976).

Results and Discussion

Seasonal Progression of Photosynthetic Capacity and Plant Canopy Development: Utilization of Solar Radiation

Photosynthetic Competency of Individual Leaves. The limited season for carbon fixation in both arctic and alpine tundras must be efficiently utilized by competitive tundra species. For several arctic and alpine species, it appears that a few leaves are typically initiated in the previous growing season but do not attain complete exsertion and development before the end of the growing season. They overwinter and assume the position of the first leaves the following spring, and maximum photosynthetic capacity of these leaves is quickly developed. With progression of the growing season photosynthetic competency of these first leaves declines and newly formed leaves somewhat higher on the plant develop and assume maximum photosynthetic capacity at about the time of full leaf expansion. This progression of leaf growth and attainment of maximum photosynthetic capacity followed by a slow senescence and loss of photosynthetic competency appeared in both the arctic and alpine habitats (Johnson and Caldwell, 1974; Tieszen and Johnson, 1975).

Ontogenetic timing of these tundra species, thus, appears geared with the surge and decline of individual leaf photosynthetic capacities so that one to several leaves operating at near maximum photosynthetic competency for exist-

ing conditions are always maintained during the growing season for each plant.
With each new leaf the active photosynthetic surface becomes progressively
elevated in the plant canopy where radiation is more available. Although such a
pattern of growth and timing of photosynthetic capacity is not confined to tundra
species (e.g., Sesták and Catský, 1962; Ludlow and Wilson, 1971), it does appear
to be remarkably similar in both pattern and timing during the short growing
seasons of the two tundra areas.

The detailed sequence of photosynthetic capacity development and leaf
growth rates are shown in Figure 1 for *Deschampsia caespitosa* in the alpine
hairgrass site and *Kobresia myosuroides* in the drier *Kobresia* meadow on Niwot
Ridge. Initially for each leaf there was a period of high relative growth rate, RGR,
associated with a period of high or increasing photosynthetic capacity. This was
followed by a reasonably long period of little or no change in length of the living
leaf. At this time, photosynthetic capacity had usually reached its maximum.
Photosynthetic capacity remained reasonably constant for a period and then

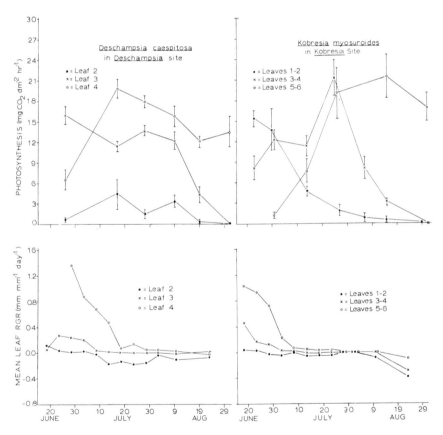

Figure 1.   Leaf photosynthesis rates and leaf relative growth rates, RGR, of *Kobresia
myosuroides* in the *Kobresia* site and *Deschampsia caespitosa* in the *Deschampsia* site.
The vertical bars represent ± one standard error. (Adapted from Johnson and Caldwell,
1974, with the permission of the Institute of Arctic and Alpine Research.)

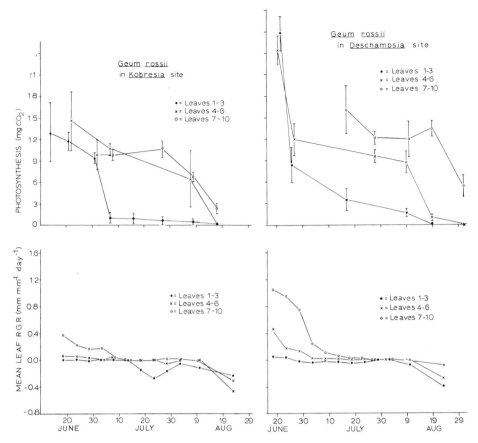

Figure 2.   Leaf photosynthesis rates and leaf relative growth rates, RGR, of *Geum rossii* in the *Kobresia* site and the *Deschampsia* site. The vertical bars represent ± one standard error. (Adapted from Johnson and Caldwell, 1974, with the permission of the Institute of Arctic and Alpine Research.)

eventually exhibited a steady decrease. Finally, there was a period of rapid senescence denoted here as negative RGR because the leaves were visibly dying back from the tips. By this time photosynthetic capacity per unit area of living leaf was rapidly decreasing. A similar depiction of photosynthetic capacity and leaf growth dynamics emerged from the arctic studies although they were not reported in such detail as for the alpine sites (Tieszen and Johnson, 1975).

   The difference in timing of the two species shown in Figure 1 is in part due to the difference in the *Kobresia* meadow and hairgrass sites. Since the *Kobresia* meadow site generally had much less snow and, in fact, was often free of snow because of winter wind patterns, plants were released much earlier in the spring as compared to the hairgrass site where snow cover persisted for several weeks longer. The differences in the sites are perhaps best illustrated by the comparable dynamics of photosynthesis and leaf relative growth rates of a species common to both sites, *Geum rossii* (Figure 2). Here the much later attainment of maximal photosynthetic capacity and a longer period of photosynthetic competency later in the season was apparent for *Geum* in the hairgrass site. Snowmelt was largely complete by 22 May in the *Kobresia* site during 1972 when these measurements were taken. Therefore, there was an earlier opportunity for growth initiation as compared with the hairgrass site where snow remained until 9 June, and the first growth was not noted until the middle of June. In the latter part of the season, the differences in water stress may have accounted for at least part of the earlier decline of photosynthetic capacity of *Geum* in the *Kobresia* site. Leaf water potentials for this species were on the order of -18 bars by 18 July in this site, whereas comparable leaf water potentials were not observed until 29 July in the hairgrass site.

It appears that not only do arctic and alpine tundra species exhibit an ontogenetic timing so as to maximize carbon gain during the short growing season, but also a certain degree of plasticity in adjustment of this timing to accommodate differences in microsites and year-to-year variations in snow cover and time of meltoff. Such flexibility was particularly evident in studies of phenological progression on the two sites of the alpine study area (Fareed and Caldwell, 1975). The progression of phenological stages for *Geum rossii* in the two sites was in some respects comparable to the shift in photosynthetic capacity and leaf growth rates. However, for the reproductive phenological stages, this shift between the two sites was not linear. Instead there was a distinct telescoping of the sequence of phenological stages in the *Deschampsia* site. This would seem to be beneficial in accommodating the less favorable weather patterns toward the end of the growing season which would, of course, occur at about the same time for these proximate sites. Thus, the successful completion of fruit formation would be ensured. In contrast, prolonged photosynthetic activity in late summer at the hairgrass site where moisture availability persists would seem more opportunistic so that these plants might fully utilize the remaining days of the short growing season for carbon gain.

Seasonal Development of the Tundra Canopy.   Although the foregoing depiction of the progression of maximum photosynthetic activity from older to newer leaves located higher in the canopy suggests that competition for solar radiation may be of significance in tundra areas, an assessment of the canopy structure and density is necessary in a consideration of radiation utilization in the entire plant community. The seasonal progression of leaf area index, the ratio of foliage area to ground area, for several height zones at the two intensive sites on Niwot Ridge and the site at Barrow are shown in Figures 3, 4, and 5 (Caldwell et al., 1974). The peak of the vegetative season as reflected in maximum living aboveground biomass was attained at all three sites in the period from 25 July to 9 August. Also, at all three sites, maximum leaf area index, F or LAI, was attained earlier in the season at the lower height zones in the canopies which corresponds with the foregoing depiction of the progress of individual leaf growth and development. At all heights in these canopies, there was less foliage in the arctic community (see Dennis et al., 1978) than in the communities of the alpine sites. For the entire canopy, leaf area index for the arctic community was about 1 while

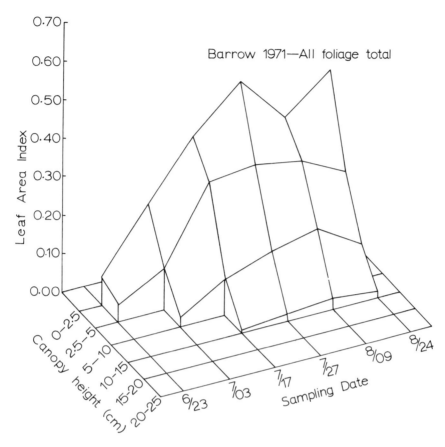

Figure 3.   Leaf area indices for different height zones in the arctic meadow community at Barrow, Alaska, during 1971. (From Caldwell et al., 1974, with permission of the Institute of Arctic and Alpine Research.)

values of 2.2 and 2.0 were reported for the *Kobresia* meadow and hairgrass sites, respectively.

Higher solar radiation intensities in alpine communities would seem to be better utilized by denser vegetative canopies when compared to the arctic tundra community where radiation intensities are much lower. When the prevailing angles of incidence of the direct beam solar radiation are considered, the difference between arctic and alpine tundras is further amplified. In alpine areas, direct beam radiation is presented to the top of the canopy at greater solar altitudes than in the Arctic resulting in shorter optical pathlengths through the alpine canopy. A comparison of the calculated radiation extinction in the Barrow community and the Niwot *Kobresia* meadow is depicted for clear (Figure 6) and overcast (Figure 7) days in late July at the peak of vegetative development (solar declination 20°). Both diffuse and direct beam radiation are incorporated in these representations (Caldwell et al., 1974). This analysis was conducted for two

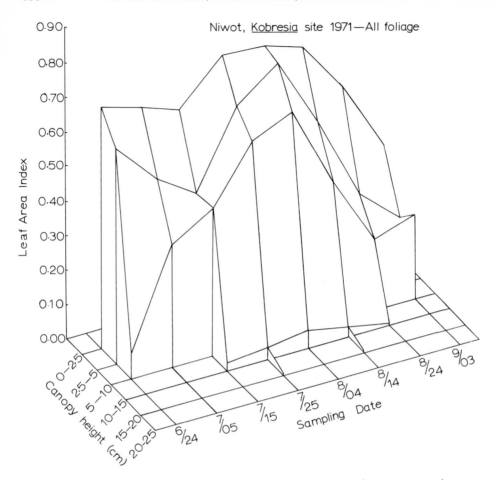

Figure 4.   Leaf area indices for different height zones in the *Kobresia* community on Niwot Ridge, Colorado, during 1971. (From Caldwell et al., 1974, with permission of the Institute of Arctic and Alpine Research.)

predominant foliage inclination angles of 60 and 85° which encompass most of the foliage in both tundra communities.

Despite the low stature of these communities, solar irradiation presented at the top of the canopy does undergo a pronounced extinction within the canopy—particularly at leaf angles near 60°. Considering that most of the daily photosynthetic carbon gain would occur during a 20-hr day at Barrow and in a 15-hr day on Niwot Ridge at this time of year, total daily irradiance of 650 and 720 cal cm^{-2} min^{-1} would be presented at the top of the canopy on clear days at Barrow and Niwot Ridge, respectively. At a foliage angle of 85°, the percentage of total daily radiation intercepted by the two communities would be nearly the same, 44% at Barrow and 51% at Niwot Ridge. At a foliage angle of 60° the interception would be 22% less at Barrow than at Niwot Ridge, i.e., 75 versus 58% of the total daily radiation presented at the top of the canopy. For totally overcast conditions

(Figure 7), a similar depiction of radiation penetration into these canopies suggests reasonably similar attenuation of the total daily available radiation in these two communities. The Niwot Ridge canopy would in this case effect a somewhat greater extinction of the radiation. However, totally overcast days on Niwot Ridge are less common than at Barrow.

Thus, although the alpine plant canopy is more than twice as dense as the Barrow community, percentage daily radiation interception is usually no more than 20% greater than in the Barrow community at comparable foliage inclina-

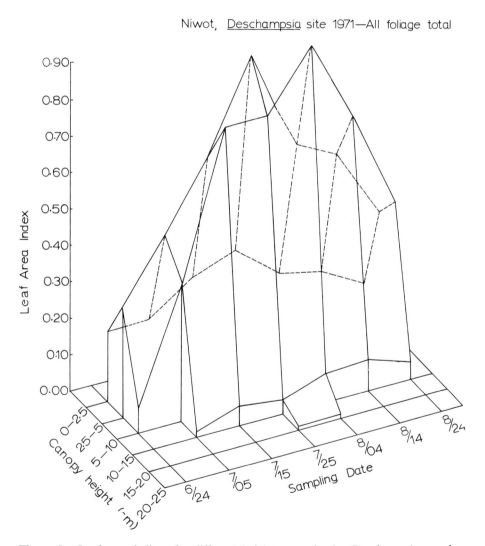

Figure 5. Leaf area indices for different height zones in the *Deschampsia* meadow community on Niwot Ridge, Colorado, during 1971. (From Caldwell et al., 1974, with permission of the Institute of Arctic and Alpine Research.)

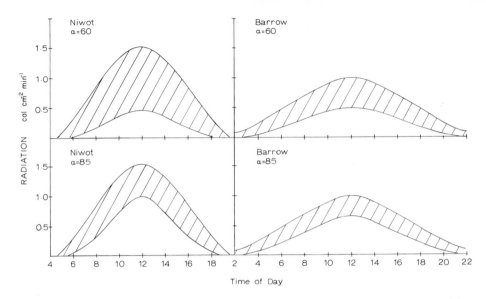

Figure 6.   Simulated radiation extinction in the Barrow community and the Niwot Ridge *Kobresia* meadow for two foliage inclination angles (60 and 85°) on clear days. In each graph the upper curve represents global solar radiation presented to the top of the canopy, the lower curve represents radiation penetrating the ground level, and the hatched area represents radiation extinction within the canopy. (From Caldwell et al., 1974, with permission of the Institute of Arctic and Alpine Research.)

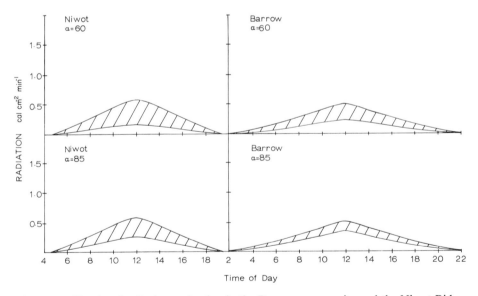

Figure 7.   Simulated radiation extinction in the Barrow community and the Niwot Ridge *Kobresia* meadow for two foliage inclination angles (60 and 85°) for overcast days. In each graph, the upper curve represents global solar radiation presented to the top of the canopy, the lower curve represents radiation penetrating to ground level, and the hatched area represents radiation extinction within the canopy. (From Caldwell et al., 1974, with permission of the Institute of Arctic and Alpine Research.)

tions for either clear or overcast days. Assuming a proportionality between radiation extinction in the canopy and radiation utilized in photosynthesis, it is apparent that radiation of greater intensities and presented at higher solar altitudes during the clear days in the alpine areas is well utilized by the denser tundra communities, whereas lower radiation intensities presented at lower solar altitudes in the longer arctic days are nearly equally utilized in a much more sparsely stocked arctic tundra community. These representations of radiation extinction in the arctic and alpine communities corroborate the implied competition for solar radiation in the foregoing discussion of the seasonal progression of individual leaf growth and photosynthetic competency.

Water Limitation and Productivity of Tundra Plants

Evidence from Field Studies.   Although tundras are not generally conceived as water-limited environments, development of plant water stress and the subsequent constraints on productivity have been suggested for a number of habitats in both arctic and alpine environments, as was discussed in the introduction. Water stress may play a key role in the segregation of tundra species along microtopographic gradients in both arctic and alpine areas. Although some species such as *Geum rossii* are prevalent in both wet and dry microsites in tundra environments, most tundra species are more restricted to certain microsites. Habitats where snow ablation takes place early in the season, which drain quickly and receive no additional water due to overland flow from other sites, constitute the drier sites in both arctic and alpine environments. In alpine areas such as the *Kobresia* meadow on Niwot Ridge, the aridity of the site depends on factors such as winter wind patterns, which control snow deposition, and exposure to solar radiation. In the arctic tundra near Barrow, which is generally considered to be wet tundra underlain with permafrost which impedes drainage of melt water, dry sites are limited to the high centers of polygons or other sites which drain more effectively.

In years of little or no summer precipitation these sites can become reasonably dry during the long arctic days. Stoner and Miller (1975) reported minimum leaf water potentials during the 1973 growing seasons approaching -15 bars for several species located in these drier sites. Summer precipitation in 1973 was 80% above the 30-yr mean for this area (Tieszen, 1978). Based on the relationship determined between stomatal diffusion resistance and leaf water potential, Stoner and Miller (1975) suggested that water potentials of several tundra species including some normally found in the wetter sites were sufficiently low to cause at least some restriction in stomatal aperture, and in some cases complete stomatal closure. A simulation model incorporating this information suggested the increased stomatal diffusion resistance to be sufficient to decrease net photosynthesis in some species. Actual measurements of stomatal diffusion resistance using a porometer in late July and early August of that year did reveal leaf diffusion resistances approaching 10 to 12 sec cm^{-1} for some species.

On Niwot Ridge, Ehleringer and Miller (1975) reported minimum seasonal leaf water potentials for some species to approach values of -30 to -40 bars for 1972. For July and August, when precipitation is most important in ameliorating water stress in the drier sites, rainfall in 1972 was within 10% of a 9-yr average for this site (Barry, personal communication). As a result, although June precipitation was 100% above the average, the 1972 growing season received nearly normal precipitation. During the same year Johnson et al., (1974) determined photosynthetic capacity as a function of leaf water potential in the field. However, it was not feasible to include the full range of leaf water potentials as determined by Ehleringer and Miller. Between −7 and −17 bars leaf water potential, *Geum rossii* exhibited a 40% decrease in photosynthetic capacity, whereas *Deschampsia caespitosa* was reduced in photosynthetic capacity by approximately 25% between −8 and −21 bars leaf water potential. Ehleringer and Miller (1975) reported minimum leaf water potentials of −23 and −31 bars for *Geum* and *Deschampsia* in 1972, respectively.

Water Relations and Photosynthetic Capacity of Selected Tundra Species. The potential for water stress to act as a constraint on tundra primary productivity and to play a significant role in the local distribution of tundra species was sufficiently well established in field studies to warrant further investigation under controlled conditions. Four tundra species were selected for more intensive investigations (Johnson and Caldwell, 1975, 1976). *Deschampsia caespitosa* and *Dupontia fisheri* are species restricted to wet sites on Niwot Ridge and at Barrow, respectively. These were compared with *Geum rossii* and *Carex aquatilis,* species which are distributed in a wider variety of habitats including drier microsites in the alpine and arctic tundras, respectively. These species were grown under controlled conditions and subjected to comparative investigations with respect to photosynthetic capacity as a function of a wide range of soil and atmospheric water stress, and other aspects of their water relations. Although these plants were grown under greenhouse conditions, their photosynthetic response to radiation and temperature corresponded closely with the responses of these species under field conditions in the arctic and alpine field sites.

These tundra species exhibited a general decline in photosynthetic capacity as atmospheric water stress was increased and particularly when soil water potential was depressed (Johnson and Caldwell, 1975). There were, however, important differences between the wet site species, *Dupontia* and *Deschampsia,* and the more widely distributed species, *Carex* and *Geum.* The photosynthetic capacities of the wet site species from both arctic and alpine areas were higher than those of the more widely distributed species under conditions of low soil water stress. This would suggest that under unstressed conditions in wet microsites, these species would exhibit higher carbon gain per unit leaf material than the wider ranging species. However, this was offset by the ability of the wider ranging species to maintain greater photosynthetic capacity as water stress increased which, of course, would be advantageous in more exposed and drier microsites and during years of limited summer precipitation.

This decline of photosynthesis with increasing water stress could be partially attributed to decreased stomatal aperture as reflected in increasing stomatal diffusion resistance (Figures 8 and 9) (Johnson and Caldwell, 1976) as had been suggested by the field studies of Ehleringer and Miller (1975) and Stoner and

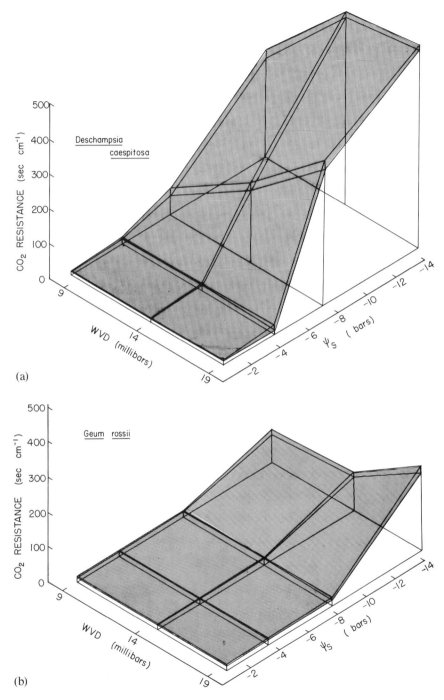

Figure 8.   Leaf (shaded top layer) and residual (unshaded bottom layer) resistances to CO_2 transfer for *Deschampsia caespitosa* (a) and *Geum rossii* (b) as a function of WVD and Ψ_s at a leaf temperature of 20°C and 900 μE m^{-2} sec^{-1} (400 to 700 nm) where photosynthesis was light saturated. (From Johnson and Caldwell, 1975)

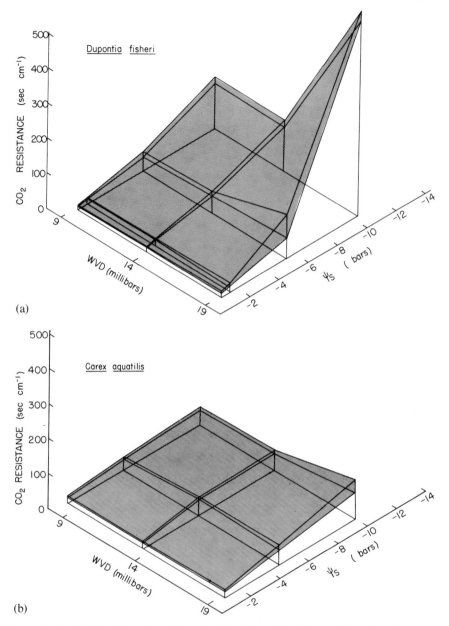

Figure 9.   Leaf (shaded top layer) and residual (unshaded bottom layer) resistances to CO_2 transfer of *Dupontia fisheri* (a) and *Carex aquatilis* (b) as a function of WVD and Ψ_s at a leaf temperature of 20°C and 900 μE m^{-2} sec^{-1} (400 to 700 nm) where photosynthesis was light saturated. (From Johnson and Caldwell, 1975.)

Miller (1975). Although both of the alpine species exhibited lower stomatal resistance, r_s, than the arctic species, species with similar habitat distributions from the two tundra areas showed similar responses. Under conditions of moist soil (high Ψ_s) all four species exhibited an increase in r_s with increasing atmospheric water stress. However, significant reductions in leaf water potential, Ψ_l, were associated with increased r_s only in *Deschampsia* and *Dupontia*, the species restricted to the wet sites. At high Ψ_s, as atmospheric stress increased from 9 to 19 mbars, Ψ_l of *Deschampsia* decreased from -19 to -21 bars, and *Dupontia* Ψ_l declined from -11 to -12 bars. In these species the increased r_s may have resulted from hydropassive reduction in stomatal aperture due to changes in bulk leaf water status as has been found in a number of mesophytic crop species. In contrast, at high Ψ_s the wider ranging species *Geum* and *Carex* exhibited no statistically significant reductions in Ψ_l with increasing WVD. Thus, the stomates may have responded directly to the WVD gradient and hence prevented changes in bulk leaf water status by partial closure as has been suggested by research on other plant species. This mechanism may have lead to a particularly significant partial stomatal occlusion in leaves of *Carex,* and to a lesser extent in *Geum,* which in turn abetted maintenance of prestress Ψ_l.

For all four species, r_s increased when soil water potential was lowered. However, this increase was not as great for the wider ranging species, particularly *Geum rossii*. Therefore, in order for a species such as *Geum* to maintain a favorable Ψ_l over a broad range of Ψ_s without a sizeable decrease in stomatal aperture, other properties of the water transport system must be involved such as lower liquid phase resistance between the soil and the leaf. Measurements of these liquid phase resistances (Johnson and Caldwell, 1976) did show *Geum* to have significantly lower resistance than the other three species.

Although stomatal diffusion resistance is the primary regulator of water loss in plants and also controls CO_2 entry into the leaf, the curtailment of photosynthetic capacity by water stress is not limited to increased stomatal diffusion resistance. Water stress also appears to affect photosynthetic capacity adversely apart from the increase in resistance to diffusion of CO_2 from the air to the intercellular spaces in the leaves, and in some circumstances this nonstomatal water stress constraint on photosynthetic capacity exceeds the effects due to decreased stomatal aperture. The nonstomatal effects may be illustrated by use of a residual resistance term, r_r', which is simply a reflection of all the reductions in photosynthetic capacity, both metabolic and diffusive, apart from leaf resistance, i.e., the diffusion resistance of CO_2 from the air to the intercellular spaces in the leaf. As soil water stress increased, the nonstomatal inhibition of photosynthesis clearly dominated, particularly for the species restricted to wet sites in both alpine and arctic tundra situations. For the wet site species, leaf water potentials corresponding to soil water potentials less than -4 bars in this study have not been encountered in field measurements of these species in their native sites.

The effects of water stress on plant gas exchange of these tundra species is also reflected in the water use efficiency, or photosynthesis/transpiration ratios. Under low water stress the species restricted to wet sites typically exhibited higher water use efficiency; however, as Ψ_s decreased, water use efficiencies of the wider ranging species exceeded those of the wet site species. Although reduced stomatal aperture increased water use efficiency for all species as water stess increased, water use efficiency for photosynthesis was higher in the wider ranging species at greater water stress. If this water use efficiency for photosynthesis corresponds to water use efficiency for growth in the field, the differences in efficiency between the species could also be an important factor in partially explaining distribution patterns of these tundra species. Although more extreme water stress has been measured in the alpine sites than in the Arctic, and *Geum* exhibits a greater tolerance of water stress than *Carex,* there is a remarkable similarity in the response pattern of the arctic and alpine wet site species and the wider ranging species from both areas.

Although water stress in arctic and alpine tundras may be severe enough to limit primary production through depression of photosynthetic rates, other physiological processes such as cell growth may be inhibited at a much lower water stress than photosynthesis (Hsiao, 1973). Since cell turgor is critical for cell expansion, the relationship between Ψ_l and Ψ_p (i.e., leaf turgor pressure) was also investigated for these four selected tundra species over a range of water stress conditions. The slope of these relationships are shown in Figure 10. The apparently negative Ψ_p shown in this figure have also been reported in a number of other studies, but may be due to unavoidable technique artifacts such as significant matric forces or a high proportion of free water in the leaves (Johnson and Caldwell, 1976). Therefore, the absolute values of Ψ_p are of less concern here than the slope of the relationship between Ψ_p and Ψ_l.

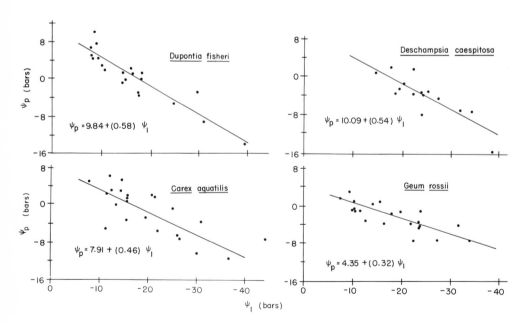

Figure 10.   Regressions of Ψ_p on Ψ_l for *Dupontia fisheri, Carex aquatilis, Deschampsia caespitosa,* and *Geum rossii.* Coefficients of determination, R^2, are 0.92, 0.84, 0.83, and 0.80, respectively. (From Johnson and Caldwell, 1976, with permission of the Scandinavian Society for Plant Physiology.)

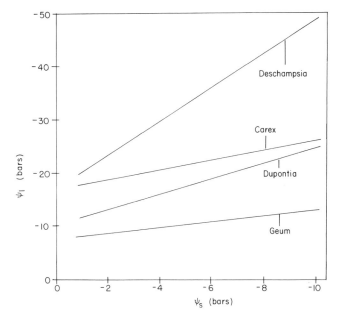

Figure 11.   Regressions of Ψ_l on Ψ_s for *Geum rossii, Dupontia fisheri, Carex aquatilis,* and *Deschampsia caespitosa.* Coefficients of determination, R^2, are 0.83, 0.98, 0.93, and 0.84, respectively. (From Johnson and Caldwell, 1976, with permission of the Scandinavian Society for Plant Physiology.)

Deschampsia, the alpine tundra species restricted mainly to a wet meadow habitat, exhibited a significantly greater slope for Ψ_p as a function of Ψ_l than did *Geum,* the wider ranging alpine species. A similar trend was suggested between the wet site and wider ranging arctic tundra species; however, because of the greater variability in the data for these arctic species, the slopes were not statistically different. This lack of significance may also reflect the smaller differences between the arctic tundra habitats where *Dupontia* and *Carex* are found. In addition, Ψ_l and Ψ_s relationships (Figure 11) show that *Deschampsia* had a significantly steeper decline in Ψ_l with decreasing Ψ_s than *Geum.*

The relationships between Ψ_l and Ψ_p when combined with the change in Ψ_l as a function of Ψ_s suggest that for the alpine species, and possibly for the arctic species, a decrease in Ψ_s would produce a larger decline in Ψ_p in the species restricted to wet meadow areas than would be the case for the wider ranging species. Because of the dependence of cell expansion on turgor, the wet site species, at least for the alpine area, would likely show earlier reduction in growth rate than the wider ranging species as soil water potential decreased. However, the ability of the wet site species to maintain higher Ψ_p at high Ψ_s would result in greater leaf expansion rates when the soil is moist than would be the case for the more widely distributed species. This is corroborated by field experiments in the alpine tundra which indicated that the wet site species, *Deschampsia,* did exhibit higher leaf relative growth rates than the wider ranging species, *Geum,* when Ψ_s was high (Johnson and Caldwell, 1974).

Table 1.   Cell Wall Elasticity, *e*, for
Deschampsia caespitosa, Geum rossii,
Dupontia fisheri, and *Carex aquatilis*

Species	*e* (bars)
Deschampsia caespitosa	52
Geum rossii	8
Dupontia fisheri	50
Carex aquatilis	27

From Johnson and Caldwell, 1976.

These differences in turgor pressure response may also be expressed as differences in cell wall elasticity. Johnson and Caldwell (1976) have calculated cell wall elasticities for these four tundra species (Table 1). The results suggest that *Geum* possessed more elastic cell walls when compared to the rather rigid inelastic cell walls of *Deschampsia*. Similarly, though the differences were not as striking for the arctic species, *Carex* had more elastic cell walls than *Dupontia*. For these wider ranging species with more elastic cell walls, a given amount of water loss from a turgid cell would result in a decrease in volume and maintenance of greater Ψ_p within the cell as opposed to the wet site species with more rigid cell walls.

Depending on the degree of cell wall elasticity, reductions in Ψ_l would be partitioned differently between Ψ_p and osmotic potential. This phenomenon may be an important attribute if the reduction in Ψ_l is diverted primarily to that water potential component which least damages the plant tissues. For example, the relatively elastic cell walls calculated for *Geum* would result in a rather small reduction in Ψ_p and, therefore, necessarily a large decrease in osmotic potential. In contrast, the rather rigid inelastic cell walls reported for the wet site species would result in a steeper decline in Ψ_p but, therefore, a small decrease of osmotic potential. This would also necessarily suggest that the protoplasm of leaf tissues of a species such as *Geum* may be more tolerant of increased solute concentrations than would be the case for the wet site species.

Stoner and Miller (1975) determined the relationship between leaf water potential and relative saturation deficit (i.e., the difference between turgid and fresh weight divided by the turgid weight and expressed as a percentage) for six tundra species at Barrow. This relationship between leaf water potential and relative saturation deficit also implies differences in cell wall elasticity. They also found relatively little difference in this relationship for *Dupontia* and *Carex*. However, when these six species are ordered as to their distribution between wet and dry microsites (Stoner et al., 1978) there was a positive correlation between apparent cell wall ridigity and occurrence of species in the wet microsites. They also report a rather strong relationship between habitat distribution and stomatal response to decreasing leaf water potentials. Species from the moist microsites exhibited stomatal closure at higher leaf water potentials than species distributed toward the drier end of the microsite spectrum.

Based on these field and laboratory studies conducted in the Tundra Biome program, the role of water stress as a secondary constraint on primary productivity in tundra areas, and as an influential factor in the segregation of tundra plant species along microtopographical gradients, is well established in both arctic and alpine tundras.

Acknowledgments. This research was supported by the National Science Foundation under Grant GV 29353 to Utah State University. It was performed under the joint sponsorship of the International Biological Programme and the Office of Polar Programs and was coordinated by the U.S. Tundra Biome. Field and laboratory activities at Barrow were supported by the Naval Arctic Research Laboratory of the Office of Naval Research.

Support provided by the Institute of Arctic and Alpine Research, University of Colorado, and the IBP Tundra Biome Center, University of Alaska, is also gratefully acknowledged. Field assistance by Kenneth Olson, Roger Hanson, Diane Hanson, Kathryn Johnson, Donald Hazlett, and Thomas Schoemaker and laboratory assistance of Harvey Neuber were invaluable in these studies.

References

Billings, W. D. (1975) Arctic and alpine vegetation: Plant adaptations to cold summer climates. In *Arctic and Alpine Environments* (J. D. Ives and R. G. Barry, Eds.). London: Methuen, pp. 403–443.

Billings, W. D., and L. C. Bliss. (1959) An alpine snowbank environment and its effects on vegetation, plant development, and productivity. *Ecology*, **40**: 388–397.

Bliss, L. C. (1956) A comparison of plant development in microenvironments of arctic and alpine tundras. *Ecol. Monogr.*, **26**: 303–337.

Bliss, L. C. (1971) Arctic and alpine plant life cycles. *Ann. Rev. Ecol. Syst.*, **2**: 405–438.

Brown, R. W. (1976) A new technique for measuring the water potential of detached leaf samples. *Agron. J.*, **68**: 432–434.

Caldwell, M. M., L. L. Tieszen, and M. Fareed. (1974) The canopy structure of tundra plant communities at Barrow, Alaska, and Niwot Ridge, Colorado. *Arct. Alp. Res.*, **6**: 151–159.

Dennis, J. G., L. L. Tieszen, and M. Vetter. (1978) Seasonal dynamics of above- and belowground production of vascular plants at Barrow, Alaska. In *Vegetation and Production Ecology of an Alaskan Arctic Tundra* (L. L. Tieszen, Ed.). New York: Springer-Verlag, Chap. 4.

Ehleringer, J. R., and P. C. Miller. (1975) Water relations of selected plant species in the alpine tundra, Colorado. *Ecology*, **56**: 370–380.

Fareed, M., and M. M. Caldwell (1975) Phenological patterns of two alpine tundra plant populations on Niwot Ridge, Colorado. *Northwest Sci.*, **49**: 17–23.

Hadley, E. B., and L. C. Bliss. (1964) Energy relationships of alpine plants on Mt. Washington, New Hampshire. *Ecol. Monogr.*, **34**: 331–357.

Hsiao, T. C. (1973) Plant responses to water stress. *Ann. Rev. Plant Physiol.*, **24**, 519–570.

Hultén, E. (1968) *Flora of Alaska and Neighboring Territories*. Stanford: Stanford University Press, 1008 pp.

Johnson, D. A., and M. M. Caldwell. (1974) Field measurements of photosynthesis and leaf growth rates of three alpine plant species. *Arct. Alp. Res.*, **6**: 245–251.

Johnson, D. A., and M. M. Caldwell. (1975) Gas exchange of four arctic and alpine tundra plant species in relation to atmospheric and soil moisture stress. *Oecologia*, **21**: 93–108.

Johnson, D. A., and M. M. Caldwell. (1976) Water potential components, stomatal function, and liquid phase water transport resistances of four arctic and alpine species in relation to moisture stress. *Physiol. Plant.*, **36**: 271–278.

Johnson, D. A., M. M. Caldwell, and L. L. Tieszen. (1974) Photosynthesis in relation to leaf water potential in three alpine plant species. In *Proceedings of the Conference on Primary Production and Production Processes, Tundra Biome, Dublin, Ireland, April 1973* (L. C. Bliss and F. E. Wielgolaski, Eds.). Stockholm: International Biological Programme Tundra Biome Steering Committee, pp. 205–210.

Koch, W., O. L. Lange, and E. D. Schulze. (1971) Ecophysiological investigations on wild and cultivated plants in the Negev Desert. I. Methods: A mobile laboratory for measuring carbon dioxide and water vapour exchange. *Oecologia,* **8**: 296–309.

Ludlow, M. M., and G. L. Wilson. (1971) Photosynthesis of tropical pasture plants. III. Leaf age. *Aust. J. Biol. Sci.,* **24**: 1077–1088.

Mayo, J. M., D. G. Despain, and E. M. Van Zinderen Bakker, Jr. (1973) CO_2 assimilation by *Dryas integrifolia* on Devon Island, Northwest Territories. *Can. J. Bot.,* **51**: 581–588.

Miller, P. C., W. A. Stoner, and J. R. Ehleringer. (1978) Some aspects of water relations of arctic and alpine regions. In *Vegetation and Production Ecology of an Alaskan Arctic Tundra* (L. L. Tieszen, Ed.). New York: Springer-Verlag, Chap. 14.

Moser, W. (1973) Licht, Temperatur und Photosynthese an der Station "Hoher Nebelkogel" (3184 m). In *Ökosystemforshung* (H. Ellenberg, Ed.). Springer-Verlag, Heidelberg, pp. 203–223.

Salisbury, F. B., and G. G. Spomer. (1964) Leaf temperatures of alpine plants in the field. *Planta,* **60**: 497–505.

Scott, D., and W. D. Billings. (1964) Effects of environmental factors on standing crop and productivity of an alpine tundra. *Ecol. Monogr.,* **34**: 243–270.

Šesták, Z., and J. Čatský. (1962) Intensity of photosynthesis and chlorophyll content as related to leaf age in *Nicotiana. Biol. Plant,* **4**: 131–140.

Stoner, W. A., and P. C. Miller. (1975) Water relations of plant species in the wet coastal tundra at Barrow, Alaska. *Arct. Alp. Res.,* **7**: 109–124.

Teeri, J. A. (1973) Polar desert adaptations of a high arctic plant species. *Science,* **179**: 496–497.

Tieszen, L. L. (1978) Photosynthesis in the principal Barrow, Alaska, species: A summary of field and laboratory responses. In *Vegetation and Production Ecology of an Alaskan Arctic Tundra* (L. L. Tieszen, Ed.). New York: Springer-Verlag, Chap. 10.

Tieszen, L. L., and D. A. Johnson. (1975) Seasonal pattern of photosynthesis in individual grass leaves and other plant parts in arctic Alaska with a portable $^{14}CO_2$ system. *Bot. Gaz.,* **136**: 99–105.

Tieszen, L. L., D. A. Johnson, and M. M. Caldwell. (1974) A portable system for the measurement of photosynthesis using $^{14}CO_2$. *Photosynthetica,* **8**: 151–160.

Warren Wilson, J. (1960) Inclined point quadrats. *New Phytol.,* **59**: 1–8.

Weber, W. A. (1967) *Rocky Mountain Flora.* Boulder: University of Colorado Press, 437pp.

14. Some Aspects of Water Relations of Arctic and Alpine Regions

P. C. MILLER, W. A. STONER, AND J. R. EHLERINGER

Introduction

The objectives of this study were to measure certain aspects of the water relations of selected plants occurring in the arctic tundra and alpine tundra and to assess the possible role of water limitation on primary production in these two ecosystems. At low leaf water potentials, water limitation of production may occur by stomatal closure restricting water loss and photosynthesis or by reduced growth because of decreased cell turgor. Jarvis and Jarvis (1963) stressed the importance of knowing the interrelationships between leaf resistance, water potential, and water content, before the water relations of a species could be understood. Hence, to study the effects of plant water relations on plant distribution and production the total soil–plant–atmosphere system should be characterized.

In the arctic there have been few studies on plant water relations. Billings and Mooney (1968) concluded that at Barrow, Alaska, plant water stress was minor with leaf water potentials above −4 to −5 bars. However, Courtin and Mayo (1975) report lower leaf water potentials (−25 to −60 bars) at Devon Island, Canada, and state that, with the exception of data from Barrow, arctic plants generally appear to have low leaf water potentials even when growing in water.

In the alpine Cox (1933) suggested that production may be limited by length of growing season, low temperatures, and moisture. Scott and Billings (1964) and Hillier (1970) considered soil moisture as a principal factor limiting plant growth in the alpine. Billings and Bliss (1959) showed that soil moisture and production were correlated in an alpine snowbank in Wyoming and that production was affected by short periods of drought even though soil water potentials were above −15 bars. Soil moisture determinations by Bliss (1956) in the top 12.5 cm of soil indicated water potentials below −15 bars in 67% of the observations on the ridgetop, in 45% on the north facing slope, and in 25% on the south facing slope. Kuramoto and Bliss (1970) reported that, in the subalpine communities of the Olympic Mountains in Washington, net photosynthesis of several species decreased as soil water potential decreased. In the Sierra Nevada alpine, Mooney et al. (1965) found that species from moist sites transpired more than species from dry sites, and transpiration of plants from moist sites decreased during midday, while transpiration of plants from dry sites remained constant throughout the day. Discussion of the arctic and alpine data presented here may be found in Stoner and Miller (1975) and Ehleringer and Miller (1975).

Conceptual Framework

The plant water relations are viewed in a simple model of the soil–plant–atmosphere continuum. Differences in water potentials between the leaf and soil provide the driving forces for water absorption across root and soil resistances. Water loss from the leaf is impeded by the resistance of the leaf, including stomates and cuticle, and leaf air boundary layer. The leaf resistance varies with the leaf water content. The model is essentially that of Honert (1948), modified by Rawlins (1963), and described in nonmathematical terms by Jarvis and Jarvis (1963). It is intended to provide a dynamic, interpretive framework for viewing leaf water potentials and is portrayed graphically in Figure 1.

The curvilinear relationship between transpiration and leaf resistance is given in the upper right quadrant. As leaf temperatures increase, transpiration will increase at any given leaf resistance. Transpiration rates at higher leaf temperatures and the same ambient water vapor density are shown by the dashed lines.

The relationship between leaf resistance and relative saturation deficit (RSD), a measure of the water lost from a leaf relative to the turgid state, is shown in the lower right quadrant. At full turgidity (an RSD of zero) leaf resistenaces are at

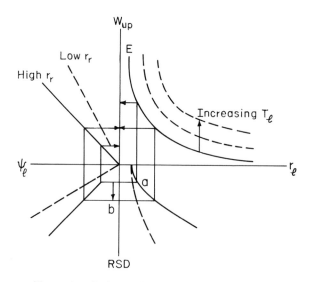

Figure 1.   Diagram illustrating the interrelations between transpiration (E), leaf resistance (r_l), relative saturation deficit (RSD), leaf water potential (Ψ_l), water uptake (W_{up}), and root resistance (r_r).

minimum values, assuming light intensities are not low enough to inhibit stomatal opening. As the RSD increases the leaf resistance increases, and the sensitivity of leaf resistance to changes in RSD will differ by species. The typical pattern is that species from wet habitats (solid line) have a sharp increase in leaf resistance as RSD increases, and species from drier habitats (dashed line) have a much lower sensitivity of leaf resistance to RSD. The relationship between leaf water potential and RSD is shown in the lower left quadrant. Species with a leaf water potential more sensitive to changes in RSD are shown by a dashed line.

The relationship of water uptake to leaf water potential, assuming a substrate water potential of zero, is shown in the upper left quadrant. Root resistance is the resistance of the entire absorbing root mass which is related to the total surface area. Since species differ in their total root biomass and their total root surface area to total root biomass ratios (i.e., fibrous versus tap root), their root resistances will also differ.

A daily course of plant water relations can be followed with these graphs. For example, starting at point "a" on the leaf resistance–RSD curve, the transpiration rate implied by that leaf resistance is given by the arrow up to the transpiration–leaf resistance curve and across to the transpiration axis. The rate of water uptake can be obtained by following from point "a" horizontally across to the water potential–RSD curve, up to the water potential–root resistance curve and across to the water uptake axis. A plant with the leaf resistance implied by the "a" arrow will have a rate of transpiration exceeding the rate of water uptake. This means that the water content of the leaf will decrease and RSD will increase. The increase in RSD is shown by the "b" arrows. At this higher RSD a higher leaf resistance occurs which produces a lower transpiration rate. The new RSD also implies a lower leaf water potential and greater water uptake. In this case transpiration and water uptake are equal, hence, the RSD will be maintained at this level. The effect of increasing leaf temperatures on the system can be explored with this diagram as can the effect of different relationships among leaf resistance, water potential, RSD, and root system resistance.

Methods

The arctic study was conducted near Barrow, Alaska, at sites 2 and 4 during the summer of 1972 and at site 1 during the summer of 1973. Leaf water potentials and the relationship between leaf water potential and relative saturation deficit were measured in *Arctophila fulva, Dupontia fisheri, Carex aquatilis, Eriophorum angustifolium, Potentilla hyparctica,* and *Salix pulchra. In situ* leaf resistances to water loss and root resistances to water absorption were measured only in *Dupontia, Carex,* and *Eriophorum*.

The alpine study was conducted on Niwot Ridge in the Front Range in central Colorado (40°02' N, 105°33' W) during the summer of 1972. The study sites were located at the 3500-m elevation in the area called the Saddle. Five sites were chosen along a moisture gradient, site 1 being the driest and site 5 the wettest.

Site 1 was on the east knoll of the Saddle in fellfield tundra where the vegetation consisted predominantly of *Silene acaulis* and *Kobresia myosuroides*. Site 2 was in an area dominated by *Kobresia*. Site 3 was exposed to wind and kept snow free throughout the winter and contained a diversity of species including *Geum rossii, Bistorta bistortoides,* and *Deschampsia caespitosa*. Site 4 occurred in the center of a trough and was dominated by *Salix nivalis*. Site 5 was protected from prevailing winds, covered by snow during the winter, and received meltwater from snow until the end of June. Vegetation consisted primarily of *Geum, Bistorta, Deschampsia, Caltha leptosepela,* and *Artemesia scopulorum*. The species studied were *Kobresia, Geum, Bistorta, Deschampsia,* and *Caltha*. *Kobresia* occurred only on sites 1, 2, and 3. *Geum* and *Bistorta* occurred on all sites. *Deschampsia* occurred on sites 3, 4, and 5. Leaf water potentials were measured in all species; while *in situ* leaf resistances to water loss, the relationship between leaf resistance and leaf water potential, and the relationship between leaf water potential and relative saturation deficit were measured only in *Bistorta* and *Caltha*.

A scholander-type pressure chamber was used to measure leaf water potentials. At Barrow leaf water potentials were measured every 2 to 3 hr in 1972 on 30 June to 2 July, 6 to 8 July, 23 to 26 July, 2 to 4 August, and 16 to 17 August; while in 1973 water potentials were measured at 3-hr intervals during two 24-hr periods, 29 to 30 July and 3 to 4 August. At Niwot, leaf water potentials were measured every 2 weeks between 15 June and 22 August, and every 3 hr throughout the day on sites 3 and 5, on 30 June, 22 July, 10 to 11 August, and 21 to 22 August.

Relative saturation deficit (RSD) (Barrs, 1968) was measured at Niwot by excising leaves and taking 0.635-cm diameter punches from a leaf. At Barrow leaves were excised and cut into 1-cm long pieces. Usually less than 2 min elapsed between leaf excision and the fresh weight measurement. Leaves were floated on distilled water at room temperature until turgid and reweighed to determine the turgid weight. RSD was calculated as the difference between the turgid and fresh weight divided by the turgid weight and expressed as a percentage.

The relationship between leaf water potential and RSD was determined at Barrow by measuring leaf water potentials while four to six adjacent leaves were taken for RSD measurements. The means of six water potentials and three RSD measurements were used for each point. At Niwot, the water potential of one leaf of a pair was measured while the other leaf was taken to measure the RSD.

At Niwot *in situ* leaf resistance to water loss was measured with a porometer (Kanemasu et al., 1969). Due to the size of the aperture on the porometer, only the broader leaves of *Bistorta* and *Caltha* could be measured. The porometer and the leaf were shaded to equalize temperatures between the porometer sensor and the leaf surface. The porometer was calibrated at 3500 m at several temperatures covering the range experienced in the field (5 to 20°) to correct for the temperature effects on the sensor (Morrow and Slatyer, 1971). Resistance of the upper and lower leaf surfaces were measured and the total leaf resistance was calculated as the mean of the parallel resistances for both surfaces. At Barrow *in situ* leaf resistances were measured with a single leaf cuvette, through which dried air was drawn.

The relationship between leaf resistance and leaf water potential was determined for *Bistorta* and *Caltha* at Niwot by measuring resistances and water potentials on groups of five leaves selected randomly from 1 m² quadrats. Measurements were taken only under high light intensities.

In order to determine the relationship between leaf resistance and RSD at Barrow, the method of measuring transpiration rates by repeatedly weighing cut leaves was modified (Hygen, 1951; Bannister, 1971; Waggoner and Turner, 1971). Groups of cut leaves were weighed every 2 min for 40 min on a Mettler balance (0.1 mg accuracy) while monitoring ambient and dew point temperatures inside the weighing chamber. Assuming that the leaves were at ambient temperature, leaf resistances were calculated from:

$$r_1 + r_a = (VS_1 - V_a)(W_t - W_{t+1})^{-1} A(\Delta t)$$

where $r_1 + r_a$ = the combined leaf and air resistance,
$\quad\quad VS_1$ = the saturation water vapor density at leaf temperature,
$\quad\quad V_a$ = the ambient vapor density,
$\quad\quad W_t$ = the weight at time t,
$\quad\quad A$ = the area of leaves in the chamber, and
$\quad\quad \Delta t$ = the time between weight measurements.
The RSD of the leaves was calculated by using the weight at each time interval as the fresh weight. The initial RSD and turgid weight were measured on other leaves collected from the same field area during the same period.

Cuticular resistances were determined for the Barrow species from the final linear section of the drying curve. Hygen (1951) designated three phases in the drying curve: a rapid linear decrease in weight (stomatal phase); a curvilinear phase, characterized by a slowing of the rate of change in weight (closing phase); and a linear decrease (cuticular phase). The slope of this final phase is the cuticular transpiration.

Root resistances were calculated using leaf resistance and leaf water potential values that were measured simultaneously on adjacent plants. By assuming that at the time of measurement transpiration and water uptake are equal:

$$r_r + r_s = (\Psi_S - \Psi_1)(r_a + r_1)(VS_1 - V_a)^{-1}$$

where Ψ_S is the soil water potential and Ψ_1 is the leaf water potential. The calculated root resistance is the resistance of the entire root mass plus the soil resistance to water flow. Most measurements were made during the mid-morning to ensure that the plants were not recovering from any water stress and hence violating the assumed equality between loss and uptake.

At Barrow, root resistances were also measured in an area where soil temperatures were raised above ambient by pipes carrying heated water which were buried 3 yr earlier.

Results

Barrow

The relationship between leaf water potential and RSD was linear throughout the range of RSD measured in the field (Figure 2). Slopes in bars RSD^{-1} were −1.82 for *Arctophila*, −1.06 for *Dupontia*, −1.05 for *Carex*, −0.57 for *Eriophorum*, −0.73 for *Potentilla*, and −0.95 for *Salix*. The responses of leaf resistance to

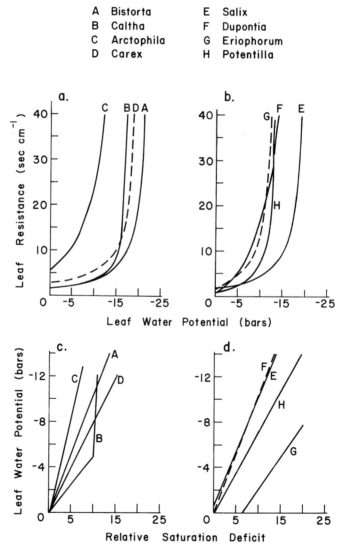

A	Bistorta	E	Salix
B	Caltha	F	Dupontia
C	Arctophila	G	Eriophorum
D	Carex	H	Potentilla

Figure 2.   Relationships between relative saturation deficit (RSD) and leaf water potential for the arctic and alpine species and between leaf water potential and leaf resistance to water loss. Data from Stoner and Miller (1975) and Ehleringer and Miller (1975).

Table 1. Cuticular Resistances, Turgid Leaf Densities, and the 95%
Confidence Limits for Turgid Leaf Densities for the Six Barrow Species
and Some Other Plants[a]

Species	Cuticular resistance (sec cm^{-1})	Turgid density (mg cm^{-2})
Arctophila	38.7	24.5 ± 2.3
Dupontia	29.4	19.0 ± 2.6
Carex	22.4	17.0 ± 1.2
Eriophorum	11.8	22.1 ± 0.9
Potentilla	37.0	19.8 ± 3.1
Salix	15.6	22.1 ± 1.0
Shade plants[b]	20.0	
Xerophytes[b]	>200.0	
Tomato[c]	20.0	
Bean[c]	20.0	
Cotton[d]	64.4	

[a] Unless noted, data are from this study.
[b] Slatyer (1967).
[c] Kuiper (1961).
[d] Slatyer and Bierhuizen (1964).

changes in RSD showed no signs of reaching a plateau, but continued to increase
with increasing RSD, similar to other studies (Kanemasu and Tanner, 1969;
Troughton, 1969; Millar et al. 1971).

By combining the relationships between leaf resistance and RSD and leaf
water potential, relationship between leaf resistance and leaf water potential is
obtained (Fig. 2). The leaf resistances of *Arctophila* and *Dupontia* increased
immediately as leaf water potential dropped below 0 bars. Resistances of the
other species increased more slowly at first, then rapidly below a water potential
which varied with each species. Leaf water potentials at which leaf resistances of
20 sec cm^{-1} occurred were -9.0 for *Arctophila*, -11.0 for *Dupontia*, -17.5 for
Carex, -11.0 for *Eriophorum*, -13.0 for *Potentilla*, and -18.0 for *Salix*.

Minimum leaf resistances, obtained by extrapolating to zero water potential,
were 6.0 sec cm^{-1} for *Arctophila*, 1.0 sec cm^{-1} for *Dupontia*, 3.0 sec cm^{-1} for
Carex, 2.0 sec cm^{-1} for *Eriophorum*, 1.5 sec cm^{-1} for *Potentilla*, and 2.0 sec
cm^{-1} for *Salix*. Cuticular resistances were between 12 and 39 sec cm^{-1}. Turgid
leaf weights were between 17 and 25 mg cm^{-2} (Table 1).

The changes of leaf resistance with either RSD or leaf water potential can be
compared among species with the "A" and "B" values of each curve. The "A"
value is the RSD or water potential at three times the minimum leaf resistance
(leaf resistance at 0 RSD or 0 water potential). The "B" value is the RSD or
water potential where maximum leaf resistances occur. A leaf water potential
less than the "A" value indicates some stomatal closure; while a water potential
lower than "B" indicates complete stomatal closure. "A" and "B" water
potential values (bars) were, respectively, -8.2 and -12.0 for *Arctophila*, -2.5
and -14.0 for *Dupontia*, -14.0 and -18.0 for *Carex*, -6.5 and -12.3 for
Eriophorum, -7.9 and -13.7 for *Potentilla*, and -12.3 and -20.0 for *Salix*.

In situ leaf resistances (Table 2) of *Dupontia* on 12 and 13 July were higher in the polygon trough than on the ridge (Mann–Whitney U test, $P < 0.05$). Resistances of *Carex* and *Eriophorum* though not significantly different between the trough and ridge, tended to be lower in the trough. Resistances of all species combined were not significantly different between the trough and the ridge (Kruskal–Wallis, $P < 0.05$). Resistances in the afternoon were higher than in the morning or at noon for all species. Resistances of *Carex* and *Eriophorum* remained similar between morning and noon readings but increased in the afternoon; those of *Dupontia* increased throughout the day. Cuvette leaf resistances were lower than the estimates of minimum leaf resistances, probably because of the higher air resistance present in the weighing chamber.

Root resistances were independent of soil temperature but were related to transpiration rates. Within a species, root resistances were not significantly different on the heated and ambient sites (U test, $P > 0.05$). However, root resistances among species on both sites were significantly different (Kruskal–Wallis, $P > 0.05$). Average root resistances (in units of 10^6 sec cm^{-1} bar^{-1}) were 10.6 for *Dupontia*, 8.4 for *Carex,* and 5.8 for *Eriophorum*. Minimum root resistances were 1.7 for *Dupontia*, 1.2 for *Carex,* and 0.6 for *Eriophorum*. These resistances occurred at transpiration rates of 198, 222, and 240 mg of H_2O cm^{-2} min^{-1}, respectively.

During the growing season of 1972, midday leaf water potentials indicated that water stress increased; while in 1973, water stress did not increase as the season progressed (Figure 3). Midday leaf water potentials were lower on sunny days, which occurred frequently before mid-July in 1972 and at various times during the season in 1973, than on cloudy days. However, the degree of water stress among species differed. A leaf water potential lower than the water potential at which the leaf resistance is three times the minimum ("A" value) was taken as an indication of water stress. In 1972 *Dupontia* was partially stressed throughout the season, with complete stomatal closure indicated in mid-season. *Carex* showed partial stress at least once during the season. In 1973 *Dupontia* was usually partially stressed, *Arctophila* and *Potentilla* were stressed at times; *Carex*, *Eriophorum*, and *Salix* were under little or no stress. Complete stomatal closure occurred during the growing season in *Arctophila* and *Potentilla*.

Table 2.   Summary of Leaf Resistances to water loss in sec cm^{-1}, in Polygon Troughs and Ridges, at Different Times of the Day on 12 and 13 June[a]

Species	Solar time		
	1000–1130	1300–1500	1530–1730
	Polygon ridge		
Dupontia	(6) 0.46	(6) 0.77	(6) 0.78
Carex	(6) 0.63	(6) 0.50	(3) 1.20
Eriophorum	(6) 0.77	(6) 0.83	(6) 1.87
	Polygon trough		
Dupontia	(7) 0.83	(6) 1.03	(6) 1.55
Carex	(6) 0.55	(6) 0.63	(6) 0.75
Eriophorum	(6) 0.58	(6) 0.85	(6) 0.78

[a]   The number of measurements is given in parentheses. All species in each site were measured on each day.

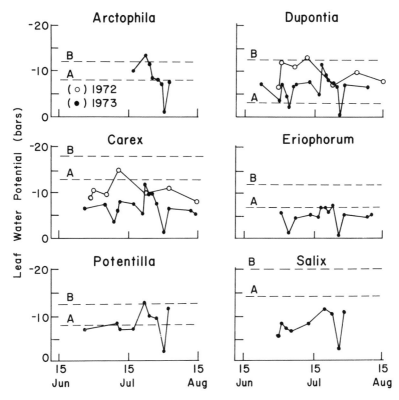

Figure 3. Seasonal course of midday leaf water potentials of the six Barrow species. Adapted from Stoner and Miller (1975) with the permission of the Institute of Arctic and Alpine Research.

Niwot

The relationships of leaf water potential and RSD of *Bisorta* and *Caltha* showed different responses (Fig. 1). *Bistorta* showed a linear response with a slope of -0.8 bars RSD^{-1}, while *Caltha* had a slope of -0.5 below 10% RSD and a slope of -10.0 above 10% RSD.

The relationship of leaf resistances to leaf water potentials differed in *Bistorta* and *Caltha*. The "A" and "B" water potentials were -14 and -22 bars for *Bistorta* and -14 and -17 bars for *Caltha*. Leaf water potentials at which leaf resistances of 20 sec cm^{-1} occurred were -20.0 and -16.0 bars for *Bistorta* and *Caltha*. By extrapolating to 0 bars, the minimum leaf resistance was 0.7 sec cm^{-1} for *Bistorta* and 1.8 sec cm^{-1} for *Caltha*. Turgid leaf weights were 30.0 and 27.0 mg cm^{-2} for *Bistorta* and *Caltha*.

Both *Bistorta* and *Caltha* had higher leaf resistances on one surface than on the other, *Bistorta* had the highest resistances on the abaxial surface. Adaxial leaf resistances were usually between 20 and 141 sec cm^{-1} for *Bistorta* and between 1 and 17 sec cm^{-1} for *Caltha*. Abaxial leaf resistances were between 0.8 and 13 sec cm^{-1} for *Bistorta* and between 15 and 35 sec cm^{-1} for *Caltha*.

The morning leaf resistances of *Bistorta* and *Caltha* did not change consistently through the season (Spearman rank order correlations, $P > 0.05$). Morning leaf resistances were between 0.8 and 4.9 sec cm^{-1} for *Bistorta* and between 1.0 and 2.7 sec cm^{-1} for *Caltha*. Although the morning values of leaf resistance did not change consistently, the daily pattern through the season did change (Figure 4). On 5 July leaf resistances of *Bistorta* decreased after 0800 hr on site 1, while *Bistorta* and *Caltha* on the other sites had low resistances into the afternoon. On 12 July leaf resistances were higher than on 5 July and tended to increase through the day. On 19 July *Bistorta* on sites 1, 2, and 3 had leaf resistances which decreased in late morning and increased in early afternoon. On the same data *Caltha* on sites 4 and 5 had leaf resistances which increased at midday and decreased in the early afternoon.

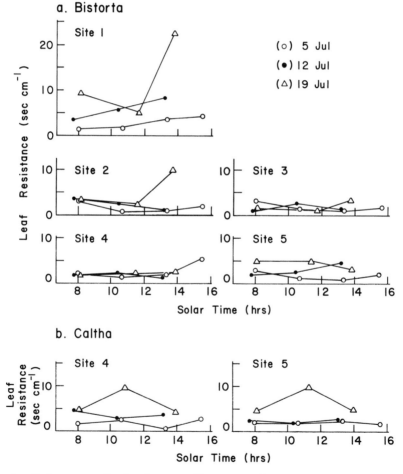

Figure 4. Daily course of leaf resistances of *Bistorta* and *Caltha* on sites 1 through 5 on 5 July (○), 12 (●), and 19 (△). Adapted from Ehleringer and Miller (1975); Copyright 1975 of the Ecological Society of America.

Table 3. Summary of Maximum and Minimum Dawn Water Potentials Measured through the 1972 Season on Niwot Ridge and Date of Occurrence on Sites 3 and 5

Species	Site 3		Site 5	
	Max Ψ_l	Min Ψ_l	Max Ψ_l	Min Ψ_l
Kobresia	−7	−22		
	7/7	8/11		
Geum	−3.5	−10	−1	−6.5
	6/30	8/22	6/20	8/1
Deschampsia	−5	−14	−3	−7
	7/7	8/22	6/20	8/1
Bistorta	−0.5	−7	−1	−5
	6/20	8/11	6/20	8/1
Caltha			−2	−4
			7/6	8/1

Root and soil resistances were measured on 2, 9, and 17 July, and on 10 August. The mean root–soil resistances were 0.89×10^6 and 0.82×10^6 sec cm^{-1} bar for *Bistorta* on sites 3 and 5, respectively, and 0.61×10^6 sec cm^{-1} bar for *Caltha*.

Minimum dawn water potentials (Table 3) for each species were lower on site 3 than on site 5. Midday leaf water potentials decreased through the season (Figure 5) and tended to be lower on site 3 than on site 5. The "A" and "B" water potential values are included on the *Bistorta* and *Caltha* graphs. As at Barrow, leaf water potentials lower than the "A" value were taken as an indication of water stress. *Caltha* appeared to be under some stress at a few points during the season; while *Bistorta* appeared to be under little stress.

Discussion and Conclusions

The relationship between leaf water potential and RSD frequently shows two stages. The first stage consists of a decrease in turgor potential and a concomitant decrease in osmotic potential and may be linear. The second stage is characterized by a rapid decrease in water potential with small increases in RSD, caused by the sorptive forces associated with the solid–liquid interfaces of various cellular components (Slatyer, 1967). Over the range of RSD found in the field, only the first stage was detectable in the six Barrow species and *Bistorta;* however, *Caltha* exhibited both phases.

The differences in the slope of the first phase of the leaf water potential–RSD curve among species is related to the sensitivity of cell turgor to water loss and is interpreted as being related to the cell wall elasticity. Plants with rigid or inelastic cell walls lose turgor more rapidly per unit of water loss than species with more elastic cell walls. Hence, plants with more elastic cell walls will tend to lose more water (develop higher RSD) before stomates close. In general, as one proceeds along a moisture gradient from wet to mesic, cell wall elasticity increases allowing a greater amount of water to be lost before stomatal closure. However,

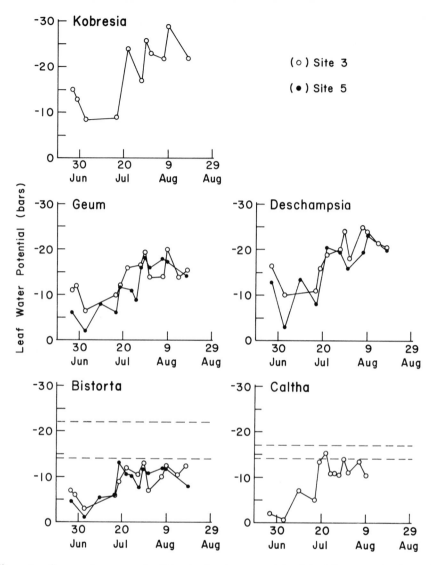

Figure 5.   Seasonal courses of midday leaf water potentials of the Niwot species from site 3 (○) and site 5 (●).

the sensitivity of stomatal opening to leaf water content may vary among species due to other factors such as guard cell construction. *Arctophila* shows the greatest rate of change of leaf water potential to RSD. *Dupontia, Carex, Salix, Bistorta,* and *Potentilla* are next. *Eriophorum* and *Caltha* showed the lowest changes in leaf water potential with RSD. However, the species do not fall out in the same order in terms of their RSD at stomatal closure (40 sec cm^{-1}, see Fig. 2). RSD at stomatal closure was 7% for *Arctophila,* 11% for *Caltha,* 13% for

Dupontia, 17% for *Carex,* 19% for *Potentilla,* 20% for *Salix,* 27.5% for *Bistorta,* and 28% for *Eriophorum.*

Perhaps a better method of comparing the species is one which takes into account the shape of the leaf resistance–leaf water potential curve. If we linearize the curve by converting to conductances and compare the first derivatives, the species in order of decreasing sensitivity of leaf resistance to leaf water potential are: (1) *Arctophila,* (2) *Dupontia,* (3) *Caltha,* (4) *Carex,* (5) *Eriophorum,* (6) *Bistorta,* (7) *Potentilla,* and (8) *Salix. Caltha* is more like the species from the wet end of the Barrow gradient; while *Bistorta* is more like those from the dry end.

At both Barrow and Niwot, lower leaf water potentials occurred than previously reported. At Barrow, leaf water potentials in the range of -15 to -20 bars occurred during both growing seasons of 1972 and 1973. At Niwot minimum leaf water potentials were -40 bars for *Kobresia,* -23 bars for *Geum,* -17 bars for *Bistorta,* -31 bars for *Deschampsia,* and -17 bars for *Caltha.*

Minimum leaf resistances and cuticular resistances are similar among arctic, alpine, and mesic species in more temperate areas. Root resistances of arctic and alpine species are similar to some agricultural species (Brouwer, 1954; Hay and Anderson, 1972; Dunham and Nye, 1973; Johnson and Caldwell, 1976). The apparent insensitivity of root resistance to soil temperature seen in the Barrow species has been seen in alpine species (McNaughton et al., 1974).

The role of water in limiting production and distributions is complex. Certainly, *Arctophila* and *Dupontia* would be under considerable stress in the alpine environment since their leaf resistances are more sensitive to leaf water potentials than *Caltha,* which is at the xeric end of its distribution at site 5. In the arctic even though some stomatal closure was evident in *Dupontia* in 1972 and in *Arctophila* and *Potentilla* in 1973, it probably did not occur frequently enough to affect the carbon balance of these species by reducing photosynthesis. However, low leaf water potentials are known to affect the mortality and expansion of new leaf tissue in agricultural species (Hsiao, 1973). At present, the effects of low leaf water potentials on these processes in tundra species are unknown.

In the alpine, no stomatal closure was evident in *Bistorta* or *Caltha,* even though 1972 was a dry year. However, the daily courses of water potential from sites 3 and 5 early in the season indicated that all species except *Kobresia* are equally adapted to the sites on which they are found since all show an afternoon recovery. The absence of this recovery in *Kobresia* may be due to its relative inactivity. Later in the season at the peak of drought stress, those species which occur on more xeric sites (*Geum* and *Bistorta* on site 5, *Bistorta* on site 3) showed a leaf water potential pattern with an afternoon recovery; and species which are at the xeric end of their distribution (*Caltha* on site 5, *Deschampsia* on site 3) showed a continual decrease in leaf water potentials throughout the day. It is expected that leaf water potentials of the species at the xeric end of its distribution will frequently be in the position of the leaf resistance–leaf water potential curve where leaf resistances increase rapidly in response to changes in leaf water and that the increased leaf resistances may also restrict production.

In general, plants in both the arctic and alpine may be limited in their distributions by water stress, but, in areas where they grow, water probably does not limit production except in very dry years.

Acknowledgments. This research was supported by the National Science Foundation under Grant GV 29345 to San Diego State University. It was performed under the joint sponsorship of the Inernational Biological Programme and the Office of Polar Programs and was coordinated by the U.S. Tundra Biome. Field and laboratory activities at Barrow were supported by the Naval Arctic Research Laboratory of the Office of Naval Research.

We especially thank Dr. Jim Mayo and Dr. L. L. Tieszen for their helpful comments during the research and Mrs. Patsy Miller for her support during and after the work was completed.

References

Bannister, P. (1971) The water relations of heath plants from open and shaded habitats. *J. Ecol.,* **59**: 51–64.

Barrs, H. D. (1968) Determination of water deficits in plant tissues. In *Water Deficits and Plant Growth* (T. T. Kozlowski, Ed.). New York: Academic Press, Vol. I, pp. 235–347.

Billings, W. D., and L. C. Bliss. (1959) An alpine snowbank environment and its effects on vegetation, plant development, and productivity. *Ecology,* **40**: 388–397.

Billings, W. D., and H. A. Mooney. (1968) The ecology of arctic and alpine plants. *Biol. Rev.,* **43**: 481–529.

Bliss, L. C. (1956) A comparison of plant development in microenvironments of arctic and alpine tundras. *Ecol. Monogr.,* **26**: 303–337.

Brouwer, R. (1954) Water absorption by the roots of *Vicia faba* at various transpiration strengths. *Proc. Kon. Ned. Akad. Wet.,* **56**: 106–115.

Courtin, G. M., and J. M. Mayo. (1975) Arctic and alpine plant water relations. In *Physiological Adaptation to the Environment* (F. J. Vernberg, Ed.). New York: Intext Educational Publishers, pp. 201–224.

Cox, C. F. (1933) Alpine plant succession on James Peak, Colorado. *Ecol. Monogr.,* **3**: 300–372.

Dunham, R. J., and P. H. Nye. (1973) The influence of soil water content on the uptake of ions by roots. I. Soil water gradients near a plane of onion roots. *J. Appl. Ecol.,* **10**: 585–598.

Ehleringer, J. R., and P. C. Miller. (1975) Water relations of selected plant species in the alpine tundra, Colorado. *Ecology,* **56**: 370–380.

Hay, R. K. M., and W. P. Anderson. (1972) Characterization of exudation from excised roots of onion, *Allium cepa.* I. Water flux. *J. Exp. Bot.,* **23**: 577–589.

Hillier, R. D. (1970) The influence of water on growth and development of alpine plants in the Medicine Bow Range, Wyoming. Ph.D. thesis, Duke University.

Honert, T. H. van den (1948) Water transport in plants as a catenary process. *Disc. Faraday Soc.,* **3**: 146.

Hsiao, T. C. (1973) Plant responses to water stress. *Ann. Rev. Plant Physiol.,* **24**: 519–570.

Hygen, G. (1951) Studies in plant transpiration I. *Physiol. Plant,* **4**: 57–183.

Jarvis, P. G., and M. S. Jarvis. (1963) The water relations of tree seedlings. IV. Some aspects of the tissue water relations and drought resistance. *Physiol. Plant.,* **16**: 501–516.

Johnson, D. A., and M. M. Caldwell. (1975) Gas exchange of four arctic and alpine tundra plant species in relation to atmospheric and soil moisture stress. *Oecologia,* **21**: 93–108.

Johnson, D. A., and M. M. Caldwell. (1976) Water potential components, stomatal function, and liquid phase water transport resistances of four arctic and alpine species in relation to moisture stress. *Physiol. Plant.,* **36**: 271–278.

Kanemasu, E. T., and C. B. Tanner. (1969) Stomatal diffusion resistance of snap beans. I. Influence of leaf water potential. *Plant Physiol.,* **44**: 1547–1552.

Kanemasu, E. T., G. W. Thurtell, and C. B. Tanner. (1969) Design, calibration, and field use of a stomatal diffusion porometer. *Plant Physiol.,* **44**: 881–885.

Kuiper, P. J. C. (1961) The effects of environmental factors on transpiration of leaves with special reference to the stomatal light response. *Meded. Landbouwhogesch, Wageningen,* **63**: 1–49.

Kuramoto, R. T., and L. C. Bliss. (1970) Ecology of subalpine meadows in the Olympic Mountains, Washington. *Ecol. Monogr.,* **40**: 317–347.

McNaughton, S. J., R. S. Campbell, R. A. Freyer, J. E. Mylroie, and K. D. Rodland. (1974) Photosynthetic properties and root chilling responses of altitudinal ecotypes of *Typha latifolia. Ecology,* **55**: 168–172.

Millar, A. A., W. R. Gardner, and S. M. Goltz. (1971) Internal water transport in seen onion plants. *Agron. J.,* **63**; 779–784.

Mooney, H. A., R. D. Hillier, and W. D. Billings. (1965) Transpiration rates of alpine plants in the Sierra Nevada of California. *Amer. Midl. Natur.,* **74**: 374–386.

Morrow, P. A., and R. O. Slatyer. (1971) Leaf resistance measurements with diffusion porometers: Precautions in calibration and use. *Agric. Meteorol.,* **8**: 223–233.

Rawlins, S. L. (1963) Resistance to water flow in the transpiration stream. In *Stomata and Water Relations in Plants* (I. Zelitch, Ed.). Connecticut Agricultural Experiment Station Bulletin 664, New Haven, pp. 69–84.

Scott, D., and W. D. Billings. (1964) Effects of environmental factors on standing crop and productivity of an alpine tundra. *Ecol. Monogr.,* **34**: 243–270.

Slatyer, R. O. (1967) *Plant Water Relationships.* New York: Academic Press, 366 pp.

Slatyer, R. O., and J. F. Bierhuizen. (1964) Transpiration from cotton leaves under a range of environmental conditions in relation to internal and external diffusive resistance. *Aust. J. Biol. Sci.,* **17**: 115–130.

Stoner, W. A., and P. C. Miller. (1975) Water relations of plant species in the wet coastal tundra at Barrow, Alaska. *Arct. Alp. Res.,* **7**: 109–124.

Troughton, J. H. (1969) Plant water status and carbon dioxide exchange of cotton leaves. *Aust. J. Biol. Sci.,* **22**: 289–302.

Waggoner, P. E., and N. C. Turner. (1971) Transpiration and its control by stomata in a pine forest. *Connecticut Agric. Exp. Sta. Bull.* 726, New Haven.

15. Radio-Tracer Measurement of Transpiration in Tundra Vegetation, Barrow, Alaska

JOHN J. KORANDA, BRUCE CLEGG, AND MARSHALL STUART

Introduction

The measurement of water fluxes in plant ecosystems has been made by several methods which typically require a considerable amount of field instrumentation and equipment. Micro-meteorological data may be used to estimate the potential evapotranspirational flux (Pruitt, 1971) and large weighing lysimeters (Pruit and Angus, 1960) have been employed in agricultural and university research facilities (Armijo et al., 1972) to make this measurement gravimetrically. Small and large cuvettes or chambers (Tieszen, 1975) are also used to obtain physiological data on water loss by plants or plant organs in semiclosed or closed experimental systems (Koller, 1970).

Recently the use of radio-tracer techniques for the measurement of tree and grassland transpiration was described (Kline et al., 1970, 1972). In the present study, radio-tracer methods similar to those used by Kline et al. (1972) were employed in a series of experiments conducted at the Tundra Biome intensive study site near Barrow, Alaska, in July 1973. We believe the use of tritiated water (THO) to label vegetation in the field permits the measurement of transpiration rates for single plants (trees and shrubs) or vegetation types under dynamic, open system conditions.

Methods

Site and Experiments

Soil and vegetation experiments were conducted near site 2 at Barrow, Alaska. Soil water use was studied in five plots, 1 × 2 m, by applying THO (10 mCi in 2 liters of HHO) to the tundra surface with a pressurized hand sprayer. The three plots at site 2 were located in the same area of the physiology studies of Tieszen (1978) and Coyne and Kelley (1978). This site was more mesic than the other study area, site 5, which was an area of low polygons along a nearly coastal slough. Samples of the thawed active layer were obtained with a 5-cm coring tool. Cores were sectioned at the plot immediately after coring. Sectioning of the core was usually at 7.6, 20.3, 28.0, and 38 cm. The core below 20 cm was usually divided in half, with the bottom half being just above the frozen layer. Because of variations in the depth of thaw, the lower core sections were not always exactly the same length.

A second type of radio-tracer experiment was conducted to determine transpiration rates in the field. This was accomplished by placing a chamber, 1 m × 1 m × 38 cm high, over the tundra plot and producing a tritiated water atmosphere in the chamber over the vegetation. Tritiated water vapor was evolved by evaporating 1 ml of THO (3.3 mCi) in the chamber by circulating the chamber air with a small fan through a gauze pad into which the THO had been pipetted. Exposures were 50 min in duration, and sampling of the standing green vegetation began 10 min after exposure termination. Between 5 and 8 g of living leaf material was collected in a random fashion from the 1-m² labeled area at each sampling interval. Samples were double wrapped in plastic bags and frozen on dry ice in the field. Samples were collected at 20-min intervals for 2 hr or more. Some limitations were experienced with this sampling method due to the low productivity of the tundra at this location.

Soil and plant tissue water was extracted from the samples by vacuum distillation or freeze drying on a specially designed vacuum manifold described by Koranda et al. (1971). Aliquots of the extracted soil or leaf tissue water were assayed for tritium by liquid scintillation counting methods. The specific activity of tritium in the extracted water sample is plotted against time after exposure, and the half-life and mean-time (T_m) of water in the soil or plant compartment are determined from the activity–time curves.

Theoretical Basis

The basic aspects of tracer theory first described by Stewart (1897) and Hamilton (1953) and later expanded by Bergner (1961, 1964, 1965, 1966), Zierler (1964), Ljunggren (1967), and Orr and Gillespie (1968) are pertinent to the analysis of these tracer data. This well-developed theory of tracer dynamics in flowing systems permits the determination of pool sizes, flow rates, and perhaps other kinetic relationships such as subpools and their turnover rates.

Transpiration in temperate latitude plants is not a steady flow of water from the leaves to the atmosphere because of photoperiod effects. The Stewart–Hamilton equation was derived under the assumption that the flow is constant during the period of measurements. Orr and Gillespie (1968), however, found that the equation is not extremely sensitive to departures from constancy in flow, and the equation can be used in systems where the flow is not steady or continuous. Kline et al. (1972) also found that the lack of continuous flow in transpiration did not obviate the use of the Stewart–Hamilton or the occupancy (Orr and Gillespie, 1968) relationships. The resulting transpiration measurement is an average estimate for the period, which, in these experiments, was 2 hr. Tieszen (1975) has shown that transpiration and CO_2 uptake occur throughout the arctic photoperiod in midsummer, and thus transpiration in arctic plants while not proceeding at steady-state rates may approximate continuous flow more so than in most temperate plant species.

It is not necessary in grassland experiments as it would be in a tree experiment to know precisely the total activity in the system at T_0, and therefore sampling is designed to provide an accurate measurement of the mean-time of the tracer in the pool. Isotope mass effects are small and are not expected to alter kinetic relationships at the level of these experiments.

Evapotranspiration is calculated by

$$F = \frac{C}{T_m} \tag{1}$$

where F = flow rate or evapotranspiration from stand (ml hr^{-1} m^{-2}),
$\quad C$ = water pool (ml m^{-2}), and
$\quad T_m$ = mean-time of water in pool (hr).

Results and Discussion

Soil Water, Evaporation, and Transpiration

Table 1 is a summary of the data obtained in the five soil water experiments. Tracer distribution profiles in the active layer (as sampled by the cores) were integrated with depth at each sampling date. Concentrations of tritium in soil water were then expressed on a surface area basis and these core inventory

Table 1.   Soil Water Experiment Summary—Barrow 1973

	Site 2 (mesic)			Site 5 (wet)	
Plot	1	2	3	1	2
ml/cm	6.31 ±0.63	7.26 ±0.98	6.76 ±0.49	7.92 ±0.69	8.21 ±0.76
$\overline{X}$	6.77 ± 0.81			8.07 ± 0.73	
Average depth to P.F.[a]	18.0	15.7	16.1	18.9	15.4
$\overline{X}$	16.6 ± 2.8			17.2 ± 0.27	
ml of HHO/core[b]	113.6	114.0	108.8	149.7	126.4
$\overline{X}$	112.1 ± 2.89			138.0 ± 16.5	
Average HHO capacity of soil (C)	89.25 mm			109.9 mm	
Percentage HHO	53.8			63.9	
Soil HHO mean time (T_m)[c]	15.8 days			23.9 days	
Flux = C/T_m =	5.65 mm-day			4.60 mm-day	

[a] Permafrost table.
[b] Core area = 12.56 cm².
[c] Determined from tracer (THO) mean-time in soil water.

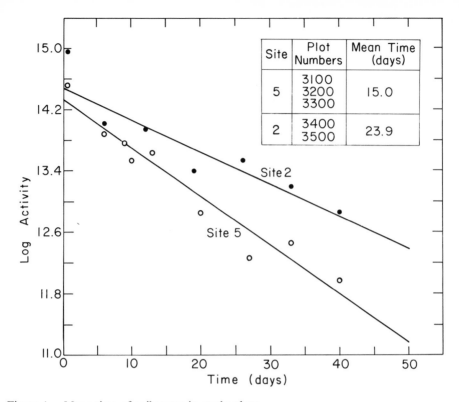

Figure 1.   Mean time of soil water in study plots.

values were then plotted against time to obtain the activity–time curve for each plot. These curves are shown for the two sites in Figure 1. The water content of the soil profile was obtained from the water extraction data and therefore the pool size was measured during sample analysis.

   Three of the soil plots (site 2) were located at the meteorology tower previously used by Weller and Cubley (1972). This site was more mesic than the second soil water study site which was an area of low polygons along a coastal slough. Plot Nos. 3400 and 3500 were established in the area of detailed plant physiological studies (Wiggins, 1951).

   The three mesic plots (site 2) had a smaller water capacity (89 vs 110 mm) than the two lower and more poorly drained plots (site 5). The labeled soil water in the mesic plots had a shorter mean-time (15.8 vs 23.9 days) which may have been due to the greater degree of slope in the area which caused the soil water to drain northward into Footprint Creek. Accumulations of water were observed on this slope during the experimental period. These measurements indicate that between 4.6 and 5.6 mm day^{-1} of water are leaving the soil water column by evaporation, transpiration, or horizontal transport. This transport could have influenced the data from the site 2 plots.

   Weller and Cubley (1972) reported mean evapotranspiration values of 4.5 mm day^{-1} for the postmeltoff period at Barrow (14 to 17 June) and during the mid-summer period, which is comparable to our measurements. Open-pan evaporation reached 3.0 mm day^{-1}. Our values of 4.6 to 5.6 mm day^{-1} may include a horizontal flow component which would not influence open-pan evaporation measurements but which would affect soil water dynamics in nature. The data obtained in the flatter and more poorly drained tundra topography (site 5) are more representative of the typical Barrow tundra.

Transpiration Measurements

Transpiration rates for sedge-grass vegetation in the Barrow area are shown in Table 2 and appear to be higher for 24 and 25 July; however, the wide range of rates obtained on 25 July embraces the ranges of all the other days. The mean value for sedge-grass transpiration for all experiments is 2.95 ± 0.76 ml g^{-1} hr^{-1}. The values for transpiration on a ml g^{-1} g^{-1} basis were converted to an area flux value by using dry biomass data from Tieszen et al. (personal communication). Tieszen's data were summarized, and a mean value of 89.65 g m^{-2} was obtained for the same general area of tundra in which our experiments were conducted. A very high value and an unusually low measurement were excluded in our

Table 2.   Transpiration Data for Sedge-Grass Experiments, Barrow Area, July 1973

Date plot	Time	T_m	HHO content[a]	Transpiration rate		
				ml-g-hr	ml-m²-hr[b]	mm-hr
7/24						
Site 2	1000–1200	44.7	2.73	3.66	329.2	0.033
	1100–1325	48.3	2.90	3.60	323.4	0.032
	1200–1400	57.2	2.97	3.12	280.3	0.028
7/25						
Site 5	1000–1200	47.9	3.42	4.30	389.3	0.039
	1100–1300	40.4	2.88	4.27	383.0	0.038
	1200–1400	41.7	2.42	3.48	312.0	0.031
	1300–1520	57.1	3.14	3.30	296.1	0.029
	1400–1620	49.6	2.26	2.73	244.2	0.024
	1600–1820	70.9	2.66	2.25	202.2	0.020
	1700–1900	81.6	2.48	1.82	163.5	0.016
	1805–2020	44.4	3.10	1.54	150.0	0.015
7/26						
Site 2	1230–1450	50.9	2.54	3.00	269.0	0.027
	1330–1550	70.7	2.43	2.01	184.4	0.018
	1400–1620	64.4	2.99	2.79	249.4	0.025
	1430–1650	86.8	2.56	1.78	158.8	0.016
	1520–1740	69.4	2.66	2.30	206.2	0.021
7/27						
Site 5	0930–1150	63.4	2.39	2.26	202.9	0.020
	1000–1220	59.4	2.94	2.96	265.7	0.026
	1200–1420	60.4	2.62	2.60	233.8	0.023
	1230–1450	57.2	2.77	2.91	260.9	0.026
	0.430–1650	55.5	2.38	2.57	230.9	0.023

[a]   g HHO-g dry wt, used as pool (C) in Eq. (1).
[b]   Assuming a mean dry biomass of 89.65 g dry wt-m^{-2}.

calculation. If the data for the 1000- to 1500-hr period for each of the 4 days of tracer experiments are compared, a reduction of transpiration rates is apparent on 26 and 27 July.

A detailed analysis of the microclimate associated with the entire period of study will be made in another report. It is appropriate to include here some basic climatic data which relate to the tracer experiments, and to use it to explain the reduction of transpiration rates observed during the latter period of tracer experiments (26 to 27 July). The partitioning of the total evapotranspirational flux is possible when actual plant transpiration data derived from tracer experiments is compared to potential evapotranspiration or latent energy values determined during the same period.

The climatic regime of the latter 2 days of tracer experiments (26 to 27 July) contrasted strongly with that of the prior 2 days (Figure 2). On the 24th and 25th of July, net radiation received was 176 and 100 cal cm^{-2}, respectively, while on the 26th and 27th of July, 254 and 308 cal cm^{-2} were recorded. Humidity gradients between 5 and 25 cm were steeper on 26 to 27 July, and wind speed more than doubled. Consequently, factors affecting evapotranspiration and tracer performance were significantly different during this 4-day period of experiments and measurements.

In Table 3, the partitioning of solar energy during the 4 days of tracer experiments is shown. In addition to the large differences in total and net energy received on the tundra during these two contrasting periods, latent energy or

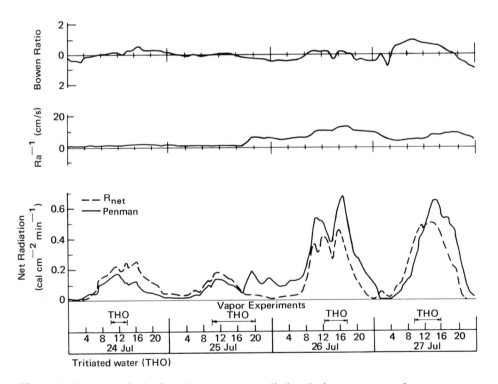

Figure 2.   Bowen ratio, leaf conductance, net radiation during vapor experiments.

Table 3. Daily energy Balance during Tritium Tracer Experiments—1973, Barrow, Alaska

Date	G^a (cal cm^{-2})	$R_n^b - G$ (cal cm^{-2})	(mm-day^{-1})	R_t^c (cal cm^{-2})	H^d (cal cm^{-2})	LE_{pen}^e (cal cm^{-2})	ET^f (mm-day^{-1})
7/24	−2.3	176	2.96	303	9.3	108	1.8
7/25	25.0	100	1.68	215	−45.0	119	2.0
7/26	3.3	254	4.28	475	−45.0	431	7.1
7/27	−17.0	308	5.18	630	116.0	374	6.2

[a] G = ground flux.
[b] R_n = Net radiation.
[c] R_t = Total incident radiation.
[d] H = Sensible heat.
[e] LE_{pen} = Latent energy by a modified Penman method.
[f] ET = Evapotranspiration.

potential evapotranspiration values on 26 to 27 July were more than three times those observed on 24 to 25 July (Figure 3). Sensible heat (H) became negative during this time, and potential evapotranspiration exceeded net energy values because of advective energy input into the Barrow tundra microclimate.

The contrasting microclimate of this 4-day period of coordinated tracer and meteorological measurements provided an opportunity to test the sensitivity of the tracer method during a period of varying evapotranspirational stress. In Table 2 it was shown that a depression of transpiration rates occurred in the latter portion of this period.

When tracer transpiration is plotted against net radiation, as in Figure 4, an almost linear relationship is apparent during the first 2 days of experimentation (24 to 25 July). Net radiation during this period did not exceed 0.3 cal cm^{-2} min^{-1}. On 26 to 27 July, this linear relationship did not pertain, and net radiation levels were approaching 0.4 cal cm^{-2} min^{-1} on 26 July, and exceeded 0.6 cal cm^{-2} min^{-1} on 27 July. Tieszen has shown that saturation for most tundra grass species in the Barrow area is usually near 0.4 cal cm^{-2} min^{-1}. It is proposed that stomatal closure occurred during this period of high radiation and turbulent diffusion conditions that occurred on 26 to 27 July, and that the tracer data represented this reduction, however subtle, in stand transpiration rates.

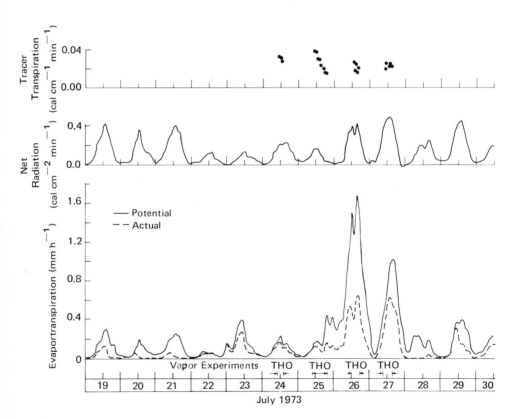

Figure 3.   Tracer transpiration, measured evapotranspiration, and net radiation during experimental periods.

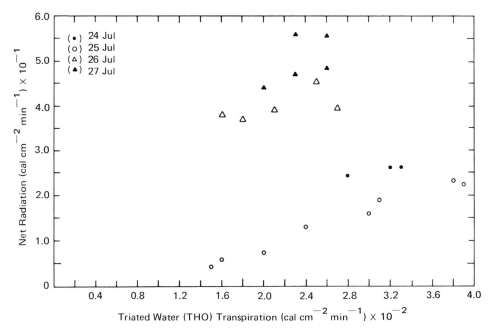

Figure 4. Net radiation and tracer transpiration relationships during experimental periods.

The mean-value for tracer transpiration rates for all sedge-grass experiments is 0.025 ± 0.007 mm hr^{-1}. This value cannot be applied to the entire 24-hr photoperiod because the experiments were conducted primarily during the high radiation and transpiration period of the day. Tieszen (1975) conducted cuvette studies on Barrow plant species and determined CO_2 uptake and water loss during the entire arctic photoperiod. Maximum rates of transpiration typically occurred in the 1000- to 1800-hr period with reduced rates throughout the night and early morning (1800 to 1000). The average transpiration rate during the night and morning hours was 24% of the maximum period value. If the mean tracer transpiration rate of 0.025 mm hr^{-1} is applied for the 8-hr period from 1000 to 1800, and 24% of that rate used for the rest of the arctic photoperiod, a daily transpiration rate of 0.296 mm, or 0.03 cm day^{-1} is obtained. This is within the estimate made by Miller et al. (1976) of 0.01 to 0.05 cm day^{-1}. The mean daily transpiration rate of 0.296 mm day^{-1} is 6.4% of the total evaporational flux estimated in the soil water experiments. A value of less than 10% was suggested by Miller et al. (1976).

Conclusions

Tracer experiments were conducted with tritiated water on the tundra near Barrow, Alaska, to determine soil water use and plant stand transpiration rates. Soil water use data obtained from five plots at two sites indicate that between 4.6

and 5.6 mm day^{-1} were disappearing from the active layer of the soil during the July to September period. The higher value may be affected by soil water interflow on a gentle slope. Information on interflow in arctic soils underlain by permafrost, especially in areas of subtle microtopographic relief such as occur at Barrow, would be very useful but cannot be obtained from these data.

Open-pan evaporation data indicate that during the midsummer period 2 to 3 mm day^{-1} of evaporation may be expected. However, Miller et al. (1976) concluded that plant transpiration of 0.1 to 0.5 mm day^{-1} was less than 10% of the open-pan evaporation for the Barrow area, which would place our values in the appropriate range.

Tracer transpiration rates are generally consistent in their diurnal pattern, in their response to varying radiation loads, and in their correlation with independent measurements such as cuvette studies. In response to higher net radiation levels, steeper vapor density gradients, and increased wind speed and turbulence, tracer transpiration rates indicated a depression of transpiration and demonstrated that a lower fraction of potential evapotranspiration was occurring as plant transpiration.

The average tracer transpiration rate for all sedge-grass experiments at Barrow was 2.95 ml g^{-1} hr^{-1} or 0.025 mm hr^{-1}. When related physiological data are employed, the daily plant transpiration rate becomes 0.03 cm day^{-1} or 6.4% of the evapotranspirational flux determined in the soil water tracer studies. These data appear to be consistent with independent measurements and simulations of the Barrow wet meadow tundra ecosystem (Miller et al., 1976).

Acknowledgments. This research was supported by the National Science Foundation under Grant GV 29342 to the University of Alaska. It was performed under the joint sponsorship of the International Biological Programme and the Office of Polar Programs and was coordinated by the U.S. Tundra Biome. Field and laboratory activities at Barrow were supported by the Naval Arctic Research Laboratory of the Office of Naval Research.

References

Armijo, J. D., J. R. Nunn, G. Twitchell, and R. D. Burman. (1972) A lysimeter for the Pawnee site. *U.S. IBP Grassland Biome Tech. Rep.* 179.

Bergner, P.-E. E. (1961) Tracer dynamics. I. A. tentative approach and definition of fundamental concepts. *J. Theor. Biol.,* 2: 120–140.

Bergner, P.-E. E. (1964) Tracer dynamics and determination of pool sizes and turnover factors in metabolic systems. *J. Theor. Biol.,* 6: 137–158.

Bergner, P.-E. E. (1965) Exchangeable mass: Determination without assumption of isotopic equilibrium. *Science,* 150: 1048–1050.

Bergner, P.-E. E. (1966) Tracer theory: A review. *Isotop. Radiat. Technol.,* 3: 245–262.

Coyne, P. I., and J. J. Kelley. (1978) Meteorological assessment of CO_2 exchange over an Alaskan arctic tundra. In *Vegetation and Production Ecology of an Alaskan Arctic Tundra* (L. L. Tieszen, Ed.). New York: Springer-Verlag, Chap. 12.

Hamilton, W. F. (1953) The physiology of the cardiac output. *Circulation,* 8: 527–543.

Kline, J. R., J. R. Martin, C. F. Jordan, and J. J. Koranda. (1970) Measurement of transpiration in tropical trees using tritiated water. *Ecology,* 5: 1068–1073.

Kline, J. R., M. L. Stewart, C. F. Jordan, and P. Kovac. (1972) Use of tritiated water for determination of plant transpiration and biomass under field conditions. In *Symposium on Isotopes and Radiation in Soil and Plant Research, Including Forestry, Vienna.* International Atomic Energy Agency, pp. 419–437.

Koller, D. (1970) Determination of fundamental plant parameters controlling carbon assimilation and transpiration by the null point compensating system. *Univ. Calif., Los Angeles, Rep.* 12.797, 26 pp.

Koranda, J. J., P. L. Phelps, L. R. Anspaugh, and G. Holladay. (1971) Sampling and analytical systems for measurement of environmental radioactivity. In *Symposium on Rapid Methods for Measuring Radioactivity in the Environment, Munich.* International Atomic Energy Agency, pp. 587–614.

Ljunggren, K. (1967) A review of the use of radioisotope tracers for evaluating parameters pertaining to the flow of materials in plant and natural systems. *Isotop. Radiat. Technol.,* **5**: 3–24.

Miller, P. C., W. A. Stoner, and L. L. Tieszen. (1976) A model of stand photosynthesis for the wet meadow tundra at Barrow, Alaska. *Ecology,* **57**: 411–430.

Orr, J. S., and F. C. Gillespie. (1968) Occupancy principle for radioactive tracers in steady state biological systems. *Science,* **162**: 138–139.

Pruitt, W. O. (1971) Factors affecting potential evapotranspiration. In *Proceedings, Third International Seminar of Hydrology Professors, July 1971,* Purdue University, Indiana.

Pruitt, W. O., and D. E. Angus. (1960) Large weighing lysimeter for measuring evapotranspiration. *Trans. Amer. Soc. Agric. Eng.,* **3**: 13–18.

Stewart, G. N. (1897) Researches on the circulation time and on the influences which affect it. *J. Physiol.,* **22**: 159–183.

Tieszen, L. L. (1975) CO_2 exchange in the Alaskan Arctic tundra: Seasonal changes in the rate of photosynthesis of four species. *Photosynthetica* **9**: 376–390.

Tieszen, L. L. (1978) Photosynthesis in the principal Barrow, Alaska, species: A summary of field and laboratory responses. In *Vegetation and Production Ecology of an Alaskan Arctic Tundra* (L. L. Tieszen, Ed.). New York: Springer-Verlag, Chap. 10.

Weller, G., and S. Cubley. (1972) The microclimates of the arctic tundra. In *Proceedings 1972 Tundra Biome Symposium, Lake Wilderness Center, University of Washington* (S. Bowen, Ed.). U.S. Tundra Biome, U.S. International Biological Program, and U.S. Arctic Research Program, pp. 5–12.

Wiggins, I. L. (1951) The distribution of vascular plants on polygonal ground near Point Barrow, Alaska. *Contrib. Dudley Herb.,* **4**: 41–56.

Zierler, K. L. (1964) Basic aspects of kinetic theory as applied to tracer distribution studies. In *Dynamic Clinical Studies with Radioisotopes. Proceedings of Symposium.* Oak Ridge, Tennessee: Oak Ridge Institute of Nuclear Studies, pp. 55–79.

16. Simulation of the Effect of the Tundra Vascular Plant Canopy on the Productivity of Four Plant Species

W. A. STONER, P. C. MILLER, AND W. C. OECHEL

Introduction

The physical environment and the structure of the vegetation are two important components which influence the nature and stability of tundra ecosystems. Low temperatures, low solar and infrared irradiances, high wind speeds, permanently frozen subsoil, and impeded water drainage restrict the development of the vegetation by affecting plant growth and development, vegetation composition, and soil faunal activity and decomposition (Bliss, 1956; Billings and Mooney, 1968; Price, 1971; McCown, 1973; Savile, 1972). The vascular plant canopy, although short, affects the solar and infrared irradiances, wind speeds, and air and surface temperatures at the moss or organic mat surface, and the depth of thaw (Drury, 1956; Price, 1971; Matveyeva, 1971; Brown, 1973; Dingman and Koutz, 1974; Miller et al., 1976; Ng and Miller, 1977). Thus, within the interactive vegetation–soil systems, the structure of the vascular plant canopy influences the production of the vascular plant and the production and distribution of moss species, by affecting the microclimate of the vascular plant canopy and moss. This chapter analyzes, by simulation models, the effects of the vascular plant canopy on microclimate and the effect of microclimate on production by vascular plants and moss.

Simulation Models and Methods

The various simulation models for the wet meadow tundra system are described in detail elsewhere (Miller et al., 1976, 1978; Ng and Miller, 1977). However, a brief overview of the submodels is included here. The simulation models are centered on the physical processes of energy exchange and on the energy budget equations for plant parts, levels in the canopy, ground or surface, and levels in the soil. Photosynthesis and transpiration follow from the solution of the energy budget equation, with the inclusion of appropriate physiological relations. The input climatic data consist of solar and infrared irradiance, air temperature, air humidity, and wind speed. The vascular canopy is composed of leaves, stems, and standing dead material of different species, each defined by inclination and by vertical profiles of area per unit area of ground. Solar and infrared radiation from the sun and sky are intercepted by the canopy and produce profiles of direct, diffuse, and reflected solar and infrared radiation within the canopy. The air temperature and humidity above the canopy and at the moss surface interact with the canopy structure, wind profile, and radiation profiles to produce profiles of air temperature, humidity, and leaf temperature.

At the moss surface the receipt of net radiation is balanced by heat exchanges due to convection, evaporation, and conduction. The convectional heat exchange occurs by turbulent exchange of air from the surface across a surface boundary layer and a bulk canopy air layer to a reference height in the canopy. Surface evaporation is related to the turbulent exchange of water vapor across the surface boundary layer and bulk canopy air layer.

The heat conduction to or from the surface is controlled by the thermal conductivity and the difference between the surface temperature and the soil temperature just below the surface. The soil profile below the surface consists of an organic soil or peat, which is underlain by a deeply extending mineral layer. Each soil layer differs in physical and thermal properties. The flow of heat into or out of the profile is calculated using an implicit finite difference solution to the one-dimensional heat conduction equation with phase change, adapted from Nakano and Brown (1972). The soil temperature model assumes the movement of heat due to water movement to be negligible.

Photosynthesis and transpiration rates of the vascular plants at different levels in the canopy are controlled by a series of feedback relations. Solar irradiance and leaf temperature affect photosynthesis. Moss photosynthesis is related to solar irradiance, tissue temperature, and water status. Solar and infrared irradiance, air temperature, humidity, and wind velocity affect the plant water status through their effect on leaf temperature and transpiration. The plant water status influences the rates of transpiration and photosynthesis through the common resistance to water and carbon dioxide diffusion. Water in the form of precipitation and dew, which is not intercepted by the vascular canopy, is added to the moss surface water film. Water flows between the green moss surface and the nongreen moss and peat layers below. Surface water can be evaporated directly or absorbed into the moss tissue, to be lost by transpiration later.

The seasonal progression of microclimate and production was simulated using a standard set of climatic conditions and deviations from the standard. The standard input climate is based on climatic and microclimatic data collected at Barrow, Alaska, in the summer of 1973 (P. C. Miller, personal observation). These data were adjusted using long-term records to produce two other sets of conditions which were ±3 standard deviations from the standard temperature conditions. The two contrived climates are hereafter referred to as HOT and COLD. The microclimatic data from 1973 have been used in other studies (Stoner and Miller, 1975; Ng and Miller, 1975; Miller et al., 1978; Ng and Miller, 1977) and were used here as the standard case to aid in interpreting the results. The HOT and COLD seasons are extreme conditions on either side of the standard case and are not typical. They are used to clarify interactions among the various components of the system.

The models were run with all three simulated growing seasons with and without vascular plant cover. Two moss species, *Dicranum angustum* and *Calliergon sarmentosum,* and two vascular species. *Dupontia fisheri,* a single shooted graminoid, and *Salix pulchra,* a deciduous dwarf shrub, were used to characterize the vascular and nonvascular components of the system. Webber (1978) has described these life form categories and the relative abundance of these species. Moss and vascular plant models were run with three substrate water potentials (-1, -5, -10 bars) at each climate to simulate a moisture gradient.

In order to simulate more fully the effect of the vascular canopy on the moss understory, a simple water interception model was used. This work has not been previously described and will be described here. The interception of rain is calculated according to the point quadrat theory (Warren Wilson, 1965) and by analogy with the interception of direct solar radiation in the canopy (Duncan et al., 1967). When the angle of incidence (α) of the drops of precipitation and dew is less than the inclination of the foliage, which is usual, the intercepted precipitation and dew (g of H_2O m_{leaf}^{-2}) is

$$INT = (precip + dew)(1 - e^{-A \cos \alpha}) \qquad (1)$$

where A is the area of plant parts in the canopy. Past studies of interception in differing vegetation types have shown linear relations between the amount of water intercepted and the amount of incoming precipitation (Reynolds and Leyton, 1963; Slatyer, 1965). Interception of low amounts of precipitation is complete, implying a storage capacity (SC) which must be filled before drip or stemflow can occur (Slatyer, 1962). Drip (DRP) and stemflow (STFL) occur when the water on the foliage surface exceeds the storage capacity (SC), which is assumed to be 200 g of H_2O m_{leaf}^{-2}, or a film 0.2 mm thick. This is consistent with Leyton et al. (1967). The fraction of the excess water diverted to each path is assumed constant. Thus,

$$STFL = k(WC_s - SC) \qquad (2)$$
$$DRP = (1-k)(WC_s - SC) \qquad (3)$$

where k is set to 0.3. Drip is added to the unintercepted precipitation and dew, and is available to the moss understory. Stemflow is absorbed at the base of the plant and is unavailable to the moss. The potential evaporation rate at Barrow is 1 to 2 kg m^{-2} day^{-1} so we assume that all stored water is evaporated in 1 day.

Two additions have been made in the moss model since it was described by Miller et al. (1977): a respiration burst following a period of desiccation and mortality due to desiccation. Respiration upon rewetting is significant when compared to the maximum photosynthetic rate (Peterson and Mayo, 1975). The sequence of the respiration burst is assumed to be the same for all species (Peterson and Mayo, 1975), while the point where mortality occurs differs among

Table 1.   Summary of Parameter Values Characterizing the Vascular Canopy, Microclimate, Vascular Production, and Moss Systems

	Canopy and vascular production submodels		
		Species	
Parameter	Dead material	*Salix pulchra*	*Dupontia fisheri*
Foliage angle (degrees)	30.0	24.0	63.0
Absorptivity (fraction)	0.40	0.50	0.50
Foliage width (cm)	0.36	0.56	0.36
Turgid weight (mg cm^{-2})	—	19.0	22.0
Root resistance (10^6 bar sec^{-1} cm^{-1})	—	10.0	9.8
Minimum leaf resistance (sec cm^{-1})	60.0	2.0	1.0
Coefficients relating leaf resistance to absorbed solar irradiance in the equation: $r_1 = b_1 (b_2 + b_3 \text{solar})^{-1}$			
b_1 (sec)		0.3	0.3
b_2 (cm)		0.007	0.007
b_3 (cm^3 min cal^{-1})		6.0	6.0
Relative water content when leaf resistance begins to increase (%)		12.0	2.5
Relative water content at maximum leaf resistance (%)		20.0	17.0

	Moss production submodel	
	Species	
Parameter	*Dicranum angustum*	*Calliergon sarmentosum*
Maximum photosynthetic rate (mg of CO_2 g dry wt^{-1} hr^{-1})	0.45	0.56
Minimum air resistance (sec cm^{-1})	35.0	13.0
Minimum leaf resistance (sec cm^{-1})	2.0	2.0
Saturated water content (g of H_2O g dry wt^{-1})	3.0	6.0
Resistance to water movement to moss surface (10^6 bar sec cm^{-1})	0.02	0.02
Resistance to water absorption into moss (10^6 bar sec cm^{-1})	0.04	0.04
Moss surface area to dry weight ratio (cm^2 g dry wt^{-1})	1360	820
Moss surface water capacity (g of H_2O g dry wt^{-1})	18.0	93.0
Percentage saturated water content below which a respiration burst occurs upon rewetting	1.0	1.5
Percentage saturated H_2O content at which 90% mortality occurs	0.25	0.38

Surface and soil temperature submodel	
Parameter	Value
Reflectance of moss surface (fraction)	0.50
Resistance of moss layer to heat and water flux (sec m^{-1})	10.0
Thickness of organic mat (m)	0.06
Thermal conductivity (cal m^{-1} sec^{-1} C^{-1})	1.61
Turbulent exchange coefficient above canopy (m^2 sec^{-1})	0.012
Extinction coefficient for turbulent exchange (foliage area index^{-1})	1.0

species (Table 1). The respiration burst sequence is such that 15, 30, 60, 300, and 480 min after rewetting, respiration or photosynthesis is −2.5, −0.8, −0.45, 0.0, and 1.0 times the normal rate. If the moss dries out while in a respiration burst sequence, the moss begins the respiration sequence at the beginning upon rewetting.

The parameters defining the physiological and structural characteristics of the species are summarized in Table 1.

Results and Discussion

The effect of the canopy on air temperature and humidity profiles is through its influence on processes of energy exchange. The canopy intercepts, reflects, transmits, and absorbs solar radiation, absorbs and emits infrared radiation, and impedes the turbulent exchange of heat and humidity in the air which directly affects rates of convection and evapotranspiration. Of the incoming direct solar beam, about 86% is intercepted in the canopy; the rest passes through to the soil or moss surface (Figure 1). The canopy is more transparent to diffuse solar

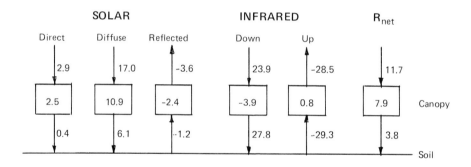

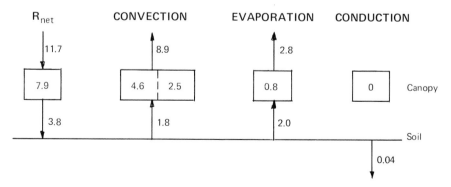

Figure 1.   Partitioning of incoming radiation into reflected solar and infrared radiation, convection, evapotranspiration, and conduction in the wet meadow ecosystem. Units are MJ m^{-2} day $^{-1}$.

radiation because of the scattering and downward reflection of direct beam radiation. The diffuse radiation reaching the soil-moss surface is 36% of the incident diffuse above the canopy because of the additional scattered components. About 18% of the incoming solar radiation is reflected back, most of this coming from the canopy rather than the soil-moss surface. Some of the absorbed solar radiation is emitted as infrared. There is a net loss of infrared radiation from the canopy, both in the upward and downward directions. However, the canopy receives more infrared from the soil-moss surface than it loses to the sky.

The net radiation in the canopy is about twice that of the soil-moss surface. The net radiation is lost by convection, evaporation, and conduction (Figure 1b). In the canopy, most of the net radiation is lost by convection, and most of this by convection from standing dead material. At the soil-moss surface, evaporation uses slightly more energy than convection. Evaporation from the soil-moss surface is 2.5 times that from the canopy. Conduction accounts for less than 1% of the incoming solar radiation.

The effect of the vascular canopy on the microclimate near the ground varied in the different simulated seasons (Table 2). The canopy intercepted 48, 48, and 44% of the incoming total solar irradiance in the standard, HOT, and COLD seasons, respectively. The decreased interception in the COLD season was the

Table 2. Summary of Microclimatic Variables above the Vascular Canopy, in the Canopy, and at the Moss Surface with and without the Vascular Canopy[a]

	Vegetation-moss configuration			
Variable	Above canopy	In the vascular canopy	Moss with canopy	Moss without canopy
Standard				
Solar incident (cal cm^{-2} day^{-1})	361	361	189	361
Average temperature (°C)	4.5	6.4	4.4	5.3
Humidity (g m^{-3})	6.12	6.33	6.46	6.33
Wind (cm sec^{-1})	200	170	61	200
Vapor density difference (g^{-3})	0.00	1.12	0.20	0.47
HOT				
Solar incident (cal cm^{-2} day^{-1})	633	633	330	633
Average temperature (°C)	15.6	16.8	12.3	13.6
Humidity (g m^{-3})	8.57	9.79	9.79	10.09
Wind (cm sec^{-1})	200	170	61	200
Vapor density difference (g^{-3})	0.00	4.52	1.07	1.68
COLD				
Solar incident (cal cm^{-2} day^{-1})	92	92	52	92
Average temperature (°C)	−3.5	−2.7	−3.0	−2.4
Humidity (g m^{-3})	3.67	3.76	3.76	3.84
Wind (cm sec^{-1})	200	170	61	200
Vapor density difference (g^{-3})	0.00	0.26	0.26	0.43

[a] The vapor density difference is between the air above the canopy and the evaporating surface. Values given are means for the growing season in the three simulated seasons. In the COLD season temperatures are above freezing during the day.

result of higher proportions of daily irradiance occurring at midday, when higher solar elevations lead to greater penetration. Average wind speeds were reduced by 23% in the middle of the canopy and by 70% at the moss surface. This difference is due largely to the concentration of standing dead material at the bottom of the canopy.

In all three simulated seasons the moss surface temperature was lower with the vascular canopy than without (Table 2). However, the differences among air temperatures, canopy temperatures, and moss surface temperatures with and without a canopy varied with season. In the standard case, the moss surface temperatures under the canopy were slightly below air temperature because of the low irradiances and high evaporation rates. Without the canopy, moss temperatures were 0.8°C above air temperatures. Vascular plant temperatures in the canopy were 1.9°C above air temperatures, due to the difference in the partitioning of the energy in the canopy from that of the surface (Miller et al., 1977). The canopy loses most of its energy through convection and reradiation, and shows low evaporative losses, while the surface loses energy equally between evaporation and convection. The lower evaporation rates within the canopy are due to the high vapor densities within the canopy when compared to the bulk air.

In the HOT season, the temperatures of the moss surface were below air temperatures both with and without the vascular canopy due to high heat losses by evaporation and conduction into the soil. The temperatures of the vascular plants were 1.2°C above air temperatures. Vascular plant and air temperatures increased as the season changed from the standard to the HOT, but vascular plant temperature increased less than air temperatures. Surface temperatures increased less than air temperatures because of the evaporational cooling, heat capacity of the ground, and heat conduction. Plant temperatures were stabilized by transpirational heat exchange. Canopy air temperatures were partially stabilized by the surface temperatures.

In the COLD season, moss temperatures were slightly above air temperatures, but vascular plant temperatures were below air temperatures. The lower vascular temperatures were caused by the low incoming irradiances which no longer compensated for radiative and convective heat losses.

In the vertical profiles humidities were always higher near the moss surface regardless of canopy cover. Humidities were higher at the moss surface without the vascular canopy because of the higher temperatures of the moss in the full sun. The differences in the vapor densities of the evaporating surface and the bulk air (Table 2) show that the vascular canopy reduced the driving force for evaporation. The vascular plants experienced greater vapor density differences than the moss. During the COLD season, the differences for moss and vascular plants were the same.

The canopy also affects the water balance of the surface by intercepting and diverting water which could have fallen directly on the moss surface. The simulations indicate that the canopy intercepts 66% of the precipitation and dew (53.8 mm) falling during the growing season. Of the total intercepted, 23% is lost from the canopy by stemflow, 54% by canopy drip, and 22% is evaporated directly. Hence, the moss under the canopy receives only 70% of the total incoming precipitation and dew. Since the canopy has a storage capacity which must be exceeded for drip and stemflow to take place, small amounts of precipitation will be completely intercepted. Thus, moss in full sun may have more available water than moss under a vascular canopy.

In summary, the canopy tends to increase the available water at the surface by reducing the vapor density differences but at the same time tends to decrease the available water by intercepting and removing 30% of the precipitation which falls during the growing season. The significance of the counteracting tendencies varies with the substrate water potential. The moss water status for the HOT season demonstrates the interaction (Figure 2). All seasons showed the same interactions; however, the effects are greater and more obvious in the HOT season due to higher evaporative stress.

Both *Calliergon* and *Dicranum* show similar seasonal courses of water content with and without the canopy and with different substrate water potentials (Figure 2). With high substrate water potentials (-1 bar), *Calliergon* had a higher water content under the canopy than in the full sun throughout the season. However, near the end of the season, a period of water stress, caused by low precipitation and high temperatures, decreased the effect of the canopy on evaporation; and the water contents of moss both with and without a canopy were similar. With increased water stress throughout the season (substrate water potentials of -5 bars), the effect of the canopy is again reduced, and the difference between moss water contents with or without a vascular canopy is less. However, moss water contents were usually greater under the canopy than in full sun. But by midseason, interception of precipitation by the canopy had reduced the moss water content, and moss in the full sun had higher water contents than did the moss under the canopy. The water contents of *Dicranum* have the same pattern as *Calliergon* although the effect of the canopy in reducing evaporation is less, due to the higher resistances of *Dicranum* to water loss.

The effect of the season, the canopy, and the substrate water potential on the seasonal carbon dioxide uptake is different for *Calliergon* and *Dicranum* (Figure 3). *Calliergon* incorporated the most CO_2 in the full sun during the standard season at -1 bar. Water stress during the HOT season reduced production by 25%, even at the high substrate potential. *Dicranum* had its maximum CO_2 incorporation in the full sun during the HOT season at -1 bar and showed no effect due to water stress. In both moss species CO_2 uptake was reduced as substrate water potential was reduced, although *Calliergon* was more sensitive than *Dicranum*. Using seasonal CO_2 incorporation as an indicator, *Calliergon* has a more optimal environment for production under the canopy than in full sun during the HOT season at -5 bars and below, and during all seasons at -10 bars. Substrate water potentials of -5 and -10 bars reduced CO_2 uptake for *Dicranum* during the HOT season, but seasonal CO_2 uptake was always lower under the canopy than in full sun. The change in seasonal CO_2 uptake per unit decrease in substrate water potential was always greatest for either moss in the full sun, with those for the hot climate being the largest (Table 3). This summary also shows that *Calliergon* was more sensitive than *Dicranum* to water stress, and that with the protection of the canopy *Calliergon* increased production.

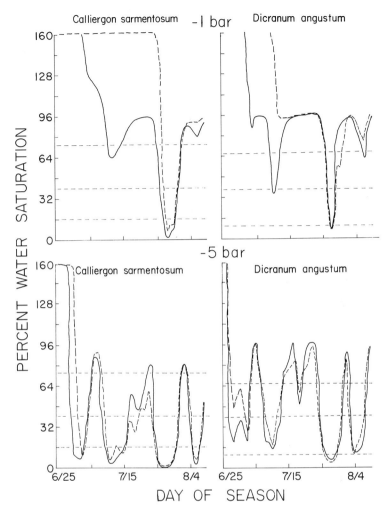

Figure 2. Seasonal course of water content (percentage of water saturation) during the HOT season for *Calliergon sarmentosum* and *Dicranum angustum* at −1 and −5 bars substrate water potential. Solid line is moss in full sun, dashed line is moss under the canopy. Horizontal dashed lines are water contents when net photosynthesis is 85, 50, and 0% (water compensation point) of maximum.

Production for the two vascular species varied during the different seasons (Figure 4). Although *Salix* has the higher absolute maximum photosynthetic rate at high substrate water potentials (−1 bar), *Dupontia* had the higher seasonal uptake because of the relative position of each in the canopy. *Salix* is only in the first 7.5 cm of the canopy, while *Dupontia* extends to 12.5 cm, well above the dwarf shrub. Even at high substrate water potentials, water stress reduced production during the HOT season for both life forms. At −5 bars under the standard season, the uptake of the shrub was slightly higher than that of the graminoid but during the HOT season the order was reversed. During the HOT

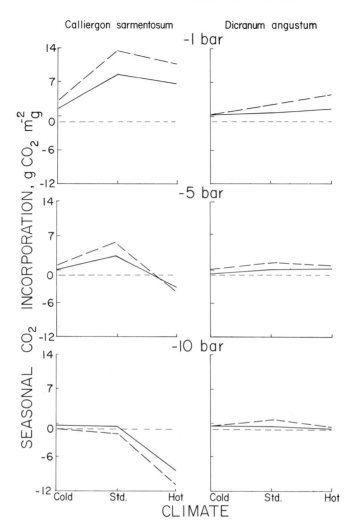

Figure 3.   Seasonal incorporation of CO_2 for *Calliergon sarmentosum* and *Dicranum angustum* at different substrate water potentials for three simulated seasons (COLD, standard, HOT) for moss in full sun (solid line) and under the canopy (dashed line).

season the stomatal resistance of the graminoid was offset by a reduction in internal resistance to CO_2 incorporation due to higher temperatures and solar irradiance. At -10 bars the favorable effect on internal resistance of the graminoid was nullified by higher stomatal resistances, and the shrub had the higher rates. Under all substrate water potentials, the magnitude of internal resistance far outweighs stomatal resistance during the COLD season for both life forms.

Carbon dioxide uptake is clearly less sensitive to substrate water potential in *Salix* than in *Dupontia* (Table 3). However, the similarity of this relation for *Dupontia* during the standard and HOT seasons was unexpected. This similarity was due to the reduction in the internal resistance to CO_2 incorporation caused by higher irradiances and temperatures. This reduction offset the increase in leaf resistance caused by the increased water stress during the HOT season.

Efficiency of water usage was greater for the vascular plants than for the mosses, although the trend with increasing water stress was the same (Table 4). As water stress increased, the efficiency of water use declined. Of the two mosses, *Calliergon* had the greatest water efficiency at high substrate water potentials and under low evaporative demand. However, under water stress *Dicranum* had the greater efficiency. During the COLD season efficiencies were at times less than during the HOT season because of the direct effect of the low temperatures on photosynthesis. *Dupontia* and *Salix* had similar water use efficiencies, although the absolute differences varied with season. During the standard season *Dupontia* was more efficient than *Salix* at high substrate water potentials while at low potentials the situation was reversed. During the HOT season the efficiency of *Dupontia* was consistantly greater than *Salix*. Finally, during the COLD season *Salix* was more efficient at high potentials while *Dupontia* was more efficient at low potentials. The actual trend is most likely the one seen in the standard season with *Salix* more efficient at low substrate water potentials. Conditions during both the HOT and COLD seasons affect photosynthesis directly through the effect of temperature and solar irradiance on internal resistance, and this masks the effects of any water stress on the water use efficiency.

Total vascular and nonvascular production was computed by summing the components and correcting for the length of the growing season (Table 5). Clearly, vascular production is greater than nonvascular. Nonvascular production using CO_2 exchange data estimated seasonal CO_2 incorporation at Barrow, Alaska, during the 1972 growing season to be 279 and 63 g of CO_2 m_g^{-2} for *Calliergon* and *Polytrichum alpinum*, respectively. Maximum rates of CO_2 incorporation for different moss species differ widely between years (Oechel and Collins, 1976; Oechel, 1976). For the simulations discussed here we used photo-

Table 3. Slopes of Seasonal CO_2 Uptake (g of CO_2 m^{-2}) in Relation to Substrate Water Potential for the Three Simulated Seasons[a]

	Season		
Species	COLD	Standard	HOT
Dicranum angustum			
With canopy	−0.02	−0.11	−0.30
Without canopy	−0.03	−0.20	−0.52
Calliergon sarmentosum			
With canopy	−0.19	−0.83	−1.54
Without canopy	−0.35	−1.53	−2.27
Salix pulchra	−1.70	−8.84	−9.94
Dupontia fisheri	−2.30	−16.75	−16.83

[a] Units are change in seasonal CO_2 uptake per bar decrease in substrate water potential. The more negative the slope the greater the depression in CO_2 uptake as substrate water potential decreases.

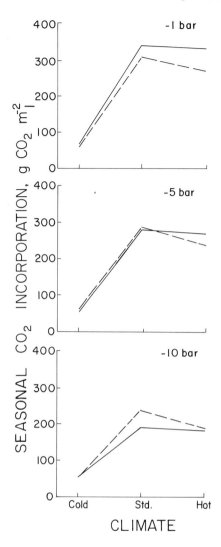

Figure 4.   Seasonal CO_2 uptake (g of CO_2 m^{-2} $season^{-1}$) for the two vascular life forms, *Dupontia* (solid line) and *Salix* (dashed line), during the three seasons at different substrate water potentials (-1, -5, and -10 bars).

synthesis rates measured in 1973 (Oechel, 1976) because they are part of the most complete data set for all physiological relations on different moss species. If the earlier photosynthesis rates had been used, moss production would have been closer to the estimates of Oechel and Collins (1976). In spite of the uncertainty in estimates of moss production, the interactions of moss production and microclimate showed *Calliergon* was consistently more sensitive to water stress than *Dicranum,* and during periods of high water stress the production of *Calliergon* was greater under the vascular canopy than in the full sun. Plant distributions reflect these results. *Calliergon* is restricted to a semiaquatic habitat in wet meadow or within polygon troughs, while *Dicranum* grows on dry elevated areas.

Table 4. Production (g of CO_2 m^{-2} Season^{-1}) and Water Use Efficiency (kg of H_2O g of CO_2^{-1}) for Two Vascular Life Forms and Two Mosses for the Different Simulated Seasons[a]

Species and configuration	Season and substrate water potential (bars)								
	COLD			Standard			HOT		
	−1	−5	−10	−1	−5	−10	−1	−5	−10
Calliergon sarmentosum									
With canopy									
Production	1.9	1.1	0.2	7.9	3.7	0.4	6.0	−2.0	−8.0
Water use efficiency	1.29	4.39	23.56	0.48	1.58	13.5	2.29	12.12	10.09
Without canopy									
Production	2.8	1.9	−0.4	12.5	6.1	−1.2	9.9	−3.0	−10.8
Water use efficiency	1.8	3.35	5.98	0.72	1.67	7.93	2.68	20.24	14.53
Dicranum angustum									
With canopy									
Production	0.7	0.6	0.5	1.4	0.8	0.4	2.3	0.7	−0.4
Water use efficiency	2.38	3.84	4.28	1.76	3.86	7.22	3.6	11.83	8.09
Without canopy									
Production	0.8	0.7	0.6	3.2	2.2	1.3	5.1	2.0	0.4
Water use efficiency	3.16	4.23	4.49	1.34	2.0	2.96	2.46	5.71	2.73
Salix pulchra									
Production	63	61	48	315	288	236	275	242	86
Water use efficiency	0.31	0.36	0.38	0.26	0.27	0.29	0.59	0.63	3.17
Dupontia fisheri									
Production	67	54	46	338	282	188	327	266	176
Water use efficiency	0.34	0.35	0.37	0.21	0.24	0.39	0.32	0.61	1.87

[a] If production was negative, numbers are water losses in kg of H_2O m^{-2} season^{-1}. In order to calculate vascular plant water use efficiency, respiration of nonphotosyn tissue was calculated from Stoner et al. (1978) and included. Water use by mosses was suppressed at high water potentials because of water evaporating from the moss surface.

Aboveground CO_2 incorporation under the standard season at high substrate potentials was similar to other estimates (Coyne and Kelley, 1975; Tieszen, 1975; Miller et al., 1976). Dennis et al. (1978) estimate seasonal aboveground vascular dry matter production to be around 100 g dry wt m^{-2} yr^{-1}. By calculating the maintenance and growth respiration from Stoner et al. (1978), we estimate the actual vascular production to be 205 g dry wt m^{-2} yr^{-1}. The difference between our estimate and that of Dennis and Tieszen is the amount used for belowground growth. This amount, 105 g dry wt m^{-2} yr^{-1}, agrees with estimates of belowground turnover (Shaver and Billings, 1975).

The pattern clarified by the simulations of the canopy-moss system is straight-forward (Figure 5). When water is amply available, i.e., the water held in the substrate is greater than the potential evaporation, which occurs with frozen subsoils and impeded drainage, interception of precipitation or suppression of evaporation by the vascular canopy is not important to the moss water balance. In this situation the moss production may be limited by either light or nutrients; in the simulation moss production under the canopy was light limited because nutrients were not included. Species with low resistances to the diffusion of water vapor and CO_2 should be favored. As available water decreases, so that water in the substrate is about equal to potential evaporation, either because of improved drainage, low precipitation, or the inability of the moss to raise its water table or conduct water up long thalli, the suppression of evaporation by the canopy becomes increasingly important, especially for species with little internal control over water loss. Because the canopy reduces water loss to about 50% while reducing water gain to about 70%, the effect of the canopy on the moss water balance is favorable. Production of moss without the moderating influence of the canopy may be suppressed by low tissue water contents. As available water decreases further, water from precipitation through the growing season is required, and the interception of precipitation by the canopy overrides the advantage of suppressed evaporation. Species growing in the open are then favored. These habitats are on the dry side of the usual moss habitat, their positions being occupied by lichens.

Table 5. Simulated CO_2 Incorporation (g of CO_2 m^{-2}) by the Vascular and Nonvascular Components for an 80-Day Growing Season

	Substrate water potential (bars)		
	-1	-5	-10
Standard season			
With canopy	597	515	378
Without canopy	28	15	0.2
HOT season			
With canopy	550	450	307
Without canopy	27	-2	-19
COLD season			
With canopy	121	105	85
Without canopy	6	5	0.4

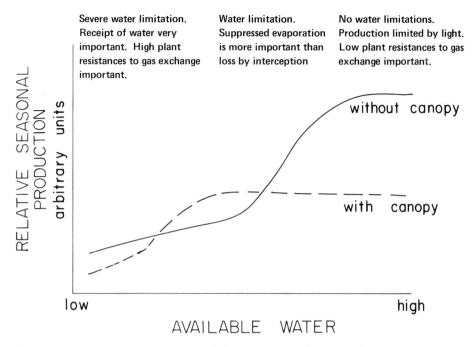

Figure 5.   Diagram of canopy-moss relations along a moisture gradient.

Acknowledgments. This research was supported by the National Science Foundation under Grant GV 29345 to San Diego State University. It was performed under the joint sponsorship of the International Biological Programme and the Office of Polar Programs and was coordinated by the U.S. Tundra Biome. Field and laboratory activities at Barrow were supported by the Naval Arctic Research Laboratory of the Office of Naval Research.

The development of these models was supported by the ERDA and CRREL. We thank Dr. L. L. Tieszen, Dr. F. S. Chapin III, Dr. B. Sveinbjörnsson, Ms. M. Poole, and Mrs. Patsy Miller for their contributions and support in this work.

References

Billings, W. D., and H. A. Mooney. (1968) The ecology of arctic and alpine plants. *Biol. Rev.*, **43**: 481–529.

Bliss, L. C. (1956) A comparison of plant development in microenvironments of arctic and alpine tundras. *Ecol. Monogr.*, **26**: 303–337.

Brown, R. J. E. (1973) Influence of climatic and terrain factors on ground temperatures at three locations in the permafrost region of Canada. *Permafrost: North American Contribution to the Second International Conference*, Washington, D.C.: National Academy of Sciences, pp. 27–34.

Coyne, P. I., and J. J. Kelley. (1975) CO_2 exchange over the Alaskan arctic tundra: Meteorological assessment by an aerodynamic method. *J. Appl. Ecol.*, **12**: 587–611.

Dennis, J. G., L. L. Tieszen, and M. Vetter. (1978) Seasonal dynamics of above- and belowground production of vascular plants at Barrow, Alaska. In *Vegetation and*

Production Ecology of an Alaskan Arctic Tundra (L. L. Tieszen, Ed.). New York: Springer-Verlag, Chap. 4.

Dingman, S. L., and R. F. Koutz. (1974) Relations among vegetation, permafrost, and potential insolation in central Alaska. *Arct. Alp. Res.,* **6**: 37–47.

Drury, W. H. (1956) Bog flats and physiographic processes in the upper Kuskokwim River region, Alaska. *Contrib. Gray Herb. Harv.,* 130 pp.

Duncan, W. C., R. S. Loomis, W. A. Williams, and R. Hanau. (1967) A model for simulating photosynthesis in plant communities. *Hilgardia,* **38**: 181–205.

Leyton, L., E. R. C. Reynolds, and F. B. Thompson. (1967) Rainfall interception in forest and moorland. In *Forest Hydrology* (W. E. Sooper and H. W. Lull, Eds.). Oxford: Pergamon Press. pp. 163–178.

McCown, B. H. (1973) Growth and survival of northern plants at low soil temperatures. *U.S. Army Cold Regions Res. Engin. Lab. Spec. Rep.,* **186**: 1–15.

Matveyeva, N. V. (1971) Dynamics of permafrost thawing in western Taimyr (Dinamika ottaivaniia merzloty v tundrakh zapadnogo Taimyr). In *Biogeocenoses of Taimyr Tundra and Their Productivity,* Vol. 1, pp. 45–56 (Biogeotsenozy Taimyrskoi tundry i ikh productivnost'). Leningrad: Nauka. (International Tundra Biome Translation 6, April 1972; Translator: P. Kuchar, 10 pp.)

Miller, P. C., W. A. Stoner, and L. L. Tieszen. (1976) A model of stand photosynthesis for the wet meadow tundra at Barrow, Alaska. *Ecology,* **57**: 411–430.

Miller, P. C., W. C. Oechel, W. A. Stoner, and B. Sveinbjörnsson. (1978) Simulation of CO_2 uptake and water relations of four arctic bryophytes at Point Barrow, Alaska. *Photosynthetica,* **12**: 7–20.

Nakano, Y., and J. Brown. (1972) Mathematical modeling and validation of the thermal regimes in tundra soils, Barrow, Alaska. *Arct. Alp. Res.,* **4**: 19–38.

Ng, E., and P. C. Miller. (1975) A model of the effect of tundra vegetation on soil temperatures. In *Climate of the Arctic. 24th Alaska Science Conference, Fairbanks, Alaska, August 15–17, 1973* (G. Weller and S. A. Bowling, Eds.). Geophysical Institute, University of Alaska, Fairbanks, Alaska, pp. 222–226.

Ng, E., and P. C. Miller. (1977) Validation of a model of the effects of tundra vegetation on soil temperatures. *Arct. Alp. Res.,* **9**: 89–104.

Oechel, W. C. (1976) Seasonal patterns of temperature response of CO_2 flux and acclimation in arctic mosses growing *in situ. Photosynthetica,* **10**: 447–456.

Oechel, W. C., and N. J. Collins. (1976) Comparative CO_2 exchange patterns in mosses from two tundra habitats at Barrow, Alaska. *Can. J. Bot.,* **54**: 1355–1369.

Peterson, W. L., and J. M. Mayo. (1975) Moisture stress and its effects on photosynthesis in *Dicranum polysetum. Can. J. Bot.,* **53**: 2897–2900.

Price, L. W. (1971) Vegetation, microtopography and depth of active layer on different exposures in subarctic alpine tundra. *Ecology,* **52**: 638–647.

Reynolds, E. R. C., and L. Leyton. (1963) Measurement and significance of throughfall in forest stands. In *The Water Relations of Plants* (A. J. Rutter and F. H. Whitehead, Eds.). New York: J. Wiley and Sons Inc., 394 pp.

Savile, D. B. O. (1972) *Arctic Adaptations in Plants.* Can. Dep. Agric. Monogr. 6, 81 pp.

Shaver, G. R., and W. D. Billings. (1975) Root production and root turnover in a wet tundra ecosystem, Barrow, Alaska. *Ecology,* **56**: 401–409.

Slatyer, R. O. (1962) Methodology of a water balance study conducted on a desert woodland (*Acacia aneura* F. Muell.). In *Symposium on Plant-water Relations in Arid and Semi-arid Conditions.* Paris: Unesco, pp. 15–26.

Slatyer, R. O. (1965) Measurements of precipitation interception by an arid zone plant community (*Acacia aneura* F. Muell.). In *Methodology of Plant Ecophysiology* (F. E. Eckardt, Ed.). Paris: Unesco, pp. 181–192.

Stoner, W. A., and P. C. Miller. (1975) Water relations of plant species in the wet coastal tundra at Barrow, Alaska. *Arct. Alp. Res.,* **7**: 109–124.

Stoner, W. A., P. C. Miller, and L. L. Tieszen. (1978) A model of plant growth and phosphorus allocation for *Dupontia fisheri* in coastal, wet, meadow tundra. In *Vegeta-*

tion and Production Ecology of an Alaskan Arctic Tundra (L. L. Tieszen, Ed.). New York: Springer-Verlag, Chap. 24.

Tieszen, L. L. (1975) CO_2 exchange in the Alaskan Arctic Tundra: Seasonal changes in the rate of photosynthesis of four species. *Photosynthetica,* **9**: 376–390.

Warren Wilson, J. (1965) Stand structure and light penetration. I. Analysis by point quadrats. *J. Appl. Ecol.,* **2**: 383–390.

Webber, P. J. (1978) Spatial and temporal variation of the vegetation and its production, Barrow, Alaska. In *Vegetation and Production Ecology of an Alaskan Arctic Tundra* (L. L. Tieszen, Ed.). New York: Springer-Verlag, Chap. 3.

Section III
Growth and the Allocation and Use of Mineral and Organic Nutrients

Section IIIa: Growth and Organic Nutrient Allocation

"Growth and the Allocation and Use of Mineral and Organic Nutrients" incorporates those aspects of plant function which are perhaps most crucial to an understanding of primary production in tundra. Yet in many ways we know the least about these processes—and these are now most important as an understanding of tundra stability, revegetation, etc. is demanded of tundra scientists. The three chapters in this subsection combine experimental and descriptive approaches to the study of the role of organic nutrients, especially in the belowground parts of the system.

Allessio and Tieszen (Chapter 17) document the ability of tundra species to translocate effectively at ambient temperatures near $0°C$. Even very early in the season when new leaves are strong sinks and when the soil is still frozen, ^{14}C is translocated to new root tips. Furthermore, the greatest extension growth of roots occurs in the first part of the summer. Secondary roots may develop and storage sites may be replenished later in the year in support of the inferences derived from the CO_2 exchange data (Section II, Chapter 10). These studies also initiated our consideration of population studies and revealed a high degree of tiller interdependence which, in combination with a generally high carbohydrate level, provides resistance to occasionally acute grazing pressure.

The detailed studies of Billings et al. (Chapter 18) dramatize the importance of a better understanding of belowground growth, physiology, and ultimately ecology. The three dominant graminoids, although coexisting and similar in photosynthetic responses (Chapter 10), illustrate strikingly different spatial and temporal patterns of growth. *Eriophorum* produces annual roots which penetrate directly to the edge of the thawing active layer. Roots of all species tested grow well at low temperatures and account for most of the ecosystem respiratory flux of CO_2.

McCown's paper (Chapter 19) combines a synoptic analysis of carbohydrate fluxes with an experimental assessment of the effect of nitrogen nutrition. He attempts to distinguish between the effects of low temperature and low nitrogen availability. Carbohydrate levels remain high throughout the seasonal cycle but are reduced somewhat when nitrogen availability increases. Ammonia is the form of nitrogen which *Dupontia* utilizes most effectively. This is also the common form in the anaerobic soils near Barrow. Again, the project points toward the necessity for population studies, since the greatest response to increased fertility or soil warming is not enhanced tiller growth but rather enhanced rhizome growth and increased plant density.

Section IIIb: Inorganic Nutrient Utilization, Fertilization, and Nitrogen Fixation

Arctic systems are often described as cold-dominated and nutrient-poor systems. Yet the nature and degree of the nutrient limitation is unclear. This subsection summarizes some of our attempts to systematically document plant nutrient status as well as seasonal allocation patterns. In addition, we explored adaptations at the level of nutrient uptake in an attempt to provide a quantitative basis for incorporating mineral nutrients into our plant production model. The response of Barrow tundra to increased nutrients is still dramatic in the experiments started by Schultz. The fertilization experiments initiated by McKendrick were set up to document more direct and physiological responses.

Ulrich and Gersper (Chapter 20) systematically surveyed tissues from a number of tundra plants to reveal the extent of mineral limitation. They describe the tundra as a "nutritional desert" and indicate that nitrogen is most often limiting but that it is closely followed by phosphorus. Among the many factors which limit production at Barrow, nutrients are among the most significant. This is of special interest since photosynthesis was not severely limited by leaf N or P concentration (Chapter 10). Thus, the control by mineral nutrients is exercised mainly at the level of growth and allocation (Chapter 19).

The low soil nutrient contents and the low temperatures place severe restrictions on normal uptake processes. Chapin's (Chapter 21) project addressed this problem directly. Tundra plants reveal a suite of adaptations. Their structure represents a compromise between expanded surface area and the need to transport oxygen in the anaerobic soils. Physiological and genetic adaptations are also significant for an exploitation of the cold, nutrient-poor substrate.

The field fertilization studies (Chapter 22) reinforce the tissue analyses of Ulrich and Gersper (Chapter 20) by indicating that nitrogen and phosphorus are clearly limiting primary production. The added nutrients stimulated growth and reduced reserve carbohydrates. The nutrient paucity of our system demands a thorough understanding of nutrient inputs, recycling, and losses. The paper by Alexander et al. (Chapter 23) documents the biotic and environmental controls over nitrogen fixation. Fixation is the dominant input of nitrogen to the system and is mainly a product of lichen and blue–green algae–moss activity.

Section IIIc: Growth, Nutrients, and Population Modeling

Our studies of primary production were guided and integrated by a mechanistic modeling approach which emphasized physiological processes. The first version of the Stand Photosynthesis model was near completion at the initiation of the field research and was published in 1972. Modifications and extensions of these simulation studies appeared in 1976 and are developed in the summary (Chapter 27). Models of plant water relations (Chapter 14 and 16) and canopy and soil thermal properties were also essential to our early studies. Near the termination of the field research it was apparent that we needed thorough and integrating frameworks to understand the role of mineral nutrients in the system, the control of plant growth, and finally the importance of reproduction strategies and population dynamics. These areas became still more crucial as we attempted to extend our understanding of growth and photosynthesis to problems of ecosystem change and development. The first chapter (24) in this section reviews our attempts to integrate our understanding of tiller growth in *Dupontia* with an incorporation of phosphorus effects. It illustrates the fact that we still need to develop a better conceptualization of this species as well as a stronger nutrient data base. This is reinforced by the more detailed (Chapter 25) model of carbohydrate, nitrogen, and phosphorus. We can now limit growth and primary production by either N or P but need a better understanding of internal physiological and biochemical controls.

The final chapter (26) emphasizes an approach to our project which was largely ignored until late in the program. The general field of plant population dynamics is now sufficiently advanced that we should be able to progress rapidly with this approach in tundra studies. This is especially important because of the high degree of vegetative reproduction in our system, the frequent periods of acute grazing pressure, and the importance of understanding revegetation processes in a system sensitive to a variety of perturbations.

17. Translocation and Allocation of ^{14}C-Photoassimilate by *Dupontia fisheri*

M. L. ALLESSIO AND L. L. TIESZEN

Introduction

A few recent studies with non-graminoid species have begun to elucidate the carbon allocation picture of tundra species. For *Polytrichum alpinum* Collins and Oechel (1974) found preferential translocation of ^{14}C-photosynthate to new belowground shoots arising from the extensive belowground stem system. In *Andromeda polifolia* translocation was to all living underground parts, especially rapidly growing rhizomes; but there was no translocation from older leaves to younger, apparently self-sufficient leaves (Johansson, 1974). Autoradiography of *Dryas integrifolia,* used to determine viability of roots, showed intensive localization in places where mycorrhizal nodules occurred (Svoboda and Bliss, 1974). In addition, Johansson (1974) showed that *Andromeda polifolia* and *Rubus chamaemorus* have different seasonal allocation patterns which are attributed to their evergreen or deciduous habit.

These studies, however, do not contribute to understanding allocation in graminoids which make up nearly 80% of the vascular plant production at Barrow, Alaska (Tieszen, 1972b). There is no information on allocation relationships between plant parts (e.g., tillers or tiller compartments), seasonal changes in allocation pattern, or effects of leaf or tiller age. During the course of our field studies with *Dupontia fisheri,* we were interested in information concerning loss of ^{14}C-photosynthate from treated leaf segments and the pattern of ^{14}C distribution and allocation to various tillers and tiller compartments as a function of the following: (1) Season—shoots were either treated at various times during the season or early in the growing season and subsampled; (2) Leaf age—three age classes (Leaf 2—senescing, Leaf 3—fully expanded, Leaf 4—young and folded) were monitored; (3) Tiller age—both short-term (1–24 hr) and long-term (1–26 days) patterns were determined.

Materials and Methods

Shoots of *Dupontia fisheri* (Figure 1, Table 1) at the Barrow, Alaska, intensive site 2 were exposed to ^{14}CO$_2$ in small transparent plastic chambers (Allessio and Tieszen, 1975b). A sodium carbonate solution of 1 μCi ml^{-1} provided the ^{14}CO$_2$

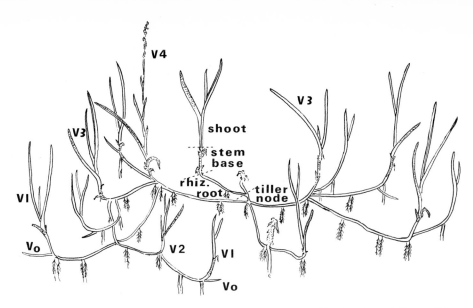

Figure 1. The *Dupontia fisheri* tillering system. Tiller classification is as follows: V0, a tiller in its first season of growth consisting only of rhizome (at Barrow, only frequently does a tiller exsert a leaf in its first season); V1, a tiller with exserted leaves in its first season of shoot production; V2, a tiller in its second year of aboveground growth; V3 and V4, tillers in their third and fourth year of growth aboveground. Flowering normally takes place in the fourth (V3) or fifth (V4) season of tiller growth. Tiller compartments are also shown. (From Allessio and Tieszen, 1975a, with the permission of the *American Journal of Botany*.)

Table 1. Summary of Field Labeling, Tundra IBP Site 2, Barrow, Alaska, Summer 1972[a]

Experiment	Labeling date	Collection date	Interval	Exposure time (hr)	$^{14}CO_2$[b] (μCi)
Seasonal	a. 7 June	8 June	1 day	0.5	1
sequence	13 June	14 June	1 day	0.5	1
	21 June	23 June	2 days	1	1
	27 June	29 June	2 days	1	1
	12 July	14 July	2 days	1	1
	27 July	29 July	2 days	1	1
	b. 27 June	29 June	2 days	1	1
		14 July	17 days	1	1
		29 July	32 days	1	1
		15 August	53 days	1	1
Tiller age	3 July	3 July	1 hr	1	1
short		3 July	6 hr	1	1
term		4 July	24 hr	1	1
(3 ages)					
Tiller age	10 July	11 July	1 day	1.5	2
long		15 July	5 days	1.5	2
term		25 July	15 days	1.5	2
(3 ages)		7 August	26 days	1.5	2
Leaf age	17 July	18 July	1 day	3 min	[b]
(3 ages)					

[a] Each collection sample consisted of five *Dupontia fisheri* plants.
[b] 2.818μCi $\cdot$ m/mole^{-1} CO_2 (flow rate: 1 ℓ $\cdot$ hr^{-1}).

upon acidification (Lovell et al., 1972; Marshall and Sagar, 1968b). Plants were harvested after varying periods of time ranging from 1 hr to 53 days (Table 1), and were removed from the soil by hand except for those collected in early June when it was necessary to use a metal soil corer because of the frozen soil. They were kept on ice until washed and pressed and then dried at about 90°C. Autoradiographs were made (5 days exposure), and the plants were prepared for liquid scintillation counting (Allessio and Tieszen, 1975a) by a procedure modified from Davidson et al. (1970) and Tieszen et al. (1974b).

The loading of 1-cm segments of a leaf to study ^{14}C loss from treated portions was accomplished using the Plexiglas chamber designed by Tieszen et al. (1974b; Allessio and Tieszen, 1975a). Monitoring of ^{14}C in the treated segment was with a General Electric Nucle-eye battery-operated counter (Tieszen et al., 1974a; Allessio and Tieszen, 1975a).

Results

Translocation Rate

The loss of radioactivity from leaves of various ages has been described by Allessio and Tieszen (1975a). The greatest rate of loss occurred during the first 2 hr; and after 24 hr, the amount remaining was 15 to 25% of original values. The youngest leaf appears to retain slightly more, but this may be a function of the relative thinness of the leaf rather than reduced translocation. The pattern of loss obtained at various times during the 1971 growing season (Tieszen et al., 1974a) was similar to that described above. Little additional loss occurred by 48 hr after treatment. The loss of radiocarbon from the treated leaf segment showed some variability, with the rate being slowest at the beginning and end of the season. The rate of ^{14}C removal appeared relatively independent of ambient temperatures and irradiances (Table 2). Export from a treated leaf was rapid even when mean daily temperatures were low (0.8 to 5.8°C). Although these data indicate that substantial amounts of radiocarbon were translocated in 24 hr, both the autoradiographs and quantitative data (Table 3) indicate high retention of radioactivity in

Table 2. Estimated Time Required for Removal of 50% of Initial Photosynthate during the 1971 Season for *Dupontia fisheri*[a]

Date	Time (hr) for 50% removal	Mean temperature (°C)	Total daily radiation (cal · cm^{-2})
18 and 24 June	11.2	about 2	—
30 June	4.8	3.5	677
13 July	8.5	2.4	360
14 July	5.6	1.1	259
21 July	8.0	5.8	590
24 July	4.8	5.0	490
7 August	12.5	0.8	418

[a] From Tieszen et al. (1974), reprinted with the permission of the National Research Council of Canada. *Can. J. Bot.,* **52**.. 2189–2193.

Table 3. Distribution of ^{14}C in Various Tillers and Tiller Compartments as Allocated from Various Age Leaves 24 hr after Treatment[a]

	Leaf 2 (senescing)		Leaf 3 (mature)		Leaf 4 (young)	
	(dpm)	(%)	(dpm)	(%)	(dpm)	(%)
Treated tiller	109134 ± 22126		96568 ± 11750		102334 ± 33845	
Treated (cm)	41170 ± 11349	37.7	47917 ± 5603	49.6	42186 ± 13251	41.2
Leaf 1	873 ± 228	0.8	560 ± 348	0.6	202 ± 96	0.2
Leaf 2	19598 ± 3481	18.0	1042 ± 417	1.1	1374 ± 478	1.4
Leaf 3	2276 ± 414	2.1	16967 ± 3926	17.6	2134 ± 456	2.2
Leaf 4	13573 ± 2656	12.4	1476 ± 327	1.5	29396 ± 7634	28.7
Leaf 5			7311 ± 3142	7.6	3295 ± 3215	3.2
Stem base	17008 ± 6959	15.6	10950 ± 3052	11.3	14559 ± 7374	14.2
Rhizome	11304 ± 3855	10.4	4912 ± 1409	5.1	3746 ± 1504	3.7
Roots	3333 ± 907	3.1	5432 ± 2010	5.6	5439 ± 4866	5.3
Nontreated	22337 ± 16225				25491 ± 21522	
Shoot	5389 ± 3957	23.1			7625 ± 7184	29.9
Stem base	5970 ± 5366	25.6			8315 ± 7844	32.6
Rhizone	6292 ± 1216	27.0			5046 ± 3001	19.8
Roots	5689 ± 5686	24.4			4506 ± 3494	17.7

[a] Radioactivity is presented as disintegrations per minute (dpm) and the amount in a tiller compartment also as percentage of the tiller total. Five replicates were used and the standard error is indicated. From Allessio and Tieszen (1975a), reprinted with the permission of *Arct. Alp. Res.*

the treated leaf segments (Allessio and Tieszen, 1975a). The treated centimeter contained more than 35% of the total radioactivity recovered from the tiller on which the treated leaf occurred.

Distribution of Radiocarbon

Seasonal Sequence. In order to determine the influence of season on the pattern of translocation, *Dupontia fisheri* tillers were treated from 7 June to 27 July 1972 (Table 1). Seasonal observations are based on results obtained from V1 or young V2 tillers (see Figure 1). It was observed that once the soil thawed, translocation of ^{14}C-photoassimilate from the treated tiller occurred. However, although nontreated tillers contained radiocarbon, the treated tiller, especially shoot and stem base parts and later in the season the rhizome, was the most densely labeled. Sites of accumulation were root and rhizome primordia and actively growing parts both on the treated tiller and on adjacent nontreated tillers (Allessio and Tieszen, 1975b).

In an effort to determine changes in the ^{14}C pattern during the season, plants were treated on 27 June and subsampled at four times. The pattern shown by plants collected after 48 hr on 29 June recurred at the later samplings. After 48 hr, much radiocarbon remained in the treated tiller. Again, translocated ^{14}C accumulated in actively growing areas, newly exserted shoots, V0 rhizomes, tips of the current year's roots, and lateral root primordia. Nontreated shoots and rhizomes typically did not contain much radiocarbon. Even after 6 weeks (15 August), the ^{14}C distribution was similar to that at 2 and 4 weeks. Redistribution was again indicated because some radiocarbon was found in newly formed root, leaf, and V0 rhizome tissue.

Plants collected more than 2 days after treatment showed some redistribution of radiocarbon. The portion of root which was actively growing at the time of treatment appeared intensely labeled and could be used as a marker; that produced subsequently was less intensely labeled. Because of this, it was possible to make a rough estimate of root and leaf growth from autoradiographs (Table 4; Allessio and Tieszen, 1975b).

Effect of Leaf Age.   Little translocation was observed between a treated leaf and other leaves on a tiller, and only the youngest leaf was consistently labeled. After 24 hr, radiocarbon fixed by Leaf 2—senescing, Leaf 3—fully expanded, and Leaf 4—young folded was distributed below ground in the pattern mentioned above. Radiocarbon was more apparent in younger tillers, accumulation was in actively growing roots and rhizomes, and there was no indication of much translocation to the aboveground parts of nontreated tillers, including flowering tillers (Allessio and Tieszen, 1975a).

Effect of Tiller Age.   Results from the short-term, 24-hr study suggested some differences between V1, V2, and V3 tillers. Young treated shoots translocated little radiocarbon to roots and rhizomes within the first hour after treatment. Even after 1 hr, young root tips and V0 rhizomes were beginning to accumulate ^{14}C-photoassimilate. After 6 hr, root tips were heavily labeled, and although some translocation to V2 tillers was evident, little radiocarbon was present beyond the tiller node.

Table 4.   Root and Leaf Growth as Determined from Autoradiographs[a]

Time interval (days)	Date labeled (collected)	Number	Range in length (cm)	Average length (cm)	Days	Growth per day during interval[b] (cm)
	Roots					
5	10 July (15)	11	1.0–2.0	1.5 ± 0.3	3	0.5
15	10 July (25)	6	2.5–6.5	4.8 ± 1.5	13	0.3
26	10 July (7 August)	9	2.8–7.5	5.8 ± 1.3	11	0.1
17	27 June (14 July)	5	3.0–5.3	4.2 ± 0.9	15	0.3
32	27 June (27 July)	4	5.7–7.5	6.6 ± 0.8	15	0.2
49	27 June (14 August)	1	—	6.0	47	0.1
	Leaves					
17	27 June (14 July)	6	3.0–5.0	4.2 ± 0.8	15	0.3
32	27 June (29 July)	7	5.5–9.0	7.1 ± 1.6	15	0.19
	new leaf	1		8.0		
49	27 June (14 August)	5	5.0–7.5	6.5 ± 2.5	47	0.16
	new leaf	3	3.5–8.5	6.2		

[a]   From Allesio and Tieszen (1975b), reprinted with the permission of the *Amer. J. Bot.*
[b]   This does not include the first 48 hr, during which autoradiograph zonation was not observed.

Translocation from older shoots was more extensive than from the young V1 tillers, especially to rhizomes of younger tillers. V1 tillers tended to retain radiocarbon. Rhizomes of the treated V2 and V3 tillers were usually less well labeled than those of the treated V1 tiller. The pattern appeared complete in 6 hr. Nevertheless, additional ^{14}C was translocated belowground as evidenced by more heavily labeled roots after 24 hr (Allessio and Tieszen, 1975b).

Long-term tiller age comparisons were made with V1, V2, and V4 (flowering) tillers treated on 10 July and collected at four dates after treatment. All of a V1 or V2 tiller was labeled after 24 hr, but little was translocated to adjacent tillers. A flowering tiller, however, was intensely labeled only in the shoot and stem base areas or on the occasional new root. Some translocate could be detected in the daughter tiller rhizome. Roots produced the previous season and those which may have been several years old on the treated V4 were also labeled. In 24 hr, small amounts of labeling appeared in rhizomes 30 cm from the treated shoot. Translocation was more extensive than observed previously, possibly because of the additional amount of ^{14}C incorporation and the age of tillers. Also, older tillers had a greater photosynthetic surface.

At 5 days after treatment, the treated tiller was most heavily labeled. V1 tillers translocated to V0 rhizomes and new roots, and very little went to other tillers. The zonation of labeling, mentioned earlier for the summer sequence series, was also present. Again, photosynthate provided by older tillers was translocated extensively, especially to younger rhizomes. The pattern of ^{14}C distribution at 5, 15, and 26 days reflected the 24-hr pattern. Although flowering tillers tended to appear independent, they were capable of supplying photosynthetically incorporated material to other actively growing portions of the plant (see Allessio and Tieszen, 1975b).

Growth. Table 4 presents the root and leaf growth data obtained from autoradiographs. Since some radiocarbon occurred in areas determined to have grown since the time of treatment, some redistribution of ^{14}C took place. This was most evident in the roots. The greatest extension root growth of *Dupontia fisheri* appeared to occur early in the season, so that by the end of July there was little further growth. This is based on both field observations and autoradiographic evidence. During this study it was observed that lateral roots were initiated in late June on roots which had been produced the year before. During the period of our observations, 1 June to 15 August, no lateral roots were produced on new (current season) roots.

Quantitative Allocation of ^{14}C-Photoassimilate

In all samples a very high percentage of the radiocarbon recovered was in the treated tiller or leaf segment except when a young V0 tiller was present. Although radiocarbon was recovered from both older and younger tillers connected to that which was treated, more was translocated to the younger.

Seasonal Sequence. The shift toward the end of the *Dupontia* growing season in the relative importance of the stem base can be seen in Figure 2A. Such

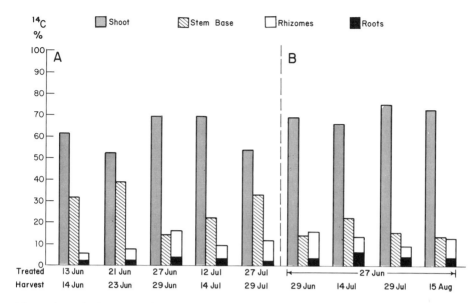

Figure 2. Distribution of ^{14}C (% dpm) in *Dupontia fisheri* tillers treated and collected at various times during the summer. (A) Tillers treated and collected at five dates during 1972. (B) Tillers treated on 27 June 1972 and subsampled at various dates. Harvest times are indicated.

a shift was not present in those plants treated on 27 June and subsampled (Figure 2B); the proportion of ^{14}C remaining in a tiller compartment appeared to become stabilized within 24 hr after treatment. The actual distribution of ^{14}C (dpm) in tiller compartments (Tables 5 and 6), and as represented by the total ^{14}C in treated tillers (Figure 3), was quite different since percentages do not reflect the actual amounts of ^{14}C incorporated. Similar trends were observed when the amount of ^{14}C was considered in terms of weight (dpm g dry wt^{-1}), so that the increased amount of radioactivity (Figure 3A) was not a function of increased photosynthetic surface, but of greater photosynthetic capacity (Tieszen, 1975). Consideration of sample variability (SE) suggests that although there was considerable variability between plants in the amount of ^{14}C incorporated, the proportion of that allocated to a given compartment was fairly constant.

During the summer there was increased incorporation at least through mid-July (Figure 3A). Although autoradiographs sometimes demonstrated extensive accumulation in roots and V0 rhizomes, usually only small concentrated amounts of radiocarbon (dpm) were involved. Translocation to nontreated tillers (Tables 5 and 6) also involved small percentages of the total radioactivity recovered.

The ^{14}C distribution ratios (Table 7) indicate that the proportion of ^{14}C apportioned above- and belowground remained relatively constant throughout the growing season with nonphotosynthetic compartments (stem base, rhizome, and roots) having less radiocarbon than the treated photosynthetic shoot. Except for 27 June (29 June), the stem base incorporated more radiocarbon than the roots and/or rhizomes. The stem base: root plus rhizome ratios were high early in the season, at the time when new shoot growth was initiated. Treated tiller ratios

Table 5. Distribution of ^{14}C in Treated and Nontreated Tillers Treated at Various Times during the Season[a]

	13 June (14 June)		21 June (23 June)		27 June (29 June)		12 July (14 July)		27 July (29 July)	
	(dpm)	(%)	(dpm)	(%)	(dpm)	(%)	(dpm)	(%)	(dpm)	(%)
Treated	18794 ± 8119		29783 ± 8965		68982 ± 10786		260004 ± 53418		19228 ± 2726	
Shoot	11606 ± 6575	61.7	15853 ± 5056	53.2	47710 ± 9026	69.1	178202 ± 39710	68.5	10402 ± 1945	54.1
Stem base	6040 ± 2429	32.1	11628 ± 3744	39.0	9955 ± 2259	14.4	58197 ± 14248	22.3	6445 ± 554	33.5
Rhizome	655 ± 242	3.4	1567 ± 656	5.2	8789 ± 2086	12.7	14884 ± 3979	5.7	1949 ± 370	10.1
Roots	493 ± 167	2.6	735 ± 382	2.4	2528 ± 756	3.6	8722 ± 3470	3.3	430 ± 201	2.2
Nontreated V3										
Shoot						20.1				
Stem base						25.3				
Rhizome						49.9				
Roots						4.5				
Nontreated V2							4182 ± 2764		1067 ± 813	
Shoot							804 ± 443	19.2	38 ± 21	3.5
Stem base							576 ± 238	13.7	257 ± 219	24.0
Rhizome							2152 ± 1754	51.4	709 ± 579	66.4
Roots							649 ± 574	15.5	62 ± 36	5.8
Nontreated V1					3123 ± 1692		7083 ± 3707			
Shoot					579 ± 252	18.5	1595 ± 1202	22.5		
Stem base					687 ± 311	22.0	629 ± 257	8.8		
Rhizome					1605 ± 1405	51.3	4233 ± 2327	59.7		
Roots					251 ± 146	8.0	618 ± 207	8.7		

[a] The harvest date is in parentheses. Radioactivity is presented as disintegrations per minute (dpm) and the amount in a tiller compartment as a percentage in the tiller total. Five replicates were used and the standard error is indicated.

Table 6. Distribution of ^{14}C (dpm) in Tillers Treated 27 June and Subsampled during the Summer[a]

	27 June (29 June)		27 June (14 July)		27 June (29 July)		27 June (15 August)	
	(dpm)	(%)	(dpm)	(%)	(dpm)	(%)	(dpm)	(%)
Treated	68982 ± 10786		107260 ± 31379		76865 ± 16689		207837 ± 54548	
Shoot	47710 ± 9026	69.3	68995 ± 22738	64.3	57787 ± 12719	75.2	152766 ± 33611	73.5
Stem base	9955 ± 2259	14.4	23673 ± 6701	22.1	12171 ± 3791	15.8	27570 ± 9377	13.3
Rhizome	8789 ± 2086	12.7	7442 ± 2057	6.9	3565 ± 1069	4.6	19209 ± 11335	9.2
Roots	2528 ± 756	3.7	7150 ± 2036	6.7	3341 ± 1518	4.4	8292 ± 3945	4.0
Nontreated V3					1377 ± 550		3710 ± 1354	
Shoot		20.1			920 ± 460	66.8	2718 ± 1114	73.3
Stem base		25.3			233 ± 93	16.9	331 + 99	8.9
Rhizome		50.0			195 ± 112	14.2	580 ± 108	15.6
Roots		4.6			29 ± 19	2.1	80 ± 41	2.2
Nontreated V2								
Shoot								
Stem base						67.8		
Rhizome		31.8				24.3		
Roots		68.2				7.9		
Nontreated V1	3123 ± 1692		37428 ± 8009				15518 ± 6614	
Shoot	579 ± 252	18.5	13435 ± 6887	35.9			4089 ± 3359	26.4
Stem base	687 ± 311	22.0	7130 ± 7130	19.1			1417 ± 720	9.1
Rhizome	1605 ± 1405	51.4	15182 ± 8210	40.6			4213 ± 3721	27.2
Roots	251 ± 146	8.0	1681 ± 42	4.5			5801 ± 5532	37.4

[a] The harvest date is indicated in parentheses. Radioactivity is presented as disintegrations per minute (dpm) and the amount in a tiller compartment also as a percentage of the total. Five replicates were used and the standard error is indicated.

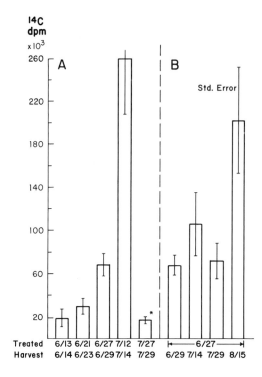

Figure 3.   [14]C (dpm) in tillers treated during summer 1972. (A) Tillers harvested 24 or 48 hr after treatment. (B) Tillers subsampled at various times after treatment. Harvest times are indicated. Standard error is indicated. (*Low incorporation can probably be attributed to technical problems.)

remained fairly constant regardless of whether tillers were treated throughout the season or treated early and subsampled (Table 7). In many cases there were considerable differences in the ^{14}C distribution ratios of treated and nontreated tiller compartments. The ratio of nonphotosynthetic:photosynthetic radioactivity was almost always greater in the nontreated tiller. This reflects the autoradiograph patterns described earlier; e.g., that little translocation occurred in shoot portions of nontreated tillers. Our study of the seasonal course of the distribution and fate of ^{14}C-photoassimilate indicates, both from analyses of autoradiographs and specific activity data, that a greater portion of the radioactivity was recovered in the treated shoot than in any other compartment.

Leaf Age.   Rate of ^{14}C loss from treated leaf segments did not change during the growing season (Tieszen et al., 1974a). However, leaf age appeared to influence rate of loss since the youngest, folded and still expanding leaf showed the slowest initial rate. After 24 hr, the proportion of [14]C (15 to 25%) remaining in a senescing leaf, fully expanded leaf, and folded young leaf was similar (Allessio and Tieszen, 1975a).

All leaf age classes translocated to belowground compartments and to other tillers (Table 3). The nontreated portion of the treated leaf contained more radiocarbon than other nontreated leaves on a shoot. Leaf 2, a senescing leaf, translocated more to belowground compartments than did the younger leaves 3 or 4. Distribution in the treated tiller was similar regardless of the age of the

Table 7. ^{14}C Distribution Ratios in Treated and Nontreated Tillers Treated on the Various Dates during the Summer and Harvested on the Dates Indicated in Parentheses

Date	Root:shoot ratio		Nonphotosynthetic: photosynthetic ratio		Stem base: root and rhizome ratio	
	Treated tiller	Nontreated tiller	Treated tiller	Nontreated tiller	Treated tiller	Nontreated tiller
13 June (14 June)	0.04		0.62		5.26	
21 June (23 June)	0.05		0.88		5.05	
27 June (29 June)	0.05	0.43	0.45	4.39	0.88	0.37
12 July (14 July)	0.05	V1 0.39	0.46	3.46	2.47	0.13
		V2 0.81		4.20		0.21
27 July (29 July)	0.04	1.63	0.85	27.05	2.71	0.33
27 June (29 June)	0.05	0.43	0.45	4.39	0.88	0.37
27 June (14 July)	0.10	0.13	0.56	1.79	1.62	0.42
27 June (29 July)	0.06	0.03	0.33	0.50	1.76	1.04
27 June (15 August)	0.05	1.42	0.36	2.80	1.03	0.14

treated leaf. In the nontreated tillers considerably more ^{14}C was recovered in photosynthetic portions than was seen previously (Tables 5 and 6). This may be due to the fact that the plants for the leaf age study were from an atypical site 2 area (plants were actively invading a moss mat and extensive vegetative growth was occurring), and under conditions of great vegetative growth greater support of the invading tiller shoot and stem base occurs.

Tiller Age.   During the first 24 hr (Figure 4), an increasing proportion of radiocarbon was found in belowground compartments and nonphotosynthetic

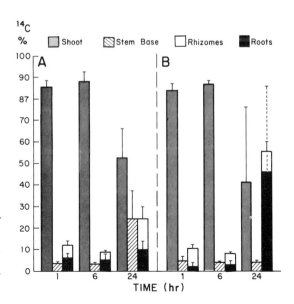

Figure 4.   Distribution of ^{14}C (% dpm) in compartments of treated V1 (A) and V3 (B) tillers. Tillers were treated on 2 July 1972 and harvested at the times indicated. Standard error is indicated.

parts. In contrast, the seasonal data (Figure 2B) demonstrated little change. It appears that when plants were treated early in the season, the distribution ratio was stabilized, while those treated during the first half of July, when photosynthesis was high and active growth was occurring, showed substantial redistribution of incorporated radiocarbon.

In contrast to the short-term (24-hr) (Figure 4A) ^{14}C allocation study, the long-term study (Figure 5A) of V1 tillers indicated a change in pattern, although in both cases the pattern at 24 hr was similar. Extensive redistribution to the rhizome occurred in mid-July, presumably at the expense of what was originally in the stem base. Since the rhizome did not retain the high proportions of radioactivity, it was not stored there, but probably used later for growth.

Over the long term, the V4 (flowering) tiller did not support, to any sizable extent, the nonphotosynthetic compartments (Figure 5B). Some material appeared to move from the shoot to the inflorescence in the first 5 days, and then the proportion recovered was stabilized. However, examination of dpm data shows that the radiocarbon recovered from the treated V4 (flowering) tillers on 7 August was only 3.7% of that recovered on 11 July. A similar decrease was not observed with V1 tillers.

The effect of tiller age on distribution to other tillers is depicted in Figure 6. The relatively greater proportion in the V1 shoot fed by the treated V4 tiller (Figure 6C) was most likely due to the fact that the V1 tillers were very young and actively growing and the translocate was used for shoot production. Figure 7 illustrates the allocation from a treated V1 and a flowering tiller.

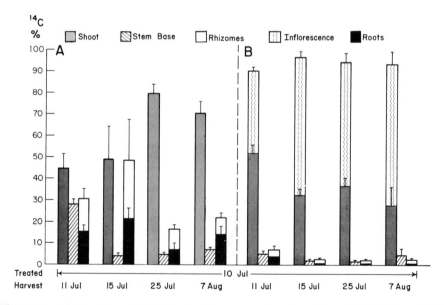

Figure 5. Distribution of ^{14}C (% dpm) in compartments of treated V1 (A) and V4 (flowering) (B) tillers. Tillers were treated on 10 July 1972. Harvest times are indicated.

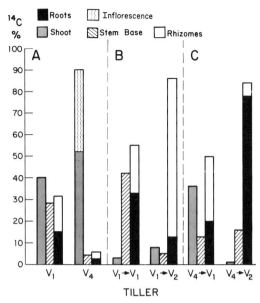

Figure 6. Distribution of ^{14}C (% dpm) in tiller compartments of treated V1 and V4 (flowering) tillers (A) and nontreated V1 and V2 tillers fed by the treated V1 (B) and V4 (flowering) tillers (C). Plants were treated on 10 July and harvested 11 July (24 hr).

Discussion

Seasonal

Our study of the seasonal course of the distribution and fate of ^{14}C-photosynthate indicates, both from analyses of autoradiographs and specific activity data, that a greater portion of the radioactivity is recovered from the treated shoot than from any other compartment. The pattern of allocation in *Dupontia fisheri* differs from those reported by Johansson (1974) for plants which were treated at various times during the season and collected 3 weeks later. In the evergreen shrub, *Andromeda polifolia,* belowground and aboveground parts varied in importance; more than 75% of recovered radioactivity was found in rhizome and roots in May, more than 25% in July, and about 50% in late August. The converse was reported for aboveground radioactivity. The deciduous shrub, *Rubus chamaemorus,* had a still different pattern. The aboveground compartment contained more than 90% of recovered radioactivity in mid-June, 55% in July and August, and 25% at the end of August. Belowground parts started the season with low incorporation but contained nearly 75% at the end of the season. The continuous production of leaves by *Dupontia fisheri* accounts for the importance of the shoot in retaining radiocarbon and the difference in seasonal pattern observed between these life forms.

Both field observations during 1972 and autoradiographic evidence (Table 4) show that the greatest extension root growth of *Dupontia fisheri* occurred early in the season. By the end of July little additional growth occurred. Based on our autoradiographic evidence and field observations, lateral roots were produced on V1 tiller roots which had been produced the year before. No lateral roots occurred on new (current season) roots between 1 June and 15 August. However, Shaver and Billings (1975) report that 90% of the lateral roots were produced by new roots, albeit late in the season. Given the warm 1972 season, lateral roots were produced after our stated observation period, or perhaps the rootbox micro-environment was sufficiently different to alter root phenology. Third, *Dupontia fisheri* may be opportunistic in its belowground growth as various tundra plants are in aboveground growth, i.e., growth is arrested at various stages of development and then continues when conditions are again favorable (Sørenson, 1941; Bliss, 1971).

During the growing season new roots appeared densely labeled, but the actual amount of radioactivity was quite small. Roots of older tillers contained radiocarbon indicating they were alive and functional. Tillers in which the shoot portion had died (V5) after flowering had radiocarbon in both roots and rhizomes. Viability of V4 and V5 tillers has also been noted by Shaver and Billings (1975) and their estimated age of up to 6 years correlates with belowground turnover rates estimated by Dennis and Johnson (1970). Our suggestion is that these living rhizomes and roots, which may acquire a fairly high percentage (2.5%) of recovered radiocarbon, may enable the plant to continue to exploit the active layer for nutrients and perhaps serve as important reserves in time of carbon stress, e.g., defoliation. We hypothesize that recovery, however, is not at the expense of root carbohydrate reserve.

It has been proposed that the fairly constant belowground biomass observed through the season by Tieszen (1972b) is possible because vegetative growth and summer blooming may depend on photosynthate of the current year (Mooney and Billings, 1960) or that the stem base serves for reserve storage and a source for new growth (Mooney and Billings, 1960; Hodgkinson, 1969). The second view is supported by McCown (1978) and Tieszen (1975, 1978). They suggest that stem base tissue may act as a storage area for carbohydrates, especially during periods of active translocation between root and leaf compartments. And although autoradiographic evidence does not indicate that the stem base accumulates large stores of radiocarbon during periods of active growth, quantitative analysis shows the stem base to be important early in the season (Figure 2A) with a decrease in mid-summer and an increase at the end of July.

Further support for the hypothesis that vegetative growth does not occur solely at the expense of reserves in rhizomes or roots comes from the following. First, translocation of newly formed (labeled) photosynthate to sites of active growth occurred even while the soil was thawing (14 June). Also, autoradiographs showing early- and late-season ^{32}P distribution in *Dupontia fisheri* indicate no accumulation early in the season, but concentration in the stem base area late in the season (Chapin, personal communication). This late-season storage would be available for growth in the spring. Utilization of stem base reserves and the current season's photosynthate would be an advantage for a plant in a regime where the aboveground environment becomes favorable at an earlier date than belowground. Localization of stored reserves in the stem base which could be used for early shoot growth would be ecologically advantageous.

Materials then would be readily available close to the point of growth, and translocation from rhizomes and roots over any distance in soil at 0°C would not be necessary; shoot growth could begin with snow melt. Tieszen (1974) points out that most Barrow plants do not photosynthesize to any extent until after snow melt and leaf exsertion. Accordingly, the ability to initiate rapid growth, as would be possible with a readily available reserve, would allow *Dupontia fisheri* to compete favorably for space and nutrients, and thus could be responsible for the high productivity and growth rates recorded by Tieszen (1972a). However, if stem base (and rhizome?) storage occurs and spring shoot regrowth necessitates translocation, a number of interesting problems arise in light of observed translocation to rhizome (root) primordia on 13 June (14 June) when the soil was frozen. This would appear to be contrary to accepted mass-flow translocation hypotheses (Crafts and Crisp, 1971). Also of note is the fact that translocation belowground occurs at very low soil temperatures (0°C). Various reports of the influence of temperature on translocation show considerable variability in low temperature effects from temporary inhibition to complete cessation of transport and diversion to other sinks (Crafts and Crisp, 1971).

Permafrost is not considered to be a primary limiting factor for the Barrow vegetation (Dennis and Johnson, 1970), although some temperature translocation studies indicate that root temperature does influence rate of accumulation and transport (Fujiwara and Suzuki, 1961; Rovira and Bowen, 1973). Low root temperature, causing slow basipetal flow, could result in accumulation in the stem and little or no loss from roots (Brouwer and Levi, 1969). Our stem base:rhizome and root ratios tend to support this latter suggestion. However, the possibility exists that the stem base during the early season, coincident with the period of greatest meristem differentiation, serves as an efficient sink for photoassimilates.

Leaf Age

According to Allessio and Tieszen (1975a), interpretation of the effect of leaf age on the rate of translocation may involve factors within the leaf, as well as assimilation and growth of other parts of the plant (Thrower, 1967; Wardlaw, 1968; Grace and Woolhouse, 1973). Photosynthetic efficiency has been shown to be affected by leaf age (Jewiss and Woledge, 1967; Larson and Gordon, 1969; Felippe and Dale, 1972; Tieszen, 1974; Tieszen and Johnson, 1975) and could determine the amount of material available for translocation. However, Nelson (1963) indicated that the amount of photosynthate does not increase the amount translocated since increased sucrose is stored in the leaf. This, incidentally, is not seen in our data (Table 3).

Our data show similar distribution patterns regardless of leaf age; Leaf 2 (the oldest) translocated a slightly higher percentage of ^{14}C to nonphotosynthetic tissue. Other investigations (Doodson et al., 1964; Wolf, 1967; Yamamoto, 1967;

Felippe and Dale, 1972; and others) have shown that leaf age affects translocation.

With *Dupontia fisheri* under arctic summer conditions, the general pattern of transport to belowground compartments from leaves of various ages was similar to that when the whole shoot was treated. Perhaps, most interesting in view of previous reports in the literature, was the export of ^{14}C, comparable to that from older leaves, by the youngest leaves and accumulation in growing roots (Allessio and Tieszen, 1975a). It appears that even young leaves may be effective sources for roots. In addition, transport from senescing leaves of *Dupontia fisheri* was documented and was extensive, unlike that of other grasses (Mayer and Porter, 1960; Williams, 1964). Senescing leaves appear to maintain themselves and belowground components. They are not parasitic and translocation continues until the leaf has reached an advanced stage of senescence as observed by Thrower (1967). Since the pattern of loss of radioactivity from leaves is similar throughout the growing season (Tieszen et al., 1974a), we can suggest an apparent maintenance of a stable sugar pool in roots and rhizomes similar to that shown by McCown (1978) for other plant parts (Allessio and Tieszen, 1975a).

Although distribution of [14]C (Table 3) in the treated tiller is similar regardless of the age of the treated leaf, in nontreated tillers considerably more ^{14}C was recovered in photosynthetic portions than was seen in other samples (Table 5 and Table 6). This may be due to the fact that these plants were actively invading a moss mat and extensive vegetative growth was occurring. Under these conditions of great vegetative growth, greater support of the invading tiller shoot and stem base might occur. This would support the contention of Mooney (1972) that great quantities of carbon are devoted to attaining or maintaining competitive advantage. The great amount of vegetative growth may be accounted for by an improved nutrient condition in the moss mat which would influence the distribution of carbon to roots and shoots (Mooney, 1972). Little flowering was observed in this site 2 area, which would correlate with the observations by Dennis et al. (1978) that in fertilized plots *Dupontia fisheri* responds to the increased nutrients by producing more vegetative growth, but not necessarily more flowering.

Tiller Age and Interrelationships between Tillers

The decline of interdependence with age as observed in *Lolium* (Sagar and Marshall, 1967) was not marked in *Dupontia fisheri*. We observed some translocation to both V1 and V2 tillers from treated V1 and V4 tillers. The greatest degree of independence was exhibited by tillers with exserted flowers which did not export much radiocarbon to daughter tillers. However, if rapidly growing sinks were present on an adjacent daughter tiller, translocation occurred. This corresponds with Wolf's report (1967) for other grasses of the lack of transport from flowering tillers to other tiller rhizomes; translocation to newly developing rhizomes occurred. Before flowers were exserted, however, translocation to other tillers took place. This is in keeping with Lupton's observation (1966) that before the wheat stem elongated, ^{14}C moved to all parts of the plant, and later little moved out of the treated shoot. There is no evidence that the flowering tiller serves as a sink for materials from a daughter tiller.

Our data show very different allocation patterns when comparing V1 and V4 (flowering) tillers at 24 hr (Figure 5) but not with V1 and V2 (V3) at 1 and 6 hr (Figure 4). Distribution to nontreated tillers of a given age class was similar (Figure 6). Although vegetative and flowering tillers have unlike distribution patterns of ^{14}C, the allocation to nontreated tillers is analogous to that observed by Marshall and Sagar (1968b) for *Lolium* where leaves on the main shoot and leaves on tillers support the main shoot. But, significant transport to parent tillers or to a major portion of the root system did not occur (Marshall, 1967). The observed allocation pattern differences in V1 and V4 treated tillers and V1 and

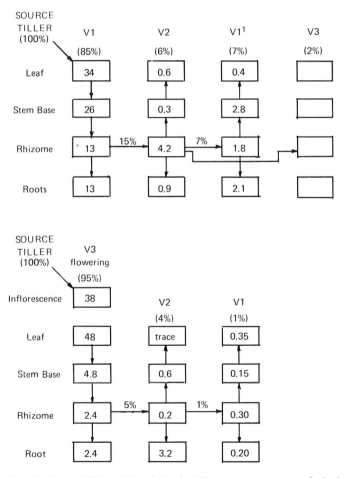

Figure 7.   Distribution of ^{14}C radioactivity in tiller compartments of single plants as influenced by tiller age. Values are percentage of the total radioactivity recovered from the tiller. Plants were treated 10 July.

V2 nontreated tillers is no doubt related to the physiological state of the tiller. For example, the V4 tiller is involved primarily with shoot processes rather than with root or rhizome processes.

Overall, we feel that the pattern observed can be interpreted as one of interdependence, at least functionally, since even flowering tillers, which usually translocated little to belowground parts or to other tillers, were observed to contribute to the carbon economy of young V1 tillers with new roots. Accordingly, it would appear that the entire *Dupontia fisheri* tiller system must be considered as a physiologically integrated unit. This is in agreement with other investigators working with grass species (Jewiss, 1965; Marshall, 1967; Marshall and Sagar, 1968a, b). Results of analysis of apportionment of radiocarbon, however, indicate that a tiller is essentially independent unless a sink (actively growing plant part such as root and rhizome primordia) is nearby. Translocation to nontreated tillers took place, but the percentage of ^{14}C recovered was usually small even when new roots and VO rhizomes was present.

The natural tillering pattern for *Dupontia fisheri* which results in a linear rhizome system may account for maintenance of the relative tiller interdependence which was observed. Growth habit may be controlled by a gradient in the C/N ratio as suggested by McIntyre (1970) for *Agropyron repens* and by McCown (personal communication) for *Dupontia*. Plant growth at Barrow is felt to be nutrient limited.

As we suggested elsewhere (Allessio and Tieszen, 1975a), this interdependence of tillers and the contribution of even young leaves to nonphotosynthetic compartments and reserves in roots and rhizomes of old tillers (V5) could prevent depletion of root reserves whenever defoliation occurs. In addition, recovery from defoliation (grazing), unless severe (Gifford and Marshall, 1973), is maximized and is not at the expense of the root reserves (Marshall and Sagar, 1965; Ryle and Powell, 1975). Root growth even under these conditions is necessary for water and nutrients to be obtained from the soil (Williams, 1969). In addition, rapid regrowth may be especially important in tundra where the growing season is short (Lewis and Callaghan, 1975), nutrients are deficient, permafrost is present, and soil and air temperatures are low. Therefore it would seem essential that plant activities should contribute to optimal nutrient absorption under field conditions. Nutrient considerations may be of considerable importance to arctic tundra plant growth (Miller and Tieszen, 1972).

This contention regarding plant activities and nutrient absorption is extended further by the observation that each season a V1 rhizome typically produces at least one V0, which may in some cases be of considerable length (20 cm), the new rhizome ultimately producing roots in places away from the parent, where space, and, more probably, a nonexhausted nutrient rhizosphere exists.

There are several questions suggested by this study. Investigation of the influence of the shoot (stem base)–rhizome temperature gradient on translocation would be of interest, since the determination of how temperature affects the rate and pattern of translocation might establish to what degree the pattern is controlled by temperature. Also, information regarding translocation and allocation patterns late in the season would ascertain the prewintering status of tillers.

In addition, knowledge of relationships between defoliation and redistribution of reserves between tillers would suggest to what degree grazing pressure is tolerated by *Dupontia fisheri*.

Acknowledgments. This research was supported by Research Corporation and the National Science Foundation under Grant GV 29343 to Augustana College, Sioux Falls, South Dakota. It was performed under the joint sponsorship of the International Biological Programme and the Office of Polar Programs and was directed under the auspices of the U.S. Tundra Biome. Field and laboratory activities at Barrow were supported by the Naval Arctic Research Laboratory of the Office of Naval Research. We would also like to thank Elizabeth Collins, Mary Vetter, Richard Mandsager, Rich Nelson, and Peggy Vance, who were our field or laboratory assistants at various times.

References

Allessio, M. L., and L. L. Tieszen. (1975a) Leaf age effect on translocation and distribution of ^{14}C-photoassimilate in *Dupontia* at Barrow, Alaska. *Arct. Alp. Res.*, 7: 3–12.

Allessio, M. L., and L. L. Tieszen. (1975b) Patterns of carbon allocation in an arctic tundra grass, *Dupontia fischeri* (Gramineae), at Barrow, Alaska. *Amer. J. Bot.*, 62: 797–807.

Bliss, L. C. (1971) Arctic and alpine plant life cycles. *Ann. Rev. Ecol. Syst.*, 2: 405–438.

Brouwer, R., and E. Levi. (1969) Responses of bean plants to root temperatures. IV. Translocation of ^{22}Na applied to leaves. *Acta Bot. Neerl.*, 18: 58–66.

Collins, N. J., and W. C. Oechel. (1974) The pattern of growth and translocation of photosynthate in a tundra moss, *Polytrichum alpinum*. *Can. J. Bot.*, 52: 355–363.

Crafts, A. S., and C. E. Crisp. (1971) *Phloem Transport in Plants*. San Francisco: W. H. Freeman and Co., 481 pp.

Davidson, J. D., V. T. Oliverio, and J. I. Peterson. (1970) Combustion of samples for liquid scintillation counting. In *The Current Status of Liquid Scintillation Counting* (E. D. Bransome, Jr., Ed.). New York: Grune and Stratton, pp. 222–235.

Dennis, J. G., and P. L. Johnson. (1970) Shoot and rhizome-root standing crops of tundra vegetation at Barrow, Alaska. *Arct. Alp. Res.*, 2: 253–266.

Dennis, J. G., L. L. Tieszen, and M. A. Vetter. (1978) Seasonal dynamics of above- and belowground production of vascular plants at Barrow, Alaska. In *Vegetation and Production Ecology of an Alaskan Arctic Tundra* (L. L. Tieszen, Ed.) New York: Springer-Verlag, Chap. 4.

Doodson, J. K., J. G. Manners, and A. Myers. (1964) The distribution pattern of ^{14}carbon assimilated by the third leaf in wheat. *J. Exp. Bot.*, 15: 96–103.

Felippe, G. M., and J. E. Dale. (1972) The uptake of $^{14}CO_2$ by developing first leaves of barley and partition of the labeled assimilates. *Ann. Bot.*, 36: 411–418.

Fujiwara, A., and M. Suzuki. (1961) Effects of temperature and light on the translocation of photosynthetic products. *Tohoku J. Agric. Res.*, 12: 363–367.

Gifford, R. M., and C. Marshall. (1973) Photosynthesis and assimilate distribution in *Lolium multiflorum* Lam. following differential tiller defoliation. *Aust. J. Biol. Sci.*, 26: 517–526.

Grace, J., and H. W. Woolhouse. (1973) A physiological and mathematical study of the growth and productivity of a *Calluna-Sphagnum* community. III. Distribution of photosynthate in *Calluna vulgaris* (L.) Hull. *J. Appl. Ecol.*, 10: 77–91.

Hodgkinson, K. C. (1969) The utilization of root organic compounds during the regeneration of lucerne. *Aust. J. Biol. Sci.*, 22: 1113–1123.

Jewiss, O. R. (1965) Morphological and physiological aspects of growth of grasses during the vegetative phase. In *Growth of Cereals and Grasses: Proceedings of Twelfth*

Easter School of Agricultural Science, University of Nottingham (F. L. Milthorpe and J. D. Ivins, Eds.). London: Butterworth.

Jewiss, O. R., and J. Woledge. (1967) The effect of age on the rate of apparent photosynthesis in leaves of tall fescue (*Festuca arundinacea* Schreb.). *Ann. Bot.*, **31**: 661–671.

Johansson, L. G. (1974) The distribution and fate of ^{14}C photoassimilated by plants on a subarctic mire at Stordalen. *Swedish International Biological Programme, Rep.* 16.

Larson, P. R., and J. C. Gordon, (1969) Leaf development, photosynthesis, and ^{14}C distribution in *Populus deltoides* seedlings. *Amer. J. Bot.*, **56**: 1058–1066.

Lewis, M. C., and T. V. Callaghan. (1975) Ecological efficiency of tundra vegetation. In *Vegetation and the Atmosphere* (J. L. Montieth, Ed.). New York: Academic Press, pp. 399–433.

Lovell, P. H., H. T. Oo, and G. R. Sagar. (1972) An investigation into the rate and control of assimilate movement from leaves in *Pisum sativum*. *J. Exp. Bot.*, **23**: 255–266.

Lupton, F. G. H. (1966) Translocation of photosynthetic assimilates in wheat. *Ann. Appl. Biol.*, **57**: 355–364.

McCown, B. H. (1978) The interactions of organic nutrients, soil nutrients, and soil temperature and plant growth and survival in the arctic environment. In *Vegetation and Production Ecology of an Alaskan Arctic Tundra* (L. L. Tieszen, Ed.). New York: Springer-Verlag, Chap. 19.

McIntyre, G. I. (1970) Studies on bud development in the rhizome of *Agropyron repens*. I. The influence of temperature, light intensity, and bud position on the pattern of development. *Can. J. Bot.*, **48**: 1903–1909.

Marshall, C. (1967) The use of radioisotopes to investigate organization in plants, with special reference to the grass plant. In *Symposium on the Use of Isotopes in Plant Nutrition and Physiology, Vienna 1966*. International Atomic Energy Agency, pp. 203–216.

Marshall, C., and G. R. Sagar. (1965) The influence of defoliation on the distribution of assimilates in *Lolium multiflorum* Lam. *Ann. Bot.*, **29**: 365–372.

Marshall, C., and G. R. Sagar. (1968a) The distribution of assimilates in *Lolium multiflorum* Lam. following differential defoliation. *Ann. Bot.*, **32**: 715–719.

Marshall, C., and G. R. Sagar. (1968b) The interdependence of tillers in *Lolium multiflorum* Lam.—A quantitative assessment. *J. Exp. Bot.*, **19**: 785–794.

Mayer, A., and H. K. Porter. (1960) The translocation from leaves of rye. *Nature*, **188**: 921–922.

Miller, P. C., and L. L. Tieszen. (1972) A preliminary model of processes affecting primary production in the arctic tundra. *Arct. Alp. Res.*, **4**: 1–18.

Mooney, H. A. (1972) The carbon balance of plants. *Ann. Rev. Ecol. Syst.*, **3**: 315–346.

Mooney, H. A., and W. D. Billings. (1960) The annual carbohydrate cycle of alpine plants as related to growth. *Amer. J. Bot.*, **47**: 594–598.

Nelson, C. D. (1963) Effect of climate on the distribution and translocation of assimilates. In *Environmental Control of Plant Growth* (L. T. Evans, Ed.). New York: Academic Press, pp. 149–174.

Rovira, A. D., and G. D. Bowen. (1973) The influence of root temperature on ^{14}C assimilate profiles in wheat roots. *Planta*, **114**: 101–107.

Ryle, G. J. D., and C. E. Powell. (1975) Defoliation and regrowth in the Graminaceous plant: The role of current assimilate. *Ann. Bot.*, **39**: 297–310.

Sagar, G. R., and C. Marshall. (1967) The grass plant as an integrated unit—Some studies of assimilate distribution in *Lolium multiflorum* Lam. Proceedings of the 9th International Grassland Congress, pp. 493–497.

Shaver, G. R., and W. D. Billings. (1975) Root production and root turnover in a wet tundra ecosystem, Barrow, Alaska. *Ecology*, **56**: 401–409.

Sørenson, T. (1941) Temperature relations and phenology of the northeast Greenland flowering plants. *Medd. Grøn.*, **125**: 1–305.

Svoboda, J., and L. C. Bliss. (1974) The use of autoradiography in determining active and inactive roots in plant production studies. *Arct. Alp. Res.*, **6**: 257–260.

Thrower, S. L. (1967) The pattern of translocation during leaf aging. In *Aspects of the Biology of Aging*. Symposium of the Society for Experimental Biology, No. XXI. London: Cambridge University Press, pp. 483–506.

Tieszen, L. L. (1972a) The seasonal course of aboveground production and chlorophyll distribution in a wet tundra meadow at Barrow, Alaska. *Arct. Alp. Res.* **4:** 307–324.

Tieszen, L. L. (1972b) Photosynthesis in relation to primary production. *Proceedings IV International Meeting on the Biological Productivity of Tundra, Leningrad, USSR.* (F. E. Wielgulaski and Th. Rosswall, Eds.). Stockholm: International Biological Programme Steering Committee. pp. 52–62.

Tieszen, L. L. (1974) Photosynthetic competence of the subnivean vegetation of an arctic tundra. *Arct. Alp. Res.*, **6:** 253–256.

Tieszen, L. L. (1975) CO_2 exchange in the Alaskan arctic tundra: Seasonal changes in the rate of photosynthesis of four species. *Photosynthetica,* **9:** 376–390.

Tieszen, L. L. (1978) Photosynthesis in the principal Barrow, Alaska, species: A summary of field and laboratory responses. In *Vegetation and Production Ecology of an Alaskan Arctic Tundra* (L. L. Tieszen, Ed.). New York: Springer-Verlag, Chap. 10.

Tieszen, L. L., and D. A. Johnson. (1975) Seasonal pattern of photosynthesis in individual grass leaves and other plant parts in arctic Alaska with a portable $^{14}CO_2$ system. *Bot. Gaz.*, **136:** 99–105.

Tieszen, L. L., D. A. Johnson, and M. L. Allessio. (1974a) Translocation of photosynthetically assimilated $^{14}CO_2$ in three arctic grasses in situ at Barrow, Alaska. *Can. J. Bot.*, **52:** 2189–2193.

Tieszen, L. L., D. A. Johnson, and M. M. Caldwell. (1974b) A portable system for the measurement of photosynthesis using $^{14}CO_2$. *Photosynthetica,* **8:**151–160.

Wardlaw, I. F. (1968) The control and pattern of movement of carbohydrates in plants. *Bot. Rev.*, **34:** 79–105.

Williams, R. D. (1964) Assimilation and translocation in perennial grasses. *Ann. Bot.*, **28:** 419–425.

Williams, T. E. (1969) Root activity of perennial grass swards. In *Root Growth: Proceedings of the Fifteenth Easter School in Agricultural Sciences, University of Nottingham* (W. J. Whittington, Ed.). New York: Plenum Press, pp. 270–279.

Wolf, D. D. (1967) Assimilation and movement of radioactive carbon in alfalfa and reed canary grass. *Crop Sci.*, **7:** 317–320.

Yamamoto, T. (1967) The distribution pattern of carbon-14 assimilated by a single leaf in tobacco plant. *Plant Cell Physiol.*, **8:** 353–362.

18. Growth, Turnover, and Respiration Rates of Roots and Tillers in Tundra Graminoids

W. D. Billings, K. M. Peterson, and G. R. Shaver

Introduction

In wet arctic tundra, by far the greater part of live vascular plant material is below the surface of the ground and above the upper level of permafrost. In this cold, narrow zone are the perennating storage and absorption systems of the dominant sedges and grasses. These over-wintering systems quickly produce and support photosynthetic leaves during the short, cold season between snowmelt in June and the first sunsets of August. An understanding of tundra productivity requires a knowledge of how much energy is allocated to roots and rhizomes, and how this energy is used in growth and reproduction of the whole plant.

Up to now, most of the work done with tundra root systems has been concerned with determining the dry matter content (biomass) and the proportion of living to dead roots (Dennis, 1968; Dennis and Johnson, 1970; Muc, 1972; Dennis and Tieszen, 1972; Wielgolaski, 1972; Tieszen, 1972). Even during summer in wet graminoid tundra, 85 to 98% of the living plant biomass is belowground. Using the data of Dennis and Tieszen (1972), dead roots and rhizomes are present in even greater amounts than those which are alive. The ratio of live to dead ranges from about 0.4 to 0.7. It is apparent that the great pool of fixed carbon, energy, and nutrients in wet tundra is belowground. However, only recently has there been anything more than scanty information on how this pool is utilized in growth and metabolism of these graminoid systems (Billings et al., 1973; Chapin, 1974; Billings et al., 1977; Peterson and Billings, 1975; Shaver and Billings, 1975). The purpose of our research is to provide information on plant growth underground in the tundra and to assess its costs to the system in terms of energy.

Methods

Field Methods

Growth and Turnover. The field location of our research is on U.S. Tundra Biome Site 2 (71°18′ N) about 3.7 km south of the Naval Arctic Research

Laboratory near Barrow, Alaska. The area consists of low-centered wet polygons, their rims and troughs, moist meadow, and some high-centered polygons. The work was concentrated in the moist meadow and wet-polygon systems. The dominants are *Eriophorum angustifolium* Honck., *Carex aquatilis* Wahlenb., and *Dupontia fisheri* R. Br.; intensive measurements of belowground growth and respiration were focused on these three graminoid species.

Root, rhizome, and tiller growth in the field was measured in two principal ways in conjunction with temperature, seasonal progression of soil thaw, and microsite. First, the growth of individual roots and rhizomes was measured *in situ* by placing natural sods of each species in wedge-shaped Plexiglas-sided boxes equipped with thermocouples. In August 1971, these sod-root boxes were inserted into the tundra soils in the same holes from which the sods were removed (Trent, 1972; Billings et al., 1976, 1977). Root growth in relation to soil temperature and depth of thaw was measured in these boxes during the growing seasons of 1972, 1973, and 1974 (Shaver and Billings, 1975; Shaver, 1976).

Second, whole tiller systems of each species were removed from the soil both in the field and from sods brought into the laboratory. These tiller systems, which included whole root systems, were dissected into roots, daughter tillers, rhizomes, stem bases, and leaves which were dried and weighed. Leaves, dead leaf bases, and nodes were counted. Using these methods, it was possible to determine dry weight of root or rhizome per unit length for each species. By using the number of leaves or leaf bases, the age of the tiller could be determined. This information also allowed us to estimate underground turnover rates (Shaver and Billings, 1975).

Nonstructural carbohydrate samples were collected from stem bases, rhizomes, and roots of the graminoid plants during the periods 3–9 June, 1–4 July, 1–5 August, and 2–5 September 1974. June samples were drilled from the frozen soil using a Haynes ice auger which took cores 7.6 cm in diameter. The cores were kept in the laboratory until they could be quickly thawed by placing them in plastic bags in hot water. The stem bases, roots, and rhizomes were then washed from the soil, sorted, and killed by the method described below. Samples collected later in the season were immediately washed from the soil and killed in the field. All tillers were sorted into a maximum of six age classes and killed in hot, 85 to 90% ethanol. Samples were killed within 45 min of thaw initiation or field collection. Details of sample collection, age classification, and killing techniques are described in Shaver and Billings (1976).

Analyses of nonstructural carbohydrate samples were done in the laboratory at Duke University. Sugars, storage polysaccharides, and total nonstructural carbohydrates were determined for each sample. The methods are described in Shaver and Billings (1976).

Dark Carbon Dioxide Flux from Soil and Turf.   The flux of CO_2 from the soil surface and turf is a rough measure of community respiration, not counting birds and mammals, of course. Dark CO_2 flux from the soil column plus the plant shoots was measured by circulating ambient air through metal chambers (Figure 3) inserted into the soil (Billings et al., 1977; Peterson and Billings, 1975). The

CO_2 concentration of the return air stream from each chamber as compared with a reference stream of ambient air was determined in a Beckman 215B differential infrared gas analyzer system with continuous recording. Automatic switching allowed six chambers to be measured within a 20-min cycle. Temperature also was continuously recorded at 0, -5, -10, and -20 cm from thermocouples installed in the soil near the chambers. Details of the dark CO_2 flux method appear in Peterson and Billings (1975).

Root respiration. Respiration rate measurements of excised live roots in the field with an infrared gas analyzer system were initiated in the 1971 summer. These were continued over a greater span of temperature (0° to 20°C) in the growing seasons of 1972 and 1973. Details of the method are in Billings et al. (1976, 1977). A laboratory method has been devised in which the respiration rates of whole intact root systems can be measured by the infrared gas analysis system. This worked well in the field laboratory as well as in the phytotron (Billings et al., 1976, 1977).

Laboratory Methods

Plants of all three graminoid species were sent to the Duke University Phytotron where they were grown under controlled environmental conditions. This made it possible to obtain more precise measurements of root growth and respiration in regard to temperature than could be obtained in the field.

Essentially the same methods were used in the Phytotron as in the field. However, the laboratory root growth boxes were completely waterproof so that they could be immersed in a cooling bath. Such a bath made it possible to manipulate the level of an ''artifical permafrost'' so that freezing effects on live roots could be studied. In the Phytotron, it was possible also to measure root growth and respiration across a wider span of temperature (Billings et al., 1977). Whole root system respiration in a gas analyzer system could be measured rather more precisely than was possible under the irregularities of field conditions. By gradually shortening the photoperiod from continuous light to 15 hr in the phytotron chamber, we could measure the effect of decreasing daylength on root and rhizome growth (Shaver and Billings, 1977).

Results

Field Results on Growth, Production, and Turnover

Root Production and Growth. Roots and rhizomes of these tundra graminoids grow relatively fast in soils whose mean daily temperature in the upper 10 cm seldom exceeds 7°C (our unpublished data). Such low temperature root growth is beyond the tolerance limits of most plants. Only those from tundra regions, alpine bogs, and also certain winter annuals and winter perennials of temperate regions can function at such low temperatures.

Even in these tundra species, however, there are pronounced differences between species in their patterns of root growth in relation to soil temperatures and thaw through the season. Figure 1 shows peak season live root distribution in relation to temperature and depth in *Eriophorum angustifolium* and *Dupontia fisheri,* respectively. *Eriophorum* produces many more roots than does *Dupontia*. The *Eriophorum* roots follow the thaw down very closely and are most abundant at temperatures below 2°C. In fact, some *Eriophorum* roots grow immediately against the top of the permafrost (Billings et al., 1976, 1977).

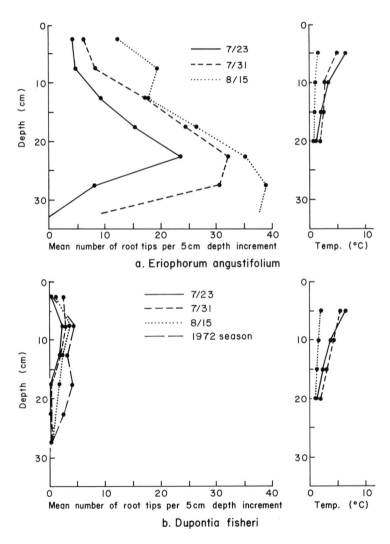

Figure 1.   The number of root tips per 5 cm increment of depth in root observation boxes at site 2, Barrow, Alaska. Temperature profiles from -5 to -20 cm are from the same site. (a) *Eriophorum angustifolium*. (b) *Dupontia fisheri*.

Table 1.   Summary of some Characteristics of Roots in Three Arctic Plant Species

	Eriophorum angustifolium	*Carex aquatilis*	*Dupontia, fisheri*
Aboveground longevity	5 to 7(8) yr	4 to 7 yr	3 to 4(5) yr
Root longevity	Annual	5 to 8(10) yr	4 to 6 yr
Elongation ability	1 yr	2 to 3 yr	1 yr
Seasonality of elongation	Continuous	Continuous	Early—Mid-season
Lateral production	None	2nd to 4th yr	90% 1st yr
Seasonality of lateral production	—	Continuous	Mid–late season
Root origins	Stem base	Stem base	Rhizome nodes

Wiggins (1953) also noted this for one or more undesignated species in the same region. Active *Carex aquatilis* root tips were not found closer than 3 to 10 cm above the frozen soil while those of *Dupontia* seldom were closer than 8 to 13 cm from the top of the frozen soil at maximum thaw.

Patterns of Root Growth.   Each of the three species has a distinct pattern of adaptation to the soil environment. Some of these characteristics are summarized in Table 1. Probably the most unusual adaptation is the production of an entire new root system each year in *Eriophorum*. Unlike the other two species, *Eriophorum* does not produce lateral roots. The significance of this remarkable adaptation is not known. It can be hypothesized that the *Eriophorum* root system plays a strong role in nutrient absorption which would be hampered if the plant depended upon deep old roots which were encased in frozen soil until late in the season.

By excavation of tiller systems (Shaver and Billings, 1975), it was possible to relate the number of live roots to the age of the tiller. Age was determined by counting the number of leaves or leaf bases per tiller. While each species differs slightly from another in number of leaves produced by an individual tiller per year, four to five leaves per year is a general figure for approximating age.

Root Turnover.   Our results indicate that root turnover rates in the Barrow tundra differ considerably according to species. One of them, *Eriophorum angustifolium,* invests a considerable amount of energy and material in the production of an annual root system. By contrast, *Carex aquatilis* and *Dupontia fisheri* produce almost all of their new roots during the first and/or second years of the lifetime of an individual (Table 1). These roots provide a return which lasts through several years. Energy invested in new roots of *Carex* and *Dupontia* presumably is utilized later in production of rhizomes and aboveground parts.

The longest lived roots, those of *Carex aquatilis,* live about 7 yr. With a total live root standing crop of 456 g m^{-2} (Dennis, 1977), the minimum ecosystem root turnover rate could be about 60 to 65 g m^{-2} yr^{-1}. This assumes that the system is in a steady state, which may not be correct.

Estimates of the annual production of aboveground vascular plant "standing dead" phytomass vary somewhat. Dennis and Johnson (1970) estimate it to be in the neighborhood of 70 g m^{-2} yr^{-1} while Dennis and Tieszen (1972) state that, at the end of the season on 4 September 1971, "current dead vascular phytomass (was) less than 28.2 g m^{-2}." Dennis et al. (1978) indicate that by the end of the growing season there is 50 to 66 g m^{-2} of dry matter present that was produced during the current year. Assuming, then, a median figure of 50 g m^{-2} yr^{-1} for annual production of "standing dead," and that decay rates in the upper 10 cm of soil range from 95 to 190 g m^{-2} yr^{-1} with a median of about 140 g (Douglas and Tedrow, 1959), maximum root turnover rates could be as high as 90 g m^{-2} yr^{-1}. These figures would indicate a root turnover time ranging from 4 to 8 yr depending upon species composition and differences in growing season temperatures. Dennis' (1977) figures of between 3 and 6 yr for live root turnover time agree fairly well with ours. Both of these rates are somewhat faster than the estimate of more than 10 yr suggested by Dennis and Johnson (1970). Part of the discrepancy may be due to the annual death and production of a wholly new root system each year by plants of *Eriophorum.*

Root Weight per Unit Length.   Table 2 shows the differences between species in root weight per unit length. *Eriophorum,* which produces large numbers of deep roots every year, invests the least amount of energy and materials in producing a millimeter of root. *Carex,* on the other hand, invests five to six times as much dry matter as does *Eriophorum* per unit length of root. But the *Carex* roots provide a return which continues for at least 5 or 6 yr. *Dupontia,* whose root longevity is intermediate, is also intermediate in root weight. Old roots of *Dupontia* weigh as much as four times that of new roots per unit length. This is due to the production of many hairlike lateral roots at the end of the first season.

Tiller Phenology.   *Carex aquatilis, Eriophorum angustifolium,* and *Dupontia fisheri* all reproduce primarily by vegetative means. Each species produces extensive networks of individual tillers connected by subsurface rhizomes. We shall define a tiller as a horizontal rhizome, stem, stem base, and leaves originating at the point where the rhizome joins the parent plant.

Each species has a distinct kind of tiller growth pattern. *Dupontia* first year tillers have less than five leaves, one or no dead leaves, and white rhizomes. Individual *Dupontia* tillers rarely live longer than 4 yr aboveground, producing 10 to 15 leaves in that time. Roots, rhizomes, and stem bases remain intact, elastic, and apparently alive for 1 to 2 yr longer. Apparently, *Dupontia* tillers depend almost entirely on roots produced during their first growing season. In

Table 2.   Root Weight per Unit Length Determinations, in mg·mm^{-1}, for Each Species, with Standard Deviations

	New live roots	Old live roots
Eriophorum angustifolium	0.0267 ± 0.0258	—
Carex aquatilis	0.1163 ± 0.0526	0.1250 ± 0.0143
Dupontia fisheri	0.0233 ± 0.0094	0.0829 ± 0.0233

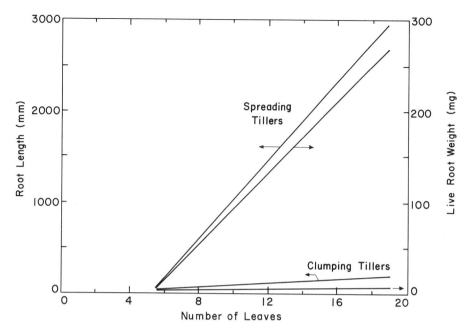

Figure 2.   Total live root length and total live root biomass versus accumulated leaf number (live plus dead) for spreading and clumping tillers of *Carex aquatilis*. Regression of length against leaf number is: $Y = -1145.3 + 218.8X$ ($r^2 = 0.62$). Regression of live root biomass against leaf number is: $Y = -107.0 + 20.3X$ ($r^2 = 0.69$). Data points are omitted for clarity.

the root boxes, none of the *Dupontia* roots which appeared in 1972 continued to elongate downward in 1973. Over 90% of these also completed growth of lateral roots in 1972.

In *Eriophorum,* roots are inserted at the stem base, the region of very short internodes at the base of the leaves. Most of the roots originate at nodes of leaves produced during the previous year, although some roots arise at nodes several years old. The oldest *Eriophorum* tillers examined had produced 20 to 40 leaves. One tiller collected in the soil cores had 27 dead leaves and four green leaves. Maximum longevity aboveground, then, is at least 6 yr and perhaps as long as 8 yr. Several very old tillers (indeterminately aged), which had no green leaves during the current season, were observed to produce new roots. Such tillers may be older than 8 yr.

Tillers and root growth patterns in *Carex aquatilis* are more complex than in *Eriophorum* or *Dupontia*. It is necessary to divide *C. aquatilis* into two tiller classes. "Spreading tillers" have long (50 to 200 mm) rhizomes and produce almost all the roots in *Carex*. "Clumping tillers" are daughters of the spreading tillers and produce relatively few roots or daughter tillers of their own. Clumping tillers have very short rhizomes (0 to 10 mm), and are bunched very tightly around the "mother" spreading tiller.

The differences in root growth between spreading and clumping tillers of *Carex* are shown in Figure 2 which is adapted from Shaver and Billings (1975).

Table 3.   Tiller Density, Mean Leaf Number (Live Plus Dead, ± 1 Standard Deviation), and Grazing Intensity in the Three Principal Plant Species on Site 2

	Tillers m^{-2}	Mean leaf number	Percentage grazed
Eriophorum angustifolium	471.12	12.00 ± 0.79	43.42
Carex aquatilis	674.46	12.54 ± 0.54	44.44
Dupontia fisheri	277.92	7.22 ± 0.47	36.11
Eriophorum scheuchzeri	467.73	—	—
Calamagrostis holmii	298.26	—	—
Total	2189	—	42.49

This demonstrates that those few roots which are produced beneath clumping tillers are much shorter and weigh less than those beneath spreading tillers. Clumping tillers may produce as many as eight to ten roots, but these roots represent a smaller investment both in terms of root length and root biomass (Figure 2) than for a spreading tiller.

In some way, clumping tillers must obtain water and nutrients. These must be coming from the spreading tiller to which a clumping tiller is attached. This spreading tiller has the only extensive root system close enough to allow for translocation into the clumping tiller.

New roots in *C. aquatilis* are rarely found beneath tillers with more than 10 to 12 leaves; such tillers would be 2 to 3 yr old. An individual spreading tiller may produce 20 to 25 (or more) leaves in 4 to 7 yr before dying above ground. Roots, rhizomes, and stem bases, though, usually do not die until 1 to 4 yr after aboveground growth stops. The belowground parts remain alive, presumably supplying water and nutrients to the still-green, rootless clumping tillers. Root longevity in *C. aquatilis*, then, is 5 to 8, or perhaps as long as 10 yr.

Results of the analysis of soil cores in regard to "tiller density" are summarized in Table 3. These are the means of 44 cores extrapolated to 1 m². The "mean leaf number" and "percentage grazed" categories are means of 22 cores. Mean leaf number is an index of the average ages of these populations; in this case, about 2 or 3 yr for *Carex* and *Eriophorum* and 2 yr for *Dupontia*.

Preliminary statistical analyses have been made of the tiller system data from 1973. It is not possible to present all this information here. However, it will be summarized briefly. For example, rhizome weight, length, weight per unit length, node number, weight per internode, and length per internode have been determined with means and standard deviations for all three species. Also, green leaf dry weight in relation to tiller age is plotted in a series of computer graphs for each species. These data show that the older a tiller is, the greater the green biomass it supports. This is in spite of extensive senescence of green leaves at the end of each summer. This is true for all three species. *Carex aquatilis* tillers reach a maximum green dry biomass of 60 to 80 mg, *Dupontia* 30 to 40 mg, and *Eriophorum angustifolium* 100 to 140 mg.

Corresponding with the production of a number of new leaves each year is the addition to the stem base of several internodes and an increase in stem base weight. Although most of the green leaf biomass dies back each year, stem base weight continues to increase with age in all three species. Maximum stem base dry weights in *Carex* range from 40 to 50 mg, in *Dupontia* from 14 to 18 mg, and in *Eriophorum* from 100 to 130 mg.

Field Results on Dark CO_2 Flux and Root Respiration

Figure 3 shows the placement of the soil CO_2 flux chambers over the tundra vegetation and soil. The tubing lines lead to the gas analyzer system. This figure also shows that most of the roots are in the upper 10 to 15 cm of the soil. This layer consists of peat which is sharply delineated from a sandy clay mineral subsoil. The chambers were left in place from 4 to 7 days at a time and then moved to a new location.

Mean daily dark CO_2 flux rates (mg of CO_2 m^{-2} hr^{-1}) from the whole vegetation–soil column were calculated and plotted against mean daily soil temperatures (Peterson and Billings, 1975). Mean daily soil temperature is defined as the average of mean daily temperatures at the surface (base of green moss layer), -5, -10, and -20 cm depths.

The lowest mean daily dark CO_2 flux rates occurred early in the season and corresponded with the lowest mean daily soil temperatures (Figures 4 and 5). The

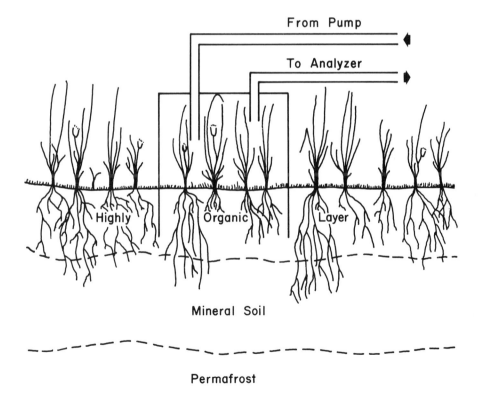

Figure 3. Soil CO_2 flux chamber placement with respect to the vegetation and soil profile. Scale is indicated by the 15-cm diameter of the chamber.

minimum flux was 52 mg of CO_2 m^{-2} hr^{-1} at a mean soil temperature of 0.1°C on 24 June 1973. Lower flux rates were observed before 19 June, but no temperature data are available for correlation before that date. During early season, mean daily CO_2 flux (72 readings per day) showed good linear regression for each day between 30 June and 7 July (Figure 4).

Late season flux rates were well above those measured during the early part of the season (Figure 5). The highest mean daily CO_2 flux measured was 367 mg of CO_2 m^{-2} hr^{-1} on 25 July 1973 at a mean daily soil temperature of 6.4°C. The data from 25 July to 2 August show relatively high dark CO_2 flux rates and correlate well with soil temperature data for the same period (Figure 5). From the data available, it appears that the seasonal course of dark CO_2 flux follows the seasonal trend in mean daily soil temperature. The seasonal trend until early August appears to be curvilinear as is the plot and regression line of all CO_2 flux data against daily mean soil temperature (Figure 6). For each *day,* however, dark CO_2 flux from the ecosystem has a linear regression on soil temperature as shown in Figure 4.

The rate of CO_2 evolution from artificially darkened tundra vegetation and soil shows a distinct diurnal pattern (Figure 7). During the entire 24-hr period, there was always measurable CO_2 output. Minimum flux rates occurred between 2300 and 0400 hr solar time; maximum rates usually were measured between 1200 and 1600 hr. Mid-day declines in rate correspond to decreases in air and soil temperatures due to fog. As Figure 7 shows, there was greater diurnal amplitude

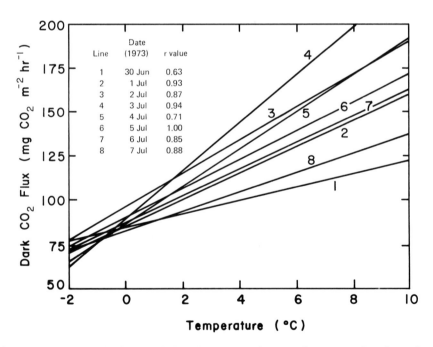

Figure 4. Carbon dioxide evolution from a tundra meadow as a function of soil temperature.

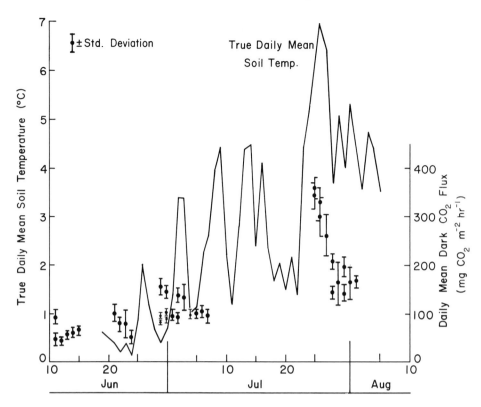

Figure 5.   Daily mean dark CO_2 flux from a tundra meadow (early and peak seasons) graphed against the true daily mean soil temperature.

in flux from two chambers in polygon troughs than from two chambers in a slightly polygonized moist meadow.

A rather rough estimate has been made of the total dark CO_2 flux for the season based upon the data in Figure 5. This indicates that for the period 9 June 1973 to 5 August 1973, about 220 to 240 g of CO_2 m^{-2} left the meadow ecosystem to the atmosphere in this flux. In view of the estimate of maximum net photosynthetic incorporation by vascular plants of 760 g of CO_2 $^{-2}$ for the season (Tieszen, 1975) and 772 g of CO_2 m^{-2} by Coyne and Kelley (1975) which includes mosses, it appears that the moist meadow ecosystem may not be in a steady state. Organic matter may be accumulating as peat and/or being exported to lakes and ponds.

An assessment of the contribution of root respiration to total carbon dioxide flux from the moist tundra meadow ecosystem yields uncertain results. Dennis and Tieszen (1972) report between 534 and 620 g m^{-2} total live subsurface standing crop. Using the data of Billings et al. (1976, 1977) derived from respiration rates of excised and intact root systems under controlled conditions, we may say, as a first approximation, that these graminoids respire about 0.2 mg of CO_2 g dry wt^{-1} hr^{-1} at 0°C and about 0.6 mg of CO_2 g dry wt^{-1} hr^{-1} at 10°C. In the tundra, roots are distributed through a range of temperatures with depth. However, if we are concerned only with a first approximation of root respiration in the field, mean soil temperature may be used. Zero to 10°C would encompass mean soil temperatures if it is remembered that 85% of the root biomass is within the top 10 cm of soil. Thus, we might predict a range of 100 to 370 mg of CO_2 m^{-2} hr^{-1} from roots alone (Peterson and Billings, 1975). Although such predictions are of the right order of magnitude, it appears that they are too high since they approximate the total dark CO_2 flux measured.

The percentage contribution of root respiration to total CO_2 flux is not known with any degree of precision. Billings et al. (1976, 1977), using our 1972 data on soil CO_2 flux from the field combined with laboratory data from intact root systems, estimated that the root and rhizome contribution to CO_2 flux from the soil ranged from 50 to 93% (see Table 4). A root and rhizome contribution of 75% would be close to average for a season in which the true daily mean soil temperature ranged between 0 and 7°C. This compares with an estimated contribution from roots of 40% within the 20 to 25°C summer soil temperature range in a tallgrass prairie in Missouri reported by Kucera and Kirkham (1971).

Peterson and Billings (1975) found that, in late June 1973, the soil contributed about 65% and aboveground living material about 35% of the total dark CO_2 flux from tundra turf. Using the rough estimate of 240 g of CO_2 m^{-2} for dark CO_2 loss from the tundra meadow for the period 9 June to 5 August 1973, it is possible from these data to estimate root and rhizome respiration for the growing season. From the soil temperature–dark CO_2 flux regression in Figure 6, one can make a rough approximation of ecosystem CO_2 loss for the entire season to 1 September.

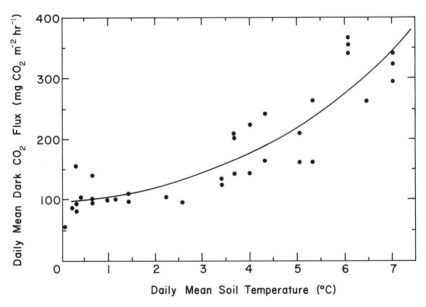

Figure 6.   Daily mean dark CO_2 flux from a tundra meadow as a function of true daily mean soil temperature. The second order quadratic equation for the regression is: $Y = 89.76 + 1.54X + 5X^2$ with a quadratic r^2 of 0.83 at $P \ll 0.001$.

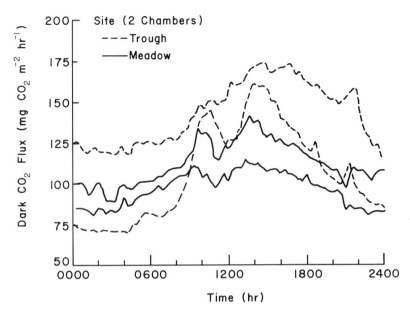

Figure 7.   Diurnal variation in CO_2 evolution into two chambers in a polygon trough site and into two chambers in a meadow site. Note the early morning minimum and the afternoon maximum which are typical of nearly all the data collected. The mid-day depression in the curves corresponds with a decline in air and soil temperatures due to fog.

This figure is about 300 g of CO_2 m^{-2} season^{-1}. Assuming that 65% of this CO_2 originates belowground, total soil respiration would be about 195 g of CO_2 m^{-2} season^{-1}. If 75% of this 195 g of CO_2 m^{-2} season^{-1} is derived from roots and rhizomes, the contribution of root and rhizome respiration to the seasonal CO_2 flux to the atmospheric pool would be about 146 g of CO_2 m^{-2}.

Compared to our estimate of 195 g of CO_2 m^{-2} season^{-1} for total soil CO_2 evolution, Kucera and Kirkham (1971) estimated that soil respiration from a

Table 4.   Estimates of the Portion of Soil CO_2 Evolution (g CO_2 · m^{-2}day^{-2}) due to Roots and Rhizomes in Moist Meadow Flat Polygons at Barrow, Alaska, at Different Times during the Summer of 1972

Time of season	Total soil respiration	Root and rhizome respiration	Percentage of total soil respiration from roots and rhizomes
Late June	2.4–3.6	1.8–2.4	66–75
Mid-July	4.8–5.4	2.4–3.2	50–59
Late July(cold day)	3.0–4.0	2.8–3.6	90–93
Late July (warm day)	5.4–6.6	4.0–4.8	73–74
Early August (cold day)	3.6–4.2	2.8–3.6	78–86
Early August (warm day)	4.2–4.8	3.6–4.3	86–90

tallgrass prairie in Missouri was about 1675 g of CO_2 m^{-2} yr^{-1}. This is about 8.6 times our estimate for the tundra growing season. Summer rates for soil respiration in the tallgrass prairie were 3.5 times those determined for the summer season in grasslands on the Gaspé Peninsula by Lieth and Ouellette (1962). Since the "soil" growing season in Missouri is somewhat longer than that in the Gaspé, it would not be surprising to find the rate per year in the tallgrass prairie would be nearer five times that of the Gaspé grasslands. Using the prairie *annual* soil respiration CO_2 yields as 100%, the other two colder grasslands would be relatively lower: Gaspé at about 20% and the Barrow tundra at about 12% of that in the prairie. In turn, soil respiration under wet and continuously warm secondary vegetation in Costa Rica as measured by Schulze (1967) was 5.5 times greater than the *summer* rates in the Missouri prairie. Since winter rates in the prairie are very low (even zero at times), it is tempting to put these four latitudinal sites in a hierarchy on the basis of CO_2 evolved per year. If in this hypothetical situation, soil CO_2 evolution in the tropics is set at 100%, then that in the prairie would be about 10%, that in the Gaspé grassland would be about 2%, and that in the Barrow tundra about 1% of the annual evolution of CO_2 from the soil in the wet tropics of Costa Rica.

As Reiners (1973) points out, direct measurements of soil CO_2 output by the direct methods used in these studies is of "unquestionable value" but fraught with methodological problems and the possible confounding of contributions by roots and mycorrhizae. As Reiners cautions, all such data available at this time probably reflect relative rather than absolute rates. It is obvious that precise respiration rates from intact root systems in the field are needed. However, even neglecting rhizosphere microorganisms, our rough estimates indicate that the belowground parts of tundra vascular plants play a very important role in the total soil CO_2 flux.

Carbohydrate Accumulation and Allocation

Shaver and Billings (1976) have measured carbohydrate contents of roots and rhizomes of the tundra graminoid plants in relation to tissue age and time of season. Tissue age is very important as a determinant of nonstructural carbohydrate content in these belowground parts. Sugars are concentrated in the growing tips of rhizomes. Polysaccharides are usually low in amount in these tips compared to the amounts in more mature tissues. Lowest levels of polysaccharides are in the oldest living tissues. Storage polysaccharide contents in such a pattern are much the same in plants of all three species. *Carex* roots and rhizomes, however, have much lower sugar contents than those of *Dupontia* and *Eriophorum*.

Storage of carbohydrate reserves in *Carex* is primarily in the rhizomes. Accumulation of carbohydrates in *Dupontia* is divided nearly equally between stem bases and rhizomes. In *Eriophorum,* almost all of the carbohydrate is stored in the stem bases.

Root Growth and Respiration under Controlled Conditions

Root Growth.   Root growth of these tundra graminoids under controlled conditions in the Phytotron has been quantified by Trent (1972), Billings et al. (1973), and Billings et al. (1976, 1977). Under these conditions where environment was steady and optimal, root growth as measured in the wedge-shaped root boxes was uniform and predictable, as opposed to field results. About 200 individual measurements of root elongation in each species were made across a temperature span from -1 to 19°C. Linear regression equations for these growth measurements (Y) against temperature (X) are:

> *Carex aquatilis, Y* = $1.061 + 0.70X$ ($r^2 = 0.896$)
> *Dupontia fisheri, Y* = $1.779 + 0.642X$ ($r^2 = 0.839$)
> *Eriophorum angustifolium, Y* = $2.531 + 0.508X$ ($r^2 = 0.829$).

These regression lines are shown in Figure 8. While linear regression equations fit the data, the point plots show a slight curvilinear trend at the lower temperatures in all three species. Root growth in *Dupontia* may be curvilinear over the whole temperature span used. Root elongation down to -0.5°C was observed in all three species. In view of the fact that *Eriophorum* roots were observed in the

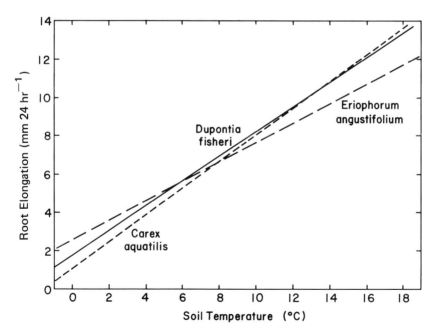

Figure 8.   Linear regression lines of root growth against soil temperature for phytotron-grown plants of *Eriophorum angustifolium* ($Y = 2.531 + 0.508X$), *Dupontia fisheri* ($Y = 1.779 + 0.642X$), and *Carex aquatilis* ($Y = 1.061 + 0.70X$).

field to be growing down against the ice front, it is not surprising that its roots appear to grow more rapidly at low temperatures than do those of the other species.

Actively growing roots were frozen by moving the "artificial permafrost" line upward in the experimental root boxes by lowering the temperature of the coolant bath. A new equilibrium was reached in about 72 hr. The roots were kept frozen for 5 days. After this time, the soil freezing line was lowered again. Resumption of root growth after freezing took place invariably in less than 12 hr. All roots exposed to temperatures between -5 and 0°C resumed growth. Of those frozen between -5 and -9°C, 80% resumed growth. Elongation rates at positive temperatures were identical to those measured before freezing.

Photoperiod and Root Elongation. Late season root elongation rates and onset of root dormancy in *Dupontia* and *Eriophorum* are controlled by decreasing daylength (Shaver and Billings, 1977). The data for *Carex* roots are inconclusive at present in regard to photoperiod; they still grow at daylengths of 15 hr but it is possible that photoperiods of between 12 and 15 hr may halt such growth.

During the growing season, root elongation is not directly limited by low temperatures. However, under controlled experimental conditions with continuous light, root elongation rate can be shown to be correlated highly with soil temperature. Plants of all three species are capable of root growth at temperatures near 0°C.

Daylength experiments under controlled conditions with soil and air temperatures above freezing show that root elongation in *Eriophorum* stops at decreasing

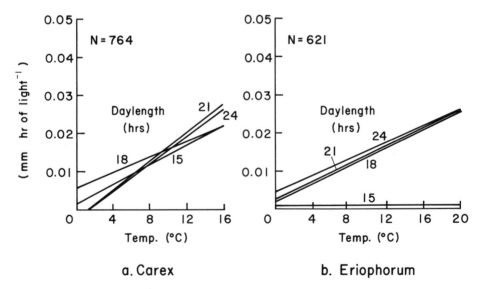

a. Carex   b. Eriophorum

Figure 9.   Root elongation of *Carex* and *Eriophorum* in the phytotron photoperiod experiment in millimeters per hour of light plotted against soil temperature. Numbers beside regression lines denote daylength in hours. N = number of observations.

photoperiods of between 18 and 15 hr (see Figure 9). However, root elongation in *Carex* still continues at daylengths of 15 hr. If the experiment had been carried down to 12-hr photoperiods, a photoperiodic effect might have been detected in *Carex*. Growth cessation in *Dupontia* roots resembles that in *Eriophorum* in that they stop elongating at photoperiods of between 18 and 15 hr. Growth rates in Figure 9 are expressed on a millimeter per hour of light basis to obviate any effect of light intensity upon the supply of available photosynthate allocated to root growth.

Root Respiration.   It has been routine to measure root respiration by excising the root tips and putting them in a respirometer or infrared gas analyzer system. We used the latter. The respiration rates of excised new roots of *Dupontia* and *Carex* at different temperatures appear in Figure 10. The curves are drawn from mean values of numerous measurements made under controlled conditions in the Duke Phytotron. At the same time, respiration rates of intact root systems of all three species were measured (Figure 10). Excised roots of both *Dupontia* and *Carex* respire at higher rates than intact root systems of the same species. Possible reasons for this were that intact whole root systems had some older roots while the measurements on excised roots were confined to young injured roots. These young roots respire from two to six times faster than old excised roots. This is an indication of where most of the respiratory activity takes place in a root system. Injury by excision may have contributed secondarily to this high respiratory rate; however, the data are from steady-state values reached at least 24 hr after excision. Whole root system data are probably more analagous to actual field situations.

The respiratory Q_{10}s of young excised roots of *Carex aquatilis* and of *Dupontia* are both about 2.5. Excised *Dupontia* roots, however, respire at higher rates than do those of *Carex*, especially at higher temperatures.

Conclusions

From our present data, we suggest the following items as working hypotheses.

1. All three of the principal graminoids at Barrow (*Eriophorum angustifolium, Carex aquatilis,* and *Dupontia fisheri*) are similar in foliar morphology and aboveground metabolism. This may be a response to the relatively severe aerial environment. Differences do exist, of course, both genetically and microenvironmentally caused, which allow each species to fit the vegetational and polygon patterns.

2. Belowground, each species is remarkably unlike either of the others. Their adaptations to depth of thaw, soil temperature, soil moisture, and soil nutrients differ considerably. Each species has a different pattern of tiller growth, and of root production and morphology. There are apparent trade-offs between root longevity and root weight per unit length.

3. Root growth under controlled conditions has been shown to be highly correlated with soil temperature. However, under field conditions, a single-factor

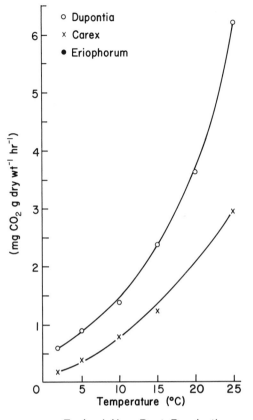

a. Excised New Root Respiration

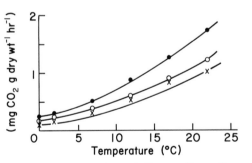

b. Respiration of Intact Root Systems

Figure 10. (a) Respiration rates at different temperatures of excised new roots from plants of *Dupontia fisheri* (circles) and *Carex aquatilis* (crosses) grown under controlled conditions. Each point is the mean of about 40 determinations in an infrared gas analyzer system. (b) Respiration rates of intact whole root systems of *Eriophorum angustifolium* (solid dots), *Dupontia fisheri* (circles), and *Carex aquatilis* (crosses) grown in coarse sand under controlled conditions. Each point, depending upon the species, is the mean of 7 to 12 determinations in an infrared gas analyzer system using the subtraction method.

explanation of plant response is not likely. It appears that other factors such as photoperiod, nutrient levels, soil oxygen tensions, soil moisture situations, permafrost, and depth of thaw are involved. Also, one cannot ignore the strong inherent controls related to the developmental stage of the whole tiller.

4. Resolution of these problems and a better picture of growth dynamics of individual tillers must await more information on root production and turnover rates. It is apparent that a variety of patterns exist in the growth, death, and

distribution of tundra graminoid root systems. The arctic soil environment may not be as restrictive as is generally thought. Rather, a number of adaptive options in root and tiller growth are available.

5. The same complex of environmental factors appears to control root respiration and soil carbon dioxide flux rates under field conditions. However, here also, soil temperature may be controlling the complex. For example, tundra CO_2 evolution is correlated strongly with mean soil temperatures up until early August. Also, soil CO_2 output closely follows diurnal air and soil temperature curves. It is possible that decreasing daylengths in late August and September may affect that part of soil CO_2 evolution originating in roots and rhizomes.

6. In view of net primary production data, the soil and dark ecosystem CO_2 evolution data do not seem high enough for the moist and wet meadow systems to be in a steady state. Rather, peat may be accumulating in place and/or organic matter may be exported to lakes and ponds.

7. The contribution of root and rhizome respiration to soil CO_2 evolution ranges, depending on site and season, from 50 to 90% of the total flux. But the results are not yet as precise as they should be.

8. Distinct patterns of nonstructural carbohydrate accumulation occur in the roots and rhizomes among the tundra graminoid species. Tissue age is an important determinant of total nonstructural carbohydrate concentrations in these belowground plant parts.

9. Decreasing daylength is an important factor in the cessation of root elongation in *Eriophorum* and *Dupontia*. The reception of this photoperiodic cue probably represents an important adaptation of tundra graminoids to the relatively rapid onset of winter.

Acknowledgments. The research reported here was supported by grants from the National Science Foundation Office of Polar Programs to the U.S. Tundra Biome of the International Biological Programme and to Duke University (GV-33850X1). This financial assistance is gratefully acknowledged. The field work was also made possible by the invaluable facilities and help of the Naval Arctic Research Laboratory and its personnel at Barrow. The laboratory work at Duke would not have been possible without the use of the Duke Unit of the Southeastern Plant Environment Laboratory which was supported by NSF grants GB19634 and GB28950-1A from General Ecology. We thank all of our many colleagues in the IBP program whose help enabled us to do the work.

References

Billings, W. D., G. R. Shaver, and A. W. Trent. (1973) Temperature effects on growth and respiration of roots and rhizomes in tundra graminoids. In *Primary Production and Production Processes* (L. C. Bliss and F. E. Wielgolaski, Eds.). IBP Tundra Biome Steering Committee, Stockholm, pp. 57–63.

Billings, W. D., G. R. Shaver, and A. W. Trent. (1976) Measurement of root growth in simulated and natural temperature gradients over permafrost. *Arct. Alp. Res.,* **8**: 247–250.

Billings, W. D., K. M. Peterson, G. R. Shaver, and A. W. Trent. (1977) Root growth, respiration, and carbon dioxide evolution in an arctic tundra soil. *Arct. Alp. Res.,* **9**: 129–137.

Chapin, F. S. III (1974) Phosphate absorption capacity and acclimation potential in plants along a latitudinal gradient. *Science,* **183**: 521–523.

Coyne, P. I., and J. J. Kelley. (1975) CO_2 exchange over the Alaskan arctic tundra: Meteorological assessment by an aerodynamic method. *J. Appl. Ecol.,* **12**: 587–611.

Dennis, J. G. (1968) Growth of tundra vegetation in relation to arctic microenvironments at Barrow, Alaska. Ph.D. dissertation, Duke University, 289 pp.

Dennis, J. G. (1977) Distribution patterns of belowground standing crop in arctic tundra at Barrow. Alaska. *Arct. Alp. Res.,* **9**: 113–127.

Dennis, J. G., and P. L. Johnson. (1970) Shoot and rhizome–root standing crops of tundra vegetation at Barrow, Alaska. *Arct. Alp. Res.,* **2**: 253–266.

Dennis, J. G., and L. L. Tieszen. (1972) Seasonal course of dry matter and chlorophyll by species at Barrow, Alaska. In *Proceedings 1972 Tundra Biome Symposium, Lake Wilderness Center, University of Washington* (S. Bowen, Ed.). U.S. Tundra Biome, pp. 16–21.

Dennis, J. G., L. L. Tieszen, and M. Vetter. (1978) Seasonal dynamics of above- and belowground production of vascular plants at Barrow, Alaska. In *Vegetation and Production Ecology of an Alaskan Arctic Tundra* (L. L. Tieszen, Ed.). New York: Springer-Verlag, Chap. 4.

Douglas, L. A., and J. C. F. Tedrow. (1959) Organic matter decomposition rates in arctic soils. *Soil Sci.,* **88**: 305–312.

Kucera, C. L., and D. R. Kirkham. (1971) Soil respiration studies in tallgrass prairie in Missouri. *Ecology,* **52**: 912–915.

Lieth, H., and R. Ouelette. (1962) Studies of vegetation of the Gaspé Peninsula. II. The soil respiration of some plant communities. *Can. J. Bot.,* **40**: 127–140.

Muc, M. (1972) Vascular plant production in the sedge meadows of the Truelove Lowland. In *Devon Island Project, High Arctic Ecosystem* (L. C. Bliss, Ed.). University of Alberta, Project Report 1970 and 1971, pp. 113–145.

Peterson, K. M., and W. D. Billings. (1975) Carbon dioxide flux from tundra soils and vegetation as related to temperature at Barrow, Alaska. *Amer. Midl. Nat.,* **94**: 88–98.

Reiners, W. A. (1973) Terrestrial detritus and the carbon cycle. In *Carbon and the Biosphere* (G. M. Woodwell and E. V. Pecan, Eds.). U.S. Atomic Energy Commission, pp. 303–327.

Schulze, E. D. (1967) Soil respiration of tropical vegetation types. *Ecology,* **48**: 652–653.

Shaver, G. R. (1976) Ecology of roots and rhizomes in graminoid plants of the Alaskan coastal tundra. Ph.D. dissertation, Duke University, 213 pp.

Shaver, G. R., and W. D. Billings. (1975) Root production and root turnover in a wet tundra ecosystem, Barrow, Alaska. *Ecology,* **56**: 401–409.

Shaver, G. R., and W. D. Billings. (1976) Carbohydrate accumulation in tundra graminoid plants as a function of season and tissue age. *Flora,* **165**: 247–267.

Shaver, G. R., and W. D. Billings. (1977) Effects of daylength and temperature on root elongation in tundra graminoids. *Oecologia,* **28**: 57–65.

Tieszen, L. L. (1972) The seasonal course of aboveground production and chlorophyll distribution in a wet arctic tundra at Barrow, Alaska. *Arct. Alp. Res.,* **4**: 307–324.

Tieszen, L. L. (1975) CO_2 exchange in the Alaskan arctic tundra: Seasonal changes in the rate of photosynthesis of four species. *Photosynthetica,* **9**: 376–390.

Trent, A. W. (1972) Measurement of root growth and respiration in arctic plants. M. A. thesis, Duke University, 75 pp.

Wielgolaski, F. E. (1972) Vegetation types and plant biomass in tundra. *Arct. Alp. Res.,* **4**: 289–306.

Wiggins, I. L. (1953) Systematic and cyto-taxonomic investigation of the vascular plants in the U.S. Naval Petroleum Reserve No. 4. Office of Naval Research and Johns Hopkins University, Final rpt. contract Nonr-248(04), 34 pp.

19.   The Interactions of Organic Nutrients, Soil Nitrogen, and Soil Temperature and Plant Growth and Survival in the Arctic Environment

BRENT H. McCOWN

Introduction

As primary producers in an ecosystem, plants act as *in situ* interfaces between the aerial environment and the soil environment. In this capacity, plants not only convert radiant energy into utilizable chemical energy, fix carbon, and absorb and incorporate soil mineral nutrients, they must also simultaneously integrate these activities between what may be two quite distinct environments which can be sharply delineated at their juncture, the ground surface. One form of this integration between the aerial and edaphic environment is the simultaneous allocation of photosynthate, mainly in the form of carbohydrate, and absorbed minerals. In arctic regions, these relationships are particularly intriguing. Studies have indicated that the lipid and carbohydrate levels and their cycling in some arctic and alpine plants may be of significance in frost tolerance and in the seasonality of growth (by allowing capitalization on the few favorable days in the relatively short growing season; Mooney and Billings, 1960; Bliss, 1962). In addition, the relatively low soil temperatures typical of these regions have been strongly indicated as major factors in limiting growth and survival of plants, especially as far as the availability and utilization of soil nitrogen is concerned (McCown, 1973, 1975).

This project was developed to address the broad question of what aspects of the organic–inorganic nutrient relationships of arctic plants may be of particular significance for their survival and success in the challenging environment of cold-dominated regions. The approach was a comparative analysis of the response of native and nonnative plant species to certain obvious stresses in arctic and alpine environments. The study was limited to graminoids, these being of prominence on the IBP sites as well as being of major importance in rehabilitation programs for disturbed soil surfaces in these regions. A second limit was the restricting of the environmental factors being considered to two soil temperature and soil nitrogen levels. Recent studies (McCown, 1973; Van Cleve, 1973) indicated that these factors, often acting in combination with each other, may be of unique importance in determining both the survival and the distribution of plant species in cold-dominated regions. Thus the original broad topic was narrowed considerably to address the question of what aspects of the organic nutrient status and soil nitrogen utilization may be of particular significance for the survival and success of plants in cold, infertile soils typical of arctic regions. The results are applicable to the practical problem of vegetation of disturbed soils in arctic and subarctic regions. Knowledge of the factors limiting growth and survival of plants in cold-dominated regions is essential in programs attempting to select and develop natural ground covers as well as those designing cultural practices for the maintenance of vegetated surfaces.

Materials and Methods

General Procedure

A field survey of the sugar, polysaccharide, and lipid levels for a range of native plant species was conducted. In addition, observations of the response of plants to a variety of stress conditions in the arctic and subarctic environments were made with the hope of isolating a limited number of environmental factors which might prove to be of major importance in determining successful growth in arctic regions. This latter activity was primarily associated with cooperating research programs throughout the north.

The second phase of research was almost completely laboratory oriented. The field research had keynoted certain environmental factors, namely, soil temperature and soil nitrogen; and further study required the control of these parameters during rather detailed observations of the plant response to changes in their status. Two major experiments were conducted. The first experiment attempted to roughly simulate the cold, infertile conditions of the arctic soil environment and to monitor the growth of a native and nonnative plant species to test if the field observations could be supported in the laboratory. The second experiment was a follow-up of the first and detailed the growth and allocation patterns of plants grown under cold, infertile conditions.

The laboratory studies were difficult in that the growth under very low soil temperature and nutrient conditions was slow and thus a long time period elapsed, usually extending over months, before the results could be utilized. This dictated the use of reliable equipment that could maintain the experimental conditions uniformly over long periods, but also negated the opportunity for conducting frequent follow-up experiments.

An analytical procedure which was rapid, sensitive, and economical was developed. It utilized solvent partitioning and estimated the following classes of compounds:
1. Total lipids.
2. Total nonstructural carbohydrates (defined as all carbohydrates readily available as a source of energy to the plant, thus excluding hemicellulose and cellulose).
   a. Total 80% alcohol-soluble carbohydrates (sugars).

   b. Total freely hydrolyzable polysaccharides (mainly storage polymers,
      e.g., starch, fructans).
With a complementary sampling and determination of dry weight/fresh weight
ratios, the above estimates were based on unit dry weight. Alternatively, the
insolubles (mostly cellulose) after extraction were estimated and used as the unit
weight basis for comparisons.

The theory of separation is based on studies conducted both in animal and
plant laboratories and has proven to be particularly adaptable. Initially, the
sample was macerated and extracted in a three-part, monophasic system containing chloroform, methanol, and water in an established ratio (1:1:0.7). Such a
solvent system offered particular advantages over the common ether extractions
for lipid components:
1. In the presence of alcohol, many of the bound, more polar lipid complexes
    are broken. Thus, such components as proteolipids and glycolipids become
    less of a problem.
2. In wholly organic solvents, some nonlipid substances will be carried into
    solution and counted as lipid material. This is particularly true when large
    amounts of phospholipids are present. The addition of water to the mixture
    prevents this problem and results in a more complete separation of
    components.
After extraction, the monophasic system was broken to form a biphasic
system with a second predetermined ratio of components (2:2:1.8). By detailing
the ratios of the solvent components in the system, an essentially contaminant-
free lipid extract can be obtained which contains close to 100% of the extracted
lipid materials in a chloroform base. The second phase, a water–alcohol mixture,
contained a large percentage of the mono- and oligosaccharides as well as amino
acids, organic acids, etc. The residue was reextracted with hot ethanol and 5%
DMSO (dimethyl-sulfoxide) to remove the remaining low molecular weight
carbohydrates. The remaining undissolved residue contained polysaccharides,
principally storage polymers (starch and fructosans), hemicellulose, and cellulose. The storage polymers are removed by mild acid hydrolysis (Smith et al.,
1964). This was the only suitable and effective extraction method employed in
this laboratory to extract polymers from a wide variety of arctic plants.

The lipid components were estimated gravimetrically after evaporating the
chloroform and drying the residue in a desiccator for 12–24 hr. The carbohydrate
components were estimated colorimetrically utilizing the phenolsulfuric acid
reaction that gave a stable colored product which was read spectrophotometri-
cally (DuBois et al., 1956).

The total procedure produced effective extraction of components (90%+).
Interferences in the lipid analyses were mainly the result of changes in lipid
chemistry during the growing season, thus influencing the ability to compare
gravimetrically estimated samples. The carbohydrate analyses overestimated the
levels since the colorimetric procedure was differentially sensitive to pentose and
hexoses, but all analyses were computed on a hexose (glucose) basis. Other
interferences may also have occurred (Ebell, 1969).

Controlled Environment Experiments

For control of root temperatures independent of air temperatures, large water baths (250 liters and larger) were utilized in walk-in environment rooms. The plant container, which was immersed most of its length in the water bath, consisted of double opaque plastic bags (15 × 20 cm). The medium surface was covered with a 8- to 10-cm layer of loose plastic foam insulation ("Polybeads") both to act as an insulator and to exclude light. This allowed unrestricted tiller production and control of root temperature to within ±1°C throughout the moist medium.

Two media were utilized. For the first preliminary experiment, a peat:perlite (1:8) mix was used (3 liters per bag). This provided an acid, light, organic soil. For the second experiment where nutrient level control was essential and where complete recovery of the root–rhizome complex was essential, a treated clay mineral (arcillite) was utilized (2 liters per bag). The medium was treated with acid (HCl) and its pH adjusted with base (NaOH) to the desired level. After repeated washings with distilled water to remove any free acid or base, the pH remained stable. Arcillite (brand name "Turface") was an ideal medium since it was heavy enough to allow immersion of the bags in the water baths, root systems could be readily and completely washed free of medium, and it contained no nitrogen.

Hoagland's nutrient solution without nitrogen (Hoagland and Arnon, 1939) was used with nitrogen added to the desired level using either ammonium sulfate or sodium nitrate. Nutrient was added only once at the beginning of the experiment. Then water was added to immerse the bags in the bath so that all the medium was at or below the bath water surface. Water was then added as required to resupply losses from transpiration. Root growth of all species was very active and freely penetrated all parts of the medium. No oxygen deficiency symptoms were observed nor were there any nutrient deficiencies seen that were not attributable to the nutrient treatments themselves.

The plants were grown in controlled environment rooms, the first experiment at the Institute of Arctic Biology, University of Alaska (24-hr photoperiods, 1000 to 3000 foot-candles of light at plant height, air temperature of 10–15°C, and relative humidity of 40 to 50%) and the second experiment at the BIOTRON, University of Wisconsin (20-hr photoperiod, 1500 foot-candles at plant height, air temperature of 15 ± 2°C, and a relative humidity of 50 to 70%). Bath temperatures were adjusted as appropriate to maintain the desired soil temperatures (±1°C).

Three selections of graminoids were utilized. *Dupontia fisheri* was used as a representative arctic grass typical of the IBP field sites and adapted to growth in low temperature, infertile soil conditions. The temperate grass used in the first experiment was a clone of brome grass, "Manchar Brome," an introduced grass utilized successfully in Alaska in warm soil areas but generally unsuccessful in cold soil regions. In the second experiment, a clone of bluegrass, "Merion Blue," was used because of its very close similarity in stature and growth habit to that of *Dupontia*. All these clones were rhizomatous under normal conditions;

however, the bluegrass during the time period of these experiments produced principally intravaginal tillers. Initial transplants were young tillers (*Dupontia*, bluegrass) or young seedlings (Brome).

Results

Seasonal and Spatial Variation of Organic Nutrient Levels

Generally, these species showed similar levels and were highly synchronous throughout the season (Figure 1) in changes in these levels. Sugar levels were relatively constant and remained at a high concentration throughout the season. However, polysaccharide levels, although consistently higher than the sugar concentrations, showed considerably more seasonal fluctuations. The highest polysaccharide concentrations were present at the earliest sampling; the lowest polysaccharide levels were found at peak season. Little indication of an accumulation of carbohydrates in leaf tissues in the fall was observed. Similar trends were apparent in the two other major graminoids at Barrow *(Carex aquatilis, Eriophorum angustifolium).*

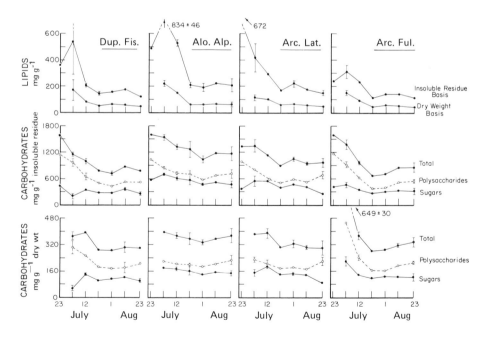

Figure 1.   Seasonal levels of organic nutrients in four species at Barrow, Alaska, during 1971. Points are presented with a ± standard error of the mean. Carbohydrate data are expressed as mg of glucose equivalent per unit weight. Dup. Fis., *Dupontia fisheri;* Alo. Alp., *Alopecurus alpinum;* Arc. Lat., *Arctagrostis latifolia;* Arc. Ful., *Arctophila fula.*

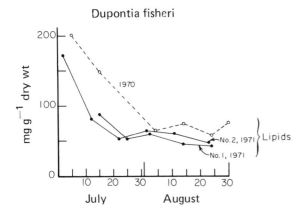

Figure 2. Seasonal levels of total sugars and polysaccharides during 1970 and 1971 at Barrow, Alaska. Separate experiments at two different sites were conducted in 1971. Data are expressed as mg of glucose equivalent per unit weight.

Lipid levels showed a surprising fluctuation early in the season which was roughly coincident with the early season polysaccharide changes. A definite peaking of levels was observed with three of the four species in early July. The air-dried lipid extracts showed a characteristic liquid residue at this time in contrast to the solid residues observed at peak season.

Comparison of yearly (1970 and 1971) and site variation at Barrow supported the constancy of sugar levels but showed significant site variation in the polysaccharide levels (Figure 2). Lipid levels (Figure 3) were nearly identical; and although the seasonal trends were similar, the time the observed changes occurred was different between seasons. Since the early season was cooler in 1970 than in 1971, the skewing of the curve to later season in 1970 may be logical and may indicate a correlation of these lipid levels with normal growth responses.

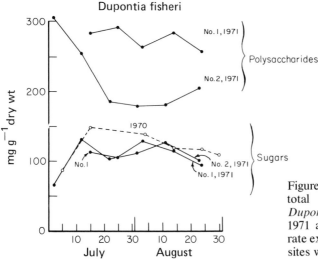

Figure 3. Seasonal levels of total lipids in the leaves of *Dupontia fisheri* during 1970 and 1971 at Barrow, Alaska. Separate experiments at two different sites were conducted in 1971.

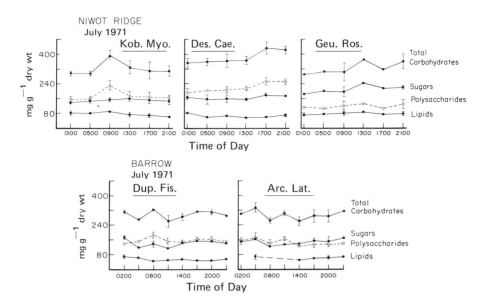

Figure 4. Diurnal levels of organic nutrients in plants at Barrow, Alaska, and Niwot Ridge, Colorado, during 1 day in 1971. Points are presented with ± standard error of the mean (N = 2 or 3). Carbohydrate data are expressed as mg of glucose equivalent per unit weight. Kob. Myo., *Kobresia myosuroides;* Des Cae., *Deschampsia caespitosa;* Geu. Ros., *Geum rossii;* Dup. Fis., *Dupontia fisheri;* Arc. Lat., *Arctagrostis latifolia.*

Diurnal Variation of Organic Nutrient Levels

Marked diurnal fluctuation of the levels of organic nutrients was not observed, either at Barrow or at Niwot Ridge, Colorado (Figure 4). These observations show that the levels observed through the season are realistic evaluations of daily means. Figure 4 also provides a comparison of monocot species, *Kobresia myosuroides* and *Deschampsia caespitosa,* with a dicot species, *Geum rossii.* Although total levels were similar, the dicot species showed lower concentrations of polysaccharides at this peak season measurement which is consistent with the sugar/starch measurement of dicots by Mooney and Billings (1960). The concentrations in the monocots from Barrow were similar to those found at Niwot Ridge.

Comparison of Organic Nutrient Levels in Different Plant Structures

Stem tissues could be obtained from only a few species, but analyses of the tissue received indicated that stems contained similar levels of sugars as leaf tissue, but higher levels of polysaccharides (Figure 5). The seasonality of these levels was similar to that observed in leaf tissue. Nongreen tissues generally had lower sugar and lipid levels but similar or higher polysaccharide levels than green tissues from the same species. In contrast to the seasonal levels of sugars in leaf tissue, sugar levels in stem bases declined to a minimum at peak season (Figure

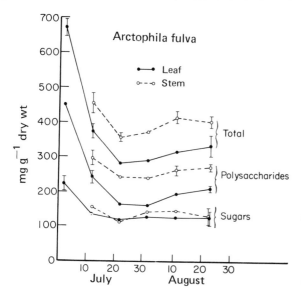

Figure 5. Seasonal levels of organic nutrients in the stem and leaf tissue of *Arctophila fulva* at Barrow, Alaska, during 1971. Points are presented with ± standard error of the mean (*N* = 2). Data are expressed as mg of glucose equivalent per unit weight.

6). Polysaccharide levels were relatively constant in stem bases until after peak season when an accumulation was observed. Lipid levels remained relatively constant with a minimum at peak season.

Organic nutrient levels in live rhizomes generally paralleled those observed in the stem bases (Figure 7). The organic nutrient content of live roots was highly variable, probably due to the difficulties in separation and sampling. Free sugars were at very low levels in roots during peak season and at soil depths greater than

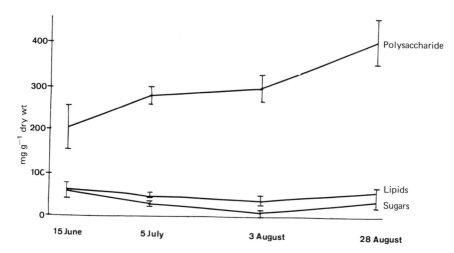

Figure 6. Seasonal levels of organic nutrients in stem bases of graminoids at Barrow, Alaska, during 1971. Samples were from cores and are a composite of all species. Points are presented with ± standard error of the mean (*N* = 3).

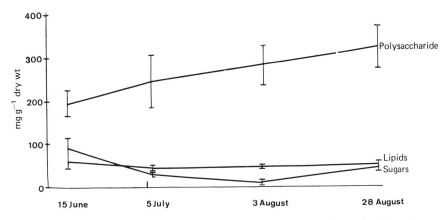

Figure 7.   Seasonal levels of organic nutrients in live rhizomes of graminoids at Barrow, Alaska, during 1971. Samples were from cores and are a composite of all species. Points are presented with ± standard error of the mean ($N = 3$).

5 cm. Inflorescences from three species did not show levels of organic nutrients markedly different from those observed in leaf tissue.

Samples of decomposing leaf tissue showed very low levels of free sugar (less than 10 mg g^{-1} dry weight). Levels of polysaccharide and lipids showed no consistent trends either over time or between species. Insolubles (principally cellulose) initially constituted 35 to 40% of the dry weight of plant parts, but after one season of decomposition, this fraction constituted 50 to 60% of the dry weight. The difficulties in sampling and the inclusion of fungal biomass probably were complicating factors.

Organic Nutrient Levels in Nonnative Plant Species Grown under Natural and Modified Arctic Environments

The organic nutrient levels in an introduced graminoid (rye) grown in pots under the natural Barrow environment were not dissimilar from those of the native graminoids (Table 1). Modifications of the natural environment, which resulted in increased environmental temperatures and reduction in desiccation stresses, produced rye plants with generally lower concentrations of carbohydrates in the leaf tissues. Plants grown at Prudhoe Bay had lower carbohydrate concentrations than those grown at Barrow, again reflecting the generally warmer environment at Prudhoe Bay.

The carbohydrate levels of an introduced dicot (pea) grown under natural Barrow conditions showed lower polysaccharide levels than the native monocots or the introduced rye, a trend remarkably similar to the comparison of arctic and alpine monocots to an alpine dicot (Figure 4). In contrast to the introduced rye, the environmental modifications produced increased concentrations of carbohydrates in the introduced dicot (Table 1). There was no apparent correlation between sugar concentrations in the leaves and the rates of respiration and

Table 1.   The Response of Nonnative Plants to Four Environmental Situations on the Arctic Coast of Alaska[a]

Species	Location	Growth (percentage of maximum biomass accumulation)	Root/shoot	Respiration rate (mg of CO_2 dm^{-2} hr^{-1}; 15°C)	Photosynthesis rate (mg of CO_2 dm^{-2} hr^{-1}) (900 μE m^{-2} sec^{-1})	Concentration of carbohydrates in leaves (mg g^{-1} dry wt)	
						Sugars	Total
Annual rye grass	Barrow Outside	1	0.46	2.5	10.6(2.9)	192(21)	379(18)
	Prudhoe Bay Outside	12	0.67	1.4	23.6(1.7)	101(8.1)	274(11)
	Barrow Outside enclosure	45	0.22	1.0	25.0(3.2)	174(8.8)	319(14)
	Barrow Greenhouse	100	0.76	1.0	24.8(0.7)	89(11)	244(29)
Pea	Barrow Outside	16	0.57	3.6	7.6(4.0)	85(2.5)	190(8.8)
	Prudhoe Bay Outside	36	0.56	1.4	22.4(1.1)	148(22)	285(44)
	Barrow Outside enclosure	100	0.23	2.2	22.5(2.1)	260(9.2)	373(26)
	Barrow Greenhouse	90	0.10	0.4	14.8(1.5)	139(5.9)	422(25)

[a] Plants were planted in pots, three to five pots per treatment. The locations at Prudhoe and in the outside enclosures had mean temperatures approximately 2°C warmer than Barrow; the greenhouse had a mean temperature of 20 to 25°C, about 20°C warmer than Barrow. Numbers in parentheses are standard errors. All but the carbohydrate analyses were conducted by Tieszen.

photosynthesis for these leaves. In both the rye and pea, the highest photosynthesis rates were associated with the higher leaf sugar levels.

Response of Arctic and Temperate Graminoids to Low Soil Temperatures and Low Soil Nitrogen Levels

Field observations of vegetation test sites along the proposed trans-Alaska pipeline route and research conducted on the Hot-Pipe test facility at Fairbanks, Alaska, provided strong evidence that low soil temperatures and infertile conditions were of major importance in determining the survival and distribution of plant species in arctic and subarctic regions. Introduced species in vegetation trials often showed classical nutrient deficiency symptoms in cold soil areas and indeed these regions were some of the more difficult to vegetate (Van Cleve, 1973). In contrast, increases in biomass production of 300 to 500% over control areas have been recorded in regions where the soils were artificially warmed as by buried, heated pipelines (McCown, 1973).

These field observations were supported by the results of the first laboratory study. A reduction of soil temperature of 18°C (to 5°C) resulted in large aboveground biomass differences, especially with the nonnative species, Brome, where the biomass production at 5°C soil temperature and high soil nitrogen levels was less than 10% that at 15°C (Figure 8). Interestingly, brome did not respond to increased levels of soil nitrogen at the 5°C soil temperature, indicating that such unadapted species were not capable of utilizing soil nutrients at low soil temperatures. In contrast, the adapted species, *Dupontia,* showed marked increases in biomass accumulation in response to increased soil nutrient levels, even at soil temperatures of 5°C. Analyses of *Dupontia* plant composition and tiller growth rates indicated that the biomass increases in response to increased soil fertility were principally the result of increased vegetative reproduction by rhizomes leading to greater stem densities under the higher nutrient levels (Figure 8). Individual tiller growth was not markedly different between the two clones under any of the test conditions (Figure 8).

The total available carbohydrate levels in the shoots of *Dupontia* were not strongly influenced by soil temperature, but did decrease with increased levels of soil nitrogen (Table 2). Considering the large increase in stem density in this species that resulted from the increased soil nitrogen levels, a greater utilization of leaf carbohydrate in the nitrogen-stimulated rhizome and leaf growth is probable.

A second non-arctic species, bluegrass, utilized in the second controlled-environment experiment, was not as inhibited by low soil temperatures as was brome. However, of the three species tested, the bluegrass clone was unique because it produced abundant new shoots via intravaginal tiller production and thus circumvented the necessity to first develop a rhizome system. Even so, the stem density of bluegrass was still reduced a greater amount by decreased soil temperature than was the arctic species (Table 3).

At higher soil temperatures (8 to 12°C), there were indications of differences in preference for nitrogen sources between the arctic and temperate grasses. The

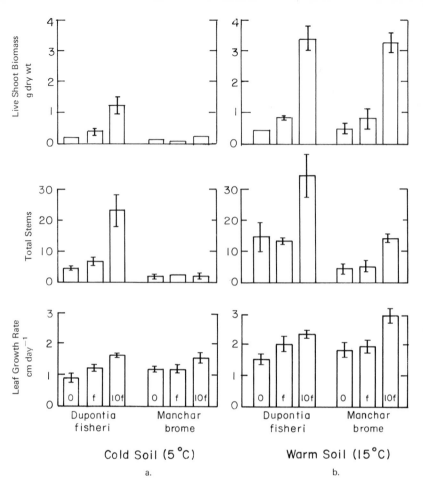

Figure 8.   The influence of soil temperature and soil ammonium concentration on the growth and vegetative reproduction of a native arctic grass, *Dupontia fisheri,* and an introduced grass, Manchar Brome. Nutrient levels: 0 = no ammonium added to medium; f = ammonium concentration approximately equal to arctic field concentrations (0.05 mg of NH_4^+ g^{-1} of medium); and 10f = 0.5 mg of NH_4^+ g^{-1} of medium.

temperate bluegrass clone showed greater biomass production under nitrate nutrition than ammonium nutrition (Table 4). As was observed earlier, this greater biomass accumulation appeared to be principally the result of an increase in number of growing shoots and not an increase in the rate of individual shoot growth (Table 5). Interestingly, any difference in growth between sources of nitrogen disappeared at the cold soil temperature of 4°C. Not nearly as dramatic as with the nitrate effects on bluegrass, *Dupontia* showed only a slight difference in biomass production with the two nitrogen sources (Table 6). However, in contrast to bluegrass, ammonium appeared to be the slightly preferred source for the arctic grass.

Table 2. The Interaction of Soil Nitrogen Levels and Soil
Temperatures with the Carbohydrate Concentration of Shoots
of a Native Arctic Grass, *Dupontia fisheri* [a]

Soil temper-ature (°C)	Soil nitrogen treatment (mg of N available per plant)	Total available carbohydrates in shoots (mg g^{-1} dry wt)
5	1	299(9)
	25	259(13)
	250	193(6)
15	1	250(7)
	25	212(11)
	250	197(23)

[a] Values in parentheses are standard errors of the mean ($N = 3$).
(Adapted from McCown, 1973.)

The relationship between the root–rhizome production and shoot production
under the various soil temperature–nutrient regimes differed between the arctic
and temperate clones. With *Dupontia,* no marked changes in plant composition
were observed with changes in soil temperature or soil nitrogen (Table 6). In fact,
the plant composition, particularly the percentage root, remained remarkably
constant. In contrast to *Dupontia,* the root/shoot ratios of the temperate bluegrass clone tended to decrease with decreased soil temperature (Table 4). As the
soil temperature decreased, a greater percentage of the bluegrass clone consisted
of shoot biomass and a lesser percentage of root biomass. However, there was a
tendency for a greater rhizome production at the lower soil temperature. As with
Dupontia, these relationships were not markedly influenced by nitrogen source
or levels.

Probably the most marked difference between the bluegrass and the *Dupontia*

Table 3. The Effect of Soil Temperature on Stem Density of Three Species of
Graminoids [a]

Differences in soil tempera-tures (°C)	Species	Percentage difference in number of shoots with leaves (warm soil–cold soil)
10	*Dupontia fisheri* (Arctic grass)	34
10	Manchar Brome (Temperate grass)	84
8	*Dupontia fisheri* (Arctic grass)	32
8	Merion Bluegrass (Temperate grass)	50

[a] Data obtained from those treatments that had soil nitrogen levels yielding the greatest
growth of each species.

Table 4. The Growth Response of a Temperate Grass Clone, Merion Bluegrass, to Alterations in the Nutrient and Temperature Status of the Soil[a]

Treatment		Biomass relationships				Composition (percentage of total biomass)		
Nitrogen source and level (mg of N plant[-1])	Soil temperature (°C)	Dry weight per plant (g)	Percentage of highest in all treatments	Percentage of highest in each treatment	Belowground/ aboveground	Shoot	Root	Rhizome
Ammonium								
50	12	4.3c	59	100	0.74a	58	42	1
	8	2.3de	32	54	0.58a	63	37	1
	4	0.9f	12	21	0.41b	71	27	2
150	12	3.1cd	42	100	0.66a	60	39	2
	8	2.0def	27	65	0.48b	68	31	1
	4	0.9f	12	29	0.41b	71	28	1
Nitrate								
50	12	5.7b	78	100	0.74a	58	42	1
	8	3.7c	51	65	0.70a	59	37	4
	4	1.0ef	14	18	0.48b	68	29	3
150	12	7.3a	100	100	0.65a	61	39	1
	8	4.2c	58	58	0.75a	57	38	5
	4	1.3ef	18	18	0.46b	69	30	1

* Plants grown at the BIOTRON. University of Wisconsin, and harvested after a minimum of 1.5 g of growth. Air temperatures were 15°C. Biomass was adjusted to a 50-day growing period. Figures followed by a common letter are not significantly different at the 0.01 level (Duncan's new multiple range test).

Table 5.   The Response of Leaf Growth Rates of a Temperate Grass (Bluegrass) and an Arctic Grass *(Dupontia fisheri)* Clone to Edaphic Conditions of Soil Temperature and Soil Nitrogen

Nutrient regime (mg of N plant^{-1})	Soil temperature during measurement (°C)	Leaf growth rate (cm extension day^{-1})	
		Bluegrass	*Dupontia*
Ammonium			
50	12	2.2	2.3
	8	0.9	1.2
	4	0.4	—
	2	0.2	0.3
150	12	2.1	2.3
	8	1.2	1.3
	4	0.5	—
	2	0.3	0.3
Nitrate			
50	12	1.6	2.2
	8	1.0	1.3
	4	0.4	—
	2	0.2	0.2
150	12	2.1	1.6
	8	1.0	1.1
	4	0.4	—
	2	0.2	0.3

clones was the amount of dead and senescing aboveground tissue, especially at the lower soil temperatures. Under all low soil temperature conditions, the amount of senescent aboveground tissue was greater with the temperate bluegrass clone than with the arctic clones. The percentage of senescent materials in the aboveground biomass in the bluegrass clone increased as soil temperature decreased (Table 7). This response was more intense under ammonium nutrition than under nitrate nutrition. With *Dupontia,* little senescent tissue was present under any condition and differences between environments and treatments were insignificant.

This difference in leaf senescence between the arctic and temperate clones was even more evident when a sudden shift to lower soil temperature occurred. With a shift of 12°C to 2°C soil temperature, 16 to 30% of the aboveground biomass of the bluegrass clone senesced within 20 days (Table 8). However, the *Dupontia* clone senesced less than 10% of its aboveground biomass during this same period.

Discussion and Conclusions

The results of these experiments and observations lend strong support to the proposition that plant growth in at least some portions of the arctic and subarctic is limited by edaphic conditions, more specifically the combined effect of low soil temperatures and low soil nitrogen levels. No evidence has been obtained that

Table 6. The Growth Response of a Native Arctic Grass, *Dupontia fisheri*, to Alterations in the Nutrient and Temperature Status of the Soil[a]

Treatment		Biomass relationships						Composition (percentage of total biomass)		
			Total live biomass		Biomass ratios					
Nitrogen source and level (mg of N plant⁻¹)	Soil temperature (°C)	Dry weight per plant (g)	Percentage of highest in all treatment	Percentage of highest in each treatment	Belowground/aboveground	Root/shoot		Shoot	Root	Rhizome
Ammonium										
50	12	3.5^{bc}	67	100	1.14^{abc}	0.71^{abc}		47	33	19
	8	2.4^{d}	46	68	1.50^{a}	0.91^{a}		41	36	24
	4	0.8^{f}	15	23	1.14^{abc}	0.75^{abc}		47	34	19
150	12	5.2^{a}	100	100	0.94^{c}	0.58^{c}		52	30	19
	8	2.6^{cd}	50	50	1.40^{ab}	0.72^{abc}		42	32	26
	4	0.8^{f}	15	15	1.07^{bc}	0.67^{bc}		49	32	19
Nitrate										
50	12	3.7^{b}	71	100	1.17^{abc}	0.73^{abc}		47	33	20
	8	2.0^{de}	38	54	1.53^{a}	0.85^{ab}		40	33	27
	4	0.8^{f}	15	22	1.20^{abc}	0.74^{abc}		47	33	20
150	12	3.6^{b}	69	100	1.14^{abc}	0.66^{abc}		47	32	21
	8	2.5^{cd}	48	69	1.61^{a}	0.90^{a}		39	35	27
	4	1.1^{ef}	21	31	1.33^{abc}	0.74^{abc}		47	34	19

[a] Plants were grown at the BIOTRON, University of Wisconsin, and harvested after a minimum of 1.5 g of growth. Aerial temperature were 15°C. Biomass was adjusted to a 50-day growth period. Figures followed by a common letter are not significantly different at the 0.01 level (Duncan's new multiple range test).

Table 7. The Relative Amount of Live Aboveground Tissue Supported by a Temperate Grass, Bluegrass, at Various Soil Temperatures and Soil Nitrogen Levels[a]

Nutrient regime (mg of N plant[-1])		Live tissue as percentage of total aboveground biomass		
		Soil temperature		
		4°C	8°C	12°C
Ammonium	50	73	84	>95
	150	73	85	>95
Nitrate	50	86	90	>95
	150	81	92	>95

[a] The arctic *Dupontia* clones showed insignificant differences between treatments (greater than 90% live tissue in all cases).

indicates the processes associated with photosynthate production and translocation are a limitation to biomass production under environments typical of the Barrow, Alaska, region. Without exception high and more-than-adequate available carbohydrate levels were observed in plants growing under arctic conditions. That these high concentrations of available carbohydrate, generally constituting more than one-fourth the dry weight of the tissues, also develop rapidly in nonnative species indicates that such levels are in themselves probably not specific adaptations to the arctic environment. Instead, they might be considered as manifestations of the limitation on photosynthetic utilization rates caused by such factors as the unavailability of inorganic nutrients.

In regard to the actual rates of photosynthesis, we have found no evidence to support the theory that high tissue concentrations of soluble carbohydrate (sugars) are inhibiting to CO_2 fixation rates (for a discussion, see Neales and Incoll, 1968). Quite to the contrary, the experiments with nonnative species growing under arctic conditions showed highest potential photosynthesis rates accompanied by the highest as well as the lowest tissue carbohydrate concentrations.

Table 8. The Increase in Senescence of Leaves of a Temperate Grass, Bluegrass, in Response to a Sudden Change in Soil Temperature[a]

Nutrient regime (mg of N plant[-1])		Percentage loss of live aboveground biomass after shift in soil temperature[b]
Ammonium	50	21 (4.5)
	150	31 (2.5)
Nitrate	50	20 (3.4)
	150	16 (1.1)

[a] The arctic *Dupontia* clones showed insignificant losses (less than 10%) during this period.
[b] Plants were grown at 12°C soil temperature for 50 days then shifted to 2°C for 20 days. Figure in parenthesis is range in values, 5 plants per treatment.

The levels and seasonal trends in carbohydrate content are quite uniform among arctic graminoid species and show less fluctuation than alpine dicots (Mooney and Billings, 1960). Monocots under arctic conditions maintain a stable sugar pool in the leaves and stems throughout the season. In addition, a large available polysaccharide component is present, but polysaccharide levels fluctuate seasonally and between sites. The sugar pool may represent the "active" pool with a high turnover and flux rate; short-term excesses and deficiencies may be overcome by exchange with the polysaccharide storage pool or by direct supply from photosynthesis. Stems, when present, appear to act as aboveground storage centers as do the stem bases and rhizomes belowground. In contrast to the aboveground tissues, the sugar pools show somewhat more intense seasonal cycles in the stem bases and rhizomes, which may be a function of active withdrawal without ready resupply from photosynthesis. However, large fluctuations in the levels of carbohydrate in these belowground tissues are not apparent and thus represent another departure from the situation present in alpine dicots. Active cycling and interchange of carbohydrate among all these pools does occur, as demonstrated by Allessio and Tieszen (1975), and, as with the aboveground tissues, at no time do the plants appear to be carbohydrate deficient.

A recurring observation is that the levels of nutrients at the beginning of the sample periods (after snowmelt) do not match those levels at the end of the sampling periods (frost). This is particularly evident with the lipid levels. Such a result may in large part be an artifact due to tissue quality changes, the early season tissue having smaller cells and less structural components than later season tissue. In any case, a considerable amount of activity is indicated during late fall, winter, and/or early spring. Allessio and Tieszen (1975) have shown that ^{14}C translocation can occur in *Dupontia* roots in frozen soils. Thus a considerable number of processes may be operating during the frozen soil periods. However, these results could also represent an actual storage of "precursors" (e.g. amino acids, fatty acids, mineral nutrients) for growth in the green overwintering leaf bases. Indeed, the quality changes in the extracted lipids during this early season period indicate that just such a process may be occurring. Such activity would allow the plant to respond with rapid growth early in the season with a minimum of biosynthesis in the otherwise inhibitively cold conditions.

Compared to the temperate graminoids investigated, the *Dupontia* plant is strongly adapted for the rhizomatous growth habit. Even under the extreme conditions of low soil fertility and low soil temperatures, heavy allocation of nutrients for rhizome formation and maintenance is apparent as evidenced by both the biomass data and the tracer studies of Allessio and Tieszen (1975). With improved conditions, as with fertilization or soil warming, *Dupontia* responds with rapid rhizome growth and eventual tiller production, culminating in an increased density of tillers. Major increases in individual tiller growth rates are not observed with changes in fertility levels, but are apparent in the response to changes in soil temperature. However, this response in itself is not atypical of northern, cool season, festucoid grasses which are known to respond to changes in the environment principally through changes in tillering rates (Mitchell, 1956). Indeed, the temperate bluegrass clone demonstrated this response, but, in contrast to *Dupontia*, increased tiller number was a result of intravaginal formation. The rhizome-dominated response may thus represent a general response of northern grasses in which allocations are made preferentially to ground level (stem bases) or belowground (rhizome) structures, thus ensuring an abundant supply of perennating organs.

In the arctic environment, tiller production via the rhizomatous habit has the obvious advantages of a more rapid exploration of new ground, a decrease in the competition between tillers, and a spacing of growing points in an environment subject to periodic heavy grazing. Since carbohydrate is not a limiting factor for arctic plant growth and since rhizomes, being nongreen tissues, demand comparatively low amounts of the scarcer mineral elements, such a habit would appear to have a high selective value.

Generally, root:shoot ratios have been observed to increase with either decreases in soil temperature or soil fertility (see Nielson and Humphries, 1966, for a review). This response has been interpreted as an acclimation mechanism compensating for the decreased nutrient uptake by an increase in the absorbing root biomass (see Chapin, 1974, for a discussion). However, the elucidation of such a response is highly complicated by the strong interaction of soil nutrient availability and soil temperature. Although the direct effect of soil nutrients, particularly nitrogen, on plant composition has been demonstrated, the effect of soil temperature has not been clearly separated from its influence on soil nutrient availability and absorption phenomena. Indeed, the high aboveground–belowground ratios often observed in tundra regions may in large part be explained by the low prevailing soil nutrient levels (Haag, 1974) and the predominant rhizomatous (or analogous) habit. The prevailing low soil temperatures in themselves have not been shown to be determinant except as they influence the availability and cycling of nutrients.

In this study, the response in plant biomass composition of both the arctic and temperate clones to the various edaphic nutrients and temperature conditions appears to be in direct conflict with the above general picture. The root:shoot ratio was either insensitive *(Dupontia)* or actually decreased with decreasing soil temperature (bluegrass); and in both species the root:shoot ratios were insensitive to nutrient level. One explanation of this seemingly contradictory response is that the factors of nutrient and temperature were essentially separate and acted rather independently in these experiments. The only significant increase in biomass in response to increased nitrogen fertility was at the warm soil temperatures (12°C) and with the preferred nutrient source (ammonium for *Dupontia;* nitrate for bluegrass). Since increases in nitrogen levels did not result in increased growth at 4 and 8°C soil temperature, it may be concluded that the soil nitrogen levels were not limiting (enough nitrogen was present to adequately supply plant requirements) or that the absorption of nitrogen was so inhibited at the lower temperatures that nutrient concentrations were unimportant. The latter conclusion is highly unlikely considering the analogous work of Chapin (1974) where the capacity to absorb a nutrient such as phosphate increased with decreasing soil temperatures. Even when increases in biomass were observed with increased fertility, the root:shoot ratios did not change significantly for either *Dupontia* or bluegrass. Thus it appears that these experiments may utilize a system that is relatively insensitive to soil nutrient concentration and therefore afford an opportunity to observe the direct effects of soil temperature on plant composition.

If the effect of soil fertility can be separated from the effect of soil temperature, then because of their higher temperature optima and their greater temperature sensitivity (Chapin, 1974), decreases in soil temperature should have a stronger inhibitory effect on the temperate grasses than on the arctic grasses. The observed decreased root content of the bluegrass at the lower soil temperature may indicate direct inhibition of root growth by temperature. In contrast, *Dupontia* showed an ability to maintain the same plant composition, regardless of soil temperature, demonstrating that root growth may not have been so strongly inhibited. However, an alternative interpretation would be that although root growth was inhibited in both species by low soil temperatures, *Dupontia* maintained a more optimal relationship between root and shoot biomass than did bluegrass. In either case, the greater senescence observed with bluegrass at the lower soil temperatures, especially after sudden shifts, lends support to the proposition that root function in bluegrass was not adequate to support shoot growth. The specific root functions involved are unknown (although nitrogen metabolism is probably not involved), however other nutrients (e.g., phosphorus, calcium, iron) and organic growth factor syntheses (Luckwill, 1960; Torrey, 1976) may be involved. The increased senescence of bluegrass under ammonium nutrition, which requires root organic syntheses, may support the latter possibility. Senescence of leaves of cuttings can be inhibited by factors produced in root systems, and this effect can at least partially be attributed to cytokinins.

In any case, the ability of *Dupontia* to maintain a seemingly optimal plant root–shoot composition under varied environments may represent an important and highly advantageous adaptation. Arctic plants may have evolved effective control systems whereby the allocation of photosynthate and nutrient is limited to the production of tissue with full potentialities for growth. Nutrient-deficient tissue may be rare, even under low soil nutrient levels. The production of metabolically insufficient tissue in an environment frequented by growth-limiting nutrient levels and extremely short growing seasons, punctuated by vegetation loss from frost and grazing, could be disastrous for the long-term survival of a species.

In comparing the laboratory studies presented here with the field and laboratory studies of other researchers, one should be highly cognizant of a number of areas of divergence which may at least in part account for the observations obtained. The medium, arcillite, appears to provide an ideal environment for root growth and function. In fact, experiments comparing this medium to the standard liquid culture indicate that plant growth appears stimulated in the presence of arcillite, and that this occurs for a wide range of species (McCown, 1974). This may be the result of a reduction in the importance of soil factors that may often be limiting to plant growth (e.g., availability of micronutrients; ion concentrations). Root separation and retrieval from the medium is complete and rapid with a minimum of washing, thus facilitating rather exact biomass and biochemical measurements. In addition to an optimal root environment, the BIOTRON provides an aerial environment with high humidities, nondesiccating light intensities, and cool temperatures, all of which might tend to optimize succulent shoot growth. Finally, the plants analyzed were generally young and in active growth. Thus the amount of structural tissue in the biomass measurements was minimal (in contrast to field experiments) and the biomass ratios presented may more closely represent activity ratios.

The edaphic environment at Barrow is primarily an ammonium-dominated system whereas the typical edaphic condition in which most temperate graminoids grow is a nitrate-dominated environment. Thus the preferences displayed by *Dupontia* for ammonium and bluegrass for nitrate are not surprising and represent adjustments to the selective pressures of the respective habitats. However, another consequence of the ammonium-dominated system of the arctic is the necessity of incorporating the ammonium into an organic molecule soon after absorption in the root to prevent ammonium toxicity. Such detoxification places added requirements on the translocation system to provide the carbon skeleton as well as on the metabolism of the root system to synthesize the organic molecule, usually glutamine. Since these activities necessarily have to occur in the edaphic environment, an intriguing hypothesis is that these processes may be an important factor in the limitation of growth resulting from low soil temperatures in the arctic environment, especially regarding nonarctic species. In addition, adjustments in the metabolism of ammonia may represent important adaptations to arctic conditions. If the hypothesis has validity, one line of evidence would be a differential response to soil temperature under ammonium or nitrate nutrition, especially for nonarctic species. The results of this project indicate that growth of a temperate grass under ammonium nutrition is not more intensely inhibited by low soil temperatures than under nitrate.

Acknowledgments. This research was supported by the U.S. Army Cold Regions Research and Engineering Laboratory under DA Task 4A061101A91D03, In-House Laboratory Independent Research, Work Unit 147, Chemical Indicators of Arctic Biological and Environmental Activities, and by the National Science Foundation under subcontract from Grant GV 29342 (University of Alaska) to the University of Wisconsin (Madison) and by Grant GV 29343 to Augustana College. It was performed under the joint NSF sponsorship of the International Biological Programme and the Office of Polar Programs and was directed under the auspices of the U.S. Tundra Biome.

The author wishes to thank many Tundra Biome associates for their cooperation, especially Dr. Larry Tieszen, and the BIOTRON staff and Directors. The assistance of Ms. Deborah McCown in the design and accomplishment of these analyses was invaluable.

References

Allessio, M. L., and L. L. Tieszen. (1975) Patterns of carbon allocation in an arctic tundra grass, *Dupontia fischeri* (Gramineae), at Barrow, Alaska. *Amer. J. Bot.*, **62**: 797–807.
Bliss, L. C. (1962) Adaptations of arctic and alpine plants to environmental conditions. *Arctic*, **15**: 117–144.

Chapin, F. S., III (1974) Morphological and physiological mechanisms of temperature compensation in phosphate absorption along a latitudinal gradient. *Ecology,* **55**: 1180–1198.

Du Bois, M., K. A. Gilles, J. K. Hamilton, P. A. Rebers, and F. Smith. (1956) Colorimetric method of determination of sugars and related substances. *Anal. Chem.,* **28**: 350.

Ebell, L. E. (1969) Variation in total sugar of conifer tissue with method of analysis. *Phytochemistry,* **8**: 227.

Haag, R. W. (1974) Nutrient limitations to plant production in two tundra communities. *Can. J. Bot.,* **52**: 103–106.

Hoagland, D. R., and D. I. Arnon. (1939) The water culture method of growing plants without soil. *Calif. Agric. Exp. Stat. Circ.,* 347 pp.

Kramer, P. J. (1942) Species differences with respect to water absorption at low soil temperatures. *Amer. J. Bot.,* **29**: 828–832.

Luckwill, L. C. (1960) The physiological relationship of root and shoot. *Sci. Hort.,* **14**: 22–26.

McCown, B. H. (1973) The influence of soil temperature on plant growth and survival in Alaska. In *Proceedings of the Symposium on Oil Resource Development and its Impact on Northern Plant Communities.* Occasional Publications on Northern Life No. 1, University of Alaska, pp. 12–33.

McCown, B. H. (1974) A new technique with high potential usefulness in root physiology and plant nutrition studies. *Hort. Sci.,* **9**: 40.

McCown, B. H. (1975) Physiological responses of root systems to stress conditions. In *Physiological Adaptation to the Environment* (F. J. Vernberg, Ed.). New York: Intext Educational Publishers, pp. 225–237.

Mitchell, K. J. (1956) Growth of pasture species under controlled environment. I. Growth at various levels of constant temperature. *N. Z. J. Sci. Tech.,* **31A**: 203–215.

Mooney, H. A., and W. D. Billings. (1960) The annual carbohydrate cycle of alpine plants as related to growth. *Amer. J. Bot.,* **47**: 594–598.

Neales, T. F., and L. D. Incoll. (1968) The control of leaf photosynthetic rate by the level of assimilate concentration in the leaf: A review of the hypothesis. *Bot. Rev.,* **37**: 107–125.

Nielson, K. F., and E. C. Humphries. (1966) Effects of root temperature on plant growth. *Soil Fert.,* **29**: 1–7.

Smith, D., G. M. Paulsen, and G. A. Raguse. (1964) Extraction of total available carbohydrates from grass and legume tissue. *Plant Physiol.,* **39**: 960–962.

Torrey, J. G. (1976) Root hormones and plant growth. *Ann. Rev. Plant Physiol.,* **27**: 435–459.

Van Cleve, K. (1973) Revegetation of disturbed tundra and taiga surfaces by introduced and native plant species. In *Proceedings of the Symposium on Oil Resource Development and its Impact on Northern Plant Communities.* Occasional Publications on Northern Life No. 1, University of Alaska, Special Report No. 2, pp. 7–11.

20. Plant Nutrient Limitations of Tundra Plant Growth

ALBERT ULRICH AND PAUL L. GERSPER

Introduction

Plant growth and primary production in tundra areas can be limited by the availability of inorganic nutrients as shown by Warren-Wilson (1957), Schultz (1964), and others. More recently, attention has focused on the experimental response of tundra plants to enrichment by N (Haag, 1974) and/or P (Chapin, 1978; Chapin et al., 1975; McKendrick et al., 1978). Rigorous attempts to determine nutrient deficiencies *in situ,* however, have not been undertaken in tundras.

Plant nutrient deficiencies can be determined by three techniques: visually, by inspection; experimentally, by field and pot tests; and chemically, by soil and plant analyses. Determinations based on visual inspection assume that specific symptoms uniquely characterize a particular nutrient deficiency. This is a quick and easy method to use by an experienced observer. Determinations based on pot and field tests depend on responses to specific treatments designed to reveal what nutrients, if any, are limiting growth. These tests are essential to the development of improved fertilizer programs, but the tests are time consuming and costly. Therefore, other tests, such as soil and plant tissue analyses, have been developed to gain comparable information.

Soil and plant tissue analyses, while much more rapid and far less costly than pot and field tests, are indirect and therefore must be calibrated to plant growth responses. This is particularly true of soil analyses, where it is assumed that what is in the soil is related directly or indirectly to plant growth. Unfortunately, however, this assumption holds only when soil nutrient levels are either very low or very high; especially where climatic factors strongly regulate nutritional supplies and plant growth rates, as on the tundra.

Plant tissue analysis, in contrast to soil analysis, assumes that what is in the plant is directly related to plant growth. Thus, when the nutrient concentration within the plant is above the critical level, growth is not limited even if soil available nutrient levels are low (Figure 1). Conversely, when the nutrient concentration within the plant decreases below the critical level, the growth rate decreases even if soil available nutrient levels are high. However, when the plant nutrient concentration is below the critical level, an increase in nutrient availabil-

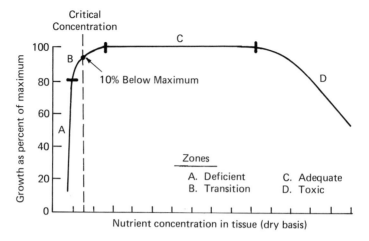

Figure 1. Plant growth in relation to tissue nutrient concentration. The critical concentration is taken at a 10% reduction on growth from the maximum. Below this concentration growth is retarded and symptoms generally appear; above this concentration growth is at a maximum and symptoms fail to appear. The transition zone (where symptoms may or may not appear) separates the zones of deficiency and adequacy. The toxic zone occurs when a nutrient concentration becomes excessive, resulting in a reduction in growth and the formation of toxicity symptoms. In practice, plant nutrient concentrations may be above the critical concentration even when soil nutrient concentrations are low, because of a low plant growth rate, and may be below the critical concentration even when soil nutrient concentrations are high, because of a high plant growth rate.

ity generally results in increased growth. The earlier in the growing season a nutrient falls below the critical level and the longer the deficiency lasts, the greater is the likelihood of a response to fertilization with that nutrient (Ulrich and Hills, 1973).

As a first step in determining the nutrient status of plants on the tundra, we conducted plant nutrient surveys of four species at numerous locations in the vicinity of Barrow, Alaska, during the summers of 1972 and 1973. The bulk of the surveys was keyed to those locations at which nutrient levels on the soil exchange complex and in soil solution were also measured during summers of 1970–73 (Challinor and Gersper, 1975; Flint and Gersper, 1974; Gersper, 1972; Gersper et al., 1974; Gersper and Challinor, 1975). Complete results of the plant nutrient surveys have been presented (Ulrich and Gersper, 1974, 1975). The results are summarized in Tables 1–5, together with critical nutrient concentrations as determined directly by growing tundra plants in solution culture in a phytotron and indirectly from information published in the literature.

Methods

Samples of four plant species were collected from 30 to 40 locations in the vicinity of Barrow, Alaska, in 1972 and 1973. In 1972 each location was sampled only once between 30 July and 8 August, but in 1973 most locations were

Table 1. Minerals in Blades of *Dupontia fisheri*, 1972 and 1973[a]

Statistic[b]	NO₃–N (ppm)	Ttl–N (%)	PO₄–P (ppm)	Ttl–P (%)	K (%)	Na (%)	Cl (%)	Ca (%)	SO₄–S (ppm)	Mg (%)	Fe (ppm)	Mn (ppm)	Zn (ppm)	Cu (ppm)
						1972								
Mean	2.9	2.38	872	2040	1.56	0.07	0.29	0.21	—	0.12	137	233	41	6
Low	0	1.41	220	990	0.52	0.01	0.04	0.07	—	0.08	42	37	18	1
High	14	3.65	2400	3720	2.75	0.48	0.69	0.51	—	0.18	570	730	80	17
CL	—	2.50	750	1500	1.00	—	—	0.1	250	0.10	50	10	10	1.0
CV	98	20	58	40	28	157	48	43	—	22	88	73	32	63
						1973								
Mean	0	3.00	1730	4010	1.63	0.06	0.50	0.22	320	0.15	119	249	49	12
Low	0	1.92	447	1240	0.60	0.01	0.23	0.13	120	0.10	46	37	22	5
High	0	4.57	6370	9240	4.44	0.33	0.92	0.46	710	0.30	364	728	70	48
CL	—	2.50	750	1500	1.00	—	—	0.1	250	0.10	50	10	10	1.0
CV	—	21	76	91	38	102	32	31	51	26	60	56	25	59

[a] N = 21–48.
[b] CL, tentative critical (deficiency) level based on best experience; CV, coefficient of variation.

Table 2.   Minerals in Stems of *Dupontia fisheri*, 1973[a]

Statistic[b]	NO$_3$-N (ppm)	Ttl-N (%)	PO$_4$-P (ppm)	Ttl-P (ppm)	K (%)	Na (%)	Cl (%)	Ca (%)	SO$_4$-S (ppm)	Mg (%)	Fe (ppm)	Mn (ppm)	Zn (ppm)	Cu (ppm)
Mean	12	2.37	2420	4160	2.17	0.09	0.70	0.08	363	0.12	88	355	42	12
Low	0	1.34	450	1520	0.85	0.01	0.16	0.03	130	0.09	28	80	30	4
High	269	4.47	5740	8080	3.56	0.42	1.42	0.20	850	0.20	296	639	61	32
CL	250	2.00	750	1500	1.00	—	—	—	—	—	—	—	—	—
CV	358	31	57	41	29	95	37	41	38	20	62	36	19	53

[a] *N* = 31–45.
[b] CL, tentative critical (deficiency) level based on best experience; CV, coefficient of variation.

Table 3. Minerals in Blades of *Carex aquatilis*, 1972 and 1973[a]

Statistic[b]	NO$_3$–N (ppm)	Ttl–N (%)	PO$_4$–P (ppm)	Ttl–P (ppm)	K (%)	Na (%)	Cl (%)	Ca (%)	SO$_4$–S (ppm)	Mg (%)	Fe (ppm)	Mn (ppm)	Zn (ppm)	Cu (ppm)
							1972							
Mean	3.3	2.90	850	2020	1.37	0.05	0.16	0.30	—	0.17	122	440	57	7
Low	0	2.54	290	1090	1.08	0.02	0.04	0.11	—	0.11	26	230	22	2
High	14	3.54	3020	4680	1.80	0.12	0.30	0.54	—	0.22	376	770	101	14
CL	—	2.50	750	1500	1.00	—	—	0.10	250	0.10	50	10	10	—
CV	31	9	79	45	16	59	54	39	—	16	62	40	40	54
							1973							
Mean	1.13	3.32	1170	2840	1.46	0.05	0.20	0.25	580	0.18	121	500	63	17
Low	0	1.67	393	940	0.74	0.01	0.04	0.05	70	0.11	70	150	31	8
High	31	4.34	4530	5210	2.95	0.22	0.40	0.51	1260	0.27	252	1000	112	38
CL	—	2.50	750	1500	1.00	—	—	0.10	250	0.10	50	10	10	—
CV	—	18	60	32	24	76	39	41	52	21	30	38	31	36

[a] N = 11–64.
[b] CL, tentative critical (deficiency) level based on best experience; CV, coefficient of variation.

Table 4. Minerals in Blades of *Petasites frigidus*, 1972 and 1973 [a]

Statistic[b]	NO$_3$-N (ppm)	Ttl-N (%)	PO$_4$-P (ppm)	Ttl-P (ppm)	K (%)	Na (%)	Cl (%)	Ca (%)	SO$_4$-S (ppm)	Mg (%)	Fe (ppm)	Mn (ppm)	Zn (ppm)	Cu (ppm)
						1972								
Mean	9	2.81	1710	2420	2.18	0.72	1.23	0.93	—	0.37	94	152	42	10
Low	0	2.03	740	1400	0.92	0.34	0.11	0.64	—	0.22	42	50	20	4
High	41	3.82	8070	8680	2.85	1.44	1.98	1.48	—	0.55	182	250	73	18
CL	—	3.0	1000	1500	1.0	—	—	0.50	250	0.10	50	10	10	—
CV	132	19	99	72	25	43	34	25	—	25	41	38	32	37
						1973								
Mean	0	3.46	1460	3060	2.63	0.35	0.77	0.81	2380	0.34	114	143	53	21
Low	0	2.64	820	1540	1.87	0.05	0.61	0.47	160	0.27	82	98	40	12
High	0	4.31	2940	5030	4.08	1.08	1.10	1.30	3680	0.45	146	250	68	32
CL	—	3.0	1000	1500	1.0	—	—	0.50	250	0.10	50	10	10	—
CV	—	20	52	38	26	86	20	35	43	16	21	37	18	26

[a] N = 9–17.
[b] CL, tentative critical (deficiency) level based on best experience; CV, coefficient of vatiation.

Table 5. Minerals in Petioles of *Petasites frigidus*, 1972 and 1973[a]

Statistic[b]	NO$_3$-N (ppm)	Ttl-N (%)	PO$_4$-P (ppm)	Ttl-P (ppm)	K (%)	Na (%)	Cl (%)	Ca (%)	SO$_4$-S (ppm)	Mg (%)	Fe (ppm)	Mn (ppm)	Zn (ppm)	Cu (ppm)
						1972								
Mean	70	1.10	1130	1620	4.09	0.66	3.78	1.10	—	0.59	80	82	104	10
Low	0	0.80	350	860	1.48	0.15	—	0.55	—	0.31	18	24	46	4
High	470	1.44	3870	2950	9.03	1.28	—	2.79	—	0.90	221	200	183	18
CL	—	1.0	750	1000	1.0	—	0.4	—	—	—	—	—	—	—
CV	199	24	84	50	45	56	—	52	—	31	72	53	38	37
						1973								
Mean	0	1.51	1600	2590	3.16	0.59	1.26	0.84	1050	0.47	119	91	94	37
Low	0	1.08	480	1130	1.38	0.07	0.87	0.41	410	0.09	17	33	42	13
High	0	2.39	4890	6320	5.10	2.35	1.88	1.33	1590	1.04	368	233	196	174
CL	—	1.0	750	1000	1.0	—	0.4	—	—	—	—	—	—	—
CV	—	34	81	64	37	119	34	40	54	53	93	61	47	122

[a] $N = 1$–16.
[b] CL, tentative critical (deficiency) level based on best experience; CV, coefficient of variation.

sampled twice: in mid-summer from 6 July to 24 July, and in late summer–early fall from 10 August to 10 September. The species and parts sampled were: *Dupontia fisheri,* first (youngest) living blade at an angle, and stem sheath tissue, exclusive of blades; *Carex aquatilis,* longest living leaf (blade); *Petasites frigidus,* blade and petiole; and *Poa arctica,* entire leaves. Each sample was a composite of 30 to 40 representative plants. Samples were dried, ground, and analyzed by conventional methods for NO_3–N, total N, PO_4–P (soluble in 2% HAc), total P, K, Na, Cl, SO_4–S, Ca, Mg, Zn, Fe, Cu, and Mn (Johnson and Ulrich, 1959). Individual plants were also tested under field conditions for NO_3–N with diphenylamine reagent (0.2 g of diphenylamine in 100 ml of NO_3-free concentrated H_2SO_4).

The calibration of nutrient concentrations within the plant to growth per unit time must be made when only the nutrient being calibrated is limiting (Figure 1). Calibration curves are obtained most effectively by growing plants in a series of culture solutions containing adequate amounts of all essential nutrients except the one being calibrated, which is varied from low to high levels over the series. In our calibrations, we used a series of eight treatments, starting with no addition of the nutrient as the control. At a time when the plants in the lower half of the series indicated a nutrient deficiency, and those in the upper half did not, they were harvested. The tops were separated from the roots, weighed, and then reweighed after subdividing the tops into appropriate fractions according to leaf age and with blades separated from petioles for *P. frigidus,* and blades separated from the stem and sheath material for grasses. Samples were then analyzed after drying and grinding.

For each combination of species, plant part, and nutrient being calibrated, the yield is plotted against the nutrient concentration over the calibration series and a line is drawn through the plotted points as shown schematically in Figure 1. The combination—of plant parts sampled and form of nutrient determined—that results in a broad range of values from deficiency to adequacy, and has a narrow transition range between these two zones, is the one used for diagnostic purposes. The critical concentration is arbitrarily set at the mid-point of the transition zone or at the point where the yield is 10% below maximal (plateau) value. Critical levels selected in this manner may be raised or lowered, depending on the investigator's experience, to indicate possible deficiency at the time of taking a field sample. Critical level calibrations are largely independent of time, location, and climatic conditions. Once the nutritional patterns have been determined over several growing seasons, then several samplings or possibly a single sampling may suffice for estimating the nutritional status of plants in an ecosystem at any future time.

Results and Discussion

The results of the plant nutrient surveys reflect the nutritional status of the plants in relation to location, climate, species, and age of the plant. This accounts for most of the variability shown in the data summaries (Tables 1 through 5). All plants, except in a few isolated places, were found to be deficient in one or more nutrients, but not necessarily in the same nutrient even when growing at the same location.

The key nutrients affecting the growth of tundra plants at Barrow are most likely N and P, with N playing the dominant role in most instances. This is in agreement with fertilization studies at Barrow (Chapin et al., 1975; McKendrick et al., 1978). Nearly all field samples were deficient in N to some extent, since nearly all of them failed to give a positive test for NO_3. However, the degree of N deficiency was only slight in many instances, as indicated by high total N concentrations in the plants. Similarly, the PO_4–P concentration was often low, indicating a deficiency of P at the time of sampling. Again, however, a high concentration of total P in many of the low PO_4–P plants indicated that P had become deficient only a short time before sampling and the response to increased available P in the soil would likely have been slight.

Nitrate–N (Field Tests)

The tests for NO_3–N under field conditions were largely negative in 1972 and entirely negative in 1973. The tests in 1972 at first corroborated the general belief that tundra plants do not contain NO_3, since tundra soils are generally cold, wet, and high in organic matter (those at Barrow are also acid), and therefore largely anaerobic. However, this view is now in doubt, because with persistent continued testing, NO_3 was found in many plants at certain locations in the vicinity of the IBP sites, near the Will Rogers Monument, at Walakpa west of Barrow Village, and elsewhere. In these places, the plants grew better and appeared greener than in most other locations sampled. Plants likely to contain NO_3 were soon more easily recognized visually and more plants containing NO_3 were located. However, had sampling been done randomly, the probability of detecting NO_3 would have been very low. On the other hand, experienced visual selection yielded strong tests for NO_3 in plants growing at locations where the soil most likely contained higher concentrations of available nutrients such as at owl mounds, drainage sloughs (Gersper et al., 1974), vehicle tracks (Challinor and Gersper, 1975), fertilized plots (Gersper et al., 1975), and areas of organic waste deposits near dwellings. At these locations the cut stems of D. fisheri and petioles of P. frigidus usually gave strong tests for NO_3. Sometimes, however, as observed for D. fisheri in a slough near Footprint Creek, the test for NO_3 was positive one day and negative a few days later, thus reflecting the delicate changing balance between soil NO_3 supply and plant NO_3 demand. Plants of all four species in the vicinity of owl mounds gave a strong, positive test for NO_3 most consistently, whereas, at other sites, C. aquatilis always gave a negative test. In 1973, none of the plants tested in the field, regardless of location, gave a positive test, although laboratory analysis showed that many of the samples contained small but significant amounts of NO_3. The much wetter and cooler conditions in 1973, compared to 1972, may have accounted for the differences in results between the 2 years.

Nitrate–N (Laboratory Analyses)

Samples of *D. fisheri* collected in 1972 showed NO_3–N levels near zero, except for 14 ppm in blades (Table 1) and 410 ppm in stems from a "green patch" in the vicinity of Will Rogers Monument. None of the blades collected in 1973 contained NO_3–N even though 5 of 45 stems ranged in content from 13 to 269 ppm (Tables 1 and 2). In 1972 blades of *C. aquatilis* had a mean NO_3–N level of 3.3 ppm (Table 3) with a high of 14 ppm. In 1973, three samples ranged from 16 to 31 ppm, with a mean of 1.1 ppm for 64 samples. Blades of *P. frigidus* contained an average of 9 ppm, with a high of 41 ppm in 1972, and none of the blades contained NO_3–N in 1973 (Table 4). Similarly, petioles contained an average of 70 ppm with a high of 470 ppm in 1972, and none of the petioles contained NO_3–N in 1973 (Table 5). The two highest levels in petioles in 1972 were from plants growing in an area of flat mesic meadow tundra at IBP site 2 (320 ppm) and from those growing on an owl mound (470 ppm). Two samples of *P. arctica* leaves growing on owl mounds near IBP sites 2 and 4 in 1972 contained 630 and 53 ppm of NO_3–N, respectively.

A particularly interesting finding in 1973 was the presence of NO_3–N in plants at Schultz's (Schultz, 1964) fertilized plots. On 27 August 1973, the stems of *D. fisheri* contained 63 ppm and the blades 0 ppm, while the blades of *C. aquatilis* contained 25 ppm. Since these plants were low in total N (see below), the presence of NO_3 was considered to be most likely a temporary condition and this hypothesis was confirmed at the next sampling, 5 days later, when all samples tested zero in NO_3–N.

Thus far, we have been unable to develop a good estimate of NO_3–N critical levels except for stem tissue of *D. fisheri*. Overall, however, levels of NO_3–N were very low when compared with those of agronomic crops.

From a diagnostic viewpoint, plants that contain NO_3 in the conducting tissues are considered to be well supplied with N at the time of sampling and, conversely, if conducting tissues do not contain NO_3 the plants are considered to be deficient in N. Since nearly all plants tested in the surveys of both years did not contain NO_3, all plants, except a very few growing in scattered spots of relatively nutrient-rich soil, were deficient in N at the time of sampling. The degree of deficiency and the response to increased availability of N would depend on the next most growth-limiting factor. If this should happen to be P (which is likely the case), then the response to increased available N would depend on the available soil P supply, which might be increased either by a faster ecosystem turnover rate of P or through P fertilization.

Results of the analyses of plants from Schultz's plots suggest that the large response to yearly fertilization from 1961 through 1964 (Schultz, 1964) was due to the addition of both N and P and that the plants in 1972 and 1973 were still abundantly supplied with P but relatively low in N (Ulrich and Gersper, 1974,

1975). Consequently, to maintain or increase growth in the fertilized plots, it would be necessary to add only N until either P or some other nutrient became deficient.

An important consequence of fertilizing tundra, as evidenced by the dramatic and long-lasting changes in tundra in Schultz's experiments (Schultz, 1964; see also Gersper et al., 1974, 1975), is the formation of a very dense, insulative vegetative cover (Figures 2 and 3; and see Discussion and Conclusions). A higher N and P supply could account for the large increase in growth of plants in landscape scars created by tracked vehicles such as Weasels (Challinor and Gersper, 1975). Blades of *D. fisheri* growing in track scars (Ulrich and Gersper, 1974, sample 327) were exceptionally green in color and high in both total N and soluble P, whereas comparable blades on adjacent plants growing outside the sphere of influence of the track scars (sample 328) were purple and green in color, very low in total N (deficient) and low, but not deficient, in P at the time of sampling (30 July 1972). In this instance, the changes in the nutrient status of the plants were affected indirectly by modified environment (Gersper and Challinor, 1975), rather than by direct fertilization. Such nutrient changes can lead to changes in plant distribution and community composition, since *D. fisheri* increased in number and in size in track scars (Challinor and Gersper, 1975), and conversely decreased in number, but not in size, after fertilization in Schultz's plots. In Schultz's plots the bryophytes increased enormously in growth and

Figure 2. Unfertilized tundra, Barrow, Alaska, Schultz's flat, sparsely vegetated by *Dupontia fisheri* and *Carex aquatilis,* resulting in a relatively firm surface.

Figure 3.   Fertilized tundra, Schultz's flat, densely vegetated by *Dupontia fisheri* and *Carex aquatilis* growing within an exceptionally dense, spongy mat of bryophytes.

formed most of the biomass of these plots, whereas they accounted for only a small part of the biomass in comparable unfertilized sites (Figures 2 and 3).

Total N

Magnitudes of total N concentrations in plants have the unique property of indicating the degree of N deficiency or sufficiency when NO_3–N values are below or above the critical NO_3–N concentration, respectively. Since practically all plants sampled in 1972 and 1973 contained none, or only very low levels of NO_3–N, they were clearly below the critical N concentration. Therefore, they were deficient to some extent, the degree of deficiency increasing with decreasing total N contained in the plant part sampled. When plants contain NO_3–N, and total N is low, this indicates a recent acquisition of NO_3–N, and, in time, total N would also increase.

Dupontia fisheri.   In 1972 total N concentrations in blades of *D. fisheri* ranged from 1.4 to 3.7% (Table 1), with no apparent relationship to the levels of other minerals contained in the plants. Sampling location had a large effect on total N; the highest average values (3.2 to 2.9%) were found in plants associated with owl mounds. Low concentrations occurred in plants growing in Schultz's (1.4 to 2.0%) and Douglas's (1.9%) unfertilized control plots. Levels in plants of Schultz's fertilized plots, even after heavy fertilization with N from 1961 through

1964, were relatively low in total N, ranging from 2.0 to 2.2%. The highest level, 3.7%, occurred in blades of lush *D. fisheri* taken from vehicle track scars in a winter haul area and the next highest, 3.4%, occurred in a lush green spot near Will Rogers Monument. In the latter plants, the stems contained 410 ppm of NO_3–N and the highest total N, 2.5%. Overall, the lowest total N level in stems was 1.5%. Levels in blades did not vary appreciably with leaf age, and in all cases blades contained more total N than corresponding stem tissues.

In 1973, 45 composite samples of *D. fisheri* blades ranged from 1.9 to 4.6% in total N content (Table 1). No sample contained detectable NO_3–N and, thus, all blades were deficient in N. Obviously, the lower the total N content, the greater the deficiency and the greater would be the response to N until at some point the yield increase would level off. The N status of the total plant, however, needs to be considered before making the final evaluation of deficiency or sufficiency. For example, a blade sample taken from an owl mound was exceptionally high in total N, 4.6%, considerably above the critical level. But the blade contained no NO_3–N at the time of sampling. However, the stem tissue also contained a very large amount of total N, 4.5%, and, moreover, contained 72 ppm of NO_3–N (Table 2). Clearly, these plants were not deficient in N at the time of sampling. On the other hand, the blades of some plants contained less than 2.5% total N, and, at the time of sampling, these plants were very deficient. The total N contents in the stems of *D. fisheri* (Table 2) ranged from 1.3 to 4.5% and, as a rule, contained considerably less than corresponding blades, particularly when the blades were in the deficiency range (Tables 1 and 2). Total N contents of blades and stems were highly correlated, $r^2 = 0.765$ (Ulrich and Gersper, 1975).

If we assume that the absence of NO_3–N in stem tissues indicates an N deficiency, then most of the plants were deficient at the time of sampling during both years. However, even when NO_3 is detected in stems, the blades may still be relatively low in total N, e.g., 1.9% at one location in 1973, thus indicating a temporary surge of NO_3, as verified at the next sampling at this location when the stems contained no NO_3–N and the level of total N was only 1.5%. In some instances, particularly late in the growing season, and especially in locations of nutrient enrichment, stems may contain NO_3–N and be high in total N concentration indicating that the plants have an ample supply of N. These findings are indicative of the changeable nature of N in plant tissues and the need for monitoring plant material regularly for NO_3 and total N in order to accurately assess the N status of plants on the tundra.

Carex aquatilis. The longest living leaf of *C. aquatilis* ranged in total N from 2.5 to 3.5% in 1972 and from 1.7 to 4.3% in 1973 (Table 3). Both the range and the mean were higher in 1973. Furthermore, some of the samples in both years contained small but significant amounts of NO_3–N. However, because of the transitory nature of NO_3–N, total N concentrations should be relied upon to evaluate the N status of this plant. Thus, the probability of a deficiency or a response to N increases directly as total N content decreases. In both years, mean total N contents were above the critical level.

Petasites frigidus. Total N concentrations in the blades of *P. frigidus* varied from 2.0 to 3.8% in 1972 and from 2.6 to 4.3% in 1973 (Table 4). The corresponding petioles contained much less, 0.8 to 1.4% in 1972 and 1.1 to 2.4% in 1973 (Table 5). Blades and petioles containing more than 3.0 and 1.0% of total N, respectively, are presently assumed to be in the adequate range of N nutrition although NO_3–N concentrations were zero or very low in both years. Nevertheless, a response to increased N availability would be unlikely for *P. frigidus* because total N concentrations were above the critical levels in both blades and petioles.

In the latter part of the growing season (10 August to 10 September), total N concentrations in *D. fisheri* declined appreciably in the blades and still more in the stems, whereas the decline in the leaves of *C. aquatilis* was relatively small (Ulrich and Gersper, 1975). Grasses in general, as represented by *D. fisheri*, showed greater changes in total N than did sedges, as represented by *C. aquatilis* (see also Chapin et al., 1975). In both species there was often a much larger decline in PO_4–P and occasionally in K than in total N, indicating that increases in more than one nutrient are required to improve plant growth in some tundra locations. Large declines in total N in the second sampling indicate that the supply of N was unable to keep up with the demand. This may result from a period of rapid growth, because the soil N supply was too low, or because of downward translocation of N, which the data of Chapin et al. (1975) indicate starts as early as 25 July. The magnitude of increased growth upon supplying N would depend on the degree of N deficiency, the proximity of the next limiting factor, and the length of growing season remaining after the deficiency became established.

Phosphate–P

Phosphate–P serves as the primary source of P for metabolic processes and for the synthesis of elaborated products within the plant, just as NO_3–N and NH_4–N serve as primary sources of N. Thus, the presence of a large reservoir of PO_4–P within a plant indicates either an ample supply of soil P or a low rate of P utilization, induced temporarily by low air temperature or by a shortage of some other nutrient, such as N. However, a relatively low PO_4–P concentration within the plant indicates that either soil P supply is low or that the rate of utilization by the tops in favorable weather has temporarily exceeded the rate of supply from the soil, especially in early summer when air temperatures are much higher than those of the soil. The converse would, of course, be true in late summer, when root temperatures may be favorable for P absorption and air temperatures too low for growth.

A low PO_4–P concentration in the plant, in conjunction with a low total P concentration, indicates deficiency associated with low soil P supply. Conversely, high concentrations would indicate a high soil P supply or another nutrient deficiency that did not influence PO_4–P absorption but did reduce growth. A low PO_4–P and high total P concentration indicates a recent depletion of P and, conversely, a high PO_4–P and low total P indicates a recent acquisition.

The earlier in the growing season a deficiency occurs and the longer the deficiency exists, the greater the likelihood of a response to increased available P, such as by fertilization, assuming of course that no other deficiency is induced after fertilization (Ulrich and Hills, 1973).

Dupontia fisheri. The blades of *D. fisheri* ranged in PO_4–P from 220 to 2400 ppm in 1972 and from 447 to 6370 ppm in 1973 (Table 1). Corresponding stem tissues ranged from 970 to 2480 ppm in 1972 (Ulrich and Gersper, 1974) and from 450 to 5740 ppm in 1973 (Table 2). In most instances stem tissues were appreciably higher in PO_4–P than corresponding blades. Tentatively, plants with blades or stems containing less than 750 ppm PO_4–P (Tables 1 and 2) are considered to be deficient in P at the time of sampling. Using this reference point, three samples from Schultz's control plots in 1972 were clearly deficient, whereas those from the fertilized plots were clearly well supplied (Ulrich and Gersper, 1974). One of the latter samples had the highest PO_4–P content of all samples collected (6370 ppm). Overall in 1973, using this reference point, only 13 of 45 blade samples and only 5 of 45 stem tissue samples were likely deficient in P. This is in sharp contrast with the 1972 results, when about half of 33 blade samples collected were deficient in P at the time of sampling and levels of PO_4–P in both stems and blades were considerably lower than in 1973. Differences in locations sampled and times of sampling may account for the higher P status of the plants in 1973, but less favorable weather for growth (smaller demand) or higher soil temperature (larger supply) could also account for the higher P status of the plants. The higher levels of PO_4–P in 1973 are all the more surprising since the second sampling took place in many instances more than a month later than the single sampling in 1972.

These differences are in agreement with those of Chapin et al. (1975). Furthermore, simulations of plant growth and P uptake rates during the 1972 and 1973 seasons indicated that the warmer air temperatures during 1972 resulted in a comparatively rapid growth rate and high P demand which greatly outstripped the comparatively smaller increase in uptake rate at slightly warmer soil temperatures.

Seasonal variations in PO_4–P concentration in the plants at various sampling locations were quite large during 1973. For example, the decrease from mid-July to mid-August in the blades of *D. fisheri* at several locations was from 2400 to 860, 3460 to 780, 3120 to 554, and 1880 to 500 ppm (Ulrich and Gersper, 1975). These rapid declines in PO_4–P over a 4-week period brought many of the plants from a zone of sufficiency into a zone of P deficiency, i.e., below 750 ppm. There were a few instances in which no change took place. Smaller but similar declines in PO_4–P took place in *C. aquatilis* and in total P levels of both plant species.

Carex aqualtilis. The PO_4–P concentrations of 64 blade smaples of *C. aquatilis* collected in 1973 ranged from 393 to 4530 ppm, with a mean of 1170 ppm (Table 3). Only 20 of the samples were below 750 ppm (Ulrich and Gersper, 1975), and since most of these were collected in mid-August, the likelihood of a response by *C. aquatilis* to increased available P would probably have been small. The 1973 findings are in sharp contrast to 1972, when the mean level was only 850 ppm (Table 3) and 11 of 15 samples collected from 30 July through 3 August from unfertilized areas contained less than 750 ppm PO_4–P (Ulrich and Gersper, 1974). Obviously, monitoring of plants for more than one season is necessary if the nutritional status of the tundra is to be adequately evaluated.

Petasites frigidus.   Phosphate–P concentrations in the blades of *P. frigidus* in 1972 ranged mainly from a relatively low value of 740 ppm to 2230 ppm (Table 4). A single high value of 8070 ppm was obtained from Schultz's fertilized plots. Correspondingly, petioles ranged in level from 860 to 2950 ppm (Table 5). In 1973 blades ranged from 1540 to 5030 ppm PO_4–P with a mean of 3060 ppm (Table 4) and petioles ranged from 480 to 6320 ppm with a mean of 2590 ppm (Table 5). Only two of 17 samples of blades in 1972 (Ulrich and Gersper, 1974) were below the tentative critical level of 1000 ppm. However, the results for corresponding petioles show that approximately half of the samples were below 750 ppm, indicating that *P. frigidus* would be likely to respond to P fertilization. The results for 1973 were similar to those of 1972 (Ulrich and Gersper, 1975).

Total P

Total P concentrations in 1972 were highly correlated with soluble PO_4–P concentrations (Ulrich and Gersper, 1974). The coefficients of determination, r^2 values, for *D. fisheri* blades, *C. aquatilis* leaves, and *P. frigidus* blades are 0.852, 0.878, and 0.968, respectively. These high values suggest that either soluble PO_4–P or total P determinations may be used to estimate the P status of the plants. However, both determinations of P have a greater diagnostic value than either one alone, and in crucial studies both should be determined for the best estimate of the P status.

Dupontia fisheri.   Total P levels in the blades of *D. fisheri* in 1972 ranged from 990 to 3720 ppm (Table 1). Assuming a critical level of 1500 ppm, based on our preliminary data, most of the blade samples appear to have been adequately supplied with P at the time of sampling, which is the opposite interpretation based on PO_4–P contents. Apparently, a blade sample can be distinctly below the critical PO_4–P level of 750 ppm and yet be distinctly above the critical total P level of 1500 ppm. What this implies is that the raw material, PO_4–P, had only recently been depleted after having been well supplied previously. However, this also implies that further growth would be limited and would occur at the expense of P reserves in the plant, unless, of course, new PO_4–P was absorbed by the roots. The excellent growth made by *D. fisheri* in a green spot near Will Rogers Monument was based on a large supply of nutrients. These plants had a total P content of 2230 ppm, which, however, was on the verge of being depleted, judging from a low soluble PO_4–P level of 620 ppm.

Total P levels in blades of *D. fisheri* in 1973 ranged from 1240 to 9240 ppm with a mean of 4010 ppm, nearly double that of 1972 (Table 1). As a rule, total P and PO_4–P varied proportionately, but, as in 1972, there were exceptions. They imply that the supply of PO_4–P had not kept up with the demand and that in time the large P reserve in the leaf blade would decline and, perhaps, go below the 1500 ppm critical level. A greater variation in difference between PO_4–P and total P in 1973 accounts for the decline in the coefficient of determination from 0.852 in 1972 to 0.581 in 1973 (Ulrich and Gersper, 1974, 1975), and supports the importance of making both determinations in assessing the P status of the plants. Total P concentrations in the stem tissue of *D. fisheri* were somewhat higher than those of corresponding blades in both 1972 and 1973 and there was good agreement between PO_4–P and total P.

Carex aquatilis. Young mature leaves of *C. aquatilis* had a mean concentration of 2020 ppm of total P and ranged from 1090 to 4680 ppm in 1972 (Table 3). In 1973 the mean was higher at 2840 ppm and ranged from 940 to 5210 ppm. Assuming that the critical concentration for total P in *C. aquatilis* is 1500 ppm, only two of 63 samples in 1973 were deficient in P (Ulrich and Gersper, 1975). This interpretation disagrees with the observation that 20 of these samples were in the zone of deficiency when based on a critical level of 750 ppm of PO_4–P. However, most of the low PO_4–P values were obtained in the second sampling near the end of the growing season and therefore a significant response to P would have been unlikely, especially in the cool, wet summer of 1973. Samples from Schultz's fertilized plots were the highest of all samples in total P.

Petasites frigidus. Total P concentrations in the blades of *P. frigidus* (Table 4) were, as a rule, much higher than those in corresponding petioles (Table 5). Again, levels in both blades and petioles were higher in 1973 than in 1972 and only a few samples (in 1972 only) were below critical levels.

Potassium

Dupontia fisheri. The blades of *D. fisheri* in 1972 ranged in K concentration from 0.52 to 2.75% (Table 1). Only three samples contained less than the tentative critical level of 1.0% (Ulrich and Ohki, 1966) at the time of sampling. The lowest value of 0.52% was for a sample collected from the basin of a low-centered polygon. The exceptionally high Na concentration of 0.48% in this sample was also indicative of severe K deficiency (Hylton et al., 1967). In 1973 the blades of *D. fisheri* ranged in K concentrations from 0.60 to 4.44% (Table 1) and the stems from 0.85 to 3.56% (Table 2). The mean of 1.63% in blades of 1973 was only slightly higher than that of 1972. Generally, the blades were much lower in K than stems (Tables 1 and 2) and concentrations in the second sampling were much lower than those of the first in 1973 (Ulrich and Gersper, 1975). Concentrations below the critical level of 1% were observed in the blades of only 5 of 44 samples collected, and, for stems, only one sample had a concentration below 1%. A response to K fertilization would most likely have taken place, but only if the fertilizer also contained P, since these plants were also deficient in this nutrient on both sampling dates.

Carex aquatilis. The leaf samples of *C. aquatilis.* collected in the 1972 survey, ranged narrowly in K concentration from 1.08 to 1.80% (Table 3), with the highest values from Schultz's fertilized plots. Concentrations of K in 1973 ranged from 0.74 to 2.95% over all sites and the mean of 1.46% was just slightly higher than that of 1972. Since only three of the 64 samples were below 1.0%, the likelihood of a response to increased K would have been small.

Petasites frigidus. Blades in 1972 ranged in K concentration from 0.92 to 2.85% (Table 4) and corresponding petioles from 1.48 to 9.03% (Table 4). A sample of *P. frigidus* from Schultz's fertilized plots contained 2.40% K in the blades and the extraordinary high level of 9.03% in the petioles. Except for one sample from site 2, none of the plants had K concentrations low enough to suggest a deficiency.

Sodium

This element is not usually considered essential to the growth of plants. Yet, in some plants it improves growth even in the presence of adequate K, and has a large K-sparing action when the plants are deficient in K (El-Sheikh et al., 1967). *D. fisheri, C. aquatilis,* and *P. arctica* are essentially Na excluders, whereas *P. frigidus* absorbs and translocates Na readily (Tables 1–5). The three excluders contained less than 0.1% Na in the plant parts analyzed in both years, whereas *P. frigidus* blades and petioles generally contained considerably more than this amount. Two blade samples of *D. fisheri* contained more than 0.4% Na, one from the basin of a low-centered polygon in Central Marsh where the plants were very low in K and the other at Will Rogers Monument where the plants were growing exceedingly well and were high in NO_3–N (Ulrich and Gersper, 1974). Mean contents in both years were about the same. Only six of the samples in 1973 contained more than 0.10% Na, and four of these were associated primarily with low K (Ulrich and Gersper, 1975). As with blades, most stems contained less than 0.10% Na, and the samples above 0.10% were associated with low K or with other exceptional conditions.

Average Na levels in *C. aquatilis* were the same in both years, with individuals ranging from 0.02 to 0.12% in 1972 and from 0.01 to 0.22% in 1973 (Table 3). Only four of the 62 samples in 1973 contained more than 0.10% Na and all of them were from Schultz's plots (Ulrich and Gersper, 1975). In *P. frigidus,* the petioles were generally higher in Na than the corresponding blades and both blades and petioles were lower in Na in 1973 than in 1972.

Chloride

Chloride concentrations in both 1972 and 1973 were largely independent of concentrations of other minerals present in a given tissue, including those of Na and SO_4–S. Concentrations differed considerably according to plant species and plant part analyzed. *P. frigidus* (Tables 4 and 5) contained more Cl than *D. fisheri* (Tables 1 and 2) and *P. arctica* (Ulrich and Gersper, 1974), and much more than *C. aquatilis* (Table 3). The petioles of *P. frigidus* and the stems of *D. fisheri* were higher in Cl than the corresponding blades. Both surveys clearly indicated that the major source of Na or Cl is from the soil and not as a spray or mist from the Arctic Ocean. This can be verified, for example, from the Cl and Na levels in the blades and stems of *D. fisheri*. In these organs the ratio of Cl to Na on a weight basis ranged from 9:1 to 30:1 instead of 1.5:1 in pure NaCl.

The blades of 36 samples of *D. fisheri* ranged in Cl concentration in 1973 from 0.23 to 0.92% (Table 1). Corresponding stems generally contained more than blades (Table 2). Chloride levels in both blades and stems were higher in 1973 than in 1972. The leaves of *C. aquatilis* in the two surveys had a very low Cl concentration, indicating that this plant is a definite Cl excluder. Chloride content of *P. frigidus* blades ranged from 0.11 to 1.98% in 1972 (Table 4) and mean contents in both blades and petioles (Table 5) were much higher in 1972 than in 1973, which is the reverse from that for *D. fisheri* and *C aquatilis*. Most plants were well above the critical level of 0.4% Cl for sugarbeet petioles. However, some of the *D. fisheri* plants and nearly all of *C. aquatilis* may have been in the range of Cl deficiency and might well have responded to increases in available Cl.

Calcium

The Ca status of the four species was about the same in both years. Significant variations in concentrations occurred according to species, plant part sampled, location, and date of sampling (Ulrich and Gersper, 1974, 1975). Blades of *D. fisheri* contained more than twice the Ca of corresponding stems (Tables 1 and 2). The blades of *P. arctica* (Ulrich and Gersper, 1975) and *C. aquatilis* (Table 3) contained slightly higher concentrations than those of *D. fisheri* (Table 1). The blades of *P. frigidus* (Table 4) were about three times higher in Ca than the blades of the other three species and the petioles of *P. frigidus* (Table 5) contained slightly higher concentrations than the corresponding blades. Moreover, concentrations increased appreciably with age (Ulrich and Gersper, 1975), especially in the blades of *D. fisheri* and in the leaves of *C. aquatilis*.

In 1973, Ca levels in the blades of *D. fisheri* ranged from 0.13 to 0.46% (Table 1) while corresponding stems ranged from only 0.03 to only 0.20% (Table 2). The leaves of *C. aquatilis* ranged from 0.05 to 0.51% (Table 3). In contrast, the blades of *P. frigidus* ranged from 0.47 to 1.30% (Table 4) and the corresponding petioles ranged from 0.41 to 1.33% (Table 5).

The blades of *D. fisheri* and *C. aquatilis* at the low end of the ranges may be low enough in Ca for these plants to respond to Ca fertilization either directly from the increased Ca or indirectly by decreasing Mn uptake and/or by decreasing soil acidity. However, the probability of decreasing Mn uptake by liming the soil at Barrow is likely to be rather low since some of the highest Mn values in the blades of *D. fisheri* and leaves of *C. aquatilis* were observed in samples with the highest Ca levels and vice versa (Ulrich and Gersper, 1974, 1975). Obviously, however, experimental treatments with lime are the only way to measure the short- and long-term effects of lime on the tundra ecosystem.

Sulfate–S

The plant samples collected at Barrow in 1973 were analyzed for SO_4–S by the Johnson–Nishita method (Johnson and Nishita, 1952) which is specific for SO_4–

S. The results revealed considerable SO_4–S in parts of the four plant species sampled (Tables 1–5; Ulrich and Gersper, 1975). This finding, together with the discovery of nitrate–N in typical tundra plants, supports our concept that the soil system of even the wet tundra is generally far more aerobic than has thus far been indicated [see nitrate–N (field tests), above].

The blades and stems of *D. fisheri* (Tables 1 and 2) ranged in SO_4–S concentrations from 120 to 710 ppm with a mean of 320 ppm and from 130 to 850 ppm with a mean of 363 ppm, respectively. *C. aquatilis* (Table 3) ranged from 70 to 1260 ppm with a mean of 580 ppm. The blades and petioles of *P. frigidus* (Tables 4 and 5) ranged from 160 to 3680 ppm with a mean of 2380 ppm and from 410 to 1590 ppm with a mean of 1050 ppm SO_4–S, respectively.

Petasites frigidus thus contained the highest concentration of SO_4–S, with *C. aquatilis* and *D. fisheri* following in that order. The blades of *P. frigidus* contained more than twice as much SO_4–S as the petioles, and there was no appreciable difference between the blade and stem tissues of *D. fisheri*. Seasonal differences between the first and second sampling were relatively small and tended to be inconsistent (Ulrich and Gersper, 1975). Responses to SO_4–S are likely when blade tissues contain less than the critical level of 250 ppm, which occurred in several samples of *D. fisheri* and *C. aquatilis* and even in a few samples of *P. frigidus*.

Magnesium

The blades of *D. fisheri* ranged in Mg concentrations from 0.08 to 0.18% in 1972 and from 0.10 to 0.30% in 1973 (Table 1). Corresponding stems ranged from 0.09 to 0.20% in 1973 (Table 2). Three samples of *P. arctica* taken from owl mounds in 1973 averaged 0.31% Mg (Ulrich and Gersper, 1975). These results indicate that *P. arctica* from owl mounds and *D. fisheri* generally were not deficient at the time of sampling if the critical level proves to be about 0.05% (a possibility). However, if the critical level proves to be as high as 0.10% (as we are inclined to think from our results to date), then most of the *D. fisheri* sampled in both 1972 and 1973 would have been deficient or nearly so. In contrast to the increase in Ca levels with time, there was little seasonal change of Mg in the blades or the stems of *D. fisheri* (Ulrich and Gersper, 1975).

The leaves of *C. aquatilis* contained slightly more Mg than the blades of *D. fisheri* (Table 3), but only about half the concentration found in *P. frigidus* (Table 4) and *P. arctica* (Ulrich and Gersper, 1974, 1975). However, response to increased available Mg would have been unlikely since the lowest concentrations were above the critical level of 0.10%. As a rule, Mg levels increased slightly with time (Ulrich and Gersper, 1975).

The blades of *P. frigidus* ranged in Mg concentration in 1972 from 0.22 to 0.55% and in 1973 from 0.27 to 0.45% (Table 4). Petioles in 1972 ranged from 0.31 to 0.90% and in 1973 from 0.09 to 1.04% (Table 5). None of the plants was deficient in Mg, judging from the high levels of the blades. However, the low level of 0.09% in the petioles of plants growing in a moderately wet plot in 1973 indicates a possible deficiency in those plants.

Iron

The Fe concentrations in the blades of *D. fisheri* ranged from 42 to 570 ppm in 1972 and 46 to 364 ppm in 1973 (Table 1). Iron deficiency would be likely in plants with concentrations below the approximate critical level of 50 ppm and, on this basis, four samples in 1972 and only one in 1973 would have been deficient at the time of sampling (Ulrich and Gersper, 1974, 1975). The stems of *D. fisheri* in 1973 ranged from 28 to 296 ppm (Table 2). Nine of the 44 samples were at or below 50 ppm indicating that some plants, in certain locations, might at times be deficient in Fe. One sample was clearly deficient, containing 46 ppm in blades and only 28 ppm in the conducting stem tissue (Ulrich and Gersper, 1975). The seasonal trend in both blades and stems was sharply downward.

Carex aquatilis ranged in Fe concentration from 26 to 376 ppm in 1972 and from 70 to 252 ppm in 1973 (Table 3). Thus, Fe deficiency was unlikely in this species except in a few locations in 1972. Iron levels in the blades of *P. frigidus* ranged from 42 to 182 ppm in 1972 and from 82 to 146 ppm in 1973 (Table 4). Mean Fe content was higher in 1973 in both blades and petioles. The data indicate that Fe deficiency did not generally occur in this species except in a few locations where the plants contained considerably less than 50 ppm in petiole tissue. There was little change in Fe content seasonally (Ulrich and Gersper, 1975).

Manganese

World-wide, Mn, except for NO_3–N, has the broadest concentration range of any nutrient in the tissues of plants. In leaf blades Mn may be as low as 4 ppm in the deficiency range and as high as 1000 ppm without toxicity. Plant tissues of the four species analyzed from Barrow ranged from about 20 to about 1000 ppm, with *C. aquatilis* generally containing the highest levels and *P. frigidus*, the lowest.

In 1972, blades of *D. fisheri* ranged in Mn concentration from 37 to 730 ppm and in 1973 from 36 to 728 ppm (Table 1). Corresponding stems ranged from 216 to 1030 ppm in 1972 (Ulrich and Gersper, 1974) and from 80 to 639 ppm in 1973 (Table 2). Average levels in both blades and stems were about the same in both years. None of the plants was deficient in Mn at the time of sampling, if the usual critical level of 10 ppm for the occurrence of grey speck in oats or barley is used as a reference.

Manganese levels in young mature leaves of *C. aquatilis* ranged from 230 to 770 ppm in 1972 and from 150 to 1000 ppm in 1973 (Table 3). Average levels were much higher in both years than in the blades of *D. fisheri* and *P. frigidus*. Quite clearly, *C. aquatilis* is not likely to be deficient in Mn anywhere in the Barrow tundra. The blades of *P. frigidus* ranged from 50 to 250 ppm Mn in 1972 and from 98 to 250 ppm in 1973 (Table 4). There was little difference between the 2 years and blades had higher average levels than petioles. Seasonal variations were not evident (Ulrich and Gersper, 1974, 1975) and none of the plants appeared to be deficient in Mn in either year.

Zinc

The plant material from the four tundra species sampled in 1972 and 1973 differed over a wide range of Zn concentrations, but with levels distinctly higher in 1973 than in 1972 (Tables 1–5). The lowest Zn levels were still considerably above the deficiency level of 10 ppm as determined for blades of nontundra plants. Since the lowest levels of Zn were about twice the critical level, it is unlikely that any of the plants were deficient in either year. Mean levels of Zn differed little among the four species or plant parts sampled except for the petioles of *P. frigidus*, which were about double those of the other samples. The blades and stems of *D. fisheri* were distinctly lower in Zn on the second sampling than on the first, but there were no consistent seasonal variations in the other species (Ulrich and Gersper, 1975).

Copper

The lowest levels of 1 to 5 ppm Cu detected in plants from the Barrow surveys, which are just within the range of determination, indicate that Cu is still adequate for growth. The highest levels of Cu occurred in *P. frigidus* and the lowest in *D. fisheri* (Tables 1–5). Levels were distinctly higher in 1973 than in 1972 and seasonal variations were inconsistent (Ulrich and Gersper, 1975).

Discussion and Conclusions

The results of the plant nutrient surveys at Barrow clearly indicate that the tundra in this area is a "nutritional desert," induced either directly or indirectly by low temperature. Nitrogen is the nutrient most frequently deficient, followed rather closely by P and then by other nutrients, either separately or in combination with N and P. In essence, it may be the supply of mineral nutrients that limits plant growth on the tundra rather than low temperature, reduced irradiance, or moisture, as often postulated heretofore. Evidence in support of our conclusion comes not only from the plant nutrient analyses but also from the vigorously growing plants in the scattered "green-spots" occurring at the Walt Disney sod houses, Will Rogers Monument, owl mounds, around dwellings in general, and in other areas of unusual nutrient availability such as tracked-vehicle scars, sloughs, and drainage channels.

The addition of nutrients by applications of artificial fertilizers, as at Schultz's plots, or in deposits of organic wastes by man, as at Disneyland North, or in natural deposits, as at owl mounds, results in dramatic increases in plant growth. As both live and dead biomass increase with time, the plant canopy becomes more dense, causing a marked decrease in thaw depth with a consequent decrease in nutrient supply. Thus, fertilization of the tundra could be self-defeating unless the increased biomass was harvested mechanically or by grazing.

Conversely, raising soil temperature, as by decreasing the albedo due to the destruction of the vegetative mat in tracked-vehicle scars (Gersper and Challinor, 1975), generally leads to an increase in nutrient supply by increasing thaw depth, enhancing root growth, and presumably accelerating mineralization. This would be followed by a new plant nutrient equilibrium that would be reestablished by the interaction of the variables of plant, climate, and micro- and macrobiological activity involved.

It is the increase in growth of plants in the "green-spots" of the tundra, along with their high nutrient status, especially in N and P, that indicates clearly that temperature, light, and moisture are not directly limiting plant growth on the tundra, but rather the supply of nutrients from the soil. This hypothesis implies, as a consequence, that tundra plant growth is based on a mineral nutrient economy, with each increment of nutrient resulting in a comparable increment in plant growth until the next limiting factor (another nutrient, temperature, irradiance, moisture, etc.) is reached. Modeling, therefore, becomes a matter of relating environmental factors to mineralization and photosynthesis, each process interacting to influence top and root growth. Thus, in the normal sequence of events, top temperatures rise sufficiently in the late spring or early summer to support photosynthesis, followed by top growth, which continues until one or more nutrients limit growth. Then, as the season progresses and temperatures rise, thaw depth and mineralization increase, and nutrient absorption and/or translocation replenishes the nutrient supply of the tops. Under these conditions top growth is again favored over root growth, until a deficiency of one or more of the nutrients recurs. At this time carbohydrates accumulate first in the tops and then in the roots, the carbohydrates supplying the energy for ion absorption and the raw material for root growth. Root growth is then favored over top growth and remains favored until mineralization and root activity replenish the nutrient supplies of the tops. At this point, top growth will again exceed root growth. In essence, top or root growth occurs whenever carbohydrates and minerals are both simultaneously adequate, and stops whenever either one becomes depleted. Later, as winter approaches, top temperatures decline faster than root temperatures and the tops cease growth, but root growth may continue until either the soluble carbohydrates are depleted or low root temperature or photoperiod inhibits growth (Billings et al., 1978). Fortunately, the factors that enhance nutrient supply, e.g., mineralization, also enhance plant growth. However, at those times when nutrient supplies exceed demand and moisture is not limiting, plant growth in arctic summers could be limited primarily by temperature. At any time when temperature becomes favorable or unfavorable for plant growth, the growth of roots and tops would proceed or recede.

Plant growth in the tundra is, therefore, time and temperature dependent, with the mineral supply the primary limiting factor. When the mineral supply is adequate and appropriately balanced, growth could depend primarily on temperature. However, a fully balanced and adequate nutrient supply under general tundra conditions at Barrow is not likely. In other words, the nutrient-supplying power of the soil under ordinary conditions does not normally meet the needs of plants, judging by the tremendous growth responses in spots of exceptionally high fertility (Figures 2 and 3) and by the generally low nutrient content of plants analyzed in the plant nutrient surveys of 1972 and 1973. Nitrogen and P were most prominently deficient, followed by K, Ca, Mg, in that order.

Results of our research have indicated so far that the nutritional status of the tundra and its plants can be monitored effectively through plant analysis. Managerial adjustments, based on such monitoring, can then be made as required for best use of the tundra as a natural resource. Managerial adjustments, however, must be made with caution.

In developing a management program for the tundra, the results of a single sampling can be used to determine the nutritional status of the plants at the time of sampling. By taking a series of samples from the same area at appropriate times, the analytical results will give the nutritional trends over an entire season. By comparing these results with the experimentally determined critical level of each nutrient, it is possible to determine which nutrients, if any, are deficient and the degree of deficiency. Then, if wanted, corrective measures could be taken to eliminate the deficiency, either by applying fertilizers or by changing management practices so as to improve mineralization and nutrient uptake. The effect of the corrective measures can be monitored by further plant sampling and analysis, and, if necessary, additional adjustments can be made in the program to achieve the nutritional and production goals planned for the tundra area in question.

Extending the plant nutrient survey to other areas of the Arctic and of Alaska and to other plant species is necessary to assess more accurately the nutrient status of cold-dominated ecosystems. Of more immediate importance and utility would be a survey of the plants along the Alaskan pipeline to gain appropriate information about their nutritional status and, consequently, their fertilizer requirements should it become desirable to increase plant growth or to reestablish vegetation along the pipeline. Hopefully, plant nutrient surveys and monitoring programs will be conducted whenever and wherever fertilizer test plots are established in Alaska in order not only to determine the effectiveness of the fertilizer applications, but to test also the validity of critical nutrient levels under field conditions.

Acknowledgments. This research was supported by NSF under Grant GV-29349. It was performed under the joint NSF sponsorship of the International Biological Programme and the Office of Polar Programs and was directed under the auspices of the US Tundra Biome. Field and Laboratory activities at Barrow were supported by the Naval Arctic Research Laboratory (NARL) of the Office of Naval Research. We thank Dr. L. Jacobson, Mr. Jeffrey Tennyson, Mr. Carlos Llano, Dr. Kwok Fong, and Mr. C. Carlson of the University of California, Berkeley, for various contributions to the research.

References

Billings, W. D., K. M. Peterson, and G. R. Shaver. (1978) Growth, turnover, and respiration rates of roots and tillers in tundra graminoids. In *Vegetation and Production Ecology of an Alaskan Arctic Tundra* (L. L. Tieszen, Ed.). New York: Springer-Verlag, Chap. 18.

Challinor, J. L., and P. L. Gersper. (1975) Vehicle perturbation effects upon a tundra soil–plant system. II. Effects on the chemical regime. *Soil Sci. Soc. Amer. Proc.*, **39**: 689–695.

Chapin, F. S. III. (1978) Phosphate uptake and nutrient utilization by Barrow tundra vegetation. In *Vegetation and Production Ecology of an Alaskan Arctic Tundra* (L. L. Tieszen, Ed.). New York: Springer-Verlag, Chap. 21.

Chapin, F. S., III, K. Van Cleve, and L. L. Tieszen. (1975) Seasonal nutrient dynamics of tundra vegetation at Barrow, Alaska. *Arct. Alp. Res.,* **7**: 209–226.

El-Sheikh, A. M., A. Ulrich, and T. C. Broyer. (1967) Sodium and rubidium as possible nutrients for sugar beet plants. *Plant Physiol.,* **42**: 1202–1208.

Flint, P. S., and P. L. Gersper. (1974) Nitrogen nutrient levels in arctic tundra soils. In *Soil Organisms and Decomposition in Tundra: Proceedings of the Microbiology, Decomposition and Invertebrate Working Groups Meetings, Fairbanks, Alaska, August 1973* (A. J. Holding et al., Eds.). Stockholm: International Biological Programme Tundra Biome Steering Committee, pp. 375–387.

Gersper, P. L. (1972) Chemical and physical soil properties and their seasonal dynamics at the Barrow intensive site. In *Proceedings of 1972 Tundra Biome Symposium, Lake Wilderness Center, University of Washington* (S. Bowen, Ed.). U.S. Tundra Biome, U.S. International Biological Programme, and U.S. Arctic Research Program, pp. 87–93.

Gersper, P. L., R. J. Arkley, R. Glauser, and P. S. Flint. (1974) Soil properties along a moisture gradient at USIBP Site Four, Barrow, Alaska. *U.S. Tundra Biome Data Rep.* 74–11, Hanover, N. H.: USACRREL, 128 pp.

Gersper, P. L., and J. L. Challinor. (1975) Vehicle perturbation effects upon a tundra soil–plant system: I. Effects on morphological and physical environmental properties of the soils. *Soil Sci. Soc. Amer. Proc.,* **39**: 737–744.

Gersper, P. L., R. Glauser, J. L. Challinor, A. M. Schultz, and R. J. Arkley. (1975) The effects of fertilizer treatment at Schultz's Flat (Site 10), Barrow, Alaska, six years following fertilization. *U.S. Tundra Biome Data Report* 75–3. Hanover, N.H.: USACRREL, 71 pp.

Haag, R. W. (1974) Nutrient limitations to plant production in two tundra communities. *Can. J. Bot.,* **52**: 103–106.

Hylton, L. O., A. Ulrich, and D. R. Cornelius. (1967) Potassium and sodium interrelations in growth and mineral content of italian ryegrass. *Agron. J.,* **59**: 311–314.

Johnson, C. M., and H. Nishita. (1952) Microestimation of sulfur in plant materials, soils and irrigation waters. *Anal. Chem.,* **24**: 736–742.

Johnson, C. M., and A. Ulrich. (1959) Analytical methods for use in plant analysis. *Calif. Agric. Exp. Sta. Bull.,* **766**: 25–78.

McKendrick, J. D., V. J. Ott, and G. A. Mitchell. (1978) Effects of nitrogen and phosphorus fertilization on carbohydrate and nutrient levels in *Dupontia fisheri* and *Arctagrostis latifolia*. In *Vegetation and Production Ecology of an Alaskan Arctic Tundra* (L. L. Tieszen, Ed.). New York: Springer-Verlag, Chap. 22.

Schultz, A. M. (1964) The nutrient recovery hypothesis for arctic microtine cycles. II. Ecosystem variables in relation to arctic microtine cycles. In *Grazing in Terrestrial and Marine Environments* (D. J. Crisp, Ed.). Oxford: Blackwell Scientific Publications, pp. 57–68.

Ulrich, A. and P. L. Gersper. (1974) Plant nutrient limitations of tundra plant growth at Barrow, Alaska: I. Plant nutrient survey, 1972. *U.S. Tundra Biome Data Rep.* 74–33. Hanover, N. H.: USACRREL, 44 pp.

Ulrich, A. and P. L. Gersper. (1975) Plant nutrient limitations of tundra plant growth at Barrow, Alaska: II. Plant nutrient survey, 1973. *U.S. Tundra Biome Data Rep.* 75–1. Hanover, N.H.: USACRREL, 64 pp.

Ulrich, A. and F. J. Hills. (1973) Plant analysis as an aid in fertilizing sugar crops: Part I. Sugar beets. In *Soil Testing and Plant Analysis* (L. M. Walsh and J. D. Beaton, Eds.). Madison, Wisc.: Soil Sci. Soc. Amer., Inc., pp. 271–288.

Ulrich, A. and K. Ohki. (1966) Potassium. In *Diagnostic Criteria for Plants and Soils* (H. Chapman, Ed.). Berkeley: Univ. Calif., Dep. Agric. Sci., pp. 263–393.

Warren-Wilson, J. (1957) Arctic plant growth. *Advance. Sci.,* **53**: 383–388.

21. Phosphate Uptake and Nutrient Utilization by Barrow Tundra Vegetation

F. Stuart Chapin, III

Introduction

Tundra soils are characterized by low nutrient availability (Hagg, 1974; Barél and Barsdate, 1978). The effect of low soil nutritional status upon tundra vegetation would be magnified by low temperatures which minimize rates of nutrient absorption. Indeed, high carbohydrate status such as is observed in tundra plants (McCown, 1978) is a feature which is frequently associated with conditions of nutrient deficiency (e.g., Leonard, 1962). The following study was undertaken to document the seasonal nutrient dynamics of tundra vegetation and to explore some of the interactions between temperature, nutrient availability, and nutrient absorption. New data are presented and the results of previous studies are summarized (Chapin, 1974a, b, 1977; Chapin et al., 1975: Chapin and Bloom, 1976).

Methods

Laboratory methods used in growing plants under controlled conditions and in the measurement of root growth parameters have been described previously (Chapin, 1974a). Briefly, plants were grown in solution culture with controlled root and air temperatures. After 6 weeks various measures of root growth and root surface area were determined. Field studies were conducted at the intensive site of the U.S. IBP Tundra Biome Program at Barrow, Alaska. The climate, site characteristics, and vegetation of this arctic coastal tundra site have been described previously (Brown et al., 1970; Miller et al., 1976; Tieszen, 1972).

Phosphate absorption studies were conducted in the field with the three principal graminoid species in the Barrow area, *Carex aquatilis* Wahlenb., *Dupontia fisheri* R. Br., and *Eriophorum angustifolium* Honck. Roots of *Dupontia* were sampled at monthly intervals for measurement of phosphate absorption rate. The polygon trough from which roots were sampled had shallow standing water during most of the growing season and was characterized by a mixed community dominated by *Carex aquatilis, Dupontia fisheri,* and *Eriophorum scheuchzeri.* An adjacent area of wet meadow tundra of similar species composition was artificially exposed to elevated soil temperatures with a heated pipe during the growing season and was also sampled at monthly intervals as described in detail elsewhere (Chapin and Bloom, 1976).

Carex aquatilis was sampled for phosphate absorption measurements from five sites along a microtopographic gradient of increasing plant production. The largest plants grew as emergents at the margin of a small pond. Here *Carex* grew in single species associations in standing water approximately 10 cm deep. The wet meadow tundra community sampled was dominated by *Carex aquatilis, Eriophorum angustifolium,* and *Dupontia fisheri* (Tieszen, 1972) and had a water table which fluctuated within 5 cm above and below the soil surface during the 1973 growing season. The polygon trough communities were similar to the wet meadow but had more *Eriophorum scheuchzeri* and less *E. angustifolium.* The basin in the middle of a low-centered polygon had the smallest plants and was populated almost entirely by widely spaced individuals of *C. aquatilis.* The basin had a moisture regime similar to that of the trough which separated adjacent polygons. The basin was separated from the trough by a ridge approximately 20 cm high and 1 m wide (Britton, 1957; Webber, 1978).

Carex aquatilis was also sampled from an area of wet meadow tundra which had been fertilized from 1960 to 1964 with a commercial fertilizer containing nitrogen, phosphorus, potassium, and calcium and from the corresponding control. The general results of this fertilization experiment are available in the literature (Schultz, 1964, 1969).

New roots of each species studied were obtained by gently removing the plant from the soil and choosing those parts of primary roots which were white and had not yet initiated secondary roots. Procedural details of field measurements of phosphate absorption have been presented elsewhere (Chapin, 1974a; Chapin and Bloom, 1976).

Nutrient dynamics of aboveground vascular plant material were examined during the 1970 growing season. Material was collected from each of 10 randomly selected quadrats at 10-day intervals between 15 June and 30 August (Chapin et al., 1975).

In order to assess the impact of increased nutrient availability upon plant production and nutrition, nitrogen–phosphorus–potassium fertilizers were applied to each of two plots immediately after snowmelt in 1970. In one plot these elements were applied in a ratio of 8:32:16, and in the other the ratio was 20:10:10. In both plots the application rate was 450 kg/ha (400 lb/acre).

Two additional treatments were employed to simulate the effect of intensive lemming grazing upon vegetation. In one plot (clip and clear treatment) all vascular vegetation was clipped at the moss surface and removed at the time of snowmelt. In the other plot (the mulch treatment) 240 g of the clipped material (largely standing dead from the previous season) was uniformly added to each square meter. This quantity was equivalent to approximately two and one-half times the annual aboveground production and was sufficient to cover the lower portion of the plant canopy. Two 0.1-m² quadrats were sampled from each treatment plot for peak season phytomass and chlorophyll and nutrient content, as described previously (Tieszen, 1972; Chapin et al., 1975). These samples were collected approximately 6 weeks after the plots were treated.

Results

Root Morphology

The root surface area exposed to the external environment by a unit weight of root was strongly influenced by root growth temperature. Root diameter decreased linearly with increasing root growth temperature in all species except *Carex aquatilis* (Figure 1). The regression equation describing this relationship was significant ($P < 0.01$) for each species of *Eriophorum* (Table 1). In these species the surface area per unit fresh weight was nearly twice as great when roots were grown at 20°C as when grown at 1°C (Figure 1). Since water content of roots was approximately the same for all species tested over a variety of root temperatures (86.7 ± 1.6% water, $N = 16$), surface area per unit dry weight biomass must also be strongly correlated with root growth temperature.

Root production was altered to a greater extent by a change in root temperature than was root diameter (Figures 1 and 2). Hence, the total root surface area of each tiller was closely correlated with root production. Change in root diameter had a minor effect upon total root surface area per tiller (Figure 2) but did appreciably alter the surface area of each gram of root biomass produced (Figure 1, Table 1).

The optimum temperature for root production was 10°C for most species tested (Figure 2). *Eriophorum angustifolium,* which was the deepest rooted

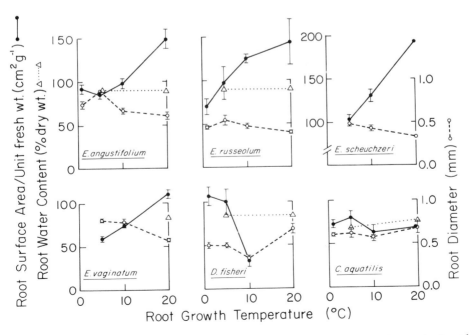

Figure 1.   Effect of root growth temperature on root surface area, root water content, and root diameter of various arctic graminoids.

Table 1. Influence of Root Growth Temperature upon Root Diameter, Surface Area, and Weight

Species	Sample size	Root diameter[a]		Surface area/weight		Surface area/length		Weight/length		Water content (% fresh weight)
		Relationship	Reg. coeff.	Relationship	Reg. coeff.	Relationship	Reg. coeff.	Relationship	Reg. coeff.	
Carex aquatilis	34	$D = 0.005T + 0.577$	0.18	$S_w = -0.33T + 74.06$	-0.11	$S_l = 0.0014T + 0.18$	0.19	$W = 0.003T + 0.268$	0.22	72.7
Dupontia fisheri	26	$D = 0.004T + 0.449$	0.20	$S_w = -1.77T + 94.24$	-0.28	$S_l = -0.0052T + 0.17$	-0.64**	$W = 0.030T + 0.113$	0.76**	84.9
Eriophorum angustifolium	40	$D = -0.009T + 0.796$	-0.52**	$S_w = 3.31T + 75.75$	0.71**	$S_l = -0.0030T + 0.25$	-0.53**	$W = -0.008T + 0.304$	-0.73**	89.2
E. russeolum	35	$D = -0.006T + 0.534$	-0.39*	$S_w = 3.50T + 78.83$	0.58**	$S_l = -0.0012T + 0.16$	-0.33*	$W = -0.005T + 0.200$	-0.58**	90.4
E. scheuchzeri	21	$D = -0.010T + 0.531$	-0.41	$S_w = 5.87T + 73.00$	0.75**	$S_l = -0.0027T + 0.16$	-0.26	$W = -0.009T + 0.194$	-0.68**	91.2
E. vaginatum	31	$D = -0.017T + 0.920$	-0.80**	$S_w = 3.47T + 40.58$	0.89**	$S_l = -0.0053T + 0.29$	-0.80**	$W = -0.018T + 0.526$	-0.94**	82.8

[a] D = root diameter (mm); T = temperature (°C); S_w = surface area/weight (cm² g⁻¹ fresh wt); S_l = surface area/length (cm² cm⁻¹); W = weight/length (mg mm⁻¹).

* Correlation is significant with a probability less than 0.05.

** Correlation is significant with a probability less than 0.01.

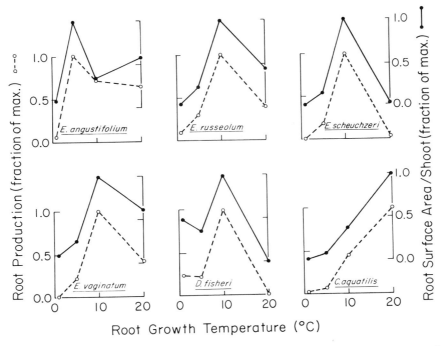

Figure 2.   Effect of root growth temperature on root production and total root surface area per shoot of various arctic graminoids.

species at Barrow, had the lowest temperature optimum for both root production and total root surface area per tiller.

Preliminary observations suggested that production of lateral roots was associated with warm root temperatures, dryness, and/or improved aeration. When roots of *Dupontia fisheri* and *Carex aquatilis* were allowed to grow in an enclosed humid atmosphere above a solution culture, root color changed from white to yellow, and abundant secondary roots were initiated. Similarly, in the field at Barrow, primary roots of *Dupontia* and *Carex* turned yellow near the base and initiated secondary roots. This occurred most prominently late in the growing season near the soil surface where soils were less saturated. Roots of *Eriophorum vaginatum* showed more pronounced yellowing at Eagle Creek (an alpine cottongrass tundra site) than in the moister Barrow site.

Temperature Response of Phosphate Absorption

Root temperature had direct effects upon the phosphate absorption process. Field-measured phosphate absorption rates increased with increasing temperature in *Dupontia fisheri* over the entire range of experimental temperatures tested from 1 to 40°C (Figure 3). During July and September, the rate of phosphate absorption increased exponentially with temperature (Figure 3). How-

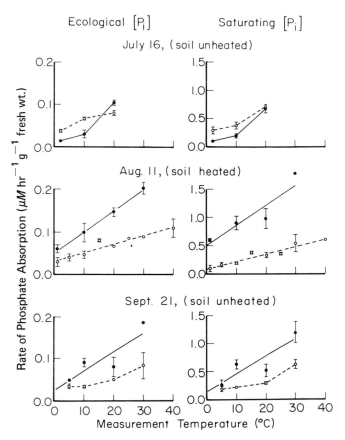

Figure 3.   Effect of measurement temperature upon phosphate absorption rate of *Dupontia fisheri* growing under natural conditions but measured under standard conditions. Measurements were made at early (16 July), mid- (11 August), and late (21 September) season on plants growing in normal permafrost-dominated soils (dashed line) and in soils which were artificially heated from 20 July till 28 August (solid line). Measurements were made at ecological (1.0 μM) and at saturating (20.0 μM) concentrations of phosphate. Each data point is the average of three or more measurements and is shown with its corresponding standard error.

ever, when phosphate absorption was measured at a greater number of temperatures (in August), the relationship between phosphate absorption rate and measurement temperature was more nearly linear. Likewise, *Carex aquatilis* measured at 5°C temperature intervals showed a linear increase of phosphate absorption rate with increasing experimental temperature. Q_{10} of phosphate absorption rate calculated between 10 and 20°C generally varied between 1.0 and 2.0.

Growth of new roots in *Dupontia fisheri* began in late June or early July. Rates of phosphate absorption measured on roots from unaltered control plots were highest when first measured in mid-July and decreased substantially by mid-August (Figure 4). This seasonal decrease in absorption rate was observed at high and low measurement temperatures. When measurements were made on 21 September, leaf bases were green, but leaf blades were largely senescent. In contrast, roots were as white and healthy looking as at any time during the growing season. Rates of phosphate absorption by these roots measured under standard conditions did not differ significantly from rates measured in early August, when shoots were actively growing (Figure 4).

Plants were grown over a heated pipe to determine the effect upon phosphate absorption of acclimation to warm soils. Before the hot pipe experiment began, individuals of *Dupontia fisheri* growing in the control area exhibited a higher rate of phosphate absorption than did those growing over the unheated pipe. After the soil had been heated for 20 days and was 14°C warmer than the control, plants growing over the heated pipe had acquired a rate of phosphate absorption three to five times greater than that of the control when measured at high phosphate concentrations. However, these warm-acclimated roots had a lower affinity for phosphate than did the control so that absorption rate measured at ecological phosphate concentrations was only twofold higher in roots from the heated soil

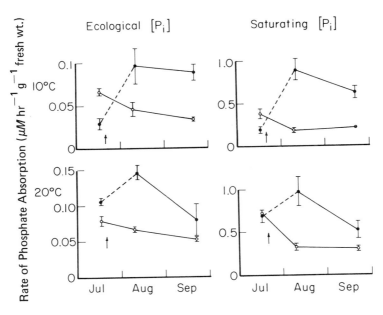

Figure 4. Seasonal course of phosphate absorption rate by *Dupontia fisheri* measured at 10 and 20°C and at ecological (1.0 μM) and at saturating (20.0 μM) concentrations of phosphate. Open circles show plants from the normal permafrost site and solid circles show the plants from the hot pipe site. Arrow indicates the date heat was first applied to soil of the hot pipe site. Each measurement is the average of three or more measurements and is shown with its corresponding standard error.

than in the control (Figure 3). There was no consistent difference in temperature dependence (Q_{10}) between plants from the two treatments (Figure 3). Roots of warm-acclimated plants had a smaller diameter and gave a visual impression of being less abundant over the heated pipe than over the control. Shoot height was clearly increased by warming the soil, although no measurements were taken.

The soil remained heated to approximately 18°C until 28 August, at which time the pipe was turned off and soil temperatures returned to control levels. When the final seasonal measurements were made on 20 September, the rate of phosphate absorption by plants in the hot pipe site was still higher than that of the unheated control. However, rates of absorption had decreased substantially in the hot pipe site from August to September, whereas rates did not change significantly in the control (Figure 4). Consequently, rate of phosphate absorption by roots in the hot pipe site was much more similar to the control in September when the soil was unheated than in August when the soil was 15°C warmer than the control.

When the three dominant species from Barrow wet meadow tundra were compared, phosphate absorption was found to respond differently to temperature in ways which complemented their different rooting strategies (Figure 5). *Carex aquatilis* and *Dupontia fisheri* frequently coexist in lowland tundra associations at Barrow, Alaska. Roots of established *C. aquatilis* plants penetrated to greater depths during their first year of growth than did those of *D. fisheri*. The *C. aquatilis* population tested had a slightly warmer average soil temperature at the average rooting depth than did *D. fisheri* (7 as opposed to 5°C), but the superficial new roots of *D. fisheri* experienced warmer maximum temperatures (as high as 15°C) than did *C. aquatilis* (10°C maximum). *D. fisheri* had a higher optimum temperature for phosphate absorption than did *C. aquatilis* (Fig. 5), and phosphate absorption by *D. fisheri* was less temperature sensitive than that by the deep-rooted *C. aquatilis*. *Dupontia fisheri* had a greater affinity for phosphate than did *C. aquatilis*, as evidenced by the high rates achieved at lower concentrations in *D. fisheri*. *Eriophorum angustifolium* grows at the greatest depth of the three species and hence experiences lowest soil temperatures. This species has the highest rate measured at 1°C of the three species and was least affected by changes in measurement temperature. The affinity of roots for phosphate was intermediate between that of the other two species.

The overall effect of temperature upon phosphate absorption rate is approximated by the following equations:

$$V_{max} \text{ (lab)} = -0.026 \ T_J + 0.71 \tag{1}$$

$$V_{max} \text{ (field)} = 0.023 \ T_M + 0.213 \tag{2}$$

$$\text{apparent } K_m = 0.492 \ T_M + 0.8 \tag{3}$$

$$V = \frac{V_{max} \ (P_i)}{K_m + (P_i)} \tag{4}$$

where V_{max} is the maximum rate of phosphate absorption at a given temperature ($\mu M \ hr^{-1} \ g^{-1}$ fresh wt), T_J is the July mean soil temperature (°C) of the habitat of

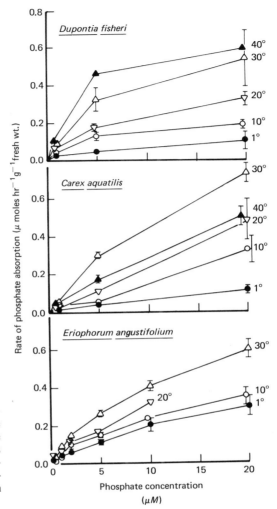

Figure 5. Effect of temperature upon phosphate absorption rate of the three principal graminoid species at Barrow. Means and standard errors for four observations are shown for each data point.

origin at the average rooting depth, T_M is the temperature (°C) at the time of measurement, apparent K_m is an inverse measure of the affinity of roots for phosphate (μM P_i), and V is the observed phosphate absorption rate (μM hr^{-1} g^{-1} fresh wt). Equation (1) accounts for differences in phosphate absorption capacity resulting from evolution at different soil temperatures. Equations (2) and (3) document the immediate effect of temperature upon the phosphate absorption process. Equation (2) was developed from field measurements at Barrow with *Dupontia fisheri*. Equation (3) resulted from field measurements at Eagle Creek with *Eriophorum vaginatum*. Equation (4) is the Michaelis–Menton equation which described the relationship between rate and phosphate concentration in all species thus far examined. Details are given elsewhere (Chapin, 1977).

Table 2.   Maximum Rate of Phosphate Absorption (V_{max}) and Affinity of Roots for Phosphate (Apparent K_m) in First Year Roots of *Carex aquatilis* Grown in Various Habitats at Barrow, Alaska, and Measured at 10°C

Treatment	V_{max} of phosphate absorption (μmoles hr^{-1} g^{-1} fresh wt)	Apparent K_m (μM)
Fertilization study		
Fertilized	0.222 ± 0.027	12.9 ± 2.6
Control	0.464 ± 0.180	26.6 ± 16.8
Gradient of plant size		
Pond margin	0.832 ± 0.195	21.6 ± 7.4
Wet meadow	0.273 ± 0.042	11.2 ± 3.0
Polygon trough	0.219 ± 0.029	11.0 ± 2.7
Polygon basin	0.117 ± 0.009	6.4 ± 1.1

Acclimation/Adaptation to Soil Phosphate

Field-measured phosphate absorption rates increased with increasing concentrations of externally supplied phosphate. *Carex aquatilis* fertilized by A.M. Schultz with a commercial fertilizer 10 yr prior to measurement had a lower V_{max} of phosphate absorption than the unfertilized control. However, fertilized plants had a greater affinity for phosphate (lower apparent K_m) than the unfertilized control such that the rate of absorption measured at low phosphate concentrations was similar between treatment and control plots (Table 2).

Carex aquatilis examined in several community types at Barrow had phosphate absorption rates which were correlated with aboveground production. Populations growing as emergents at the margins of small ponds produced the largest shoots and had the highest rates of phosphate absorption (Table 2). Plants growing in polygon basins were quite stunted, having shoots less than half as tall as pond populations. These basin plants exhibited a correspondingly low rate of phosphate absorption, when measured under standard conditions. Populations of *C. aquatilis* from wet meadow and polygon trough populations exhibited an intermediate shoot size in the field and had intermediate phosphate absorption rates when measured under standard conditions. Affinity of roots for phosphate was inversely correlated with production in that roots of plants from the polygon basin showed the greatest affinity for phosphate (lowest apparent K_m), whereas roots of the more productive pond margin plants had the lowest affinity for phosphate (Table 2).

Nutrient Manipulations

The addition of nitrogen, phosphorus, and potassium in 1970 consistently increased production of the Barrow wet meadow tundra, even just 6 weeks after application. The increase was particularly pronounced with the high phosphorus fertilizer (Table 3). With only two biomass samples per treatment, any significance of these changes could not be demonstrated. Nitrogen concentration of the

Table 3. Response of Barrow Wet Meadow Tundra Vegetation to Treatments of Added Nutrients, Clipping, and Added Litter

Parameter	Control	N fertilized[a]	P fertilized[b]	Clip and clear[c]	Mulch[d]
Peak season biomass (g m^{-2})	102 ± 10	110 (100–119)[e]	146 (141–150)	61 (57–65)	73 (66–79)
Chlorophyll content (mg g^{-1} dry wt)	5.7 ± 0.3	3.9 (3.4–4.5)	3.8 (3.7–3.9)	4.3 (3.8–4.9)	4.5 (4.2–4.8)
(g m^{-2})	455 ± 35	439 (339–539)	555 (523–588)	268 (216–319)	325 (319–330)
Average monocot nutrient content (%)					
Nitrogen	1.86 ± 0.07	1.89 ± 0.14	1.77 ± 0.12	2.15 ± 0.12	1.87 ± 0.16
Phosphorus	0.129 ± 0.007	0.216 ± 0.033***	0.287 ± 0.034***	0.235 ± 0.027***	0.183 ± 0.018
Potassium	0.94 ± 0.04	1.15 ± 0.12***	1.47 ± 0.13***	1.28 ± 0.07*	1.24 ± 0.06***
Calcium	0.159 ± 0.014	0.183 ± 0.022	0.159 ± 0.013	0.179 ± 0.018	0.146 ± 0.020
Magnesium	0.196 ± 0.016	0.130 ± 0.011	0.154 ± 0.012	0.138 ± 0.013	0.136 ± 0.008
Total aboveground nutrient content					
Nitrogen (g m^{-2})	1.79 ± 0.20	1.98 (1.88–2.09)[e]	2.38 (2.33–2.44)	1.23 (1.08–1.38)	1.34 (1.33–1.35)
Phosphorus (mg m^{-2})	116 ± 14	222 (212–233)	334 (332–337)	143 (136–150)	120 (119–121)
Potassium (g m^{-2})	0.90 ± 0.09	1.19 (1.04–1.35)	2.09 (1.71–2.47)	0.79 (0.74–0.83)	0.83 (0.82–0.84)
Calcium (mg m^{-2})	167 ± 24	223 (201–245)	232 (227–238)	114 (99–129)	124 (112–135)
Magnesium (mg m^{-2})	197 ± 23	158 (155–160)	225 (212–239)	88 (87–88)	105 (91–118)

[a] 20–10–10 commercial fertilizer applied at 450 kg/ha 6 weeks before sampling.
[b] 8–32–16 commercial fertilizer applied at 450 kg/ha 6 weeks before sampling.
[c] All vascular vegetation clipped at the moss surface and removed 6 weeks before sampling.
[d] 240 g m^{-2} of standing dead added to plot 6 weeks before sampling.
[e] Range of duplicate samples.
* Treatment differs significantly from the control at 0.05 level of probability.
*** Treatment differs significantly from the control at 0.001 level of probability.

monocot plant tissue did not change significantly with fertilization, but both phosphorus and potassium concentrations were significantly higher ($P < 0.001$) in plants which had received added nutrients. Of the three main species the shallow-rooted *Dupontia* showed the greatest increase in phosphorus concentration (240%) and production (130%) in response to application of the high phosphorus fertilizer.

Clipping and removal of vascular vegetation (simulated lemming grazing) decreased peak season standing crop (Table 3). This treatment also resulted in a significant increase in phosphorus ($P < 0.001$) and potassium ($P < 0.05$) concentrations in leaves above the control, although there was no change in nitrogen. Addition of excess standing dead from outside the plot decreased production. Only potassium concentration was significantly altered ($P < 0.001$) by this treatment.

Seasonal Nutrient Dynamics

In each of the three major monocot species at Barrow, *Dupontia fisheri, Carex aquatilis,* and *Eriophorum angustifolium,* the percentage nitrogen of current year's aboveground material reached its 1970 maximum on 25 June, 10 days after

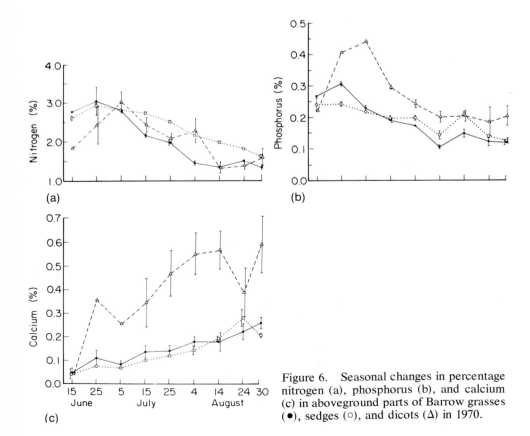

Figure 6.   Seasonal changes in percentage nitrogen (a), phosphorus (b), and calcium (c) in aboveground parts of Barrow grasses (●), sedges (○), and dicots (Δ) in 1970.

snowmelt (Figure 6, Table 4). Grasses as a group tended to have a higher nitrogen concentration than sedges. However, after 25 June, the percentage nitrogen of grass shoots decreased exponentially, reaching a minimum on 4 August, coincident with the time of peak standing crop of biomass (Figure 6). In contrast, the percentage nitrogen in the current year's sedge shoots decreased gradually through the growing season and reached no stable minimum during the period of sampling. Percentage nitrogen in the current year's aboveground production peaked later in the season in dicots than in monocots reaching a maximum intermediate between that of grasses and sedges. In both monocots and dicots the percentage nitrogen dropped by about 50% in the course of the growing season. All grasses and sedges and the principal dicot species (*Petasites frigidus*) continued to produce new leaves throughout the growing season. More and more of these leaves senesced as the season progressed. Therefore, early in the growing season, most of the leaves sampled were newly produced. However, as the season progressed, a larger percentage of the biomass sampled consisted of mature or senescent leaves. This coincided with a decreased nitrogen concentration.

The total quantity of nitrogen accumulated in aboveground plant material more rapidly than did the biomass itself in each of the three major Barrow monocots and in the foliage as a whole (Figure 7). In *Carex aquatilis* and *Dupontia fisheri* and in the aboveground biomass as a whole, the peak standing crop of nitrogen was achieved at least 10 days before the time of peak standing crop of biomass (Table 4). Consequently, at mid-season the quantity of aboveground nitrogen was decreasing at a time when biomass was still increasing. By 24 August, 40% of the aboveground nitrogen had disappeared from the shoots and by the following spring another 20% had disappeared from the standing dead (Table 5).

The seasonal changes in percentage phosphorus and total phosphorus in aboveground biomass followed a pattern similar to that of nitrogen. There was no consistent difference in the seasonal course of phosphorus content between grasses and sedges. As with nitrogen, phosphorus percentage peaked later in the season in dicots than in monocots (Figure 6). However, in marked contrast to nitrogen, the percentage phosphorus was much greater in dicots than in monocots. Half of the maximum standing stock of phosphorus was translocated belowground (or leached) by 24 August and of that remaining, 40% disappeared by spring (Table 5).

The seasonal pattern of calcium concentration was quite different from those of the soluble elements previously discussed (nitrogen and phosphorus). In all species calcium concentration of aboveground phytomass increased progressively through the growing season. The calcium concentration of dicots was three to four times greater than that of monocots (Figure 6). Consequently, although dicots constituted only 8% of the average plant biomass and contained only 6 to 9% of the average standing stock of nitrogen, phosphorus, and potassium, they accounted for over 16% of the aboveground plant calcium. The standing stock of aboveground calcium increased exponentially through the early part of the growing season. Late in the season, when the standing crop of biomass began decreasing, the quantity of aboveground calcium ceased increas-

Table 4. Concentration and Standing Stock of Nutrients in Aboveground Vascular Plant Material

Parameter	N	P	K	Ca	Mg	Fe	Mn	Sample size
Maximum nutrient concentration[a] (%)								
Monocots	2.98 ± 0.11	0.25 ± 0.01	1.47 ± 0.09	0.08 ± 0.01	0.12 ± 0.01	0.008 ± 0.001	0.019 ± 0.003	16
Dicots	3.02 ± 0.28	0.44[c]	1.16	0.26	0.17	0.016	0.024	2–5
Total foliage	2.90 ± 0.12	0.26 ± 0.02	1.43 ± 0.09	0.10 ± 0.02	0.12 ± 0.01	0.008 ± 0.001	0.018 ± 0.003	21
Concentration at peak biomass[b] (%)								
Monocot	1.86 ± 0.07	0.13 ± 0.01	0.94 ± 0.04	0.16 ± 0.01	0.20 ± 0.02	0.014 ± 0.002	0.018 ± 0.002	37
Dicot	2.26 ± 0.31	0.20 ± 0.02	1.57 ± 0.29	0.55 ± 0.09	0.53 ± 0.07	0.016	0.013	5
Total foliage	1.90 ± 0.07	0.13 ± 0.01	1.01 ± 0.06	0.20 ± 0.03	0.23 ± 0.02	0.015 ± 0.002	0.018 ± 0.002	42
Standing stock at peak biomass[b] (mg m^{-2})								
Monocot	1699 ± 190	109.3 ± 14.4	834 ± 86	145.6 ± 20.2	177.9 ± 20.3	8.84 ± 1.88	18.50 ± 4.21	8
Dicot	94 ± 28	6.3 ± 2.3	68 ± 22	21.6 ± 6.9	18.8 ± 6.0	0.47 ± 0.19	0.15 ± 0.11	8
Total foliage	1793 ± 204	115.6 ± 14.3	902 ± 90	167.2 ± 23.6	196.7 ± 22.9	9.32 ± 1.93	18.65 ± 4.31	8

[a] Values for monocots and total foliage were determined 6/25/70, the date when most nutrients showed their highest concentration in most monocot species. Dicot values are for 7/5/70, the date of peak concentration.
[b] Determined 8/4/70.
[c] Values without standard error are averages of duplicate determinations.

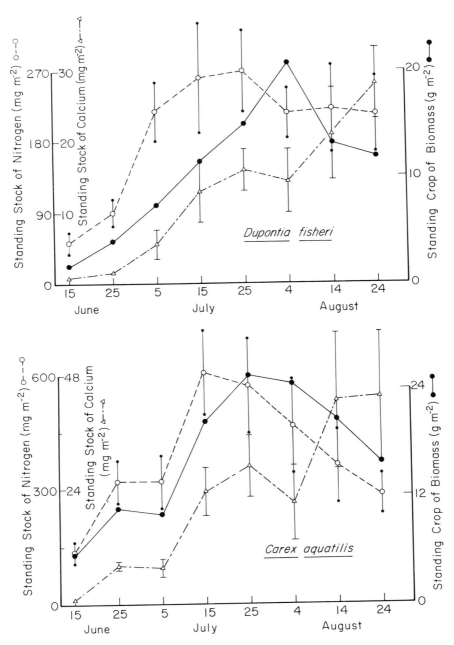

Figure 7. Seasonal progression of biomass, standing stock of nitrogen, and standing stock of calcium in the three principal Barrow species and in the total foliage during 1970. (Parts adapted from Chapin et al., 1975, with the permission of the Institute of Arctic and Alpine Research.)

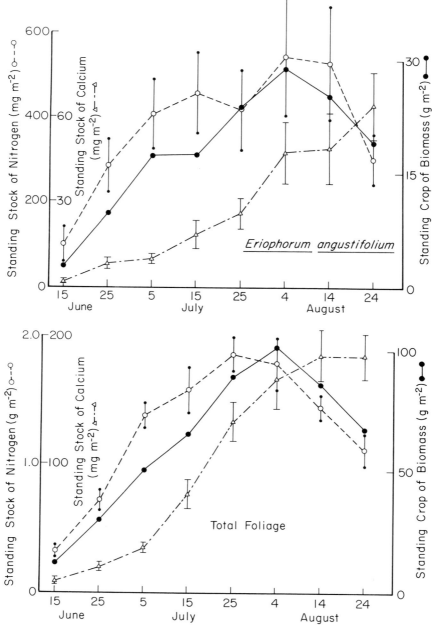

Figure 7 (continued).

Table 5. Estimated Nutrient Loss from Shoots through Leaching and through
Retranslocation Belowground

	N	P	K	Ca	Biomass[a]
Nutrient retranslocation[b] (percentage maximum standing stock)					
Carex aquatilis	53	55	64	0	38
Dupontia fisheri	21	44	25	0	43
Eriophorum angustifolium	45	61	48	0	34
Monocot	48	45	42	0	35
Dicot	65	78	76	—	33
Total	40	49	44	0	34
Nutrient leaching[c]					
Carex aquatilis	26	54	84	−45	--
Dupontia fisheri	27	54	83	−26	--
Eriophorum angustifolium	22	38	82	−115	--
Monocot	19	41	82	−19	--

[a] Calculated from Tieszen (1972).
[b] Assumes all nutrient disappearance from time of maximum standing stock of nutrients until August 24 results from retranslocation.
[c] Assumes all nutrient loss from shoots between 8/24/70 and 6/15/71 is due to leaching; standing dead determinations were made 6/15/70; end of season values were determined 8/30/70; 1971 standing dead values for percentage nitrogen were within 5% of the 1970 values but were not used in these calculations because data were unavailable for other elements.

ing and remained constant (Figure 7). The percentage calcium of standing dead material at the beginning of the growing season was 20% greater than the end of season values for live material. This contrasted with the more mobile elements (nitrogen, phosphorus, and potassium) which decreased in concentration.

Discussion

Low temperatures can limit nutrient absorption by plants in the tundra, just as elsewhere, but there are important means by which tundra plants maximize nutrient gain and minimize nutrient loss and thereby compensate for low nutrient availability and slow rates of nutrient absorption.

Root morphology is strongly influenced by root growth temperature. The large root diameter observed at low root growth temperatures results in a small surface-to-volume ratio. Consequently, roots which grow at low temperatures have less nutrient-absorbing surface per unit biomass than do roots grown at warm temperatures. However, tundra plants have a smaller root diameter than do temperate plants at any given growth temperature (Chapin, 1974a). This difference between species results in similar root diameters in tundra and temperate marsh plants at their respective environmental temperatures and compensates for the temperature influence upon root diameter. Moreover, the effect of root temperature on root diameter is less pronounced in tundra than in temperate species (Chapin, 1977).

Theoretically, nutrient absorption rates are a function of the total surface area of root cortical cells and not of the surface area of the root per se. However, wet meadow tundra plants are characterized by low transpiration rates (Miller et al, 1976) and high root-to-shoot ratios (Dennis and Johnson, 1970). Both of these features would result in slow rates of water movement (and hence nutrient transport) through the root cortex. Therefore, in tundra plants peripheral cells of the cortex may be more active in nutrient absorption than would be those cells toward the interior of the cortex. Hence, root diameter may exert a more important influence upon nutrient absorption rates in tundra plants than in plants which experience high rates of nutrient movement through the intercellular spaces of the root cortex. Root diameter definitely is important in the sense that a small diameter root can exploit a larger soil volume per gram of biomass invested than can a large diameter root.

The anaerobic soil environment of the tundra places certain restrictions upon root diameter. The virtual absence of secondary roots below the water table and their production after the rhizosphere becomes more aerobic either in the field or in the laboratory suggest that thin roots function most effectively when they receive an adequate oxygen supply from the soil. The large diameter which characterizes the primary roots of *Dupontia, Carex,* and *Eriophorum* may represent a balance between selection for thick roots which contain abundant aerenchyma for oxygen transport (Armstrong, 1972) and thin roots with a high surface-to-volume ratio for more effective nutrient absorption.

There appear to be two alternative rooting strategies used by all Barrow graminoids. In submerged soils thick primary roots are produced which have adequate aerenchyma to support root metabolism. The white color of these roots and the maintenance of a substantial phosphate absorption capability along the entire root length of one such species, *Eriophorum vaginatum* (Chapin, 1974a), suggest that these roots are relatively unsuberized and retain their nutrient-absorptive capacities for quite a distance behind the root tip. As the water table lowers later in the growing season, the thick primary roots become suberized and are probably no longer as effective in nutrient absorption throughout their length. In such soils thin secondary roots are produced. These roots may be quite effective in nutrient absorption because of their large surface-to-volume ratio. Thin roots are presumably a viable alternative rooting strategy only when the soil has adequate aeration.

Although root diameter may influence the effectiveness in nutrient absorption of biomass and protein which are allocated to roots, root-to-shoot ratio appears to be more important in determining the total supply of and demand for nutrients. Root production is, of course, temperature dependent (see also Shaver and Billings, 1975; Billings et al., 1978), but tundra plants have a lower optimum temperature for root production than do temperate marsh plants. This results in part from lower temperature optima for both root initiation and root elongation (Chapin, 1974a). Hence, tundra plants can produce a substantial quantity of roots at temperatures which would prevent root growth by most temperate species (Richards et al., 1952).

Nutrient absorption processes are themselves temperature dependent. The increase in phosphate absorption rate with increasing temperature to at least 40°C in *Dupontia fisheri* and *Eriophorum angustifolium* is similar to the temperature response of temperate graminoids (Carter and Lathwell, 1967). Tundra plants are apparently not adapted to low soil temperatures through any change in the temperature optimum for the phosphate absorption process. Similarly, Tieszen and Sigurdson (1973) found that the optimum temperature for RuDP carboxylase activity in tundra graminoids was comparable to values for temperate zone grasses.

The linearity of the relationship between phosphate absorption rate and temperature, even down to 1°C, indicates that roots of tundra plants are quite capable of absorbing phosphate from cold soils whenever soils are thawed. Temperate plants are unable to absorb phosphate at these low temperatures (Sutton, 1969; Rathore et al., 1970). The relatively slight temperature sensitivity of the phosphate absorption process (low Q_{10}) in both field (Chapin and Bloom, 1976) and laboratory (Chapin, 1974a) measurements would suggest that the tundra plants investigated do not depend upon daily or seasonal warming of the soil for acquisition of phosphate but instead actively absorb phosphate from cold soils.

The seasonal changes in phosphate absorption capacity must be viewed within the context of seasonal patterns of nutrient supply and demand. Nutrient absorptive processes at Barrow have a season which lags considerably behind that of shoot growth. Roots retain a substantial capacity to absorb phosphate and to translocate that phosphate to rhizomes even in late September (Chapin and Bloom, 1976). At this time, aboveground portions of plants are senescent. Such nutrient absorption and translocation to storage organs probably continues until the soil begins freezing from the surface downward in late September or October. In contrast, shoot growth proceeds rapidly in mid-June (Tieszen, 1972), while the ground is largely frozen. Production of new roots in *Dupontia* begins approximately 2 weeks after the onset of shoot growth. The asynchrony of aboveground and belowground plant functions is a necessary feature of arctic plants. Snowmelt precedes the summer solstice and concomitant favorable radiation regime by only a few days, whereas the maximum depth of thaw and the warmest root temperatures come in late July or early August.

The differences in temperature response of phosphate absorption in *Eriophorum angustifolium, Carex aquatilis,* and *Dupontia fisheri* are consistent with their ecological differentiation. Although the species differ in optimum temperature, this is unlikely to have ecological importance, because the optimum for all three species far exceeds temperatures ever encountered in nature. Of greater importance is the low sensitivity to temperature change. The species which experiences the lowest root temperatures *(E. angustifolium)* is least affected by temperature in its absorption. Similarly, in laboratory studies, tundra species as a group showed less temperature sensitivity of phosphate absorption than did temperate species (Chapin 1974a). This suggests that, within a single tundra community as well as over a broad latitudinal gradient, extreme low temperature selects for low sensitivity to temperature. This may contribute to the notable success of *E. angustifolium* at exploiting low soil temperature.

The increase in phosphate absorption capacity following warm acclimation over the heated pipe conflicts with laboratory results (Chapin, 1974a, b) which demonstrate a consistent decrease in phosphate absorption capacity following warm acclimation. Moreover, the observed increase in phosphate absorption capacity over the heated pipe probably did not result from the improved soil phosphate status relative to the control (Chapin and Bloom, 1976). Agricultural studies show that improved soil phosphate status, such as seen in the heated soils, generally leads to a decline in phosphate uptake capacity (Hoagland and Broyer, 1936). However, changes in soil nitrogen may have been important. Cole et al. (1963) report that improved nitrogen status of barley increases its capacity to absorb phosphate. McKendrick et al. (1978) showed that nitrogen is a prime factor limiting growth of *Dupontia* at Barrow. Heating the tundra soil resulted in higher *in situ* microbial respiration rates (Benoit, personal communication) and a threefold increase in the concentration of ammonia in the soil solution (Barsdate and Barél, personal communication). This would presumably lead to an improved soil and plant nitrogen status. Regardless of the cause of the increase in phosphate absorption capacity by warm-acclimated roots, it is clear that heating the soil results in a complex interaction between the roots and soil which extends beyond a simple temperature acclimation response. The implications of this field acclimatization experiment have been more thoroughly discussed elsewhere (Chapin and Bloom, 1976).

Individuals of *Carex aquatilis* growing on soils fertilized by A. M. Schultz in 1960–64 had a higher phosphate content (Ulrich and Gersper, 1978) than control plants. Dennis (1977) found a similar root-to-shoot ratio in fertilized and unfertilized plots, indicating that there was little compensatory shift in biomass allocation as a result of long-term fertilization. However, at the physiological level the lower capacities for phosphate absorption by fertilized than by unfertilized plants presumably reflected a reallocation of protein from the phosphate absorption process to other more limiting processes. These findings support laboratory studies which show that low phosphate status increases the phosphate absorption capacity of plants (Hoagland and Broyer, 1936; Humphries, 1951).

The correlation between phosphate absorption capacity in *Carex aquatilis* and shoot size is consistent with the hypothesis that phosphate is a critical nutritional element in the Barrow ecosystem and that plants capable of absorbing more phosphate will be more productive and will retain a selective advantage. The high affinity for phosphate shown by the least productive populations would tend to offset the effect of their low phosphate absorption capacity. The observed differences in V_{max} and apparent K_m are both genetically and environmentally controlled (Shaver et al., in press).

Clearly, tundra plants are capable of exploiting the cold, low phosphate environment in which they live. Their seasonal pattern of nutrient utilization indicates that they have been equally successful at retaining and recycling those nutrients which they acquire.

Barrow monocots begin the growing season with a small amount of overwintering green material. The high concentrations of nitrogen, phosphorus, and potassium in this material suggest that even in mid-June the plants are quite active metabolically. This is necessary for the rapid shoot growth observed at snowmelt (Tieszen, 1972). In mid-June, part of the tissue sampled consisted of dead leaf tips of overwintering live leaves. Growth of monocots is most rapid and senescence is minimal early in the season (Tieszen, 1972), so that by the end of June the shoots consist almost entirely of new leaves and therefore exhibit peak concentrations of nitrogen, phosphorus, and potassium. The pronounced decrease in concentration of these elements over the following month is accompanied by an increase in their total standing stock, and therefore clearly does not represent any net nutrient translocation out of the shoot. Rather, as the shoots age and contain a larger percentage of old leaves, shoot weight is substantially increased by the synthesis of structural carbohydrates as secondary cell walls. The increase in leaf weight as the growing season progresses would dilute the nutrient content and thereby decrease the percentage concentration.

The fact that total aboveground standing stocks of nitrogen and phosphorus begin decreasing before peak biomass is achieved indicates that there is a net translocation of these nutrients to belowground organs even as shoot growth continues. This retranslocation of soluble nutrients out of leaves begins before the onset of obvious leaf senescence. The substantial amount of translocation of mobile nutrients out of shoots to belowground organs is consistent with previous observations in tundra-like moorlands (Goodman and Perkins, 1959). Clearly, tundra plants are quite effective in retrieving nitrogen, phosphorus, and potassium which would otherwise be lost to the plant at the end of the season through leaf senescence.

The very high percentage retranslocation of phosphorus from shoots at the end of the growing season (at least 50% of maximum standing stock) suggests that in the fall plants actively hydrolyze or transport organic phosphate compounds belowground. The decrease in aboveground phosphorus does not correlate with periods of rainfall, suggesting that translocation plays a larger role than leaching in this removal. Nitrogen, which has an even larger organic component (proteins) than phosphorus, is less readily retranslocated from aboveground biomass and is much less readily leached from standing dead than is phosphorus. Barrow graminoids contain approximately 30 to 40% nonstructural carbohydrate at time of maximum aboveground biomass (McCown, 1978). The fact that aboveground biomass decreases up to 35% between time of maximum biomass and the end of the growing season suggests that these tundra plants are quite efficient in extracting nonstructural carbohydrates as well as nutrients from senescing shoots.

The continued increase in calcium, magnesium, and iron concentrations through the growing season is consistent with their low mobilities. Calcium in particular is not readily absorbed and translocated by phloem cells except under conditions of calcium deficiency (Biddulph et al., 1959). Hence, its concentration in a tissue is a function of the rate at which it is transported to that tissue in the xylem. In short, calcium concentration should be largely a function of cumulative transpiration by each tissue. This would explain the increase in calcium concentration as the season progresses. Such calcium would then be available for synthesis of secondary cell walls.

Differences in seasonal patterns of nutrient concentrations between monocots and dicots are consistent with observed phenology (Tieszen, 1972). Dicots begin growth later in the season than do monocots and therefore achieve their peak nutrient concentration later than do monocots. The peak standing stock of nutrients in dicots is achieved only a month after they begin growth. The higher phosphorus concentrations in dicots than in monocots may result from their observed mycorrhizal association. Dicots at Barrow are consistently mycorrhizal, whereas the major monocot species which are lower in phosphorus are generally nonmycorrhizal (Miller and Laursen, 1978). Only the few secondary roots of *Carex aquatilis* which occur in the litter layer are mycorrhizal among the three principal graminoid species. Mycorrhizae are critical to the phosphorus nutrition of many plants. The higher potassium, calcium, and magnesium concentrations of dicots than of monocots may result from the tendency of dicots to be in drier, less acid soils where these elements are more available (Gersper, personal communication). Differences in soil chemistry would not explain differences in phosphorus content since dry areas tend to be phosphorus deficient (Barél and Barsdate, 1978). It is unclear to what extent differences in distribution between monocots and dicots are caused by differences in soil nutrients and to what extent the differences in their nutrient contents result from distribution patterns along a moisture-nutrient gradient that might be caused by other factors. The implication for herbivores of seasonal variation and interspecific variation in plant nutrient content has been discussed elsewhere (Chapin et al., 1975).

The Barrow wet meadow study site responds to nutrient additions after only 6 weeks, indicating that it is indeed a nutrient-limited system, as previously demonstrated in other tundra sites (Warren Wilson, 1957; Schultz, 1969; Haag, 1974; Ulrich and Gersper, 1978). The accumulation of potassium and phosphorus but not of nitrogen could result from differential movement to or uptake by the plant during the short period between application and sampling or could suggest that nitrogen is more strongly limiting than phosphorus in the study site. *Dupontia,* which responded particularly strongly to nutrient addition, occurs most abundantly in polygon troughs which are heavily grazed and fertilized by lemmings and which have high concentrations of available phosphorus (Barél and Barsdate, 1978). The ability to exploit soil nutrients when readily available rather than to depend entirely upon internal stores may be an important criterion for success in heavily grazed portions of the Barrow tundra such as troughs where *Dupontia* grows.

The clipping experiment suggests that removal of all leaves by grazers would decrease aboveground standing crop in that year but that the leaves produced following grazing have higher nitrogen, phosphorus, and potassium concentrations and a calcium concentration similar to those which are not stressed by grazing. These data conflict with the nutrient recovery hypothesis (Schultz, 1964, 1969), which explains the crash of tundra lemming populations by assuming that

N, P, and K concentrations of vegetation decline following heavy grazing. Improvement in nutritional content of leaves following grazing might increase photosynthetic capacity and could encourage grazers to regraze the same areas rather than to continually search for ungrazed areas. Such a grazing pattern would constitute a strong selective force and lead to a patterning of vegetation according to capability for regrowth following grazing.

The mulching experiment suggests that the increased shading and insulation provided by an enlarged canopy of standing dead (as might occur in absence of lemmings) decreases standing crop in the fashion shown by long-term exclosures and predicted by simulation modeling (Miller et al., 1976).

Acknowledgments. This research was supported by the National Science Foundation under Grants 29342 to the University of Alaska and 29343 to Augustana College. It was performed under the joint NSF sponsorship of the International Biological Programme and the Office of Polar Programs and was directed under the auspices of the U.S. Tundra Biome. Considerable support was provided by the Institute of Arctic Biology, University of Alaska. Larry Tieszen, Keith Van Cleve, and Arnold Bloom cooperated actively in this research. The assistance of Robert Porter, Frances Randall, and Laraine Noonan in data and sample analysis is gratefully acknowledged.

References

Armstrong, W. (1972) A re-examination of the functional significance of aerenchyma. *Plant Physiol.*, **27**: 173–177.

Barél, D., and R. J. Barsdate. (1978) Phosphorus dynamics of wet coastal tundra soils near Barrow, Alaska. In *Environmental Chemistry and Cycling Processes* (D. C. Adriano and I. L. Brisbin, Eds.). U.S. Dept. of Energy Symposium Series.

Biddulph, O., R. Cory, and S. Biddulph. (1959) Translocation of calcium in the bean plant. *Plant Physiol.*, **34**: 512–519.

Billings, W. D., K. M. Peterson, and G. R. Shaver. (1978) Growth, turnover, and respiration rates of roots and tillers in tundra graminoids. In *Vegetation and Production Ecology of an Alaskan Arctic Tundra* (L. L. Tieszen, Ed.). New York: Springer-Verlag, Chap. 18.

Britton, M. E. (1957) Vegetation of the arctic tundra. In *Arctic Biology: 18th Biology Colloquium* (H. P. Hansen, Ed.). Corvallis, Oregon: Oregon State University Press, pp. 67–130.

Brown, J., H. Coulombe, and F. Pitelka. (1970) Structure and function of the tundra ecosystem at Barrow, Alaska. In *Proceedings of Conference on Productivity and Conservation in Northern Circumpolar Lands, Edmonton, Alberta, 15 to 17 October 1969* (W. A. Fuller and P. G. Kevan, Eds.). Int. Union Conserv. Natur., Morges, Switzerland, Publ. 16, N.S., pp. 41–71.

Carter, O. G., and D. J. Lathwell. (1967) Effects of temperature on orthophosphate absorption by excised corn roots. *Plant Physiol.*, **42**: 1407–1412.

Chapin, F. S., III (1974a) Morphological and physiological mechanisms of temperature compensation in phosphate absorption along a latitudinal gradient. *Ecology*, **55**: 1180–1198.

Chapin, F. S., III (1974b) Phosphate absorption capacity and acclimation potential in plants along a latitudinal gradient. *Science*, **183**: 521–523.

Chapin, F. S., III (1977) Temperature compensation in phosphate absorption occurring over diverse time scales. *Arct. Alp. Res.*, **9**: 137–146.

Chapin, F. S., III, and A. Bloom. (1976) Phosphate absorption: Adaptation of tundra graminoids to a low temperature, low phosphorus environment. *Oikos,* **26**:111–121.

Chapin, F. S., III, K. Van Cleve, and L. L. Tieszen. (1975) Seasonal nutrient dynamics of tundra vegetation at Barrow, Alaska. *Arct. Alp. Res.,* **7**: 209–226.

Cole, C. V., D. L. Grunes, L. K. Porter, and S. R. Olsen. (1963) The effects of nitrogen on short-term phosphorus absorption and translocation in corn *(Zea mays). Soil Sci. Soc. Amer. Proc.,* **27**: 671–674.

Dennis, J. G. (1977) Distribution patterns of belowground standing crop in arctic tundra at Barrow, Alaska. *Arct. Alp. Res.,* **9**: 113–137.

Dennis, J. G., and P. L. Johnson. (1970) Shoot and rhizome–root standing crops of tundra vegetation at Barrow, Alaska. *Arct. Alp. Res.,* **2**: 253–266.

Goodman, G. T., and D. F. Perkins. (1959) Mineral uptake and retention in cottongrass *(Eriophorum vaginatum* L.). *Nature,* **184**: 467–468.

Haag, R. W. (1974) Nutrient limitations to plant production in two tundra communities. *Can. J. Bot.,* **52**: 103–106.

Hoagland, D. R., and T. C. Broyer. (1936) General nature of the process of salt accumulation by roots with description of experimental methods. *Plant Physiol.,* **11**: 471–507.

Humphries, E. C. (1951) The absorption of ions by excised root systems. II. Observations on roots of barley grown in solutions deficient in phosphorus, nitrogen and potassium. *J. Exp. Bot.,* **2**: 344–379.

Leonard, E. R. (1962) Interrelations of vegetative and reproductive growth, with special reference to indeterminate plants. *Bot. Rev.,* **28**: 353–410.

McCown, B. H. (1978) The interaction of organic nutrients, soil nitrogen, and soil temperature on plant growth and survival in the arctic environment. In *Vegetation and Production Ecology of an Alaskan Arctic Tundra* (L. L. Tieszen, Ed.). New York: Springer-Verlag, Chap. 19.

McKendrick, J. D., V. J. Ott, and G. A. Mitchell. (1978) Effects of nitrogen and phosphorus fertilization on the carbohydrate and nutrient levels in *Dupontia fisheri* and *Arctagrositis latifolia.* In *Vegetation and Production Ecology of an Alaskan Arctic Tundra* (L. L. Tieszen, Ed.). New York: Springer-Verlag, Chap. 22.

Miller, O. K., Jr., and G. A. Laursen. (1978) Ecto- and endomycorrhizae of Arctic plants at Barrow, Alaska. In *Vegetation and Production Ecology of an Alaskan Arctic Tundra* (L. L. Tieszen, Ed.). New York: Springer-Verlag, Chap. 9.

Miller, P. C., W. A. Stoner, and L. L. Tieszen, 1976. A model of stand photosynthesis for the wet meadow tundra at Barrow, Alaska. *Ecol.* **57**: 411–430.

Rathore, V. S., S. G. Wittwer, W. H. Jyung, Y. P. S. Bajaj, and M. W. Adams. (1970) Mechanism of zinc uptake in bean *(Phaseolus vulgaris)* tissues. *Physiol. Plant.,* **23**: 908–919.

Richards, S. J., R. M. Hagan, and T. M. McCalla. (1952) Soil temperature and plant growth. In *Soil Physical Conditions and Plant Growth* (B. T. Shaw, Ed.). New York: Academic Press, pp. 303–408.

Schultz, A. M. (1964) The nutrient recovery hypothesis for arctic microtine cycles. II. Ecosystem variables in relation to arctic microtine cycles. In *Grazing in Terrestrial and Marine Environments* (D. J. Crisp, Ed.). Oxford: Blackwell Scientific Publications, pp. 57–68.

Schultz, A. M. (1969) A study of an ecosystem: The arctic tundra. In *The Ecosystem Concept in Natural Resource Management* (G. M. Van Dyne, Ed.). New York: Academic Press, pp. 77–93.

Shaver, G. R., and W. D. Billings. (1975) Root production and root turnover in a wet tundra ecosystem, Barrow, Alaska. *Ecology,* **56**: 401–409.

Shaver, G. R., F. S. Chapin, III, and W. D. Billings. (in press). Ecotypic differentiation of *Carex aquatilis* in relation to ice-wedge polygonization in the Alaskan coastal tundra. *J. Ecol.*

Sutton, C. D. (1969) Effect of low soil temperature on phosphate nutrition of plants—a review. *J. Sci. Food Agric.,* **20**: 1–3.

Tieszen, L. L. (1972) The seasonal course of aboveground production and chlorophyll distribution in a wet arctic tundra at Barrow, Alaska. *Arct. Alp. Res.*, **4**: 307–324.

Tieszen, L. L., and D. C. Sigurdson. (1973) Effect of temperature on carboxylase activity and stability in some Calvin cycle grasses from the arctic. *Arct. Alp. Res.*, **5**: 59–66.

Ulrich, A., and P. L. Gersper. (1978) Plant nutrient limitations of tundra plant growth. In *Vegetation and Production Ecology of an Alaskan Arctic Tundra* (L. L. Tieszen, Ed.). New York: Springer-Verlag, Chap. 20.

Warren Wilson, J. (1957) Arctic plant growth. *Adv. Sci.*, **13**: 383–388.

Webber, P. J. (1978) Spatial and temporal variation of the vegetation and its production, Barrow, Alaska. In *Vegetation and Production Ecology of an Alaskan Arctic Tundra* (L. L. Tieszen, Ed.). New York: Springer-Verlag, Chap. 3.

22. Effects of Nitrogen and Phosphorus Fertilization on Carbohydrate and Nutrient Levels in *Dupontia fisheri* and *Arctagrostis latifolia*

J. D. McKendrick, V. J. Ott, and G. A. Mitchell

Introduction

Increasing activities by man in the North American Arctic require concomitant responsibilities for land management. In some instances revegetation of disturbed sites is needed to maintain stability of the soil and the integrity of the tundra biological system. Recent explorations in arctic tundra revegetation by staff of this station have shown that fertilization to improve the nutrient availability of some arctic soils is necessary for establishing seedlings of native as well as introduced grasses. It is known, however, from experiences with turf and forage grass management systems in temperate regions, that winter hardiness of certain grasses is significantly reduced by excessive and late-season fertilization with nitrogen. Theoretically, the fertilizer stimulates additional vegetative growth late in the growing season causing carbohydrates that would normally remain as metabolizable reserves to be allocated for structural tissues, thus reducing the carbohydrate reserves below critical levels.

Arctic grasses are subjected to the climatic extremes of winter severity and duration; hence, their ability for winter survival is possibly paramount among grass species. Also, freezing temperatures periodically interrupt the growing season in the Arctic. Since management and cultural practices introduced into the Arctic must preserve the natural adaptation of indigenous species, the effects of fertilization on the overwintering of indigenous grasses must be understood. Our project objectives were to determine the annual cycles of carbohydrate and nitrogen reserves and the effects of heavy applications of nitrogen (N) and phosphorus (P) fertilizers on those cycles in *Dupontia fisheri* and *Arctagrostis latifolia*.

Nitrogen and phosphorus were chosen because both were being considered in arctic revegetation programs. Nitrogen was known to affect winter hardiness adversely when improperly applied, and phosphorus was known to be essential for carbohydrate synthesis, translocation, and metabolism. Also, both nitrogen (McKendrick et al., 1975) and phosphorus are conserved as inorganic reserves, i.e., remigrate to storage organs prior to dormancy periods (Weinmann, 1942).

Some tundra plants are known to release phosphorus sparingly as one means of survival in the nutrient-poor arctic environment (Goodman and Perkins, 1959).

Dupontia fisheri is an indigenous arctic grass whose distribution is confined to the Arctic. *Arctagrostis latifolia* is an indigenous arctic grass whose distribution extends beyond the Arctic into the northern boreal regions. *Arctagrostis* also increases in abundance in mesic disturbed sites in the Arctic and is being scrutinized by Canadian and United States agronomists as a possible candidate for some arctic revegetation programs. Commercial seed production of *Arctagrostis* began recently in the Matanuska Valley, Alaska (Mitchell, 1976).

Methods

Site Description

The well-drained site (2) was situated on the bank of Footprint Creek with *A. latifolia* and *Alopecurus alpinus* as the dominant grasses. These species had increased in abundance apparently in response to previous disturbances by track vehicles. However, such traffic had been restricted from the area for some time. A moist site on the opposite side of the drainage about 30 m south and slightly west of the *A. latifolia* test site was chosen as the *Dupontia fisheri* site. *Carex aquatilis* and *Eriophorum angustifolium* were also common in this area.

The two sites had very obvious microclimatic differences. The *Arctagrostis* site was nearly level and accordant with the uppermost elevations in the vicinity. The *Dupontia* site sloped to the north and lay about 0.5 to 1.0 m below the *Arctagrostis* site. During the winter, snow depth on the *Arctagrostis* plots ranged between 6 and 15 cm. In the spring, the *Arctagrostis* plots were snow free about 2 weeks earlier than *Dupontia* plots. Due to the *Dupontia* site's lower position on the terrain, snow drifted into that area and accumulated to depths of 0.3 to 1 m. These greater snow depths compared to those on the *Arctagrostis* site and the slope aspect were two primary factors delaying spring snowmelt on the *Dupontia* site. Desiccation during winter was probably less on the *Dupontia* site than on the more exposed *Arctagrostis* site.

The soils from the two sites differed chemically in that pH tended to decrease with depth on the *Dupontia* site and to increase with depth on the *Arctagrostis* site (Table 1). According to soil tests, the 2.5- to 12.5-cm zone in the *Arctagrostis* soil contained more available N, P, and K (1.7, 1.6, and 1.6 times, respectively) than in the *Dupontia* soil. The surface zones (0 to 2.5 cm) were comparable except in the Bray P-1 available P which in all treatments averaged higher in the *Arctagrostis* plot. Of the two soils, that on the *Dupontia* site obviously contained more organic matter, which was probably indicative of greater plant productivity due to higher moisture. A shorter summer season for soil microbiological activity possibly depressed decomposition on the *Dupontia* site compared to that on the *Arctagrostis* site.

Table 1. Soil Test Results[a] for N, P, K, and pH from the *Dupontia* and *Arctagrostis* Plots Sampled about 1 yr (5 September 1973) following Fertilization (25 August 1972)

| Sampling depths (cm) | Fertilizer treatment | | | | | | | | | | | | | | | |
| | Control | | | | Phosphorus | | | | Nitrogen | | | | Nitrogen and phosphorus | | | |
	pH	N	P	K	pH	N	P	K	pH	N	P	K	pH	N	P	K
Dupontia plots																
0–2.5	6.2	68	22	375	6.2	49	202	371	5.9	91	34	415	6.2	94	111	341
2.5–12.5	5.9	21	5	103	6.0	19	8	104	5.4	26	10	100	6.1	19	8	104
Arctagrostis plots																
0–2.5	6.0	58	33	400	5.9	43	>280	346	5.7	112	42	450	5.7	72	>280	390
2.5–12.5	6.3	55	13	166	6.0	29	13	154	6.1	32	6	166	5.9	35	17	169

[a] N as $NH_3 + NO_3$ by micro-Kjeldahl; P as Bray P-1 extractable; K as N ammonium acetate extractable (pH 7.0); and pH 1:1 soil:water. N, P, and K reported in ppm for air-dry soil.

Field Collections

On 25 August 1972, four adjacent rectangular plots 1.5 × 6.1 m were established. Four fertilizer treatments were assigned and applied that day: 207 kg of N/ha (183 lb of N/acre), 147 kg of P/ha (131 lb of P/acre), these two combined treatments, and an untreated control plot. Nitrogen was applied as ammonium nitrate (34–0–0) and P as treble superphosphate (0–45–0). From the N and P analyses of soil samples collected a year later (Table 1), neither N nor P had moved significantly below the surface 2.5 cm. Weathered fertilizer prills were still visible in the surface mat of vegetation during September 1973.

During the 1972–73 winter, preliminary tissue collections were taken from the control plots. Between 20 and 120 tillers (a tiller was an individual grass shoot with its underground stem axis, rhizome, and attached roots) were collected weekly from each plot from the last week in June through the first week in September 1973.

Laboratory Procedures

Tillers were washed in cold tap water. Roots were removed and discarded, and the shoot was separated from the rhizome at the base of the first fully expanded leaf. Rhizomes and shoots were counted and then oven-dried for 24 hr at 65°C. Total nonstructural carbohydrates (TNC) were extracted from ground samples with 0.2 N H_2SO_4 and measured by the iodometric method (Smith et al., 1964; Smith, 1969). Nitrogen was determined by micro-Kjeldahl. Phosphorus was measured via the spectrophotometer with the ammonium molybdate method on wet ashed 100-mg subsamples.

Results and Discussion

Main Effects on the *Dupontia* Shoot

All fertilizer additions increased *Dupontia* shoot weights (Table 2). The combined N and P treatment produced the largest season-long average increase of 25% followed by the N treatment with a 20% increase. Near peak season (15 August), the average shoot weights were about 60 mg for the N and N + P plots and 44 mg for the control and P treatments, respectively (Figure 1).

Dry Matter. The seasonal trends in average dry matter accumulation for *Dupontia* shoots generally followed the sigmoid growth curve between 11 July and 15 August. There were depressions in shoot weights occurring between 26 June and 11 July and 15 August and 29 August, about 11 and 7 mg, respectively. The early season weight loss was probably due to a number of new tillers (and hence light-weight individuals) being included in the collections. If the late-August weight loss was real, it could have been partially due to several factors: (1) translocation of TNC from shoots to rhizomes for winter hardening, (2) losses

Table 2.   Effects of N, P, and N + P Treatments on the Weight, Percentage TNC (Total Nonstructural Carbohydrates), Percentage N, and Percentage P of *Dupontia fisheri* Shoots and Rhizomes during the 1973 Growing Season near Barrow, Alaska

Treatments	Weight (mg)	Percentage TNC	Percentage N	Percentage P
	Shoots			
Control	32 a[a]	23.4 b	1.87 a	0.30 a
207 kg of N ha^{-1}	39 ab	23.7 c	2.33 b	0.28 a
207 kg of N ha^{-1} + 149 kg of P ha^{-1}	40 b	21.5 a	2.79 c	0.48 b
149 kg of P ha^{-1}	33 ab	23.0 b	1.88 a	0.44 b
	Rhizome			
Control	36 a	38.0 c	1.01 a	0.28 b
207 kg of N ha^{-1}	36 a	35.4 b	1.66 b	0.22 a
207 kg of N ha^{-1} + 149 kg of P ha^{-1}	34 a	32.1 a	2.13 c	0.35 c
149 kg of P ha^{-1}	34 a	37.4 c	1.02 a	0.35 c

[a]   Within each column, shoot and rhizome means followed by the same letter were not significantly different at the 0.01 level according to Duncan's new multiple range test.

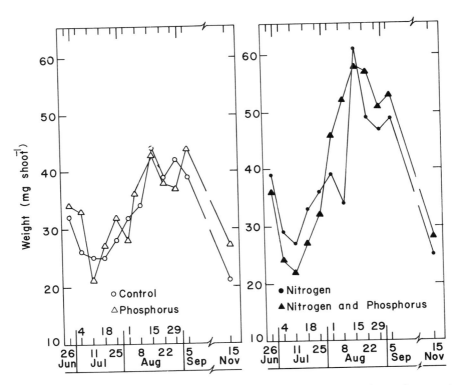

Figure 1.   Average shoot weights for weekly collections of *Dupontia* shoots from control and N, P, and N + P fertilized plots during the 1973 growing season, Barrow, Alaska.

of aerial plant parts due to senescence and weathering, and (3) carbohydrates being mobilized to meet demands by extensive root growth resulting as the soil temperature and the thickness of the active layer reached their annual maxima. The end-of-season loss in dry matter from *Dupontia* shoots was mainly due to senescence and weathering of leaves and inflorescences.

Billings et al. (1978) indicate, however, that elongation ceases at photoperiods between 15 and 18 hr (Shaver and Billings, 1977). Root biomass, and perhaps the production of laterals, may still have increased.

TNC.   Annual average levels of TNC in *Dupontia* shoots were significantly reduced by the combined N and P treatment (Table 2). Judging from the seasonal cycle of TNC levels in *Dupontia* shoots (Figure 2), the fertilizers' influences occurred primarily during the rapid growth period (11 July to 15 August). During senescence, the effects of N + P on TNC levels in the shoot essentially disappeared.

According to data cited by Tieszen et al. (1974), there was little change in the levels of soluble carbohydrate pools in aboveground portions of plants at Barrow. This appears contrary to our data. This inconsistency may have arisen through differences in carbohydrate extraction methods between the two studies. Also our shoot samples consisted of stem bases and leaves while their samples contained primarily leaves. Later analyses of graminoid stem bases (McCown, 1978) in fact, do show changes in soluble sugars which are similar in magnitude to those we are describing.

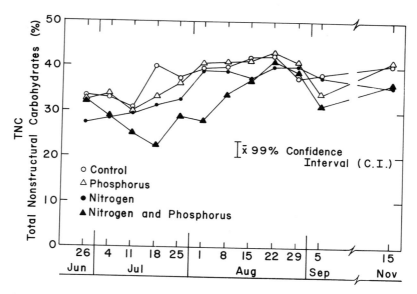

Figure 2.   Total nonstructural carbohydrate (TNC) percentages in weekly collections of *Dupontia* shoots from control and N, P, and N + P fertilized plots during the 1973 growing season, Barrow, Alaska.

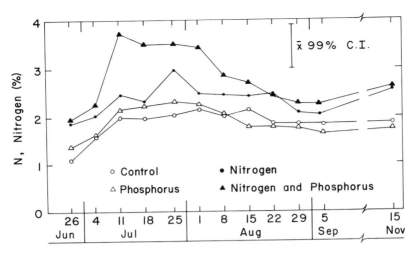

Figure 3.   N percentages in weekly collections of *Dupontia* shoots from control and N, P, and N + P fertilized plots during the 1973 growing season, Barrow, Alaska.

Reductions of TNC levels induced by fertilization provide evidence that the added nutrients were stimulating the allocation of carbon for structural development. Both N and P were either limiting or nearly limiting in the natural soil plant system.

Nitrogen.   The peak-season N levels for *Dupontia* shoots (Figure 3) occurred during early July, which seemed quite unusual for a grass and more typical of the pattern described by Leopold (1964) for seasonal N levels in the xylem sap of trees. Usually the peak N concentration in grass foliage occurs as the bud emerges. During the remainder of the growing season, N levels decline until the seasonal minimum is reached, as noted for *Arctagrostis* (Figure 19). However, Chapin (1978) also describes monocots in the Barrow vicinity whose N levels peaked during late June. He attributed that to dilution of the early spring foliage with overwintering leaves. That could also explain our findings since the *Arctagrostis* shoots sampled early in the growing season were primarily new leaves, while the *Dupontia* shoots consisted of both new leaves and leaves from the previous season.

Nitrogen levels in *Dupontia* shoots significantly increased with N and N + P fertilization (Table 2). Nitrogen alone increased the mean seasonal N level by about 64%, and N + P similarly increased that level by about 109%. The influence of N fertilization was apparent in the N contents of *Dupontia* shoots even at snowmelt (25 June for these plots), indicating that N was either being absorbed from the frozen soil or had entered the plant during the previous autumn. Absorption of N probably occurred during the previous autumn when the active layer was near its maximum depth (Chapin, 1978). However, growth and development occur at near freezing temperatures in some montane species

(Kimball et al., 1973). Tieszen et al. (1974; Allessio and Tieszen, 1978) also reported that translocation occurred at near freezing temperatures in *Dupontia*.

The soil test data indicated (Table 1) that after 1 yr N and P applied to the soil surface had not penetrated significantly below the upper 2.5 cm. This would support Chapin's (1978) observation that shallow-rooted species were affected more by fertilization than deeper rooted species. Other 1973 data (unpublished) from the Prudhoe Bay area of the Arctic indicated that P applied to a bare soil surface remained in the upper 2 cm (which was also the maximum depth of cracks in the soil surface) for the entire growing season. At Prudhoe Bay, it was likely that P penetrated the soil profile only as the fertilizer pellets fell into the cracks in the soil's surface.

Even though the added fertilizer remained at the soil surface for at least a year and the *Dupontia* rhizomes occur in the uppermost soil layer, nutrient uptake by *Dupontia* during the snowmelt period must have occurred by plant parts other than roots and rhizomes. These structures were still frozen during snowmelt.

Phosphorus.   Phosphorus levels in *Dupontia* shoots increased significantly with the application of P alone and P combined with N (Table 2). There is a trend

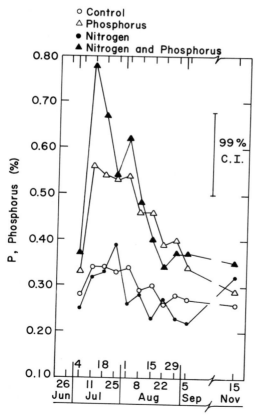

Figure 4.   P percentages in weekly collections of *Dupontia* shoots from control and N, P, and N + P fertilized plots during the 1973 growing season, Barrow, Alaska.

in the data indicating that N fertilization alone may have depressed P levels and that N combined with P may have had an even greater influence on P levels than P alone (Figure 4). It is obvious from these data that the influences of P applications were greatest during the early growth period. Also, the influence of P was apparent even at snowmelt, at least in the rhizomes (see results later).

The seasonal trend for amount of P in *Dupontia* shoots increased from snowmelt until about mid-July and declined thereafter. The pattern appeared similar to that found in N levels. Since P and N are considered to be closely associated during plant development, that similarity in seasonal trends might have been expected. Others have reported seasonal remigration of 50 to 60% of P in certain tundra species (Chapin, 1978; Dowding, 1974).

Main Effects on the *Dupontia* Rhizome

Dry Matter. We detected no effect from fertilizer treatments on *Dupontia* rhizome weights (Table 2). However, average rhizome weights may have changed relatively little annually compared to shoots which produced and lost leaves seasonally. Based on weight variance for *Dupontia* rhizomes from Allessio and Tieszen's data (personal communication), detecting 10 and 5 mg differences at the 99% probability level would have required samples of at least 111 and 444 rhizomes, respectively. Therefore, our sampling intensity was too low to detect subtle rhizome weight changes induced by fertilization. Furthermore, it is well known that fertilization influences aerial growth more than underground development in grasses.

TNC. TNC levels in *Dupontia* rhizomes were significantly reduced in plots receiving N fertilization (Table 2). Nitrogen alone reduced the average seasonal TNC levels an average of 2.6% and N + P reduced the average seasonal TNC levels 5.9%. As with the shoots, the influence of fertilization on rhizome TNC levels was primarily during the rapid growth period (Figure 5). By 15 November, the effects of fertilization had disappeared.

The balance of N and P available to *Dupontia* in the unfertilized condition must have been least favorable for N, because N applied alone induced a TNC depression in the rhizome while P applied alone failed to affect TNC levels (Table 2). However, available P in the soil must have been very near the critical level, because where N and P were applied in combination, the demand on TNC for structural growth was greater (less TNC was present) than where only N was applied. These findings concur with data of Chapin (1978) and Ulrich and Gersper (1978) which suggest that N was more limiting than P on wet tundra sites at Barrow.

The rhizome TNC pool was more sensitive to demands for structural carbon than the shoot's TNC pool, because shoot TNC levels were affected less by the fertilizer treatments even though shoot weights increased with fertilization (Table 2). The sensitivity of the rhizome's TNC pool to fertilization was anticipated. Interruptions of carbohydrate supply within grasses by clipping, grazing, and shading, etc. of leaves generally affects carbohydrate supplies to underground tissues sooner and to a greater degree than it affects carbohydrates in aerial

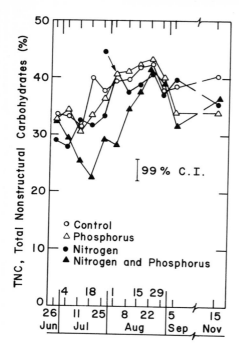

Figure 5. Total nonstructural carbohydrate (TNC) percentages in weekly collections of *Dupontia* rhizomes from control and N, P, and N + P fertilized plots during the 1973 growing season, Barrow, Alaska.

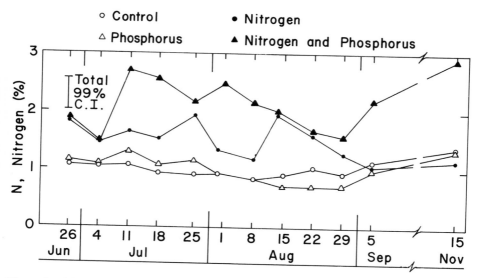

Figure 6. N percentages in weekly collections of *Dupontia* rhizomes from control and N, P, and N + P fertilized plots during the 1973 growing season, Barrow, Alaska.

tissues (Jameson, 1963). This has usually been observed by the cessation of root growth (Brouwer, 1966; Milthorpe and Davidson, 1966). Logically we could extend that theory to the *Dupontia* rhizome. The simulated grazing data of Mattheis, Tieszen, and Lewis (1976) do show a slight decrease in rhizome TNC following a single defoliation. The TNC level, however, is still 50% that of the control even after six chronic clipping events.

Nitrogen.   Nitrogen levels in *Dupontia* rhizomes increased significantly with N and N + P fertilization (Table 2). Since rhizome weights were unaffected by fertilization (as far as our sampling intensity could detect), the differences in N levels were accepted as indices of actual N accumulations rather than dilutions (or concentrations) resulting from carbohydrate fluctuations. The seasonally nonparallel patterns in N levels were apparent as an N × P interaction in the data (Figure 6). The combined N and P treatment permanently raised the N levels in *Dupontia* rhizomes during the sample period, while effects of the N treatment alone disappeared by early September, suggesting that given additions of either or both elements, this plant would benefit more if both elements were added.

Phosphorus.   Phosphorus levels in *Dupontia* rhizomes were depressed by N fertilization and increased by the P and N + P applications (Table 2). Since the N application stimulated shoot growth and since available soil P was probably at or near a limiting level, it seems reasonable that P in rhizomes might be mobilized and incorporated into the expanding dry matter of the shoot. Thus, the P levels in rhizomes would have been lowered.

Seasonal trends in *Dupontia* rhizome P levels in the control plot were relatively constant about the mean of 0.20%, except during the autumn senescence

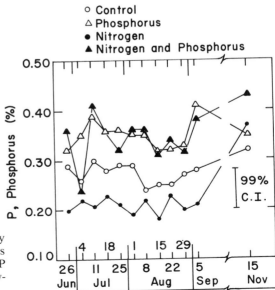

Figure 7.   P percentages in weekly collections of *Dupontia* rhizomes from control and N, P, and N + P fertilized plots during the 1973 growing season, Barrow, Alaska.

period when a sharp increase occurred (Figure 7). In plots treated with P fertilizer, there seemed to be a depression in the rhizome P levels during August just prior to the time when plants began translocating P into the rhizomes. The N + P treatment produced a higher rhizome P level by 15 November than any other treatment. Further study would be needed to verify that response and to elucidate a plausible explanation.

Seasonality of the *Dupontia* Shoot and Rhizome

TNC. The average TNC levels in shoots and rhizomes tended to parallel each other (Figure 8). With the exception of the (presumably) anomalous peak that occurred 15 August, the highest seasonal level occurred in the rhizomes before freeze-up. The aerial shoot retained relatively high TNC levels (compared to temperate grasses) into the dormant season. This, undoubtedly, would have significant forage value to grazing animals. Overall, average TNC levels ranged between 18 and 30% in the shoots; while in the rhizomes, the range was between 28 and 40%. Considering the TNC ranges in just the control plots, the levels were between 17 and 30% and 31 and 42% for shoots and rhizomes, respectively. Thus, the maximum fluctuations were about 30% of the peak TNC level. McCarty and Price (1942) reported a seasonal fluctuation of from 4 to 15% in the carbohydrates of mountain brome. Sampson and McCarty (1930) reported a seasonal fluctuation of 4 to 16% in *Stipa pulchra*. McCarty (1938) also reported a seasonal fluctuation from 4 to 20% in mountain brome. Thus, in those studies

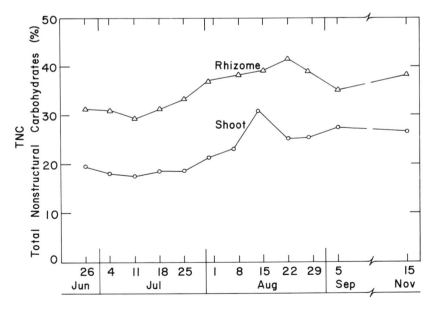

Figure 8.   Average TNC levels in weekly collections of *Dupontia* shoots and rhizomes during the 1973 growing season, Barrow, Alaska.

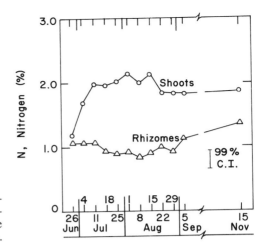

Figure 9. N percentages in weekly collections of *Dupontia* shoots and rhizomes from the control plot during the 1973 growing season, Barrow, Alaska.

with temperate-zone cool-season grasses, the seasonal loss in carbohydrates was about 75 to 80% of the maximum pool. Weinmann (1952) cites work by Hall et al. wherein a warm season grass *(Cynodon dactylon)* stored very high quantities of starch reserves and lost relatively little of these during normal commencement of spring growth. However, when *Cynodon* rhizomes were forced to expend reserves in darkness, they lost about 80% of their reserves. It appears that *Dupontia* behaved in a manner more similar to the descriptions of the warm-season *Cynodon* species in its annual TNC consumption than the cool-season temperate-zone grasses which utilized greater amounts of reserves under normal conditions.

The tendency for shoot and rhizome TNC levels to parallel each other was unusual (Figure 8). Normally, the shoot TNC levels would be expected to remain relatively constant with a possible drop during senescence. However, as mentioned above, our "shoots" included the stem base which was defined as the shortened-internode portion of the stem axis upon which the basal leaves were inserted. That tissue was found to be relatively rich in organic and inorganic nutrients (unpublished data; McCown, 1978) and accumulates carbon during senescence (Allessio and Tieszen, 1978). Therefore, the shoot TNC could have been enriched in our samples compared to TNC levels of strictly aerial leaf and elongated culm tissues. Thus, TNC levels we obtained from shoots probably reflected the general reserve carbohydrate status for the entire tiller, even though the shoot is not normally considered a permanent repository for TNC. Although the TNC levels in *Dupontia* shoots and rhizomes paralleled each other during the growing season, the patterns for N levels in rhizomes and shoots were opposite each other. The shoots generally gained N during the initiation of rapid growth and lost N during senescence (Figure 9). Rhizomes lost N during the shoot's rapid growth period and gained N during shoot senescence.

According to observations of temperate and boreal zone grass species by the senior author, buds of perennial grasses generally contain relatively high concen-

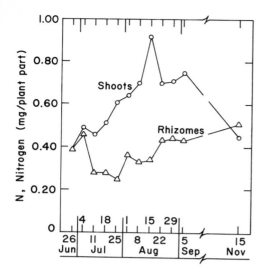

Figure 10. Average milligrams of N in weekly collections of *Dupontia* shoots and rhizomes from the control plot during the 1973 growing season, Barrow, Alaska.

trations of N and P. Since such buds originate on the rhizome rather than on the aerial stem of *Dupontia* and *Arctagrostis,* those N and P reserves would have been included with the rhizome samples in this study. Hence, autumn rhizome collections contained N- and P-rich buds. Late-spring and summer rhizomes probably had fewer N- and P-rich buds because at that time buds were enlongating and emerging as shoots with relatively lower N and P levels compared to buds.

Nitrogen.   From the percentage nitrogen, rhizome weight, and shoot weight data for the non-N fertilized plots, we calculated the mg of N contained in the shoots and rhizomes during the growing season (Figure 10). The general trend was to increase from 0.40 mg of N rhizome^{-1} and shoot to a maximum of about 0.90 mg of N shoot^{-1} at peak season (Table 3). During that rapid growth period the rhizome lost an estimated 0.05 mg of N. During senescence the shoot lost about 0.47 mg of N, and the rhizome gained about 0.17 mg of N. During the dormant period, N was lost from both the rhizome and the shoot, with the shoot's loss being one-half that of the rhizome's.

Seasonal budget estimates should be interpreted cautiously in calculating absolute quantities of N utilized by the average tiller. During the early portion of rapid growth, 11 June to 25 July, the net N loss from the rhizome was 0.14 mg (Figure 10), while our budget sheet (Table 3) only showed a net loss of 0.05 mg for the entire rapid growth period. Obviously, plant and rhizome N was in a dynamic condition and discrete observations could not account for continual losses and gains to the entire plant.

However, the trends we found in *Dupontia's* N budget are comparable to those reported by Chapin (1978) and to certain South African grasses (Weinmann, 1940, 1942) and prairie grasses (McKendrick et al., 1975) wherein autumnal accumulations occurred in underground tissues. Accumulations were presumed available for initiating spring growth. More recently, Milthorpe and Davidson (1966) and Perry and Moser (1974) also speculated that protein-like substances served as carbon and energy sources during regrowth. Studies by Hanway (1962) indicated that significant seasonal redistribution occurred in the corn plant (an annual shoot) with 60 to 70% of the N from the aerial shoot moved to the grain at maturity.

Figure 11 illustrates the growing season N budget for the *Dupontia* rhizome and shoot combined. The general upward trend for the rapid growth period is followed by a more gradual decline during senescence. The 4 July sample was unusually high in N as was the 15 August collection. Apparently, *Dupontia* acquired its N coincidentally with the progressive thawing of the active layer and rapid growth period (Figure 10). That would partially explain why effects of autumn fertilization increased spring N levels in tissue at snow melt in the fertilized plots. The quantity of mineral and organic nutrient reserves available to the rapidly developing tundra grass shoot in spring would seem to be critically important especially since there is very little soil thawed at that time.

Phosphorus.   P levels in *Dupontia* shoots and rhizomes tended to parallel each other as did the TNC levels, except in the autumn. During the autumn when shoots senesced, rhizomes gained in P while the shoot's P levels declined (Figure 12). This finding is in keeping with Hanway's (1962) results with corn, wherein the senescing shoot mobilized 72 to 82% of its P.

The P budget (Table 4) was based on the estimated amounts of P in an average *Dupontia* rhizome and shoot. According to those estimates, the shoot gained 0.04 mg of P during the rapid growth period while the rhizome lost slightly.

Table 3.   An Estimate of the Seasonal mg of N Requirements per *Dupontia* Rhizome and Shoot Based on Unfertilized Plot Data, Barrow, Alaska

	Rapid growth Period (mg of N plant part^{-1})		Net change during period
	Snowmelt	Peak season	
Rhizome	0.39	0.34	−0.05
Shoot	0.39	0.92	+0.53
Rhizome and shoot	0.78	1.26	+0.48
	Senescence period (mg of N plant part^{-1})		Net change during period
	Peak season	Freezeup	
Rhizome	0.34	0.51	+0.17
Shoot	0.92	0.45	−0.47
Rhizome and shoot	1.26	0.96	−0.30
	Dormant period (mg of N plant part^{-1})		Net change during period
	Freeze-up	Snowmelt	
Rhizome	0.51	0.39	−0.12
Shoot	0.45	0.39	−0.06
Rhizome and shoot	0.96	0.78	−0.18

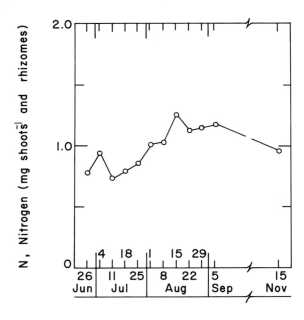

Figure 11. Average milligrams of N in weekly collection of *Dupontia* tillers (shoots and rhizomes with roots removed) from the control plot during the 1973 growing season, Barrow, Alaska.

During senescence, the rhizome gained 0.03 mg of P and the shoot lost 0.07 mg of P. Essentially, the budget pattern for P was similar to that for N. We estimate that during the dormant period there was a net gain in the shoot and a net loss in the rhizome P contents. This may have been a result of P being closely associated with carbohydrate translocation. If carbohydrates moved from dying tiller's rhizomes into a tiller offspring during the early spring before snowmelt, there could have been a net gain in P for the live tillers. Since some translocation

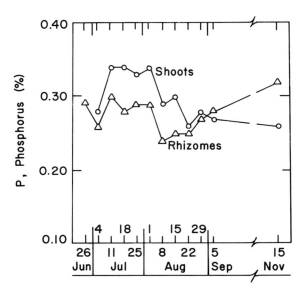

Figure 12. P percentages in weekly collections of *Dupontia* shoots and rhizomes from the control plot during the 1973 growing season, Barrow, Alaska.

Table 4. An Estimate of the Seasonal mg of P Requirements per *Dupontia* Rhizome and Shoot Based on Unfertilized Plot Data, Barrow, Alaska (1973)

	Rapid growth Period (mg of Plant part^{-1})		Net change during period
	Snowmelt	Peak season	
Rhizome	0.10	0.09	−0.01
Shoot	0.09	0.13	+0.04
Rhizome and shoot	0.19	0.22	+0.03
	Senescence period (mg of P plant part^{-1})		Net change during period
	Peak season	Freezeup	
Rhizome	0.09	0.12	+0.03
Shoot	0.13	0.06	−0.07
Rhizome and shoot	0.22	0.18	−0.04
	Dormant period (mg of P plant part^{-1})		Net change during period
	Freeze-up	Snowmelt	
Rhizome	0.12	0.10	−0.02
Shoot	0.06	0.09	+0.03
Rhizome and shoot	0.18	0.19	+0.01

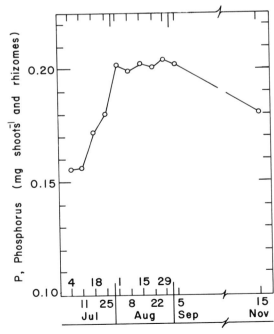

Figure 13. Average milligrams of P in weekly collections of *Dupontia* tillers (shoots and rhizomes with roots removed) from the control plot during the 1973 growing season, Barrow, Alaska.

occurs in the Arctic under the snow during the early spring (Allessio and Tieszen, 1975), this could explain a net gain in P for live *Dupontia* tillers during the "dormant" period. Allessio and Tieszen (1978) also indicate that roots appear to remain active for 1 yr after the shoot has senesced.

The seasonal rhizome and shoot P contents increased rapidly from 0.15 to 0.20 mg before the shoot reached its peak season dry matter stage (Figure 13). Then a rather steady state predominated until some time between 5 September and 15 November, when about 0.04 mg of P was lost from the shoot–rhizome (tiller) system.

Main Effects on *Arctagrostis* Shoots

Dry Matter. N and P applications significantly increased the average shoot weight in *Arctagrostis* (Table 5). Nitrogen alone induced the largest increase, 148%, followed by the N + P and P treatments with 70 and 35%, respectively. It was significant that the N + P treatment produced an intermediate response between those induced by the single nutrient treatments. Both nutrients appeared to be limiting growth at this site according to the weight data. *Arctagrostis* plants appeared greener, grew taller, and produced more inflorescences in the N-treated plots. These responses were still apparent during the second year (1974). By August 1977 these visible effects had disappeared. The influence of P seemed to be confined to a slight increase in plant height and stimulation of inflorescence production.

Seasonally, the *Arctagrostis* shoot reached its peak dry-matter accumulation about mid-July (Figure 14), remained relatively constant for about 4 weeks, and then lost weight during the senescence period, 15 August through 5 September.

TNC. Fertilization reduced TNC levels in *Arctagrostis* shoots significantly (Table 5). TNC reductions coincided with growth, i.e., greater TNC reductions

Table 5. Effects of N, P, and N + P Treatments on the Weight, Percentage TNC (Total Nonstructural Carbohydrates), Percentage N, and Percentage P of *Arctagrostis latifolia* Shoots and Rhizomes during the 1973 Growing Season near Barrow, Alaska

Treatments	Weight (mg)	Percentage TNC	Percentage N	Percentage P
	Shoots			
Control	40.1 a [a]	19.5 a	2.25 a	0.402 a
207 kg of N ha^{-1}	99.5 d	14.5 d	3.53 b	0.416 a
207 kg of N ha^{-1} + 149 kg of P ha^{-1}	68.4 c	15.1 c	3.40 b	0.587 b
149 kg of P ha^{-1}	54.2 b	17.9 b	2.37 a	0.445 a
	Rhizome			
Control	70.3 a	35.5 a	0.75 a	0.343 a
207 kg of N ha^{-1}	52.6 b	26.4 c	1.71 b	0.321 a
207 kg of N ha^{-1} + 149 kg of P ha^{-1}	59.6 a	26.6 c	1.83 b	0.454 b
149 kg of P ha^{-1}	68.2 a	34.0 b	0.90 a	0.428 b

[a] Within each column shoot and rhizome means followed by the same letter are not different at the 0.05 level according to Duncan's new multiple range test.

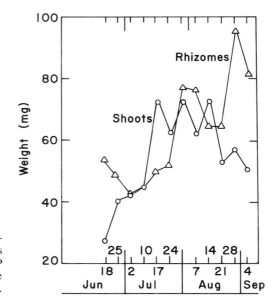

Figure 14. Average shoots and rhizome weights for weekly collections from control and N, P, and N + P fertilized *Arctagrostis* plots during the 1973 growing season, Barrow, Alaska.

occurred with increases in growth. The descending order of influence was N, N + P, and P. The effects of N and N + P occurred during the rapid growth period (Figure 15). The influence of N alone was most noticeable during the senescent period and persisted into freeze-up (Figure 15).

There were two depressions (Figure 15) in the seasonal TNC curves, one during the rapid growth period, and one during senescence. The latter depression corresponded to an increase in dry matter and was probably indicative of carbohydrate allocation into structural compounds. During late August, there were several warm days wherein vegetative tillers could have grown. On 29

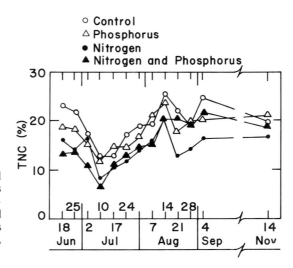

Figure 15. Total nonstructural carbohydrate (TNC) percentages in weekly collections of *Arctagrostis* shoots from control and N, P, and N + P fertilized plots during the 1973 growing season, Barrow, Alaska.

August 1974, such conditions occurred at Prudhoe Bay and some grass leaves elongated as much as 5 mm during a mid-day 5-hr period.

Nitrogen.   Nitrogen and N + P fertilization increased N levels equally in *Arctagrostis* shoots (Table 5). This contrasted to *Dupontia's* response wherein the N + P treatment increased N levels more than the N only treatment. According to the soil test data, the native levels of P in the *Arctagrostis* site were higher than those of the *Dupontia* site, particularly in the 2.5- to 12.5-cm zone (Table 1). Therefore, it is possible that indigenous P was less limiting in the *Arctagrostis* plots than in the *Dupontia* plots. Chemical evaluations of soil P available to plants are difficult and require considerable correlation with plant responses, however.

Phosphorus.   Phosphorus levels in *Arctagrostis* shoots were unaffected by P fertilization except where N was also added (Table 5). That would indicate P levels in the soil were adequate but near the critical level during the 1973 growing season.

Main Effects on *Arctagrostis* Rhizomes

Dry Matter.   N fertilization increased dry matter of the *Arctagrostis* rhizomes (Table 5), indicating available soil nitrogen was limiting growth of that grass. Since sample size was too small to adequately measure slight changes in rhizome weights, there may have been smaller responses in other treatments that we were unable to detect.

TNC.   Only N depressed TNC reserves significantly over the entire season (Table 5), supporting the hypothesis that N was more limiting than P on this site. However, from the graph (Figure 16), both the N and N + P treatments depressed TNC levels during the rapid growth period. But the N + P plots recovered more during senescence than did the N treatment alone; thus, the seasonal average TNC value of N was low enough to be significant statistically. However, the plants recovered their TNC losses and contained sufficient TNC for winter survival. Observations during subsequent years confirmed that these plants survived over winter well.

Nitrogen.   Nitrogen levels in *Arctagrostis* rhizomes increased significantly in the N and N + P treatments (Figure 17). P added alone tended to increase N levels after mid-July (Table 5), although the seasonal average was not statistically different from the control.

Phosphorus.   Phosphorus levels in *Arctagrostis* rhizomes increased with the addition of P (Table 5). Those increases may have reflected luxury uptake. Even though N was probably the most limiting element, P was near the critical level of availability judging from the slight depression in rhizome P levels resulting when N was applied alone.

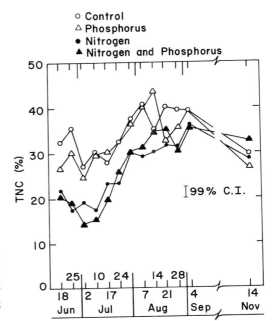

Figure 16.   Total nonstructural carbohydrate (TNC) percentages in weekly
collections of *Arctagrostis* rhizomes
from control and N, P, and N + P
fertilized plots during the 1973 growing
season, Barrow, Alaska.

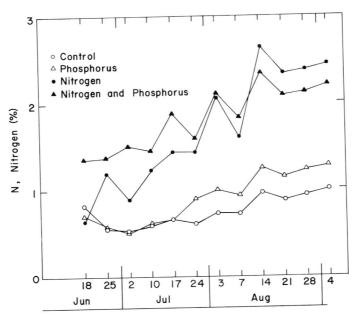

Figure 17.   N percentages in weekly collections of *Arctagrostis* rhizomes from control
and N, P, and N + P fertilized plots during the 1973 growing season, Barrow, Alaska.

Seasonality of the *Arctagrostis* Shoots and Rhizomes

Dry Matter. Rhizome and shoot dry matter accumulations were generally upward during the growing season. Rhizome dry matter increases lagged shoot dry matter increases (Figure 14). The variability of rhizome weights was high enough to limit precision of weight estimates for individual collections as noted earlier for the *Dupontia* data.

TNC. TNC levels in rhizomes and shoots paralleled each other over time, and were greatest in the rhizome (Figure 18). TNC levels in the untreated *Arctagrostis* rhizomes ranged between 27 and 41% for a fluctuation of 34% of the maximum pool. In the shoots that range was from 12 to 24% for a 50% fluctuation of the maximum pool.

Nitrogen. *Arctagrostis* rhizomes increased in N levels while shoots decreased in N levels throughout the June and September period (Figure 19). These N level patterns in *Arctagrostis* shoots were similar to those of cool-season grasses in the temperate zone and differed from those found in *Dupontia* (Figure 9). The difference between *Arctagrostis* and *Dupontia's* seasonal N levels may have been due to dissimilarities in the growth pattern of tillers between the two species. Green leaves were found on *Dupontia* tillers throughout the winter (see Chapin, 1978), while no green leaves were found during the

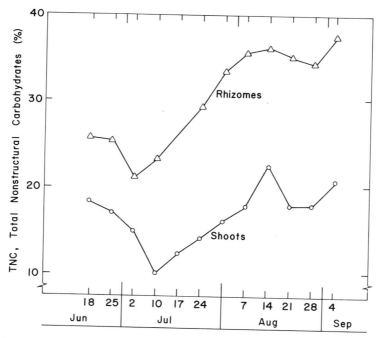

Figure 18. Average TNC percentages from all plots in weekly collections of *Arctagrostis* shoots and rhizomes during the 1973 growing season, Barrow, Alaska.

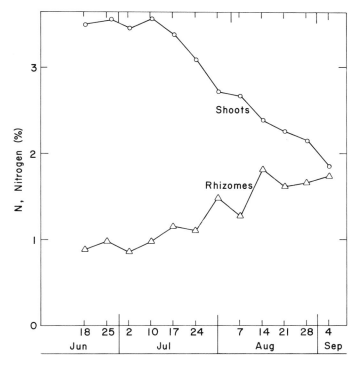

Figure 19.   Average N percentages from all plots in weekly collections of *Arctagrostis* shoots and rhizomes during the 1973 growing season, Barrow, Alaska.

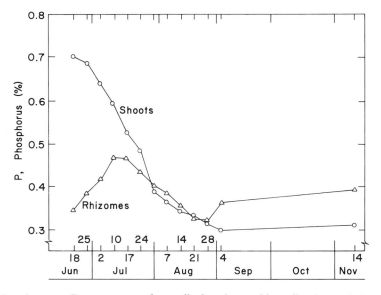

Figure 20.   Average P percentages from all plots in weekly collections of *Arctagrostis* shoots and rhizomes during the 1973 growing season.

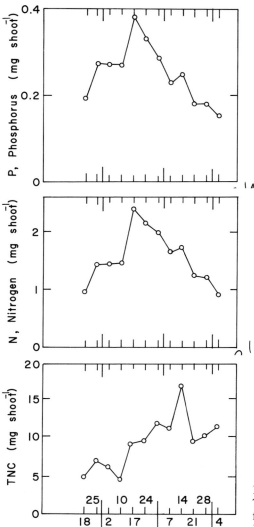

Figure 21. Average milligrams of TNC, N, and P per *Arctagrostis* shoot from weekly collections (all plots) during the 1973 growing season, Barrow, Alaska.

winter on *Arctagrostis* tillers. If those green leaves on *Dupontia* senesced after spring thaw, the N level in *Dupontia* shoots would have declined temporarily. In contrast *Arctagrostis* had no overwintering green leaves, hence all green tissues in the spring were newly developed. New tissues usually contain high N concentration and decline in N thereafter. The seasonal N levels in *Dupontia* and *Arctagrostis* rhizomes reflected the withdrawal and storage of N reserves as affected by tissue development and senescence of the respective species' shoots.

Phosphorus.   Seasonal P levels in *Arctagrostis* shoots were downward from June through August. In contrast P levels in *Arctagrostis* rhizomes increased

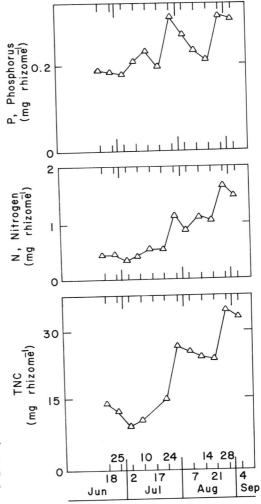

Figure 22. Average milligrams of TNC, N, and P per *Arctagrostis* rhizome from weekly collections (all plots) during the 1973 growing season, Barrow, Alaska.

during the late-June to early-July period, then declined through August and increased again in early September (Figure 20). The seasonal pattern for absolute P per shoot was upward from June until mid-July and then downward for the rest of the growing season (Figure 21). The seasonal pattern for absolute P per rhizome was generally upward except for a decline during August, which may have reflected the development of new (lighter weight) rhizomes in the stand (Figure 22).

Seasonal Budget for Nitrogen.   According to our seasonal budget estimates for *Arctagrostis* shoots, about 50% of the peak-season shoot N was lost during the senescence and dormant periods (Table 6). That estimate is slightly lower, but comparable with our estimate for *Dupontia* shoots. More N entered the rhizome compartment during senescence than was removed from the shoot

Table 6.   An Estimate of the Seasonal mg of N Requirements per *Arctagrostis* Rhizome and Shoot Based on Unfertilized Plot Data, Barrow, Alaska (1973)

| | Rapid growth Period (mg of N plant part⁻¹) | | Net change during period |
	Post-snowmelt	Peak season	
Rhizome	0.29	0.33	+0.04
Shoot	0.87	1.73	+0.86
Rhizome and shoot	1.16	2.06	+0.90
	Senescence period (mg of N plant part⁻¹)		Net change during period
	Peak season	Freeze-up	
Rhizome	0.33	1.48	+1.15
Shoot	1.73	1.26	−0.47
Rhizome and shoot	2.06	2.74	+0.68
	Dormant period (mg of N plant part⁻¹)		Net change during period
	Freeze-up	Post-snowmelt	
Rhizome	1.48	0.29	−1.19
Shoot	1.26	0.87	−0.39
Rhizome and shoot	2.74	1.16	−1.58

Table 7.   An Estimate of the Seasonal mg of P Requirement per *Arctagrostis* Rhizome and Shoot Based on Unfertilized Plot Data, Barrow, Alaska (1973)

| | Rapid growth Period (mg of P plant part⁻¹) | | Net change during period |
	Post-snowmelt	Peak season	
Rhizome	0.15	0.20	+0.05
Shoot	0.18	0.29	+0.11
Rhizome and shoot	0.33	0.49	+0.16
	Senescence period (mg of P plant part⁻¹)		Net change during period
	Peak season	Freeze-up	
Rhizome	0.20	0.25	+0.05
Shoot	0.29	0.08	−0.21
Rhizome and shoot	0.49	0.33	−0.16
	Dormant period (mg of P plant part⁻¹)		Net change during period
	Freeze-up	Post-snowmelt	
Rhizome	0.25	0.15	−0.10
Shoot	0.08	0.18	+0.10
Rhizome and shoot	0.33	0.33	0

during that period, indicating two phenomena: (1) some losses of shoot N were probably physical weathering of exposed tissues and physical movement into new buds, and (2) there was a substantial amount of N being taken up from the soil and being stored in the rhizome during the autumn. This corresponds to findings by Chapin (1978).

Seasonal Budget for Phosphorus. Seasonal budget estimates for *Arctagrostis* P requirements (Table 7) indicated P was slightly more mobile within the plant than N. Chapin (1978) also noted that difference between P and N. Seventy percent of the P in *Arctagrostis* shoots was lost during senescences, concurring with Chapin's (1978) report for *Dupontia*. Only 17% of the P from *Arctagrostis* shoots appeared in the rhizome compartment. That indicated there must have been substantial P movement into new rhizomes and P losses through weathering. Our data indicated all of the P absorption for *Arctagrostis* occurred during the rapid growth period. That was contrary to Chapin's (1978) findings for other arctic grasses. At present we cannot explain this discrepancy except through possible species differences. An overriding factor was the lumping of all tiller age classes together in our study, thereby masking the significant but relatively low level of growth of young tillers during the autumn period.

Conclusions

The influence of fertilization on TNC levels in both *Dupontia* and *Arctagrostis* was only transitory except for the effect of N on TNC levels in *Arctagrostis* shoots which was most evident during senescence. The TNC levels in these two arctic grasses fluctuated about 30% seasonally, except in *Arctagrostis* shoots which lost up to 50% of their maximum TNC level during senescence.

The soils at these two locations were limiting or nearly limiting in both N and P with respect to these two grasses. N was the most limiting of the two nutrients. Available soil P was greater on the *Arctagrostis* site than on the *Dupontia* site.

Both *Dupontia* and *Arctagrostis* mobilize significant quantities of N, P, and TNC seasonally. Relatively more mineral nutrients were mobilized than organic nutrients. This indicates that mineral nutrients are marginally adequate to these grasses in this ecosystem, and are conserved by these plants through recycling.

Acknowledgments. This research was supported by the National Science Foundation under Grant GV 29342 to the University of Alaska. It was performed under the joint sponsorship of the International Biological Programme and the Office of Polar Programs and was coordinated by the U.S. Tundra Biome. Field and laboratory activities at Barrow were supported by the Naval Arctic Research Laboratory of the Office of Naval Research.

We also wish to thank Dr. L. L. Tieszen for encouraging the study, editing our reports, and reviewing our progress. Dr. Jerry Brown of USACRREL loaned a semi-micro Wiley mill. Typing and figure drafting contributions of USACRREL, Hanover, N.H., are also appreciated. Several people at the Alaska Agricultural Experiment Station, Palmer Research Center, also deserve credit: Dr. W. W. Mitchell for assisting with the original field plot development in 1972; Mrs. Kathy Moffitt for typing, accounting, and coordinating communications between the Barrow field and Palmer laboratory facilities; and Mr.

Pete Scorup for helping collect the 1973 soil samples. The Alaska Agricultural Experiment Station, Palmer Research Center, provided field tools, fertilizer, chemicals, and analytical laboratory facilities.

References

Allessio, M. L., and L. L. Tieszen. (1975) Patterns of carbon allocation in an arctic tundra grass, *Dupontia fischeri* (Gramineae), at Barrow, Alaska. *Amer. J. Bot.*, **62**: 797–807.

Allessio, M. L., and L. L. Tieszen. (1978) Translocation and allocation of ^{14}C-photoassimilate by *Dupontia fisheri*. In *Vegetation and Production Ecology of an Alaskan Arctic Tundra* (L. L. Tieszen, Ed.). New York: Springer-Verlag, Chap. 17.

Billings, W. D., K. M. Peterson, and G. R. Shaver. (1978) Growth, turnover, and respiration rates of roots and tillers in tundra graminoids. In *Vegetation and Production Ecology of the Alaskan Arctic Tundra* (L. L. Tieszen, Ed.). New York: Springer-Verlag, Chap. 18.

Brouwer, R. (1966) Root growth of grasses and cereals. In *The Growth of Cereals and Grasses, Proceedings of Twelfth Easter School of Agricultural Science, University of Nottingham* (F. L. Milthorpe and J. D. Ivins, Eds.). London: Butterworth, pp. 163–166.

Chapin, F. S. III (1978) Phosphate uptake and nutrient utilization by Barrow tundra vegetation. In *Vegetation and Production Ecology of an Alaskan Arctic Tundra* (L. L. Tieszen, Ed.). New York: Springer-Verlag, Chap. 21.

Dowding, P. (1974) Nutrient losses from litter on IBP tundra sites. In *Soil Organisms and Decomposition in Tundra: Proceedings of the Microbiology, Decomposition and Invertebrate Working Groups Meeting, Fairbanks, Alaska, August 1973* (A. J. Holding et al., Eds.). Stockholm: International Biological Programme, Tundra Biome Steering Committee, pp. 363–373.

Goodman, G. T., and D. F. Perkins. (1959) Mineral uptake and retention in cottongrass (*Eriophorum vaginatum* L.). *Nature*, **184**: 467–468.

Hanway, J. J. (1962) Corn growth and composition in relation to soil fertility. II. Uptake of N, P and K and their distribution in different plant parts during the growing season. *Agron. J.*, **54**: 217–222.

Jameson, D. A. (1963) Responses of individual plants to harvesting. *Bot. Rev.*, **29**: 532–594.

Kimball, S. L., B. D. Bennett, and F. B. Salisbury. (1973) The growth and development of montane species at near-freezing temperatures. *Ecology*, **54**: 168–173.

Leopold, A. C. (1964) *Plant Growth and Development*. New York: McGraw–Hill, 58 pp.

Mattheis, P. J., L. L. Tieszen, and M. C. Lewis. (1976) Responses of *Dupontia fisheri* to simulated lemming grazing in an Alaskan arctic tundra *Ann. Bot.*, **40**: 179–197.

McCarty, E. C. (1938) The relation of growth to the varying carbohydrate content in mountain-brome. *U.S. Dep. Agric. Tech. Bull.* 598, 24 pp.

McCarty, E. C., and R. Price. (1942) Growth and carbohydrate content of important mountain forage plants in central Utah as affected by clipping and grazing. *U.S. Dep. Agric. Tech. Bull.* 818, 51 pp.

McCown, B. H. (1978) The interactions of organic nutrients, soil nitrogen, and soil temperature and plant growth and survival in the arctic environment. In *Vegetation and Production Ecology of an Alaskan Arctic Tundra* (L. L. Tieszen, Ed.). New York: Springer-Verlag, Chap. 19.

McKendrick, J. D., C. E. Owensby, and R. M. Hyde. (1975) Big bluestem and indiangrass vegetative reproduction and annual reserve carbohydrate and nitrogen cycles. *Agro-Ecosystem*, **2**: 75–93.

Milthorpe, F. L., and J. L. Davidson. (1966) Physiological aspects of regrowth following defoliation. In *The Growth of Cereals and Grasses: Proceedings of Twelfth Easter*

School of Agricultural Science, University of Nottingham (F. L. Milthorpe and J. D. Ivins, Eds.). London: Butterworth, pp. 153–166.

Mitchell, W. W. (1976) Native grass seed enters commercial production. *Agroboreal,* **8**: 19–21.

Perry, L. J., Jr., and L. E. Moser. (1974) Carbohydrate and organic nitrogen concentrations within range grass parts at maturity. *J. Range Manag.,* **27**: 276–278.

Sampson, A. W., and E. C. McCarty. (1930) The carbohydrate metabolism of *Stipa pulchra. Hilgardia,* **5**: 62–100.

Shaver, G. R., and W. D. Billings. (1977) Effects of daylength and temperature on root elongation in tundra graminoids. *Oecologia,* **28**: 57–65.

Smith, D. (1969) Removing and analyzing total nonstructural carbohydrates from plant tissue. Research Division, College of Agriculture and Life Sciences, University of Wisconsin, *Res. Rep.* 41, 11 pp.

Smith, D., G. M. Paulsen, and G. A. Raguse. (1964) Extraction of total available carbohydrates from grass and legume tissue. *Plant Physiol.,* **39**: 960–962.

Tieszen, L. L., D. A. Johnson, and M. L. Allessio. (1974) Translocation of photosynthetically assimilated $^{14}CO_2$ in three arctic grasses in situ at Barrow, Alaska. *Can. J. Bot.,* **52**: 2189–2193.

Ulrich, A., and P. L. Gersper. (1978) Plant nutrient limitations of tundra plant growth. In *Vegetation and Production Ecology of an Alaskan Arctic Tundra* (L. L. Tieszen, Ed.). New York: Springer-Verlag, Chap. 20.

Weinmann, H. (1940) Seasonal chemical changes in the roots of some South African highveld grasses. *J. South Afric. Bot.,* **6**: 131–145.

Weinmann, H. (1942) On the autumnal remigration of nitrogen and phosphorus in *Trachypogon plumosa. J. South Afric. Bot.,* **8**: 179–196.

Weinmann, H. (1952) Carbohydrate reserves in grasses. *Proceedings, 6th International Grassland Congress,* Part I, pp. 655–660.

23. Nitrogen Fixation in Arctic and Alpine Tundra

Vera Alexander, Margaret Billington, and Donald M. Schell

Introduction

The role of biological nitrogen fixation in tundra ecosystems was studied in order to determine the quantitative input of nitrogen from this source, the organisms involved, and the ecological interactions between these organisms and the tundra vegetative community. We were especially concerned with the environmental factors which influence nitrogen fixation rates. Very early in the study it became apparent that although the total amount of nitrogen entering the system through fixation was rather low, nevertheless this represented the major input, especially in low, mossy areas. Possibly the ecological significance is greater than would be assumed from the amount of nitrogen involved, since this nitrogen is in a mobile pool and subject to short-term cycling, whereas the bulk of the nitrogen in tundra soils is tied up in relatively unavailable forms.

Our approach was threefold. Initially, a survey was conducted to identify nitrogen-fixing organisms and the general range of activity to be expected. This was done in the arctic tundra at Barrow, in the Alaska Range alpine tundra, and at the Eagle Summit alpine tundra sites. The second approach, which continued throughout the entire program, was a mapping exercise whereby variations in input rate were determined along transects or in an area with multipoint stations. Finally, experiments were done using individual plants to determine the effects of light, temperature, moisture, oxygen tension, nutrients, and trace metals.

Methods

We used the acetylene reduction method for assaying nitrogenase activity (Stewart et al., 1967). Our modifications (Schell and Alexander, 1970) included adding 10% by volume freshly generated acetylene to the gas phase in the samples, and following incubation we collected the gases in Vacutainers® for subsequent gas chromatographic measurement of ethylene. For the mapping work, we collected cores with an area of 18 cm², placed the top 3 cm in a small glass jar, added acetylene, and carried out *in situ* incubation with incubation periods ranging from 6 to 24 hr. Small glass vials (25 or 36 ml) were used for the experiments on individual nitrogen-fixing organisms, except that for assays of nitrogenase activity in larger plants of the alpine tundra, plastic bags were used for the incubations. Nitrogen-15 experiments were used for two major types of work. First we confirmed many of our early acetylene reduction assays with nitrogen-15 gas uptake, in order to be certain that the ethylene production was indeed a measure of nitrogen fixation activity. In the second type of experiment, we looked at the uptake of nitrogen-15 gas by algae and lichens, and looked for transfer of the newly fixed nitrogen to other plants, in particular *Eriophorum* and mosses. We also carried out some cross-calibration experiments between acetylene reduction and nitrogen-15 uptake.

In this chapter, results will be presented in terms of ethylene produced in all cases where relative activity is the primary interest. Where absolute nitrogen fixation activity must be documented, we have assumed a 1.5 molar ratio between ethylene and nitrogen produced, and therefore multiplied the ethylene results by $14/1.5 = 9.33$. The reader may at any time convert results reported in terms of μmol ethylene to μg nitrogen in this same way.

Results and Discussion

The General Features of Nitrogen Fixation in Arctic and Alpine Tundra in Alaska

Our survey studies indicated that the major nitrogen-fixing components of the tundra ecosystem were blue–green algae, either free living or symbiotic in lichens. In the alpine tundra of the Alaska Range, we found some legume activity [*Oxytropis scammaniana* Hult. and *Oxytropis nigrescens* (Pall.) Fisch.] and also activity in connection with *Dryas octopetala* L. This activity was quite significant ($\sim$2 μg of N mg plant dry wt^{-1} day^{-1}) on a plant dry weight basis. However, these plants contributed little to total biomass and existed only as scattered individuals. Lichens were quite abundant, and blue–green algae were extremely abundant in the damper areas. The nitrogenase activity associated with some lichen genera was very high, as was the activity of the blue–green algal genera *Nostoc* and *Anabaena*. During our first alpine tundra visit (1970) we observed that *Stereocaulon* sp. was an active form among the lichens, and during the second visit (1972) we found *Peltigera* sp., *Nephroma* sp., and *Solorina crocea* (L.) Ach. all abundant and active. In the Eagle Summit alpine tundra, *Nostoc*, *Oxytropis nigrescens*, and mosses were active. We had already observed in the Alaska Range alpine tundra that nitrogenase activity was associated with mosses, and this turned out to be a very important feature in Barrow.

A large proportion of the surface of the Arctic Coastal Plain is covered with small ponds and lakes during the summer, and in spring an even larger area is covered with meltwater. Precipitation is low, but the water content is strongly influenced by the presence of permafrost near the soil surface throughout the growing season. Blue–green algae were very abundant and closely associated with mosses, an observation which has turned out to be common to all the Tundra Biome circumpolar sites except those that are essentially moorland rather than tundra, and to Devon Island. Such mossy areas were often overlain by a nearly continuous cover of *Drepanocladus* sp. with *Carex aquatilis* Wahlenb. as the canopy. Interpolygonal ponds in troughs usually contained a continuous layer of the mosses *Calliergon* sp. and *Mnium* sp. with *Dupontia fisheri* R. Br. the emergent plant. Nitrogen fixation was especially significant in these environments, but even on a relatively dry grassy meadow site there was distinct acetylene reduction activity. On higher polygon tops, nitrogenase activity associated with lichens became relatively more important. These relationships are discussed more fully in Alexander and Schell (1973) and Schell and Alexander (1973).

Nitrogen Fixation as a Nitrogen Input in Arctic Tundra at Barrow—The Results of Microvegetational Mapping

The first year's seasonal survey at site 2, a wet meadow site, showed very little activity in mid-June, a rapid increase throughout the summer, and a maximum in late July or early August. We estimate that the annual input during the summer of 1971 (Table 1) was about 28 mg of N m^{-2}. Subsequent work showed that this site was one of the areas with the lowest amounts of nitrogen fixation.

The detailed site (site 12, see Introduction) selected for nitrogen fixation mapping was located within 500 m of site 2. Photographic flight lines were made over the area of interest and from the photographs a site was selected which encompassed a transition zone from high-centered to low-centered polygonal tundra. A concentric distribution of sampling sites was then laid out at 10-, 20-, and 30-m radii. Points 10 m apart were marked off on the 30-m circle and on radial intersections with the 20- and 10-m circles. The 52 points of the plot were sampled three times during the summer of 1971 (24 and 28 July and 2 August). During the summer of 1972, sampling was done at 10-day intervals. We divided the area into three broad categories—high-centered polygons; polygonal troughs, marsh, and interpolygonal ponds; and low-centered polygons. Monthly means for 1972 are shown in Figure 1. More recently we have divided the site into five different areas based in part on relative dryness, but also on the vegetation. These divisions are shown, with the 1973 mean results for the

Table 1. Summary of Acetylene Reduction Assays on IBP Terrestrial Intensive Site 2, Summer 1971[a]

	June 25	July 5	July 15	July 25	August 3	August 14
Plot 6	0.6 (5.60)	1.47 (13.72)	3.27 (30.52)	2.84 (26.51)	0.61 (5.69)	3.69 (34.44)
Plot 7	0.62 (5.78)	2.71 (25.29)	1.85 (17.27)	1.81 (16.89)	1.56 (14.56)	5.22 (48.72)
Plot 8	0.20 (1.87)	2.54 (23.71)	3.60 (33.60)	2.34 (21.84)	1.12 (10.45)	.23 (2.15)
Plot 9	0.70 (6.53)	—	0.38 (3.55)	0.84 (7.84)	—	1.42 (13.25)
Plot 10	0.15 (1.40)	—	0.43 (4.01)	1.53 (14.28)	3.26 (30.43)	1.47 (13.72)
Average	0.46 (4.31)	2.24 (20.91)	2.00 (18.67)	1.87 (17.47)	1.64 (15.31)	2.41 (22.49)
	July 10	August 3				
Plot 32	0.04 (0.37)	0.20 (1.87)				
Plot 33	0.40 (3.73)	0.22 (2.05)				

*   Results are expressed in μmol of ethylene m^{-2} hr^{-1} with approximate conversion to μg of N $m^2 hr^{-1}$ in parentheses.

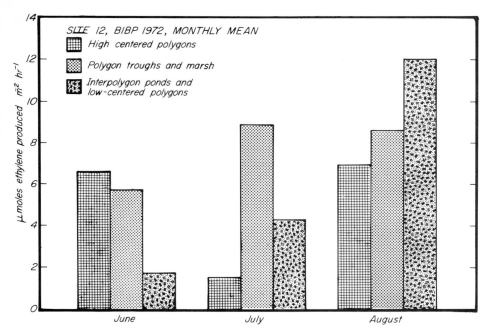

Figure 1. Site 12 acetylene reduction experimental results—1972. (From Alexander et al., 1974, with the permission of the Kevo Subarct. Res. Sta.)

summer, in Figure 2. Table 2 gives a breakdown of the site and summarizes the principal vegetation. Dry areas are dominated by grasses, with some *Salix* sp. and some mosses as the principal vegetation. Intermediate areas also have grasses, mosses, and some *Salix* sp. as the major vegetation. The dry areas had a higher proportion (10) of sites with nitrogen-fixing lichens present than did the intermediate areas (8). The low wet stations had grasses and mosses predominant, with six stations having nitrogen-fixing lichens also present. Areas designated marsh were primarily soil covered with standing water throughout the season; no nitrogen-fixing lichens were found, but clumps of *Nostoc* were present at all of these stations. Finally, areas designated as pool were primarily moss covered and also inundated, with *Nostoc* clumps present at one station. However, *Nostoc* associated with moss was extremely prevalent in this environment. During the 1973 summer, the site was divided into quadrats for sampling purposes with two quadrats sampled per week. Each station was sampled in duplicate for acetylene reduction assay, so that each value given for an area is the mean of several stations, each run in duplicate.

Based on the 1971 data, we concluded that acetylene reduction activity is highest in the wet polygon troughs and marsh, and lowest in the high-centered polygons. Rates ranged from 2.8 to 10.5 μmol of ethylene m^{-2} hr^{-1}, which is equivalent to a nitrogen input of 25 to 100 μg of nitrogen m^{-2} hr^{-1}. The 1972 data are shown as monthly means in Figure 1, such that seasonal trends as well as averages for the summer are available. High-centered polygons show a distinct minimum at midsummer, with maxima in June and August, a pattern which may be related to the seasonal distribution of moisture and/or temperature. During June, there is considerable moisture on the ground, with plenty of sunlight, and the high-centered polygons are the first areas to become clear of snow cover. July is dry, but relatively warm. For the polygon troughs and marsh, snow cover and low ground temperatures persist during much of June, but during the warmer months of July and August, sufficient moisture is present for active nitrogen fixation, resulting in a high mean input throughout the summer. Such low interpolygonal ponds and low-centered polygons tend to increase in acetylene reduction activity as the summer progresses. These relationships can be seen more clearly in the results obtained during the summer of 1973 (Figure 3). Acetylene reduction activity appears related to clear sky and high temperatures. Although temperature is probably more important, we cannot exclude light as an important influencing agent, since all the nitrogen fixation in tundra appears attributable to photosynthetic organisms. Intermediate areas have similar patterns, except that the June peak is much more spectacular, possibly because of longer exposure after snowmelt. Dry sites tended to have lower overall rates with less pronounced peaks, whereas moss pools had a mid-summer maximum,

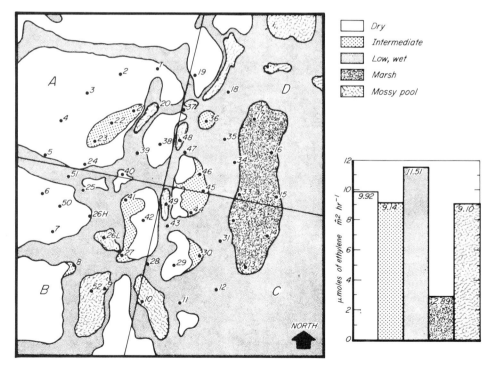

Figure 2.   Map of site 12, showing mean 1973 summer ethylene production rates. (From Alexander et al., 1974, with the permission of Kevo Subarct. Res. Sta.)

Table 2.   Station Designations for Site 12, 1973

Site subdivision	Number of stations	Dominant vegetation per site	Stations with N-fixing lichens	Stations with *Nostoc* sp. in mats
Quadrat A and D				
Dry	9	Grass (6), *Salix* sp. (1), moss (1), *Salix* sp. and grass (1)	7	0
Intermediate	6	Grass (2), *Salix* sp. (1), moss (2), grass and *Salix* sp. (1)	5	0
Low, wet	6	Grass (2), moss (4)	3	0[a]
Marsh	4	Soil (4)	0	4
Mossy pool	2	Moss (1), moss and grass (1)	0	0
Quadrat B and C				
Dry	6	Grass (3), *Salix* sp. (1), moss (1), moss and grass (1)	3	0
Intermediate	5	Grass (2), moss (3)	3	0
Low, wet	7	Grass (3), moss (2), moss and grass (2)	1	0[a]
Marsh	3	Soil (3)	0	3
Mossy pool	5	Moss (5)	0	1

[a]   *Nostoc* present, associated with moss.

possibly temperature related. Marshes had the lowest rates. These observations differ from those reported above for 1971 and 1972, but, in the earlier results, no distinction had been made between moss pools and marshes (primarily devoid of moss cover). The highest activities on site 12 were associated with low wet areas where both blue–green algae and lichens occured. The importance of clear skies seems to be supported by the low late-summer values (after 14 August), which are associated with continuous cloud cover. Light limitation is a reasonable hypothesis at this time.

In terms of annual input, rather than seasonal variations, we can calculate overall values for site 12 for 1971, 1972, and 1973. The data for the latter 2 years are more reliable, since the frequency of experimental sampling was much greater than during the 1971 summer. The 1973 seasonal average was 9.1 μmol of ethylene m^{-2} hr^{-1} (85.12 μg of N m^{-2} hr^{-1}) for a total of 788 samples, whereas for 1972 the mean was 6.4 (59.7) for 141 samples, and the 1971 mean was 6.7 μmol of ethylene m^{-2} hr^{-1} (62.5). The total input, converted to nitrogen using a 1.5 molar ratio between ethylene produced and ammonia produced, and assuming a 24-hr fixation day and a 60-day season with a linear decline to zero during a 10-day period at each end of the season, was 143 mg of N m^{-2} for 1973, 100 mg of N m^{-2} for 1972, and 105 mg of N m^{-2} for 1971. The higher value for 1973 possibly reflects an unusually wet summer.

Additional mapping exercises included transects across Nuwuk, a sparsely vegetated sandspit north of Barrow. Site selection was made from an aerial

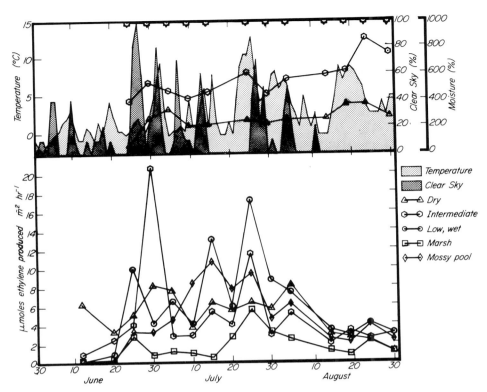

Figure 3.   Site 12 results for 1973, showing moisture, temperature, and percentage clear sky. (From Alexander et al., 1974, with the permission of Kevo Subarct. Res. Sta.)

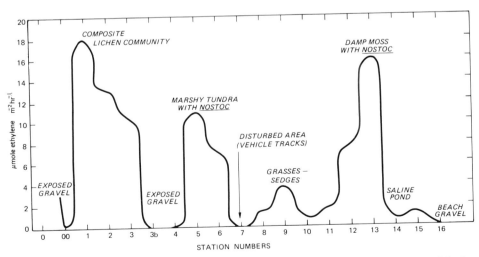

Figure 4.   Transect across Nuwuk, 1972. (From Schell and Alexander, 1973, with the permission of the Arctic Institute of North America.)

photograph and a transect was laid out with sampling points at 10-m intervals. The first point near the crest of the spit was in gravel and the final point at the beach was also in gravel. The vegetation on the slope between the beach and crest ranged from lichen crusts at the upper edge through grassy bogs to submerged grasses coated with attached algae near the beach. Mats of *Nostoc* were present in the wetter areas except near the beach. For details of the location of this transect see Schell and Alexander (1973). The results in terms of ethylene production are shown in Figure 4. Clearly, small scale variations in ground cover are very significant in influencing rates of nitrogenase activity in tundra. Two additional transects were conducted during July 1973, with moisture data taken in addition to notes on the principal vegetation type. Once again, considerable variations were found, related clearly to vegetation type and moisture; the overall rates, however, were low during this experiment, probably because temperatures were low (1.1°C).

The mapping exercise was extended to yet one more site in 1972 and 1973. This was a site selected for microbiological intensive work related to moisture tension (site 4). We used a portion of the site encompassing one polygon with a diameter of approximately 20 m and a rim height of about 30 cm above the basin and 25 cm above the trough. A diagram of this polygon is shown in Figure 5.

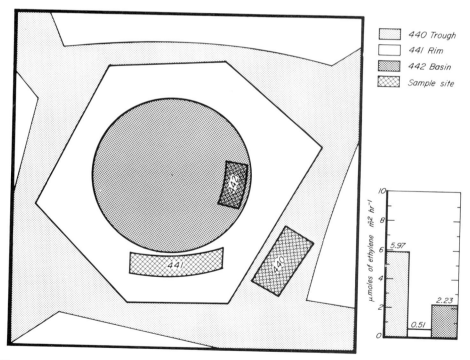

Figure 5.   A diagrammatic representation of the site 4 polygon showing mean summer ethylene production rates. (From Alexander et al., 1974, with the permission of Kevo Subarct. Res. Sta.)

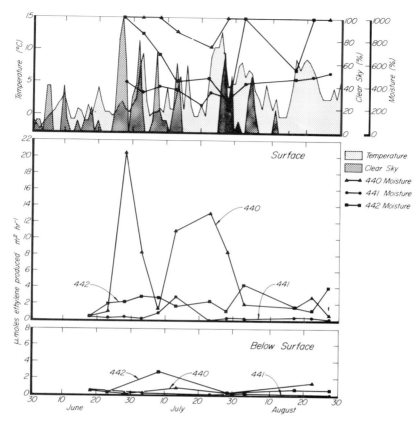

Figure 6.   Ethylene production at site 4, 1973. (From Alexander et al., 1974, with the permission of Kevo Subarct. Res. Sta.)

Attempts were made to prevent more than minimal disturbance to the sampling areas, which involved the surrounding trough, the relatively dry polygon rim, and the enclosed basin. The trough appeared saturated throughout the summer, whereas the rim showed noticeable drying and the basin developed a crust over the top at mid-season. Cores were taken every 5 days, and each of the stations was sampled to a depth of 10 cm approximately every 15 days. The results obtained here are shown in Figure 6. The highest rates were associated with the polygon trough (Sta. 440). This portion of the polygon had extreme fluctuations in activity, with two main peaks during the summer, one in June and the second in July. The first peak correlates well with clear skies, and the second with a temperature maximum. The next in activity was the polygon basin (Sta. 442), with much lower acetylene reduction rates. Possibly the correlation with moisture in August is significant. The polygon rim, the driest part of the polygon, had the lowest nitrogenase activity. In general, rates of acetylene reduction below the ground surface were not significant except on one occasion, 8 July, when the rates in the basin were constant down to 10 cm. Acetylene reduction rates on all

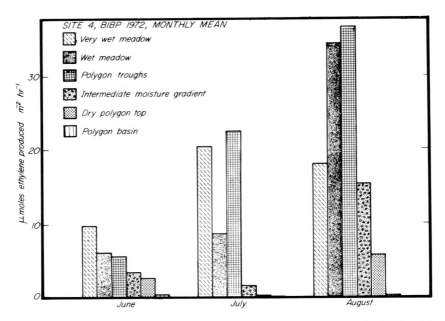

Figure 7.   Ethylene production at site 4, 1972. (From Alexander et al., 1974, with the permission of Kevo Subarct. Res. Sta.)

parts of the polygon system were much lower than those found in either the damp mossy or the lichen-rich portions of site 12. However, the 1972 site 4 results (Figure 7) confirm the observations from site 12, with lowest activity on dry polygon tops and basins, and high rates in areas with adequate moisture (wet meadow and polygon trough).

Factors Influencing Nitrogen Fixation Rates in the Major Nitrogen-fixing Organisms, and their Role in the Natural System

The work described above, as well as our earlier work regarding environmental influences, had suggested an important role for temperature and for moisture. In particular, the moisture regime of a particular site seemed to influence greatly the total annual nitrogen fixation input. However, within a particular site, given adequate moisture, the seasonal trend was related to temperature. Our previous work had shown a strong temperature dependence (Q_{10} of 3.7) and an importance for moisture (see Alexander and Schell, 1973). We have also looked at the role of oxygen in influencing nitrogen fixation in *Nostoc* and *Peltigera,* by placing these organisms in test tubes, adding a gas of predetermined composition to a known volume, and inverting the tubes in a water bath under light for incubation. For both *Nostoc* sp. and *Peltigera aphthosa* (L.) Willd., a strong response of acetylene reduction to low oxygen tensions is evident, such that at 20% O_2, the activity is only 50% of the potential for *Nostoc commune*. The response of *Peltigera aphthosa* is not as easy to interpret, but there does appear to be a depression of activity at high oxygen tensions. The tendency of *Nostoc* sp. to exist in wet mossy areas where low oxygen tensions occur (as evidenced by the presence of methane in gas samples) suggests that enhancement of nitrogen fixation may occur in the natural environment in connection with low oxygen tensions. Our diurnal oxygen tension measurements made in water surrounding moss support significant reductions in oxygen tension in this environment. On the other hand, *Peltigera* occurs frequently in less waterlogged environments, and apparently would be able to fix nitrogen effectively at the ambient atmospheric oxygen content.

We have run a series of experiments involving light and temperature simultaneously, which yielded results shown in Figures 8 and 9. Experiments designed to study the rate of recovery of *Peltigera aphthosa* after storage at −18 and 0°C showed that recovery is rapid, and within 2 hours maximum fixation rates are attained at 15°C. The optimal temperature for *Peltigera* and *Nostoc* nitrogenase activity lies between 15 and 20°C. Temperature appears to be an important

Table 3.   Seasonal Average Acetylene Reduction Rates for the Major Study Sites, 1972 and 1973

	μmol of C_2H_2 m^{-2}hr^{-1}	μg of N m^{-2}hr^{-1}	N
Site 12			
1973			
Dry	9.69	90.44	227
Intermediate	9.36	87.36	165
Low	11.69	109.10	190
Marsh	3.18	29.68	100
Mossy pool	8.85	82.60	100
$\bar{x}$ (12 June–1 September)	9.13	85.21	788
1972			
High-centered polygon	3.98	37.15	44
Polygon troughs and marsh	7.85	73.26	65
Interpolygon ponds and low-centered polygon	6.96	64.96	32
$\bar{x}$ (16 June–27 August)	6.40	59.73	141
Site 4			
1973			
Trough (440)	5.97	55.72	24
Rim (441)	0.51	4.76	24
Basin (442)	2.23	20.81	24
$\bar{x}$ (18 June–28 August)	2.90	27.07	72
Site 2 (meadow)			
1973 (23 June–23 August)	2.66	24.83	12
1971 (24 June–14 August)	1.72	16.05	27

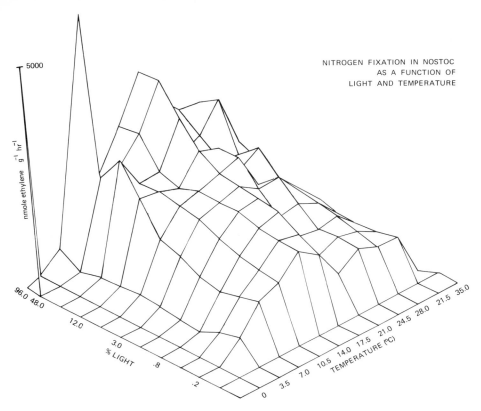

Figure 8.   The response of *Nostoc* acetylene reduction to light and temperature.

factor, and although diel studies showed no intrinsic rhythmicity in either *Nostoc* or *Peltigera,* an experiment with *Peltigera aphthosa* under *in situ* conditions has shown a distinct diurnal pattern in acetylene reduction, which appears to be related to light levels and temperature (Figure 10).

1970 site 2 studies showed that fertilization eliminates nitrogenase activity, and almost any disturbance to the system seems to have a negative response by acetylene reduction (see the site 2 data, Table 1, and the effects of vehicle tracks in the Nuwuk transect, Figure 4). In 1973, an experiment was conducted to determine the effect of phosphorus addition on acetylene reduction by *Peltigera* and *Nostoc.* The rationale was the possibility that low nutrient levels may be limiting nitrogenase activity in tundra, in particular, phosphorus. Phosphorus limitation was believed to occur in the tundra ponds. Therefore a similar situation could exist in the interstitial and standing waters of the terrestrial environment, especially in relation to nitrogen-fixing organisms, where nitrogen limitation is excluded. Stewart and Alexander (1971) have reported on the relationship between acetylene reduction and phosphorus availability in blue–green algae, and in fact have shown that it is possible to use nitrogenase activity as a form of bioassay for biologically available phosphorus. Therefore, the rationale was

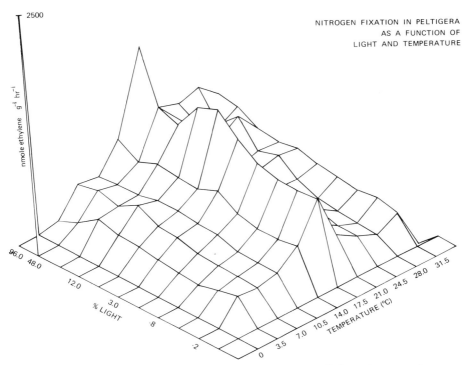

Figure 9.   The response of *Peltigera* acetylene reduction to light and temperature.

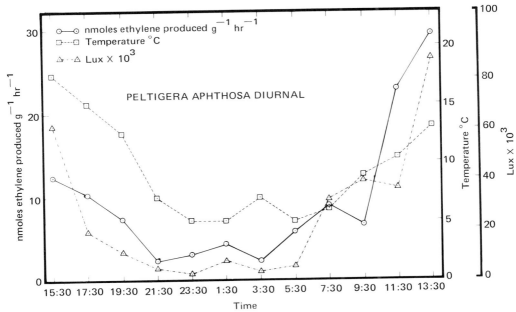

Figure 10.   Diel curves for *Peltigera* acetylene reduction. (From Alexander et al., 1974, with the permission of Kevo Subract. Res. Sta.)

Table 4. The Effect of Phosphorus Addition on Nitrogenase Activity

P added (ng)	ng of P added mg dry wt^{-1} sample	nmol of C_2H_4 g dry wt^{-1}		
		Rate 1 (before)	Rate 2 (after)	Percentage change
Nostoc commune				
3000	72.12	409.1	762.4	+86.3
450	18.44	475.8	816.6	+71.6
150	5.90	527.8	930.7	+76.3
0	0.00	368.8	833.7	+126.0
Peltigera aphthosa				
3000	51.11	990.1	691.6	−30.0
450	8.82	899.5	750.8	−16.5
150	2.56	686.5	1035.2	+59.3
0	0.00	1506.1	1115.5	−26.0

adopted that if, indeed, the tundra is a low phosphorus environment, then a positive response in acetylene reduction should result from adding phosphorus to the organisms. In the case of *Nostoc,* the addition of phosphorus at three different levels caused less of an increase in the acetylene reduction rate than adding a similar amount of distilled water (Table 4). With *Peltigera,* the rate decreased with distilled water and increased slightly at moderate phosphorus levels.

We looked into the possibility that molybdenum or cobalt might be limiting to nitrogen fixation in the Barrow area. It was presumed that an enhancement of nitrogenase activity by addition of either element would indicate suboptimal environmental levels. Fresh *Peltigera aphthosa* was collected, cleaned, and cut into approximately 18-mm circles. The samples were allowed to soak in tap water at room temperature prior to the beginning of the experiment. One milliliter of tap water was added to each vial, the circles inserted, the vials capped, and 5 ml of freshly generated acetylene added (with venting). There were four replicates for each set. The samples were then incubated for 1 hr at 15°C after which the gases were collected with Vacutainers®, the stoppers removed, and the water decanted. At this time the cobalt and molybdenum compounds were added in varying concentrations. Three different compounds were utilized: $CoCl_2 \cdot 6H_2O$, $Co(NO_3)_2 \cdot 6H_2O$, and $H_2MoO_4 \cdot H_2O$, separately and in combination. Once again, acetylene was added with further incubation for an hour, whereupon gas samples were collected as before. A water blank was also run to determine changes between the first and second incubation period without trace element addition. A definite increase in nitrogenase activity was noted for the majority of concentrations (Figure 11). A major exception was with the 1000 ng (45.5 ng·mg dry wt^{-1}) addition of $CoCl_2 \cdot 6H_2O$. Presumably the cobalt at this concentration caused a toxic or inhibitory reaction. It may also be significant to note that the baseline nitrogenase activity of the replicates for the 1000-ng level was much higher than the other replicates. There were also negative reactions to the Co^{2+} as nitrate solution + $H_2MoO_4 \cdot H_2O$ additions at all concentrations. The

CoCl$_2$·6H$_2$O additions showed the highest activity enhancement. The nitrate in Co(NO$_3$)$_2$·6H$_2$O was at concentrations which would tend to dampen fixation rates. This would explain the lesser differences in activities of the baseline incubation and the nutrient addition run for this compound. There may be two explanations for the reactions to the H$_2$MoO$_4$·H$_2$O treatments. Either the molybdenum is not limiting in the environment or the concentrations used in the experiment were too low to give a maximal effect.

The rate of acetylene reduction by *Peltigera aphthosa* declined with time in total darkness. After 1 hr in the dark, the rate was reduced by 60%, after the second hour by another 45%, and after a third hour by another 69%. The total loss in activity over a 3-hr period in the dark, when compared with a light control, was 93%. Normally a reduction in activity occurs with incubation time, but amounts to less than 30%. Light requirement is obviously connected to the photosynthetic nature of the blue–green algae, and thus to an oxidizable carbon

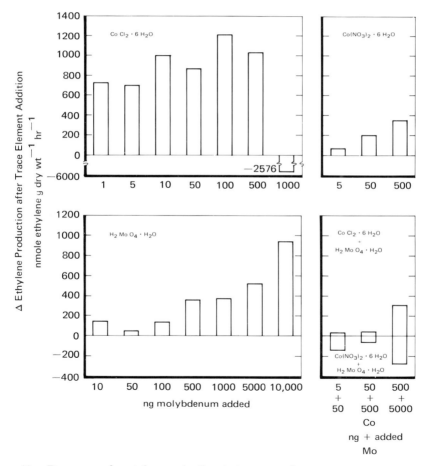

Figure 11.   Response of acetylene reduction to trace metals.

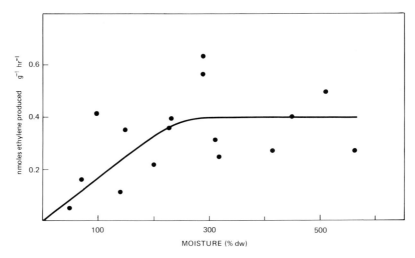

Figure 12.   The response of *Peltigera aphthosa* to moisture.

source. However, oxygen production may be involved directly. For example, Kallio (1973) found that light was important in allowing fixation by *Stereocaulon paschale* (L.) Hoffm. under anaerobic conditions. We therefore tested for nitrogenase activity in *Peltigera aphthosa* in the light and dark under anaerobic and aerobic conditions, and also in light under anaerobic conditions with pyrogallol added. We hoped that removal of some of the photosynthetically produced oxygen might lead to clarification of this relationship. Dark anaerobic fixation was 99% lower than light aerobic, and dark aerobic was 44% lower than light aerobic. Light anaerobic was 19.5% lower than light aerobic, a somewhat surprising result in view of our earlier observations with oxygen response. Light anaerobic with pyrogallol was 44% lower than light aerobic, very similar to dark aerobic. This suggests that oxygen does in fact play a role, but further experimentation is needed.

We carried out experiments to determine the moisture response of *Peltigera aphthosa,* and found that 300% of dry weight moisture was sufficient for maximum nitrogenase activity (Figure 12).

Nitrogen Fixation in Other Tundra Biome Circumpolar Sites, in Relation to our Alaska Observations

Nitrogen fixation by free-living or symbiotic blue–green algae appeared to be important in all the subpolar sites. The total input is of the same order of magnitude in many of the sites, although the moss–blue–green algal complex at Stordalen, Sweden, had activities somewhat higher (Granhall and Selander, 1973) than our Barrow associations. Environmental factors, too, seem to correlate well with our findings. At Kevo, Finland, moisture is the overriding problem, and a low pH apparently precludes much blue–green algal activity, except

Table 5. Nitrogenase Activity in Aerobic and Anaerobic Conditions in *Peltigera aphthosa*

Light aerobic	713.25	nmol of ethylene mg dry wt^{-1}
Light anaerobic	573.95	nmol of ethylene mg dry wt^{-1}
Dark aerobic	397.60	nmol of ethylene mg dry wt^{-1}
Light anaerobic, pyrogallol	396.51	nmol of ethylene mg dry wt^{-1}

perhaps in mires where moss beds occur. pH does seem to determine the occurrence of blue–green algae, but possibly the buffering effect of moss is able to produce a tolerable environment. Temperature is universally accepted as an important factor, whereas in Stordalen, light was also attributed a major role. We feel now, based on our results discussed above, that for blue–green algae moist environments, moss communities with adequate light, and low oxygen tensions provide the optimal environment for acetylene reduction activity. Bacterial activity appears minimal at Barrow, and all our depth sampling of cores has shown little activity below the surface, in the dark, whether under aerobic or anaerobic conditions. Only Devon Island has implicated bacteria as important in the nitrogen fixation regime, but this was based on indirect evidence rather than on observations relating to the microbial community (Stutz, 1973). Granhall and Selander (1973) have found that bacteria do, in fact, contribute on the high and dry areas, and we have not completely eliminated this possibility for the Alaskan arctic tundra.

Principal organisms involved in nitrogenase activity in Alaska tundra areas and their nitrogen fixation efficiencies measured under field conditions (except in some cases where optimal conditions were used) are shown in Table 6. The predominance of photosynthetic microorganisms is not surprising. It is logical that nitrogenase activity, which is highly temperature sensitive, should occur at maximum rates in the zone of greatest solar heating, which lies at or slightly above the ground surface. Dark-colored mosses provide a particularly warm environment. This, coupled with a relatively abundant energy source in the form of light, enables such algal fixation to proceed at rates sufficiently high to be of ecological importance. Experiments with nitrogen-15 have shown that this newly

Table 6. Nitrogenase Activity in Blue–Green Algae and Lichens at 15°C in the Barrow, Alaska, Area

	μmol of C_2H_4 mg dry wt^{-1}	
	13 July	30 August
Peltigera aphthosa	1.36	1.00
Peltigera canina (L.) Willd.	0.32	2.77
Lobaria linita (Ach.) Rabh.	0.50	1.67
Peltigera scabrosa Th. Fr.	0.98	0.94
Stereocaulon tomentosum Fr.		0.48
Nostoc commune		4.45

Also present: *Nephroma expallidum, Peltigera malacea* (Ach.) Funk, *Peltigera rufescens* (Weiss) Humb., *Solorina crocea, Stereocaulon alpinum, Stereocaulon paschale* (L.) Hoffm.

Table 7.   The Transfer of Newly Fixed Nitrogen from *Nostoc* to Other Components of Tundra Vegetation (22 July–4 August 1972)

Sample	$^{15}N_2$-N uptake (ng mg^{-1}hr^{-1})
Nostoc commune	3.4
Nostoc commune	15.4
Eriophorum scheuchzeri Hoppe (adjacent to *Nostoc*)	
Main root	0.21
Main root	0.56
Offshoot roots	0.02
Offshoot roots	5.31
Main stem	0.00
Main stem	0.00
Offshoot stems and leaves	0.02
Offshoot stems and leaves	0.58
Flower	0.28
Flower	0.42

fixed nitrogen is transferred to mosses and other plants rather rapidly (Table 7), but further work is needed to clarify the fate and rate of cycling of this newly fixed nitrogen. The existence of oxidizing and reducing conditions in tundra systems suggests that some of this nitrogen may be lost through denitrification, such that the net addition to the system on a long-term basis may be less than suggested by the inputs discussed above.

Nitrogen Fixation and the Nitrogen Budget of Alaskan Arctic Tundra

Barsdate and Alexander (1975) have discussed the overall nitrogen budget of arctic tundra, and have concluded that nitrogen fixation is the larger of the two major nitrogen inputs to the system. The other input, precipitation, was estimated to add 20.3, 9.1, and 2.7 mg of N m^{-2} yr^{-1} as ammonia, nitrite, and nitrate respectively. Nitrogen fixation on the average was estimated to supply 69.3 mg of N m^{-2} yr^{-1}. Obviously the rates of fixation vary extensively between microenvironments, and in some situations the input may be very significant whereas in others it may be trivial. Clearly, the tundra system operates with a relatively small nitrogen input.

Total loss of nitrogen to the system through denitrification and runoff (14.5 mg m^{-2} yr^{-1}) is also relatively small. The rate of accumulation of nitrogen stored by the system is, however, not exceptional in comparison with other environments. The significance of nitrogen fixation as a nitrogen source may rest in part in its relatively rapid availablility to other ecosystem components in the form of ammonia, and the tendency of such ammonia to be readily available and resistant to leaching in the mossy wet areas where high rates of nitrogen fixation predominate. Therefore, the newly fixed nitrogen may be particularly important in the short-term nitrogen turnover during the growing season.

Conclusions

Although much work yet remains to be done, we now have an idea of the mode of nitrogen fixation in Alaskan tundra environments. The high rates found in wet mossy areas are related to the large populations of blue–green algae of the genus *Nostoc,* which benefit from the low oxygen conditions and the rather warm environment provided by the moss. The moss may also play a role in buffering the pH. Such damp mossy environments apparently provide ideal conditions for nitrogen-fixing blue–green algae, since they occur and are active in such environments at all tundra sites (Granhall and Selander, 1973; Schell and Alexander, 1973; Kallio, personal communication). This relationship between mosses and algae deserves further investigation.

Another area deserving further study is the role of oxygen in tundra systems. The different responses to oxygen of *Nostoc* and *Peltigera* suggests that these responses are appropriate to their microenvironments. The role of oxygen in natural systems has been studied in relation to nitrogen fixation, but primarily in connection with soils (Brouzes et al., 1971).

Further study is also recommended in the area of distribution and abundance of nitrogen-fixing lichens in arctic tundra. In particular, the potential impact of human activities on nitrogen fixation and thus on the nitrogen regime of tundra needs further exploration.

Acknowledgments. This research was supported by the National Science Foundation under Grants GB 25024 and GV 29342 to the University of Alaska. It was performed under the joint sponsorship of the International Biological Programme and the Office of Polar Programs and was coordinated by the U.S. Tundra Biome. Field and laboratory activities at Barrow were supported by the Naval Arctic Research Laboratory of the Office of Naval Research.

We appreciate the invaluable assistance provided by Mrs. G.B. Threlkeld (botanical illustration), Dr. David Murray (alpine site selection and identification of plants), Mrs. Barbara Murray (lichen identification), and Mrs. Joanne Flock (lichen identification and collection). We also thank Mr. John Chang for field assistance during the 1971 season, and Dr. Sinikka Kallio for assistance with experiments during summer, 1973. We are grateful to Messrs. Hubert Sager, Sr. and Jr. for the use of their Alaska Range hunting camp as a base site for our alpine tundra studies. We are also grateful for the computer assistance and advice by Dr. Fred Bunnell, University of British Columbia, especially for the preparation of Figures 13 and 14.

References

Alexander, V., and D.M. Schell. (1973) Seasonal and spatial variation of nitrogen fixation in the Barrow, Alaska, tundra. *Arct. Alp. Res., 5*: 77–88.

Alexander, V., M. Billington, and D.M. Schell. (1974) The influence of abiotic factors on nitrogen fixation rates in the Barrow, Alaska, arctic tundra. *Rep. Kevo Subarct. Res. Sta., 11*: 3–15.

Barsdate, R.J., and V. Alexander. (1975) The nitrogen balance of arctic tundra: pathways, rates and environmental implications. *J. Environ. Qual., 4*: 111–117.

Brouzes, R., C.I. Mayfield, and R. Knowles. (1971) Effect of oxygen partial pressure on nitrogen fixation and acetylene reduction in a sandy loam soil amended with glucose. *Plant Soil,* Spec. Vol., 481–494.

Granhall, U., and H. Selander. (1973) Nitrogen fixation in a subarctic mire. *Oikos,* **20**: 175–178.

Kallio, S. (1973) The ecology of nitrogen fixation in *Stereocaulon paschale. Rep. Kevo Subarct. Res. Sta.,* **10**: 34–42.

Schell, D.M., and V. Alexander. (1970) Improved incubation and gas sampling techniques for nitrogen fixation studies. *Limnol. Oceanogr.,* **15**: 961–962.

Schell, D.M., and V. Alexander. (1973) Nitrogen fixation in arctic coastal tundra in relation to vegetation and micro-relief. *Arctic,* **26**: 130–137.

Stewart, W.D.P. and G. Alexander. (1971) Phosphorus availability and nitrogenase activity in aquatic blue–green algae. *Freshwater Biol.,* **1**: 389–406.

Stewart, W.D.P., G.P. Fitzgerald, and R.H. Burris. (1967) In situ studies on N_2 fixation using the acetylene reduction technique. *Proc. Nat. Acad. Sci. (Washington),* **58**: 2070–2078.

Stutz, R.C. (1973) Nitrogen fixation in a high arctic ecosystem. Ph.D. dissertation, University of Alberta, 62 pp.

24. A Model of Plant Growth and Phosphorus Allocation for *Dupontia fisheri* in Coastal, Wet-Meadow Tundra

WAYNE A. STONER, PHILIP C. MILLER, AND LARRY L. TIESZEN

Introduction

Primary production in the arctic tundra has been measured and summarized by diverse techniques (Bliss, 1966; Tieszen, 1972; Coyne and Kelley, 1975; Miller et al., 1976) and related to various environmental factors, such as low temperature, length of the growing season, light, water, and inorganic nutrients (Tieszen, 1972; Miller et al., 1975; Stoner and Miller, 1975; Miller and Tieszen, 1972). Usually primary production is related to one or a few factors qualitatively, and indeed a full quantitative understanding is not yet possible. The effects of the totality of environmental factors and of their interactions on primary production are difficult to study in the field or laboratory and impossible to conceptualize in detail. Computer simulation models can make a contribution by clarifying aspects of the system being considered, integrating data and concepts on these diverse aspects, and allowing explorations into the quantitative consequences of the data and concepts. The objective of this chapter is to describe a simulation model which summarizes much of the current understanding of processes affecting primary production in the wet meadow tundra of the Arctic Coastal Plain at a level of resolution appropriate to seasonal and multiyear simulations. The model incorporates the effects of light, water, phosphorus, and plant characteristics.

The potential importance of inorganic nutrients, especially phosphorus, in affecting primary production and herbivore population fluctuations in the Arctic was emphasized by Pitelka et al. (1955), Pitelka (1958a, b), and Schultz (1964, 1969). These ideas have forced a preliminary concept of the role of inorganic nutrients in arctic plant production. Most models of primary production processes (Monsi and Saeki, 1953; Davidson and Philip, 1958; de Wit, 1965; Duncan et al., 1967; Miller, 1972; Miller and Tieszen, 1972; Lemon et al., 1971; Miller et al., 1976) have included only aboveground, within canopy, micrometeorological, and physiological processes. Models providing an integrative framework of plant growth processes are considerably less sophisticated and more rudimentary (Loomis, 1970), even for plants with agricultural importance. Monsi and Murata (1970) showed by simple calculations how the allocation of carbon to productive tissues, rather than to absorptive tissues, can affect production. Brouwer and de Wit (1969) proposed a model of plant growth in which allocation to shoots or roots was controlled by the availability of carbohydrate reserves and water. De Wit et al. (1970) elaborated upon this model. The model was developed for agricultural crops and compared against data on corn in particular. Thornley (1972a, b) proposed a general model describing the growth of shoots and roots by logistic equations. Botkin (1975) developed a model of the growth of several forest trees to describe the process of succession in northeastern hardwood forest using as simple formulations as possible to describe the responses to light and water. Fick et al. (1975) describe a model for the growth and carbohydrate allocation in sugar beets. Penning de Vries et al. (1976) present a model of nitrogen dynamics and growth of loblolly pine over a multiyear time interval. Van Keulen (1975) describes a model of the growth of annual grasses in an arid region, which was extended to include nitrogen (van Keulen et al., 1975). Harpaz (1975) included the nitrogen dynamics for an arid region in a model similar to van Keulen's. However, these models include only a superficial treatment of nitrogen dynamics in the plant.

   The lack of plant growth models which include the effects of minerals may reflect a current lack of a consistent set of hypotheses and data regarding the role of minerals in plant development and the mechanisms of their interactions. Approaches to this problem have been neither detailed and mechanistic (Horowitz, 1958; Dainty, 1965) nor descriptive (Warren Wilson, 1964, 1966; Blackman, 1968; Květ et al., 1971). Emphasis on description is valuable for comparing the behavior of different plant life forms and understanding ecosystem behavior, while emphasis on mechanisms is valuable for pressing the "state of the art" in plant physiological research. The model described here is somewhere in between but should contribute both to physiologically oriented process studies and to management decisions which require a prediction capability in the face of simultaneous changes in a variety of interacting environmental factors.

Description of the Model

The objective of this model is to provide, in a simple framework, means of describing the patterns of carbohydrate and nutrient allocation, of exploring the interrelations between the carbohydrate and nutrient allocation patterns within a single tiller and a whole tiller system, and of exploring the possible effects of grazing. Although the model includes both nitrogen and phosphorus, only phosphorus is considered in detail because of the early emphasis on phosphorus in field research. The model was developed for *Dupontia fisheri,* an arctic grass which is common in the coastal wet meadow near Barrow, Alaska. The model uses daily time steps through the season, beginning at melt-off and ending at freeze-up.

General Description of Plant Compartments

The basic plant unit of *Dupontia,* in the field, consists of five to six age classes of tillers connected by rhizomes (see Allessio and Tieszen, 1978). Four age classes have aboveground parts (V1 · · · V4) and one contains only newly initiated, subterranean tillers (V0). We have modeled a system consisting of three aboveground tillers, one newly initiated belowground tiller, and one old tiller without aboveground parts. Each of the three aboveground tillers includes leaves, stem base, rhizome, and roots. A tiller during its life produces about 12 leaves before the shoot meristem differentiates into an inflorescence. After differentiation, no more leaves are produced. In its first and second years aboveground (V1 and V2) a tiller will initiate four to five leaves each growing season. In the third year (V3) the tiller normally produces three to four leaves and an inflorescence. In the year following flowering, the shoot senesces, but the rhizome and roots remain active. In the next year the rhizome and roots senesce and die. Through the life cycle, leaves grow from the stem base and roots from the rhizome. Rhizomes serve as storage and translocation organs.

Each plant part is subdivided into five chemical compartments: total nonstructural carbohydrates (TNC), noncarbohydrate cell contents (NCC), cell wall material (CW), and labile and nonlabile phosphorus (Table 1). The TNC is divided into sugars and storage polymers. The NCC includes components which are in the protoplasm enclosed within the cell wall such as protein, amino acids, lipids, etc., but not sugars and storage polymers. In the leaves, the cell wall materials of blades and sheaths are distinguished. Photosynthesis produces sugars which are added to the sugar compartment of the leaf. Sugars are used for synthesizing NCC, CW, and storage polymers. Nitrogen and phosphorus are required for the synthesis of NCC. However, only phosphorus uptake is included here.

Sugar and phosphorus translocation occur between the stem base and its leaves, the stem base and its connecting rhizome, the rhizome and its roots, and adjacent rhizomes (intertiller translocation). Sugar and phosphorus are translocated among leaves, stem base, rhizome, and roots according to concentration equilibria. Translocation is thus a response to growth priorities of the plant.

Growth and Senescence

Growth is defined as the sum of NCC, CW, and phosphorus added to a structure, and is calculated in g dry wt (tiller part)$^{-1}$ day^{-1}. Senescence is considered as a conversion of NCC back to sugar and labile phosphorus.

For leaves we assume that there is an inherent pattern of growth and senescence of individual leaves, which is genetically based, but which can be altered

Table 1. Average Composition of Newly Formed Plant Parts

	Noncarbohydrate cell content (%)	Cell wall material (%)	Total P %
Leaf	26.1	73.5	0.4
Stem base	8.4	91.2	0.4
Rhizome	8.6	91.2	0.2
Root	6.6	93.2	0.2

by temperature and the availability of phosphorus and sugars. We assume that leaves are initiated every 10 days, until the oldest mature leaf starts senescence. The first leaf begins elongating in the sheath at snowmelt (Tieszen, 1972; Tieszen, 1978). The initiation of the $n + 1^{th}$ leaf coincides with the initiation of the sheath of the nth leaf, and also with the maximum rate of growth of the blade of the nth leaf. The life cycle of a leaf has three stages: growth, maturity, and senescence, followed by death. The period of growth is 22 days. The blade grows through the first 20 days and the sheath through the last 12 days, overlapping by 10 days. Following initiation, the growth rates of both the blade and the sheath increase to a maximum and then decrease to zero, following

$$B = C_1 t - C_2 t^2$$

where t is the time in days from the initiation of the growth of this leaf, $C_1 = 4B_{max}/t_{max}$, B_{max} is the maximum growth rate, t_{max} is the total length of the growth phase, and $C_2 = C_1/t_{max}$. New NCC, CW, and total phosphorus make up 26.1, 73.5, and 0.4% of the new material, respectively.

In the first 10 days of leaf growth, sugars and phosphorus used in growth come from the stem base. After photosynthesis begins, when the blade emerges from the sheath, the blade supports its own growth, having developed a sugar and phosphorus reserve through translocation and photosynthesis. Materials for sheath growth come from the stem base.

Senescence proceeds as a linear decline in NCC biomass with time, so that 80% of the total NCC is converted to sugars and phosphorus in 15 days. No respiratory cost is associated with the breakdown of NCC. A leaf is considered dead when its NCC drops to 20% of the contents during the mature phase. This amount of NCC will eventually be lost from the plant to litter. During senescence cell wall materials are unaffected and at death the sheath cellulose is added to the protective layer of old sheaths. Death of leaves may be caused also by low temperatures. Exposed leaf blades are killed by air temperatures below $-4°C$. Leaves developing within the sheath are not killed by these temperatures. In the late growing season all leaves start to senesce when minimum air temperatures fall below $-2°C$. When air temperatures below $-4°C$ occur, all leaf material not covered by sheath is killed and all aboveground metabolic activity stops. The parts of leaves in the growth phase which are covered by old sheath become dormant, but will continue to grow in the next growing season.

Stem bases grow in all tiller age classes. In the V0 the stem is initiated when the leading rhizome is 50 mm long and then grows at a fixed rate equal to the rate of growth of the V0 rhizome. In other age classes the stem base grows at 10% of the sheath growth rates to represent the growth of the stem base internodes. New growth is 8.4% NCC, 91.2% CW, and 0.4% total phosphorus.

Rhizomes grow only in the V0 age class. Rhizome growth is stimulated by reduced phosphorus availability and decreased by reduced sugar availability. Thus,

$$B = C_3 f(\% \text{ sugar}) f(\% \text{ phosphorus})$$

where C_3 is growth rate with optimal concentrations of sugar and phosphorus, and $f(\%\ \text{sugar})$ and $f(\%\ \text{phosphorus})$ are functions of sugar and phosphorus concentrations, respectively. Rhizome growth is not limited by carbon above 20% TNC, is zero at 5% TNC, and is linearly related to intermediate concentrations. Growth is not affected by phosphorus above 1% and is 20% above normal at 0% phosphorus. The high level of phosphorus is arbitrary, so that under normal situations growth is always limited by phosphorus. New growth in rhizomes consists of 8.6% NCC, 91.2% CW, and 0.2% total phosphorus.

Root growth is initiated in the V0 when the rhizome is 25 mm long and in the V1 and V2 in early July. Older tillers do not grow roots. Root growth is affected by sugar and phosphorus in the same manner as rhizome growth and by soil temperature. Thus,

$$B = C_4 f(\%\ \text{sugar}) f(\%\ \text{phosphorus}(\ f(T_s))$$

where C_4 is the growth rate with optimal conditions, and $f(\%\ \text{sugar})$ and $f(\%\ \text{phosphorus})$ are the same as for rhizome growth. The $f(T_s)$ is the regression relationship between root growth and soil temperature presented by Billings et al. (1977). The roots grow in the V1 and V2 at 50% of the V0 root growth rate to represent the growth of secondary roots. New growth in roots is 6.6% NCC, 93.2% CW, and 0.2% total phosphorus.

The inflorescence is initiated when the 12th leaf is initiated. The inflorescence is a sink for carbon and phosphorus and is important in the carbon and nutrient balance of the tiller system. However, we are not following the pollination and development of seeds, or seed viability and germination. Growth of the inflorescence is calculated in a fashion similar to that of other plant parts except that low levels of available phosphorus in the stem base and rhizome retard growth.

$$B = C_5 f(\%\ \text{sugar}) f(\%\ \text{phosphorus}).$$

Growth of the inflorescence is not affected if the available phosphorus concentration in the stem base and rhizome is greater than 0.2% and is zero at concentrations below 0.04%.

In addition to the effects of carbon and nutrients on growth no more than 80% of the available sugar and nutrients in a structure may be used for growth during one time period. The remaining 20% represents an estimate of the threshold levels required for maintenance of existing metabolic machinery.

Photosynthesis and Respiration

A response surface for the photosynthetic rate of leaf blades in relation to different microclimatic conditions was generated with a canopy photosynthesis model (Miller et al., 1976) run with different microclimatic conditions and different amounts of live and dead leaf area. This surface was used to interpolate the rate under any set of conditions. Photosynthesis rates in the sheath and inflorescence are 60 and 20% respectively of blade photosynthesis (Tieszen and

Johnson, 1975). The photosynthesis rate of leaves is multiplied by the exposed phytosynthetic surface area of the leaf. The exposed surface area is calculated by multiplying the blade and sheath cell wall compartments by their area:biomass ratios and subtracting the area of surrounding dead sheaths. If the leaf is senescing, its photosynthetic competence is modified according to its NCC. A linear decline in photosynthetic competence is used, varying from the full competence at normal NCC level to zero at 20% of the NCC of the mature leaf.

Respiration in g of TNC day^{-1} is calculated as the sum of three separately controlled respiratory processes: growth, maintenance, and translocation. Growth respiration is calculated from the biochemical efficiencies involved in converting glucose to different cell constituents (Penning de Vries, 1972) and from the average biochemical composition of the new tissues. Considering the proportion of NCC and CW, the growth respiration is about 0.3 times the growth rates of each part.

Maintenance respiration depends on the protein content of the plant part (Penning de Vries, 1972). In leaves, maintenance respiration is included in the calculated net photosynthesis rate. In the stem base, rhizomes, and roots, maintenance respiration is defined as the amount of sugar needed by the NCC to carry on all metabolic activity related to maintenance of present conditions in the plant. Maintenance respiration is calculated by

$$R_m = \phi(\text{NCC})Q_{10}^{(0.1T_s)}$$

where ϕ is g of sugar respired per g of NCC, $Q_{10} = 2.0$, and T_s is the soil temperature. Based on the protein content of stem bases, rhizomes, and roots, indicated by the measured nitrogen contents, the maintenance respiration coefficients (ϕ) are 0.075 for stem bases and 0.035 g g^{-1} day^{-1} for rhizomes and roots.

Translocation respiration is assumed to be 10% of the weight of phosphorus and TNC translocated and is shared equally by the source and sink structures.

Storage Polymers—Sugar Reactions

An equilibrium between the concentrations of storage polymers and sugars is assumed such that 63% of the carbohydrate in any structure is in the form of storage material. However, storage polymers are not allowed to exceed 30% of the total biomass, which is visualized as internal resistance to growth of starch grains to prevent injury from membrane deformation.

Phosphorus Uptake

Phosphorus uptake (UP) by roots is described by a Michaelis–Menton relationship (Chapin, 1974). Thus:

$$\text{UP} = (V_x C_s)/(K_m + C_s)B_r$$

where V_x is the maximum uptake rate, K_m is the concentration of phosphorus at half V_x, C_s is the soil solution concentration of phosphorus, and B_r is the root biomass. The maximum uptake rate (V_x) and K_m are calculated from Chapin (1974) and are:

$$V_x = (7.13T_s + 66)10^{-6}$$
$$K_m = (0.015T_s + 0.024)31 \times 10^{-6}$$

where T_s is soil temperature in °C.

Labile and Nonlabile Phosphorus Reactions

Labile phosphorus is defined as phosphorus in forms which may be translocated or utilized in growth instantaneously. Phosphorus required for growth is removed only from the labile compartment. The nonlabile pool contains a large amount of phosphorus which may be converted to the labile form without any impairment of physiological activity. Only about 5% of the total phosphorus in the plant is truly fixed in compounds such as DNA and RNA (Bieleski, 1973). Labile and nonlabile phosphorus pools are in equilibrium, with the nonlabile pool acting as a buffer for the labile pool. At present, it appears that this reaction takes place only in leaves (Ulrich, personal communication). In the active cell there is more phosphorus in the nonlabile state than in the labile state (Figure 1), but during leaf senescence the ratio changes to 1.

During the growth of a leaf 0.4% of the total biomass must be phosphorus for proper physiological activity (Lessander and Garz, 1975). However, after a leaf

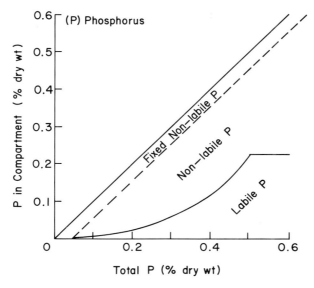

Figure 1.   Relationship between labile and nonlabile phosphorus pools in the leaf with different plant phosphorus contents.

is mature much of the phosphorus is no longer needed and may be removed. If phosphorus is limiting leaf growth, the required amount is removed from the labile pool of the oldest mature leaf. A mature leaf will go into early senescence when its total phosphorus is below 0.08% and will die when its total phosphorus reaches 0.06%. The level at which senescence begins (0.08%) was chosen since this may be the level at which photosynthetic rates are affected (Tieszen, 1978; Lessander and Garz, 1975).

Translocation

The translocation of sugars and phosphorus is simulated by differences in the concentrations of the substances between a plant structure and the equilibrium concentration of the plant as a whole. The equilibrium concentration is:

$$C = \Sigma S_i / \Sigma B_i$$

where S_i is the translocatable substance (sugar or labile nutrient) in a plant part, and B_i is the total biomass of the plant part. Translocation to or from the ith plant structure (T_i) is:

$$T_i = -(S_i - B_i C).$$

A positive value of T_i indicates translocation into and a negative value indicates translocation out of the ith plant structure. This formulation is not realistic, but no better formulation has appeared. Brouwer and de Wit (1969), de Wit et al. (1970), and Fick et al. (1975) bypass the problem by pooling all reserves into one plant pool which is drawn upon for growth by each plant part. Thornley (1972a, b) considered translocation as driven by a concentration gradient impeded by resistances. All formulations ignore the possibility of sugars being transported against a gradient, which occurs in the arctic graminoids (Shaver, personal communication) as well as other plants.

There are two constraints placed on the translocation reactions between the leaves and the rest of the tiller: a mature leaf translocates at least 70% of its daily net photosynthesis and cannot import carbon and phosphorus.

Results

The total production and percentages of the total production allocated to roots, rhizomes, stem bases, leaves, and inflorescences through the season were calcu-lated for each of the four tillers in the model tiller system (Figure 2). The leading rhizome develops roots and stem base in mid-season. The stem base remains approximately constant through the remaining 3 yr. In the first-year tiller, carbohydrate reserves in the rhizome are drawn upon to develop the leaves, leading rhizome, and root material. Rhizome reserves recover late in the growing

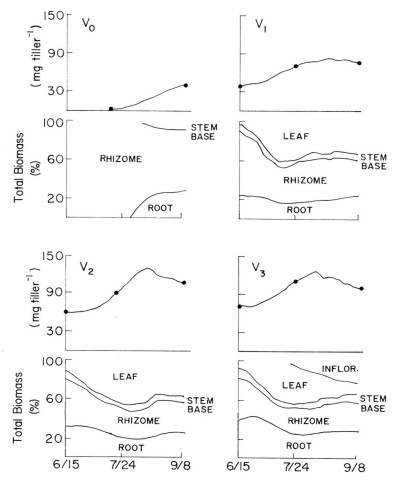

Figure 2.   Seasonal course of total dry weight and percentage of the total weight in root, rhizome, stem base, leaf, and inflorescence compartments in a tiller system. The V0 is the leading rhizome. The V4 (not shown) is the old, senescing rhizome.

season. In the second-year tiller, rhizome reserves are drawn upon to grow leaves, and supply the deficits of the first-year tiller. In the final year, the production of leaves ceases and the inflorescence is initiated.

An annual, steady-state carbon balance of a four-tiller system of *Dupontia*, consisting of the leading rhizome (V0), first-year tiller (V1), second-year tiller (V2), and third-year tiller (V3) with an inflorescence, shows that the leading rhizome is the largest carbon sink in the system (Figure 3a). The rhizome produces no carbohydrate by photosynthesis, grows to about 42 mg, and loses about 16 mg of dry weight in respiration. Thus the leading rhizome requires 58 mg of dry weight per tiller. The first-year tiller accumulates 116 mg of dry weight

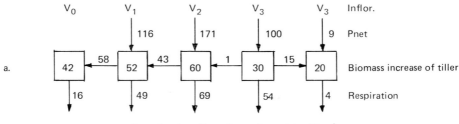

a.

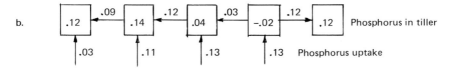

b.

Figure 3.   (a) Annual fluxes of carbon in a four-tiller system with an inflorescence as a steady state. Values are mg of dry weight per tiller and growing season. Upper downward arrows represent carbon gained by net photosynthesis. Lower downward arrows represent carbon lost by growth respiration, maintenance respiration of nonphotosynthetic parts, and translocation respiration. Horizontal arrows represent carbon fluxes between tillers. Numbers at the intersection are the biomass increase of the tiller. (b) Annual fluxes of phosphorus in the same tiller system. Values are mg of phosphorus per tiller and growing season.

through photosynthesis, respires 49 g of dry weight, and accumulates 52 g of dry weight in growing leaves and roots. It contributes 58 mg of dry weight to the leading rhizome. These exchanges develop a deficit of 43 mg which comes from the attached second-year tiller. The second-year tiller incorporates 171 mg of dry weight (compared with 154 mg season^{-1} reported by Tieszen et al., personal communication, 1977), respires 69 mg and accumulates 60 mg in growth. The V2 contributes 43 mg to the V1 and develops a deficit of 1 mg. The third-year tiller incorporates 100 mg of dry weight from leaves, loses 58 mg in respiration, and accumulates 30 mg in growth. The inflorescence is photosynthetic and accumulates about 9 mg of dry weight by photosynthesis, but the growth of the inflorescence requires 20 mg of new biomass and 4 mg of respiratory loss. The V3 tiller then must supply the inflorescence with 15 mg for its growth and contributes 1 mg to the V2.

The annual phosphorus budget for the same tiller system shows a different relationship (Figure 3b). The V0, V1, and the inflorescence accumulate similar amounts of phosphorus, the V2 accumulates 70% less, and the V3 has a negative phosphorus balance in part due to its support of the inflorescence.

The biomass of the four-tiller system was converted to grams per square meter assuming 1500 tillers per m^2 so that the simulations could be compared with field

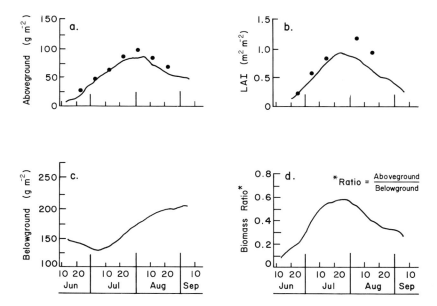

Figure 4.   Seasonal progressions of (a) aboveground biomass and (b) leaf area index, measured (•) and calculated (line), (c) belowground biomass, and (d) aboveground:belowground biomass ratio.

data on aboveground biomass and leaf area index, which were on a m² basis and collected in a mixed sward rather than in pure *Dupontia*. The simulated values follow the observed values better for biomass than for leaf area index (Figure 4), perhaps due to not including changing area to biomass ratios in the model although indicated in the data of Tieszen (1972). The model predicts an increase of about 50 g m^{-2} season^{-1} in the belowground biomass and a peak season above:below biomass ratio of 0.5.

Comparisons of observed and simulated TNC levels for leaves and rhizomes show general agreement (Figure 5), although simulated TNC levels were usually higher than observed. In the simulations the V3 rhizome gained TNC at the end of the season unlike the observed. There may be a developmentally controlled redistribution of TNC in the V3 after flowering. However, the inability of the V3 to regain lost TNC is probably not due to lack of photosynthate or demands from the inflorescence. It may be important that Allessio and Tieszen (1975) have shown that roots appeared active even 1 yr after their associated tiller had produced an inflorescence.

In order to look at the effects of reduced soil phosphorus on the tiller system, 10-yr runs were made at different phosphorus levels (Figure 6). Little or no effect on leaf area index and biomass or on the above:below ratio was seen at soil phosphorus concentrations above 4 ppb. Luxury uptake of phosphorus in the aboveground tissue occurred at about 4 ppb. At 1 ppb biomass was reduced

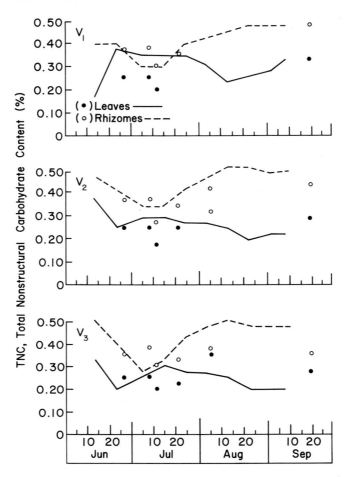

Figure 5. Seasonal progression of the total nonstructural carbohydrate content (TNC) in the V1, V2, and V3 leaves (——) and rhizomes (---). Data are given by circles (● leaves, ○ rhizomes) and simulations by lines.

because the growth of the V0 was not completed in a single season at that phosphorus level.

The model's behavior following grazing was simulated by following as closely as possible the clipping regime of experiments performed in the field (Mattheis et al., 1976). All aboveground material 5 mm above the surface was removed from the whole system on 7 July. The simulated TNC levels in the rhizome were more sensitive and usually lower following grazing than the observed (Figure 7). Physiological responses to grazing such as increased photosynthesis or reduced activity of belowground plant parts are not included in the model and may explain the greater sensitivity of the model to the grazing process.

The effects of changes in the length of the growing season were simulated by increasing or decreasing temperatures at the beginning of the season as might occur with changes in the time of melt-off. As the length of the growing season increased, seasonal net photosynthesis and total system biomass increased (Figure 8). As the biomass increased respiratory losses increased and net photosynthesis contributed less to new biomass. CO_2 uptake and biomass increased at about 0.10 g of CO_2 day^{-1} and 0.005 g of dry wt day^{-1} respectively for the tiller system, and 5 g of CO_2 m^{-2} day^{-1} and 2.55 g of dry wt m^{-2} day^{-1} respectively on an area basis. The latter are higher than estimates of Miller et al. (1976), due either to overestimating the tiller density used to convert the present model production estimates to a m^2 basis or to increased CO_2 incorporation caused by growth of new photosynthetic tissue included in this model. The latter is suspected, since the seasonal courses of biomasses were reasonable (Figure 4).

Lower air temperatures during the growing season have been observed to correspond with higher phosphorus contents (Chapin et al., 1975). This possibility was simulated by decreasing the air temperatures by 2°C. This change decreased seasonal net CO_2 incorporation by 0.08 g of CO_2 per tiller, total biomass by 0.02 g of dry wt tiller^{-1}, and leaf area index by 0.13 m^2 m^{-2}. The phosphorus contents increased in the leaves of the V1, V2, and V3 tillers by 0.11, 0.17, and 0.00%, respectively, in the inflorescence by 0.02%, and in the root by 0.04% in all age classes. In the V0, phosphorus increased 0.03%. This model response is somewhat less than the observed increase of about 0.3% P, and may be related to higher soil solution levels of phosphate in the simulated standard case.

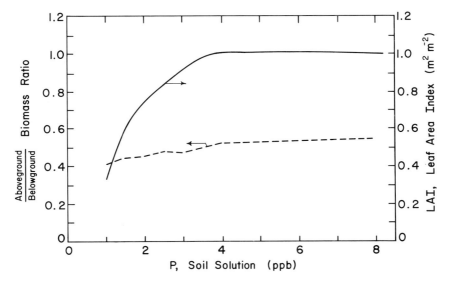

Figure 6.   Peak season leaf area index and aboveground:belowground biomass ratio at different soil phosphorus concentrations.

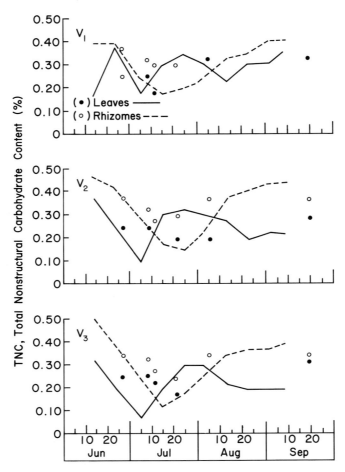

Figure 7.   Seasonal progression of the total nonstructural carbohydrate content (TNC) in the V1, V2, and V3 leaves (———) and rhizomes (---) with removal of plant parts about 5 mm above the ground surface on 7 July. Data are given by circles (● leaves, ○ rhizomes) and simulations by lines.

Discussion

The model of growth and carbon–phosphorus allocation summarizes much of what is known about the dynamics of the tiller system of *Dupontia fisheri* and points up new conclusions and hypotheses. The simulations indicate that the second-year tiller has the largest carbon surplus of the tillers in the system. Rapid mobilization of nutrients and carbon early in the season increases production. In order for rapid leaf growth to occur, nutrient levels must be high in the plant. For an average year, the fraction of the seasonal requirements of phosphorus coming from internal reserves is 0.82. The carbon budget indicates 15 mg of dry wt and 0.12 mg of P are used to produce a floral structure plus seeds for sexual reproduction, and about 58 mg of dry wt and 0.09 mg of P are used to produce a vegetative rhizome. However, when the relative survival of seeds and seedlings vs rhizomes is considered (see Lawrence et al., 1978), it is about 30 times more costly in terms of g of dry wt and 200 times more costly in terms of g of phosphorus to produce an individual via sexual means than via vegetative rhizomes. This may help explain, in quantitative terms, the preponderance of vegetative reproduction in tundra.

The total CO_2 uptake per m² through the season which was simulated in this model was about 400 g of CO_2 m^{-2} yr^{-1} and is lower than the 500 to 600 g of CO_2 m^{-2} yr^{-1} simulated by Miller et al. (1976) or the 600 to 700 g of CO_2 m^{-2} yr^{-1} estimated by Tieszen (1975). This difference may be due to the environmental simulations used here or to an underestimation of tiller density.

One of the major deficiencies in the model is the lack of experimental estimates of growth rates for various plant parts without phosphorus limitation. Without such experiments it is difficult to explain the fact that there is no limitation of plant growth above soil phosphorus concentrations of 4 ppb. This may be caused by underestimating the maximum growth rates and hence the phosphorus required for growth. However, the problem may be related to reduction in phosphorus concentration from the bulk soil solution to the root surface (Bieleski, 1973). Phosphorus concentrations in the soil solution at Barrow range from 10 to 50 ppb and an order of magnitude decrease in concentration from bulk soil solution to root surface is reasonable (Bieleski, 1973).

Since the model is more sensitive to grazing than field observations indicate, it seems reasonable that the effect of grazing is not simply a loss of photosynthetic tissue; but other processes are affected as well. The reduction of root growth which has been observed in pasture and clipping experiments should increase the

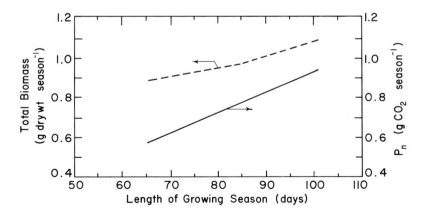

Figure 8.   Seasonal CO_2 incorporation and seasonal biomass increase with different lengths of the growing season.

carbohydrate available for leaf growth and densensitize the plant to the effect of grazing. The functional balance between roots and leaves, and the control of this balance, needs further work.

Growth is a complex process, influenced by many internal and external variables. The simulation model described is preliminary, and needs further validation and development to increase confidence in the prediction made. However, the success of the model in predicting reasonable responses of the tiller system to several treatments indicates a consistency in the several data sets involved in the model construction. Critical areas needing experimental data are: the functional balance of roots and leaves and its control, the response of growth to different nutrient levels, and the dynamics of phosphorus in the root–soil system.

Acknowledgments. This research was supported by the National Science Foundation under Grant GV 29345 to San Diego State University. It was performed under the joint sponsorship of the International Biological Programme and the Office of Polar Programs and was coordinated by the U.S. Tundra Biome. Field and laboratory activities at Barrow were supported by the Naval Arctic Research Laboratory of the Office of Naval Research.

We thank Drs. F.S. Chapin III, G. Shaver, B. McCown, and D. Johnson for their helpful criticisms.

References

Allessio, M.L., and L.L. Tieszen. (1975) Patterns of carbon allocation in an arctic tundra grass, *Dupontia fischeri* (Gramineae), at Barrow, Alaska. *Amer. J. Bot.,* **62**: 797–807.

Allessio, M. L., and L. L. Tieszen. (1978) Translocation and allocation of ^{14}C-photoassimilate by *Dupontia fisheri.* In *Vegetation and Production Ecology of an Alaskan Arctic Tundra* (L. L. Tieszen, Ed.). New York: Springer-Verlag, Chap. 17.

Bieleski, R.L. (1973) Phosphate pools, phosphate transport and phosphate availability. *Ann. Rev. Plant Physiol.,* **24**: 225–252.

Billings, W.D., K.M. Peterson, G.R. Shaver, and A.W. Trent. (1977) Root growth, respiration, and carbon dioxide evolution in an arctic tundra soil. *Arct. Alp. Res.,* **9**: 127–135.

Blackman, G.E. (1968) The application of the concepts of growth analysis to the assessment of productivity. In *Functioning of Terrestrial Ecosystems at the Primary Production Level* (F.E. Eckardt, Ed.). Paris: Unesco.

Bliss, L.C. (1966) Plant productivity in alpine microenvironments on Mt. Washington, New Hampshire. *Ecol. Monogr.,* **36**: 125–155.

Botkin, D.B. (1975) A functional approach to the niche concept in model communities. Paper delivered at National Science Foundation Conference on the Future of Systems Ecology, Utah State University, 19–21 February 1975.

Brouwer, R., and C.T. de Wit. (1969) A simulation model of plant growth with special attention to root growth and its consequences. In *Root Growth: Proceedings of the Fifteenth Easter School of Agricultural Science, University of Nottingham* (W.J. Whittington, Ed.). London: Butterworth, pp. 224–244.

Chapin, F.S., III (1974) Morphological and physiological mechanisms of temperature compensation in phosphate absorption along a latitudinal gradient. *Ecology,* **55**: 1180–1198.

Coyne, P.I., and J.J. Kelley. (1975) CO_2 exchange over the Alaskan arctic tundra: Meteorological assessment by an aerodynamic method. *J. Appl. Ecol.,* **12**: 587–611.

Dainty, J. (1965) Osmotic Flow in the State and Movement of Water in Living Organisms. New York: Academic Press.

Davidson, J.L., and J.R. Phillip. (1958) Light and pasture growth. In *Climatology and Microclimatology.* Paris: Unesco, pp. 181–187.

Duncan, W.C., R.S. Loomis, W.A. Williams, and R. Hanau. (1967) A model for simulating photosynthesis in plant communities. *Hilgardia,* **38**: 181–205.

Fick, G.W., R.L. Loomis, and W. Williams. (1975) Sugar beet. In *Crop Physiology* (L.T. Evans, Ed.). London: Cambridge University Press, pp. 259–295.

Harpaz, Y. (1975) Simulation of the nitrogen balance in semi-arid regions. Ph.D. thesis, Hebrew University, Jerusalem.

Horowitz, L. (1958) Some simplified mathematical treatments of translocation in plants. *Plant Physiol.,* **33**: 81–93.

Květ, J., J.P. Ondok, J. Nečas, and P.G. Jarvis. (1971) Methods of growth analysis. In *Plant Photosynthetic Production: Manual of Methods* (Z. Šesták, J. Čatský, and P.G. Jarvis, Eds.). The Hague: Junk, pp. 343–411.

Lawrence, B.A., M.C. Lewis, and P.C. Miller. (1978) A simulation model of population processes of arctic tundra graminoids. In *Vegetation and Production Ecology of an Alaskan Arctic Tundra* (L.L. Tieszen, Ed.). New York: Springer-Verlag, Chap. 26.

Lemon, E., D.W. Stewart, and R.W. Shawcroft. (1971) The sun's work in a cornfield. *Science,* **174**: 371–378.

Lessander, R., and J. Garz. (1975) Photosynthesis and Wachstum du Clatter von jungan Zuchurrubenpflanzen in Abhangigkeit von der P. Ernahrung. *Arch. Acker. Pflanz. Bodenk., Berlin,* **19**: 545–554.

Loomis, R.S. (1970) Summary: Dynamics and development of photosynthetic systems. In *Prediction and Measurement of Photosynthetic Productivity: Proceedings of the IBP/ PP Technical Meeting, Trebon, 14–21 September 1969.* Wageningen: Center for Agricultural Publishing and Documentation, pp. 137–141.

Mattheis, P.J., L.L. Tieszen, and M.C. Lewis. (1976) Responses of *Dupontia fischeri* to simulated lemming grazing in an Alaskan arctic tundra. *Ann. Bot.,* **40**: 179–197.

Miller, P.C. (1972) Bioclimate, leaf temperature, and primary production in red mangrove canopies in south Florida. *Ecology,* **53**: 22–45.

Miller, P.C., B.D. Collier, and F.L. Bunnell. (1975) Development of ecosystem modeling in the Tundra Biome. In *Systems Analysis and Simulation in Ecology* (B. Patten, Ed.). New York: Academic Press, Vol. III, pp. 95–115.

Miller, P.C., W.A. Stoner, and L.L. Tieszen. (1976) A model of stand photosynthesis for the wet meadow tundra at Barrow, Alaska. *Ecology,* **57**: 411–430.

Miller, P.C., and L.L. Tieszen. (1972) A preliminary model of processes affecting primary production in the arctic tundra. *Arct. Alp. Res.,* **4**: 1–18.

Monsi, M., and T. Saeki. (1953) Über den Lichtfaktor in den Pflanzengesellschaften und seine Bedeutung für die Stoffproduction. *Japan. J. Bot.,* **14**: 22–52.

Monsi, M. and Y. Murata (1970) Development of photosynthetic systems as influenced by distribution of matter. In *Prediction and Measurement of Photosynthetic Productivity: Proceedings of IBP/PP Technical meeting, Trebon, 14–21 September 1969.* Wageningen: Center for Agricultural Publishing and Documentation, pp. 115–129.

Penning de Vries, F.W.T. (1972) Respiration and growth. In *Crop Processes in Controlled Environments* (A.R. Rees, K.E. Cockshull, D.W. Hand, and R.G. Hurd, Eds.). London: Academic Press, pp. 327–347.

Penning de Vries, F.W.T., C.E. Murphy, Jr., C.G. Wells, and J.R. Jorgensen. (1976) Simulation of nitrogen distribution in time and space in even-agral loblolly pine plantations and its effect on productivity. IBP Contribution No. 156. Paper presented at Symposium on Mineral Cycling in Southern Ecosystems, Augusta, Georgia, 1–3 May 1974.

Pitelka, F.A. (1958a) Some characteristics of microtine cycles in the arctic. *Arctic Biol.*, **18**: 73–78.

Pitelka, F.A. (1958b) Some aspects of population structure in the short-term cycle of the brown lemming in northern Alaska. In *Cold Spring Harbor Symp. Quant. Biol.* The Biological Laboratory, Cold Spring Harbor, New York, pp. 237–251.

Pitelka, F.A., P.Q. Tomich, and G.W. Treichel. (1955) Ecological relations of jaegers and owls as lemming predators near Barrow, Alaska. *Ecol. Monogr.*, **25**: 85–117.

Schultz, A.M. (1964) The nutrient recovery hypothesis for arctic microtine cycles. II. Ecosystem variables in relation to arctic microtine cycles. In *Grazing in Terrestrial and Marine Environments* (D.J. Crisp, Ed.). Oxford: Blackwell Scientific Publications, pp. 57–68.

Schultz, A.M. (1969) A study of an ecosystem: The arctic tundra. In *The Ecosystem Concept in Natural Resource Management* (G.M. Van Dyne, Ed.). New York: Academic Press, pp. 77–93.

Stoner, W.A., and P.C. Miller. (1975) Water relations of plant species in the wet coastal tundra at Barrow, Alaska. *Arct. Alp. Res.*, **7**: 109–124.

Thornley, J.H.M. (1972a) A balanced quantitative model for root:shoot ratios in vegetative plants. *Ann. Bot.*, **36**: 431–441.

Thornley, J.H.M. (1972b) A model to describe the partitioning of photosynthate during vegetative plant growth. *Ann. Bot.*, **36**: 419–430.

Tieszen, L.L. (1972) The seasonal course of aboveground production and chlorophyll distribution in a wet arctic tundra at Barrow, Alaska. *Arct. Alp. Res.*, **4**: 307–324.

Tieszen, L.L. (1975) CO_2 exchange in the Alaskan Arctic Tundra: Seasonal changes in the rate of photosynthesis of four species. *Photosynthetica*, **9**: 376–390.

Tieszen, L.L. (1978) Photosynthesis in the principal Barrow, Alaska, species: A summary of field and laboratory responses. In *Vegetation and Production Ecology of an Alaskan Arctic Tundra* (L.L. Tieszen, Ed.). New York: Springer-Verlag, Chap. 10.

Tieszen, L.L., and D.A. Johnson, (1975) Seasonal pattern of photosynthesis in individual grass leaves and other plant parts in arctic Alaska with a portable $^{14}CO_2$ system. *Bot. Gaz.*, **136**: 99–105.

van Keulen, H. (1975) *Simulation for Water Use and Herbage Growth in Arid Regions*. Wageningen: Center for Agricultural Publishing and Documentation, 176 pp.

van Keulen, H., N.G. Seligman, and J. Goudriaan. (1975) Availability of anions in the growth medium of roots of an actively growing plant. *Neth. J. Agric. Sci.*, **23**: 131–138.

Warren Wilson, J. (1964) Annual growth of *Salix arctica* in the high-arctic. *Ann. Bot.*, **28**: 71–76.

Warren Wilson, J. (1966) An analysis of plant growth and its control in arctic environments. *Ann. Bot.*, **30**: 383–402.

de Wit, C.T. (1965) Photosynthesis of leaf canopies. Institute of Biol. Chem. Res. on Field Crops and Herbage, Wageningen. *Agric. Res. Rep.* 663.

de Wit, C.T., R. Brouwer, and F.W.T. Penning de Vries. (1970) The simulation of photosynthetic systems. In *Prediction and Measurement of Photosynthetic Productivity: Proceedings of the IBP/PP Technical Meeting, Trebon, 14–21 September 1969*. Wageningen: Center for Agricultural Publishing and Documentation, pp. 47–70.

25. A Model of Carbohydrate, Nitrogen, Phosphorus Allocation and Growth in Tundra Production

P. C. MILLER, W. A. STONER, L. L. TIESZEN, M. ALLESSIO, B. McCOWN, F. S. CHAPIN, G. SHAVER

Introduction

Most models of plant production processes have been models of primary production and aboveground canopy processes and have included various environmental and physiological parameters in an attempt to predict photosynthesis and growth (Davidson and Philip, 1958; de Wit, 1965; Duncan et al., 1967; de Wit et al., 1970; Miller and Tieszen, 1972; Monsi and Saeki, 1953; Murphy and Knoerr, 1972; Paltridge, 1970; Lemon et al., 1971). Many have assumed that temperatures were optimal, most have assumed that water was not limiting growth, and all have assumed that minerals were not limiting growth. Horowitz (1958) developed mathematical formulations for alternative concepts of the mechanism of translocation. Thornley (1972) presented a primary production model which incorporated carbon and nitrogen to simulate the seasonal development of the root:shoot ratios. The model consisted of only two compartments, roots and shoots, composed of carbon and nitrogen. Although the model is fairly simple, it has the basic feedback mechanisms which have been associated with the carbon–nitrogen flow in plants. The lack of plant models which include minerals may reflect a current lack of a consistent set of hypotheses and data regarding the role of minerals in plant development and the mechanisms of their interaction.

The development of the current concepts of the dynamics of the Barrow tundra ecosystem has included major inputs from Pitelka et al. (1955) and Pitelka (1958a, b) regarding the importance of the lemming–vegetation interactions in causing lemming population fluctuations at Barrow. He did not regard these interactions necessarily as the causes of population fluctuations of other microtines in other arctic areas. Schultz (1964, 1969) developed the concept of the interaction between lemmings and their food supply in the nutrient-recovery hypothesis. He postulated that the food quality of the vegetation, the amount of standing crop insulating the soil surface, the depth of the active layer, and the size of the lemming population were interrelated and showed correlations between the size of the lemming population and the phosphorus content of the vegetation. Plant reproduction was assumed to be limited by the availability of minerals in the soil and was increased by the addition of minerals in a field experiment. A conspicuous feature of arctic plants is a high sucrose content, which is considered significant in cold resistance and can be produced by nutrient deficiencies.

The potential importance of minerals in affecting primary production and herbivore population fluctuations in the arctic forced a preliminary concept of the role of minerals in plant production. A model which incorporates possible effects of minerals on primary production and plant growth in the arctic tundra at Barrow, Alaska, was developed to clarify the understanding of the possible mineral limitation of production in the tundra and to provide a framework for interaction with consumer and decomposer processes. The model was to include the possibilities of high sucrose content because of mineral limitations, of separate processes of photosynthesis and aboveground production because Tieszen (1978) has shown that the rate of photosynthesis in leaves of fertilized plants was hardly different from control plants although the aboveground production was higher, and of a changing allocation of growth to above- or belowground parts depending upon the mineral availability. The model was to simulate the seasonal courses of leaf area index and plant biomass in different biologically and ecologically important plant parts and of at least phosphorus, nitrogen, and energy contents of the different plant parts. Tieszen (1975) has shown that net photosynthesis in the arctic remained positive 24 hr a day until late in the season. Respiration rates remained low because of the low air and soil temperatures. When nutrients were added to an area of wet meadow, production increased (Schultz, 1969). These findings taken together form the basis for the assumption that production is nutrient limited, probably by nitrogen (McCown, unpublished data, 1978) or phosphorus (Chapin, 1978; McKendrick et al., 1978). A shortage of carbohydrates, as might occur when CO_2 fixation rates are low, acts indirectly to reduce growth by reducing amino acid synthesis and ion uptake.

The model was developed for an idealized monocot, relying primarily on field and laboratory data which were available for *Dupontia fisheri,* an arctic grass which occurs on wet meadow sites near the coast. The model is being adapted to apply to other monocots and to dicots. The specific objectives of the model were to provide a framework in which to view detailed mechanisms of plant mineral relations, alternative plant strategies for the tundra, and the hypothesis of mineral limitation of primary production. In addition it provides a means of interacting conceptually with herbivores and decomposers by considering carbon to nitrogen ratios, energy content, protein content, etc., through the growing season. The model is an early attempt by the primary producer group to synthesize the current understanding of tundra plant physiological processes. On many aspects additional data are now available and ideas may have changed. The fundamental framework is still viable, however.

In the model all standing crops have the units of g dry weight m^{-2} and all transfers have the units g m^{-2} day^{-1}. "Standing crop" is used in a general sense to refer to the amount of substance present. It may be g of sucrose, g of dry organic matter, g of protein, etc., per m^2.

Plant Description

The monocot plant, for which *Dupontia fisheri* is an example, is considered to consist of six parts:

1. Leaf blade—that part of the leaf above the ligule.

2. Leaf sheath—that part of the leaf below the ligule but above the meristem and stem base. This is taken to include the aboveground portion of the sheath.

3. Stem base complex—belowground part of the leaf sheath to the point where it terminates at a node, stem base, and associated meristems. The stem tissue may be distinguished from the rhizome by its characteristically compressed internodes in nonreproductive tillers and by the presence of expanded leaf sheaths. The stem base complex functions both meristematically and as a storage site during winter.

4. Rhizome—the underground stem from which most of the tillers and roots are produced. The rhizome is responsible for the vegetative extension of the plant horizontally underground.

5. Roots—the roots are the major absorptive structures of the plant.

6. Aboveground reproductive structures—inflorescences, fruits, and seeds.

Justification for the separation of the plant into these parts lies in the ability to visually and physically distinguish them (Dennis and Johnson, 1970). Relative biomass values and discrete samples of each plant part can be obtained for analysis throughout the year. In addition, each component has a relatively distinct biological function in the plant and interacts with other components of the ecosystem in different ways.

Plant Part Components

Each plant part consists of eight biochemical components:

1. Cell wall polymers—principally cellulose, but including hemicelluloses, lignins when present, and some associated proteinaceous compounds.

2. Sugars—sucrose, fructose, and glucose, and other low molecular weight, freely mobile saccharides.

3. Polysaccharides—storage polymers, such as glucans and fructans, as well as some more readily hydrolyzed molecules such as pectins and lower molecular weight glucans.

4. Protein—proteins and other complex nitrogenous compounds including nucleic acids.

5. Amino acids and amides.

6. Lipids and fatty acids—compounds freely soluble in chloroform.

7. Phosphate—occasionally a sugar phosphate, but for the sake of simplicity considered to be inorganic phosphate.

8. Phytic acid—a molecule serving as a phosphate storage reserve.

These biochemical components can be quantitatively determined, and they represent more than 95% of the plant biomass (McCown, personal communication). It is also important to note that with separation into these biochemical classes, it is possible to introduce mechanistic limitations on plant growth and to

allow for precise transfer of minerals and energy in forms with different digestibility to other components of the ecosystem, such as herbivores and
decomposers.

Transfers within Plant Parts

Transfers between the biochemical components considered in each plant part are
(Figure 1): amino acids to and from proteins, amides to and from amino acids,
sugars to and from fatty acids, fatty acids to and from lipids, inorganic phosphates to and from lipids, inorganic phosphates to and from phytic acids, sugars
to and from phytic acids, inorganic phosphates to and from proteins, sugars to
cell wall material, sugars to and from polysaccharides, and sugars to amino acids.
In the roots only, sugars and nitrogen are transferred to amino acids. Associated
with each transfer is a biochemical efficiency, such that organic matter is lost in
each transfer (Penning de Vries, 1974; Penning de Vries et al., 1974). The transfer
which controls plant growth in each plant part is the conversion of amino acids to
proteins.

The transfer of amino acids (AA) to proteins (Prot) proceeds as a fraction (Φ_s)
of the amino acid pool which is modified by temperature (TEMP), the size of the
phosphate pool, and the time of year (t). Thus,

$$T_{AA \rightarrow Prot} = \Phi_s AA f(TEMP) f(t). \tag{1}$$

In the simulation Φ_s varies in each plant part. In the sheaths, lamina, and

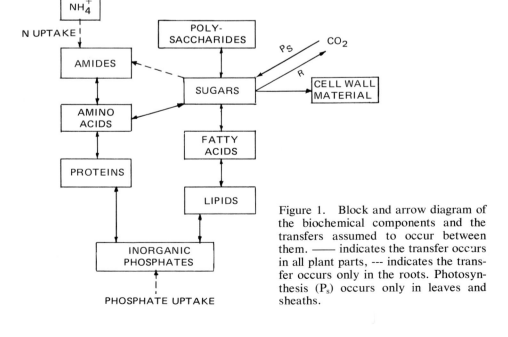

Figure 1.   Block and arrow diagram of
the biochemical components and the
transfers assumed to occur between
them. —— indicates the transfer occurs
in all plant parts, --- indicates the transfer occurs only in the roots. Photosynthesis (P_s) occurs only in leaves and
sheaths.

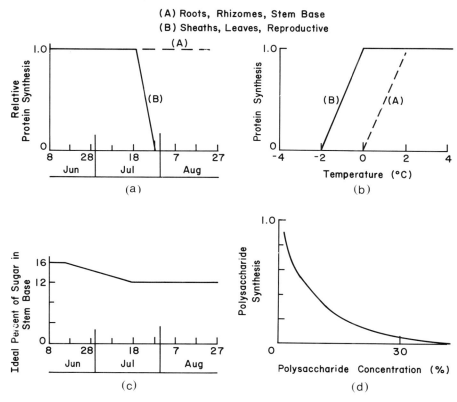

Figure 2.   (a) Seasonal course of the modification of the synthesis of proteins from amino acids due to effects of senescence. (b) Effect of temperature of the air or ground on the rates of protein synthesis and breakdown and of lipid breakdown. (c) Seasonal course of the percentage of sugars in the stem base complex under ideal conditions. (d) Effect of polysaccharide concentration on polysaccharide synthesis.

reproductive tissues, protein synthesis does not occur at temperatures below −2°C and occurs without temperature limitations above 0°C. Between −2 and 0°C protein synthesis varies linearly with temperature. In roots, rhizomes, and meristem complex, protein synthesis does not occur below 0°C and occurs without temperature limitations above 2°C. It varies linearly with temperature between 0 and 2°C. New protein is assumed to be associated with 0.25% phosphorus. If the size of the phosphate pool is inadequate to supply the required amount, protein synthesis is reduced to use only the available phosphate. The time of year modification incorporates the effects of senescence on plant growth (Figure 2a). The increase of aboveground biomass ceases and aboveground biomass begins to decrease in early August at Barrow (Dennis and Johnson, 1970; Dennis et al., 1978; Tieszen, 1972) at a time when environmental conditions are still favorable for growth (Tieszen, 1975; Miller and Tieszen, 1972). Such decreases in biomass may be caused by internal hormonal relationships affecting protein synthesis. The senescence effect is such that by early August the rate of protein synthesis is less than the rate of protein hydrolysis in leaves.

The hydrolysis of protein into amino acids and phosphate is considered to be a constant fraction (Φ_b) of the amount of proteins, modified by temperature. Thus,

$$T_{\text{Prot}\rightarrow\text{AA}} = \Phi_b \,\text{Prot} f(\text{TEMP}). \tag{2}$$

In the simulation Φ_b equaled 0.05. Protein hydrolysis is assumed to have the same temperature relations as protein synthesis. That is, in the sheaths, lamina, and reproductive tissues protein synthesis does not occur below $-2°C$, occurs without temperature limitation above $0°C$, and varies linearly between -2 and $0°C$.

Amides, containing two nitrogen groups, are created when the amino acid pool exceeds 5% of the total biomass in any plant part. Conversion of amides to and from amino acids involving the incorporation or release of sugars occurs in all plant parts. The synthesis of amino acids from sugars and inorganic nitrogen occurs mainly in the roots (Thomas, 1927; Bollard, 1956, 1957). Ammonium is absorbed by the roots and synthesized immediately to amino acids to prevent toxification by ammonia. When sugars are scarce, amino acids can be converted into amides.

The transfer of inorganic phosphate (P_i) to phytic acid (PA) is assumed to occur significantly only in storage tissue (rhizome and stem base complex) (Bieleski, 1973) and to proceed as a fraction (Φ_s) of the inorganic phosphate concentration modified by temperature (T):

$$T_{P_i\rightarrow PA} = \Phi_s Q_{10}^{-1(\text{TEMP})}$$

where $\Phi_s = 0.95$ g if $P_i > 0.13\%$ dry wt, and $\Phi_s = 0$, if $P_i < 0.0013\%$ dry wt. In the model this transfer occurs only after the phosphorus concentration reaches a threshold of 0.13% of dry weight (Bieleski, 1973), presumably as a result of induction of phytic acid synthetase. These high levels of inorganic phosphate accumulate only in the fall when there are no actively growing tissues to compete as sinks for inorganic phosphate. Phytic acid requires one hexose sugar for six phosphate groups; synthesis does not occur if there is insufficient sugar.

The hydrolysis of phytic acid to sugar and inorganic phosphate by phytase (Bieleski, 1973) is assumed to proceed as a fraction of the phytic acid concentration as modified by temperature and inorganic phosphate concentration. Inorganic phosphate inhibits activity (Sartirana and Bianchetti, 1967) as well as synthesis of phytase (Bianchetti and Sartirana, 1967). The transfer from phytic acid to inorganic phosphate is assumed to occur only when inorganic phosphate concentrations are less than 0.13% dry wt

$$T_{PA\rightarrow P_i} = \frac{\Phi_s Q_{10}^{0.1(\text{TEMP})}}{f(P_i)}$$

where $\Phi_s = 0$, if $P_i > 0.0013$. This transfer is particularly important in the spring

Table 1. Composition of New Material in Each Monocot Plant Part

Box	Root (%)	Rhizome (%)	Stem (%)	Sheath (%)	Lamina (%)
Cell (1)	76.8	69.8	49.5	42.5	42.0
Sugar (2)	1.6	5.3	12.0	12.5	13.0
Polysaccharide (3)	10.0	16.0	25.0	20.0	20.0
Lipid (4)	2.4	2.8	3.0	7.0	7.0
Protein (5)	8.0	5.0	10.0	17.0	17.0
Amino acid (6)	1.2	1.1	1.0	0.9	0.8

when rapid shoot growth depletes inorganic phosphate supplies; the transfer ceases in the fall when inorganic phosphate concentrations increase.

The synthesis of fatty acids from sugars is proportional to the transfer of amino acids to proteins and is expressed by

$$T_{\text{sugar}\rightarrow\text{fatty acids}} = (\%L/\%\text{Prot})T_{\text{AA}\rightarrow\text{Prot}} \qquad (3)$$

where %L and %Prot are the percentages of lipids and proteins, respectively, in new plant tissue (Table 1). Fatty acids may be transferred back to sugar, following Eq. (3), when proteins are broken down into amino acids.

The production of lipids from fatty acids is a function of fatty acid synthesis, the amount of fatty acids, and the amount of phosphates present. The transfer is expressed by

$$T_{\text{fatty acids}\rightarrow\text{lipids}} = \Omega_s \text{ fatty acids } T_{\text{sugar}\rightarrow\text{fatty acids}} \qquad (4)$$

where Ω_s is a constant, equal to 0.85. Lipids are assumed to contain 0.5% of phosphorus by weight. Lipid synthesis is reduced if the phosphate pool is inadequate.

Lipids break down into fatty acids and phosphates at a rate proportional to the breakdown of proteins. Thus

$$T_{\text{lipids}\rightarrow\text{fatty acids}} = \Omega_b(\%L/\%\text{Prot})\text{lipid } T_{\text{Prot}\rightarrow\text{AA}} \qquad (5)$$

where Ω_b is a constant, equal to 0.5.

Cell wall material is synthesized as a function of protein synthesis.

$$T_{\text{sugar}\rightarrow\text{cell wall}} = (\%\text{cell}/\%\text{Prot})T_{\text{amino acids}\rightarrow\text{protein}} \qquad (6)$$

where %cell is the percentage of cellulose in the new plant part (Table 1). Cell wall synthesis ceases in aboveground parts when the canopy reaches the daily light compensation point. The daily compensation point (Θ) is calculated by:

$$\Theta = \{-\ln(\lambda/Q_o)/([K(1 - M)]\} - \text{LAI} \qquad (7)$$

where λ is the light compensation point equal to 0.015 cal cm^{-2} min^{-1}, Q_0 is the solar radiation outside of the canopy, K is the extinction coefficient for radiation in the canopy (calculated after Miller and Tieszen, 1972), M is the transmittance of the canopy, and LAI is the total leaf area of the canopy on a specific day. If Θ is less than zero, the LAI is considered greater than an optimal size and material is translocated out of the leaf cells. Cellulose is not translocated.

Polysaccharide synthesis is controlled by the availability of sugar remaining after all other cellular components have been synthesized or translocated. The rate of polysaccharide synthesis is inhibited by product accumulation in a curvilinear fashion (Figure 2d). This function may be visualized as an internal resistance to growth of starch grains which prevents injury to chloroplasts from membrane deformation. Below $-2°C$ no transfer occurs. This relationship is expressed by:

If $\%S_A < \%S_1$

$$T_{starch \to sugar} = (\%S_1 - \%S_A)W_1 exp(-W_2\%STAR),$$

If $\%S_A < \%S_1$

$$T_{sugar \to starch} = (\%S_A - \%S_1)(1.0 - W_1 exp(-W_2\%STAR)) \tag{8}$$

where $\%S_A$ and $\%S_1$ are the actual and ideal sugar concentrations in each plant part, $\%STAR$ is the actual starch concentration of the plant part, and W_1 and W_2 are constants. Values of 4.0 and 10.0 for W_1 and W_2 were used to reduce starch synthesis to 0.05 of its maximum rate when the starch concentration reached 20%.

The ideal sugar concentration of the stem base complex is allowed to vary with the season (Figure 2c) while those of the other plant parts remain constant. Accordingly, in the spring the equilibrium tends toward sugar formation in the stem complex. Sugar moves into the leaf prior to growth. In the fall, sugar released from the senescing leaf will accumulate as polysaccharide in the stem complex. This tissue becomes a reservoir through the winter for stored organic nutrients which are utilized for spring growth.

Translocation between Plant Parts

Translocation of materials occurs between roots and rhizomes, stem base, sheaths, laminae, and reproductive structures. Sugars, amino acids, amides, phosphates, and fatty acids are translocated (Figure 3).

Translocation of sugars, amino acids, amides, inorganic phosphates, and fatty acids between the plant parts is calculated as a transfer along a concentration gradient, which is donor controlled and follows a Q_{10} relationship (Mason and Maskell, 1928). Translocation ceases below $-2°C$. Thus, if $\%\Psi_I > \%\Psi_J$

$$T_{I,J} = (\%\Psi_I - \%\Psi_J)\epsilon_{I,J}\Psi_I Q_{10}^{0.1(TEMP)},$$

If $\%\Psi_I < \%\Psi_J$

$$T_{J,I} = (\%\Psi_J - \%\Psi_I)\epsilon_{J,I}\Psi_J Q_{10}^{0.1(\text{TEMP})}$$

where $T_{I,J}$ is the translocation from the Ith to the Jth plant part, $\%\Psi_I$ is the percentage concentration in the Ith plant part, Ψ_I is the standing crop of organic matter (g m^{-2}), and $\epsilon_{I,J}$ is the transfer coefficient from the Ith to the Jth plant part (Table 2). If the concentration in a plant part drops below a minimal level, translocation out of that part ceases. Thus, each plant part maintains a certain minimum concentration against a gradient.

Phosphate is translocated upward from the root primarily as inorganic phosphate (Crossett and Loughman, 1966; Morrison, 1965; Selvehdian, 1970) in the xylem (Stout and Hoagland, 1939; Loughman, 1968; Bieleski, 1973). Rate of translocation is assumed to be independent of transpiration rate, to depend upon root temperature (Bieleski, 1973), and to have a Q_{10} only slightly greater than 1.0. Redistribution of inorganic phosphate to other plant parts occurs via the phloem (Bieleski, 1973) and is assumed to have a higher Q_{10} than does xylem translocation. The transfer coefficient (E_{IJ}) is a sigmoidally increasing function of the inorganic phosphate content of the donor plant part. The shape of this function (Figure 4) is suggested by studies of translocation in phosphate-starved and phosphate-supplied plants (Russell and Martin, 1953; Bieleski, 1973); the values used are taken from inorganic phosphate contents of phosphate-starved and phosphate-supplied plants (Bieleski, 1973).

The principal source of nitrogen in the humus soils at Barrow (Gersper, 1972) and generally in soils with high humus content and low pH (Kirby, 1967) is ammonium. It is assumed that absorbed ammonium is converted immediately in the roots to amino acids and amides which are then available for the synthesis of

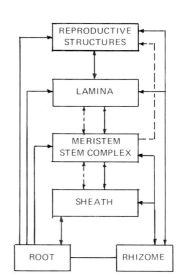

Figure 3. Block and arrow diagram of translocation paths between plant parts. —— indicates a translocation in xylem or phloem, --- indicates a transfer of materials by differentiation.

Table 2.   Transfer Coefficients for Translocation between Plant Parts

| Donor | Monocot plant part—recipient | | | | |
	(1) Root	(2) Rhizome	(3) Stem	(4) Sheath	(5) Lamina
(1) Root	0	1	1	0	0
(2) Rhizome	1	0	1	1	0
(3) Stem	1	2	0	2	0
(4) Sheath	0	1	1	0	2
(5) Leaf	0	0	0	2	0

protein in the roots or for translocation to other plant parts (Thomas, 1927; Bollard, 1956, 1957). The sugars required for the synthesis of amino acids are supplied from photosynthesizing tissues or from storage tissues early in the season. Since the growth of photosynthetic tissues requires the synthesis of amino acids in the roots and the synthesis depends on the amount of sugar translocated, a feedback between the amounts of photosynthesizing and root tissue occurs.

Another feedback control on growth is established because some nitrogen is translocated as amide and the plant tissues must synthesize 50% of their amino acids from amides. Utilization of amide nitrogen will not occur unless an adequate supply of sugars is available to prevent the effects of ammonium toxicity.

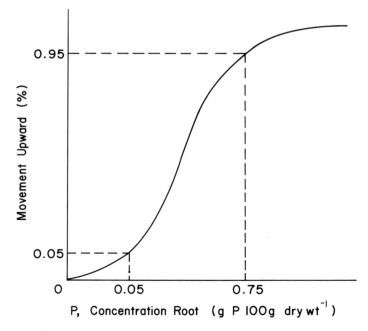

Figure 4.   Fraction of root phosphate translocated upwards.

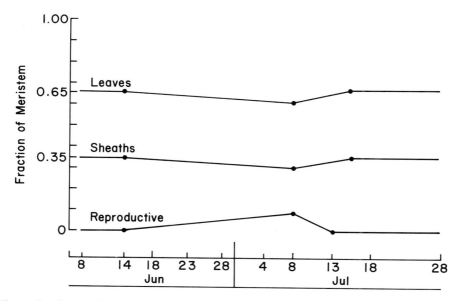

Figure 5.   Seasonal course of the fractions of meristem differentiating into reproductive structures, laminae, and sheaths.

The meristematic portion of the stem base differentiates into sheaths, lamina, and reproductive structures. In this process a full complement of all stem base chemical compartments is transferred to sheaths, laminae, and reproductive structures. The fraction of the stem base which differentiates is 0.0005 until day 170, increases linearly to 0.0050 at day 180, then decreases linearly to 0.000 by day 185. This differentiated material is converted to laminae, sheath, or reproductive material. Of the differentiated material, about 0.63 goes to laminae, 0.33 goes to sheaths, and 0.02 goes to reproductive tissue, but the fractions vary through the year (Figure 5).

Photosynthesis

The model incorporates light and dark respiration in net photosynthesis. Net photosynthesis is calculated from a light curve determined for young expanded leaves of *D. fisheri* (Figure 6a) by

$$P_n = \frac{B_1 Q}{B_2 + Q} \qquad (10)$$

where Q is the solar radiation in the canopy, B_1 is the maximum photosynthesis rate of the leaves at optimum temperatures and light, and B_2 is the solar radiation at $\frac{1}{2} B_1$. Q is calculated by

$$Q = Q_0 \int_0^{LAI} e^{K(M-1)F} \, dF. \qquad (11)$$

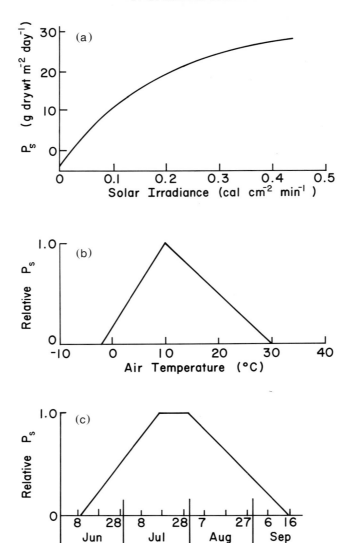

Figure 6.   (a) Net photosynthesis in relation to mean daily intensity. (b) Modification of daily net photosynthesis by air temperature. (c) Modification of daily net photosynthesis by time of year or stage of development.

The use of data derived from this specific class of expanded leaves to represent all leaves in the canopy overestimates CO_2 uptake by some unknown amount because the young leaves are more efficient. However, the overestimation is assumed to be small since older, senescent leaves occupy the more shaded parts of the canopy and contribute less to the total photosynthetic output of the canopy.

The effects of temperature are incorporated by using a linear approximation to the actual measured reduction in photosynthesis (Figure 6b). The effect of temperature is assumed to be equally exerted at all light intensities. Temperatures in the field at Barrow are almost always below the optimum temperatures for photosynthesis in *Dupontia fisheri* (Miller and Tieszen, 1972; Weller and Holmgren, 1974; Tieszen, 1975). The use of mean daily temperatures will also overestimate photosynthesis since CO_2 uptake is reduced at both the high and low ends of the temperature range. If the model is modified to use hourly temperature data, or to incorporate water relations, a function relating CO_2 uptake to these parameters will need to be included. Presently, the use of mean temperature data on a daily basis results in temperature conditions where observed phenomena like midday depressions in photosynthesis are not attained.

Seasonal changes in CO_2 uptake are included to account for the development of photosynthetic competence and the effects of leaf senescence (Figure 6c). Preliminary data (Tieszen, 1975) suggested that photosynthetic efficiency started at a high level and decreased after 1 August, approaching zero near 15 September. More recent data (Tieszen, 1978) indicate that the development of photosynthetic competence was not complete until near 15 July and varied as a function of leaf position.

Net photosynthesis for stems, leaves, and reproductive structures is computed by

$$P_{NI} = \{P_N f(TEMP) f(YEAR)\} LAI_l A_p \tag{12}$$

where LAI_l is the area of each plant part, and A_p is the relative photosynthetic ability of that part as compared to leaves. At present, stems and reproductive structures are assumed to have a 60% efficiency (Tieszen and Johnson, 1975). Net photosynthesis is not affected by phosphorus or nitrogen concentrations in the range which occurs in the field.

Respiration

Maintenance respiration in photosynthetic parts is assumed to be included in net photosynthesis. Maintenance respiration of belowground parts was calculated from Trent (1972). Respiration begins at $-2°C$ and proceeds as a Q_{10} reaction with Q_{10} equal to 2. At $0°C$, belowground parts respire at a rate of 0.02 mg of sugar $g^{-1} hr^{-1}$. At present no data are available on the relative rates of respiration in roots, rhizomes, and stem complexes, nor is any distinction made between new or old roots. Thus, for all these parts

$$R = R_0 Q_{10}^{0.1(TEMP - T_0)} \tag{13}$$

with the parameters defined as above.

Growth respiration is accounted for by including inefficiencies in the biochemical transfers. Thus, 1.62 g of sugar are required to make 1 g of amino acid when the nitrogen source is NH_3, 1.17 g of sugar are required to make 1 g of cellulose, and 3.0 g of sugar are required to make 1 g of fatty acid or lipid.

Mineral Uptake

The uptake of nitrogen and phosphorus is controlled by soil temperatures, their concentrations in the soil, the content of amino acids and phosphorus in the absorbing roots, and the efficiency of absorption as expressed by a Michaelis–Menton relationship (Becking, 1956; Fried et al., 1965; Carter and Lathwell, 1967; Chapin, 1978).

Specifically, for nitrogen uptake the equation is

$$\text{Uptake} = \frac{V_{max} N_s}{K_m + N_s} Q_{10}^{0.1(\text{TEMP}-T_0)}(1 - (1/\%\text{Ideal AA})\%\text{AA}) \text{ Roots} \quad (14)$$

where V_{max} is the maximum velocity of uptake, K_m is the NH_4 concentration at one-half V_{max}, N_s is the soil ammonium concentration, %Ideal AA and %AA are the ideal and actual concentrations of amino acids in the roots, and Roots is the biomass of absorbing roots. Ammonium uptake is reduced if the sugar concentration and the synthesis of amino acids and amides in the roots are low in order to prevent the accumulation of ammonium and toxicity. Exudation of ammonium is not allowed so that if uptake is less than zero, it is set to zero.

For phosphorus uptake the equation is

$$\text{Uptake} = \frac{V_{max} P_s}{K_m + P_s} \frac{Q_{10}^{0.1(\text{TEMP}-T_0)}}{AP_{10}^{0.1(\text{TEMP}-T_0)}} (1 - (1/\%\text{Ideal P})\%\text{P}) \text{ Roots} \quad (15)$$

where P_s is the phosphorus concentration in the soil and %Ideal P and %P are the ideal and actual phosphorus concentrations of the roots. As with nitrogen, no exudation is allowed.

For *Dupontia fisheri*, V_{max} equals 1.5×10^{-4} g of phosphate g dry wt^{-1} hr^{-1} at 5°C and 4.6×10^{-4} g of phosphate g dry wt^{-1} hr^{-1} at 20°C; K_m equals 4.2 (TEMP) + 78, a linear function of temperature; Q_{10} is 2.0; and AP_{10} is 1.0. The variable AP_{10} allows for an acclimation potential. The present data indicate that *Dupontia* does not acclimate, although *Carex* and *Eriophorum* do.

Results and Discussion

The value of this model is its structure more than its output, since the limited data base available when the assumptions were made make the output suggestive but not final. The model structure constitutes a consistent set of hypotheses regarding the essential variables constituting the plant, the form of the relation between these variables, and the quantitative values associated with these relations. Therefore, the structure and output can provide a focal point for discussion on fundamental physiological processes.

Simulations of the seasonal growth of the plant parts and the biochemical compartments were run using soil phosphate and soil nitrogen at concentrations

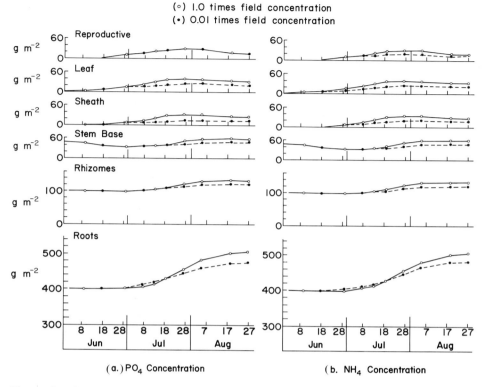

(a.) PO₄ Concentration   (b. NH₄ Concentration)

Figure 7. Seasonal course of biomass (g m⁻²) of different plant parts at two soil ammonium and phosphate concentrations. 1 × (○) and 0.01 × (●) the concentrations measured by Gersper (1972).

measured by Gersper (1972) in 1970, 0.1 times these concentrations and 0.01 times these concentrations (Figures 7 through 13). No change in the seasonal courses occurred between measured concentration and 0.1 times the concentration. In the simulation, deficiencies of nitrogen and phosphorus have about the same magnitude of effect on biomass production. Low nitrogen and low phosphorus each increased lipid concentration in leaves and sheaths at the end of the season. In addition low nitrogen and low phosphorus decreased the concentrations of sugars and storage polymers in the stem base and rhizomes. Low phosphorus had a greater effect than low nitrogen on the sugar and storage polymer concentrations.

Leaf and sheath biomass decreases after mid-season because of decreases in the sugars and storage polymers. At the same time the stem base biomass increased as the sugars and storage polymers were translocated into the stem base.

Several changes in model structures are suggested by data made available since this model was developed. The maximum net photosynthesis is too high.

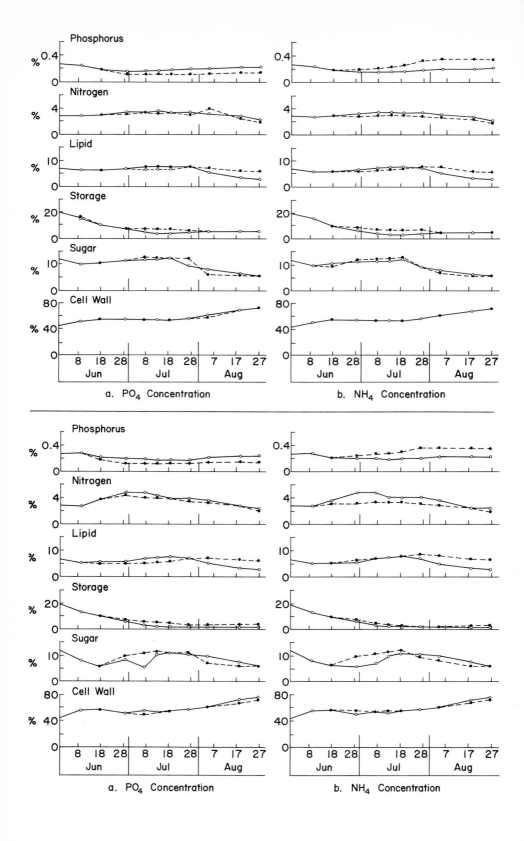

a. PO₄ Concentration

b. NH₄ Concentration

a. PO₄ Concentration

b. NH₄ Concentration

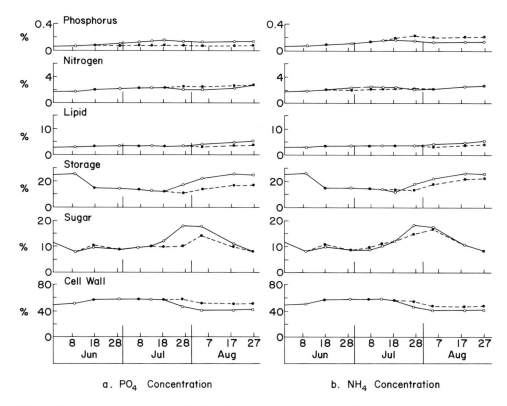

a. PO$_4$ Concentration   b. NH$_4$ Concentration

Figure 10. Seasonal course of the biochemical composition of the stem base complex at two soil ammonium and phosphate concentrations. 1 × (○) and 0.01 × (●) the concentrations measured by Gersper (1972).

The temperature for maximum photosynthesis is 10 to 15°C; but photosynthesis decreases with temperature less abruptly so that at 0°C, photosynthesis is 40% of the maximum rate. Internal limitations on polysaccharide synthesis may not exist, since bean plants can become turgid with starch grains if maintained under continuous light. Protein degradation may occur at 0.1 g g^{-1} day^{-1}, although the range of breakdown rates varies widely (Penning de Vries et al., 1974). Soil solution nitrogen and phosphorus concentrations were later found to be lower than used in these simulations, although the processes of the movement of nitrogen (or NH$^+_4$) and phosphorus from the bulk soil to the roots need clarification. Also, the fundamental pathways of conversion, the biochemical efficiencies

◀──

Figure 8. (Top) Seasonal course of the biochemical composition of the lamina at two soil ammonium and phosphate concentrations, 1 × (○) and 0.01 × (●) the concentrations measured by Gersper (1972).
Figure 9. (Bottom) Seasonal course of the biochemical composition of the leaf sheaths at two soil ammonium and phosphate concentrations, 1 × (○) and 0.01 × (●) the concentrations measured by Gersper (1972).

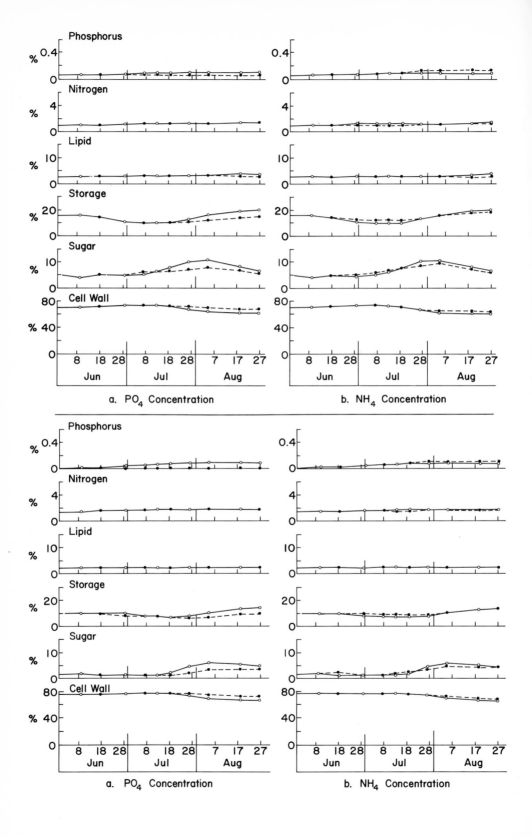

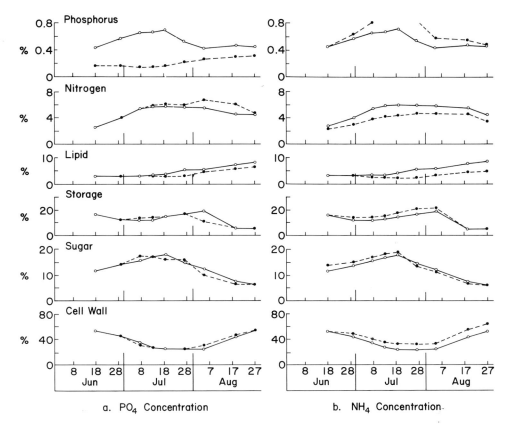

a. PO$_4$ Concentration   b.   NH$_4$ Concentration.

Figure 13.   Seasonal course of the biochemical composition of the reproductive structures at two soil ammonium and phosphate concentrations, $1 \times$ (○) and $0.01 \times$ (●) the
concentrations measured by Gersper (1972).

of the steps considered, and the effects of temperatures and water stress on the
rates still need to be clarified.

Acknowledgments. This research was supported by the National Science Foundation under the joint sponsorship of the International Biological Programme and the Office
of Polar Programs and was coordinated by the U.S. Tundra Biome. Field and laboratory
activities at Barrow were supported by the Naval Arctic Research Laboratory of the
Office of Naval Research.

Figure 11.   Seasonal course of the biochemical composition of the rhizomes at two soil
ammonium and phosphate concentrations, $1\times$ (○) and $0.01\times$ (●) the concentrations
measured by Gersper (1972).
Figure 12.   Seasonal course of the biochemical composition of the roots at two soil
ammonium and phosphate concentrations, $1\times$ (○) and $0.01\times$ (●) the concentrations
measured by Gersper (1972).

References

Becking, J. H. (1956) On the mechanism of ammonium uptake by maize roots. *Acta Bot. Neerl.*, **5**: 1–79.

Bianchetti, R., and M. L. Sartirana. (1967) The mechanism of the repression by inorganic phosphate of phytase synthesis in the germinating wheat embryo. *Biochim. Biophys. Acta*, **145**: 484–490.

Bieleski, R. L. (1973) Phosphate pools, phosphate transport and phosphate availability. *Ann. Rev. Plant Physiol.*, **24**: 225–252.

Bollard, E. G. (1956) Nitrogeneous compounds in plant xylem sap. *Nature*, **178**: 1189–1190.

Bollard, E. G. (1957) Composition of the nitrogen fraction of Apple Tracheal sap, *Aust. J. Biol. Sci.*, **10**: 279–287.

Carter, O. G., and D. J. Lathwell. (1967) Effects of temperature on orthophosphate absorption by excised corn roots. *Plant Physiol.*, **42**: 1407–1412.

Chapin, F. S. III (1978) Phosphate uptake and nutrient utilization by Barrow tundra vegetation. In *Vegetation and Production Ecology of an Alaskan Arctic Tundra* (L. L. Tieszen, Ed.). New York: Springer-Verlag, Chap. 21.

Crossett, R. N., and B. C. Loughman. (1966) The absorption and translocation of phosphorus by seedlings of *Hordeum vulgare* (L.). *New Phytol.*, **65**: 459–468.

Davidson, J. L., and J. R. Phillip. (1958) Light and pasture growth. In *Climatology and Microclimatology*. Paris: Unesco, pp. 181–187.

Dennis, J. G., and P. L. Johnson. (1970) Shoot and rhizome-root standing crops of tundra vegetation at Barrow, Alaska. *Arct. Alp. Res.*, **2**: 253–266.

Dennis, J. G., L. L. Tieszen, and M. Vetter. (1978) Seasonal dynamics of above- and belowground production of vascular plants at Barrow, Alaska. In *Vegetation and Production Ecology of an Alaskan Arctic Tundra* (L. L. Tieszen, Ed.). New York: Springer-Verlag, Chap. 4.

Duncan, W. C., R. S. Loomis, W. A. Williams, and R. Hanau. (1967) A model for simulating photosynthesis in plant communities. *Hilgardia*, **38**: 181–205.

Fried, M., F. Zsoldos, P. B. Vose, and I. L. Shatokin. (1965) Characterizing the NO_3 and NH_4 uptake process of rice roots by use of ^{15}N labeled NH_4NO_3. *Physiol. Plant*, **18**: 313–320.

Gersper, P. L. (1972) Chemical and physical soil properties and their seasonal dynamics at the Barrow intensive site. In *Proceedings of 1972 Tundra Biome Symposium, Lake Wilderness Center, University of Washington* (S. Bowen, Ed.). U.S. Tundra Biome, U.S. International Biological Program, and U.S. Arctic Research Program, pp. 87–93.

Horowitz, L. (1958) Some simplified mathematical treatments of translocation in plants. *Plant Physiol.*, **33**: 81–93.

Kirkby, E. A. (1967) A note on the utilization of nitrate, urea, and ammonium nitrogen by *Chenopodium album. Z. Pflanzenernahr. Bodenkd*, **117**: 204–209.

Lemon, E., D. W. Stewart, and R. W. Shawcroft. (1971) The sun's work in a cornfield. *Science*, **174**: 371–378.

Loughman, B. C. (1968) The selective effect of organic metabolites and metabolic inhibitors on the processes of absorption of phosphate by whole plants. *Argo. Chim.*, **12**: 130–139.

McCown, B. H. (1978) The interactions of organic nutrients, soil nitrogen, and soil temperature and plant growth and survival in the arctic environment. In *Vegetation and Production Ecology of an Alaskan Arctic Tundra* (L. L. Tieszen, Ed.). New York: Springer-Verlag, Chap. 19.

McKendrick, J. D., V. J. Ott, and G. A. Mitchell. (1978) Effects of nitrogen and phosphorus fertilization on carbohydrate and nutrient levels in *Dupontia fisheri* and *Arctagrostis latifolia*. In *Vegetation and Production Ecology of an Alaskan Arctic Tundra* (L. L. Tieszen, Ed.). New York: Springer-Verlag, Chap. 22.

Mason, T. G., and E. J. Maskell. (1928) Studies on the transport of carbohydrates in cotton plants. II. The factors determining the rate and direction of the movement of sugars. *Ann. Bot.*, **42**: 571–636.

Miller, P. C., and L. L. Tieszen. (1972) A preliminary model of processes affecting primary production in the arctic tundra. *Arct. Alp. Res.*, **4**: 1–18.

Monsi, M., and T. Saeki. (1953) Über den Lichtfaktor in den Pflanzengesellschaften und seine Bedeutung für die Stoffproduction. *Japan. J. Bot.*, **14**: 22–52.

Morrison, T. M. (1965) Xylem sap composition in woody plants. *Nature*, **205**: 1027.

Murphy, C. E., and K. R. Knoerr. (1972) Modeling the energy balance processes of natural ecosystems. *U.S. IBP Analysis of Ecosystems, Eastern Deciduous Forest Biome Subproject Research Report EDFB-IBP 72-10, Oak Ridge National Laboratory*, 141 pp.

Paltridge, G. W. (1970) A model of a growing pasture. *Agric. Meteorol.*, **7**: 93–130.

Penning de·Vries, F. W. T. (1974) Substrate utilization and respiration in relation to growth and maintenance in higher plants. *Neth. J. Agric. Sci.*, **22**: 40–44.

Penning de Vries, F. W. T., A. H. M. Brunsting, and H. H. van Laar. (1974) Products, requirements and efficiency of biosynthesis: A quantitative approach. *J. Theoret. Biol.*, **45**: 339–377.

Pitelka, F. A. (1958a) Some characteristics of microtine cycles in the arctic. *Arctic Biol.*, **18**: 73–78.

Pitelka, F. A. (1958b) Some aspects of population structure in the short-term cycle of the brown lemming in northern Alaska. In *Cold Spring Harbor Symp. Quant. Biol.* The Biological Laboratory, Cold Spring Harbor, New York, pp. 237–251.

Pitelka, F. A., P. Q. Tomich, and G. W. Treichel. (1955) Ecological relations of jaegers and owls as lemming predators near Barrow, Alaska. *Ecol. Monogr.*, **25**: 85–117.

Russell, R. S., and R. P. Martin. (1953) A study of the absorption and utilization of phosphate by young barley plants. I. The effect of external concentration on the distribution of absorbed phosphate between roots and shoots. *J. Exp. Bot.*, **4**: 108–127.

Sartirana, M. L., and R. Bianchetti. (1967) Effects of phosphate on development of phytase in wheat embryos. *Physiol. Plant.* **20**: 1066–1075.

Schultz, A. M. (1964) The nutrient recovery hypothesis for arctic microtine cycles. II. Ecosystem variables in relation to arctic microtine cycles. In *Grazing in Terrestrial and Marine Environments* (D. J. Crisp, Ed.). Oxford: Blackwell Scientific Publications, pp. 57–68.

Schultz, A. M. (1969) A study of an ecosystem: The arctic tundra. In *The Ecosystem Concept in Natural Resource Management* (G. M. Van Dyne, Ed.). New York: Academic Press, pp. 77–93.

Selvendian, R. R. (1970) Changes in the composition of the xylem exudata of tea plants (*Camellia sinensis* L.) during recovery from pruning. *Ann. Bot.*, **34**: 825–833.

Stout, P. R., and D. R. Hoagland. (1939) Upward and lateral movement of salt in certain plants as indicated by radioactive isotopes of potassium, sodium and phosphorus absorbed by roots. *Amer. J. Bot.*, **26**: 320–324.

Thomas, W. (1927) The seat of formation of amino acids in *Bryus malus* L. *Science*, **66**: 115–116.

Thornley, J. H. M. (1972) A balanced quantitative model for root:shoot ratios in vegetative plants. *Ann. Bot.*, **36**: 431–441.

Tieszen, L. L. (1972) The seasonal course of aboveground production and chlorophyll distribution in a wet arctic tundra at Barrow, Alaska. *Arct. Alp. Res.*, **4**: 307–324.

Tieszen, L. L. (1975) CO_2 exchange in the Alaskan arctic tundra: Seasonal changes in the rate of photosynthesis of four species. *Photosynthetica*, **9**: 376–390.

Tieszen, L. L. (1978) Photosynthesis in the principal Barrow, Alaska, species: A summary of field and laboratory responses. In *Vegetation and Production Ecology of an Alaskan Arctic Tundra* (L. L. Tieszen, Ed.). New York: Springer-Verlag, Chap. 10.

Tieszen, L. L., and D. A. Johnson. (1975) Seasonal patterns of photosynthesis in individual grass leaves and other plant parts in Arctic Alaska with a portable $^{14}CO_2$ system. *Bot. Gaz.*, **136**: 99–105.

Trent, A. W. (1972) Measurement of root growth and respiration in arctic plants. M.A. thesis, Duke University, 75 pp.

Weller, G., and B. Holmgren. (1974) The microclimate of the arctic tundra. *J. Appl. Meteorol.*, **13**: 854–862.

de Wit, C. T. (1965) Photosynthesis of leaf canopies. Institute of Biol. Chem. Res. on Field Crops and Herbage, Wageningen. *Agric. Res. Rep.* 663.

de Wit, C. T., R. Brouwer, and F. W. T. Penning de Vries. (1970) The simulation of photosynthetic systems. In *Prediction and Measurement of Photosynthetic Productivity: Proceedings of the IBP/PP Technical Meeting, Trebon, 14–21 September 1969.* Wageningen: Center for Agricultural Publishing and Documentation, pp. 47–70.

26. A Simulation Model of Population Processes of Arctic Tundra Graminoids

B. A. Lawrence, M. C. Lewis, and P. C. Miller

Introduction

Wet meadow tundra predominates along the Arctic Coastal Plain of Alaska and has been studied intensively near Barrow, Alaska. Of the three species of monocotyledons that account for nearly 80% of the vascular plant production (Tieszen, 1972), *Dupontia fisheri,* a grass, has been studied in detail to characterize the basic patterns of photosynthesis, water relations, nutrition, growth, and reproduction found in graminoid species in the tundra (Allessio and Tieszen, 1975a, b; Tieszen et al., 1974; Lewis, unpublished data; McCown, 1975; Tieszen, 1975; Stoner and Miller, 1975; Chapin, 1977). These data assisted the development of a simulation model of population processes in perennial graminoid species that integrates physiological and population processes found in the graminoid plant form in the Arctic. The model forms a framework to deal precisely with broad questions about tundra processes.

Various models have been proposed for plant growth (Thornley, 1972a, b; Grace and Woolhouse, 1973; de Wit et al., 1970), canopy photosynthesis (Verhager et al., 1963; de Wit, 1965; Monteith, 1965; Duncan et al., 1967; Miller, 1972; Miller and Tieszen, 1972), and the production or biomass yield for annual crop plants (Holliday, 1960; Baeumer and de Wit, 1968; Harper, 1960). However, these models do not combine perennial plant population processes with physiological and environmental processes such as reproduction, mortality, and dispersion. The model constructed in this study is unique in that it uses physically connected individuals (tillers) as the simulation unit, in order to assess the effects of different physical and chemical environments, allocation patterns, patterns of reproduction, grazing, population growth, and spatial arrangement. Some of these effects will be assessed in individual case studies in this chapter.

The main objectives of this study were to: (1) quantify the present understanding of processes involved in the growth and maintenance of populations of tundra graminoids growing at Barrow, Alaska; (2) develop a general model of plant population dynamics which incorporates vegetative propagation and reallocation of reserves between members of vegetative systems, both of which have proved problematic when attempting to apply the concepts developed for animal populations to populations of perennial plants; and (3) simulate through yearly time periods the effects of seed reproduction, lemming grazing, early season allocation to aboveground parts, vegetative reproduction, and aspects of the plant and environment on the structure and dynamics of tundra graminoid populations.

The model relies heavily on data collected near Barrow during the 1973 growing season and was developed with the cooperation of researchers who provided estimates of parameter values used. Many of these values were "educated" guesses because no quantitative data were available; other values were based on preliminary experimental results; and some were based on field data from temperate zone crops or from other tundra regions. Thus, one of the implicit purposes of this chapter is to provide a framework for future plant population and community studies of tillering or clonal plants. The model is not meant to represent all the details of the real world, but rather to isolate and provide a mechanism to explore the essential interaction of plant population and environment. The model is a set of assumptions about the behavior of the population-environmental system, and the simulation output and the logical deductions from these assumptions.

The Model

The Plant

The basic plant unit of *Dupontia* is a tiller system comprised of innovating rhizomes (V0) and several tillers from 1 to 5 yr old (V1···V5) which can be connected by common rhizomes (Figure 1; see Allessio and Tieszen, 1978). A tiller consists of leaves (consisting of blades and sheaths), stem base, rhizome, and associated roots. Fifth and sixth year tillers are in the process of senescing and consist only of belowground parts containing stored carbohydrates and nutrients. Leaves grow from the stem base and roots grow from the rhizomes. Rhizomes serve as storage and translocation organs. During the first 3 years of aboveground growth a tiller will grow four or five leaves each growing season (Mattheis et al., 1976). At maturity, in the third or fourth year, the tiller will

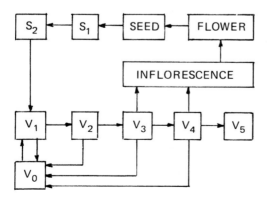

Figure 1. Diagram of state variables and transfers included in the model. Transfer between V0, V1, V2, V3, V4, and V5 inflorescence flowers and seeds are included only within tiller systems. Seedling age classes (S_1 and S_2) are independent of the tiller system and expressed on a m^2 basis.

initiate an inflorescence during the latter part of the growing season. After floral initiation no further aboveground growth occurs on that tiller.

Belowground growth occurs primarily in the rhizome and roots of the V0. V0's grow from any aged tiller but are more common on younger tillers. The parent tiller supplies carbohydrates and nutrients to the rhizome, stem base, and roots of the V0. In the second and third years, tillers supply some carbohydrates and minerals to the rest of the tiller system, while the roots and rhizomes of the 5- and 6-yr-old tillers contribute carbohydrates and minerals as they senesce.

The tiller is the basic population unit in the model and all information is cataloged by tiller. Dry weights of the rhizome, leaf, root, and inflorescence compartments are recorded for each tiller in the simulation. In the rhizome the dry weight is divided into carbohydrate reserves and the remaining organic constituents. The reserve carbohydrates of the rhizome represent the total reserves of the tiller. The status of these reserves determines most tiller activities.

Spatial relationships of the tillers are treated by arranging the tillers on a two-dimensional grid. Each tiller is located by a pair of coordinates. The grid is subdivided into cells to calculate the availability of light, nutrients, and water to each tiller. The processes of growth and reproduction of tillers through the season involve many calculations including production of photosynthate, allocation of the photosynthate to the plant compartments, determination of the requirements of and the allocation to inflorescences and V0's, and the death of individuals. In a natural system these processes are continuous and simultaneous, but in a complex simulation model they are treated discretely in numerical solutions of the model. In the discussion which follows a parameter n will be identified, P_n, and an intermediate algebraic variable will be identified, A_n. The parameters are listed in Table 1.

Photosynthesis and Respiration

The production of photosynthesis by a tiller involves the interception of solar radiation and the conversion of the intercepted radiation into carbohydrate. The irradiance intercepted by a tiller in a grid square is calculated by first calculating the irradiance intercepted by live leaves and standing dead material in the grid square and then multiplying the intercepted irradiance by the fraction of the total foliage area (live plus dead) which is live leaves. Thus,

$$I_t = (A_1/A_t)I_0(1 - e^{-CKA_t})A_g \tag{1}$$

where I_t is the solar irradiance intercepted by the tiller in joules tiller^{-1}, I_0 is the incoming solar irradiance (A3), A_g is the area of the grid square (0.0160 m^2), A_1 is the live leaf area for the individual tiller, A_t is the total foliage area comprised of live leaves and standing dead material for the grid square, K is the extinction coefficient for solar irradiance (P7), and C is a leaf clumping factor.

The clumping factor is included because all leaves on a tiller originate from the same meristematic region. As leaves become more clumped, the canopy

Table 1. Description of Parameters and Values Used in the Standard Case

Parameter	Description	Units	Value	Source[a]
1	Semiannual amplitude of mean daily temperature at the ground surface	°C	14.0	1
2	Mean daily temperature at ground surface over a year	°C	−12.0	1
4	Semiannual amplitude of mean daily solar radiation	$cal \cdot cm^{-2} \cdot day^{-1}$	265.0	1
5	Mean daily solar radiation received over a year	$cal \cdot cm^{-2} \cdot day^{-1}$	265.0	1
7	Extinction coefficient of leaf canopy at Barrow	$m^2 \cdot m^{-2}$	0.8	2
8	Maximum relative growth rate of inflorescence	$g \cdot g^{-1} \cdot day^{-1}$	0.05	3
9	Maximum inflorescence biomass for plants under normal field conditions	g	0.04	4
10	Maximum relative growth rate of V0 tiller	$g \cdot g^{-1} \, day^{-1}$	0.2	4
11	Maximum biomass of V0 tillers for plants under normal field conditions	g	0.04	4
12	Maximum fraction of parent rhizome carbohydrate reserves which are usable by V0		0.75	5
13	Number of days in a simulation time step	days	10.0	
14	Minimum size of reserves required for floral initiation	g	0.01	5
15	Minimum weight of inflorescence required for flower opening	g	0.020	3
16	Minimum daily mean temperature required for flower opening	°C	3.0	5
17	Minimum daily mean temperature required for pollination	°C	3.0	5
18	Pollination rate of inflorescences	fraction pollinated $\cdot$ day^{-1}	0.10	5
19	Mean weight of mature seed	g	0.0008	3
20	Fractional germinability of mature seeds		0.0	
21	Maximum distance from parent plant to which a seed may be dispersed	m	1.0	5
22	Maximum age attained by any tiller	yr	5.0	3

26	Specific leaf area at end of season just prior to leaf death	cm²·g⁻¹	200.0	5
27	Fraction of reserves in V0 rhizome		0.30	4
28	Fraction of leaf weight retrieved during senescence		0.40	5
30	Earliest day at which seeds may be dispersed	days	230.0	
31	Minimum daily mean temperature required for seed dispersal	°C	1.0	5
32	Maximum gross photosynthetic efficiency under nonlimiting conditions	g·cal⁻¹	0.026	11
33	Minimum nutrient availability in arbitrary units required for P32 (based on nonlimiting soil nutrient of 1 arbitrary unit and 500 plants per m²)	Arbitrary	0.2	
35	Calorific value of *Dupontia* plant tissue	cal·g⁻¹	4500.0	7
36	Minimum temperature required for P32. (This value should have been 15°C) Tieszen, 1978	°C	10.0	
37	Number of years of a simulation	yr	20.0	
38	Maximum inflorescence weight attainable before pollination	g	0.018	3
39	Minimum water availability required for P32 (based on limiting soil water potential of −5 bars and a mean root biomass of 0.003 g)	bars g·root⁻¹	−1666.0	
41	Fraction reserves in rhizome at initialization		0.4	8
42	Length to weight ratio of rhizome at initiation	m·g⁻¹	3.0	3
43	Slope of rhizome length to weight ratio (with nutrient level on the abscissa)	m·g⁻¹·arbitrary⁻¹	−0.05	3
44	Length to weight ratio for rhizome at 0 nutrient level	m·g⁻¹	5.0	5
45	Minimum rhizome length to weight ratio	m·g⁻¹	2.0	5
46	Minimum length of rhizome required for negative geotropism	m	0.03	3
47	Maximum azimuth angle of rhizome relative to parent	degrees	30.0	3
48	Fraction of rhizome reserves allocated to leaves at leaf initiation		0.6	5

Table 1.   Description of Parameters and Values Used in the Standard Case (*Continued*)

Parameter	Description	Units	Value	Source [a]
49	Mean leaf area index of standing dead at initialization	$m^2 \cdot m^{-2}$	1.0	9
51	Day number at snowmelt	days	160.0	6
52	Day number at freeze-up	days	270.0	6
53	Minimum temperature for photosynthesis	°C	−4.0	7
54	Water availability for $E_{water} = 0$ (based on limiting soil water potential of −10 bars and a mean root biomass of 0.003 g)	bars g·root^{-1}	−3333.0	
55	Fraction of rhizome which is incorporated into reserves for V1 tillers and seedlings		0.30	5
56	Fraction of rhizome allocation which is incorporated into reserves for V2, V3, and V4 tillers		0.9	5
58	Minimum level of reserves as fraction of rhizome biomass required for new V0 initiation		0.25	5
59–64	Minimum level of reserves as a fraction of rhizome biomass required for survival of V0–V5 tillers		0.05	5
68	Nutrient levels	Arbitrary	1.0	
73	Light to dry weight conversion factor, with no respiration	g cal^{-1}	0.04	10
74	Constant in sine-wave distortion used in calculating solar radiation		0.485	1
75	Photosynthetic efficiency for conversion of solar radiation to primary photosynthetic products, not adjusted for maintenance respiration		0.027	10

82	Maximum percentage modification of normal allocation per time step (P13)	%	20.0	5
85	Minimum amount of photosynthate and reserves expressed as a fraction of total biomass required for survival of any plant during growing season		0.001	5
86	Maximum weight of average aboveground green material in a 100-cm^2 grid (W) for which $C = 1$ (no clumping effect)	g	0.02	5
87	Value of W for which $C = 0.5$	g	0.1	5
90–93	Probability of any plant being grazed in the 4 yr of a lemming cycle		0.0, 0.036, 0.18, 0.54	4
95	Minimum level of reserves as a function of rhizome biomass which is required for V0 initiation when the plant has been grazed		0.015	5

[a]
1 Timin et al. (1973)
2 Miller and Tieszen (1972)
3 Lewis unpublished data
4 Chapin et al. (1975)
5 guess based on best estimates from available field data
6 1973 growing season field observation (M. C. Lewis, pers. obs.)
7 Tieszen et al. (1974)
8 McCown 1975
9 Dennis and Tieszen (1972)
10 Warren Wilson (1967)
11 Tieszen et al. (1975)

becomes less efficient at intercepting radiation (Acock et al., 1970). Clumping of leaves is assumed to be proportional to the mean weight of leaves on a tiller and assumed to vary from a random ($C = 1$) to a clumped ($C < 1$) arrangement. Thus,

$$C = e^{-c_1 \overline{W}} \tag{2}$$

where c_1 is a constant equal to ln $2/(P87 - P86)$ and $\overline{W}$ is the mean leaf weight per tiller on the grid square.

Net photosynthesis, in gram dry weight per tiller, is calculated by converting the solar irradiance intercepted per tiller first to energy stored in the tiller and then to organic matter produced minus the maintenance respiration. The organic matter produced (P_g) is given by

$$P_g = (E_m I_t)/C_2 \tag{3}$$

where C_2 is the average energy content of plant material in joules $\cdot$ gram dry weight^{-1} and E_m is a minimum efficiency of conversion. The efficiency of conversion is potentially influenced by four factors: irradiance, temperature, soil nutrients, and soil water potential. An efficiency is calculated for each factor assuming that the other factors are not limiting, and the minimum efficiency is used in Eq. (3). The limiting factor approach is similar to one used by Paltridge (1970) and indicated by Blackman (1905). The efficiency for irradiance is assumed constant and maximal (P32). Separate efficiencies for temperature, soil nutrients, and soil water content are calculated by modifying this maximum efficiency, using linear relations with temperature, with available soil nutrients, and with soil water. The efficiency is modified up to $\pm 20\%$ over the range of each factor. In each grid square the availabilities of soil nutrients and water are calculated by partitioning the amount of each substance present in the square among each tiller, according to the proportion of the total root biomass in the square belonging to each tiller. This method of partitioning soil resources is indicated in results of Donald (1961) who showed that one plant in a pot secured all the available nitrogen in the pot, but 50 shared the nitrogen with plants of similar size gaining equal shares.

Maintenance respiration depends on the tiller weight, a maintenance respiration cost, and an influence of light, temperature, soil nutrients, and water, and is calculated from

$$R_m = (E_m/E_0)C_3 W_T \tag{4}$$

where W_T is the tiller weight in grams, C_3 is the fraction of total biomass required to maintain a tiller per day (= 0.015 g g^{-1} day^{-1}) and (E_m/E_0) adjusts for the limiting factors. E_m is the minimum efficiency selected above and E_0 is the maximum or optimal efficiency. Thus, the ratio, E_m/E_0, becomes smaller as temperature, soil nutrients, and soil water become limiting. The reduction of respiration with temperature is well recognized. The reduction of respiration because of soil nutrients follows from a relationship of the maintenance cost to protein content (Penning de Vries, 1972) and protein content to nitrogen availability (Darwinkle, 1975).

Growth and Allocation

The simulated plants begin growth immediately after thaw. The initiation of growth involves the translocation of a percentage of reserves (P48) from the rhizomes into the leaf compartment in the first time step of the model. The percentage of reserves translocated is not known, but rapid shoot elongation and occasional depletion of reserves at the beginning of the season has been reported by Billings and Mooney (1968), McCown (1978), and Chapin et al. (1975).

The photosynthate produced [Eq. (3)] is allocated to leaf, root, rhizome, and inflorescence depending on the time of year and either the level of a limiting factor, in the case of leaf, root, and rhizome, or the size of the sink strength, in the case of the inflorescence. A time-dependent allocation pattern (Fig. 1) for a tiller is summarized by a fourth-degree polynomial, based on data from the 1973 field season (Lewis, unpublished data; Chapin et al., 1975), using, as the single independent variable, days since the start of the growing season. The polynomial calculates the fraction of photosynthate produced in each time step allocated to leaf, root, rhizome, and inflorescence in the same time step (Figure 2). A different polynomial is used for each age class (V1 through V4) and reproductive mode (i.e., vegetative or sexually reproductive). Seedlings are assumed to allocate their photosynthate similarly to a V1.

The calculated fractions of photosynthate allocated to leaves, roots, and rhizomes are then modified so that the effect of the predominant limiting factor is decreased in the next time step. For example, if the tiller is limited by nutrients, a larger proportion of carbohydrates is allocated to roots and less to leaves. The assumption of this modification scheme is supported by Aspinall (1960) who found that in grasses the response to light limitation was an increased shoot:root ratio while the response to decreased nutrients was a decreased shoot:root ratio. The degree of modification depends on (1) the availabilities of the various potential limiting factors, (2) the modification by the most limiting factor, and (3) the maximum modification possible. The modification is linearly related to the limiting factor availability, with a slope defined by the maximum modification possible (P82) and the minimum value of the limiting factor. The maximum modification possible is 20%, which (although chosen independently) is similar to the simulated effects of phosphorus limitation of growth given in Stoner et al. (1978). The fraction calculated by this function is added to the fraction generated by the polynomials.

Thus, the fraction of photosynthate allocated to leaves, roots, and rhizomes is calculated by equations of the form:

$$P_i = (P82)\frac{X_0 - X_A}{X_0 - X_M} + b_0 + b_1 D^1 + b_2 D^2 + b_3 D^3 + b_4 D^4 \qquad (5)$$

where P_i is the fraction allocated to the ith plant part, X_0 is the level of the most limiting factor with optimal conditions, X_A is the actual level of the most limiting

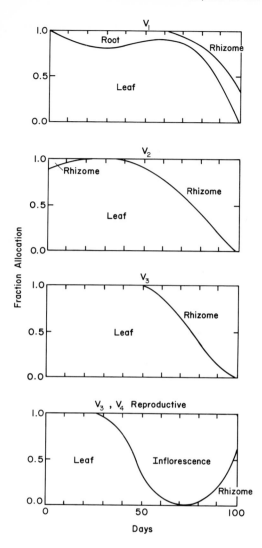

Figure 2.   The allocation of photosynthate to different plant parts of tillers of different ages, where days is the number of days since melt-off.

factor, X_m is the minimal level of the most limiting factor, b_0 is the fraction allocated to the ith plant part on day zero, i.e., the beginning of the growing season, $b_1 \cdots b_4$ are coefficients, and D is the days since the beginning of the growing season.

The growth of an inflorescence is assumed to differ from the growth of leaves, roots, and rhizomes in that the inflorescence has a carbohydrate requirement or sink strength which must be satisfied regardless of the amount of photosynthate produced by the tiller. This assumption is made because of the low variability in inflorescence size observed in tillers varying greatly in size. The sink strength is a method to ensure this low variability. The sink strength can be satisfied by the fraction of photosynthate allocated to the inflorescence in the normal allocation scheme. However, if this amount is not sufficient, the rhizome reserves of the parent tiller are used to complete the carbohydrate requirement. If these reserves are still insufficient, reserves of any directly attached tiller are used.

In the field, not all inflorescences which are initiated complete growth. The assumptions made concerning the abortion of inflorescences are that: (1) the inflorescence will continue growing only if it is the largest inflorescence in the immediate system, (2) the flowers of an inflorescence will open only if temperature conditions are favorable (P16) and the inflorescence is greater than a minimum weight (P15), and (3) pollination of an inflorescence can occur only if temperatures are favorable (P17) but that is randomly determined (P18). If an inflorescence is aborted, the carbohydrate which would have been allocated to it is assigned to rhizome reserves.

For those tillers whose inflorescences are the largest of their system, the sink strength is calculated by:

$$B_p = r_m(1 - B_i/B_{im})B_i \qquad (6)$$

where B_p is the potential growth increment, r_m is the maximum related growth rate of the inflorescence (P8), B_i is the inflorescence biomass, and B_{im} is the mean maximum inflorescence biomass (P6). Photosynthate, allocated to the inflorescence but not used because the inflorescence has attained the maximum allowable size (P38) before being pollinated, is diverted to reserves.

The growth of a V0 is exactly analogous to the growth of an inflorescence; however, the potential growth increment of the V0 can be satisfied only by the reserves of the parent tiller to which it is attached. The length of the growing rhizome is calculated from the product of its weight and the length:weight ratio (A8). The length:weight ratio (A8) decreases to a minimum ratio (P45) as soil nutrient concentration (A5) increases, and is calculated by

$$A8 = (P44) - (P43)(A5). \qquad (7)$$

This function is based on the observation that rhizomes are longer in areas of low phosphorus (Chapin et al., 1975).

Initiation and Dispersion of Plant Parts

A new V0 is initiated if the reserves in the rhizome of a tiller are above a critical level (P58), and fewer than five living V0's exist on the tiller. We assume that a healthy plant will tiller and that the best measure of plant health is the status of the carbohydrate reserves.

Sexual reproduction requires several steps. Any V3 with sufficient carbohydrate reserves (P14) and any V4 that has not yet done so will initiate an inflorescence. The necessary level of carbohydrate reserves is not known, but that inflorescence initiation depends on these reserves seems reasonable. Allsopp (1965, cited in Sachs, 1972) suggested that in some species the carbohydrate level influences inflorescence induction. Lewis (unpublished data) reported field observations which showed that all V4's will initiate an inflorescence. However, only a few of the tillers which have initiated an inflorescence will complete their development and produce seed. Maturity of an inflorescence and seed dispersal are synchronous and determined by temperature conditions (P31) and the time of the season (P30). *Dupontia* is considered to be self-pollinated as are most arctic species (Bliss, 1962). A dispersed seed has a germination probability (P20). If successfully germinated it is then located on the grid as a seedling. At germination the seed weight is divided equally to form the leaf and root biomasses. The seedling must survive two growing seasons to become a V1. These mechanisms are not based on field data, but are believed to be appropriate for the environmental constraints on flowering and seed production.

Dispersal is by the elongation of rhizomes and the dissemination of seeds. The innovating rhizome (V0) has associated with it a direction, randomly selected within 60° of the azimuth of the parent tiller, and a maximum length. After the V0 attains a required length (P_x), it turns up, produces leaves, and becomes a V1. The new V1 has associated with it the grid coordinates giving its location.

Mature seeds are dispersed individually in random directions and to a distance calculated by

$$X = X_m \exp - 5\phi \tag{8}$$

where X is the distance from parent, X_m is the maximum distance from parent, and ϕ is a random variable between zero and 1. An exponential decrease in seed density away from the plant is discussed by Harper (1968).

Death of Plant Parts

Death of a tiller (either V1, V2, V3, V4, V5) may occur at any time during the growing season if the sum of new photosynthate and carbohydrate reserves falls below a critical percentage of tiller plant biomass (P85). At the end of the growing season, a fraction of the leaf compartment (P28) is transferred to the rhizome reserve compartment and all remaining live leaf material is converted to standing dead. A tiller is considered to die during the winter if the rhizome reserves of any tiller fall below a specified critical level, which is expressed as an age-specific fraction of rhizome biomass (P59 to P63), required for survival through the winter and leaf initiation in the spring. All V0 tillers younger than 2 yr survive the winter. Older V0's die.

Grazing

A simulator is used to determine which tillers are grazed. The simulator compares a random variable to the probability of a plant being grazed in the current year (Z). If the random variable is less than Z, the plant will be grazed. Z is calculated from assumptions about the population cycle and metabolic requirements of lemmings. Lemmings were assumed to require 25 kcal day^{-1} individual^{-1} and to have a 4-yr population cycle of 0, 5, 25, and 70 lemmings per hectare

(S. MacLean, personal communication). We assumed 4.5 kcal g^{-1} as the caloric value of aboveground plant tissue (Dennis et al., 1978), a 35% digestive efficiency (Melchior, 1972), a tiller density of 2000 tillers m^{-2}, and an average leaf dry weight of 0.035 g. Thus, the probability (Z) of a tiller being grazed in each of the 4 years of a lemming cycle is 0, 0.036, 0.18, and 0.54.

If the tiller is grazed, leaves and standing dead material associated with the tiller are removed and the apical meristem may be lost. The grazed tillers are assumed to be clipped to the ground level (Thompson, 1955; Bunnell, 1973). Grazing has a more adverse effect on older tillers than on younger tillers, because the apical meristem is closer to the ground surface in the older tillers and is even above the surface in the V4 tillers which have initiated an inflorescence (Mattheis et al., 1976). This increasingly adverse effect was modeled by increasing the probability of removal of the apical meristem by grazing for increasing age classes. It may be, however, that the meristems are generally damaged only during very high population densities of the lemming.

Methods of Simulations

In most simulations initial populations consisted of 75 tillers arranged in several six-tiller systems (two V1, two V2, one V3, and one V4 tillers). The tiller systems were located at random on the grid. The initial spatial arrangement was the same in all experiments in order to eliminate unnecessary variation between experiments. The initial, overwintering biomasses differed for each age class (Table 2). Seed germination was not allowed, although carbohydrate was allocated to the inflorescence to produce seeds, because allocation to seeds is a major carbohydrate sink. The driving variables of solar radiation and temperature were generated by modified sine curves. Soil water potential was constant throughout all experiments. Nutrient availability was a constant arbitrary value (P68).

The model operates with an arbitrary, variable time step of multiples of 1 day, and in these simulations 10-day increments were used. All experiments presented simulated a time period of 20 yr.

Simulations were performed using a standard set of parameter values (Table 1) and varying one or more parameter values at a time. Summaries of the average biomasses of all tiller compartments for each age class were calculated for each year and for the total 20-yr simulation in each simulation experiment. Parameter values from the literature, field experiments, or the best estimate available were used in the standard case simulation which was used to validate the model

Table 2. The Initial Weights of Roots and Rhizomes of the Various Age Classes[a]

Age class	Root (g)	Rhizome (g)
V1	0.002	0.018
V2	0.004	0.027
V3	0.005	0.026
V4	0.001	0.037

[a] Based on Lewis (unpublished data).

against literature and field data. Parameter values were not adjusted or timed to increase the agreement with observed data. Because random variables were included, simulations were replicated and resulting mean values calculated.

Validation

A preliminary validation was obtained by comparing results simulated in the standard case with values observed in the field. In the standard case, simulations were run through a 20-yr period beginning with a tiller density of 470 tillers m^{-2} (Figures 3 and 4).

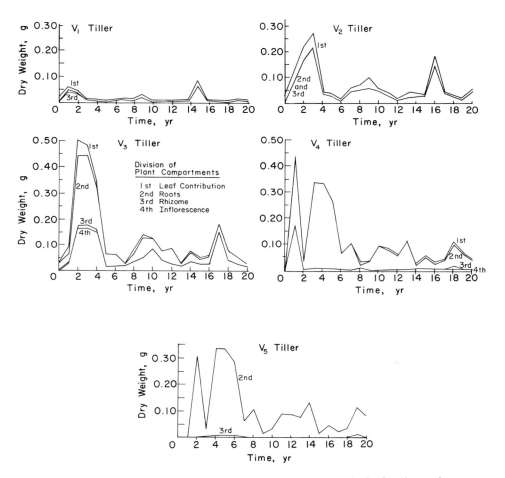

Figure 3.   The weight of plant compartments plotted cumulatively for the various age classes for the 20 yr of the standard case simulation. The division of plant compartments is most clearly seen in the V3 age class where the first line (from the abscissa) represents leaf contribution, the second represents root, the third rhizome, and the fourth inflorescence. The V5 age class has no leaf and the V1, V2, and V5 age classes have no inflorescence.

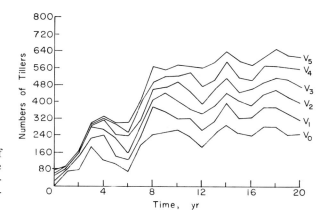

Figure 4. The number of tillers of the various age classes (plotted cumulatively) for 20 yr of the standard case simulation.

The simulated mean biomass of all tillers was similar to observed results, but the simulated and observed biomasses of tillers of each age class differed. Simulated tillers grew too slowly in the younger age classes and continued to grow rather than decline in the older age classes, so that young plants were too small and old plants were too large (Table 3). However, the agreement between the results for the total population and observed data was good on several aspects, especially considering the large number of imprecise estimates of parameters included in the model. Observed values were generally bracketed by the mean and maximum values simulated.

The resulting mean density was 1600 tillers m^{-2} and maximum density was 2500 tillers m^{-2}. Dennis et al. (1978) estimated a tiller density of just under 2000 tillers m^{-2} and Lewis (unpublished data) estimated a density of 2400 to 2800 tillers m^{-2}. A mean aboveground biomass of 47 g m^{-2} and a maximum of 82 g m^{-2} was simulated, compared to measured aboveground biomasses of 102 and 87 g m^{-2} (Tieszen, 1972; Dennis et al., 1978) and 82 g m^{-2} (Johnson, 1969). A mean belowground biomass of 73 g m^{-2} and a maximum of 120 g m^{-2} was simulated, compared with observations ranging substantially higher (Dennis, 1977; Dennis et al., 1978). A mean belowground:aboveground ratio of 1.3 and a maximum of 13 were simulated, compared to 0.9 to 1.6 estimated by McCown (1973). A mean

Table 3. Simulated and Observed Mean Tiller Weights for Different Age Classes and the Total Population

Age class	Mean tiller weight (mg)		Percentage error
	Simulated	Observed[a]	
V0	6	25	−76
V1	20	70	−71
V2	71	97	−26
V3	118	71	+66
V4	120	67	+79
Population	67	66	+ 1

[a] Lewis (unpublished data).

LAI of 0.79 and a maximum of 1.48 simulated compares with values around 1 reported by Caldwell et al. (1974). Leaf area index varies greatly at Barrow (Dennis et al., 1978) but our simulated maximum of 1.48 is in close agreement with the maximum attained at site 2. The model predicted that about 18% of the tillers would be flowering on the average, whereas Dennis (Dennis et al., 1978) found about 5% of all shoots flowering but noted that this increased to 16.5% in the plot fertilized by Schultz. A 14% total nonstructural carbohydrate content was calculated whereas McCown (1978) reports mid-season values for *Dupontia* from nearly 20% to nearly 30%.

Simulation Experiments

Seed Reproduction

In the simulated standard case, seeds were not allowed to germinate or establish, to be consistent with observations on tundra plants in which reproduction is primarily vegetative (Billings and Mooney, 1968). Savile (1972) did not find any *Dupontia* seedlings in what he described as a favorable year. To explore the effect of seed reproduction on total plant production (g m^{-2}), aboveground plant production (g m^{-2}), and density of tillers (number m^{-2}), simulations were run with different probabilities for seed germination and establishment.

Increasing the probability of seed germination and establishment from zero to 2.5, 10, and 50% increased all three population indicators and particularly the total numbers of tillers in the population after 20 simulated years. With 2.5, 10, and 50% seed establishment, the number of tillers increased by 40, 120, and 200%, respectively. Total plant production increased less. With 2.5, 10, and 50% establishment, biomass increased 1, 5, and 15%. Aboveground production increased more than total production, increasing 5, 20, and 28% at the respective establishment percentages. The simulations indicate that after a series of years with favorable conditions for seedling establishment, the number of tillers should increase noticeably. However, the increase in total plant biomass is less and the individual plant size decreases. In the simulations, the greater plant density increased competition for light and thus aboveground growth. This increased biomass in aboveground parts at the expense of belowground reserves may result in greater plant sensitivity to grazing and winter killing, and may also alter the nutrient balance.

Lemming Grazing

Grazing was not included in the standard case. Including the simulated grazing regime, aboveground biomass decreased 33% after 20 yr, which compares with the 26% decrease measured over 4 yr of an exclosure study (Thompson, 1955). The grazing regime caused a 35% decrease in the total number of tillers, but little change in the age structure of the population and in the age specific mortality rates (Table 4). Mortality rates of V2's increased, while mortality rates of the other age classes decreased. Grazing shifted the percentage of total biomass in

Table 4. Percentages of Tillers of Different Ages, Age Specific Mortality Rates (q_x), and Percentage of the Total Biomass in Different Aged Tillers in the Standard and Grazed Simulations[a]

Age class	Percentage of tillers in each age class		Age specific mortality rates		Percentage of total biomass in each age class	
	Standard	Grazed	Standard	Grazed	Standard	Grazed
V0	42.3	42.0	—	—	5	5
V1	15.5	15.0	21.0	19.5	6	5
V2	12.1	12.0	9.9	11.3	16	14
V3	10.8	10.0	8.1	7.4	30	25
V4	9.9	9.0	8.4	6.9	23	27
V5	9.1	9.0	9.0	100.0	20	25

[a] The age specific mortality rate is the number of tillers entering age class X minus the number of tillers entering age class $X + 1$, divided by the number of tillers entering age class X.

different age classes. The biomass of the V0's was unaffected, while the biomass of the V1's, V2's, and V3's decreased and that of the V4's and V5's increased with grazing. Thus, the long-term grazing simulations suggest that the primary effects of grazing are decreased foliage and total number of tillers. The effects on age class structure and age specific mortality are minimal.

Early Season Allocation

A percentage of all carbohydrates stored in the rhizome is allocated to the leaf compartment at the beginning of each growing season, and rapidity of translocation thus has been regarded as important in regions with a short growing season. In the standard case, 35% of the stored carbohydrate was allocated. Simulations were also run using 20 and 50%, both with and without grazing. With a 50% allocation, the total number of plants and the total biomass m^{-2} each decreased 9%, and with a 20% allocation, the total number of plants decreased 9% and the total biomass m^{-2} decreased 30%. However, with grazing and the 50% allocation, the total number of plants decreased 16%, but the total biomass m^{-2} decreased only 1%. With grazing and a 20% allocation, the total number of plants decreased 42% and the total biomass m^{-2} decreased 67%. While allocation percentages greater or less than the standard case and the grazing regime both had a negative effect on the population, the negative effect of grazing was reduced by higher allocation percentages and increased by lower allocation percentages. The simulations indicate that the large early season allocation observed in arctic plants, in addition to the physiological advantage suggested by Billings and Mooney (1968) and Miller and Tieszen (1972), may confer an advantage under the influence of grazing.

Percentage Reserves Required for New V0 and Rhizome Length

Vegetative reproduction is affected by the minimum percentage of stored carbohydrates required to initiate a new V0 (P58). Increasing this percentage by 100%

decreased the number of tillers m^{-2} by 32% and the total biomass m^{-2} by 4%. Decreasing the minimum percentage by 40% increased the number of plants m^{-2} by 4%, but decreased the total biomass m^{-2} by 9%. The model was less sensitive to this parameter than to other parameters.

When the minimum length of the V0 before negative geotropism occurred (P46) was increased to 6 cm, the population failed after 20 yr. The model proved very sensitive to the parameter and more study is warranted to find a good field estimate of P46.

Environmental Conditions and Photosynthetic Efficiency

Several experiments were performed to evaluate potential effects of changes in the arctic environment on the population. A 13% increase in incoming solar irradiance increased the total biomass m^{-2} by 12% but decreased the total number of tillers by 15%. Similarly when the maximum efficiency of conversion of the incoming solar irradiance into photosynthetic products (P32) was increased from 0.026 to 0.035, the total biomass m^{-2} increased 38% and the total number of tillers decreased 23%. Decreasing solar irradiance by 13% decreased the total biomass m^{-2} by 32% and decreased the total number of tillers by 25%. The sensitivity to solar irradiance is in the same direction but about an order of magnitude greater than reported in peak season diurnal simulations by Miller et al. (1976), which is partially due to the mixed sward and standing dead included in their simulation and the lack of a light saturation response assumed here. Increasing temperature by 29% increased the total biomass m^{-2} by 2% and decreased the total number of tillers by 51%. Decreasing temperature by 29% caused the population to fail after 8 years. The effect of increased temperature was in the same direction but less than given by Miller et al. (1976). Increasing nutrients by 50% increased the total biomass m^{-2} by 2% and increased the total number of tillers m^{-2} by 40%. Decreasing nutrients by 50% decreased the total biomass m^{-2} by 33% and decreased the total number of tillers by 55%. Decreases in the environmental variables decreased both the biomass m^{-2} and the number of tillers. Increases in solar irradiance and temperature increased biomass m^{-2} but decreased the number of tillers. Increases in nutrients increased both the biomass m^{-2} and the number of tillers.

Summary

Results of several simulations showed that light was the main limiting factor in the standard run. Large increases in both mean temperature and soil nutrient concentration caused smaller increases in total biomass per unit area than a small increase in radiation. These results are consistent with the general view that light is often a limiting factor in the tundra. The photosynthesis data of Tieszen (1973, 1975) suggest, for example, that the *Dupontia* system is often light limited at Barrow.

The role of modeling is partly to locate counter intuitive conclusions (Forester, 1971). Several such conclusions were derived from the model. The interaction between early season allocation of stored material to aboveground parts and grazing intensity was not expected. The generally inverse relationship between standing crop and density was not expected and probably relates to the limitation by light which appears in the model. These interactions were not explicitly built into the model and were not predictable before simulation runs because of the complexity of the model and the number of processes and parameters included. All the simulations should be validated further by field experiments oriented toward model output, model structure, and parameters before generalizing in detail from the model to the field situation. The model is a hypothesis, but a more complex hypothesis than can usually be derived from single experiments or data sets, and this is a vehicle for making logical deductions from our current understanding of tundra processes. As the model is used and improved, it can provide insight into the complexities of the real situation.

Acknowledgments. This research was supported by the National Science Foundation under Grant GV 29345 to the San Diego State University. It was performed under the joint sponsorship of the International Biological Programme and the Office of Polar Programs and was coordinated by the U.S. Tundra Biome. Field and laboratory activities at Barrow were supported by the Naval Arctic Research Laboratory of the Office of Naval Research.

We thank Dr. Edward Britton for his preliminary discussions and work developing the spatial bookkeeping parts of the model. The help of Drs. Larry Tieszen, F. Stuart Chapin, III, Steve MacLean, and Brent McCown in providing estimates of various model parameters is greatly appreciated.

References

Acock, B. J., H. M. Thornley, and J. Warren Wilson. (1970) Spatial variation of light in the canopy. In *Prediction and Measurement of Photosynthetic Productivity: Proceedings of the IBP/PP Technical Meeting, Trebon, 14–21 September 1969.* Wageningen: Center for Agricultural Publishing and Documentation, pp. 91–102.

Allessio, M. L., and L. L. Tieszen. (1975a) Patterns of carbon allocation in an arctic tundra grass, *Dupontia fischeri* (Gramineae), at Barrow, Alaska. *Amer. J. Bot.,* **62**: 797–807.

Allessio, M. L., and L. L. Tieszen. (1975b) Leaf age effect on translocation and distribution of ^{14}C-photoassimilate in *Dupontia* at Barrow, Alaska. *Arct. Alp. Res.,* **7**: 3–12.

Allessio, M. L., and L. L. Tieszen. (1978) Translocation and allocation of ^{14}C-photoassimilate by *Dupontia fisheri*. In *Vegetation and Production Ecology of an Alaskan Arctic Tundra* (L. L. Tieszen, Ed.). New York: Springer-Verlag, Chap. 17.

Allsopp, A. (1965) The significance for development of water supply, osmotic relations and nutrition. In *Handbuch der Pflanzen.* (W. Ruhland, Ed.). Berlin: Springer-Verlag, Vol. 15, part 2, pp. 504–522.

Aspinall, D. (1960) An analysis of competition between barley and white persicaria. II. Factors determining the course of competition. *Ann. Appl. Biol.,* **48**: 637–654.

Baeumer, K., and C. T. de Wit. (1968) Competitive interference of plant species in monocultures and in mixed stands. *Neth. J. Agric. Sci.,* **16**: 103–122.

Billings, W. D., and H. A. Mooney. (1968) The ecology of arctic and alpine plants. *Biol. Rev.,* **43**: 481–529.

Blackman, F. F. (1905) Optimal and limiting factors. *Ann. Bot.,* **19**: 281–295.

Bliss, L. C. (1962) Adaptations of arctic and alpine plants to environmental conditions. *Arctic,* **15**: 117–144.

Bunnell, F. L. (1973) Theological ecology or models and the real world. *For. Chron.,* **49**: 167–171.

Caldwell, M. M., L. L. Tieszen, and M. Fareed. (1974) The canopy structure of tundra plant communities at Barrow, Alaska, and Niwot Ridge, Colorado. *Arct. Alp. Res.,* **6**: 151–159.

Chapin, F. S. III (1977) Temperature compensation in phosphate absorption occurring over diverse time scales. *Arctic and Alpine Res.* **9**: 139–148.

Chapin, F. S., III, K. Van Cleve, and L. L. Tieszen. (1975) Seasonal nutrient dynamics of tundra vegetation at Barrow, Alaska. *Arct. Alp. Res.,* **7**: 209–226.

Darwinkle, A. (1975) Aspects of assimilation and accumulation of nitrate in some cultivated plants. Wageningen: Centre for Agricultural Publishing and Documentation. 64 pp.

Dennis, J. G. (1977) Distribution patterns of belowground standing crop in Arctic Tundra at Barrow, Alaska. *Arct. Alp. Res.* **9**: 113–127.

Dennis, J. G., and P. L. Johnson. (1970) Shoot and rhizome-root standing crops of tundra vegetation at Barrow, Alaska. *Arct. Alp. Res.,* **2**: 253–266.

Dennis, J. G., L. L. Tieszen, and M. Vetter. (1978) Seasonal dynamics of above- and belowground production of vascular plants at Barrow, Alaska. In *Vegetation and Production Ecology of an Alaskan Arctic Tundra* (L. L. Tieszen, Ed.). New York: Springer-Verlag, Chap. 4.

Donald, C. M. (1961) Competition for light in crops and pastures. In *Society for Experimental Biology Symposium XV*. New York: Academic Press, pp. 282–313.

Duncan, W. C., R. S. Loomis, W. A. Williams, and R. Hanau. (1967) A model for simulating photosynthesis in plant communities. *Hilgardia,* **38**: 181–205.

Forrester, J. W. (1971) *World Dynamics*. Cambridge, Mass.: Wright Allen Press, 142 pp.

Grace, J., and H. W. Woolhouse. (1973) A physiological and mathematical study of the growth and productivity of a *Calluna-Sphagnum* community. III. Distribution of photosynthate in *Calluna vulgaris* (L.) Hull. *J. Appl. Ecol.,* **10**: 77–91.

Harper, J. L. (1960) Factors controlling plant numbers. In *Biology of Weeds* (J. L. Harper, Ed.). Oxford: Blackwell Scientific Publications, pp. 119–132.

Harper, J. L. (1968) The regulation of numbers and mass on plant populations. In *Population Biology and Evolution* (R. C. Lewontin, Ed.). Syracuse, N.Y.: Syracuse University Press, pp. 138–158.

Holliday, R. (1960) Plant population and crop yield. *Nature,* **186**: 22–24.

Johnson, P. L. (1969) Arctic plants, ecosystem and strategies. *Arctic,* **22**: 341–355.

McCown, B. H. (1973) The influence of soil temperature on plant growth and survival in Alaska. In *Proceedings of the Symposium on Oil Resource Development and its Impact on Northern Plant Communities*. Occasional Publications on Northern Life No. 1, University of Alaska, pp. 12–33.

McCown, B. H. (1975) Physiological responses of root systems to stress conditions. In *Physiological Adaptation to the Environment* (F. J. Vernberg, Ed.). New York: Intext Educational Publishers, pp. 225–237.

McCown, B. H. (1978) The interactions of organic nutrients, soil notrogen, and soil temperature and plant growth and survival in the arctic environment. In *Vegetation and Production Ecology of an Alaskan Arctic Tundra* (L. L. Tieszen, Ed.). New York: Springer-Verlag, Chap. 19.

Mattheis, P. J., L. L. Tieszen, and M. C. Lewis. (1976) Responses of *Dupontia fischeri* to simulated lemming grazing in an Alaskan arctic tundra. *Ann. Bot.,* **40**: 179–197.

Melchior, H. R. (1972) Summer herbivory by the brown lemming in Barrow, Alaska. In *Proceedings 1972 Tundra Biome Symposium, Lake Wilderness Center, University of Washington* (S. Bowen, Ed.). U.S. Tundra Biome, U.S. International Biological Program and U.S. Arctic Research Program, pp. 136–138.

Miller, P. C. (1972) Bioclimate, leaf temperature, and primary production in red mangrove canopies in south Florida. *Ecology*, **53**: 22–45.

Miller, P. C., and L. L. Tieszen. (1972) A preliminary model of processes affecting primary production in the arctic tundra. *Arct. Alp. Res.*, **4**: 1–18.

Miller, P. C., W. A. Stoner, and L. L. Tieszen. (1976) A model of stand photosynthesis for the wet meadow tundra at Barrow, Alaska. *Ecology*, **57**: 411–430.

Monteith, J. L. (1965) Light distribution and photosynthesis in field crops. *Ann. Bot.*, **29**: 17–37.

Paltridge, G. W. (1970) A model of a growing pasture. *Agric. Meteorol.*, **7**: 93–130.

Penning de Vries, F. W. T. (1972) Respiration and growth. In *Crop Processes in Controlled Environments* (A. R. Rees, K. E. Cockshull, D. W. Hand, and R. G. Hurd, Eds.). London: Academic Press, pp. 327–347.

Sachs, R. M. (1972) Inflorescence induction and initiation. In *The Biology and Utilization of Grasses* (V. B. Younger and M. C. McKell, Eds.). New York: Academic Press, pp. 348–364.

Savile, D. B. O. (1972) *Arctic Adaptations in Plants*. Can. Dep. Agric. Monogr. 6, 81 pp.

Stoner, W. A., and P. C. Miller. (1975) Water relations of plant species in the wet coastal tundra at Barrow, Alaska. *Arct. Alp. Res.*, **7**: 109–124.

Stoner, W. A., P. C. Miller, and L. L. Tieszen. (1978) A model of plant growth and phosphorus allocation for *Dupontia fisheri* in coastal, wet-meadow tundra. In *Vegetation and Production Ecology of an Alaskan Arctic Tundra* (L. L. Tieszen, Ed.). New York: Springer-Verlag, Chap. 24.

Thompson, D. Q. (1955) The role of food and cover in population fluctuations of brown lemming at Point Barrow, Alaska. In *Transactions of the 20th North American Wildlife Conference*, pp. 166–176.

Thornley, J. H. M. (1972a) A balanced quantitative model for root:shoot ratios in vegetative plants. *Ann. Bot.*, **36**: 431–441.

Thornley, J. H. M. (1972b) A model to describe the partitioning of photosynthate during vegetative plant growth. *Ann. Bot.*, **36**: 419–430.

Tieszen, L. L. (1972) The seasonal course of aboveground production and chlorophyll distribution in a wet arctic tundra at Barrow, Alaska. *Arct. Alp. Res.*, **4**: 307–324.

Tieszen, L. L. (1973) Photosynthesis and respiration in arctic tundra grasses: field light intensity and temperature responses. *Arct. Alp. Res.* **5**: 239–251.

Tieszen, L. L. (1975) CO_2 exchange in the Alaskan Arctic Tundra: Seasonal changes in the rate of photosynthesis of four species. *Photosynthetica*, **9**: 376–390.

Tieszen, L. L., D. A. Johnson, and M. L. Allessio. (1974) Translocation of photosynthetically assimilated $^{14}CO_2$ in three arctic grasses in situ at Barrow, Alaska. *Can. J. Bot.*, **52**: 2189–2193.

Tieszen, L. L. (1978). Photosynthesis in the principal Barrow, Alaska species: A summary of field and laboratory responses. In *Vegetation and Production Ecology of an Alaskan Arctic Tundra* (L. L. Tieszen, Ed.). New York: Springer-Verlag, Chap. 10.

Verhager, A. M. W., J. H. Wilson, and E. J. Britton. (1963) Plant production in relation to foliage illumination. *Ann. Bot.*, **27**: 627–640.

Warren Wilson, J. (1967) Ecological data on dry-matter production by plants and plant communities. In *The Collection and Processing of Field Data* (E. M. Bradley and O. T. Denmead, Eds.). C.S.I.R.O. Symposium, New York, p. 77.

de Wit, C. T. (1965) Photosynthesis of leaf canopies. Institute of Biol. Chem. Res. on Field Crops and Herbage, Wageningen. *Agric. Res. Rep.* 663.

de Wit, C. T., R. Brouwer, and F. W. T. Penning de Vries. (1970) The simulation of photosynthetic systems. In *Prediction and Measurement of Photosynthetic Productivity: Proceedings of the IBP/PP Technical Meeting, Trebon, 14–21 September 1969*. Wageningen: Center for Agricultural Publishing and Documentation, pp. 47–70.

27. Summary

LARRY L. TIESZEN

Introduction

The productivity of a plant or a system can be limited by a large variety of resources and environmental factors. At any given point in time several of these may have a significant and simultaneous impact. Presumably, through evolutionary time the selection of species within ecological systems and the adaptations of individuals or populations within species result in individuals which minimize the impact of any single component in this complex. Thus in a highly adapted system an individual should be responisve to perhaps small changes in a large number of niche components. Ideally, we want to quantify the responses to these niche components in some way which allows us to make quantitative comparisons both among species and among components.

We have attempted to incorporate many of these components which we have judged to be significant and important in a simulation model of primary productivity. The details of these relationships are presented in Miller and Tieszen (1972) and refined in Miller et al. (1976). This chapter is based to a large extent on the insights derived during the iterative modifications of this framework, on a study of the output, and on the data available from Barrow, Alaska. It is my intent, in this summary chapter, to discuss the significant proximate and ultimate features which structure the plant component of this ecological system. Explicit in my approach is the acceptance of the individual plant as a functional unit and as the unit which responds to the environment directly. I will use as a measure of this plant's success its ability to accumulate energy in the form of organic material. Perhaps a more useful criterion would include some measure of reproductive success. However, at this stage of our understanding of plant population systems—especially in arctic regions—too little is known to examine this quantitatively.

Historical and Floristic Relationships

A. Species Richness

The increased impoverishment of the vegetation and flora is readily apparent as latitude increases on the Arctic Slope. This has also been described quantitatively in the USSR by Alexandrova (1970) and more recently by Tikhomirov et al. (in prep.). They indicate that the living aboveground biomass increased by an order of magnitude from the subarctic tundra to the arctic tundra and again to polar desert. In Northern Alaska this reduction in vegetation is associated with a reduced flora. At Barrow, for example, we have around 125 (124) vascular species, 95 bryophytes, and 75 (73) lichens (see Murray and Murray, 1978). This is a significant localized reduction from the pool of about 470 bryophytes and 435 vascular plants on the Arctic Slope (Wiggins and Thomas, 1962).

This reduced flora is a consequence of the local accentuation of an already highly selective wet and cold climatic regime, although as Murray (1978) and Webber (1978) indicate the lack of local topographical relief and geological diversity at Barrow is also important. The effect of Pleistocene events on tundra elements is not clear. The imposed migrations and constrictions of ranges likely reduced the flora in a manner analogous to the reduction in the flora of the Northern Great Plains.

The result of these selective forces is a limited flora, the members of which have a number of structural and physiological features of ecological importance. They are perennial plants with perennating structures generally at or below the soil surface. The species either possess well developed root and belowground storage systems (vascular plants) or possess none at all (lichens, bryophytes). Associated with these features is a well-developed system of vegetative reproduction in species of generally long life cycles. This highly developed vegetative reproduction is accompanied by a generally reduced level of sexual reproductive success. Although quantitative data are still lacking, it is generally accepted that seed set and seedling establishment are only marginally successful on the Coastal Plain. This reduced sexual success must have important consequences for speciation in the Arctic as well as for the migrations to, colonization of, and succession on disturbed habitats.

The mere existence of arctic species indicates that they possess metabolic systems that allow them to function well at low temperatures. It is my opinion that this pervasive adaptation to very low temperatures is more a reflection of altered lipid components (see also McCown, 1978) than a large number of changes at the enzyme level. Tieszen and Sigurdson (1973) suggested early that all species were C_3 types, a finding supported by the latitudinal trend in the distribution of C_3 and C_4 plants which has been summarized for North America by Teeri and Stowe (1976). The existence of only this photosynthetic type probably represents the clearest example of the selectivity for specific metabolic systems adapted to low temperatures. Associated with this is a clear possession of frost hardiness in all species which seems to be retained even during periods of active growth. This is further expressed in the apparent ability to enter dormancy (perhaps quiescence) in nearly any stage of development.

B. Growth Forms

Another characteristic feature associated with an increase in lattitude is a progressive change in the spectrum of growth forms. In fact, "Treeless Northern Plains" is used occasionally to describe arctic tundras. Even dwarf shrubs

Table 1. Growth Form Compositions of the
Barrow Vascular Flora[a]

Growth form	Composition (%)
Graminoids	27
Forbs	64
Dwarf deciduous shrubs	7
Dwarf evergreen shrubs	2

[a] Compiled from Murray and Murray (1978).

become uncommon; and in polar deserts lichens clearly dominate and only a few grasses and forbs represent the vascular plant group. The Barrow vascular plant assemblage is composed (Table 1) mainly of forbs and graminoids with only a few shrubs. The wet habitats are generally rich in bryophytes and rhizomatous monocots and poor in dicots and lichens. At the drier end of the spectrum there is a greater diversity of species and life forms, especially dicots and lichens, but an overall lower biomass and productivity.

The latitudinal trend is one of decreasing allocation to supporting structures. Thus, the energy costs of maintaining existing tissues relative to the costs of growth and other processes are lowered, since forms with high maintenance costs are selected against. At Barrow (Dennis et al., 1978) the dominant vascular plants (graminoids and forbs) are characterized by high photosynthetic capacities as are the deciduous shrubs (Johnson and Tieszen, 1976; Tieszen, 1978). In contrast, however, dwarf evergreen shrubs have lower capacities although still much higher than those of lichens and bryophytes. The increasing dominance of lichens and bryophytes under the most severe conditions apparently results because nearly all tissue is potentially productive and activity is very opportunistic.

The biomass and productivity relationship of a community reflect the growth form composition to a large extent (Webber, 1978). Productivity is highest in communities allocating large proportions of organic matter to highly productive tissue which here means communities dominated by graminoids and forbs. One must, however, recognize the importance of nutrient availability. Strong nutrient limitations may give a competitive advantage to lichens, bryophytes, and dwarf evergreen shrubs which, however, will be characterized by low primary productivity. The common xeromorphic nature of a number of tundra plants, in fact, may be in response to a general nutrient limitation (Haag, 1974).

The growth form spectrum of a community is also an important determinant of other ecosystem features. The spectrum, for example, is responsible for vertical structure and thereby may affect the diversity of higher trophic levels. In our Barrow system the relationship with herbivores is perhaps more directly important. The graminoid growth form of sequential leaf production with high photosynthetic capacities is obviously adapted to grazing situations, and at Barrow graminoids are heavily utilized by the brown lemming. Although photosynthetic rates and productivities of deciduous shrubs may be equally high, they provide less usable energy for herbivores, and their more synchronous leaf exsertions make them more susceptible to early season grazing events. Dwarf evergreen shrubs are still more susceptible to grazing because of their high synthetic costs and low photosynthetic capacities. Leaves must, therefore, be retained for long periods to repay the investment. The aboveground parts are selected by the direct action of the physical environment such that growth forms based on aboveground parts segregate into different habitats (Webber, 1978). We would also expect a higher investment in chemical materials to discourage herbivory in these forms. General support for these observations are found in the grazing studies at Barrow and unpublished observations at Meade River.

Abiotic Relationships

A. Temperature

The high rates of photosynthesis described before suggest that this process is well adapted to the low ambient temperatures. However, the temperature optima for *Dupontia* leaves as well as tundra graminoids in general are about 15°C or well above usual leaf and air temperatures (Tieszen and Wieland, 1975). The response curves for both graminoids and mosses (Oechel and Collins, 1976) are strongly positive at 0°C and *Dupontia* can photosynthesize at temperatures below −4°C.

The significance of these responses has been studied with the stand photosynthesis simulations (Miller et al., 1976). We determined the daily uptake of CO_2 for *Dupontia fisheri* with the entire response curve shifted to higher or lower temperatures. The results of the temperature simulations under a standard Barrow day (Figures 1 and 2) suggest that a plant with an optimum of 15°C performs nearly as well at the standard temperature (4.2°C) as do plants with 10 or 5°C optima. Furthermore, a plant with a 15°C optimum photosynthesizes at a high rate under all but the extreme lowest temperatures where plants with lower optima possess higher daily rates. A plant with an optimum temperature (5°C) near the mean Barrow temperature has high rates at the low temperatures but lower rates at ambient temperatures of 10 to 15°C. The simulations suggest clearly that the *Dupontia* response curve is well adapted to the seasonal range of temperatures, that there would be a strong selection pressure against plants with optima near 25°C, and that relatively little would be gained by a seasonal acclimation in the position of the response curve. The lack of strong seasonal acclimation in photosynthesis was previously indicated by Oechel (1976) and Oechel and Sveinbjörnsson (1978).

These interesting temperature relationships occur because the response curve encompasses the occasional very low temperatures and because of canopy relationships. A plant with a 15°C optimum attains high rates near the base of the canopy where most of the leaf area is located (Figure 1). High rates for plants with lower optima are attained only higher in the canopy—where less leaf area is present. These relationships may help explain the absence of tall growth forms in tundras, since still lower optima would be required.

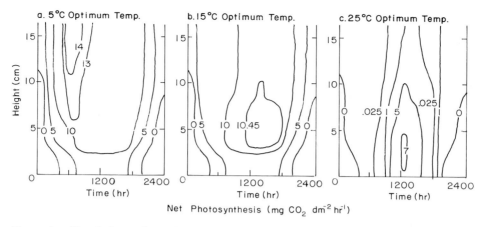

Net Photosynthesis (mg CO_2 dm^{-2} hr^{-1})

Figure 1. Simulations of net photosynthesis using the entire stand photosynthesis model and photosynthesis temperature optima of 5, 15, and 25°C. Lines represent isopleths through the canopy and through the day . Height = canopy height.

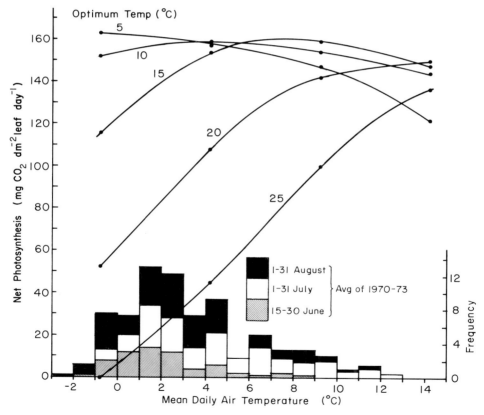

Figure 2. Simulation of the daily photosynthetic response of *Dupontia fisheri* leaves as a function of temperature, and the frequency distribution of mean daily temperatures (1970 through 1973) for 15–30 June, July, and August. (Other input = standard day.)

In a further attempt to analyze the temperature effects, we simulated a standard *Dupontia* response through the season and varied the ambient temperature by one standard deviation around the long-term mean. The seasonal course of CO_2 uptake was surprisingly insensitive to these temperature changes (Figure 3) providing further support for the generalization that photosynthesis per se is not strongly reduced by the low arctic temperatures. The species present are active at temperatures which would reduce photosynthesis in most other plants.

Field observations indicate that growth and primary production are more sensitive to temperature changes than the above simulations of photosynthesis indicate. The data of Wieland and Tieszen indicate that in the three dominant graminoids, weight accumulation is maximal at 15°C and substantially reduced at lower temperatures. This suggests that leaf growth or the allocation for leaf production is temperature sensitive and may be much more significant than the temperature sensitivity of photosynthesis in limiting primary production. The rate of leaf emergence and the rate of elongations are only slightly reduced at 5°C relative to 15°C; however, the rate at which new tillers are produced is 2.5 to 4 times greater at 15°C than at 5°C. Thus, tiller production appears to be very temperature sensitive, and a change in this rate would have a large impact on total production as well as on nutrient demands.

Other plant processes also occur efficiently at the low temperatures as shown by the accumulation of new photoysnthate by primordia in frozen soil (Allessio and Tieszen, 1975). And certainly root growth occurs better at the low temperatures (Chapin, 1974; Billings et al., 1976) than in temperate zone forms. Shaver and Billings (1977) even indicate that root elongation in the field is independent of soil temperature, although it is positively related to temperature for at least two species in phytotron studies.

The sensitivity of some other processes to temperature is apparent in the plant nutrient relationships. There is apparently a greater stimulation of leaf growth (perhaps by more tillering) than nutrient uptake when air temperatures are increased. This greater nutrient demand in warmer years results in greater nutrient "deficiencies" (Chapin, personal communication; Ulrich and Gersper, 1978; Wielgolaski et al., 1975) on both control and fertilized tundra plots. This extreme sensitivity resulted in large differences in nutrient content when mean ambient temperatures differed by only 1°C.

The field data (Webber, 1978) do not show a clear temperature response. Leaf growth is restricted by low temperature. However, during the years of study aboveground production did not increase in warmer seasons. This may result from an intraseasonal temperature variability or a delayed initiation of growth (e.g., 1971) (Dennis et al., 1978), either of which could override a seasonal temperature difference. The temperature relationships in this system are certainly complex, highly interactive, and merit more attention.

B. Irradiance and Photoperiod

Most graminoid leaves light saturate around 0.4 cal cm^{-2} min^{-1} (400 to 700 nm) (Tieszen, 1973). The large diurnal variation in irradiance means that often leaf

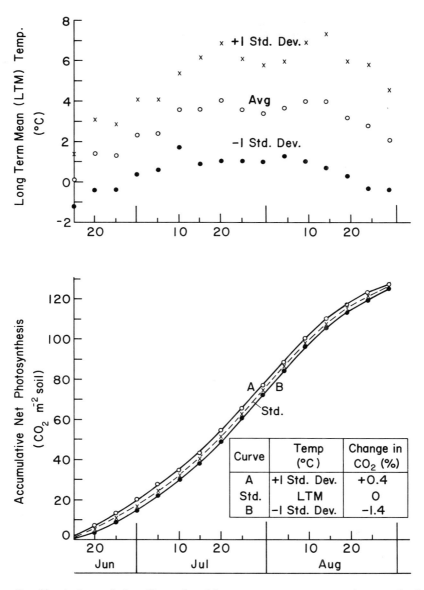

Figure 3. Simulations of the effect of ambient temperature on net photosynthesis of *Dupontia fisheri*. Long-term temperature was varied by ± 1 standard deviation; standard seasonal course is also given. Percentages indicate percentage differences from standard season net photosynthesis. (Other input = standard season.)

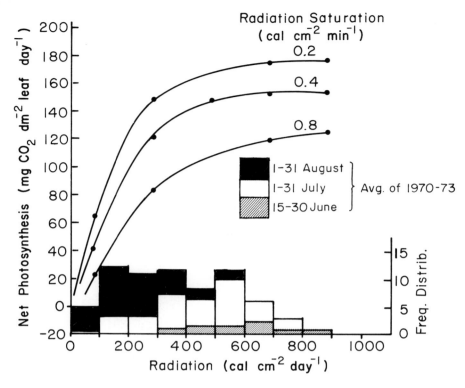

Figure 4. Simulated response of photosynthesis to daily irradiance, and the frequency distribution of daily irradiances (1970 through 1973) for 15–30 June, July, and August. (Other input = standard day.) Simulations of the effects of varying requirements for light saturation of leaf photosynthesis are also shown.

CO_2 uptake will be light limited. Similarly, the seasonal variation in daily receipt is high and varies from less than 100 to around 900 cal cm^{-2} (300 to 3000 nm). The simulations (Figure 4) suggest that on a daily basis the canopy becomes saturated with a daily rate around 600 cal cm^{-2}. On a clear day with high irradiance the leaves at the top are saturated and become water stressed. Leaf resistances increase, and the photosynthetic rates of the leaves with highest absorbed irradiances decrease. Thus, during the course of the season the canopy is light limited on a daily basis nearly half the time. It is also occasionally saturated. An increased irradiance should result in an increased seasonal uptake of CO_2 as is in fact observed. A simulated season with the addition of one standard deviation of irradiance resulted in an increase in CO_2 uptake due only to greater photosynthesis.

The continuous photoperiod during much of the season is of obvious significance, since it allows continuously positive CO_2 balances for most plants until the sun sets on 2 August. The extent to which flowering, senescence, or other developmental phenomena are controlled is not known; however, Shaver and

Billings (1977) show that root and rhizome elongation are inhibited by short photoperiods. Interestingly, late in the season, when inhibition would be effective, photosynthesis is still substantial. In *Dupontia* this coincides with the formation of lateral roots which may continue exploring the horizon as nutrients are accumulated.

C. Nutrients

It is evident that one of the key factors limiting primary production in this arctic system is the availability of nutrients. This has been shown at Barrow by the nutrient enrichment experiments of Schultz (1964), Chapin et al. (1975), and McKendrick et al. (1978) and in the Canadian North by Bliss (1966) and Haag (1974). Although it is impossible at this time to rank these limiting nutrients for even one area due to insufficient data, McKendrick et al. (1978) certainly show a stimulation of production with nitrogen and phosphorus. Webber (1978) also indicates that phosphorus concentration is correlated with the distribution of communities. Ulrich and Gersper (1978) suggest from plant analyses that nearly all plants are deficient in N. Their critical level studies indicate, however, that nitrogen and phosphorus act among a complex of limiting factors and probably never as the sole limiting factors.

They emphatically assert that the Barrow system is a "nutritional desert," and that the system is limited by the availability of soil nutrients. This nutrient-poor status results primarily from the low energy input to the system rather than from inherently impoverished parent materials. The low energy input results in a slow weathering of parent material, a root zone restricted within the narrow zone between the surface and the bottom of the active layer, and a slow rate of decomposition or mineralization. This, in combination with water-logged soils and poor soil aeration, facilitates an accumulation of organic matter with a further loss of nutrients from the available pool. This results in a highly conservative nutrient system, but the conserved chemicals are largely not exploitable. Hence, the inverse relationship between primary production and organic matter accumulation as shown by Webber (1978).

Tundra plants have responded to this resource limitation in a number of ways. Belowground biomass is high (Dennis et al., 1978) and provides a large amount of absorptive area even in the absence of root hairs. Chapin (1974) has also shown in the graminoids he has studied that phosphate absorption capacities are high especially at the low temperatures characteristic of these tundra soils. However, in all species except *Carex aquatilis,* root diameter decreased linearly as temperature increased. Thus, at 1°C root surface area per unit mass was only half that at 20°C. This is partially overcome because uptake activity is retained throughout the length of the root.

Different rooting strategies are apparent (Billings et al., 1978) and are suggestive of niche differentiation belowground. Thus, rooting systems of species in the same habitat tend to diverge to exploit the root profile. *Eriophorum angustifolium* produces annual roots which grow adjacent to the front of the thawing active layer and are most abundant at 2°C. *Dupontia* roots on the other hand were located nearer the surface with *Carex* intermediate. When physical disturbance is intense, for example, on frost boils, the requirements for anchorage override the requirements for nutrients, and the rooting systems are more convergent. Primary production, however, is decreased.

It has often been suggested that the large belowground (especially roots) biomass results from low soil temperatures and/or limiting nutrients (McCown, 1978). In support of the relationship with nutrient availability is the observation of Chapin (1978) that *Carex* had fewer and shorter roots in fertilized plots than in controls.

The one feature which is apparent is that the plants exercise control over allocation such that photosynthesis (Tieszen, 1978) under field conditions is not strongly affected across a large gradient of available soil nutrients or even by natural variations in leaf phosphate levels of mature leaves. Apparently Barrow graminoids produce new tissue only if nutrient reserves adequate to support the photosynthetic system are available. Allocation of additional productive tissue, however, is very sensitive to nutrient levels, and when allocation is limited it is responsible for lower productivity on a land area basis. The response of plants to changed nutrient availability is through density-dependent mechanisms rather than by alterations of the efficiency of different plant compartments.

D. Water

The abundance and availability of soil moisture has an overriding influence on the distribution of species. Associated with this is an increase in productivity with an increase in soil moisture (Webber, 1978). This apparently results from the fact that the species selected near the wetter end of the gradient possess higher photosynthetic capacities and may also have greater access to limiting nutrients as a result of more rapid nutrient cycling. Both vascular and nonvascular species possess different responses with respect to water stress depending upon their position along the gradient.

Photosynthetic capacity is more quickly curtailed in vascular species restricted to wet sites than in those which range more widely (Caldwell et al., 1978). This is due in part to a greater sensitivity of leaf resistance to leaf water potential (Miller et al., 1978) in plants from wet habitats. Under existing field conditions, however, stomatal closure was relatively uncommon (Tieszen, 1978) in a number of vascular plants. Thus, the field gas exchange data support the contention of Miller et al. (1978) that the productivity at one site is probably not limited by water stress except in very dry years. The moisture relationships are complex, however, and are reflected in plant distributions.

In a further attempt to explore the importance of water stress, we simulated seasonal production under varying degrees of atmosphere moisture stress. Varying atmospheric water vapor density by ±40% resulted in only a 9% change in CO_2 uptake. Similarly, increasing root resistance by 50% resulted in a reduction of CO_2 uptake by only 6% for the growing season. Thus, production at one site is relatively insensitive to increased water stress.

In a further attempt to clarify some of the environmental controls over

Table 2. Simulated Response of the Barrow System to Two Extreme Growing Seasons with and without Changes in Allocation for Leaf Area

Simulation	Percentage departure in seasonal CO_2 incroporation
Standard	0
Warm and bright	+6
Warm and bright +20% allocation	+45
Cold and dark −20% allocation	−29
Cold and dark	−56

production, we simulated production for an actual season which was warm and bright as well as for one that was exceptionally cold and dark. The warm season was 2.5°C warmer on the average than the standard season and the cold season was 2.5°C colder. Irradiance was 12% higher during the warm season and 30% lower during the cold season. The results of these simulations suggest that the effect of the warmer and brighter season on CO_2 uptake was minimal (Table 2) since it resulted in only a 6% increase. The cooler season departed more from the standard season mainly because irradiance was lower and more often limiting photosynthesis.

These simulations, however, include no control over allocation, and allocation may be markedly altered by the different environments and the different carbohydrate budgets. In an attempt to examine the effect of changing allocation patterns during these seasons, we increased the photosynthetic leaf area by 20% each period in the warm season and decreased it by 20% during the cold season. The effect of these allocation patterns was much more dramatic and resulted in a 45% increase and a 56% decrease respectively in these two seasons. The simulations therefore support the hypothesis that production is limited by low temperature through its greater effects on leaf growth and allocation than on photosynthesis per se.

E. Season Length and Leaf Allocation

The shortness of the growing season is impressed upon any field worker in the Arctic, and the photosynthesis data (Tieszen, 1978; Coyne and Kelley, 1978; Oechel and Sveinbjörnsson, 1978) clearly suggest that this is a major factor limiting production. Aboveground production begins after melt-off and biomass usually decreases after the beginning of August (Caldwell et al., 1978) although CO_2 incorporation occurs for about another month, and nutrient uptake perphaps still longer (Chapin, 1978). Thus, the effective season length is different for various processes but is constrained to a short period. Since production appears to be very sensitive to the allocation of leaf area, and since only small amounts of leaf material are produced before the summer solstice, length of season may be highly significant. The earlier initiation of growth in 1970 than in 1971 may account for the greater production even though 1970 was colder.

Table 3.   Simulation of the Effect of the Length of the
Growing Season on Accumulative CO_2 Uptake. LAI
Changes at the Same Rate as the Standard Run

Simulation	Percentage departure in seasonal CO_2 incorporation
Standard (80 days)	0
Long season (100 days)[a]	+55
Long season + LAI[b]	+77
Short season (65 days)[c]	−27
Short season[d]	−39

[a]   LAI plateaus at maximum value of standard run from 7 July to 4 August.
[b]   LAI is allowed to increase at the same rate as standard case *and* exceed the standard case until 4 August.
[c]   LAI is not allowed to senesce until 19 August.
[d]   LAI changes are identical to standard case.

Estimates of effective season lengths are absent as are comparable estimates of primary production over a long time period. Lewellen (1972), however, has compiled the thaw seasons for 1922–1967 and indicates that it has a mean length of 87 days, usually extending from 12 June to 6 September. Our standard simulations were run for 80 days and still possessed some productive tissue at the end of this period. We varied the season length to encompass most of the variability that has occurred. If we only lengthened the season to 100 days (Table 3) by initiating growth on 26 May and by increasing leaf area at the same rate as the standard case, production was stimulated by 55%.

Allowing leaf area to continue to increase until the normal season peak (4 August) resulted in 30% greater LAI at this time and a still higher uptake of CO_2. Similarly, shortening the season by starting it 15 days later substantially decreased CO_2 uptake especially if senescence was initiated at the normal time. These simulations clearly show the sensitivity to the amount of leaf area and to the onset of the development of the canopy. Physical extension of the season in fall was not studied since the low incoming solar radiation makes this less likely and of lesser quantitative importance.

Since these simulations suggest that the amount of leaf area exerts a major effect on seasonal CO_2 incorporation, we increased the allocation to leaf area (Figure 5). Seasonal uptake was very sensitive to a change in leaf area (allocation to leaves). Increasing the rate of leaf production by 25%, for example, resulted in a 45% increase in CO_2 incorporation. Thus, those factors which alter season length and allocation pattern are among the most important factors limiting primary production which need to be clarified and quantified.

F. Grazing

One of the characteristics of the Barrow system is a very low diversity of herbivores. This is obvious from the near absence of damaged leaves from either macro herbivores or plant pathogens. This apparent simplicity, however, is confounded by the presence of two herbivores, *Dicrostonyx groenlandicus* and *Lemmus sibericus*, which vary widely in population densities and therefore in their impact on the vegetation. Batzli (1975), for example, suggests that brown lemmings (*L. sibericus*) range in density in Northern Alaska from 0.1/ha to 75–200/ha. These high densities are not stable and are shortly followed by rapid declines in density. During the period of maximum densities, their effects are dramatic and can even result in a physical alteration of the soil surface. These effects may often be localized because of the moist habitat preference of the brown lemmings, and because high densities are often attained in early spring when the available vegetation is at a minimum.

Little is known about the nutritional requirements of the brown lemming, the effect on the physiology of the preferred monocot species, or the ecological consequences of the variable densities. Observations of old exclosures suggest that in the absence of lemmings some species or growth forms (e.g., lichens) increase and that a thicker vegatative mat (including moss) develops. The effect is to increase insulation and therefore reduce the depth of the active layer. Long-term consequences of these changes could be highly significant.

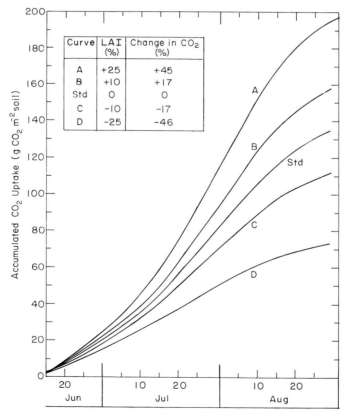

Figure 5.   Simulation of the effect of changes in the rate of production of leaf area on seasonal CO_2 incorporation.

The physiological response to grazing pressure is only now becoming known. The brown lemming selects monocots (Batzli, 1975), and in the wetter part of the moisture gradient those species have generally high inorganic contents (Chapin et al., 1975; Chapin 1978) and TNC levels (McCown, 1978). Yet Batzli (1975) indicates that these species show a low digestibility by *Lemmus*. During low densities observations indicate that *Dupontia* and other graminoids are clipped near the moss surface. Single leaves, or, perhaps more commonly, whole tillers are severed and then eaten. The apical meristem, however, is positioned below the moss surface (Mattheis et al., 1976) and is usually not damaged directly unless the inflorescence was developing and the culm had elongated.

Dupontia responds to this type of grazing event by continuing to produce leaves at about the same rate as control plants except in very young tillers where leaf production is slightly depressed. The stored reserves in stem bases and rhizomes in these multitiller systems are apparently very important to sustain this regrowth. In the field, a single grazing event results in only a slight depression of TNC levels. In fact, after chronic regrazing (six times at 10-day intervals) TNC levels are only reduced to 50% that of the controls, and the plants continue to produce new leaf material. This buffering capability must be a result of the large belowground biomass, since laboratory tests show a more drastic response to even single grazing events (Tieszen, unpublished).

In a *Dupontia*-dominated system then, the effect of low level grazing is mainly to modify the canopy. This will have significant effect on energy balance at the surface and nutrient cycling as well as on carbon uptake and the replenishment of reserves. Our photosynthesis simulations can at least examine the direct effect of the removal of this photosynthetic material. Acute grazing, especially in late winter or early spring, removes the canopy including standing dead and live material. Thus, the microenvironment surrounding the photosynthetic leaves is altered dramatically. Leaves even near the moss surface are now exposed to more intense radiation and generally have lower temperatures since convectional losses are greater.

The net result of this early season grazing is a temporary depression of CO_2 uptake and a later stimulation (Figure 6). This results because the canopy has been opened and leaves are more fully illuminated. Thus, over the season 10% more CO_2 is incorporated. Grazing at any time other than winter or early spring has a greater effect because more photosynthetic tissue is removed. In fact these simulations suggest that mid-summer grazing is the most damaging since mature and highly productive tissues are removed at a time when much photosynthesis but little leaf production is left to occur. Thus, the CO_2 taken up is only 49% of the controls. The more typical pattern of high lemming densities in early spring is least detrimental in terms of the total carbon budget.

The time of the grazing event will be important in determining the amount of dry matter that will become available that season for plant reserves or successive grazing events. In this context the effect of grazing becomes more important as the season progresses (Table 4), and only little production occurs if grazing occurs after the canopy has been exserted. Plants cannot be expected to replenish reserves, grow roots, or undertake other expensive processes if reserves are not restored before this time.

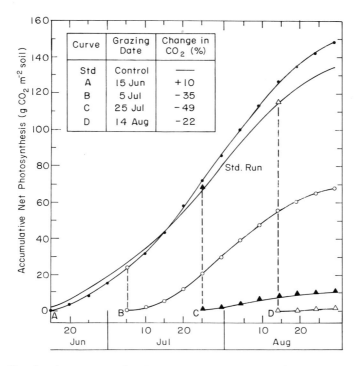

Figure 6. Simulated response of a *Dupontia* system to complete canopy removal by grazing at different times of the season. Regrowth of leaves is assumed to occur at the same rate as growth occurs at that time in the ungrazed situation.

The effects of grazing during a high density of lemmings are much different. At a density of 100 lemmings per hectare (0.01 animals m^{-2}) the animals not only clip tillers but also grub through the surface and apparently eat rhizomes and stem bases. If we use the consumption estimate of Batzli (1975), the population could require 0.85 kcal m^{-2} day^{-1} in June. This is equivalent to slightly less than 0.2 g of dry matter. At the time of snowmelt there is something more than 7 g of green material present aboveground (Dennis et al., 1978). If all of this were eaten, it

Table 4. Carbon Dioxide Taken Up after a Grazing Event Expressed as a Percentage of the Control

Date of grazing	Percentage CO_2 incorporated after grazing
None	100
15 June	110
5 July	59
20 July	14
14 August	3

could sustain the population for at least 35 days. Thus, these high populations cannot be supported for long periods beneath the snow unless alternative foods, for example mosses and rhizomes, are exploited. After snowmelt, however, the rate of production (around 1.45 to 1.85 g m^{-2} day^{-1}) appears high enough to support this high density. The long-term effects of this pressure are not known. However, from a physiological basis, only plants with good regrowth capabilities and inexpensive leaves could perform well. Thus, shrubs and species with expensive reproductive strategies are incompatible with a *Lemmus* grazing strategy. They may, however, be more compatible with a *Dicrostonyx* strategy of more continuous, but less intensive, grazing.

Spatial and Temporal

A. Spatial Patterns of Primary Production

It is apparent that primary production varies greatly over short distances on the Coastal Plain. However, nearly all systems are characterized as being carbon rich. In the intensive site, for example, the large standing crop of carbon is bound mainly (96%) in the form of peat, whereas only 1.7% of the carbon is in living organisms. Even most of this is belowground since the mean plant biomass belowground is three times greater than that aboveground at the time of its maximum.

Webber's study of 43 stands within the study area indicates that the minimum overall estimate for aboveground production is 41.6 g m^{-2} yr^{-1}. This estimate ranges from a low of 18.1 g m^{-2} yr^{-1} in dry heaths to a high of 118.5 g m^{-2} yr^{-1} in *Arctophila fulva* pond margins. There is a clear increase in production with an increase in soil moisture. The increase is not clearly related to species diversity (Figure 7) but is associated with an increased frequency of monocots and dramatic decreases in woody dicots and especially lichens. Thus, in the broad

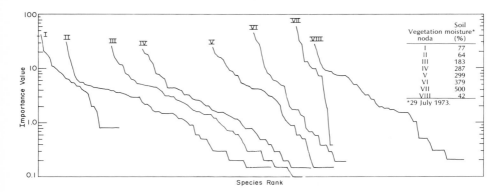

Figure 7. Dominance-diversity curves for eight vegetation noda recognized at Barrow, 1971–1974. Vascular and nonvascular plants are included. Data and importance value are from Webber (1978).

pattern the most productive communities consist of growth forms which are characterized by generally high rates of photosynthesis. Single-shooted graminoids, for example, increase markedly from Nodum I through Noda VI and VII. Species diversity is greatest in the more mesic communities, and there is less dominance by individual species.

Johnson and Tieszen (1976), indicate that arctic monocots and deciduous shrubs possess rates of maximum carbon uptake that are two to eight times greater than those of dwarf evergreen shrubs. This pattern is verified by the data available from all tundra regions. Furthermore, there is an inverse relationship between maximum photosynthetic rates and leaf longevity. Leaves of long lifetimes also make large commitments to chemical defenses against herbivores. Thus, they are expensive to synthesize as well as to maintain on a year-round basis (Penning de Vries, 1974). Net carbon gain by leaves of all growth forms may in fact be similar on a lifetime basis, but longer lived leaves provide less organic matter for herbivores at any point in time.

The different productivities are therefore due in part to the different species selected in each nodum. The productivities, however, may also be causally related to other factors. For example, the continuum from the aquatic pond margin to dry polygon tops or other areas may also reflect a reduction in the rate of nutrient cycling. Decomposition and mineralization are more rapid in the wetter sites. This in itself may select for those species of high photosynthetic rates and leaf turnover. In support of this relationship with available nutrients is the obvious positive association between accumulated soil organic matter and low rates of primary production, or, for that matter, secondary production as well.

It thus appears that in the terrestrial system organic matter accumulates as primary production exceeds decomposition and consequently restricts primary production as nutrients become unavailable. Nutrient replenishment now becomes very important in this system and operates at several levels and time frames. Decompoistion is facilitated at the wet end of the gradient on a continuous basis. Occasionally, however, the total felling of vegetation by lemmings provides a rapid pulse of nutrients for the system. Further nutrient replenishment is achieved by direct physical distrubance to the organic surface. This can be caused by extreme herbivore activity, vehicular disturbance, or by other means. It can result in an apparently rapid decomposition and can account for the very productive, but less diverse, communities Dennis et al. (1978) describe for some disturbed sites. In the long run, however, it may be the thaw lake cycle which replenishes nutrients by a thorough decomposition of the accumulated organic matter. This would help account for the apparently high productivities at Barrow (Tieszen, 1972) relative to inland sites (see Tieszen and Johnson, 1968).

B. Temporal Sequence of Primary Production

Arctic systems are generally characterized by a markedly distinct seasonality, especially for primary producer organisms. This is especially apparent in the Barrow system where the plants are either quiescent or dormant for 8 to 9 months of each year, and where there is an abrupt transition from the inactivity of winter to the onset of spring growth and production. This onset is initiated immediately after snowmelt as indicated by Tieszen (1974). Prior to this time, the insulative effect of the snow maintains low plant temperatures because of the heat sink of the underlying permafrost. Although as much as 10 to 15 g of dry matter may be present in vascular stem bases and unexpanded leaves, photosynthesis does not occur because of low irradiance, low temperature, and an incompletely developed photosynthetic apparatus. Mosses on the other hand are more competent and are highly active as soon as they are exposed.

The onset of the productive season is therefore tightly controlled by the amount of snow accumulation and meteorological phenomena. It usually occurs before the summer solstice and after 5 June. The termination of the productive season is less clearly defined, since dormancy is established in many species well before permanent snow is received. Peak aboveground production, however, occurs around the beginning of August (Tieszen, 1972; Webber, 1978; Dennis et al., 1978). Tieszen (1975) suggests that on the basis of physiological responses, photosynthesis would occur well beyond this early August period, and the potentially productive season should extend into early September. Thus, the maximum limits of the growing season are around 3 months even though nutrient uptake occurs for a slightly longer period (see Chapin 1978).

The extent to which photoperiodic cues are used to control seasonal events at this latitude are not known. However, Shaver and Billings (1977) have shown that root and rhizome elongation are stopped by the short photoperiods which develop in late August. The photoperiod may also help control leaf and shoot senescence, but personal observations indicate that unseasonally mild temperatures in August will delay senescence.

The direct harvests provide good estimates of shoot production but are not very suitable for quantitative assessments of new growth in roots or other belowground components (Dennis et al., 1978). This assessment is especially important in wet tundra because of the concentration of live organic matter belowground (Billings et al., 1978), the general accumulation of organic matter in tundra systems, and the high degree of invertebrate activity this organic matter can support (MacLean, personal communication). Therefore, I will attempt to utilize our gas exchange studies and simulations to analyze the seasonal kinetics of primary production.

Estimates of net CO_2 uptake by cuvette (Tieszen, 1975), simulation (Miller et al., 1976), and aerodynamic (Coyne and Kelley, 1975, 1978) approaches are very comparable as is indicated by the two diurnal courses shown in Figure 8. This figure also illustrates the positive uptake throughout the day by whole shoots as well as a significant ecosystem respiratory component which results in negative aerodynamic values for nearly half the 24-hr periods. When these ecosystem respiration components are included in the aerodynamic estimate of photosynthesis, all three methods are in close agreement. An integration of these approaches for the entire season (Table 5) again provides similar estimates for CO_2 m^{-2}.

In addition to this net uptake by vascular plants, mosses are responsible for the incorporation of from 41 g of CO_2 m^{-2} (Rastorfer, 1978) based on harvest data

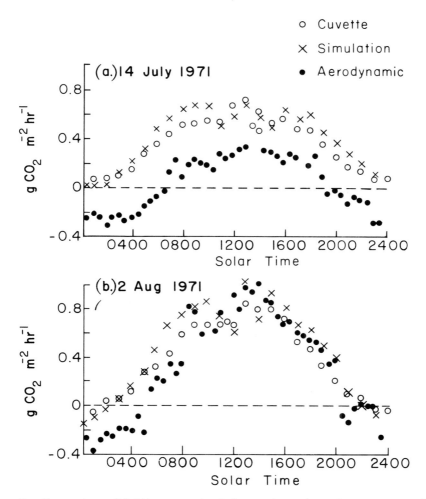

Figure 8.   Comparison of field cuvette, simulation, and aerodynamic assessment of net CO_2 exchange. Aerodynamic data represent flux from the atmosphere only. Cuvette and simulation data are for vascular plants. (Cuvette data from Tieszen, 1975; aerodynamic data derived from Coyne and Kelley, 1975.)

to 44 g of CO_2 m^{-2} (Oechel and Sveinbjörnsson, 1978) based on gas exchange data. Thus, although mosses may often represent up to 40 to 50% of the biomass, the contribution to primary production is much less.

An analysis of all available gas exchange data and biomass estimates provides an annual partitioning of carbon. Around 19% of the gross amount fixed is lost as aboveground respiration. Aboveground production accounts for another 22%. More than half, 59%, is translocated belowground where half is used in respiration and half to produce new tissue. During the course of our studies the system did not appear to be in a steady state but was accumulating carbon. The general

Table 5.   Estimate of the Seasonal Gross Uptake of
CO_2 by the Vascular Components of the *Dupontia*-
Dominated Site 2

Method	CO_2 flux (g of CO_2 m^{-2} season^{-1})
Cuvette [a,c]	742
Simulator model [b,c]	750
Aerodynamic [d]	733

[a]   Based on Tieszen (1975, 1978) corrected for self-shading.
[b]   Based on Miller et al. (1976)
[c]   Includes direct estimate of net uptake plus 140 g of CO_2
estimates for top dark respiration.
[d]   Based on Coyne and Kelley (1978).

occurrence of peat indicates that this accretion has been common ever since the
area became tundra.

Previous harvest estimates of Tieszen (1972) and Dennis et al. (1978), indicated that the rate of aboveground increase in biomass was constant until peak
season. This differs markedly from the present analysis of CO_2 incorporation
(Figure 9) which accelerated until late July when reates as high as 10 to 12 g of
CO_2 m^{-2} day^{-1} were attained. After the period of peak biomass, both net
photosynthesis and net aboveground production decreased. However, net photosynthesis remained positive until September. These data all suggest that
substantial carbon incorporation occurs after aboveground production ceases. It
is during this period, then, that belowground reserves are replenished or that new
growth is occurring. Allessio (Allessio and Tieszen, 1978) has shown increased
labeling intensity in stem bases late in the season; and the data of Mattheis
(Mattheis et al., 1976) clearly show a replenishment of dry weight and TNC in
Dupontia rhizomes and stem bases beginning mid-July and continuing to early
September. Thus, the inferences from gas analysis studies are reinforced by
experimental studies as well as the harvest studies of Dennis (Dennis et al.,
1978). The last third of the CO_2 taken up in the summer is used belowground for
the replenishment of reserves, expansion of new tissue, or maintenance, but
apparently not for root elongation (Shaver and Billings, 1977).

The marked seasonality of photosynthesis and primary production occurs in
the absence of major temperature changes and in the presence of a continuously
decreasing irradiance. Figure 10 clearly illustrates that net community photosynthesis is limited strongly by the availability of photosynthetic tissue throughout
most of the season. The proportion of total irradiance incident on green tissue is
quite low early in the season and does not reach maximum values until the end of
July. Thus, vascular plant photosynthesis occurs at low effficiencies early in the
season because of limiting leaf area and photosynthetic competence (Tieszen,
1978) rather than limiting irradiance.

Interception efficiency (I_g/I_o) and absorption by green tissue approaches a
maximum as LAI approaches 1.0. The dependency of the rate of community
photosynthesis on interception and absorption is shown by the linear and vir-

tually constant relationship between P_N and I_g (Figure 10). The conversion efficiency of intercepted radiation is maintained at a high level (between 2.0 and 2.6%) characteristic of temperate zone systems. This also suggests that the main limitation to photosynthesis and productivity is the allocation pattern and the rate of foliage production rather than the photosynthetic process.

Summary

The Barrow system is therefore characterized by a generally small number of species which are segregated into a complex array of communities. The production in these communities varies over a large range, and near the wet end of the moisture gradient attains high values on a daily basis. Factors which control the extent of carbon allocation to new tillers or leaves appear most important in limiting primary production. Foremost among these are the length of the growing season and the availability of soil nutrients.

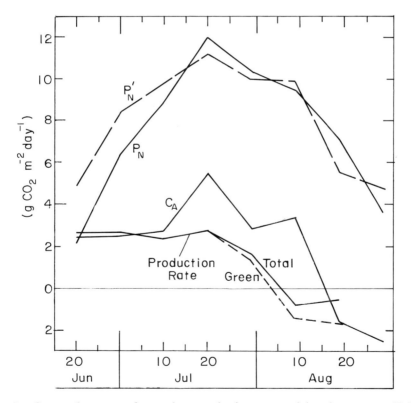

Figure 9. Seasonal course of net photosynthesis measured by the cuvette (P_N) and simulated by the canopy model ($P_N{}^1$), atmospheric CO_2 flux (C_A) estimated by the aerodynamic method, and aboveground production rate (CO_2 equivalent) determined by the harvest method in 1971.

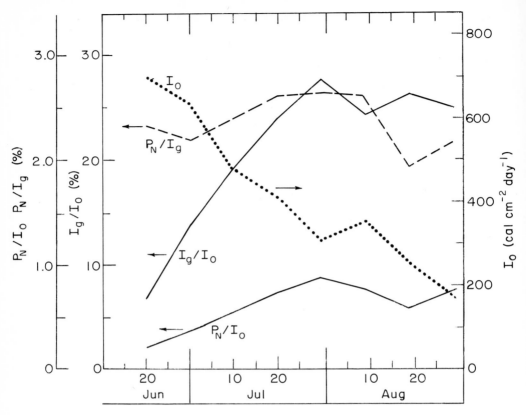

Figure 10. Seasonal courses of incident irradiation (I_o, 300 to 3000 nm), interception efficiency by green tissue (I_g/I_o), and the ratios of simulated net photosynthesis (P_N) to I_o and I_g in 1971.

Perturbations which affect the cycling of inorganic nutrients or the thermal and moisture status of the community will have significant effects on primary production and community stability. Removal of the canopy and moss layer, shearing of vascular plant roots, or the initiation of thermal erosive forces appear to have the most detrimental long-term effects. An awareness of the perennial nature of the flora of the Coastal Plain, the sensitivity of the tundra to slight changes in the thermal radiation regime, the natural limitations on the release of inorganic nutrients, and the limited dispersal of sexual propagules, however, should allow man to plan effectively for activities in the tundra.

References

Alexandrova, V. D. (1970) The vegetation of the tundra zones in the USSR and data about its productivity. In *Productivity and Conservation in Northern Circumpolar Lands* (W. A. Fuller and P. G. Kevan, Eds.). Int. Union Conserv. Natur., Morges, Switzerland, Publ. 16, N.S., pp. 93–114.

Allessio, M. L., and L. L. Tieszen. (1975) Patterns of carbon allocation in an arctic tundra grass, *Dupontia fisheri* (Gramineae), at Barrow, Alaska. *Amer. J. Bot.,* **62**: 797–807.

Allessio, M. L., and L. L. Tieszen. (1978) Translocation and allocation of ^{14}C-photoassimilate by *Dupontia fisheri*. In *Vegetation and Production Ecology of an Alaskan Arctic Tundra* (L. L. Tieszen, Ed.). New York: Springer-Verlag, Chap. 17.

Batzli, G. O. (1975) The role of small mammals in arctic ecosystems. In *Small Mammals: Productivity and Dynamics of Populations* (F. Golley, et al., Eds.). IBP Series, Cambridge University Press, Chap. 11, pp. 245–268.

Billings, W. D., G. R. Shaver, and A. W. Trent. (1976) Measurement of root growth in simulated and natural temperature gradients over permafrost. *Arct. Alp. Res.,* **8**: 247–250.

Billings, W. D., K. M. Peterson, and G. R. Shaver. (1978) Growth, turnover, and respiration rates of roots and tillers in tundra graminoids. In *Vegetation and Production Ecology of an Alaskan Arctic Tundra* (L. L. Tieszen, Ed.). New York: Springer-Verlag, Chap. 18.

Bliss, L. C. (1966) Plant productivity in alpine microenvironments on Mt. Washington, New Hampshire. *Ecol. Monogr.,* **36**: 125–155.

Caldwell, M. M., D. A. Johnson, and M. Fareed. (1978) Constraints on tundra productivity: Photosynthetic capacity in relation to solar radiation utilization and water stress in arctic and alpine tundras. In *Vegetation and Production Ecology of an Alaskan Arctic Tundra* (L. L. Tieszen, Ed.). New York: Springer-Verlag, Chap. 13.

Chapin, F. S. III (1974) Phsophate absorption capacity and acclimation potential in plants along a latitudinal gradient. *Science,* **183**: 521–523.

Chapin, F. S. III (1978) Phosphate uptake and nutrient utilization by Barrow tundra vegetation. In *Vegetation and Production Ecology of an Alaskan Arctic Tundra* (L. L. Tieszen, Ed.). New York: Springer-Verlag, Chap. 21.

Chapin, F. S. III, K. Can Cleve, and L. L. Tieszen. (1975) Seasonal nutrient dynamics of tundra vegetation at Barrow, Alaska. *Arct. Alp. Res.,* **7**: 209–226.

Coyne, P. I., and J. J. Kelley. (1975) CO_2 exchange over the Alaskan arctic tundra: Meteorological assessment by an aerodynamic method. *J. Appl. Ecol.,* **12**: 587–611.

Coyne, P. I., and J. J. Kelley. (1978) Meteorological assessment of CO_2 exchange over an Alaskan arctic tundra. In *Vegetation and Production Ecology of an Alaskan Arctic Tundra* (L. L. Tieszen, Ed.). New York: Springer-Verlag, Chap. 12.

Dennis, J. G., L. L. Tieszen, and M. Vetter. (1978) Seasonal dynamics of above- and belowground production of vascular plants at Barrow, Alaska. In *Vegetation and Production Ecology of an Alaskan Arctic Tundra* (L. L. Tieszen, Ed.). New York: Springer-Verlag, Chap. 4.

Haag, R. W. (1974) Nutrient limitations to plant production in two tundra communities. *Can. J. Bot.,* **52**: 103–106.

Johnson, D. A., and L. L. Tieszen. (1976) Aboveground biomass allocation, leaf growth, and photosynthesis patterns in tundra plant forms in arctic Alaska. *Oecologia,* **24**: 159–173.

Lewellen, R. I. (1972) *Studies on the Fluvial Environment Arctic Coastal Plain Province, Northern Alaska*. Published by the author (P.O. Box 1068, Littleton, Colorado 80120). 2 volumes, 282 pp.

McCown, B. H. (1978) The interaction of organic nutrients, soil nitrogen, and soil temperature on plant growth and survival in the arctic environment. In *Vegetation and Production Ecology of an Alaskan Arctic Tundra* (L. L. Tieszen, Ed.). New York: Springer-Verlag, Chap. 19.

McKendrick, J. D., V. J. Ott, and G. A. Mitchell. (1978) Effects of nitrogen and phosphorus fertilization on carbohydrate and nutrient levels in *Dupontia fisheri* and *Arctagrostis latifolia*. In *Vegetation and Production Ecology of an Alaskan Arctic Tundra* (L. L. Tieszen, Ed.). New York: Springer-Verlag, Chap. 22.

Mattheis, P. J., L. L. Tieszen, and M. C. Lewis. (1976) Responses of *Dupontia fisheri* to simulated lemming grazing in an Alaskan arctic tundra. *Ann. Bot.,* **40**: 179–197.

Miller, P. C., W. A. Stoner, and L. L. Tieszen. (1976) A model of stand photosynthesis for the wet meadow tundra at Barrow, Alaska. *Ecology, 57*:411–430.

Miller, P. C., and L. L. Tieszen. (1972) A preliminary model of processes affecting primary production in the arctic tundra. *Arct. Alp. Res., 4*: 1–18.

Miller, P. C., W. A. Stoner, and J. R. Ehleringer. (1978) Some aspects of water relations of arctic and alpine regions. In *Vegetation and Production Ecology of an Alaskan Arctic Tundra* (L. L. Tieszen, Ed.). New York: Springer-Verlag, Chap. 14.

Murray, D. F. (1978) Ecology, floristics, and phytogeography of northern Alaska. In *Vegetation and Production Ecology of an Alaskan Arctic Tundra* (L. L. Tieszen, Ed.). New York: Springer-Verlag, Chap. 2.

Murray, B. M., and D. F. Murray. (1978) Checklists of vascular plants, bryophytes, and lichens for the Alaskan U.S. IBP Tundra Biome study areas—Barrow, Prudhoe Bay, Eagle Summit. In *Vegetation and Production Ecology of an Alaskan Arctic Tundra* (L. L. Tieszen, Ed.). New York: Springer-Verlag, Appendix 1.

Oechel, W. C. (1976) Seasonal patterns of temperature response of CO_2 flux and acclimation in arctic mosses growing *in situ*. *Photosynthetica, 10*: 447–456.

Oechel, W. C., and N. J. Collins. (1976) Comparative CO_2 exchange patterns in mosses from two tundra habitats at Barrow, Alaska. *Can. J. Bot., 54*: 1355–1369.

Oechel, W. C., and B. Sveinbjörnsson. (1978) Primary production processes in arctic bryophytes at Barrow, Alaska. In *Vegetation and Production Ecology of an Alaskan Arctic Tundra* (L. L. Tieszen, Ed.). New York: Springer-Verlag, Chap. 11.

Penning de Vries, F. W. T. (1974) Substrate utilization and respiration in relation to growth and maintenance in higher plants. *Neth. J. Agric. Sci., 22*: 40–44.

Rastorfer, J. R. (1978) Composition and bryomass of the moss layers of two wet-meadow tundra communities near Barrow, Alaska. In *Vegetation and Production Ecology of an Alaskan Arctic Tundra* (L. L. Tieszen, Ed.). New York: Springer-Verlag, Chap. 6.

Schultz, A. M. (1964) The nutrient recovery hypothesis for arctic microtine cycles. II. Ecosystem variables in relation to arctic microtine cycles. In *Grazing in Terrestrial and Marine Environments* (D. J. Crisp, Ed.). Oxford: Blackwell Scientific Publications, pp. 57–68.

Shaver, G. R., and W. D. Billings. (1977) Effects of daylength and temperature on root elongation in tundra graminoids. *Oecologia, 28*: 57–65.

Teeri, J. A., and L. G. Stowe. (1976) Climatic patterns and the distribution of C_4 grasses in North America. *Oecologia, 23*: 1–12.

Tieszen, L. L. (1972) The seasonal course of aboveground production and chlorophyll distribution in a wet arctic tundra at Barrow, Alaska. *Arct. Alp. Res., 4*: 307–324.

Tieszen, L. L. (1973) Photosynthesis and respiration in arctic tundra grasses: Field light intensity and temperature responses. *Arct. Alp. Res., 5*: 239–251.

Tieszen, L. L. (1974) Photosynthetic competence of the subnivean vegetation of an arctic tundra. *Arct. Alp. Res., 6*: 253–256.

Tieszen, L. L. (1975) CO_2 exchange in the Alaskan arctic tundra: Seasonal changes in the rate of photosynthesis of four species. *Photosynthetica, 9*: 376–390.

Tieszen, L. L. (1978) Photosynthesis in the principal Barrow, Alaska, species: A summary of field and laboratory responses. In *Vegetation and Production Ecology of an Alaskan Arctic Tundra* (L. L. Tieszen, Ed.). New York: Springer-Verlag, Chap. 10.

Tieszen, L. L., and P. L. Johnson. (1968) Pigment structure of some arctic tundra communities. *Ecology, 49*: 370–373.

Tieszen, L. L., and D. C. Sigurdson. (1973) Effect of temperature on carboxylase activity and stability in some Clavin cycle grasses from the arctic. *Arct. Alp. Res., 5*: 59–66.

Tieszen, L. L., and N. K. Wieland. (1975) Physiological ecology of arctic and alpine photosynthesis and respiration. In *Physiological Adaptation to the Environment*, pp. 157–200.

Tikhomirov, B. A., V. F. Shamurin, and V. D. Alexandrova. (In prep.) Primary production of tundra communities. In *Botanical Institute of the USSR Academy of Sciences*, pp. 1–19.

Ulrich, A., and P. L. Gersper. (1978) Plant nutrient limitations of tundra plant growth. In *Vegetation and Production Ecology of an Alaskan Arctic Tundra* (L. L. Tieszen, Ed.). New York: Springer-Verlag, Chap. 20.

Webber, P. J. (1978) Spatial and temporal variation of the vegetation and its production, Barrow, Alaska. In *Vegetation and Production Ecology of an Alaskan Arctic Tundra* (L. L. Tieszen, Ed.). New York: Springer-Verlag, Chap. 3.

Wielgolaski, F. E., S. Kjelvik, and P. Kallio. (1975) Mineral content of tundra and forest tundra plants in Fennoscandia. In *Fennoscandian Tundra Ecosystems, Vol. 1, Plants and Microorganisms* (F. E. Wielgolaski, Ed.). Berlin: Springer-Verlag.

Wiggins, I. L., and J. H. Thomas. (1962) *A Flora of the Alaskan Arctic Slope*. Arctic Institute of North America Spec. Publ. 4. Toronto: University of Toronto Press, 425 pp.

Appendix
Checklists of Vascular Plants, Bryophytes, and Lichens for the Alaskan U.S. IBP Tundra Biome Study Areas— Barrow, Prudhoe Bay, Eagle Summit

BARBARA M. MURRAY AND DAVID F. MURRAY

Introduction

The checklists to vascular plants, bryophytes, and lichens for the three Alaskan study areas are based on primary literature sources (books, journals, dissertations, and processed publications) or unpublished records of herbarium specimens in cases for which no published reports are available. For the documentation of the known flora at each study area we thought it essential that there be provision for checking determinations and making corrections or adjustments in nomenclature. Therefore, sources were used in which there was a declaration of the voucher specimen and the herbarium at which it is housed. All such literature records known to us are cited, but when a report is based on unpublished material only one herbarium specimen is used.

Some ecological papers (Britton, 1967; Brown and Johnson, 1965; Hanson, 1950; Rastorfer, Webster, and Smith, 1973; Weeden, 1968; Wiggins, 1951) and graduate theses (Clebsch, 1957; Dennis, 1968; Koranda, 1954; Shacklette, 1962) as well as several papers on soils and physiological ecology cited species, but in many cases these are not referrable to specimens. Such references are included in these lists only when they constitute the sole record for a taxon. These reports may be valid, but they can not be confirmed at present, and they are, therefore, set off by brackets. Records for Barrow by Hultén (1968) are based on his distribution maps.

Hultén's comprehensive treatments of the vascular flora (1941–1950, 1967, 1968, 1973) give us a nomenclatural standard to follow. Synonyms are given when we have departed from Hultén (1968) to accommodate the changes proposed by him (1973) and by others. When subspecific status was not indicated in the literature record, we attempted to assign subspecies on the basis of distribution according to the concepts of Hultén, but only if the biogeographic distinction was obvious. We are following the treatment of Argus (1973) for the genus *Salix*.

Nomenclature of mosses follows Crum, Steere, and Anderson (1973); for Mniaceae we are using the treatment of Koponen (1968, 1971, 1972, 1974); and for hepatics nomenclature follows Worley (1970). Lichen nomenclature follows Hale and Culberson (1970). For the cryptogams, synonyms are given if they were used in the literature, by the determiners of collections, or in instances when more than one name is commonly in use.

Unless otherwise stated, specimens cited are in the University of Alaska Herbarium (ALA) and have been determined by the collectors. We have made no effort to verify determinations of literature reports unless duplicate specimens are in our herbarium. Other herbaria housing specimens cited in this report are University of Alberta Herbarium (ALTA); University of Calgary Herbarium (UAC); National Herbarium of Canada, Bryophyte Section (CANM); herbarium at the Naval Arctic Research Laboratory, Barrow (NARL); New York Botanical Garden (NY); and University of Wisconsin Herbarium (WIS).

The taxa are listed alphabetically by genus and species for convenience to Tundra Biome participants. Corrections and additions are requested.

Alaska Study Areas

The diversity of plants at any one area, as manifested in the checklists, is a function of several variables. Aside from the obvious differences in the natural features peculiar to each area that affect diversity, it should be noted that the areal extent covered by our inventory and the physical setting represented are not strictly comparable, certainly not equivalent. Furthermore, as one would expect for Alaska, attention to botanical inventory in the past has varied, and background information on floristics was relatively good (Barrow), sketchy (Eagle Summit), or absent (Prudhoe Bay).

The flora at Barrow has been investigated by a number of botanists over a span of almost 25 yr. All of the work there has been conducted under the auspices of the Naval Arctic Research Laboratory, and their logistic base has supported much of the work for the entire Arctic Slope. Ira Wiggins, with his students and colleagues, explored the environs of Barrow and numerous localities to the east and southwest. Their work of several years culminated in a manual to the vascular flora for the Alaskan Arctic Slope (Wiggins and Thomas, 1962). Eric Hultén visited several points in northern Alaska and drew up the first list of plants for the immediate vicinity of the Laboratory at Barrow. His botanical discoveries and the finds of others have been reported (1967) and incorporated in this exemplary flora of Alaska (1968).

Hildur Krog published a study of Alaskan macrolichens (1968) based primarily on her own collections, some of which came from Barrow. John W. Thomson and colleagues were in the field at Barrow as well as at several localities to the east during one summer. His large collection of macro- and microlichens is the basis for a manuscript that he recently completed. Not only is this a major study of the lichen flora for northern Alaska, but also it has keys to genera and species which, when available, will make it easier for more people to learn about lichens. Records of Barrow lichens were extracted from this manuscript which he kindly made available to us.

For 10 field seasons, William C. Steere has studied the bryoflora of the Arctic Slope. He and several other bryologists have published on the taxonomy and distribution of various species, and he is now preparing a comprehensive treatment of Arctic Slope bryophytes. We have included many of Dr. Steere's unpublished records in the checklist. Rastorfer (1972) and Smith (1974) have compiled lists of bryophytes from Barrow.

The Prudhoe Bay flora was unknown until Tundra Biome studies began there in 1970, and the first collections were made in 1971. Numerous habitats are available via the local road system that links drilling pads and related facilities in the oil field. The only published reports are Halliday (1977), Murray and Murray (1975), and Rastorfer et al. (1973). Other Tundra Biome personnel and Drs. Steere and Zennoske Iwatsuki provided us with records.

Eagle Summit has been visited by botanists frequently for a number of years since it is such a beautiful area of alpine tundra and accessible by the Steese Highway from Fairbanks. Many small collections of vascular plants are scattered among a number of herbaria, but most of these have been evaluated by Hultén and included in his treatments of the Alaskan vascular flora (1941–1950, 1968). An important study is that by Gjaerevoll (1958–1967) who not only added vascular plant records, but also brought back bryophytes that were identified by Persson (Persson and Gjaerevoll 1957). The only published records of the lichen flora (Krog 1962, 1968) provided a list of macrolichens, but we also have some unpublished records of Drs. Thomson and Teuvo Ahti. Our list includes records of vascular plants from tundra adjacent to the Eagle Creek Camp occupied by Tundra Biome personnel. Since there has been no intensive study of the Eagle Summit cryptogamic flora, substantial additions to the list can be made in the future.

Acknowledgments. We wish to acknowledge the aid of those Tundra Biome participants who responded to our requests and sent specimens to us and those people cited as determiners. We are grateful to John W. Thomson, University of Wisconsin (Madison), for many determinations and verifications of lichens, for permitting us to cite specimens of lichens from Barrow listed in his manuscript (Thomson, no date) and unpublished records for Eagle Summit collected by him and Teuvo Ahti, and for the use of facilities at the University of Wisconsin Herbarium. We also thank William C. Steere, The New York Botanical Garden, for providing help with bryophyte identifications and for making available to us his lists of specimens collected at the study areas and permitting us to cite these unpublished records. The New York Botanical Garden and its staff generously provided office space and help during the final preparation of this report.

This project was supported through grants and contributions by the National Science Foundation under Grant GV 29342 to the University of Alaska, the State of Alaska, and the Prudhoe Bay Environmental Subcommittee. It was performed under the joint NSF sponsorship of the International Biological Programme and the Office of Polar Programs and was directed under the auspices of the U.S. Tundra Biome.

Checklists

B = Barrow   ES = Eagle Summit   PB = Prudhoe Bay

Vascular Plants

Aconitum delphinifolium DC.
  ES: Scamman, 1940
Agropyron boreale (Turcz.) Drobov ssp. *hyper-arcticum* (Pol.) Meld.
  PB: *D. Murray 4575*
Alopecurus alpinus Sm. ssp. *alpinus*
  B: Hultén, 1968; Packer and McPherson, 1974; Wiggins and Thomas, 1962
  PB: *D. Murray 3410*
Andromeda polifolia L.
  ES: Scamman, 1940
Androsace chamaejasme Host ssp. *lehmanniana* (Spreng.) Hult.
  B: Hultén, 1968; Wiggins and Thomas, 1962
  PB: *D. Murray 3387*
  ES: Gjaerevoll, 1958–1967
Androsace septentrionalis L.
  B: Hultén, 1968; Wiggins and Thomas, 1962
  PB: *D. Murray 4505*
[*Anemone drummondii* S. Wats.
  ES: Weeden, 1968]
Anemone narcissiflora L. ssp. *interior* Hult.
  ES: Gjaerevoll, 1958–1967; Hämet-Ahti, 1971; Hultén, 1941–1950; Scamman, 1940
Anemone parviflora Michx.
  PB: *D. Murray 4531*
  ES: Gjaerevoll, 1958–1967; Scamman, 1940
Anemone richardsonii Hook.
  PB: *D. Murray 4537*
  ES: *Kessel s.n.* (ALA 22881), det. V. L. Harms
Antennaria friesiana (Trautv.) Ekman ssp. *alaskana* (Malte) Hult.
  B: *McPherson and Galeski 398*
  PB: *D. Murray 4571*
  ES: Scamman, 1940
Antennaria monocephala DC.
  B: Packer and McPherson, 1974
  ssp. *philonipha* (Pors.) Hult.
    ES: Gjaerevoll, 1958–1967; Hultén 1941–1950
Arabis lyrata L. ssp. *kamchatica* (Fisch.) Hult.
  PB: *D. Murray 3359*
Arctagrostis latifolia (R. Br.) Griseb.
  var. *arundinacea* (Trin.) Griseb.
    ES: Gjaerevoll, 1958–1967; Hämet-Ahti, 1971
  var. *latifolia*

    B: Hultén, 1968; Packer and McPherson, 1974; Wiggins and Thomas, 1962
  PB: *D. Murray 4554*
Arctophila fulva (Trin.) Anderss.
  B. Hultén, 1968; Packer and McPherson, 1974; Wiggins and Thomas, 1962
  PB: *D. Murray 4555*
Arctostaphylos alpina (L.) Spreng.
  ES: Hultén, 1941–1950
Arctostaphylos rubra (Rehd. & Wils.) Fern.
  PB: *D. Murray 4561*
Armeria maritima (Mill.) Willd. ssp. *arctica* (Cham.) Hult.
  B: Hultén, 1968
  PB: *D. Murray 3356*
Arnica angustifolia Vahl ssp. *attenuata* (Greene) Dougl. & Ruyle-Dougl. (= *A. alpina* ssp. *attenuata,* Douglas and Ruyle-Douglas, 1978)
  ES: *Cross 50,* det. *A. Batten*
Arnica frigida C. A. Mey.
  ES: Gjaerevoll, 1958–1967; Hämet-Ahti, 1971; Scamman 1940
Arnica lessingii Greene ssp. *lessingii*
  ES: Gjaerevoll, 1958–1967; Hämet-Ahti, 1971; Scamman, 1940
Artemisia arctica Less. ssp. *arctica*
  PB: *D. Murray 4568*
  ES: Hämet-Ahti, 1971; Hultén, 1941–1950; Scamman, 1940
Artemisia borealis Pall.
  PB: *D. Murray 3356*
Artemisia glomerata Ledeb.
  PB: *D. Murray 4532*
Artemisia tilesii Ledeb. ssp. *tilesii*
  B: Hultén, 1968
  PB: *D. Murray 4569*
Aster sibiricus L.
  PB: *D. Murray 4574*
Astragalus aboriginorum Richards.
  ES: Gjaerevoll, 1958–1967
Astragalus alpinus L.
  B: Packer and McPherson, 1974; Wiggins and Thomas, 1962
  PB: *D. Murray 4540*
  ssp. *alpinus*

B: Hultén, 1968
ssp. *arcticus* (Bunge) Hult.
B: Hultén, 1968
Astragalus umbellatus Bunge
B: Hultén, 1968; Wiggins and Thomas, 1962
PB: *D. Murray 4517*
ES: Hämet-Ahti, 1971; Hultén, 1941–1950
Betula nana L. ssp. *exilis* (Sukatsch.) Hult.
ES: *Harms 6261*
Boykinia richardsonii (Hook.) Gray
PB: *Walker 503*
Braya pilosa Hook.
PB: *D. Murray 3383*
Braya purpurascens (R. Br.) Bunge
PB: *D. Murray 3385*
Bromus pumpellianus Scribn. var. *arcticus*
(Shear) Pors.
PB: *Walker 570*
Bupleurum triradiatum Adams ssp. *arcticum*
(Regel) Hult.
ES: *Anderson and Densmore 2231*
Calamagrostis holmii Lange
B: Hultén, 1968; Packer and McPherson, 1974;
Wiggins and Thomas, 1962
Calamagrostis lapponica (Wahlenb.) Hartm.
ES: Gjaerevoll, 1958–1967
Caltha palustris L. ssp. *arctica* (R. Br.) Hult.
B: Hultén, 1968; Packer and McPherson, 1974;
Wiggins and Thomas, 1962
PB: *D. Murray 4512*
Campanula lasiocarpa Cham. ssp. *lasiocarpa*
ES: Gjaerevoll, 1958–1967; Hämet-Ahti, 1971;
Hultén, 1941–1950; Scamman, 1940
Campanula uniflora L.
PB: *D. Murray 3398*
ES: Scamman, 1940
Cardamine bellidifolia L.
B: Hultén, 1968; Packer and McPherson, 1974;
Wiggins and Thomas, 1962
ES: Gjaerevoll, 1958–1967, Hämet-Ahti, 1971;
Hultén, 1941–1950; Scamman, 1940
Cardamine digitata Richards. (= *C. hyperborea*
O. E. Schulz)
B: Hultén, 1968; Wiggins and Thomas, 1962
PB: *D. Murray 3399*
Cardamine pratensis L. ssp. *angustifolia*
(Hook.) O. E. Schulz
B: Hultén, 1968; Packer and McPherson, 1974;
Wiggins and Thomas, 1962
PB: *Walker 549*
Cardamine purpurea Cham. & Schlecht.
ES: Gjaerevoll, 1958–1967; Hämet-Ahti, 1971;
Hultén, 1941–1950; Scamman, 1940
Carex aquatilis Wahlenb. (including *C. stans*
Drej.)

B: Hultén, 1968; Packer and McPherson, 1974;
Wiggins and Thomas, 1962
PB: *D. Murray 3586*
Carex atrofusca Schkuhr
PB: *D. Murray 3370*
ES: Gjaerevoll, 1958–1967; Hämet-Ahti, 1971
Carex bigelowii Torr. (including *C. lugens*
Holm, *C. consimilis* Holm)
PB: *D. Murray 3416*
ES: Gjaerevoll, 1958–1967; Hämet-Ahti, 1971
Carex capillaris L.
ES: Gjaerevoll, 1958–1967; Hämet-Ahti, 1971
Carex chordorrhiza Ehrh.
PB: *Walker 288*, det. D. Murray
Carex glareosa Wahlenb. ssp. *glareosa*
B: Hultén, 1968; Wiggins and Thomas, 1962
Carex lachenalii Schkuhr
ES: Gjaerevoll, 1958–1967, Hämet-Ahti,
1971
Carex marina Dewey (= *C. amblyorhyncha*
Krecz., Halliday and Chater, 1969, *C. glar-
eosa* sensu Hultén in part)
PB: *Walker 4*, det. A. Batten
Carex maritima Gunn.
PB: *D. Murray 4514*
Carex media R. Br.
ES: *Anderson and Densmore 2212*
Carex membranacea Hook.
PB: *Walker s.n.*, 5 August 1974, det. D. Murray
ES: Gjaerevoll, 1958–1967
Carex microchaeta Holm. ssp. *microchaeta*
ES: Hämet-Ahti, 1971
Carex misandra R. Br.
PB: *D. Murray 3384*
ES: Gjaerevoll 1958–1967; Hämet-Ahti, 1971
Carex podocarpa R. Br.
ES: Gjaerevoll, 1958–1967; Hultén, 1941–1950;
Lepage, 1951; Scamman, 1940
Carex ramenskii Kom.
B: Hultén, 1968; Wiggins and Thomas, 1962
Carex rariflora (Wahlenb.) J. E. Sm.
PB: *D. Murray 3364*
Carex rupestris All.
PB: *D. Murray 4583*
ES: *Anderson and Densmore 2216*
Carex saxatilis L. ssp. *laxa* (Trautv.) Kalela
PB: *Walker s.n.*, 5 August 1974
Carex scirpoidea Michx.
PB: *D. Murray 4519*
ES: Gjaerevoll, 1958–1967; Hämet-Ahti,
1971
Carex subspathacea Wormsk.
B: Hultén, 1968; Packer and McPherson, 1974;
Wiggins and Thomas, 1962
PB: *Walker and Batty PB039*

Carex ursina Dew.
  B: Hultén, 1968; Packer and McPherson, 1974;
    Wiggins and Thomas, 1962
  PB: *D. Murray 3406*
Carex vaginata Tausch
  PB: *Walker 526*
  ES: *Anderson and Densmore 2182,* det. D.
    Murray
Cassiope tetragona (L.) D. Don ssp. *tetragona*
  B: Hultén, 1968; Packer and McPherson, 1974;
    Wiggins and Thomas, 1962
  PB: *D. Murray 4539*
  ES: Gjaerevoll, 1958–1967; Hultén, 1941–1950;
    Scamman, 1940
Castilleja hyperborea Pennell
  ES: Gjaerevoll, 1958–1967; Scamman, 1940
Cerastium beeringianum Cham. & Schlecht.
  ES: Hämet-Ahti, 1971; Scamman, 1940
  var. *beeringianum*
    B: Hultén, 1968; Wiggins and Thomas, 1962
    PB: *D. Murray 4538*
  var. *grandiflorum* (Fenzl) Hult.
    ES: Scamman, 1940 (Hulten, 1965)
Cerastium fontanum Baumg. ssp. *triviale* (Link)
  Jalas
  ES: Hämet-Ahti, 1971
Cerastium jenisejense Hult.
  B: Hultén, 1968; Packer and McPherson, 1974
Chrysanthemum bipinnatum L. ssp. *bipinnatum*
  PB: Halliday, 1977
Chrysanthemum integrifolium Richards.
  PB: *D. Murray 3394*
Chrysosplenium tetrandrum (Lund) Th. Fr.
  B: Hultén, 1968; Packer and McPherson, 1974;
    Wiggins and Thomas, 1962
  PB: *D. Murray 4525*
  ES: Gjaerevoll, 1958–1967
Claytonia sarmentosa C. A. Mey.
  ES: Gjaerevoll, 1958–1967; Hämet-Ahti, 1971;
    Hultén, 1941–1950; Scamman, 1940
Claytonia scammaniana Hult.
  ES: *Cross s.n.,* 10 July 1971 (type locality on
    nearby Porcupine Dome)
Claytonia tuberosa Pall.
  ES: Gjaerevoll, 1958–1967; Scamman, 1940
Cochlearia officinalis L. ssp. *arctica* (Schlecht.)
  Hult.
  B: Hultén, 1968; Packer and McPherson, 1974;
    Wiggins and Thomas, 1962
  PB: *D. Murray 4511*
Colpodium vahlianum (Liebm.) Nevski
  PB: Halliday, 1977
Cornus canadensis L.
  ES: *Herrick s.n.* (ALA 18519)

Corydalis pauciflora (Steph.) Pers.
  ES: Gjaerevoll, 1958–1967; Scamman, 1940
Crepis nana Richards.
  ES: *Cross 53*
Cystopteris fragilis (L.) Bernh.
  ES: Hultén, 1941–1950 (in Suppl. p. 1697)
Cystopteris montana (Lam.) Bernh.
  ES: *Anderson and Densmore 2220*
Delphinium chamissonis Pritz (= *D. brachycen-*
  trum of authors)
  ES: Gjaerevoll, 1958–1967; Hämet-Ahti, 1971;
    Scamman, 1940
Deschampsia cespitosa (L.) Beauv. ssp. *orien-*
  talis Hult.
  B: Hultén, 1968; Wiggins and Thomas, 1962
  PB: *D. Murray 3412*
Deschampsia pumila (Trin.) Ostenf.
  B: Hultén, 1968
Descurainia sophioides (Fisch.) O. E. Schulz
  PB: *J. McKendrick s.n.,* 11 September 1976
Diapensia lapponica L. ssp. *obovata* (F. Schm.)
  Hult.
  ES: Gjaerevoll, 1958–1967
Dodecatheon frigidum Cham. & Schlecht.
  PB: *Walker 308*
  ES: Hämet-Ahti, 1971; Scamman, 1940
Douglasia arctica Hook.
  ES: Gjaerevoll, 1958–1967
Draba adamsii Ledebour (= *D. oblongata* R.
  Br., Mulligan, 1974)
  B: Hultén, 1968; Mulligan, 1974; Packer and
    McPherson, 1974
Draba alpina L.
  B: Hultén, 1968; Packer and McPherson, 1974;
    Wiggins and Thomas, 1962
  PB: *D. Murray 3381,* det. G. A. Mulligan
Draba chamissonis G. Don
  B: Hultén, 1968; Wiggins and Thomas, 1962
Draba cinerea Adams
  B: Hultén, 1968; Wiggins and Thomas, 1962
  PB: *D. Murray 3402,* det. G. A. Mulligan
Draba corymbosa R. Br. *ex* DC. (= *D. bellii*
  Holm, *D. macrocarpa* Adams; Mulligan,
  1974)
  B: Hultén, 1968; Wiggins and Thomas, 1962
  PB: *D. Murray 3371,* det. G. A. Mulligan
Draba fladnizensis Wulf.
  B: Hultén, 1968
  ES: Gjaerevoll, 1958–1967 mentions Scamman
    specimens, but not cited by Scamman, 1940
Draba glabella Pursh (= *D. hirta* L., Mulligan,
  1970)
  PB: *Walker 272,* det. D. Murray
  ES: *Anderson and Densmore 2294*

Draba lactea Adams
  B: Hultén, 1968; Packer and McPherson, 1974;
    Wiggins and Thomas, 1962
  PB: *D. Murray 3382,* det. G. A. Mulligan
Draba longipes Raup
  PB: *Walker 241,* det. D. Murray
Draba nivalis Liljebl.
  B: Hultén, 1968
[*Dryas "alaskana"*
  ES: Shacklette, 1962]
Dryas integrifolia M. Vahl ssp. *integrifolia*
  B: Hultén, 1968; Packer and McPherson, 1974;
    Wiggins and Thomas, 1962
  PB: *D. Murray 4533*
Dryas octopetala L.
  ES: Hämet-Ahti, 1971; Scamman, 1940
  ssp. *alaskensis* (Pors.) Hult.
  ES: *Kessel 6,* det. A. Batten
Dryopteris fragrans (L.) Schott
  ES: Hultén, 1941–1950
Dupontia fisheri R. Br.
  ssp. *fisheri*
  B: Hultén, 1968; Packer and McPherson, 1974;
    Wiggins and Thomas, 1962
  ssp. *psilosantha* (Rupr.) Hult.
  B: Wiggins and Thomas, 1962
  PB: *D. Murray 4563*
Elymus arenarius L. ssp. *mollis* (Trin.) Hult.
  B: Packer and McPherson, 1974; Wiggins and
    Thomas, 1962
  var. *villosissimus* (Scribn.) Hult.
  B: Hultén, 1968
  PB: *D. Murray 3411*
Empetrum nigrum L. ssp. *hermaphroditum*
  (Lange) Böcher
  ES: Gjaerevoll, 1958–1967
Epilobium anagallidifolium Lam.
  ES: *Anderson and Densmore 2295*
Epilobium angustifolium L. ssp. *angustifolium*
  ES: *Hatler 42*
Epilobium davuricum Fisch. var. *arcticum*
  (Sam.) Polunin
  PB: Halliday, 1977
Epilobium latifolium L.
  PB: *Walker 557*
  ES: Hultén, 1941–1950
Equisetum arvense L.
  B: Hultén, 1968; Packer and McPherson, 1974;
    Wiggins and Thomas, 1962
  PB: *D. Murray 4515*
  ES: *Trent 21*
Equisetum scirpoides Michx.
  PB: *D. Murray 3380*
  ES: *Anderson and Densmore 2247*

Equisetum variegatum Schleich.
  PB: *D. Murray 4565*
Erigeron eriocephalus J. Vahl
  B: *McPherson and Galeski 330*
  PB: *D. Murray 4545*
Erigeron humilis Grah.
  PB: *D. Murray 3378*
Erigeron purpuratus Greene
  ES: Hultén, 1941–1950; Scamman, 1940
Eriophorum angustifolium Honck.
  B: Packer and McPherson, 1974; Wiggins and
    Thomas, 1962
  [ES: Hanson, 1950]
  ssp. *subarcticum* (Vassil.) Hult.
  B: *Keeley 1920*
  PB: *Walker s.n.,* 5 August 1974
Eriophorum callitrix Cham.
  PB: *Walker 327,* det. D. Murray
Eriophorum russeolum Fr.
  B: Packer and McPherson, 1974
  var. *albidum* Nyl.
  B: Hultén, 1968; Wiggins and Thomas, 1962
  ssp. *rufescens* (Anders.) Hyl.
  B: Hultén, 1968; Wiggins and Thomas, 1962
Eriophorum scheuchzeri Hoppe var.
  scheuchzeri
  B: Hultén, 1968; Wiggins and Thomas, 1962
  PB: *D. Murray 3405*
Eriophorum triste (Th. Fr.) Hadac and Löve
  (= *E. angustifolium* ssp. *triste*)
  B: Hultén, 1968
  PB: *D. Murray 3375*
  ES: Gjaerevoll, 1958–1967; Hämet-Ahti,
    1971
Eriophorum vaginatum L.
  B: Packer and McPherson, 1974
  PB: *D. Murray 4550*
  ssp. *spissum* (Fern.) Hult.
  B: Wiggins and Thomas, 1962
  ES: Hämet-Ahti, 1971
Eritrichum chamissonis DC. (= *E. aretioides* of
  authors)
  ES: Gjaerevoll, 1958–1967; Hämet-Ahti, 1971;
    Hultén, 1941–1950; Scamman, 1940
Eutrema edwardsii R. Br.
  B: Hultén, 1968; Packer and McPherson, 1974;
    Wiggins and Thomas, 1962
  PB: *D. Murray 3368*
  ES: Gjaerevoll, 1958–1967; Hämet-Ahti, 1971;
    Scamman, 1940
Festuca altaica Trin.
  ES: Gjaerevoll, 1958–1967
Festuca baffinensis Polunin
  PB: *D. Murray 3417*

Festuca brachyphylla Schult.
  B: Hultén, 1968; Packer and McPherson, 1974;
    Wiggins and Thomas, 1962
  PB: *D. Murray 4564*
  ES: Hämet-Ahti, 1971
Festuca rubra L.
  PB: *D. Murray 3415*
Festuca saximontana Rydb.
  ES: Gjaerevoll, 1958–1967
Gentiana algida Pall.
  ES: *Trent 134-1965*
Gentiana glauca Pall.
  ES: Hämet-Ahti, 1971; Scamman, 1940
Gentiana prostrata Haenke
  PB: *D. Murray 4556*
  ES: Gjaerevoll, 1958–1967
Gentianella propinqua (Richards.) J. M. Gillett
  ssp. *propinqua* (= *Gentiana propinqua*)
  PB: *D. Murray 3407*
Geum rossii (R. Br.) Ser. (= *Acomastylis rossii*
  (R. Br.) Greene)
  ES: Gjaerevoll, 1958–1967; Hämet-Ahti, 1971;
    Hultén, 1941–1950; Scamman, 1940
Hedysarum mackenzii Richards.
  B: Hultén, 1968
Hierochloe alpina (Sw.) Roem. & Schult.
  B: Hultén, 1968; Packer and McPherson, 1974;
    Wiggins and Thomas, 1962
  ES: Hämet-Ahti, 1971; Scamman, 1940
Hierochloe pauciflora R. Br.
  B: Hultén, 1968; Packer and McPherson, 1974;
    Wiggins and Thomas, 1962
  PB: *Walker 1*
Hippuris vulgaris L.
  B: Hultén, 1968; Packer and McPherson, 1974;
    Wiggins and Thomas, 1962
  PB: *D. Murray 4552*
Honckenya peploides (L.) Ehrh. ssp. *peploides*
  B: Hultén, 1968; Packer and McPherson, 1974;
    Wiggins and Thomas, 1962
Juncus arcticus Willd. ssp. *alaskanus* Hult.
  PB: *D. Murray 4553*
Juncus biglumis L.
  B: Hultén, 1968; Packer and McPherson, 1974;
    Wiggins and Thomas, 1962
  PB: *D. Murray 4560*
  ES: Gjaerevoll, 1958–1967; Hämet-Ahti, 1971
Juncus castaneus Sm. ssp. *castaneus*
  PB: *D. Murray 3404*
  ES: Hämet-Ahti, 1971
Juncus triglumis L. ssp. *albescens* (Lange) Hult.
  PB: *Walker and Batty s.n.,* August 1974
  ES: Gjaerevoll, 1958–1967; Hämet-Ahti, 1971
Kobresia myosuroides (Vill.) Fiori & Paol.
  PB: *D. Murray 4557*

Kobresia sibirica Turcz.
  PB: *D. Murray 3352*
Koenigia islandica L.
  PB: *D. Murray and Johnson 6207*
Lagotis glauca Gaertn. ssp. *minor* (Willd.) Hult.
  PB: *D. Murray 4526*
  ES: Hämet-Ahti, 1971; Scamman, 1940
Lappula myosotis Moench
  ES: Gjaerevoll, 1958–1967
Ledum palustre L. ssp. *decumbens* (Ait.)
  Hult.
  ES: Hämet-Ahti, 1971; Scamman, 1940
Lesquerella arctica (Wormsk.) Wats. ssp.
  arctica
  PB: *D. Murray 3395*
Ligusticum mutellinoides (Grantz) Willar
  ssp. *alpinum* (Ledeb.) Thell.
  ES: Hämet-Ahti, 1971
Linnaea borealis L.
  ES: *Anderson and Densmore 2282*
Lloydia serotina (L.) Rchb.
  PB: *D. Murray 3390*
  ES: Gjaerevoll, 1958–1967; Scamman, 1940
Loiseleuria procumbens (L.) Desv.
  ES: Gjaerevoll, 1958–1967; Hämet-Ahti, 1971;
    Hultén, 1941–1950; Scamman, 1940
Lupinus arcticus S. Wats.
  ES: Hultén, 1941–1950
Luzula arctica Blytt
  B: Hultén, 1968; Packer and McPherson, 1974;
    Wiggins and Thomas, 1962
  PB: *D. Murray 4580*
  ES: Gjaerevoll, 1958–1967
Luzula arcuata (Wahlenb.) Sw.
  ES: *Anderson and Densmore 2296*
Luzula confusa Lindeb.
  B. Hultén, 1968; Packer and McPherson, 1974;
    Wiggins and Thomas, 1962
  ES: Hämet-Ahti, 1971
Luzula kjellmaniana Miyabe & Kudo (= *L. tun-
  dricola* Gorodk., Hämet-Ahti and Virrankoski,
  1971)
  B: Hultén, 1968
  ES: Hämet-Ahti, 1971; Hämet-Ahti and Virran-
    koski, 1971
Luzula multiflora (Retz.) Lej. ssp. *multiflora*
  var. *frigida* (Buchenau) Sam.
  ES: Hämet-Ahti, 1971
Luzula wahlenbergii Rupr.
  ES: Gjaerevoll, 1958–1967
Lycopodium alpinum L.
  ES: *Anderson and Densmore 2222*
Lycopodium annotinum L. ssp. *pungens* (La
  Pyl.) Hult.
  ES: Hämet-Ahti, 1971; Scamman, 1940

Lycopodium complanatum L.
  ES: *Anderson and Densmore 2259*
Lycopodium selago L.
  ES: Scamman, 1940
  ssp. *appressum* (Desv.) Hult.
    ES: Hämet-Ahti, 1971
Mertensia maritima (L.) S. F. Gray ssp.
    maritima
  B: Hultén, 1968; Packer and McPherson, 1974;
    Wiggins and Thomas, 1962
Mertensia paniculata (Ait.) G. Don
  ES: Gjaerevoll, 1958–1967
Minuartia arctica (Stev.) Aschers, & Graebn.
  PB: *D. Murray 3379*
Minuartia dawsonensis (Britt.) Mattf.
  ES: Scamman, 1940
Minuartia macrocarpa (Pursh) Ostenf.
  ES: Gjaerevoll, 1958–1967; Hämet-Ahti, 1971;
    Scamman, 1940
[*Minuartia obtusiloba* (Rydb.) House
  ES: Weeden, 1968]
Minuartia rossii (R. Br.) Graebn.
  PB: *Schofield and Williams P-G16*
  var. *elegans* (Cham. & Schlecht.) Hult.
    ES: Gjaerevoll, 1958–1967; Scamman,
    1940
Minuartia rubella (Wahlenb.) Graebn.
  B: Hultén, 1968; Wiggins and Thomas, 1962
  PB: *D. Murray 3403*
Minuartia stricta (Sw.) Hiern
  ES: Hämet-Ahti, 1971
Myosotis alpestris F. W. Schmidt ssp. *asiatica*
    Vestergr.
  ES: *Trent 95-1965*
Orthilia secunda (L.) House ssp. *obtusata*
    (Turcz.) Böcher (= *Pyrola secunda* ssp.
    obtusata)
  PB: *Walker and Batty PB005*
Oxyria digyna (L.) Hill
  B: Hultén, 1968; Packer and McPherson, 1974;
    Wiggins and Thomas, 1962
  PB: *D. Murray 4520*
  ES: Hämet-Ahti, 1971; Hultén, 1941–1950;
    Scamman, 1940
Oxytropis arctica R. Br.
  PB: *D. Murray 3396*
Oxytropis borealis DC.
  PB: *D. Murray 4559*
Oxytropis deflexa (Pall.) DC. var. *foliolosa*
    (Hook.) Barneby
  PB: *D. Murray 4584*
Oxytropis maydelliana Trautv.
  PB: *D. Murray 4513*
  ES: Hämet-Ahti, 1971; Hultén, 1941–1950;
    Scamman, 1940

Oxytropis mertensiana Turcz.
  ES: Gjaerevoll, 1958–1967; Hämet-Ahti, 1971;
    Hultén, 1941–1950; Scamman, 1940; Welsh,
    1967
Oxytropis nigrescens (Pall.) Fisch. ssp. *bry-*
    ophila (Greene) Hult.
  PB: *D. Murray 4541*
  ES: Hämet-Ahti, 1971; Weeden, 1968; Welsh,
    1967; Scamman, 1940 cited *O. pygmaea* (Pall.)
    Fern. which is probably this taxon
Oxytropis scammaniana Hult.
  ES: (type locality) Gjaerevoll, 1958–1967; Hul-
    tén, 1941–1950; Scamman, 1940; Welsh, 1967
Papaver hultenii Knaben
  B: Hultén, 1968; Packer and McPherson, 1974
Papaver lapponicum (Tolm.) Nordh. ssp. *occi-*
    dentale (Lundstr.) Knaben
  B: Hultén, 1968; Wiggins and Thomas, 1962
  PB: *D. Murray 4521*
  ES: Gjaerevoll, 1958–1967
Papaver macounii Greene
  B: Hultén, 1968; Packer and McPherson, 1974;
    Wiggins and Thomas, 1962
  PB: *D. Murray 3377*
  ES: Gjaerevoll, 1958–1967; Hämet-Ahti, 1971;
    Scamman, 1940
Parnassia kotzebuei Cham. & Schlecht.
  PB: *D. Murray 4570*
  ES: Scamman, 1940
Parrya nudicaulis (L.) Regel.
  ES: Scamman, 1940
  cf. ssp. *interior* Hult.
    ES: Hämet-Ahti, 1971
  ssp. *nudicaulis*
    PB: *D. Murray 3408*
  ssp. *nudicaulis* var. *grandiflora* Hult.
    ES: Hultén, 1968
Pedicularis capitata Adams
  PB: *D. Murray 3386*
  ES: Scamman, 1940
Pedicularis hirsuta L.
  PB: Halliday, 1977
Pedicularis labradorica Wirsing
  ES: *Harms 6267*
Pedicularis lanata Cham. & Schlecht. (= *P.*
    kanei Durand)
  B: Hultén, 1968; Packer and McPherson, 1974;
    Wiggins and Thomas, 1962
  PB: *D. Murray 3556*
  ES: Scamman, 1940
Pedicularis langsdorffii Fisch. ssp. *arctica* (R.
    Br.) Pennell
  B: Hultén, 1968; Wiggins and Thomas, 1962
  PB: *D. Murray 3362*
  ES: Scamman, 1940

Pedicularis oederi M. Vahl
 ES: Gjaerevoll, 1958–1967; Hämet-Ahti, 1971;
 Scamman, 1940
Pedicularis sudetica Willd.
 B: Packer and McPherson, 1974; Wiggins and
 Thomas, 1962
 PB: *D. Murray 3391*
 ssp. *albolabiata*
 B: Hultén, 1968
 PB: *D. Murray 3372*
 ssp. *interior* Hult.
 ES: Gjaerevoll, 1958–1967; Scamman, 1940
Pedicularis verticillata L.
 ES: *Hatler 61*
Petasites frigidus (L.) Franch.
 B: Hultén, 1968; Packer and McPherson, 1974;
 Wiggins and Thomas, 1962
 PB: *D. Murray 4582*
 ES: *Hatler 5*
Phippsia algida (Soland.) R. Br.
 B: Hultén, 1968; Packer and McPherson, 1974;
 Wiggins and Thomas, 1962
 PB: Halliday, 1977
Pleuropogon sabinei R. Br.
 PB: Halliday, 1977
Poa alpigena (Fr.) Lindm.
 B: Packer and McPherson, 1974; Wiggins and
 Thomas, 1962
 PB: *D. Murray 4576*
Poa arctica R. Br.
 B: Packer and McPherson, 1974
 PB: *Walker 330,* det. D. Murray
 ES: Scamman, 1940
 ssp. *arctica*
 B: Hultén, 1968; Wiggins and Thomas, 1962
 ES: Gjaerevoll, 1958–1967; Hämet-Ahti, 1971
 ssp. *caespitans* (Simmons) Nannf.
 B: Hultén, 1968
 ssp. *williamsii* (Nash) Hult.
 ES: Gjaerevoll, 1958–1967
Poa glauca M. Vahl
 PB: *D. Murray 3419*
Poa malacantha Kom.
 B: Hultén, 1968; Wiggins and Thomas, 1962
Poa paucispicula Scribn. & Merr. (including *P.
 leptocoma* Trin.)
 ES: Gjaerevoll, 1958–1967; Hämet-Ahti, 1971
Poa vaseyochloa Scribn.
 ES: Hämet-Ahti, 1971
Polemonium acutiflorum Willd.
 B: Hultén, 1968
 PB: Halliday, 1977
 ES: Scamman, 1940
Polemonium boreale Adams
 PB: *D. Murray 3353*

Polygonum bistorta L. ssp. *plumosum* (Small)
 Hult. (= *Bistorta plumosa* (Small) Greene)
 PB: *Walker 528*
 ES: Scamman, 1940
Polygonum viviparum L. (= *Bistorta vivipara*
 (L.) S. F. Gray)
 B: Hultén, 1968; Packer and McPherson, 1974;
 Wiggins and Thomas, 1962
 PB: *D. Murray 3389*
 ES: Gjaerevoll, 1958–1967; Scamman, 1940
Potentilla biflora Willd.
 ES: *Trent 81-1965,* det. V. Harms
Potentilla fruticosa Willd.
 ES: Gjaerevoll, 1958–1967
Potentilla hookeriana Lehm. ssp. *hookeriana*
 PB: *D. Murray 3401*
Potentilla hyparctica Malte
 B: Hultén, 1968; Packer and McPherson, 1974;
 Wiggins and Thomas, 1962
Potentilla pulchella R. Br.
 B: Hultén, 1968; Packer and McPherson, 1974;
 Wiggins and Thomas, 1962
 PB: *D. Murray 3358*
Potentilla uniflora Ledeb.
 PB: *D. Murray 4529*
 ES: Gjaerevoll, 1958–1967
Primula borealis Duby
 PB: *D. Murray 4510*
Puccinellia andersonii Swallen
 PB: *D. Murray 3414,* ver. A. E. Porsild
Puccinellia angustata (R. Br.) Rand & Redf.
 PB: Halliday, 1977
Puccinellia langeana (Berl.) Sørens.
 B: Hultén, 1968; Wiggins and Thomas, 1962
Puccinellia phryganodes (Trin.) Scribn. & Merr.
 B: Hultén, 1968; Packer and McPherson, 1974;
 Wiggins and Thomas, 1962
 PB: *D. Murray 4567*
Pyrola asarifolia Michx. var. *purpurea* (Bunge)
 Fern.
 ES: Hämet-Ahti, 1971
Pyrola grandiflora Radius
 PB: *Walker 545*
 ES: Gjaerevoll, 1958–1967; Hämet-Ahti, 1971;
 Scamman, 1940
Ranunculus gmelinii DC. ssp. *gmelinii*
 B: Hultén, 1968; Packer and McPherson, 1974;
 Wiggins and Thomas, 1962
Ranunculus hyperboreus Rottb. ssp.
 hyperboreus
 B: Hultén, 1968; Wiggins and Thomas, 1962
 PB: Halliday, 1977
Ranunculus nivalis L.
 B: Hultén, 1968; Packer and McPherson, 1974;
 Wiggins and Thomas, 1962

PB: *D. Murray 3555*
ES: Gjaerevoll, 1958–1967; Hämet-Ahti, 1971; Scamman, 1940
Ranunculus pallasii Schlecht.
B: Hultén, 1968; Packer and McPherson, 1974; Wiggins and Thomas, 1962
PB: Halliday, 1977
Ranunculus pedatifidus Sm. ssp. *affinis* (R. Br.) Hult.
PB: *D. Murray 4536*
Ranunculus pygmaeus Wahlenb.
B: Packer and McPherson, 1974
ssp. *pygmaeus*
B: Hultén, 1968; Wiggins and Thomas, 1962
ES: Gjaerevoll, 1958–1967; Hultén, 1941–1950; Scamman, 1940
ssp. *sabinei* (R. Br.) Hult.
B: Hultén, 1968; Wiggins and Thomas, 1962
Ranunculus sulphureus Soland. var. *sulphureus*
B: Hultén, 1968; Packer and McPherson, 1974; Wiggins and Thomas, 1962
Ranunculus trichophyllus Chaix. ssp. *eradicatus* (Laest.) Cook
PB: *Walker 532*
Rhododendron lapponicum (L.) Wahlenb.
ES: Gjaerevoll, 1958–1967; Hultén, 1941–1950
Rubus arcticus L.
ES: *Hatler 47*
Rubus chamaemorus L.
B: Hultén, 1968; Wiggins and Thomas, 1962
ES: *Trent 21-1965*
Rumex arcticus Trautv.
B: Hultén, 1968; Packer and McPherson, 1974; Wiggins and Thomas, 1962
Sagina intermedia Fenzl
B: Hultén, 1968; Packer and McPherson, 1974; Wiggins and Thomas, 1962
PB: *D. Murray 4562*
Salix alaxensis (Anderss.) Cov. var. *alaxensis*
PB: *D. Murray 4566*
Salix arctica Pall.
PB: *D. Murray 4523*
ES: Argus, 1973; Gjaerevoll, 1958–1967; Hämet-Ahti, 1971; Hultén, 1941–1950
Salix chamissonis Anderss.
ES: Argus, 1973; Gjaerevoll, 1958–1967
Salix fuscescens Anderss.
B: Argus, 1973; Hultén, 1968; Wiggins and Thomas, 1962
Salix hastata L.
ES: Argus, 1973
Salix lanata L. ssp. *richardsonii* (Hook.) A. Skvortz.
B: Argus, 1973
PB: *D. Murray 3351*

ES: *Anderson and Densmore 2184*
Salix ovalifolia Trautv.
B: Hultén, 1968; Packer and McPherson, 1974; probably Wiggins and Thomas, 1962 as *S. pseudopolaris*
var. *arctolitoralis* (Hult.) Argus (= *S. arctolitoralis* Hult.)
B: Hultén, 1968; Wiggins and Thomas, 1962; not cited for Barrow by Argus 1973
var. *glacialis* (Anderss.) Argus *(= S. arctica × ovalifolia)*
B: Argus, 1973
var. *ovalifolia*
B: Argus, 1973
PB: *D. Murray 3366*
Salix phlebophylla Anderss.
B: Argus, 1973; Hultén, 1968; Packer and McPherson, 1974; Wiggins and Thomas, 1962
ES: Argus, 1973; Gjaerevoll, 1958–1967; Hultén, 1941–1950; Scamman, 1940, incorrectly cited as *S. rotundifolia* according to Hultén, 1941–1950
Salix planifolia Pursh ssp. *pulchra* (Cham.) Argus var. *pulchra*
B: Argus, 1973; Hultén, 1968; Wiggins and Thomas, 1962
PB: *D. Murray 4522*
ES: Gjaerevoll, 1958–1967; Hultén, 1941–1950
Salix polaris Wahlenb.
ES: Argus, 1973; Gjaerevoll, 1958–1967; Scamman, 1940
Salix pseudomonticola Ball (= *S. monticola* Bebb, Dorn 1975)
ES: Argus, 1973
Salix reticulata L. ssp. *reticulata*
B: Argus, 1973; Hultén, 1968; Packer and McPherson, 1974; Wiggins and Thomas, 1962
PB: *D. Murray 4534*
ES: Argus, 1973; Gjaerevoll, 1958–1967; Hämet-Ahti, 1971
Salix rotundifolia Trautv. ssp. *rotundifolia*
B: Argus, 1973; Hultén, 1968; Packer and McPherson, 1974; Wiggins and Thomas, 1962
PB: *D. Murray 4548*
ES: Argus, 1973
Saussurea angustifolia (Willd.) DC.
PB: *Walker 555*
ES: Gjaerevoll, 1958–1967; Scamman, 1940, as *S. densa* Hook.
Saussurea viscida Hult. var. *yukonensis* (Pors.) Hult.
B: Hultén, 1968
ES: Hämet-Ahti, 1971; Scamman, 1940, as *S. densa*

Saxifraga adscendens L. ssp. *oregonensis* (Raf.)
  Bacigalupi
  ES: Scamman, 1940
Saxifraga bronchialis L. ssp. *funstonii* (Small)
  Hult.
  B: Hultén, 1968
  ES: Hämet-Ahti, 1971; Hultén, 1941–1950;
  Scamman, 1940
Saxifraga caespitosa L.
  B: Hultén, 1968; Packer and McPherson, 1974;
  Wiggins and Thomas, 1962
  PB: *D. Murray 4546*
Saxifraga cernua L.
  B: Hultén, 1968; Packer and McPherson, 1974;
  Wiggins and Thomas, 1962
  PB: *D. Murray 4547*
  ES: Gjaerevoll, 1958–1967
Saxifraga calycina Sternb. (= *S. davurica*
  Willd. ssp. *grandipetala* (Engler & Irmsch.)
  Hult.
  ES: Gjaerevoll, 1958–1967
Saxifraga flagellaris Willd.
  ssp. *platysepala* (Trautv.) Pors.
  B: Hultén, 1968; Packer and McPherson, 1974;
  Wiggins and Thomas, 1962
  ssp. *setigera* (Pursh) Tolm.
  ES: Hämet-Ahti, 1971
Saxifraga foliolosa R. Br.
  B: Packer and McPherson, 1974
  PB: Halliday, 1977
  ES: Scamman, 1940
  var. *foliolosa*
  B: Hultén, 1968; Wiggins and Thomas, 1962
Saxifraga hieracifolia Waldst. & Kit.
  B: Hultén, 1968; Packer and McPherson, 1974;
  Wiggins and Thomas, 1962
  PB: *D. Murray 4543*
  ES: Gjaerevoll, 1958–1967; Hämet-Ahti, 1971;
  Scamman, 1940
Saxifraga hirculus L.
  B: Hultén, 1968; Packer and McPherson, 1974;
  Wiggins and Thomas, 1962
  PB: *D. Murray 3369*
  ES: Gjaerevoll, 1958–1967; Hämet-Ahti, 1971;
  Scamman, 1940
Saxifraga nelsoniana D. Don (= *S. punctata*
  L. ssp. *nelsoniana*)
  B: Hultén, 1968; Packer and McPherson, 1974;
  Wiggins and Thomas, 1962
  ES: Calder and Savile, 1960; Hämet-Ahti, 1971;
  Hultén 1941–1950; Scamman, 1940
Saxifraga nivalis L.
  B: Hultén, 1968; Packer and McPherson, 1974;
  Wiggins and Thomas, 1962

  var. *rufopilosa* Hult.
  ES: Hultén (1967) wrote that *S. ?eriophora* S.
  Wats. cited by Porsild (1965) is this taxon;
  Weeden, 1968 as *Saxifraga* sp. Treated by
  Porsild (1974) as *S. rufopilosa*
Saxifraga oppositifolia L. ssp. *oppositifolia*
  B: Hultén, 1968; Wiggins and Thomas, 1962
  PB: *D. Murray 4524*
Saxifraga reflexa Hook.
  B: *McPherson and Galeski 336* (ALTA)
  ES: Gjaerevoll, 1958–1967
Saxifraga rivularis L. (including *S. hyperborea*
  R. Br.)
  B: Packer and McPherson, 1974
  PB: Halliday, 1977
  var. *flexuosa* (Sternb.) Engler & Irmsch.
  ES: Gjaerevoll, 1958–1967; Scamman, 1940
  var. *rivularis*
  B: Hultén, 1968; Wiggins and Thomas, 1962
Saxifraga tricuspidata Rottb.
  PB: *D. Murray 4544*
Sedum rosea (L.) Scop. ssp. *integrifolium* (Raf.)
  Hult.
  PB: *D. Murray 3354*
  ES: Scamman, 1940
Selaginella sibirica (Milde) Hieron.
  ES: *Anderson and Densmore 2260*
Senecio atropurpureus (Ledeb.) Fedtsch.
  ssp. *frigidus* (Richards.) Hult.
  B: Hultén, 1968; Packer and McPherson, 1974;
  Wiggins and Thomas, 1962
  PB: *D. Murray 3365*
  ES: Gjaerevoll, 1958–1967; Scamman, 1940
  ssp. *tomentosus* (Kjellm.) Hult.
  ES: Gjaerevoll, 1958–1967; Hämet-Ahti, 1971;
  Hultén, 1941–1950; Porsild, 1974; Scamman,
  1940
Senecio congestus (R. Br.) DC.
  B: Hultén, 1968; Packer and McPherson, 1974;
  Wiggins and Thomas, 1962
  PB: *Walker s.n.*, July 1974
Senecio fuscatus (Jord. & Fourr.) Hayek
  ES: Hämet-Ahti, 1971; Hultén, 1941–1950
[*Senecio lugens* Richards.
  ES: Weeden, 1968]
Senecio resedifolius Less.
  PB: *D. Murray 3376*
  ES: Scamman, 1940
Senecio yukonensis Pors.
  ES: Gjaerevoll, 1958–1967; Hämet-Ahti, 1971
Silene acaulis L.
  PB: *D. Murray 4535*
  ssp. *acaulis*
  ES: Scamman, 1940

ssp. *subacaulescens* (F. N. Williams) Hult.
  ES: Hämet-Ahti, 1971
Silene involucrata (Cham. & Schlecht.) Bocq.
  (= *Melandrium affine* J. Vahl)
  PB: *D. Murray 3373*
Silene macrosperma (Pors.) Hult. (= *Melandrium macrospermum* Pors.)
  ES: *Scamman 764*
  Hultén (1941–1950) referred with reservation the report of *Melandrium taylorae* (Robins.) Tolm. by Scamman (1940) to *M. macrospermum*. See also note under *Silene wahlbergella*.
Silene repens Patrin
  ES: Scamman, 1940
Silene uralensis (Rupr.) Bocquet (= *Melandrium apetalum* (L.) Fenzl ssp. *arcticum* (Fr.) Hult., McNeill, 1978)
  B: Hultén, 1968; Packer and McPherson, 1974; Wiggins and Thomas, 1962
  PB: *D. Murray 3363*
  ES: Gjaerevoll, 1958–1967; Hämet-Ahti, 1971; Scamman, 1940. Gjaerevoll (1958–1967) also reported *Melandrium soczavianum* Schischk. which, according to Hultén (1968), is a synonym of *M. apetalum*. Gjaerevoll (1958–1967) said his specimen might be *M. macrospermum*, which is placed in the Eagle Summit area by Hultén (1968).
Solidago multiradiata Ait.
  ES: *Anderson and Densmore 2299*
Spiraea stevenii (Schneid.) Rydb. (= *S. beauverdiana* Schneid., Uttal, 1973)
  ES: Gjaerevoll, 1958–1967; Hämet-Ahti, 1971
Stellaria calycantha (Ledeb.) Bong.
  ES: Gjaerevoll, 1958–1967
Stellaria crassifolia Ehrh.
  B: Hultén, 1968; Wiggins and Thomas, 1962
Stellaria edwardsii R. Br.
  B: Hultén, 1968; Wiggins and Thomas, 1962
Stellaria humifusa Rottb.
  B: Hultén, 1968; Packer and McPherson, 1974; Wiggins and Thomas, 1962
  PB: *Walker and Batty PB037*
Stellaria laeta Richards.
  B: Hultén, 1968; Packer and McPherson, 1974; Wiggins and Thomas, 1962
  PB: *D. Murray 4549*
  ES: Gjaerevoll, 1958–1967; Hämet-Ahti, 1971
Stellaria "laxmanii"
  ES: Hämet-Ahti, 1971
Stellaria longipes Goldie
  ES: Hämet-Ahti, 1971; Scamman, 1940

Stellaria umbellata Turcz.
  ES: Gjaerevoll, 1958–1967
[*Syntheris borealis* Pennell
  ES: Weeden, 1968]
Taraxacum alaskanum Rydb.
  B: Hultén, 1968; Packer and McPherson, 1974; Wiggins and Thomas, 1962
Taraxacum ceratophorum (Ledeb.) DC.
  B: Hultén, 1968; Packer and McPherson, 1974; Wiggins and Thomas, 1962
  PB: *D. Murray 3355*
Taraxacum lateritium Dahlstedt
  B: Hultén, 1968
Taraxacum phymatocarpum J. Vahl
  PB: *D. Murray 3397*
Thalictrum alpinum L.
  PB: *D. Murray 3392*
  ES: Gjaerevoll, 1958–1967
Thlaspi arcticum Pors.
  PB: *D. Murray 4530*
Tofieldia coccinea Richards.
  ES: Hultén, 1941–1950; Scamman, 1940
Tofieldia pusilla (Michx.) Pers.
  PB: *Walker and Batty PB028*
  ES: Hämet-Ahti, 1971; Scamman, 1940
Tripleurospermum phaeocephalum (Rupr.) Pobad.
  B: Hultén, 1968
Trisetum spicatum (L.) Richter
  B: Packer and McPherson, 1974
  PB: *D. Murray 3413*
  ES: *Anderson and Densmore 2298*
  ssp. *spicatum*
  B: Hultén, 1968
Utricularia vulgaris L. ssp. *macrorhiza* (Le Conte) Clausen
  PB: *Walker s.n.* 5 August 1974, det. D. Murray
Vaccinium uliginosum L. ssp. *alpinum* (Bigel.) Hult.
  ES: Gjaerevoll, 1958–1967
Vaccinium vitis-idaea L. ssp. *minus* (Lodd.) Hult.
  B: Hultén, 1968; Packer and McPherson, 1974; Wiggins and Thomas, 1962
  ES: Gjaerevoll, 1958–1967; Hämet-Ahti, 1971
Valeriana capitata Pall.
  B: Hultén, 1968; Wiggins and Thomas, 1962
  PB: *D. Murray 4551*
  [ES: Weeden, 1968]
Viola epipsila Ledeb.
  ES: *Hatler 58*
Wilhelmsia physodes (Fisch.) McNeill
  PB: *D. Murray 4528*

Woodsia glabella R. Br.
  ES: Hultén, 1941–1950, in Suppl. p. 1697
Zygadenus elegans Pursh
  ES: Gjaerevoll, 1958–1967

Rejected Vascular Plant Taxa

Barrow:
Salix polaris Wahlenb. (= *S. pseudopolaris*
  Flod.) Cited by Hultén (1968), Wiggins and
  Thomas (1962). Argus (1973) said that he has
  not seen authentic *S. polaris* from the Arctic
  Slope of Alaska and that many of the specimens cited by Wiggins and Thomas as *S.
  pseudopolaris* are *S. ovalifolia* var. *glacialis*
  or *S. ovalifolia* var. *ovalifolia*.

Prudhoe Bay:
Poa abbreviata R. Br.
  Cited by Murray and Murray (1975) is *Colpodium vahlianum*.
Eagle Summit:
Douglasia gormanii Constance cited by Weeden
  (1968) is *D. arctica*.
Salix stolonifera Cov.
  The report by Scamman (1940) was considered
  doubtful by Hultén, according to Gjaerevoll
  (1958–1967). However, Hultén (1968) did map
  a disjunct locality in the vicinity of Eagle Summit. Argus (1973) showed the species as limited to south coastal Alaska and doubted the
  existence of this species in the interior.
Saussurea densa Hook.
  Incorrectly cited by Scamman (1940), according to Hultén (1941–1950) who determined her
  plants as *S. viscida* var. *yukonensis* and *S.
  angustifolia*.

Bryophytes

Hepatics[1]

Anastrophyllum minutum (Schreb.) Schust.
  B: Persson, 1963 in discussion of *Scapania
  mucronata*
  ES: *Gravesen A1 130,* det. K. Damsholt
Anastrophyllum saxicolus (Schrad.) Schust. *(=
  Sphenolobus saxicolus)*
  ES: Persson and Gjaerevoll, 1957
Aneura pinguis (L.) Dum. *(= Riccardia pinguis)*
  B: Smith, 1974
  PB: Rastorfer, Webster, and Smith 1973
  ES: Persson and Gjaerevoll, 1957
Anthelia juratzkana (Limpr.) Trev. *(= A.
  nivalis)*
  B: Smith, 1974
  ES: *Gravesen A1 803* (P. Gravesen personal
  herbarium, Copenhagen), det. K. Damsholt
Arnellia fennica (Gott.) Lindb.
  PB: *Walker 52,* det. W. C. Steere
Blepharostoma trichophyllum (L.) Dum.
  B: Smith, 1974
  PB: Rastorfer, Webster, and Smith, 1973

  ES: Persson and Gjaerevoll, 1957
  var. *brevirete* Bryhn & Kaalaas
    B: *Gravesen A1 1120,* det. K. Damsholt
    PB: Rastorfer, Webster, and Smith, 1973
Cephalozia bicuspidata (L.) Dum.
  B: Persson, 1963 in discussion of *Scapania
  mucronata*
Cephalozia pleniceps (Aust.) Lindb. *(= Calypo-
  geia pleniceps)*
  B: Smith, 1974
  ES: Persson, 1952a
Cephaloziella arctica Bryhn & Douin
  B: Persson, 1962
  [PB: Rastorfer, Webster, and Smith, 1973]
Chandonanthus setiformis (Ehrh.) Lindb.
  ES: *Gravesen A1 97,* det. K. Damsholt
Chiloscyphus pallescens (Ehrh.) Dum.
  B: Persson, 1962
Chiloscyphus polyanthus (L.) Corda
  B: *Gravesen A1 1119,* det. K. Damsholt
  [var. *fragilis* (Roth) K. Muell. (= *C. fragilis*)
  B: Britton, 1967, p. 48]

[1]Since this manuscript was submitted, a complete account of arctic Alaskan hepatics has appeared (Steere, W. C. and H. Inoue, 1978, *J. Hattori Bot. Lab.* No. **44**: 251–345). It includes the following hepatics not reported here for Barrow: *Calycularia laxa* Lindb. & Arn., *Cephalozia ambigua* Mass., *Cladopodiella fluitans* (Nees) Joerg., *Lophozia groenlandica* (Nees) Macoun, *L. laxa* (Lindb.) Grolle, *L. pellucida* var. *minor* Schust., *L. polaris* (Schust.) Schust. & Damsholt, *Scapania apiculata* Spruce, *S. obcordata* (Berggr.) S. Arn., *S. paludosa* (K. Muell.) K. Muell. or for Prudhoe Bay: *Lophozia pellucida*. It does not include *Lophozia wenzelii* for Barrow which is listed here based on a Steere collection.

Clevea hyalina (Sommerf.) Lindb.
  PB: *B. Murray 6215,* det. W. C. Steere
Diplophyllum taxifolium (Wahlenb.) Dum.
  B: Persson, 1963 in discussion of *Scapania mucronata*
  ES: Persson and Gjaerevoll, 1957
Gymnocolea inflata (Huds.) Dum.
  B: Smith, 1974
Gymnomitrion coralloides Nees
  B: Smith, 1974
Lepicolea fryei Persson (= *Pseudolepicolea fryei* (Perss.) Grolle & Ando, Ando, 1963)
  B: Smith, 1974
Lophozia alpestris (Schleich.) Evans
  B: Smith, 1974
Lophozia binsteadii (Kaal.) Evans (= *Orthocaulis binsteadii*)
  B: Smith, 1974
Lophozia heterocolpa (Thed.) M. A. Howe (= *Leiocolea heterocolpa*)
  B: Smith, 1974
  var. *harpanthoides* (Bryhn & Kaal.) Schust.
  B: *Steere 15131* (NY)
Lophozia incisa (Schrad.) Dum.
  B: *Battrum 330* (UAC), det. K. Damsholt
  ES: Persson and Gjaerevoll, 1957; Steere, 1975
Lophozia kunzeana (Hueb.) Evans (= *Orthocaulis kunzeana*)
  B: Arnell and Persson, 1961; Smith, 1974
Lophozia longidens (Lindb.) Mac.
  ES: Persson and Gjaerevoll, 1957
Lophozia lycopodioides (Wallr.) Cogn.
  ES: *Gravesen A1 134,* det. K. Damsholt
Lophozia opacifolia Culm. (= *Massula opacifolia*)
  B: Smith, 1974
  ES: *Gravesen A1 913* (Gravesen personal herbarium), det. K. Damsholt
Lophozia pellucida Schust.
  B: Arnell and Persson, 1961
Lophozia quadriloba (Lindb.) Evans (= *Orthocaulis quadrilobus*)
  B: Persson, 1952a as *O. quadrilobus* var. *heterophylla* Kaal. & Bryhn
  ES: Persson and Gjaerevoll, 1957
Lophozia rutheana (Limpr.) M. A. Howe
  B: *Steere 72-605* (mainly *Mesoptychia sahlbergii*) (NY)
Lophozia ventricosa (Dicks.) Dum.
  B: Persson, 1962
  ES: *Gravesen A1 104,* det. K. Damsholt
Lophozia wenzelii (Nees) Steph.
  B: *Steere 15765* (NY)
  ES: Persson and Gjaerevoll, 1957

Macrodiplophyllum plicatum (Lindb.) Perss.
  ES: *Steere 72-868* (NY)
Marchantia alpestris Nees
  PB: *B. Murray 4417,* det. K. Damsholt
Marchantia polymorpha L.
  B: Persson, 1962
  PB: *B. Murray 4428*
Marsupella apiculata Schiffn.
  ES: Hermann, 1973, as new to Alaska
Mesoptychia sahlbergii (Lindb. & Arn.) Evans
  B: *Steere 72-605* (NY)
  PB: *Steere 72-700a* (NY)
Moerckia flowtowiana (Nees) Schiffn.
  B: Smith, 1974
  ES: *Gravesen A1 913* (Gravesen personal herbarium), det. K. Damsholt
Moerckia hibernica (Hook.) Gott.
  B: *Battrum 331* (UAC), det. K. Damsholt
  According to Damsholt (written communication to Battrum) this specimen may possibly be *M. flowtowiana.*
Mylia anomala (Hook.) S. Gray
  B: Smith, 1974
Odontoschisma elongata (Lindb.) Evans
  B: Smith, 1974
Odontoschisma macounii (Aust.) Und.
  B: Smith, 1974
  PB: Rastorfer, Webster, and Smith, 1973
Pellia neesiana (Gott.) Limpr.
  ES: Persson and Gjaerevoll, 1957
Plagiochila arctica Bryhn & Kaal.
  B: Persson, 1962
  PB: *Walker 71,* det. W. C. Steere
  ES: *Steere 72-880* (mainly *Ditrichum flexicaule*) (NY)
Preissia quadrata (Scop.) Nees
  PB: *B. Murray 6243*
Ptilidium ciliare (Web.) Hampe
  B: Smith, 1974
  PB: Rastorfer, Webster, and Smith, 1973
  var. *ciliare*
  ES: *Gravesen A1 121,* det. K. Damsholt
Radula prolifera H. Arnell
  PB: Rastorfer, Webster, and Smith, 1973
Scapania curta (Mårt.) Dum.
  ES: Persson and Gjaerevoll, 1957
Scapania cuspiduligera (Nees) K. Muell.
  ES: *Gravesen A1 121,* det. K. Damsholt
Scapania degenii Schiffn.
  B: *Gravesen A1 1122,* det. K. Damsholt
  var. *dubia* Schust.
  B: *Gravesen A1 1123,* det. K. Damsholt
Scapania irrigua (Nees) Dum.
  B: Persson, 1962

[PB: Rastorfer, Webster, and Smith, 1973]
var. *rufescens* Buch
  B: *Battrum 274* in part (UAC filed under *Chiloscyphus pallescens*), det. K. Damsholt
Scapania mucronata Buch
  B: Persson, 1963
Scapania paludicola Loeske & K. Muell.
  B: Persson, 1962
Scapania parvifolia Warnst.
  B: *Steere 16-359* (NY)
  ES: *Gravesen A1 913* (Gravesen personal herbarium), det. K. Damsholt
  New to Alaska. Not in Worley's checklist (1970).
Scapania simmonsii Bryhn & Kaal.
  B: Smith, 1974
  ES: *Steere 72-891*
Scapania subalpina (Nees) Dum.
  ES: Persson and Gjaerevoll, 1957
Scapania uliginosa (Sw.) Dum.
  B: *Gravesen A1 1503,* det. K. Damsholt
Solenostoma pumilum (With.) K. Muell. ssp.
  polaris (Berggr.) Schust.
  B: Smith, 1974
Tritomaria quinquedentata (Huds.) Buch
  B: Persson, 1962
  PB: Rastorfer, Webster, and Smith, 1973
  ES: Persson and Gjaerevoll, 1957

Mosses[2]

Aloina brevirostris (Hook. & Grev.) Kindb.
  B: Smith, 1974
  PB: *B. Murray 6231*
Andreaea rupestris Hedw.
  ES: Persson and Gjaerevoll, 1957
Aplodon wormskjoldii (Hornem.) R. Br. *(= Haplodon wormskjoldii)*
  B: Persson, 1962; Steere, 1954, 1973; Steere, Holmen, and Mogenson, 1976
  PB: Rastorfer, Webster, and Smith, 1973
Arctoa fulvella (Dicks.) B.S.G.
  ES: *Steere 72-817* (NY)
Aulacomnium acuminatum (Lindb. & Arnell) Kindb.
  B: Steere written communication, 1977
  PB: Rastorfer, Webster, and Smith, 1973
  ES: Persson and Gjaerevoll, 1957

Aulacomnium palustre (Hedw.) Schwaegr.
  B: Persson, 1962
  PB: Rastorfer, Webster, and Smith, 1973
  ES: Persson and Gjaerevoll, 1957
Aulacomnium turgidum (Wahlenb.) Schwaegr.
  B: Harvill, 1948; Persson, 1962
  PB: Rastorfer, Webster, and Smith, 1973
  ES: Persson, 1954; Persson and Gjaerevoll, 1957
[*Barbula icmadophila* Schimp. *ex* C. Muell.
  PB: Rastorfer, Webster, and Smith, 1973]
Bartramia ithyphylla Brid.
  B: Ochi, 1962; Persson, 1954, 1962; Steere, 1954
  ES: Persson and Gjaerevoll, 1957
Blindia acuta (Hedw.) B.S.G.
  B: Harvill, 1948
  ES: Persson and Gjaerevoll, 1957
Brachythecium albicans (Hedw.) B.S.G.
  B: Steere written communication, 1977
Brachythecium groenlandicum (C. Jens.) Schljak.
  B: *Steere 19192* (NY), det. K. Holmen
Brachythecium nelsonii Grout *(= B. latifolium)*
  B: Persson, 1962
  ES: Persson, 1946; Persson and Gjaerevoll, 1957
Brachythecium salebrosum (Web. & Mohr) B.S.G.
  ES: Persson and Gjaerevoll, 1957
Brachythecium turgidum (C. J. Hartm.) Kindb.
  B: Persson, 1962, 1963
  [PB: Rastorfer, Webster, and Smith, 1973]
  ES: Persson, 1952b; Persson and Gjaerevoll 1957
Bryobrittonia longipes (Mitt.) Horton (= *B. pellucida*)
  PB: *B. Murray 6247*
Bryoerythrophyllum recurvirostrum (Hedw.) Chen *(= Didymodon recurvirostris)*
  B: Persson, 1962; Steere, 1954
  PB: Rastorfer, Webster, and Smith, 1973
Bryum arcticum (R. Br.) B.S.G.
  B: Persson, 1963; Steere, 1954
  PB: Rastorfer, Webster, and Smith, 1973
Bryum argenteum Hedw.
  B: Persson, 1962
  PB: *B. Murray 6249*

[2]Since this manuscript was submitted, *The Mosses of Arctic Alaska* by W. C. Steere has been published by J. Cramer (*Bryophytorum Bibliotheca* **14.** 1978). It includes the following mosses not reported here for Barrow or Prudhoe Bay: *Brachythecium turgidum* (B), *Bryum purpurascens* (PB), *Calliergon trifarium* (B), *Cyrtomnium hymenophylloides* (B), *Drepanocladus revolvens* var. *intermedius* (Lindb.) Cheney *ex* Wils. (B), *Fissidens viridulus* (Sw.) Wahlenb. (B), *Funaria hygrometrica* (PB), and *Hypnum hamulosum* (B).

Bryum cf. *caespiticium* Hedw.
  PB: *Walker 24,* det. W. C. Steere
Bryum calophyllum R. Br.
  B: Persson, 1962; Steere, 1954
  PB: Rastorfer, Webster, and Smith, 1973;
  Steere, written communication, 1977
Bryum creberrimum Tayl. *(= B. cuspidatum)*
  B: *Wiggins 12614A* (NY)
Bryum cryophilum Mårt. *(= B. obtusifolium)*
  B: Smith, 1974
  PB: *B. Murray 6248*
Bryum knowltonii Barnes
  B: Steere written communication, 1977
Bryum lonchocaulon C. Muell. *(= B. cirratum)*
  ES: Persson and Gjaerevoll, 1957
Bryum nitidulum Lindb.
  B: Steere, 1954
Bryum pallens (Brid.) Sw. *ex* Roehl.
  B: Persson, 1962
Bryum pallescens Schleich. *ex* Schwaegr.
  PB: Rastorfer, Webster, and Smith, 1973
Bryum pendulum (Hornsch.) Schimp.
  B: Steere written communication, 1977
  PB: Steere written communication, 1977
Bryum pseudotriquetrum (Hedw.) Gaertn.,
  Meyer & Scherb.
  B: Smith, 1974
  ES: Persson and Gjaerevoll, 1973
Bryum purpurascens (R. Br.) B.S.G.
  B: Steere written communication, 1977
Bryum salinum I. Hag. *ex* Limpr.
  B: Nyholm and Crundwell, 1958; Persson, 1962
Bryum stenotrichum C. Muell. *(= B. inclinatum)*
  B: Steere, 1954
  PB: Rastorfer, Webster, and Smith, 1973
  ES: Persson and Gjaerevoll, 1957
Bryum tortifolium Funck. *ex* Brid. *(= B.
  cyclophyllum)*
  B: *Steere 19115* (NY)
Bryum weigelii Spreng.
  ES: *Gravesen A1 116,* det. V. B. Lauridsen
Bryum wrightii Sull. & Lesq.
  B: Smith, 1974; Steere and Murray, 1974
  PB: Steere and Murray, 1974
Calliergidium pseudostramineum (C. Muell.)
  Grout
  B: Smith, 1974
Calliergon cordifolium (Hedw.) Kindb.
  B: *Shanks 9-C* (NY)
  ES: Persson and Gjaerevoll, 1957
Calliergon giganteum (Schimp.) Kindb.
  B: Persson, 1962
  PB: *Steere 72-718* (NY)
Calliergon orbicularicordatum (Ren. & Card.)
  Broth.

  B: Karczmarz, 1971
  PB: *Steere 72-665* (NY)
Calliergon richardsonii (Mitt.) Kindb. *ex*
  Warnst.
  B: Smith, 1974
  PB: Rastorfer, Webster, and Smith, 1973
  ES: Persson and Gjaerevoll, 1957
  var. *robustum* (Lindb. & Arn.) Broth. em. Kar.
    PB: *B. Murray 6220*
Calliergon sarmentosum (Wahlenb.) Kindb.
  B: Persson 1962, 1963 (in discussion of *Drepan-
    ocladus purpurascens*)
  PB: Rastorfer, Webster, and Smith, 1973,
    Steere written communication, 1977
  ES: Persson and Gjaerevoll, 1957
  var. *beringianum* (Card. & Ther.) Grout
    ES: *Steere 72-775* (NY)
Calliergon stramineum (Brid.) Kindb.
  B: Arnell and Persson, 1961; Smith, 1974
  ES: Persson and Gjaerevoll, 1957
  var. *laxifolium* (Kindb.) Kar.
    B: Karczmarz, 1971
Calliergon trifarium (Web. & Mohr) Kindb.
  PB: Steere written communication, 1974
Campylium stellatum (Hedw.) C. Jens.
  B: Persson, 1962; Rastorfer, 1974
  PB: Rastorfer, Webster, and Smith, 1973
  ES: Persson and Gjaerevoll, 1957
  var. *arcticum* (Williams) Sav.-Ljub. *(= C.
    arcticum)*
    B: *McPherson 587,* det. D. H. Vitt
    PB: *B. Murray 517*
Catoscopium nigritum (Hedw.) Brid.
  B: Steere written communication, 1977
  PB: Rastorfer, Webster, and Smith, 1973
  ES: *Steere 72-848*
Ceratodon heterophyllus Kindb. *ex* J. M. Mac.
  B: Steere, 1954, 1955
Ceratodon purpureus (Hedw.) Brid.
  B: Persson, 1962; Steere, 1954
  PB: Rastorfer, Webster, and Smith, 1973
  ES: *Gravesen A1 11*
Cinclidium arcticum (B.S.G.) Schimp.
  B: Morgensen, 1973a, based on distribution
    map; Persson, 1963
  PB: Rastorfer, Webster, and Smith, 1973
  ES: *Gravesen A1 904* (Gravesen personal
    herbarium)
Cinclidium latifolium Lindb.
  B: Mogenson 1973a, based on distribution map,
    1973b; Smith, 1974
  PB: Rastorfer, Webster and Smith 1973
Cinclidium stygium Sw.
  B: Smith, 1974
  ES: Persson and Gjaerevoll, 1957

Cinclidium subrotundum Lindb.
  B: Mogensen, 1973a, based on distribution
    map, 1973b; Smith, 1974
Cirriphyllum cirrosum (Schwaegr. *ex* Schultes)
    Grout
  B: Persson, 1952b
  PB: Rastorfer, Webster, and Smith, 1973
  ES: Persson and Gjaerevoll, 1957
Climacium dendroides (Hedw.) Web. & Mohr
  ES: *Steere and B. Murray BMM 4632*
Conostomum tetragonum (Hedw.) Lindb.
  B: Persson, 1963, in discussion of *Scapania
    mucronata;* Steere, 1954
  ES: *Steere and B. Murray BMM 4635*
Cratoneuron arcticum Steere
  PB: *Walker 49,* det. W. C. Steere
Cratoneuron filicinum (Hedw.) Spruce
  PB: *Steere 72-739* (NY)
Ctenidium molluscum (Hedw.) Mitt.
  PB: *Walker 29,* det. W. C. Steere
Cynodontium alpestre (Wahlenb.) Milde
  B: Steere, 1954
Cynodontium strumiferum (Hedw.) Lindb.
  ES: *Steere 72-857* (NY)
[*Cyrtomnium hymenophylloides* (Hueb.) Kop.
  PB: Rastorfer, Webster, and Smith, 1973]
Cyrtomnium hymenophyllum (B.S.G.)
    Holmen
  B: Smith, 1974
  PB: Rastorfer, Webster, and Smith, 1973
  ES: *Steere 72-816*
Desmatodon heimii (Hedw.) Mitt. *(= Pottia
    heimii)*
  B: *Steere 15-267*
  PB: *B. Murray 4472*
  var. *arctica* (Lindb.) Crum (= *P. heimii* var.
    obtusifolia)
    B: Persson, 1962 as *Pottia obtusa* (R. Br.) C.
      Muell.; Steere 1954, 1955; Steere, Holmen,
      and Mogenson, 1976, as *Pottia obtusifolia*
Desmatodon latifolius (Hedw.) Brid.
  B: Steere written communication, 1977
  PB: Steere written communication, 1977
Desmatodon leucostoma (R. Br.) Berggr. *(= D.
    suberectus)*
  B: Smith, 1974
  PB: Rastorfer, Webster, and Smith, 1973
Dichodontium pellucidum (Hedw.) Schimp.
  B: Smith, 1974
Dicranella cerviculata (Hedw.) Schimp.
  B: Steere, 1954
Dicranella crispa (Hedw.) Schimp. *(= Anisothe-
    cium crispum)*
  B: Persson, 1962; Steere, 1954
  [PB: Rastorfer, Webster, and Smith, 1973]

Dicranella grevilleana (Brid.) Schimp.
  B: Smith, 1974
  ES: *Steere 72-856* (NY)
Dicranella subulata (Hedw.) Schimp.
  B: Smith, 1974
  ES: Persson and Gjaerevoll, 1957
Dicranoweisia crispula (Hedw.) Lindb.
  B: *Steere and Crum 19190* (NY), det. K.
    Holmen
  ES: *Hermann 21036*
Dicranum acutifolium (Lindb. & Arnell) C. Jens.
  B: Steere written communication, 1977
Dicranum angustum Lindb.
  B: Persson, 1962; Steere and Holmen, 1975
  PB: Rastorfer, Webster, and Smith, 1973
Dicranum elongatum Schleich. *ex* Schwaegr.
  B: Arnell and Persson, 1961; Persson, 1962
  PB: Rastorfer, Webster, and Smith, 1973
  ES: Persson and Gjaerevoll, 1957
Dicranum fuscescens Turn.
  B: Harvill, 1948
  ES: Persson and Gjaerevoll, 1957
Dicranum groenlandicum Brid.
  B: *Steere and Crum 19121* (NY), det. K.
    Holmen
Dicranum majus Sm.
  B: Harvill, 1948; Persson, 1962
  ES: Persson and Gjaerevoll, 1957
Dicranum muehlenbeckii B.S.G.
  B: Smith, 1974
  ES: Persson and Gjaerevoll, 1957
Dicranum scoparium Hedw.
  B: Steere written communication, 1977
  ES: *Steere 72-771*
Dicranum spadiceum Zett.
  B: Smith, 1974
  ES: *Hermann 21058* (CANM)
Didymodon asperifolius (Mitt.) Crum, Steere &
    Anderson
  PB: *B. Murray 4446*
  ES: *Steere 72-888*
Diphyscium foliosum (Hedw.) Mohr
  ES: *Steere 72-806* (NY)
  This is the first report of this species for Alaska
    (Steere written communication, 1974)
Distichium capillaceum (Hedw.) B.S.G.
  B: Steere, 1954
  PB: Rastorfer, Webster, and Smith, 1973
  ES: Persson and Gjaerevoll, 1957
Distichium hagenii Ryan *ex* Philib.
  B: Steere, 1954
  PB: Rastorfer, Webster, and Smith, 1973
Distichium inclinatum (Hedw.) B.S.G.
  B: Persson, 1962
  PB: Rastorfer, Webster, and Smith, 1973

Ditrichum flexicaule (Schwaegr.) Hampe
 B: Harvill, 1948; Persson, 1962
 PB: Rastorfer, Webster, and Smith, 1973
 ES: *Steere 72-880*
Drepanocladus aduncus (Hedw.) Warnst.
 B: Persson, 1962
Drepanocladus badius (C. J. Hartm.) Roth
 B: Smith, 1974
 ES: Persson and Gjaerevoll, 1957
Drepanocladus exannulatus (B.S.G.) Warnst.
 (= D. purpurascens)
 B: Persson, 1963 as *D. purpurascens,* Steere
 written communication, 1977
 PB: *Steere and Iwatsuki 74-317* (NY)
 ES: *Steere 72-885*
Drepanocladus fluitans (Hedw.) Warnst.
 B: *Steere 15171*
Drepanocladus lycopodioides (Brid.) Warnst.
 B: Smith, 1974
 PB: *Steere 72-731*
 var. *brevifolius* (Lindb.) Moenk *(= D.
 brevifolius)*
 B: *Steere 19185* (NY)
 PB: Rastorfer, Webster, and Smith, 1973
 ES: Persson and Gjaerevoll, 1957
Drepanocladus revolvens (Sw.) Warnst.
 B: Persson, 1962
 PB: Rastorfer, Webster, and Smith, 1973
 ES: Persson and Gjaerevoll, 1957
Drepanocladus uncinatus (Hedw.) Warnst.
 B: Arnell and Persson, 1961; Smith, 1974
 PB: *Rastorfer, Webster and Smith 28,* det. B.
 Murray
 ES: Persson and Gjaerevoll, 1957
Encalypta alpina Sm.
 B: Steere, 1954
 PB: *Steere 72-707* (NY)
Encalypta brevicolla (B.S.G.) Bruch *ex* Ångstr.
 ES: *Steere and B. Murray BMM 4638*
Encalypta ciliata Hedw.
 B: Smith, 1974
Encalypta mutica Hag.
 PB: *B. Murray 6214*
Encalypta procera Bruch
 B: Steere, 1954
 PB: *B. Murray 6244*
Encalypta rhaptocarpa Schwaegr. *(= E. vul-
 garis* var. *rhaptocarpa)*
 B: Steere, 1954
 PB: Rastorfer, Webster, and Smith, 1973
Eurhynchium pulchellum (Hedw.) Jenn.
 B: *Steere 15252*
 ES: Persson and Gjaerevoll, 1957
Fissidens adiantoides Hedw.
 PB: *Steere and Iwatsuki 74-318* (NY)

Fissidens arcticus Bryhn
 B: Brassard, 1971, based on distribution map;
 Smith, 1974; Steere and Brassard, 1974
Fissidens osmundoides Hedw.
 B: Smith, 1974
 PB: Rastorfer, Webster, and Smith, 1973
 ES: Persson and Gjaerevoll, 1957
Funaria arctica (Berggr.) Kindb. *(= F. hygro-
 metrica* var. *arctica, F. microstoma* var.
 obtusifolia)
 B: Smith, 1974
 PB: *B. Murray 6251*
Funaria hygrometrica Hedw.
 B: Smith, 1974
Funaria polaris Bryhn
 PB: Rastorfer, Webster, and Smith,
 1973
Grimmia apocarpa Hedw. *(= Schistidium
 apocarpum)*
 B: *Steere and Crum 19171* (NY)
 PB: *Battrum 304*
 ES: *Gravesen A1 806* (Gravesen personal
 herbarium)
 var. *stricta* (Turn.) Hook. & Tayl. *(= G. apo-
 carpa* var. *gracilis, Schistidium gracile)*
 B: *Steere and Crum 19189* (NY)
 ES: *Hermann 21035* (CANM)
Grimmia donniana Sm.
 ES: *Steere 72-895*
Grimmia ovalis (Hedw.) Lindb.
 ES: Persson and Gjaerevoll, 1957
Grimmia tenera Zett. *(= Schistidium
 tenerum)*
 B: *Steere and Crum 19172* (NY)
Hylocomium pyrenaicum (Spruce) Lindb.
 ES: Persson and Gjaerevoll, 1957
Hylocomium splendens (Hedw.) B.S.G.
 B: Harvill, 1948
 var. *obtusifolium* (Geh.) Par. *(= H.
 alaskanum)*
 B: *Rastorfer, Webster and Smith 107*
 PB: Rastorfer, Webster, and Smith, 1973
 ES: *B. Murray 4582*
Hypnum bambergeri Schimp.
 B: Smith, 1974
 PB: Rastorfer, Webster, and Smith, 1973
 ES: Persson and Gjaerevoll, 1957
Hypnum callichroum Funck *ex* Brid.
 B: Steere written communication, 1977
 ES: *Steere 72-787* (NY)
Hypnum cupressiforme Hedw.
 B: *Steere 15002*
 [PB: Rastorfer, Webster, and Smith, 1973]
 var. *subjulaceum* Mol.
 ES: Hermann, 1973

Hypnum dieckii Ren. & Card. *ex* Roell.
 ES: *Steere 72-869*
Hypnum hamulosum B.S.G.
 ES: Persson, 1947
Hypnum lindbergii Mitt.
 B: Steere written communication, 1977
 ES: Persson and Gjaerevoll, 1957
Hypnum plicatulum (Lindb.) Jaeg. & Sauerb. *(=*
 H. subplicatile)
 B: *Steere 15191* (NY)
 ES: Persson, 1947; Persson and Gjaerevoll,
 1957
Hypnum pratense Koch *ex* Brid.
 B: Steere written communciation, 1977
Hypnum procerrimum Mol.
 B: Steere written communication, 1977
 PB: *B. Murray 4440*
Hypnum revolutum (Mitt.) Lindb.
 B: Steere written communication, 1977
 PB: *Walker 55,* det. W. C. Steere
 ES: Persson and Gjaerevoll, 1957
Hypnum vaucheri Lesq.
 PB: Rastorfer, Webster, and Smith, 1973
Isopterygiopsis muelleriana (Schimp.) Iwats.
 ES: *Steere 72-878* (NY)
Isopterygium pulchellum (Hedw.) Jaeg. &
 Sauerb.
 B: Persson 1954, 1962
Kiaeria glacialis (Berggr.) I. Hag.
 ES: Persson and Gjaerevoll, 1957
Leptobryum pyriforme (Hedw.) Wils.
 B: Persson, 1962
 PB: *B. Murray 4412*
Meesia longiseta Hedw.
 B: Smith, 1974
Meesia triquetra (Richt.) Ångstr.
 B: Smith, 1974
 PB: Rastorfer, Webster, and Smith, 1973
 ES: Persson and Gjaerevoll, 1957
Meesia uliginosa Hedw.
 B: Smith, 1974
 PB: Rastorfer, Webster, and Smith, 1973
 ES: *Gravesen A1 26*
Mnium blyttii B.S.G.
 B: *Steere 15124* (NY), det. T. Koponen
 PB: *B. Murray 4426*
 ES: *Gravesen A1 4*
Mnium marginatum (With.) Brid. *ex* P.-Beauv.
 B: Smith, 1974
Mnium thomsonii Schimp. *(= M.*
 orthorrhynchum)
 B: *Steere 15256* (NY)
 PB: *Steere 72-683* (NY)
Myurella julacea (Schwaegr.) B.S.G.
 B: Persson, 1962; Steere and Holmen,
 1975

PB: Rastorfer, Webster, and Smith, 1973
 ES: *Steere and B. Murray BMM 4636* in part
Myurella tenerrima (Brid.) Lindb.
 B: Smith, 1974
 PB: Rastorfer, Webster, and Smith, 1973
 ES: *Steere and B. Murray BMM 4636* in part
Oncophorus virens (Hedw.) Brid.
 B: Steere, 1954
 ES: *Hermann 21048*
Oncophorus wahlenbergii Brid.
 B: Harvill, 1948; Smith, 1974
 PB: Rastorfer, Webster, and Smith, 1973
 ES: *Steere 72-777*
Orthothecium chryseum (Schwaegr. *ex*
 Schultes) B.S.G.
 B: Persson, 1962
 PB: Rastorfer, Webster, and Smith, 1973
 ES: Persson and Gjaerevoll, 1957
Orthothecium intricatum (C. J. Hartm.) B.S.G.
 PB: *B. Murray 6234,* det. W. C. Steere
Orthothecium rufescens (Brid.) B.S.G.
 B: *Steere and Crum 19184* (NY)
 PB: *Steere 72-715* (NY)
Orthothecium strictum Lor.
 B: Smith, 1974
 PB: Steere written communication, 1977
Paludella squarrosa (Hedw.) Brid.
 B: Smith, 1974
Paraleucobryum enerve (Thed. *ex.* C. J. Hartm.)
 Loeske
 ES: Persson and Gjaerevoll, 1957
Philonotis fontana (Hedw.) Brid.
 ES: *Gravesen A1 22*
 var. *pumila* (Turn.) Brid. *(= P. tomentella)*
 B: *Sharp A-58171E* (NY), det. W. Zales
 PB: *Steere 72-679* (NY)
Philonotis sp.
 PB: Rastorfer, Webster, and Smith, 1973, as
 possibly *P. fontana*
Plagiomnium ellipticum (Brid.) Kop. *(= P. rugi-*
 cum (Laur.) Kop.)
 B: Persson, 1962; Smith, 1974
 PB: Rastorfer, Webster, and Smith, 1973;
 Steere written communication, 1977
 ES: Persson and Gjaerevoll, 1957
Plagiomnium medium (B.S.G.) Kop.
 ssp. *curvatulum* (Lindb.) Kop.
 B: Steere written communication, 1977
Plagiothecium cavifolium (Brid.) Iwats.
 B: *Steere, Inoue, and Iwatsuki 73-1* (NY)
Plagiothecium denticulatum (Hedw.) B.S.G.
 B: Persson, 1962
 ES: Persson, 1952a
Plagiothecium piliferum (Sw. *ex* C. J. Hartm.)
 B.S.G.
 B: Smith, 1974

Platydictya jungermannoides (Brid.) Crum *(= Amblystegiella jungermannoides)*
  B: *Wiggins 12448* (NY)
  PB: *Walker 51*, det. W. C. Steere
Pleurozium schreberi (Brid.) Mitt.
  ES: Persson and Gjaerevoll, 1957
Pogonatum alpinum (Hedw.) Roehl. *(= Polytrichastrum alpinum, Polytrichum alpinum)*
  B: Smith, 1974
  ES: Persson and Gjaerevoll, 1957
  var. *arcticum* (Brid.) Brid.
    B: *Rastorfer, Webster and Smith 129*, det. H. Webster
  var. *fragile* (Bryhn) Crum
    B: *Gravesen 1302*, det. V. B. Lauridsen
  var. *septentrionale* (Brid.) Brid. (= var. *brevifolium*)
    B: Steere, 1954
    PB: Rastorfer, Webster, and Smith, 1973
Pogonatum dentatum (Brid.) Brid. *(= P. capillare)*
  B: Steere, 1954
  ES: Persson and Gjaerevoll, 1957
Pogonatum urnigerum (Hedw.) P.-Beauv.
  B: *Steere 15231* (NY)
Pohlia annotina (Hedw.) Lindb.
  B: Smith, 1974
Pohlia bulbifera (Warnst.) Warnst.
  B: Smith, 1974
Pohlia cruda (Hedw.) Lindb.
  B: Persson, 1954; 1962; Steere, 1954
  PB: Rastorfer, Webster, and Smith, 1973
  ES: Persson and Gjaerevoll, 1957
Pohlia crudoides (Sull. & Lesq.) Broth.
  B: *Steere and Crum 19142* (NY)
  ES: *Steere 72-803* (NY)
Pohlia drummondii (C. Muell.) Andr.
  ES: Persson and Gjaerevoll, 1957
Pohlia nutans (Hedw.) Lindb.
  B: Arnell and Persson, 1961; Persson, 1962; Steere, 1954
  [PB: Rastorfer, Webster, and Smith, 1973]
  ES: Persson and Gjaerevoll, 1957
Pohlia proligera (Kindb. *ex* Limpr.) Lindb. *ex* H. Arnell
  B: Smith, 1974
Pohlia rothii (Corr. *ex* Limpr.) Broth.
  ES: *Steere 72-803*
Polytrichum commune Hedw.
  var. *maximoviczii* Lindb. (= var. *jensenii*)
    B: Rastorfer, 1974; Smith, 1974
  var. *nigrescens* Warnst. *(= P. swartzii)*
    B: Smith, 1974
  var. *yukonense* (Card. & Ther.) Frye
    B: Smith, 1974

Polytrichum juniperinum Hedw.
  B: Smith, 1974
  [ES: Shacklette, 1962]
Polytrichum piliferum Hedw.
  B: Smith, 1974
  ES: Persson and Gjaerevoll, 1957
  var. *hyperboreum* (R. Br.) C. Muell. *(= P. hyperboreum)*
    B: Rastorfer, 1974; Smith, 1974
    ES: *Steere 72-805* (NY)
Polytrichum sexangulare Brid. *(= P. norvegicum)*
  ES: Persson and Gjaerevoll, 1957
Polytrichum strictum Brid. (= *P. juniperinum* var. *affine, P. juniperinum* var. *gracilius*)
  B: Smith, 1974
  ES: Persson and Gjaerevoll, 1957
Pseudobryum cinclidioides (Hueb.) Kop. *(= Mnium cinclidioides)*
  B: Koponen, 1974; Smith, 1974
Psilopilum cavifolium (Wils.) I. Hag.
  B: Persson, 1962; Steere 1954, 1955; Steere and Holmen, 1975
Psilopilum laevigatum (Wahlen.) Lindb.
  B: Harvill, 1948
Pterygoneurum lamellatum (Lindb.) Jur. (= *P. arcticum*, Steere, 1976)
  B: Steere written communication, 1977
Ptilium crista-castrensis (Hedw.) De Not.
  ES: *Steere 72-774* (NY)
Rhacomitrium canescens (Hedw.) Brid.
  ES: Persson and Gjaerevoll, 1957
Rhacomitrium heterostichum Hedw.
  B: Smith, 1974
  ES: *Gravesen A1 128*
  cfr. var. *microcarpon* (Hedw.) Boul.
    ES: *Gravesen A1 12*
Rhacomitrium lanuginosum (Hedw.) Brid.
  B: Smith, 1974
  PB: Rastorfer, Webster, and Smith, 1973
  ES: Persson, 1954; Persson and Gjaerevoll, 1957
Rhizomnium andrewsianum (Steere) Kop. *(= Mnium andrewsianum)*
  B: Persson, 1963; Steere, 1958
  PB: Steere written communication, 1974
Rhytidiadelphus triquetrus (Hedw.) Warnst.
  ES: *Gravesen A1 15*
Rhytidium rugosum (Hedw.) Kindb.
  PB: Rastorfer, Webster, and Smith, 1973
  ES: Persson, 1954
Scorpidium scorpioides (Hedw.) Limpr.
  B: Smith, 1974
  PB: Rastorfer, Webster, and Smith 1973
  ES: *Gravesen A1 112*

Scorpidium turgescens (T. Jens.) Loeske
  (= Calliergon turgescens)
  B: Karczmarz, 1971; Persson, 1962
  PB: Rastorfer, Webster, and Smith, 1973
  ES: Persson and Gjaerevoll, 1957
Sphagnum balticum (Russ.) Russ. *ex* C. Jens
  B: Steere written communication, 1977
Sphagnum capillaceum (Weiss) Schrank *(= S. nemoreum)*
  B: Smith, 1974
  var. *tenellum* (Schimp.) Andr. *(= S. rubellum, S. subtile)*
  B: Arnell and Persson, 1961; Persson, 1962 in discussion of *Dicranum angustum;* Smith, 1974
  ES: *Gravesen A1 24,* det. V. B. Lauridsen
Sphagnum cuspidatum Ehrh. *ex* Hoffm.
  B: Smith, 1974
Sphagnum fallax (Klinggr.) Kinggr. (= *S. recurvum* auct., non P.-Beauv.)
  B: Steere written communication, 1977
Sphagnum fimbriatum Wils. *ex* J. Hook.
  B: Smith, 1974; Steere and Holmen, 1975
Sphanum fuscum (Schimp.) Klinggr.
  B: Brown and Johnson, 1965; Steere written communication, 1977
Sphagnum girgensohnii Russ.
  B: Persson, 1962
  ES: *Gravesen A1 18,* det. B. Lange
Sphagnum imbricatum Hornsch. *ex* Russ.
  B: Arnell and Persson, 1961
Sphagnum lenense H. Lindb. *ex* Pohle
  B: *McPherson 588-b,* det. D. H. Vitt
Sphagnum obtusum Warnst.
  B: Steere written communication, 1977
Sphagnum squarrosum Crome
  B: Persson, 1962; Steere and Holmen, 1975
Sphagnum subsecundum Nees *ex* Sturm
  B: Brown and Johnson, 1965; Steere written communication, 1977
Sphagnum teres (Schimp.) Ångstr. *ex* C. J. Hartm.
  B: Steere written communication, 1977
  ES: Persson and Gjaerevoll, 1957
Splachnum sphaericum Hedw.
  PB: Rastorfer, Webster, and Smith, 1973
Splachnum vasculosum Hedw.
  B: Persson, 1962; Steere, 1954, 1973; Steere and Holmen, 1975
  PB: *B. Murray 4415*
Stegonia latifolia (Schwaegr. *ex* Schultes) Vent. *ex* Broth. var. *pilifera* (Brid.) Broth.
  PB: *B. Murray 6246*
Tayloria acuminata Hornsch.
  PB: *B. Murray 6249*

Tayloria lingulata (Dicks.) Lindb.
  PB: Steere written communication, 1977
Tetraplodon mnioides (Hedw.) B.S.G.
  B: Harvill, 1948; Steere, 1954
  PB: *B. Murray 4448*
  ES: *Gravesen A1 28*
  Many specimens collected in these localities are var. *cavifolius* Schimp. *(= T. urceolatus).*
Tetraplodon pallidus Hag.
  B: Steere written communication, 1977
  PB: Steere written communication, 1977
Tetraplodon paradoxus (R. Br.) Hag.
  B: *Steere 19030* (NY)
  PB: Steere written communication, 1977
  ES: *Steere and B. Murray BMM 4633*
Thuidium abietinum (Hedw.) B.S.G. *(= Abietinella abietina)*
  PB: Rastorfer, Webster, and Smith, 1973
  ES: *Gravesen A1 917* (Gravesen personal herbarium)
Timmia austriaca Hedw.
  B: Smith, 1974
  PB: *B. Murray 4431,* det. V. B. Lauridsen
  ES: *Steere and B. Murray BMM 4637*
Timmia megapolitana Hedw. var. *bavarica* (Hessl.) Brid.
  PB: *B. Murray 6235*
Timmia norvegica Zett.
  B: Smith, 1974
  PB: Rastorfer, Webster, and Smith, 1973
Tomenthypnum nitens (Hedw.) Loeske
  (= Homalothecium nitens)
  B: Smith, 1974; Steere and Holmen, 1975
  PB: Rastorfer, Webster, and Smith, 1973
  ES: Persson and Gjaerevoll, 1957
Tortella arctica (Arn.) Crundw. & Nyh.
  B: *Steere 15185*
  PB: *Rastorfer, Webster and Smith 68,* det. B. Murray
  ES: *Steere 72-872*
Tortella fragilis (Drumm.) Limpr.
  B: Smith, 1974
  PB: *B. Murray 6242*
  ES: *Steere 72-897*
Tortella tortuosa (Hedw.) Limpr.
  ES: Persson, 1952a; Persson and Gjaerevoll, 1957
Tortula mucronifolia Schwaegr.
  B: Steere, 1954
  PB: *B. Murray 6250*
Tortula ruralis (Hedw.) Gaertn., Meyer, & Scherb.
  B: Persson, 1962
  PB: Rastorfer, Webster, and Smith, 1973
  ES: Persson and Gjaerevoll, 1957

Trichodon cylindricus (Hedw.) Schimp.
  B: Steere, 1954
Trichostomum arcticum Kaal. (= *T. cuspidatissimum*)
  PB: *Walker s.n.*, 20 July 1974, det. B. Murray
Voitia hyperborea Grev. & Arnott
  PB: Rastorfer, Webster, and Smith, 1973, as *V. nivalis* Hornsch.; Steere, 1974

Rejected Moss Taxa

Barrow:
Desmatodon obtusifolius (Schwaegr.) Schimp.
  Steere 15095 cited by Smith (1974) is *D. heimii* var. *arctica* according to W. C. Steere.
Tortella tortuosa (Hedw.) Limpr. *Steere 15185* cited by Smith (1974) is *T. arctica* according to W. C. Steere.

Prudhoe Bay:
Encalypta vulgaris Hedw.
  cited by Murray and Murray (1975) is *E. mutica.*
Hypnum callichroum Funck *ex* Brid.
  cited by Rastorfer, Webster, and Smith (1973) is *Drepanocladus uncinatus.*
Oncophorus virens (Hedw.) Brid.
  cited by Rastorfer, Webster, and Smith (1973) is *O. wahlenbergii.*
Voitia nivalis Hornsch.
  Rastorfer, Webster, and Smith (1973) cited their collections numbered 14, 73, and 85 as this species. Steere (1974) discussed *V. hyperborea* and *V. nivalis* and cited the Rastorfer, Webster, and Smith material as *V. hyperborea.*
Eagle Summit:
Fossombronia foveolata Lindb.
  The specimen cited by Hermann (1973) is *Lophozia incisa* according to W. C. Steere (1975).

Lichens

Alectoria chalybeiformis (L.) S. Gray
  B: *Thomson and Shushan 5644* (WIS)
Alectoria lanea (Ehrh.) Vain.
  ES: Krog, 1968
Alectoria minuscula Nyl.
  ES: Krog, 1968
Alectoria nigricans (Ach.) Nyl.
  B: Krog, 1962
  PB: *Schofield Ak-86,* det. M. E. Williams
  ES: Krog, 1968
Alectoria ochroleuca (Hoffm.) Mass.
  B: *Flock Fl-270*
  PB: *B. Murray 6219*
  var. *ochroleuca*
  ES: Krog, 1968
Alectoria pubescens (L.) R. H. Howe
  ES: Krog, 1968
Arthrothapis citrinella (Ach.) Poelt var. *alpina* (Schaer.) Poelt (= *Bacidia alpina*)
  B: *Flock Fl-224*
  ES: *Thomson and Ahti 17921* (WIS)
Asahinea chrysantha (Tuck.) W. Culb. & C. Culb.
  B: Culberson and Culberson, 1965
  PB: *Williams Ak-652,* det. B. Murray
  ES: Culberson and Culberson, 1965; Krog, 1968
Asahinea scholanderi (Llano) W. Culb. & C. Culb.
  ES: Krog, 1968

Bacidia bagliettoana (Mass. & De Not.) Jatta
  B: *Thomson, Shushan, and Sharp 11591* (WIS)
Bacidia beckhausii Koerb.
  B: *Thomson, Shushan, and Sharp 11591* (WIS)
Bacidia microcarpa (Th. Fr.) Lett.
  B: *Sharp 11445* (WIS), det. J. W. Thomson
Bacidia sphaeroides (Dicks.) Zahlbr.
  B: *Battrum s.n.,* 1971, det. B. Murray
Baeomyces placophyllus Ach.
  ES: *Thomson and Ahti 18096* (WIS)
Buellia alboatra (Hoffm.) Branth. & Rostr.
  PB: *Battrum 325A* (UAC), det. C. D. Bird
Buellia papillata (Somm.) Tuck.
  B: *Thomson, Shushan, and Sharp 6423* (WIS)
  PB: *B. Murray 4355,* det. J. W. Thomson
Buellia punctata (Hoffm.) Mass.
  B: *Flock Fl-225*
Caloplaca cinnamomea (Th. Fr.) Oliv.
  B: *Thomson and Shushan, Lich. Arct. 1.,* as *C. livida*
  PB: *B. Murray 6245,* det. J. W. Thomson
Caloplaca discolor (Will.) Fink
  PB: *B. Murray 6227,* det. J. W. Thomson
Caloplaca stillicidiorum (Vahl) Lynge
  B: *Thomson, Shushan, and Sharp 11589* (WIS)
  PB: *B. Murray 6245,* det. J. W. Thomson
Caloplaca tiroliensis Zahlbr.
  B: *Thomson, Shushan, and Sharp 11587* (WIS)

Candelariella aurella (Hoffm.) Zahlbr.
 PB: *B. Murray 6228,* det. J. W. Thomson
Candelariella xanthostigma (Pers.) Lett.
 PB: *B. Murray 6241,* det. J. W. Thomson
Cetraria commixta (Nyl.) Th. Fr.
 ES: Krog, 1968
Cetraria cucullata (Bell.) Ach.
 B: Krog, 1962
 PB: *B. Murray 4331*
 ES: Krog, 1968
Cetraria delisei (Bory *ex* Schaer.) Th. Fr.
 B: Krog, 1962
 PB: *B. Murray 4345*
Cetraria ericetorum Opiz
 [B: Brown and Johnson, 1965]
 ES: Krog, 1968
Cetraria hepatizon (Ach.) Vain.
 ES: Krog, 1968
Cetraria inermis (Nyl.) Krog
 B: Krog, 1973
Cetraria islandica (L.) Ach.
 B: *Flock Fl-277*
 PB: *B. Murray 4335*
 ES: Krog, 1968
Cetraria kamczatica Sav.
 B: Krog, 1962
Cetraria laevigata Rass.
 B: *Thomson and Shushan 5825* (WIS)
 ES: Krog, 1962
Cetraria nigricans (Retz.) Nyl.
 ES: Krog, 1968
Cetraria nivalis (L.) Ach.
 B: *Battrum 201*
 PB: *B. Murray 4330*
 ES: Krog, 1968
Cetraria richardsonii Hook.
 B: *Battrum 199*
 PB: *B. Murray 4332*
 ES: Krog, 1968
Cetraria simmonsii Krog. var. *simmonsii*
 B: Krog, 1968
Cetraria subalpina Imsh.
 B: Imshaug, 1950; Krog, 1968
Cetraria tilesii Ach.
 PB: *B. Murray 4349*
 ES: Krog, 1968
Cladina arbuscula (Wallr.) Hale & W. Culb. ssp.
 beringiana (Ahti) Bird (= *Cladonia arbuscula*
 ssp. *beringiana*)
 ES: Krog, 1968
Cladina mitis (Sanst.) Hale & W. Culb. (= *Cla-*
 donia mitis)
 B: *Williams Ak-549,* det. E. D. Rudolph
 ES: *Thomson and Ahti 17940* (WIS)

Cladina rangiferina (L.) Harm. *(= Cladonia*
 rangiferina)
 B: *Thomson and Shushan 6469* (WIS)
 ssp. *rangiferina*
  ES: Krog, 1962
Cladonia amaurocraea (Floerke) Schaer.
 B: *Thomson, Shushan, and Sharp, Lich. Arct.*
 24
 PB: *Williams Ak-655,* det. J. W. Thomson
 ES: Krog, 1968
Cladonia bellidiflora (Ach.) Schaer.
 B: Krog, 1962
 ES: Krog, 1968
Cladonia carneola
 B: *Williams Ak-606,* det. E. D. Rudolph
[*Cladonia chlorophaea* (Floerke *ex* Somm.)
 Spreng.
 B: Brown and Johnson, 1965]
Cladonia coccifera (L.) Willd.
 B: *Williams Ak-608,* det. B. Murray
 var. *coccifera*
  ES: Krog, 1968 as barbatic acid strain
 var. *stemmatina* Ach.
  B: Krog, 1962
Cladonia ecmocyna (Ach.) Nyl.
 B: *Schofield Ak-15*
Cladonia gonecha (Ach.) Asah. *(= C. deformis*
 var. *gonecha)*
 ES: Krog, 1968
Cladonia gracilis (L.) Willd. var. *gracilis*
 B: *Thomson and Shushan 6445* (WIS)
 ES: Krog, 1968
 f. *amaura* (Floerke) Aigr.
  B: *Thomson, Shushan, and Sharp, Lich.*
  Arct. 20
Cladonia kanewskii Oxn. (= *C. nipponica* Asah.
 var. *aculeata* Asah., Ahti, 1973)
 ES: Krog, 1973
Cladonia lepidota Nyl.
 B: *Thomson, Shushan, and Sharp, Lich. Arct.*
 34
Cladonia macrophylla (Schaer.) Stenham.
 B: *Thomson and Shushan 6427* (WIS)
 ES: Krog, 1968
Cladonia major (Hag.) Sandst.
 B: *Schofield Ak-49,* det. J. W. Thomson
Cladonia metacorallifera Asah.
 B: Evans, 1955
Cladonia phyllophora Hoffm.
 B: *Thomson and Shushan 6125* (WIS)
[*Cladonia pleurota* (Floerke) Schaer.
 B: Brown and Johnson, 1965]
Cladonia pocillum (Ach.) O. Rich.
 B: *Gravesen s.n.,* 1971, det. B. Murray

PB: *B. Murray 4350*
ES: Krog, 1962
Cladonia pyxidata (L.) Hoffm. ssp. *pyxidata*
ES: Krog, 1968
Cladonia squamosa (Scop.) Hoffm.
 PB: *Schofield Ak-91,* det. M. E. Williams
Cladonia subfurcata (Nyl.) Arn.
 B: *Schofield Ak-17,* det. J. W. Thomson
 PB: *Schofield Ak-90,* det. J. W. Thomson
 ES: *Thomson and Ahti 18199* (WIS)
Cladonia transcendens (Vain.) Vain.
 B: *Koranda SA-5818* (WIS), det. J. W.
 Thomson
Cladonia uncialis (L.) Wigg.
 B: *Flock Fl-257*
 ES: Krog, 1968
Cladonia sp.[3]
 B: Thomson 1967 (1968) as *C. wainii* Sav.,
 according to Ahti (1973).
 Ahti said that he and other authors (Yoshimura,
 1968, Krog, 1968, Ahti, 1968, Ahti, Scotter,
 and Vänskä (1973) regard this material as a
 different, probably undescribed species.
Collema bachmanianum (Fink) Degel. (= *C.
 tenax* var. *bachmanianum*)
 PB: *B. Murray 4328,* det. J. W. Thomson
 var. *millegranum* Degel.
 PB: *B. Murray 4387,* det. J. W. Thomson
Collema ceraniscum Nyl.
 B: *Koranda A1-1* (WIS), det. J. W. Thomson
Collema tuniforme (Ach.) Ach. em. Degel.
 PB: *B. Murray 4342,* det. J. W. Thomson
Cornicularia aculeata (Schreb.) Ach.
 B: Krog, 1962
 PB: *B. Murray 4326*
Cornicularia divergens Ach.
 B: Krog, 1962
 PB: *B. Murray 6237*
 ES: Krog, 1962
Cornicularia odontella (Ach.) Roehl.
 ES: Krog, 1968
Dactylina arctica (Hook.) Nyl.
 B: Krog, 1962
 PB: *B. Murray 4333*
 ES: Krog, 1968
Dactylina madreporiformis (Wulf.) Tuck.
 B: *Thomson and Shushan 5650* (Thomson
 ms)
Dactylina ramulosa (Hook.) Tuck.
 B: *Thomson and Shushan 6442* (WIS)
 PB: *B. Murray 4329*
 ES: Krog, 1968

Evernia perfragilis Llano
 PB: *B. Murray 4344,* det. J. W. Thomson
Fulgensia bracteata (Hoffm.) Raes.
 PB: *B. Murray 4363*
Geisleria sychnogonioides Nitschke
 B: *Thomson and Shushan 9182* (WIS),
 Thomson (ms) as new to North America
Gyalecta foveolaris (Ach.) Schaer.
 PB: *B. Murray 4364,* det. J. W. Thomson
Haematomma lapponicum Raes.
 ES: Culberson, 1963; Thomson, 1968
Hypogymnia bitteri (Lynge) Ahti
 ES: Krog, 1962 as *Parmelia obscurata* Bitt.
Hypogymnia intestiniformis (Vill.) Raes.
 ES: Krog, 1968
Hypogymnia physodes (L.) W. Wats.
 PB: *B. Murray 4400,* det. J. W. Thomson,
 esorediate
 ES: Krog, 1968
Hypogymnia subobscura (Vain.) Poelt
 B: *Gravesen s.n.,* 1971, det. B. Murray
 PB: *B. Murray 4327*
 ES: *Viereck 7406*
Lecanora beringii Nyl.
 PB: *Richardson s.n.* (ALA 61974), det. J. W.
 Thomson
Lecanora castanea (Hepp) Th. Fr.
 B: *Thomson, Shushan, and Sharp 6428* (WIS)
Lecanora epibryon (Ach.) Ach.
 B: *Thomson, Shushan, and Carson 9186* (WIS)
 PB: *B. Murray 4337*
 ES: *Thomson and Ahti 17894* (WIS)
Lecanora leptacina Somm.
 B: *Thomson, Shushan, and Carson 9186* (WIS)
 Not in Hale and Culberson (1970)
Lecanora polytropa (Ehrh.) Rabenh.
 ES: *Thomson and Ahti 18049* (WIS)
Lecanora verrucosa Ach.
 PB: *B. Murray 4339*
Lecidea assimilata Nyl.
 B: *Thomson and Shushan 6432* (WIS)
 PB: *Richardson s.n.* (ALA 61975), det. J. W.
 Thomson
Lecidea berengeriana (Mass.) Nyl.
 B: *Thomson and Shushan 6411* (WIS)
Lecidea crustulata (Ach.) Spreng.
 B: *Thomson and Shushan 6400* (WIS), det. H.
 Hertel
Lecidea demissa (Rutstr.) Ach.
 B: *Thomson and Shushan 9178* (WIS)
Lecidea granulosa (Ehrh.) Ach.
 B: *Thomson and Shushan 6408* (WIS)

---

[3]*Cladonia thomsonii* Ahti (*Bryologist* **81:** 334. 1978).

Lecidea ramulosa Th. Fr.
  B: *Thomson and Shushan 6427* (WIS)
Lecidea tornoensis Nyl.
  B: *Thomson, Shushan, and Carson 6404*
  (WIS)
Lecidea uliginosa (Schrad.) Ach.
  B: *Thomson and Shushan 6414* (WIS)
Lecidea vernalis (L.) Ach.
  PB: *B. Murray 4361*, det. J. W. Thomson
Lecidella wulfenii (Hepp) Koerb.
  B: *Thomson and Shushan 9045* (WIS)
Lepraria membranacea (Dicks.) Vain.
  PB: *B. Murray 6240*, det. J. W. Thomson
Leptogium saturninum (Dicks.) Nyl.
  ES: Krog, 1968
Leptogium tenuissimum (Dicks.) Fr.
  PB: *B. Murray 4347*, det. J. W. Thomson
Lobaria linita (Ach.) Rabenh.
  B: Krog, 1962
  ES: Krog, 1962
Lopadium coralloideum (Nyl.) Lynge
  B: *Miller and Laursen 10652*, det. B. Murray
Lopadium fecundum Th. Fr.
  PB: *B. Murray 4396*, det. J. W. Thomson
  This collection is the first record of this species
  for Alaska (Thomson, ms).
Lopadium pezizoideum (Ach.) Koerb.
  B: *Thomson, Shushan, and Sharp 6120* (WIS)
  ES: *Thomson and Ahti 18191* (WIS)
Nephroma expallidum (Nyl.) Nyl.
  B: Wetmore, 1960
  ES: Krog, 1968
Ochrolechia androgyna (Hoffm.) Arn.
  ES: *Thomson and Ahti 18191* in part (WIS)
Ochrolechia frigida (Sw.) Lynge
  B: Howard, 1970
  PB: *B. Murray 4396*
  f. *thelephoroides* (Ach.) Lynge, including *P.*
  pterulina (Nyl.) Howard
    B: Howard, 1970
    PB: *B. Murray 4395*, det. J. W. Thomson
Ochrolechia grimmiae Lynge
  ES: *Thomson and Ahti 17891* (WIS)
Ochrolechia inaequatula (Nyl.) Zahlbr.
  B: *Battrum 233*, det. M. J. Dibben
Ochrolechia upsaliensis (L.) Mass.
  PB: *B. Murray 4360*, det. J. W. Thomson
  ES: *Thomson and Ahti 17892* (WIS)
Omphalina sp. (= *Coriscium viride* (Ach.)
  Vain.)
  B: *Flock Fl-233*
Pannaria pezizoides (G. Web.) Trev.
  B: *Thomson, Shushan, and Sharp 6156* (WIS)
Parmelia almquistii Vain.
  ES: Esslinger, 1974; Krog, 1963, 1968

Parmelia alpicola Th. Fr.
  ES: Krog, 1968
Parmelia centrifuga (L.) Ach.
  ES: Krog, 1968
Parmelia incurva (Pers.) Fr.
  ES: *Thomson and Ahti 18025* (WIS)
Parmelia omphalodes (L.) Ach.
  B: *Thomson, Shushan, and Sharp, Lich. Arct.*
  25
  PB: *B. Murray 4392*
  ES: Krog, 1968
Parmelia separata Th. Fr.
  ES: Krog, 1968
Parmelia stygia (L.) Ach.
  ES: Krog, 1968
Parmelia sulcata Tayl.
  ES: Krog, 1962
Parmeliella praetermissa (Nyl.) P. James
  PB: *Richardson s.n.* (ALA 61971), det. J. W.
  Thomson
Peltigera aphthosa (L.) Willd.
  B: Krog, 1962
  PB: *B. Murray 4397*
Peltigera canina (L.) Willd.
  B: Krog, 1962
  PB: *B. Murray 4382*
Peltigera leucophlebia (Nyl.) Gyeln. (= *P.*
  aphthosa var. *leucophlebia*)
  B: *Thomson and Shushan 9179* (WIS)
  ES: Krog, 1962
Peltigera malacea (Ach.) Funck
  B: *Flock Fl-242*
  PB: *B. Murray 4325*, det. J. W. Thomson
  ES: *Viereck 7382*
Peltigera polydactyla (Neck.) Hoffm.
  PB: *Richardson s.n.* (ALA 61970), det. B.
  Murray
Peltigera rufescens (Weis.) Humb. (= *P. canina*
  var. *rufescens*)
  B: *Thomson and Shushan 6162* (WIS)
  PB: *B. Murray 4374*, det. J. W. Thomson
Peltigera scabrosa Th. Fr.
  B: Krog, 1962
  ES: Krog, 1968
Peltigera spuria (Ach.) DC. (= *P. canina* var.
  spuria)
  B: *Thomson and Shushan 6124* (WIS)
  f. *sorediata* Schaer.
    B: *Thomson and Shushan 6118* (WIS)
    PB: *B. Murray 4340*, det. J. W. Thomson
Peltigera venosa (L.) Baumg.
  ES: *B. Murray 6210*
Pertusaria dactylina (Ach.) Nyl.
  B: Dibben, 1974
  ES: *Thomson and Ahti 17898* (WIS)

Pertusaria glomerata (Ach.) Schaer.
 B: Dibben 1974
Pertusaria octomela (Norm.) Erichs.
 PB: *B. Murray 4394*
Pertusaria oculata (Dicks.) Th. Fr.
 B: Dibben, 1974
 ES: *Thomson and Ahti 17942* (WIS)
Pertusaria panyrga (Ach.) Mass.
 PB: *B. Murray 4358*, det. J. W. Thomson
 ES: *Thomson and Ahti 18044* (WIS)
Pertusaria subobducens Nyl.
 PB: *B. Murray 4338*, det. J. W. Thomson—K
 + yellow, but no stictic acid in GAoT
Physcia dubia (Hoffm.) Lett.
 PB: *B. Murray 6218*
Physconia muscigena (Ach.) Poelt *(= Physcia muscigena)*
 B: *Thomson and Shushan 6462* (WIS)
 PB: *B. Murray 4348*
Polyblastia bryophila Lönnr.
 PB: *B. Murray 6227*, det. J. W. Thomson
 This is apparently the first record of this species
 for Alaska.
Polyblastia gelatinosa (Ach.) Th. Fr.
 B: *Thomson, Shushan, and Sharp 6149* (WIS)
 Not listed in Hale and Culberson (1970).
 According to Thomson (ms) the specimen
 cited here is the first Alaskan record.
Polyblastia sendtneri Kremph.
 PB: *B. Murray 4386*, det. J. W. Thomson
 Not listed in Hale and Culberson (1970). Thom-
 son (written communication) says that the spe-
 cies has been reported from Ellesmere Island
 by Darbishire, and that he has it in Canada
 only from the Firth River, Yukon Territory,
 and Baffin Island.
Psoroma hypnorum (Vahl) S. Gray
 B: *Flock Fl-246*
 ES: *Thomson and Ahti 17893* (WIS)
Ramalina almquistii Vain.
 B: *Koranda s.n.*, 14 Aug 1956 (WIS)
 PB: *B. Murray 4346*
Rhizocarpon disporum (Naeg. *ex* Hepp) Muell.
 Arg.
 PB: *B. Murray 4351*
Rhizocarpon geographicum (L.) DC.
 B: *Flock Fl-250* (NARL), det. J. W.
 Thomson
Rinodina roscida (Somm.) Arn.
 PB: *B. Murray 4341*
Rinodina turfacea (Wahlenb.) Koerb.
 B: *Thomson and Shushan 5654* (WIS)
 PB: *B. Murray 6238*, det. J. W. Thomson
Siphula ceratites (Wahlenb.) Fr.
 B: *Thomson and Shushan, Lich. Arct. 23*

Solorina bispora Nyl.
 ES: Krog, 1968
Solorina crocea (L.) Ach.
 B: *Williams Ak-104*
 ES: Krog, 1968
Solorina saccata (L.) Ach.
 B: *Thomson and Shushan 6461* (WIS)
 PB: *B. Murray 4362*
Solorina spongiosa (Sm.) Anzi
 PB: *B. Murray 4356*
Sphaerophorus fragilis (L.) Pers.
 B: *Thomson and Shushan, Lich. Arct. 31*
 ES: Krog, 1962
Sphaerophorus globosus (Huds.) Vain.
 B: Krog, 1962
 PB: *Walker s.n.*, 22 Aug 1974, det. B. Murray
 var. *globosus*
 ES: Krog, 1968
Stereocaulon alpinum Laur.
 PB: *B. Murray 4375*, det. I. M. Lamb
 var. *alpinum*
 B: *Shanks and Sharp 37*, det. I. M.
 Lamb
 ES: Krog, 1968
Stereocaulon glareosum (Sav.) Magn.
 ES: *Thomson and Ahti 17911* (WIS)
 var. *brachyphylloides* Lamb
 B: *Thomson and Shushan 6428* (WIS)
Stereocaulon paschale (L.) Hoffm.
 [B: Brown and Johnson, 1965]
 ES: Krog, 1968
Stereocaulon rivulorum Magn.
 PB: *?B. Murray 4365*, det. I. M. Lamb, too
 scanty and depauperate to determine with
 certainty.
Stereocaulon saxatile Magn.
 B: *Thomson and Shushan 6164* (WIS)
Stereocaulon tomentosum Fr.
 B: *Flock Fl-254*
Sticta arctica Degel.
 ES: Krog, 1968
Thamnolia subuliformis (Ehrh.) W. Culb.
 B: Sato, 1959
 PB: *B. Murray 4336*
 ES: Krog, 1968 as baeomycic acid strain of *T. vermicularis*
Thamnolia vermicularis (Sw.) Ach. *ex* Schaer.
 B: Krog, 1962
 ES: Krog, 1968
Toninia lobulata (Somm.) Lynge
 B: *Flock Fl-248* (NARL), det. J. W.
 Thomson
 PB: *B. Murray 6216*, det. J. W. Thomson
Umbilicaria arctica (Ach.) Nyl.
 ES: Krog, 1968

Umbilicaria caroliniana Tuck.
  ES: *Viereck 7405,* det. B. Murray
Umbilicaria hyperborea (Ach.) Ach.
  ES: Krog, 1968
Umbilicaria proboscidea (L.) Schrad.
  ES: Krog, 1968
Umbilicaria torrefacta (Lightf.) Schrad.
  ES: Krog, 1968
Verrucaria devergens Nyl.
  PB: *B. Murray 4354,* det. J. W. Thomson
Xanthoria candelaria (L.) Th. Fr.
  B: *Webster Ak-589,* det. E. D. Rudolph

Xanthoria elegans (Link) Th. Fr.
  PB: *B. Murray 4353*

Rejected Lichen Taxon from Barrow

Cladonia wainii Sav.
  This species was cited for Barrow by Thomson
  (1968). Ahti (1973) said that he and other
  authors (Yoshimura, 1968, Krog, 1968, Ahti,
  1968, Ahti, Scotter, and Vänskä, 1973) regard
  this material as a different, undescribed
  species.

References

Ahti, T. (1968) (Review) The lichen genus *Cladonia* in North America. By J. W. Thomson. *Bryologist,* **71**: 292–294.

Ahti, T. (1973) Taxonomic notes on some species of *Cladonia,* subsect. *Unciales. Ann. Bot. Fenn.,* **10**: 163–184.

Ahti, T., G. W. Scotter, and H. Vänskä. (1973) Lichens of the Reindeer Preserve, Northwest Territories, Canada. *Bryologist,* **76**: 48–76.

Ando, H. (1963) A *Pseudolepicolea* found in the middle Honshu of Japan. *Hikobia,* **3**: 177–183.

Argus, G. W. (1973) The genus *Salix* in Alaska and the Yukon. *Nat. Mus. Canada, Publ. Bot.* No. 2, 279 pp.

Arnell, S., and H. Persson. (1961) Notes on *Lophozia pellucida* Schuster. *Sv. Bot. Tidskr.,* **55**: 376–378.

Brassard, G. R. (1971) The mosses of northern Ellesmere Island, arctic Canada. I. Ecology and phytogeography, with an analysis for the Queen Elizabeth Islands. *Bryologist,* **74**: 233–281.

Britton, M. E. (1967) Vegetation of the arctic tundra. In *Arctic Biology* (H. P. Hansen, Ed.). Corvallis: Oregon State University Press, 2nd ed., pp. 67–130.

Brown, J., and P. L. Johnson. (1965) Pedo-ecological investigations, Barrow, Alaska. U.S. Army Cold Regions Research and Engineering Laboratory (CRREL) Technical Report 159, 32 pp.

Calder, J. A., and D. B. O. Savile. (1960) Studies in Saxifragaceae III. *Can. J. Bot.,* **38**: 409–435.

Clebsch, E. E. C. (1957) The summer season climatic and vegetational gradient between Point Barrow and Meade River, Alaska. M.S. thesis, University of Tennessee, Knoxville.

Crum, H., W. C. Steere, and L. E. Anderson. (1973) A new list of mosses of North America north of Mexico. *Bryologist,* **76**: 85–130.

Culberson, W. L. (1963) A summary of the genus *Haematomma* in North America. *Bryologist,* **66**: 224–236.

Culberson, W. L., and C. F. Culberson. (1965) *Asahinea,* a new genus in the Parmeliaceae. *Brittonia,* **17**: 182–190.

Dennis, J. G. (1968) Growth of tundra vegetation in relation to arctic microenvironments at Barrow, Alaska. Ph.D. thesis, Duke University.

Dibben, M. J. (1974) The chemosystematics of the lichen genus *Pertusaria* in North America, north of Mexico. Ph.D. thesis, Duke University, 587 pp.

Dorn, R. D. (1975) A systematic study of *Salix* section *Cordatae* in North America. *Can. J. Bot.,* **53**: 1491–1522.

Douglas, G. W., and G. Ruyle-Douglas. (1978) Nomenclatural changes in the Asteraceae of British Columbia. I. Senecioneae. *Can. J. Bot.,* **56:** 1710–1711.

Esslinger, T. L. (1974) A chemosystematic revision of the brown *Parmeliae*. Ph.D. thesis, Duke University, 669 pp.

Evans, A. W. (1955) Notes on North American *Cladoniae*. *Bryologist, 58*: 93–112.

Gjaerevoll, O. (1958–1967) Botanical investigations in central Alaska, especially in the White Mountains, *Kgl. Norske Vidensk, Selsk. Skr. Trondheim.* 1958, No. 5, 74 pp: 1963, No. 4, 115 pp; 1967, No. 10, 63 pp.

Hale, M. E., Jr., and W. L. Culberson. (1970) A fourth checklist of the lichens of the continental United States and Canada. *Bryologist, 73*: 499–543.

Halliday, G. (1977) Vascular plants new to Alaska. *Can. J. Bot.*

Halliday, G., and A. O. Chater. (1969) *Carex marina* Dewey, an earlier name for *C. amblyorhyncha* Krecz. *Feddes Rep., 80*: 103–106.

Hämet-Ahti, L. (1971) List of vascular plants collected in Alaska, the Yukon, northern British Columbia, and Alberta by Leena Hämet-Ahti and Teuvo Ahti in 1967. *Mimeogr. Papers Bot. Mus. Univ. Helsinki, 3*: 1–17.

Hämet-Ahti, L., and V. Virrankoski. (1971) Cytotaxonomic notes on some monocotyledons of Alaska and northern British Columbia. *Ann. Bot. Fenn., 8*: 156–159.

Hanson, H. C. (1950) Vegetation and soil profiles in some solifluction and mound areas in Alaska. *Ecology, 31*: 606–630.

Harvill, A. M., Jr. (1948) A phytogeographic study of Alaskan mosses. Ph.D. thesis, University of Michigan, Ann Arbor, 337 pp.

Hermann, F. J. (1973) Additions to the bryophyte flora of Alaska. *Bryologist, 76*: 442–446.

Howard, G. E. (1970) The lichen genus *Ochrolechia* in North America north of Mexico. *Bryologist, 73*: 93–130.

Hultén, E. (1941–1950) Flora of Alaska and Yukon, Parts 1–10. Lund: C. W. K. Gleerup, 1902 pp.

Hultén, E. (1965) The *Cerastium alpinum* complex. A case of worldwide introgressive hybridization. *Sv. Bot. Tidskr., 50*: 411–495.

Hultén, E. (1967) Comments on the flora of Alaska and Yukon. *Arkiv Bot., 2*(7): 1–147.

Hultén, E. (1968) *Flora of Alaska and Neighboring Territories.* Stanford: Stanford University Press, 1008 pp.

Hultén, E. (1973) Supplement to ''Flora of Alaska and neighboring territories.'' *Bot. Not., 126*: 459–512.

Imshaug, H. A. (1950) New and noteworthy lichens from Mt. Rainier National Park. *Mycologia, 42*: 743–752.

Karczmarz, K. (1971) A monograph of the genus *Calliergon* (Sull.) Kindb. *Polskie Towarzystwo Bot. Monogr. Bot., 34*: 209 pp., Pl. I–XX.

Koponen, T. (1968) Generic revision of Mniaceae Mitt. (Bryophyta). *Ann. Bot. Fenn., 5*: 117–151.

Koponen, T. (1971) A monograph of *Plagiomnium* sect. *Rosulata* (Mniaceae). *Ann. Bot. Fenn., 8*: 305–367.

Koponen, T. (1972) Notes on *Mnium arizonicum* and *M. thomsonii. Lindbergia, 1*: 161–165.

Koponen, T. (1974) A guide to the Mniaceae in Canada. *Lindbergia, 2*: 160–184.

Koranda, J. J. (1954) A phytosociological study of an uplifted marine beach ridge, near Point Barrow, Alaska. M.S. thesis, Michigan State College, Lansing, 132 pp.

Krog, H. (1962) A contribution to the lichen flora of Alaska. *Arkiv Bot.*, Ser. 2, *4*(16): 489–513.

Krog, H. (1963) *Parmelia almquistii* Vain. and its distribution. *Bryologist, 66*: 28–31.

Krog, H. (1968) The macrolichens of Alaska. *Norsk. Polarinst. Skr., (144)*: 180 pp.

Krog, H. (1973) *Cetraria inermis* (Nyl.) Krog, a new lichen species in the Amphi-Beringian flora element. *Bryologist, 76*: 299–300.

Lepage, E. (1951) New or noteworthy plants in the flora of Alaska. *Amer. Midl. Nat., 46*: 754–759.

McNeill, J. (1978) *Silene alba* and *S. dioica* in North America and the generic delimitation of *Lychnis, Melandrium,* and *Silene* (Caryophyllaceae). *Can. J. Bot.* **56**: 297–308.

Mogensen, G. S. (1973a) A revision of the moss genus *Cinclidium* Sw. (Mniaceae Mitt.). *Lindbergia,* **2**: 49–80.

Mogensen, G. S. (1973b) List of specimens studied in G. S. Mogensen: A revision of the moss genus *Cinclidium* Sw. (Mniaceae Mitt.). *Lindbergia,* **2**: 49–80, 1973. Processed publication without pagination.

Mulligan, G. A. (1970) Cytotaxonomic studies of *Draba glabella* and its close allies in Canada and Alaska. *Can. J. Bot.,* **48**: 1431–1443.

Mulligan, G. A. (1974) Confusion in the names of three *Draba* species of the Arctic: *D. adamsii, D. oblongata,* and *D. corymbosa. Can. J. Bot.,* **52**: 791–793.

Murray, B. M., and D. F. Murray. (1975) Provisional checklist to the vascular, bryophyte, and lichen flora of Prudhoe Bay, Alaska. In *Ecological Investigations of the Tundra Biome in the Prudhoe Bay Region, Alaska* (J. Brown, Ed.). Biol. Pap. Univ. Alaska, Special Report No. 2, pp. 203–212.

Nyholm, E., and A. C. Crundwell. (1958) *Bryum salinum* Hagen *ex* Limpr. in Britain and in America. *Trans. Brit. Bryol. Soc.,* **3**: 373–377.

Ochi, A. (1962) Contribution to the mosses of Bartramiaceae in Japan and the adjacent regions I. *Nova Hedwigia,* **4**: 87–108.

Packer, J. G., and G. D. McPherson. (1974) Chromosome numbers in some vascular plants from northern Alaska. *Can. J. Bot.,* **52**: 1095–1099.

Persson, H. (1946) Some Alaskan and Yukon bryophytes. *Bryologist,* **49**: 41–58.

Persson, H. (1947) Further notes on Alaskan–Yukon bryophytes. *Bryologist,* **50**: 279–310.

Persson, H. (1952a) Critical or otherwise interesting bryophytes from Alaska–Yukon. *Bryologist,* **55**: 1–25, 88–116.

Persson, H. (1952b) Additional list of Alaska–Yukon mosses. *Bryologist,* **55**: 261–279.

Persson, H. (1954) Mosses of Alaska–Yukon. *Bryologist,* **57**: 189–217.

Persson, H. (1962) Bryophytes from Alaska collected by E. Hultén and others. *Sv. Bot. Tidskr.,* **56**: 1–35.

Persson, H. (1963) Bryophytes of Alaska and Yukon Territory collected by Hansford T. Shacklette. *Bryologist,* **66**: 1–26.

Persson, H., and O. Gjaerevoll. (1957) Bryophytes from the interior of Alaska. *Kgl. Norske Vidensk. Selsk. Skr. Trondheim,* 1957, No. 5, 74 pp.

Porsild, A. E. (1965) Some new or critical vascular plants of Alaska and Yukon. *Can. Field-Nat.,* **79**: 79–90.

Porsild, A. E. 1974 (1975) Materials for a flora of central Yukon Territory. *Nat. Mus. Canada, Publ. Bot.,* No. 4, 77 pp.

Rastorfer, J. R. (1972) Bryophyte taxa lists of the high Alaskan Arctic. Institute of Polar Studies, Ohio State University, 54 pp.

Rastorfer, J. R. (1974) Element contents of three Alaskan-arctic mosses. *Ohio J. Sci.,* **74**: 55–59.

Rastorfer, J. R., H. J. Webster, and D. K. Smith. (1973) Floristic and ecologic studies of bryophytes of selected habitats at Prudhoe Bay, Alaska. Institute of Polar Studies, Ohio State University Rpt. No. 49, 20 pp.

Sato, M. (1959) Mixture ratio of the lichen genus *Thamnolia* collected in Japan and adjacent regions. *Misc. Bryol. Lich.,* **22**: 1–6.

Scamman, E. (1940) A list of plants from interior Alaska. *Rhodora,* **42**: 309–343.

Shacklette, H. T. (1962) Influences of the soil on boreal and arctic plant communities. Ph.D. thesis, University of Michigan, Ann Arbor, 349 pp.

Smith, D. K. (1974) Floristic, ecologic, and phytogeographic studies of the bryophytes in the tundra around Barrow, Alaska. Ph.D. thesis, University of Tennessee, Knoxville, 191 pp.

Steere, W. C. (1954) Chromosome number and behavior in arctic mosses. *Bot. Gaz.,* **116**: 93–133.

Steere, W. C. (1955) Bryophyta of arctic America. VI. A collection from Prince Patrick Island. *Amer, Midl. Nat.,* **53**: 231–241.

Steere, W. C. (1958) *Mnium andrewsianum,* a new subarctic and arctic moss. *Bryologist,* **61**: 173–182.

Steere, W. C. (1973) Observations on the genus *Aplodon. Bryologist,* **76**: 347–355.

Steere, W. C. (1974) The status and geographical distribution of *Voitia hyperborea* in North America (Musci: Splachnaceae). *Bull. Torrey Bot. Club,* **101**: 55–63.

Steere, W. C. (1975) *Lophozia incisa:* A curious reaction to desiccation. *Bryologist,* **78**: 368–369.

Steere, W. C. (1976) Identity of *Pterygoneurum arcticum* with *P. lamellatum. Bryologist,* **79**: 221–222.

Steere, W. C., and G. R. Brassard. (1974) The systematic position and geographical distribution of *Fissidens arcticus. Bryologist,* **77**: 195–202.

Steere, W. C., and K. A. Holmen. (1975) Bryophyta Arctica Exsiccata. Fasc. I (No. 1–50), New York and Copenhagen, 16 pp.

Steere, W. C., and B. M. Murray. (1974) The geographical distribution of *Bryum wrightii* in arctic and boreal North America. *Bryologist,* **77**: 172–178.

Steere, W. C., K. A. Holmen, and G. S. Mogensen. (1976) Bryophyta Arctica Exsiccata. Fasc. II (No. 51–100), New York and Copenhagen.

Thomson, J. W. 1967 (1968) *The lichen genus Cladonia in North America.* Toronto: University of Toronto Press, 172 pp.

Thomson, J. W. (1968) *Haematomma lapponicum* Räs. in North America. *J. Jap. Bot.,* **43**: 305–310.

Thomson, J. W. (No date) Lichens of the Alaskan Arctic Slope. Manuscript, 504 pp.

Uttal, L. J. (1973) The scientific name of the Alaska *Spiraea. Bull. Torrey Bot. Club,* **100**: 236–237.

Weeden, R. B. (1968) Dates of first flowers of alpine plants at Eagle Creek, central Alaska. *Can. Field-Nat.,* **82**: 24–31.

Welsh, S. L. (1967) Legumes of Alaska II: *Oxytropis* DC. *Iowa State J. Sci.,* **41**: 277–303.

Wetmore, C. M. (1960) The lichen genus *Nephroma* in North and Middle America. *Publ. Mus. Mich. State Univ., Biol. Ser.,* **1**: 369–452.

Wiggins, I. L. (1951) The distribution of vascular plants on polygonal ground near Point Barrow, Alaska. *Contr. Dudley Herb.,* **4**: 41–56.

Wiggins, I. L., and J. H. Thomas. (1962) *A Flora of the Alaskan Arctic Slope.* Toronto: University of Toronto Press, 425 pp.

Worley, I. A. (1970) A checklist of the Hepaticae of Alaska. *Bryologist,* **73**: 32–38.

Yoshimura, I. (1968) Lichenological notes. 1. Some species of *Cladonia* with taxonomic problems. *J. Hattori Bot. Lab.,* **31**: 109–204.

Index